T0213426

Lecture Notes in Artificial Intelligence 9992

Subseries of Lecture Notes in Computer Science

More information about this series at http://www.springer.com/series/1244

Byeong Ho Kang · Quan Bai (Eds.)

AI 2016: Advances in Artificial Intelligence

29th Australasian Joint Conference
Hobart, TAS, Australia, December 5–8, 2016
Proceedings

 Springer

Editors
Byeong Ho Kang
University of Tasmania
Hobart
Australia

Quan Bai
Auckland University of Technology
Auckland
New Zealand

ISSN 0302-9743   ISSN 1611-3349   (electronic)
Lecture Notes in Artificial Intelligence
ISBN 978-3-319-50126-0   ISBN 978-3-319-50127-7   (eBook)
DOI 10.1007/978-3-319-50127-7

Library of Congress Control Number: 2016958512

LNCS Sublibrary: SL7 – Artificial Intelligence

This Springer imprint is published by the registered company Springer Nature Switzerland AG
The registered company address is: Gewerbestrasse 11, 6330 Cham, Switzerland

Preface

This volume contains the papers presented at the 29th Australasian Joint Conference on Artificial Intelligence 2016 (AI 2016), which was held in Hobart, Australia, December 5–8, 2016.

The conference is the premier event for artificial intelligence in Australasia and provides a forum for researchers and practitioners across all subfields of artificial intelligence to meet and discuss recent advances. This year, we were co-located with the 2016 International Conference on Smart Media and Applications (SMA 2016), with which we shared the keynote speech session, a joint session, and a social event.

The technical program of AI 2016 comprised a number of high-quality papers that were selected in a thorough, single-blind reviewing process with at least two expert reviews per paper. Out of 121 submissions, our senior Program Committee with the help of an experienced international Program Committee selected 40 long papers and 18 short papers for presentation at the conference and inclusion in these proceedings. In addition to the technical program, we also selected two papers from the doctoral consortium and six papers from SMA 2016 to be included in the proceedings as invited papers.

Papers were submitted by authors from 28 countries, demonstrating the broad international appeal of our conference. In addition to the 58 paper presentations, we had four keynote talks by high-profile speakers:

- Prof. Rayid Ghani, University of Chicago, USA
- Prof. Takayuki Ito, Nagoya Institute of Technology, Japan
- Prof. Maurice Pagnucco, UNSW, Australia
- Prof. Zhi-Hua Zhou, Nanjing University, China

AI 2016 was complemented by a doctoral consortium, and featured an exciting selection of four workshops and a tutorial that were free for all conference participants to attend. The four workshops were:

- The 8th International Workshop on Collaborative Agents Research and Development
- The Workshop on Time Series Analytics and Applications
- The Workshop on Experiential Knowledge Platform Development Research
- Korean Academy of Scientists and Engineers in Australasian Annual Workshop

The tutorial were on:

- Deep Learning and Applications in Non-Cognitive Domains by Truyen Tran (Deakin University)

A large number of people and organizations helped make AI 2016 a success. First and foremost, we would like to thank the authors for contributing and presenting their latest work at the conference. Without their contribution this conference would not have been possible. The same is true for the members of the conference organization. We also thank our senior Program Committee members, the members of our Program Committee, as well as additional reviewers who were all very dedicated and timely in their contributions to selecting the best papers for presentation at AI 2016.

We are grateful for support and sponsorship from the Australian Computer Society for AI (ACS-AI) Meeting, the School of Engineering and ICT, the University of Tasmania (UTAS), and the Science and Engineering Research Support Society. In particular, we appreciate the ACS-AI Meeting for student scholarships and the School of Engineering and ICT of UTAS for the administration and technical supports for this conference. We also appreciate the free conference management system EasyChair, which was used for putting together this volume. Last but not the least, we thank Springer for their sponsorship and their support in preparing and publishing this volume in the *Lecture Notes in Computer Science* series.

October 2016

Quan Bai
Byeong Ho Kang

Organization

General Chairs

Geoff Webb — Monash University, Australia
Craig Lindley — CSIRO, Australia

Program Chairs

Byeong Ho Kang — University of Tasmania, Australia
Quan Bai — Auckland University of Technology, New Zealand

Workshop Chair

Ashfaqur Rahman — CSIRO, Australia

Tutorial Chair

Charlotte Sennersten — CSIRO, Australia

Doctoral Consortium Chairs

Alan Blair — University of New South Wales, Australia
James Montgomery — University of Tasmania, Australia

Publicity Chair

Soyeon Caren Han — University of Tasmania, Australia

Senior Program Committee

Wray Lindsay Buntine — Monash University, Australia
Stephen Cranefield — University of Otago, New Zealand
Reinhard Klette — Auckland University of Technology, New Zealand
Jimmy Lee — The Chinese University of Hong Kong, Hong Kong, SAR China
Michael Maher — UNSW, Australia
Thomas Meyer — University of Cape Town and CAIR, South Africa
Abhaya Nayak — Macquarie University, Australia
Fatih Porikli — ANU, Australia
Mikhail Prokopenko — University of Sydney, Australia
Fabio Ramos — University of Sydney, Australia

Jussi Rintanen	Aalto University, Finland
Michael Thielscher	UNSW, Australia
Dianhui Wang	La Trobe University, Australia
Ian Watson	University of Auckland, New Zealand
Chengqi Zhang	University of Technology, Sydney, Australia
Dongmo Zhang	University of Western Sydney, Australia
Mengjie Zhang	Victoria University of Wellington, New Zealand

Program Committee

Ayse A. Bilgin	Macquarie University, Australia
Ivan Bindoff	University of Tasmania, Australia
Rafael H. Bordini	PUCRS, Brazil
Xiongcai Cai	UNSW, Australia
Lawrence Cavedon	RMIT University, Australia
Jeffrey Chan	RMIT University, Australia
Songcan Chen	Nanjing University of Aeronautics and Astronautics, China
Winyu Chinthammit	University of Tasmania, Australia
Sung-Bae Cho	Yonsei University, Korea
Michael Cree	University of Waikato, New Zealand
Hepu Deng	RMIT University, Australia
Peter Eklund	IT University of Copenhagen, Denmark
Atilla Elci	Aksaray University, Turkey
Daryl Essam	UNSW Canberra, Australia
Cèsar Ferri	Universitat Politècnica de València, Spain
Marcus Gallagher	University of Queensland, Australia
Xiaoying Gao	Victoria University of Wellington, New Zealand
Edel Garcia	CENATAV, Cuba
Saurabh Kumar Garg	University of Tasmania, Australia
Manolis Gergatsoulis	Ionian University, Greece
Chi Keong Goh	Advanced Technology Centre, Rolls Royce, Singapore
Hans W. Guesgen	Massey University, New Zealand
Christian Guttmann	UNSW, Australia
Nader Hanna	Macquarie University, Australia
Jose Hernandez-Orallo	Universitat Politècnica de València, Spain
Weidong Huang	University of Tasmania, Australia
Paul Kennedy	University of Technology, Sydney, Australia
Philip Kilby	Data61, CSIRO and ANU, Australia
Myunghee Kim	Defence Science Technology (DST), Department of Defence, Australia
Yang Sok Kim	Keimyung University, Korea
Kevin Korb	Monash University, Australia
Rudolf Kruse	University of Magdeburg, Germany
Jérôme Lang	LAMSADE, Université Paris-Dauphine, France
Nung Kion Lee	Universiti Malaysia Sarawak, Malaysia

Additional Reviewers

Arun Anand
Flaulles Bergamaschi
Aidan Bindoff
Zied Bouraoui
Khalil Bouzekri
Christian Braune
Nathan Brewer
Kinzang Chhogyal
Maisa Daoud
Dave De Jonge
Alexander Dockhorn
Christoph Doell
Alex Feng

Matthew Gibson
Vitor Guizilini
Aaron Hunter
Eleftherios Kalogeros
Luke Lake
Weihua Li
Zhidong Li
Rodrigo A. Lima
Craig Lindley
Qinxue Meng
Stefano Moretti
Doan Tung Nguyen
Lei Niu

Diogo Patrão
Rivindu Perera
Gavin Rens
Sobia Saleem
Ransalu Senanayake
Upul Senanayake
Darren Shen
Damiano Spina
Xishun Wang
Mohammad Yousef
Jihang Zhang

Contents

Big Data

Constraint Satisfaction, Search and Optimisation

Knowledge Representation and Reasoning

Machine Learning and Data Mining

Social Intelligence

Text mining and NLP

Selected Papers from AI 2016 Doctoral Consortium

Selected Papers from SMA 2016

Agents and Multiagent Systems

Lifted Backward Search for General Game Playing

Dave de Jonge[✉] and Dongmo Zhang

Western Sydney University, Werrington, Australia
{d.dejonge,d.zhang}@westernsydney.edu.au

Abstract. A General Game player is a computer program that can play games of which the rules are only known at run-time. These rules are usually given as a logic program. General Game players commonly apply a tree search over the state space, which is time consuming. In this paper we therefore present a new method that allows a player to detect that a future state satisfies some beneficial properties, without having to explicitly generate that state in the search tree. This may lead to faster algorithms and hence to better performance. Our method employs a search algorithm that searches backwards through formula space rather than state space.

1 Introduction

The development of programs that can play specific games such as Chess and Go, has long been an important field in AI research. However, such game players are limited in the sense that they are only able to play one specific game and often rely on game-specific heuristics invented by humans. Therefore, recently more attention has been given to the concept of *General* Game Playing (GGP). A General Game Playing agent is able to interpret the rules of a game at run-time and devise a strategy for it without any human intervention. Since 2005 General Game Playing competitions have been held annually at the AAAI Conference [6]. A language called Game Description Language (GDL) [11] was invented to write down game-rules in a standardized, machine-readable way. GDL specifies games as a logic program, and is similar to other logic-based languages such as ASP [4] and Datalog [3].

Most GGP players apply a generic search algorithm such as minimax search [12], alpha-beta pruning [9] and, most importantly, Monte Carlo Tree Search (MCTS) [7,10]. These techniques are based on the principle of forward searching: they start with the initial state, determine the set of legal moves in that state, then determine for each of these moves the next state that would result if that move were executed, evaluate this hypothetical future state, and then repeat the procedure.

In order to be able to quickly generate a future state given the current state and a legal move, most GGP players translate a GDL-specified game into a state transition machine. A disadvantage of this technique is that it requires *grounding*

B.H. Kang and Q. Bai (Eds.): AI 2016, LNAI 9992, pp. 3–16, 2016.
DOI: 10.1007/978-3-319-50127-7_1

to get rid of any variables in the rules. Every rule is replaced by several variable-free copies of that rule by replacing every variable with a possible ground term for that variable. Grounding makes logical reasoning much simpler, but it also causes the size of the search space to explode, and may cause explicit symmetries in the game rules to get lost.

In this paper we propose a technique that may improve the speed of tree-search algorithms. The idea is that one may be able to evaluate a game state before it is generated in the tree. We achieve this by applying a backward search that does not require grounding. It starts with a formula that describes states satisfying some desired property (such as being a terminal winning state) and then searches for formulas that describe states which are one move away from a state with that property, and so on. This backward search is applied before the forward search is started.

In the field of planning such a technique is known as Lifted-Backward-Search [8]. The difference between planning and GGP, however, is that planning is normally concerned with a single agent, or a group of agents with a common goal. In GGP on the other hand the agent needs to deal with adversaries. Therefore, our approach combines Lifted-Backward-Search with a min-max strategy. We show that our approach is sound and compete, in the sense that, given any desired property every state satisfying that property can be found by our algorithm, and that every state found by our algorithm indeed satisfies that property.

The rest of the paper is organized as follows: In Sect. 2 we give a short overview of existing work. In Sect. 3 we explain how our technique can be used in a GGP agent. In Sect. 4 we will formalize the language in which we present our algorithm. In Sect. 5 we will present the algorithm itself and prove its main properties. Finally, in Sect. 6 we present our conclusions.

2   Related Work

FluxPlayer [13], the winner of the 2006 AAAI GGP competition is a player that applies an iterated deepening depth-first search method with alpha-beta pruning, and uses Fuzzy logic to determine how close a given state is to the goal state. Cadia Player [5], the winner in 2007, 2008, and 2012, is based on MCTS, extended with several heuristics to guide the rollouts so that they are better informed and hence give more realistic results, and also the winner of the 2014 competition, Sancho,[1] as well as the winner of 2015, Galvanise,[2] apply variants of MCTS.

A technique similar to ours is described in [14]. The main difference however is that their backward search is grounded. Furthermore, their algorithm only goes 1 step back, whereas our algorithm may take any number of backward steps.

---

[1] http://sanchoggp.blogspot.co.uk/2014/05/what-is-sancho.html.
[2] https://bitbucket.org/rxe/galvanise_v2.

In [2] the authors implement a heuristic backward search algorithm called HSPr for single-agent planning domains. They do not find much benefit in backward search however, because it generates many 'spurious states': states that are impossible to reach. However, in [1] the authors manage to improve HSPr by extrapolating several common techniques from forward search to backward search and thus creating a new regression planner called FDr. A general overview of planning algorithms, including Lifted Backward Search, can be found in [8]. To the best of our knowledge lifted backward search has never been used in domains with adversarial agents.

3   A Player Based on Backward Search

A standard approach to General Game Playing is to apply Monte Carlo Tree Search. Key to this approach is that one evaluates a game state w by simulating a game in which random moves are played starting at state w until a terminal state is reached, and then evaluating the player's utility for that terminal state. This is called a *rollout*. Repeating this many times and averaging the utility values returned by these rollouts yields an estimation of the true utility value of the state w. Unfortunately, since these rollouts are random, one needs to perform a lot of them before the result becomes accurate.

We propose to increase the accuracy by combining MCTS with Lifted-Backward-Search. The idea is that, before applying MCTS, the player will apply backward search for a given amount of time, say 10 s for example. Assuming the game is a turn-taking game for two players, which we refer to as A and B, the backward search generates two sequences of formulas $\alpha_0, \alpha_1, \alpha_2 \ldots$ and $\beta_0, \beta_1, \beta_2 \ldots$ which are stored in memory and will be used by the rollouts of the MCTS. Here, α_0 represents the set of winning states for our player A, while α_1 represents the set of all states for which player A has a move that will lead to winning state. The formula α_2 represents the set of states for which the opponent cannot avoid the next state to satisfy α_1. That is, in general, if a state satisfies α_i it means that A has a strategy that can guarantee that the next state will satisfy α_{i-1} (either because A has a move that leads to such a state or because B does not have any move that can lead to a state that does not satisfy it). The formulas β_i have the opposite interpretation: β_0 represents a winning state for the opponent B, and if a state satisfies β_i our opponent can enforce the next state to satisfy β_{i-1}. After the backward search our player applies MCTS. For each game state explored during a rollout, the algorithm determines whether it satisfies any of the α_i or β_i. If it does satisfy any of these formulas the rollout can be stopped and return either the value 100 (if the state satisfies any α_i), or 0 (if it satisfies some β_i).

Not only does this allow us to terminate the rollout earlier, it also yields a much more accurate result. Say for example that we have a state w in which player A has 20 possible moves, of which only one leads to a winning state yielding 100 points, and all others lead to a draw, yielding 50 points. Of course, any rational player would always pick that winning move, so the non-terminal state w itself can be assigned a value of 100. However, since rollouts are random, a rollout will only pick the correct winning move and return 100 once in every 20 times it visits w. With our algorithm on the other hand, the rollout will detect that w satisfies α_1 and therefore always return the correct utility value of 100.

Our backward search works with non-grounded formulas, which means that many different states can be described by a single formula. Of course, generating those formulas may take a lot of time, and in the worst case the size of the formulas α_n may grow exponentially with n. However we expect that in many cases the formulas remain relatively compact because they contain variables.

The fact that α_0 and β_0 represent terminal winning states for the respective players is just an example. One could also give them the interpretation of some other important game property. For example, in chess, α_0 could represent those states in which A just captured the opponent's queen. In that case, whenever the rollout function finds a state that satisfies any of the α_i's it will return some heuristic value.

4   Formal Definitions

In general, GGP deals with games that take place over one or more discrete rounds. In each round the game is in a certain state, and each player chooses an action (also referred to as a *move*) to take. The game description specifies the initial state, which actions each player is allowed to choose given any state, and how in any state the chosen actions determine the next state.

In this paper we assume the game is a two-player turn-taking game that can end in three possible ways: a win for the first player, a win for the second player, or a draw. Furthermore, we assume that if the game has a winner, then the winner is always the player who made the last move. Examples of such games are Chess, Checkers and Tic-Tac-Toe. In fact, it seems that most of the games available on the GGP website[3] satisfy these criteria. Furthermore, we also assume that the game description does not contain any cycles (see Definition 4). This is a restriction, because GDL only requires rules to be *stratified* (see [11]), which is a weaker assumption than being cycle-free. We do expect however that the assumption of being cycle-free can be dropped, but we leave it as future work to investigate this.

We will refer to the two players by A and B, where A is the player that applies our algorithm and B is our opponent. Since we assume turn-taking games in each round there is only one player that has an action to choose. We call this player the *active* player of that round.

4.1   State Transition Model

We first define the notion of a *game* in terms of a state transition model.

Definition 1. *A **game** $\mathcal{G}$ is a tuple $\langle P, \mathcal{A}, W, w_1, T, L, u, U \rangle$, where:*

---

[3] http://www.ggp.org/view/tiltyard/games/.

- *P is the set of* players: $P = \{A, B\}$
- *A is the set of* actions.
- *W is a non-empty set of* states.
- $w_1 \in W$ *is the* initial *state.*
- $T \subset W$ *is the set of* terminal *states.*
- $L : W \backslash T \rightarrow 2^A$ *is the* legality *function, which describes for each non-terminal state which actions are legal.*
- $u : W \times A \rightarrow W$ *is the* update function *that maps each state and action to a new state.*
- $U : T \times P \rightarrow [0, 100]$ *is the* utility function *that assigns a utility value to each terminal state and player.*

The idea is that the players A and B both have their own set of actions, denoted $\mathcal{A}_A$ and $\mathcal{A}_B$ respectively, with $\mathcal{A} = \mathcal{A}_A \cup \mathcal{A}_B$ and $\mathcal{A}_A \cap \mathcal{A}_B = \emptyset$. The game consists of discrete rounds, in which the players alternately pick an action a from their respective action sets and hence generate a sequence of actions $a_1, a_2, \ldots a_n$. Given a state w and an action a the update function u defines a new state $u(w, a)$. Therefore, given the initial state w_1 and a sequence of actions the update function defines a sequence of states $w_1, w_2, \ldots$ where each w_{i+1} equals $u(w_i, a_i)$. If the current state is w_i then the players may only pick their action from the set $L(w_i)$. Since we are assuming turn-taking games it is always the case that either $L(w) \subseteq \mathcal{A}_A$ or $L(w) \subseteq \mathcal{A}_B$.

4.2 Syntax

In this paper we will use our own language $\mathcal{L}$, which is a segment of first-order logic, because it is easier to describe our algorithm in $\mathcal{L}$ than in GDL. Our language borrows its basic components from GDL so that any game description given in GDL can be translated easily into $\mathcal{L}$. Given a game $\mathcal{G}$ we define $\mathcal{L}$ to be a language with a finite set of constants, function symbols and relation symbols which are specific to the game $\mathcal{G}$, a finite set of variables and logical connectives ($\neg, \vee, \wedge, \rightarrow$ and $\exists$).[4] $\mathcal{L}$ inherits the following game-independent relation symbols from GDL: *distinct, true, does, legal, goal, terminal,* and *next.* Their semantics will be given in the next subsection.[5] As a segment of first-order logic, we do not employ the full syntax of first-order logic. Instead we impose heavy restrictions on the structure of terms and formulas. A *formula* in $\mathcal{L}$ can either be a *complex formula,*[6] which is a combination of atoms using $\neg, \vee, \wedge,$ or $\exists$, or a *rule* defined as follows:

Definition 2. *A **rule** is an expression in $\mathcal{L}$ of the form: $p_1 \wedge p_2 \wedge \ldots p_n \rightarrow q$ where each p_i is a positive or negative literal, and q is a positive literal. The atom q is called the **head** of the rule and the p_i's are called the **subgoals** of the rule. The conjunction of subgoals is called the **body** of the rule.*

---

[4] The universal quantifier $\forall$ is not included in the language.

[5] GDL defines more relations, but these are not relevant for this paper.

[6] The implication $\rightarrow$ is not allowed in a complex formula.

The body of a rule may be the empty conjunction (in which case the rule is also called a *fact*), and the subgoals and the head of a rule may contain variables. For clarity we will always denote variables with a question mark, e.g. $?x$. Similar to the restrictions on rules in GDL, the relation symbols *distinct*, *true* and *does* cannot appear in the head of any rule, while *legal*, *goal*, *terminal* and *next* cannot appear in the body of any rule. Apart from these key words rules may also contain user-defined relation symbols.

Definition 3. *There is a certain subset of the constants of $\mathcal{L}$ that we call **action-constants**, and a certain subset of the function symbols that we call **action-functions**. An action-function can only contain action-constants or variables that range over the action-constants. A term containing an action-function is called an **action-term**.*

The relation symbols *does* and *legal* can only contain action terms. A ground atom of the form *does*(t) is called an *action-proposition*. The size of the set of ground action terms must be equal to the size of $\mathcal{A}$.

Definition 4. *The **dependency graph** of a set of rules $\mathcal{R}$ is a directed graph that contains a vertex v_p for each relation symbol p that appears in any rule in $\mathcal{R}$, and there is an edge from v_p to v_q if there is a rule in which p appears in the body and q appears in the head. A set of rules $\mathcal{R}$ is **cycle-free** if its dependency graph does not contain any cycles.*

Definition 5. *A proposition of the form true(t), where t is a ground term of $\mathcal{L}$, is called a **base proposition**. The (finite) set of all base propositions is denoted as $\mathcal{B}$.*

Definition 6. *A formula is called **unwound** if it only contains the relation symbols 'distinct', 'true' and 'does', and it is called **wound** otherwise.*

We will see below that for an unwound formula we can determine whether it is satisfied or not in a straightforward manner, whereas for wound formulas we first need to rewrite ('unwind') the formula as an unwound formula, using the rules of the game.

Definition 7. *Let φ be an atom and p be the relation symbol of φ. Then φ is called **statewise** if none of the paths in the dependency graph of $\mathcal{R}$ that go through v_p pass through the vertex corresponding to the relation symbol 'does'. A formula ϕ is called statewise if all its atoms are statewise. A formula is called **non-statewise** otherwise.*

As we will see below, the satisfaction of a statewise formula only depends on the game state, while for non-statewise formulas it depends on the state as well as on the chosen action. Just like in GDL, we pose the restriction on $\mathcal{L}$ that any atom containing *legal*, *goal*, or *terminal* must be statewise.

4.3   Semantics

Given a game $\mathcal{G}$ and a language $\mathcal{L}$ for that game, let V be a valuation function. That is: V is an injective function $V : W \rightarrow 2^{\mathcal{B}}$ that maps each world state of $\mathcal{G}$ to a different set of base-propositions. The interpretation of V is that when the game is in a state w the propositions in $V(w)$ are considered true and all other base-propositions are considered false. Furthermore, we define a bijective map μ that maps each action $a \in \mathcal{A}$ to a ground action-term t_a. In the following we will assume that $\mathcal{G}$, $\mathcal{L}$, $\mathcal{R}$, V, and μ are fixed.

Definition 8. *Let t and s be ground terms, $w \in W$, and $a, b \in \mathcal{A}$, then we define:*

- $V, \mathcal{R} \vDash_{(w,a)} true(t)$ iff $true(t) \in V(w)$
- $V, \mathcal{R} \vDash_{(w,a)} does(\mu(b))$ iff $a = b$ and $a \in L(w)$
- $V, \mathcal{R} \vDash_{(w,a)} distinct(t, s)$ iff $t \neq s$

Here $a = b$ means that a and b are syntactically equal, and $t \neq s$ that t and s are syntactically different. Note that the entailment of $true(t)$ and $distinct(t, s)$ does not depend on the action a.

Definition 9. *Let ϕ be any (non-ground) formula. Then we define $V, \mathcal{R} \vDash_{(w,a)} \exists \phi$ to hold iff there exists a substitution θ such that $\phi[\theta]$ is ground and $V, \mathcal{R} \vDash_{(w,a)} \phi[\theta]$ holds.*

Definition 10. *Let $\mathcal{R}$ be a set of rules, q an atom, r a rule in $\mathcal{R}$ and θ the most general substitution that unifies q with the head of r. Then we say the body of $r[\theta]$ is a **premise** of q. Furthermore, we say the disjunction of all premises of q is the **complete premise** of q, and is denoted as $Pr(q)$.*

For example, if we have an atom $q(t)$ where t is a ground term, and we have the following two rules: $p_1(?x) \rightarrow q(?x)$ and $p_2(?y) \rightarrow q(?y)$ then $Pr(q(t)) = p_1(t) \vee p_2(t)$. Note that this definition implies that if there is no rule r such that q can be unified with the head of r then $Pr(q) = \bot$, and that if q can be unified with a fact then $Pr(q) = \top$.

Definition 11. *Let q be any ground wound atom. Then we define:*
$V, \mathcal{R} \vDash_{(w,a)} q$ iff $V, \mathcal{R} \vDash_{(w,a)} \exists Pr(q)$.

Definition 12. *Let ϕ be any ground formula. Then $V, \mathcal{R} \vDash_{(w,a)} \phi$ is defined by the standard interpretation of the connectives of propositional logic, and Definitions 8 and 11.*

Definitions 8, 9, 11 and 12 together recursively define satisfaction. The fact that $\mathcal{R}$ is finite and cycle-free guarantees that the recursion terminates. If for some formula ϕ we have $V, \mathcal{R} \vDash_{(w,a)} \exists \phi$ then we say that (w, a) satisfies ϕ. If ϕ is statewise we may also say that w satisfies ϕ, and we may denote this as $V, \mathcal{R} \vDash_w \exists \phi$.

For example, if $\phi = goal(A, 100)$, and $\mathcal{R}$ contains a rule $true(t) \rightarrow goal(A, 100)$. Then w satisfies ϕ if $true(t) \in V(w)$.

Definition 13. *A game description G for a game $\mathcal{G}$ is a tuple $\langle \mathcal{L}, V, \mu, \mathcal{R} \rangle$ where $\mathcal{R}$ is finite and cycle-free. Furthermore, all of the following must hold:*

- *$V, \mathcal{R} \vDash_{(w,a)} terminal \quad iff \quad w \in T$*
- *$V, \mathcal{R} \vDash_{(w,a)} legal(\mu(a)) \quad iff \quad a \in L(w)$*
- *$V, \mathcal{R} \vDash_{(w,a)} next(t) \quad iff \quad V \vDash_{u(w,a)} true(t)$*
- *$V, \mathcal{R} \vDash_{(w,a)} goal(p, x) \quad iff \quad U(w, p) = x$*

5   Backward Search

The goal of player A is to bring about a state in which both *terminal* and *goal(A, 100)* are satisfied. For a given state it is easy to verify whether these are satisfied or not. However, we would like our player to determine in advance whether it can enforce these propositions to be true in any future state. We therefore apply an algorithm that determines a sequence of unwound statewise formulas $\alpha_0, \alpha_1, \alpha_2 \ldots$ until time runs out. Their interpretation is that if a state w satisfies α_i then A can enforce a victory in i rounds. Note that our assumption that the winner always makes the last move implies that if the game is in a state that satisfies α_i and i is an odd number, then A is the active player, whereas if i is an even number then B is the active player. In order to explain how these formulas are calculated we need to define two operators, which we call the C-operator and the N-operator.

5.1   The C-Operator

In our language any formula can be rewritten as an equivalent unwound formula. Therefore, we here define an operator that takes a formula ϕ as input and outputs an equivalent unwound formula. In Sect. 5.2 we will see that this is important because the N-operator can only operate on unwound formulas.

Definition 14. *We define the C-operator as follows:*

- *For any terms t, s: $C(true(t)) = true(t)$, $C(distinct(t, s)) = distinct(t, s)$.*
- *For any action-term a: $C(does(a)) = does(a) \wedge legal(a)$*
- *For any wound atom q: $C(q) = \exists Pr(q)$*
- *For any non-atomic ϕ we obtain $C(\phi)$ by replacing each atom p in ϕ by $C(p)$.*

Thus, if we have: $\phi = q(t) \wedge does(a)$ with t a ground term, and the only rules in $\mathcal{R}$ of which the head can be unified with $q(t)$ are the following two: $p_1(?x) \to q(?x)$ and $p_2(?y) \to q(?y)$, with $?x$ and $?y$ variables, then:
$C(\phi) = (p_1(t) \vee p_2(t)) \wedge does(a) \wedge legal(a)$.

We use the notation $C^2(\phi)$ to denote $C(C(\phi))$, and $C^n(\phi)$ for $C(C^{n-1}(\phi))$. Since we require that $\mathcal{R}$ is finite and cycle-free, we have that for any formula ϕ there is always some n for which $C^n(\phi) = C^{n-1}(\phi)$. In other words, the sequence $C^1(\phi), C^2(\phi) \ldots$ always converges. Therefore, we can define $C^\infty(\phi)$ to be the limit of this sequence.

Definition 15. *We define C^∞ as follows: $C^\infty(\phi) = C^n(\phi)$ iff $C^n(\phi) = C^{n-1}(\phi)$*

Lemma 1. *For any formula ϕ the formula $C^\infty(\phi)$ is unwound.*

Proof. Suppose that the formula $C^n(\phi)$ contains an atom p of which the relation symbol is not *distinct*, *true*, or *does*. Then $C^{n+1}(\phi)$ contains $\exists Pr(p)$ instead of p, but $\exists Pr(p)$ is not equal to p because $\mathcal{R}$ is cycle-free. This means $C^n(\phi)$ is not equal to $C^{n+1}(\phi)$, which means that $C^n(\phi)$ is not $C^\infty(\phi)$. The conclusion is that if $C^n(\phi) = C^\infty(\phi)$ then $C^\infty(\phi)$ cannot contain any such atom, and thus is unwound.

Lemma 2. *For any formula ϕ, any state w and any action a we have:*
$V, \mathcal{R} \vDash_{(w,a)} \exists C^\infty(\phi)$ *iff* $V, \mathcal{R} \vDash_{(w,a)} \exists \phi$

Proof. It follows directly from Definitions 9, 11, 12 and 14, that $V, \mathcal{R} \vDash_{(w,a)} \exists C(\phi)$ iff $V, \mathcal{R} \vDash_{(w,a)} \exists \phi$. This argument can be repeated to prove the lemma.

Let η_A denote some wound statewise formula that player A desires to be satisfied, for example: $\eta_A = terminal \wedge goal(100, A)$ (this example formula describes the terminal states in which A wins the game). Then we define: $\alpha_0 := C^\infty(\eta_A)$. Note that α_0 describes exactly the same property as η_A, but α_0 is unwound.

5.2   The N-Operator

Now that we have defined α_0 we want to define the other formulas α_i. For this we need an operator that 'translates' a formula ϕ into a new formula ϕ' such that we know that if ϕ' is true in the current state then ϕ will be true in the next state.

Definition 16. *Given any unwound statewise formula ϕ, the formula $N(\phi)$ is obtained by replacing every occurrence of the relation symbol 'true' by the relation symbol 'next'. The resulting formula $N(\phi)$ is wound and non-statewise.*

For example, if $\phi = true(t_1) \vee true(t_2)$ then $N(\phi) = next(t_1) \vee next(t_2)$.

Suppose we have a game state w and we want to pick an action a such that the next state $u(w, a)$ will satisfy some formula ϕ. The action we need to pick is then the action a for which (w, a) satisfies $N(\phi)$. Specifically, if we apply this to α_0 it means that if a state-action pair (w, a) satisfies $N(\alpha_0)$ then A can guarantee the property α_0 to be satisfied in the next state by playing action a in state w.

Lemma 3. *Let ϕ be an unwound statewise formula. Then we have:*
$V, \mathcal{R} \vDash_{(w,a)} \exists N(\phi)$ *iff* $V, \mathcal{R} \vDash_{u(w,a)} \exists \phi$

Proof. We will only prove this for a specific example, but it can be generalized easily. Suppose that $\phi = true(t_1) \wedge true(t_2)$. We have that (w, a) satisfies $N(\phi) = next(t_1) \wedge next(t_2)$ iff (w, a) satisfies $next(t_1)$ and (w, a) satisfies $next(t_2)$. But, according to Definition 13 this is true iff $u(w, a)$ satisfies $true(t_1)$ and $u(w, a)$ satisfies $true(t_2)$, which means that $u(w, a)$ satisfies $true(t_1) \wedge true(t_2)$, which is ϕ.

For example, in Tic-Tac-Toe, suppose we have: $\phi = true(cell(1,1,X))$ which states that ϕ is satisfied if the top left cell contains an X. Then we have:

$$N(\phi) = next(cell(1,1,X)).$$

By itself this formula is not very useful, but by using the C^∞-operator we can transform it into an unwound formula, for which we can easily check whether it is satisfied or not. The game description of Tic-Tac-Toe contains the following three rules:

$$true(cell(1,1,X)) \rightarrow next(cell(1,1,X))$$
$$does(mark(1,1,X)) \rightarrow next(cell(1,1,X))$$
$$true(cell(1,1,b)) \rightarrow legal(mark(1,1,X))$$

The first rule states that if the upper left cell is marked with an X then it will also be marked with an X in the next state. The second rule states that if A makes the move $mark(1,1,X)$ then the upper left cell will be marked with an X in the next state. The third rule states that it is legal for player A to make a mark in the upper left cell if that cell is currently empty. Using these rules we obtain:

$$C^\infty(N(\phi)) \;\; = \; true(cell(1,1,X)) \;\; \vee \;\; (does(mark(1,1,X)) \wedge true(cell(1,1,b)))$$

which states that if in the current state the upper left cell contains an X, or if the upper left cell is blank and player A marks it with an X, then in the next turn ϕ will be satisfied.

We now know that if the game is in a state w that contains the base-proposition $true(cell(1,1,X))$ or it contains the base-proposition $true(cell(1,1,b))$ and A chooses the action $mark(1,1,X)$ then the next state $u(w,a)$ will (also) satisfy $true(cell(1,1,X))$.

5.3   Action Normal Form

We have seen in the previous section that if we want some formula ϕ to be true in the next state, then we want $N(\phi)$ to be satisfied in the current state. However, $N(\phi)$ is non-statewise, so the satisfaction of $N(\phi)$ depends on the action chosen by the active player. Therefore, what we want is to determine, given a state w, which action a the active player needs to choose in order to satisfy $N(\phi)$. To make this easier, we transform the formula into *Action Normal Form*, as defined in [15].

Definition 17. *We say a formula ϕ is in **Action Normal Form** (ANF) for player A if it is written in the following form: $\phi = \bigvee_{t \in S} \mathcal{X}_t^\phi \wedge does(t)$ where S is a set of action-terms such that for each action $a \in \mathcal{A}$ there is a term $t \in S$ that can be unified with $\mu(a)$, and all $\mathcal{X}_t^\phi$ are statewise.*

The idea of ANF is that it explicitly separates the action-propositions from the other types of atoms. It gives us a clear recipe to determine which action to choose in a state w if we want ϕ to be satisfied.

Lemma 4. *For any non-statewise formula ϕ there is a formula $\overline{\phi}$ which is in ANF, such that: $V, \mathcal{R} \vDash_{(w,a)} \exists \overline{\phi}$ iff $V, \mathcal{R} \vDash_{(w,a)} \exists \phi$.*

Proof. We only prove this for the case that ϕ is ground. For each action a we can generate a formula $\mathcal{X}^{\phi}_{\mu(a)}$ by replacing every occurrence of $does(\mu(a))$ in ϕ with $\top$ and replacing all other action-propositions with $\bot$. Note that we then have that, for every $a \in \mathcal{A}$, (w, a) satisfies $\mathcal{X}^{\phi}_{\mu(a)} \wedge does(a)$ iff (w, a) satisfies ϕ.

Now suppose that for some state w and some formula ϕ we want to choose an action a such that (w, a) satisfies ϕ. We can achieve this by first generating an equivalent formula $\overline{\phi}$ which is in ANF. We can then check for every $\mathcal{X}^{\phi}_t$ in the expression whether it is satisfied by w or not. If for some term $t \in S$ we have that $\mathcal{X}^{\phi}_t$ is indeed satisfied by w then there is an action a such that $\mu(a)$ is unifiable with t and such that (w, a) satisfies $\mathcal{X}^{\phi}_t \wedge does(t)$ and therefore we have that (w, a) satisfies ϕ.

From now on, the notation $\overline{\phi}$ will denote any formula that is in ANF and that is equivalent to ϕ. Furthermore, we will use $\overline{\phi}^+$ to denote $\overline{\phi}$ in which all action-propositions have been replaced with $\top$. Thus, if: $\overline{\phi} = \bigvee_{t \in S} \mathcal{X}^{\phi}_t \wedge does(t)$ then: $\overline{\phi}^+ := \bigvee_{t \in S} \mathcal{X}^{\phi}_t$.

Lemma 5. *If a state w satisfies $\overline{\phi}^+$ then there exists an action a such that (w, a) satisfies $\overline{\phi}$, and hence also ϕ.*

Proof. A state w satisfies $\overline{\phi}^+$ iff there is some $\mathcal{X}^{\phi}_t$ in the ANF of ϕ that is satisfied, which means there is a substitution θ such that w satisfies $\mathcal{X}^{\phi}_t[\theta]$. We then have that $(w, t[\theta])$ satisfies $\mathcal{X}^{\phi}_t[\theta] \wedge does(t[\theta])$, which means that (w, a) satisfies $\mathcal{X}^{\phi}_t \wedge does(t)$, with $a = \mu^{-1}(t[\theta])$, and therefore (w, a) satisfies $\overline{\phi}$, and because of Lemma 4 (w, a) satisfies ϕ.

Note that $\overline{\phi}$ is not uniquely defined. However, from now on we will simply assume we have some algorithm that outputs a unique $\overline{\phi}$ for any given ϕ. Then, for any odd positive integer n we can define:

$$\alpha_n = \overline{C^{\infty}(N(\alpha_{n-1}) \wedge \neg terminal)}^+ \qquad (1)$$

Lemma 6. *Let n be any odd positive integer. A state w satisfies α_n iff w is non-terminal and there is some action a for A that is legal in state w and for which the resulting state $u(w, a)$ satisfies α_{n-1}.*

Proof. If we combine Eq. (1) with Lemma 5 we conclude that w satisfies α_n iff there exists an action a such that (w, a) satisfies $C^{\infty}(N(\alpha_{n-1}) \wedge \neg terminal)$. According to Lemma 2 this holds iff (w, a) satisfies $N(\alpha_{n-1}) \wedge \neg terminal$, and then using Lemma 3 we conclude that this holds iff $u(w, a)$ satisfies α_{n-1} and w is non-terminal.

We now still need to define α_n for even n. Note that if n is even and a state w satisfies α_n then it means that in state w player B is the active player. However, since B is an adversary, the existence of an action for B that leads to a state satisfying α_{n-1} is not enough to guarantee that α_{n-1} will be satisfied in the next state. After all, B may choose a different action in order to prevent this. Therefore, for states in which B is the active player we demand that *all* actions of B lead to a state satisfying α_{n-1}. Let us first define a formula α'_n as follows: $\alpha'_n = C^\infty(N(\alpha_{n-1} \vee \alpha_{n-2} \vee \ldots \alpha_0))$. Then, for n is even, we can define:

$$\alpha_n = C^\infty\left(\neg terminal \wedge \neg \bigvee_{t \in S} \exists (legal(t) \wedge \neg \mathcal{X}_t^{\alpha'_n}) \right) \tag{2}$$

where S is the same set of action-terms as the one that appears in the ANF of α'_n and the $\mathcal{X}_t^{\alpha'_n}$ are also obtained from the ANF of α'_n.

Lemma 7. *Let n be an even number. A state w satisfies α_n iff for every move a of B that is legal in state w there is an integer $m < n$ such that $u(w, a)$ satisfies α_m.*

Proof. Equation (2) states that w satisfies α_n iff w is non-terminal and there is no action-term t and no substitution θ such that w satisfies $legal(t[\theta]) \wedge \neg \mathcal{X}_t^{\alpha'_n}[\theta]$. This means that for every legal action $a = \mu^{-1}(t[\theta])$ we have that w satisfies $\mathcal{X}_t^{\alpha'_n}[\theta]$ and hence that (w, a) satisfies $\mathcal{X}_t^{\alpha'_n}[\theta] \wedge does(a)$, which means that $(w, t[\theta])$ satisfies α'_n, which is $C^\infty(N(\alpha_{n-1} \vee \alpha_{n-2} \vee \ldots \alpha_0))$. Again, using Lemmas 2 and 3 this means that $u(w, a)$ satisfies $\alpha_{n-1} \vee \alpha_{n-2} \vee \ldots \alpha_0$, which means there is some $m < n$ for which $u(w, a)$ satisfies α_m.

Lemma 8. *If a state w satisfies α_n and $n > 0$ then w is a non-terminal state.*

Proof. This follows directly by applying Lemma 2 to Eqs. 1 and 2.

Theorem 1. *If a state w satisfies α_n for some n then there exists a strategy for A that guarantees that a state satisfying α_0 can be reached in at most n steps whenever the game is in the state w.*

Proof. We know from Lemmas 6 and 7 that if w satisfies α_n and A plays optimally, then the next state will satisfy some α_m with $m < n$. This means that no matter what strategy is played by B, in every round that follows some α_m will be satisfied, and m will be decreasing with every new round. Moreover, since every α_m is non-terminal (except for $m = 0$) this means the game will eventually reach a state that satisfies α_0.

Theorem 2. *Let w be a state such that A has a strategy that guarantees that some state w' that satisfies α_0 can be reached then there is an integer n such that w satisfies α_n.*

Proof. The proof goes by induction. If w already satisfies α_0 then the theorem clearly holds. Our induction hypothesis says that for any state in which A has such a strategy in $n-1$ steps, α_{n-1} is satisfied. We now need to prove that if w is non-terminal and A has a strategy that guarantees α_0 in n steps then α_n is satisfied. If n is odd, then A is the active player, so we know A has a move a that leads to a state $u(w, a)$ which is only $n-1$ steps away from w'. According to the induction hypothesis we then know that $u(w, a)$ satisfies α_{n-1}. From Lemma 6 it then follows that w satisfies α_n. If n is even, then it means that all legal moves of B lead to a state that satisfies α_m for some $m < n$. According to Lemma 7 this means that w satisfies α_n.

6   Conclusions

We have presented a backward search algorithm for turn-taking games in General Game Playing. Given any desired property encoded by some formula η_A it generates a sequence of formulas $\alpha_0, \alpha_1, \ldots$ with the interpretation that if a game state w satisfies some α_i then player A has a strategy that guarantees that η_A will be satisfied in i steps. We expect this method to be useful as an addition to MCTS because it allows one to quickly evaluate whether a target state can be reached without having to generate all the states that lead to it. The effectiveness of this method depends on whether it is possible to keep the formulas compact. Therefore, an empirical evaluation is left as future work.

Acknowledgments. This work was sponsored by an Endeavour Research Fellowship awarded by the Australian Government Department of Education.

References

1. Alcázar, V., Borrajo, D., Fernández, S., Fuentetaja, R.: Revisiting regression in planning. In: IJCAI 2013, Beijing, China, 3–9 August 2013 (2013)
2. Bonet, B., Geffner, H.: Planning as heuristic search. Artif. Intell. **129**(1–2), 5–33 (2001)
3. Ceri, S., Gottlob, G., Tanca, L.: What you always wanted to know about datalog (and never dared to ask). IEEE Trans. Knowl. Data Eng. **1**(1), 146–166 (1989)
4. Eiter, T., Ianni, G., Krennwallner, T.: Answer set programming: a primer. In: Tessaris, S., Franconi, E., Eiter, T., Gutierrez, C., Handschuh, S., Rousset, M.-C., Schmidt, R.A. (eds.) Reasoning Web 2009. LNCS, vol. 5689, pp. 40–110. Springer, Heidelberg (2009). doi:10.1007/978-3-642-03754-2_2
5. Finnsson, H.: Simulation-based general game playing. Ph.D. thesis, School of Computer Science, Reykjavik University (2012)
6. Genesereth, M., Love, N., Pell, B.: General game playing: overview of the AAAI competition. AI Mag. **26**(2), 62–72 (2005)
7. Genesereth, M.R., Thielscher, M.: General Game Playing. Synthesis Lectures on Artificial Intelligence and Machine Learning. Morgan & Claypool Publishers, San Rafael (2014)
8. Ghallab, M., Nau, D., Traverso, P.: Automated Planning: Theory & Practice. Morgan Kaufmann Publishers Inc., San Francisco (2004)
9. Knuth, D.E., Moore, R.W.: An analysis of alpha-beta pruning. Artif. Intell. **6**(4), 293–326 (1975)

10. Kocsis, L., Szepesvári, C.: Bandit based Monte-Carlo planning. In: Fürnkranz, J., Scheffer, T., Spiliopoulou, M. (eds.) ECML 2006. LNCS (LNAI), vol. 4212, pp. 282–293. Springer, Heidelberg (2006). doi:10.1007/11871842_29

11. Love, N., Genesereth, M., Hinrichs, T.: General game playing: game description language specification. Technical report. LG-2006-01, Stanford University, Stanford, CA (2006)

12. von Neumann, J.: On the theory of games of strategy. In: Tucker, A., Luce, R. (eds.) Contributions to the Theory of Games, pp. 13–42. Princeton University Press, New Jersey (1959)

13. Schiffel, S., Thielscher, M.: M.: Fluxplayer: a successful general game player. In: Proceedings of AAAI 2007, pp. 1191–1196. AAAI Press (2007)

14. Trutman, M., Schiffel, S.: Creating action heuristics for general game playing agents. In: Cazenave, T., Winands, M.H.M., Edelkamp, S., Schiffel, S., Thielscher, M., Togelius, J. (eds.) CGW/GIGA -2015. CCIS, vol. 614, pp. 149–164. Springer, Heidelberg (2016). doi:10.1007/978-3-319-39402-2_11

15. Zhang, D., Thielscher, M.: A logic for reasoning about game strategies. In: Proceedings of the Twenty-Ninth AAAI Conference on Artificial Intelligence (AAAI-15), pp. 1671–1677 (2015)

Corrupt Strategic Argumentation:
The Ideal and the Naive

Michael J. Maher[✉]

School of Engineering and Information Technology,
University of New South Wales, Canberra, ACT 2600, Australia
michael.maher@unsw.edu.au

Abstract. Previous work introduced a model of corruption within strategic argumentation, and showed that some forms of strategic argumentation are resistant to two forms of corruption: collusion and espionage. Such a result provides a (limited) basis on which to trust agents acting on our behalf. That work addressed several argumentation semantics, all built on the notion of admissibility. Here we continue this work to three other well-motivated semantics: the ideal, naive, and stage semantics. The latter two are not admissibility-based. We show that the naive semantics does not support strategic argumentation, in the sense that the outcome of the game is determined by the initial state, if the players are not corrupt. As a result, the semantics is corruption-proof. We show that the ideal semantics is resistant to both collusion and espionage. The stage semantics is resistant to espionage, but its resistance to collusion depends on the strategic aims of the players.

1 Introduction

Strategic argumentation provides a simple model of disputation and negotiation among agents. Agents are intended to act on our behalf but – whether they are human or software – we cannot be sure that they are acting in our best interests. Social structures, including criminal sanctions, are used to discourage corruption by human agents. For computational agents, [2] introduced another way corruption can be discouraged: if the computational requirements to take advantage of the corruption are too great then there is no incentive to act corruptly. [13–15] adapted this idea to strategic argumentation. That work formulated a notion of resistance to corruption, where corruption may be collusion of the two nominally opposed agents, or espionage by one agent in gaining illicit knowledge of her opponent's arsenal of arguments.

[14,15] addressed the problem in terms of Dung's abstract argumentation [6]. Abstract argumentation abstracts away from the structure of arguments and the conflicts between them; the resulting argumentation framework can be interpreted by many different semantics. Each semantics expresses compatibility criteria that a set of arguments must satisfy to be jointly accepted. The different semantics reflect different intuitions and principles of how conflicting arguments should be resolved.

© Springer International Publishing AG 2016
B.H. Kang and Q. Bai (Eds.): AI 2016, LNAI 9992, pp. 17–28, 2016.
DOI: 10.1007/978-3-319-50127-7_2

While [14,15] addressed many semantics for argumentation, attention was focussed on semantics based on admissibility: that for a set of arguments to be compatible it must defend itself against any attack. In this paper we address one admissibility-based semantics (the ideal semantics) that was not addressed in previous work, and two semantics that are not based on admissibility (the naive and stage semantics). It has been argued ([1,17], among others) that admissibility-based semantics do not accord with intuitions, for some argumentation frameworks.

The ideal semantics [7] offers a form of scepticism that is not as severe as that provided by the grounded semantics. Starting from the preferred semantics, it provides a single maximal coherent set of arguments that are accepted in every preferred extension, and defends itself against attack from any other arguments. It is claimed to express "justifiably accepted skeptical belief" [9].

The other semantics define a collection of sets of arguments, rather than a single set, and are not based on self-defence. The naive semantics [4] is quite natural: it consists of maximal conflict-free sets of arguments, where conflict-free entails that no argument in the set is incompatible with another. It reflects a desire to accept as many arguments as possible that are consistent with each other.

The stage semantics [17] is, like the naive semantics, based on conflict-free sets. It consists of those conflict-free sets that maximize the arguments that are decided – accepted or rejected. It reflects the intuition that as many arguments as possible should have their status decided. It has a close relation to some admissible semantics [17].

This paper is structured as follows. The next section provides necessary background on abstract argumentation and computational complexity. The following sections introduce strategic argumentation, and the computation problems arising in playing strategic argumentation games, including problems that arise when exploiting corruption. Then these problem are examined for, respectively, the ideal, naive, and stage semantics. Their computational complexity is established, which provides the basis for identifying resistance to corruption.

2   Background

2.1   Abstract Argumentation

This work is based on abstract argumentation in the sense of [6], which addresses the evaluation of a static set of arguments. An *argumentation framework* $\mathcal{A} = (S, \gg)$ consists of a finite set of arguments S and a binary relation $\gg$ over S, called the attack (sometimes, defeat) relation. If $(a, b) \in \gg$ we write $a \gg b$ and say that a attacks b. The semantics of an argumentation framework is given in terms of *extensions*, which are subsets of S.

Given an argumentation framework, an argument a is said to be *accepted* in an extension E if $a \in E$, and said to be *rejected* in E if some $b \in E$ attacks a. The set of rejected arguments in E is denoted by E^-. An argument that is neither accepted nor rejected in E is said to be *undecided* in E. We say an argument a is *self-defeating* if it attacks itself (that is, $a \gg a$). We say there is a *conflict* between arguments a_1 and a_2 if either $a_1 \gg a_2$ or $a_2 \gg a_1$. An extension E is *conflict-free* if the restriction of $\gg$ to E is empty. The *naive* extensions are the maximal conflict-free extensions. The *stage* extensions are the conflict-free extensions that are maximal under the containment ordering of $E \cup E^-$. Alternatively, the stage extensions are the conflict-free extensions that minimize the set of undecided arguments.

An argument a is *defended* by E if every argument that attacks a is attacked by some argument in E. An extension E of $\mathcal{A}$ is *admissible* if it is conflict-free and, for every argument $a \in E$ is defended by E. An extension E of $\mathcal{A}$ is *complete* if it is conflict-free and, $a \in E$ iff a is defended by E. The least complete extension under the containment ordering exists and is called the *grounded* extension. It reflects a strongly sceptical attitude towards accepting arguments. The *preferred* extensions are the maximal admissible extensions under the containment ordering. The *ideal* extension is the maximal admissible extension contained in all preferred extensions. A semantics is *unitary* if every argumentation framework has a single extension under the semantics. The grounded and ideal semantics are unitary; the naive and stage semantics are not.

A semantics is defined to be a set of extensions: the ideal semantics consists only of the ideal extension, the naive semantics is the set of naive extensions, the stage semantics is the set of stage extensions, etc. Each semantics expresses a criterion for what arguments can coherently be accepted together, given an argumentation framework. Each extension in the semantics represents a "reasonable" adjudication, according to that criterion, of the arguments in the argumentation framework.

2.2 Computational Complexity

We can view a complexity class as a set of decision problems. We assume the reader has knowledge of the polynomial complexity hierarchy (see, for example, [12]). We use PTIME to refer to the class of problems solvable in polynomial time. PSPACE is the class of decision problems solvable in polynomial space. It contains the entire polynomial hierarchy PH. As usual, the notation $\mathcal{C}^{\mathcal{D}}$, where $\mathcal{C}$ and $\mathcal{D}$ are complexity classes, refers to the class of problems that can be decided by an algorithm of complexity $\mathcal{C}$ with calls to a $\mathcal{D}$ oracle.

There are some additional complexity classes within the hierarchy that we need. D_2^p is the class of problems that can be expressed as the conjunction of a problem in Σ_2^p and a problem in Π_2^p. Θ_2^p is the class of decision problems solvable by a deterministic polynomial algorithm with $O(\log n)$ calls to an NP oracle. It is equal to $\text{PTIME}_{||}^{\text{NP}}$, the class of problems solvable by a deterministic polynomial algorithm with non-adaptive calls to an NP oracle. Non-adaptive refers to the restriction that oracle calls cannot depend on the outcome of previous calls. We have

$$NP, coNP \subseteq D^p \subseteq \Theta_2^p \subseteq \Delta_2^p \subseteq \Sigma_2^p, \Pi_2^p \subseteq D_2^p \subseteq \Sigma_3^p, \Pi_3^p$$

with $\text{NP}^{\Theta_2^p} = \Sigma_2^p$ and $\text{NP}^{D_2^p} = \Sigma_3^p$. Also, $\text{PTIME}^{\Sigma_i^p} = \Delta_{i+1}^p$.

Table 1. Complexity of several argumentation reasoning problems under selected semantics

	Credulous acceptance	Sceptical acceptance	Verification	Non-emptiness
Ideal	in Θ_2^p	in Θ_2^p	in Θ_2^p	in Θ_2^p
Naive	in PTIME	in PTIME	in PTIME	in PTIME
Stage	Σ_2^p-c	Π_2^p-c	coNP-c	in PTIME

There are several prominent decision problems in argumentation. For any semantics σ:

- The *Verification problem* asks, given an argumentation framework $\mathcal{A}$ and a set of arguments S, is S a σ-extension?
- The *Credulous Acceptance problem* asks, given $\mathcal{A}$ and an argument a, is there a σ-extension containing a?
- The *Sceptical Acceptance problem* asks, given $\mathcal{A}$ and an argument a, do all σ-extensions contain a?
- The *Non-emptiness problem* asks, is there a σ-extension of $\mathcal{A}$ that is non-empty?

Table 1 summarizes complexity results for these problems under the semantics of interest in this paper, drawn from [8–10]. For a complexity class $\mathcal{C}$, $\mathcal{C}$-c denotes $\mathcal{C}$-completeness.

3   Strategic Argumentation

Strategic argumentation provides a simple model of dynamic argumentation. Originally [11] it was formulated for a concrete argumentation system based in a defeasible logic, but we will use the model of [15] which is defined in terms of abstract argumentation. In strategic abstract argumentation, players take turns to add arguments to an argumentation framework. At each turn, the player adds arguments so that the argumentation framework is in a desired state. We refer to such states interchangeably as *desired outcomes* or *strategic aims* of the player. A player loses the strategic argumentation game when she is unable to achieve her desired outcome. In general, both players can win if the argumentation reaches a state that is desired by both players, but in this paper we consider an adversarial setting where the players' aims are mutually exclusive.

Strategic abstract argumentation is formalized as follows [15]. We assume there are two players, a proponent P and her opponent O. A *split argumentation framework* $(\mathcal{A}_{Com}, \mathcal{A}_P, \mathcal{A}_O, \gg)$ consists of three sets of arguments: $\mathcal{A}_{Com}$ the arguments that are common knowledge to P and O; $\mathcal{A}_P$ the arguments available to P, and $\mathcal{A}_O$ the arguments available to O; and an attack relation $\gg$ over $\mathcal{A}_{Com} \cup \mathcal{A}_P \cup \mathcal{A}_O$. $\mathcal{A}_P$ is assumed to be unknown to O, and $\mathcal{A}_O$ is unknown to P. Each player is aware of $\gg$ restricted to the arguments they know. We assume that P's desired outcome is that a distinguished argument a is accepted, in some sense, while O's aim is to prevent this. Starting with P, the players take turns in adding sets of arguments to $\mathcal{A}_{Com}$ from their available arguments, ensuring that their desired outcome is a consequence of the resulting argumentation framework[1]. As play continues, the set of arguments that are common knowledge $\mathcal{A}_{Com}$ becomes larger. When a player is unable to achieve her aim when it is her turn to play, she loses. We say that a player is *honest* if she plays rationally in trying to win. In particular, an honest player does not abandon a game when she has a play that achieves her aim, and does not play arguments that are unnecessary to achieve her aim.

[15] identifies several plausible strategic aims that the proponent P might have under an argumentation semantics σ. In this paper we focus on the following four:

1. **Existential:** a is accepted in at least one σ-extension
2. **Universal:** a is accepted in all σ-extensions
3. **Unrejected:** a is not rejected in any σ-extension
4. **Uncontested:** a is accepted in at least one σ-extension and is not rejected in any σ-extension

The existential and universal aims are credulous and sceptical acceptance. [15] also identifies some "counting aims". They will not be addressed in this paper.

In addition to these aims, a player may wish to "spoil" or prevent such aims from being achieved. Such aims are the negation of the above aims. For example, the negation of the uncontested aim aims to have a not accepted in any extension or have a rejected in some extension. In this paper, player O's aim is to prevent P's desired outcome; thus O's aim is the negation of P's aim.

In general, all these aims are distinct. However, for a unitary semantics σ (such as the ideal semantics) this variety of aims collapses: all the above aims – except the unrejected aim – collapse into one, that a is accepted in the σ-extension. For a unitary semantics σ there are six possible aims: (1) a is accepted in the σ-extension; (2) a is rejected; (3) a is undecided; (4) a is not accepted; (5) a is not rejected; and (6) a is not undecided. Each of these aims can be expressed as a disjunction of (some of) the three properties: a is accepted; a is rejected; and a is undecided. There are, in theory, two other aims. One is the empty disjunction, which represents an aim that can never be satisfied in a unitary semantics[2]. The other is the disjunction of all three properties, but this is always satisfied in any non-empty semantics. Of the six possible aims, the second three are the negations of the first three.

---

[1] Each player's move is a normal expansion [3].

[2] This possibility is not so outré in general: the stable semantics can be empty, and this is a sensible aim when a player wants to sabotage the game (that is, prevent any conclusion about the status of a).

4   Corruption in Strategic Argumentation

[13–15] presents a model of corruption within strategic argumentation. Two corrupt behaviours are defined: espionage, where one player (say P) violates the privacy of $\mathcal{A}_O$, and collusion, where P and O arrange for one of them to win, in violation of the best interests of the other's client.

Resistance to corruption under this model, adapting the notion of resistance of voting systems to manipulation [2], arises when the computational problems that arise when exploiting corruption and hiding the corruption from view require greater computational resources than playing the game honestly. This greater computational cost can act as a disincentive to corruption, since a player might be unable to exploit the results of corruption. The computational problems are formulated as decision problems, rather than functional problems, to avoid less familiar complexity classes.

The problem of verifying that an aim is satisfied by some state of strategic argumentation is a fundamental part of each move in a game, and of the exploitation of corrupt behaviour. However, its main interest is as a component of other problems.

The Aim Verification Problem

Instance An argumentation framework $(\mathcal{A}_{Com}, \gg)$, an argumentation semantics, and an aim.
Question Is the aim satisfied under the given semantics by the given argumentation framework?

The Desired Outcome problem [15] is the problem that a player must solve at each step of a strategic abstract argumentation game. It involves identifying that the player has a legal move.

The Desired Outcome Problem for P

Instance A split argumentation framework $(\mathcal{A}_{Com}, \mathcal{A}_P, \mathcal{A}_O, \gg)$ and a desired outcome for P.
Question Is there a set $I \subseteq \mathcal{A}_P$ such that P's desired outcome is achieved in the argumentation framework $(\mathcal{A}_{Com} \cup I, \gg)$?

It is not difficult to see that this problem can be solved by a non-deterministic algorithm with an oracle for the Aim Verification problem.

Playing strategic argumentation involves solving the desired outcome problem at each turn. We can formulate this as a deterministic polynomially bounded algorithm with an oracle for the player's desired outcome problem. Consequently, we can identify the complexity of playing strategic argumentation as PTIME^{DO}, where DO is the complexity of the desired outcome problem.

We now turn to corruption, and the computational problems that must be solved to exploit corruption.

In the case of collusion between P and O to ensure that (say) P wins, the players must arrange a sequence of moves that satisfy the rules of the game and leads to P winning. This sequence must give the appearance of being normal play. In particular, O cannot simply "give up" and fail to make a move – such behaviour would open her to charges of incompetence or corruption. Instead, she must exhaust her possible moves.

The Winning Sequence Problem for P

Instance A split argumentation framework $(\mathcal{A}_{Com}, \mathcal{A}_P, \mathcal{A}_O, \gg)$ and a desired outcome for P.
Question Is there a sequence of moves such that P wins?

This problem can be solved by a non-deterministic algorithm that guesses moves for P and O and uses oracles for the aim verification problem for P and O and the (complement of) the desired outcome problem for O.

In the case of espionage, one player, say P, illicitly learns her opponent's arguments $\mathcal{A}_O$ and desires a strategy that will ensure P wins, no matter what moves O makes. A *strategy* for P in a split argumentation framework $(\mathcal{A}_{Com}, \mathcal{A}_P, \mathcal{A}_O, \gg)$ is a function s_P from a set of common arguments and a set of playable arguments to the set of arguments to be played in the next move. A sequence of moves $S_1, T_1, S_2, T_2, \ldots$ resulting in common arguments $\mathcal{A}_{Com}^{P,1}, \mathcal{A}_{Com}^{O,1}, \mathcal{A}_{Com}^{P,2}, \mathcal{A}_{Com}^{O,2}, \ldots$ is *consistent with* a strategy s for P if, for every j, $S_{j+1} = s_P(\mathcal{A}_{Com}^{O,j}, \mathcal{A}_P)$. A strategy for P is *winning* if every valid sequence of moves consistent with the strategy is won by P.

The Winning Strategy Problem for P

Instance A split argumentation framework $(\mathcal{A}_{Com}, \mathcal{A}_P, \mathcal{A}_O, \gg)$ and a desired outcome for P.
Question Is there a winning strategy for P?

Strategic argumentation is said to be *resistant to collusion (espionage)* if the complexity of the Winning Sequence (Winning Strategy) problem is greater than the complexity of playing the strategic argumentation game, under the widely-believed complexity-theoretic assumption that the polynomial hierarchy does not collapse. In that case, the computational work needed to exploit the corrupt behaviour is greater than that required to simply play the argumentation game.

In the next sections we investigate the complexity of the problems defined above for the three semantics of abstract argumentation under investigation.

5 Strategic Argumentation Under the Ideal Semantics

Building on the work of [8], and the previous analysis of the aims under a unitary semantics, we have an upper bound on the complexity of aim verification under the ideal semantics.

Theorem 1. *The Aim Verification problem for P with any of the six strategic aims under the ideal semantics is in Θ_2^p. The corresponding aim verification problem for O is also in Θ_2^p.*

On the other hand, for the remaining problems we can give a tight analysis of their complexity. The ideal semantics is amenable to the techniques developed in [14] for other admissibility-based semantics. By combining constructions in [8, 14] we obtain the following results.

Theorem 2. *The Desired Outcome problem for P with any of the aims under the ideal semantics is Σ_2^p-complete. The same complexity holds for O.*

Theorem 3. *The Winning Sequence problem for P with any of the aims under the ideal semantics is Σ_3^p-complete.*

Theorem 4. *The Winning Strategy problem for P with any of the aims under the ideal semantics is PSPACE-complete.*

The complexity of honestly playing strategic argumentation is $\text{PTIME}^{\Sigma_2^p} = \Delta_3^p$, using Theorem 2. Consequently, we see that strategic argumentation under the ideal semantics is resistant to both collusion and espionage.

6   Argumentation Under the Naive Semantics

We can characterize the aims under the naive semantics.

Lemma 1. *Consider an argument a in an argumentation framework and the naive semantics.*

1. *a is in at least one naive extension iff a is not self-defeating*
2. *a is in every naive extension iff the only arguments that attack or are attacked by a are self-defeating*
3. *a is unrejected in every naive extension iff the only arguments that attack a are self-defeating*
4. *a is uncontested iff a is not self-defeating, and the only arguments that attack a are self-defeating*

It follows from this lemma that the aim verification problem under the naive semantics can be solved in polynomial time for each of the aims.

However, using the above characterization we find a more surprising result: under the naive semantics, if the players are honest, the outcome of the strategic argumentation game is determined by the initial split argumentation framework $\mathcal{A}$. There is no strategy involved.

For example, for the unrejected aim, if $\mathcal{A}_O$ contains an argument b that attacks a and is not self-defeating then O simply has to play b in order to win. Furthermore, additional arguments do not affect the existence of b, so it is sufficient for O to play her entire set of arguments $\mathcal{A}_O$ at her first move. She does not even need to know what the focal argument is!

Theorem 5. *Consider strategic argumentation under the naive semantics where P and O are honest. Suppose that P can make an initial move that includes the focal argument a.*

1. *If P has the existential aim, then P wins.*
2. *If P has the universal aim, then P wins iff all the arguments in $\mathcal{A}_O$ that conflict with a are self-defeating*
3. *If P has the unrejected or uncontested aim, then P wins iff all the arguments in $\mathcal{A}_O$ that attack a are self-defeating*

In each case, if P does not win then O wins, and if P cannot make an initial play then O wins.

The characterization for the uncontested aim is the same as the characterization for the unrejected aim because the assumption that P can make an initial move ensures that there is a naive extension where a is accepted.

Consequently, for each of these aims, the outcome is totally predictable. Any deviation from that result is a sign that one of the players is corrupt. Thus strategic argumentation under naive semantics is more than resistant to corruption: any collusive behaviour cannot be hidden, and the results of espionage cannot be used to affect the outcome. We say that strategic argumentation under the naive semantics is *proof against corruption* or *corruption-proof*.

This result contrasts markedly with a result of [16]. That work characterized strategy-proof games in similar argumentation games under the grounded semantics. That characterization suggests that only rarely is a game under the grounded semantics not strategic.

7   Strategic Argumentation Under the Stage Semantics

Unlike the ideal semantics, but like many other semantics [14], the complexity of the problems under the stage semantics varies with the players' strategic aim.

Proposition 1. *Consider the Aim Verification problem for P under the stage semantics.*

1. *The complexity for the existential aim is Σ_2^p-complete*
2. *The complexity for the universal aim is Π_2^p-complete*
3. *The complexity for the unrejected aim is Π_2^p-complete*
4. *The complexity for the uncontested aim is D_2^p-complete*

The complexity of Aim Verification for O, assuming O's aim is to prevent P from achieving her aim, is the complement of the complexity of Aim Verification for P.

As mentioned earlier, the Desired Outcome problem is in NP^{AV}, where AV is the Aim Verification problem. Building on constructions of [10,15], we can establish hardness results.

Theorem 6. *Consider the Desired Outcome problem for P under the stage semantics.*

1. *The problem with the existential aim is Σ_2^p-complete*
2. *The problem with the universal aim is Σ_3^p-complete*
3. *The problem with the unrejected aim is Σ_3^p-complete*
4. *The problem with the uncontested aim is Σ_3^p-complete*

Hence the complexity of P honestly playing strategic argumentation is Δ_3^p for the existential aim, and Δ_4^p for the other aims.

Theorem 7. *Consider the Winning Sequence problem for P under the stage semantics.*

1. *The problem with the existential aim is Σ_4^p-complete*
2. *The problem with the universal aim is Σ_3^p-complete*
3. *The problem with the unrejected aim is Σ_3^p-complete*
4. *The problem with the uncontested aim is Σ_4^p-complete*

Like all the admissibility-based semantics, and unlike the naive semantics, under the stage semantics the Winning Strategy problem is PSPACE-complete.

Theorem 8. *The Winning Strategy problem under the stage semantics is PSPACE-complete for each of the aims addressed in this paper.*

Thus we see that there is resistance to espionage under the stage semantics, but resistant to collusion only for the existential and uncontested aims.

Table 2. Resistance to collusion to ensure P wins, for several aims and semantics.

	Grounded	Preferred	Ideal	Naive	Stage
Existential	Resistant	Resistant	Resistant	Proof	Resistant
Universal	Resistant		Resistant	Proof	
Unrejected	Resistant		Resistant	Proof	
Uncontested	Resistant	Resistant	Resistant	Proof	Resistant

8   Conclusion

We have investigated the resistance to corruption of strategic argumentation under three semantics for argumentation. The naive semantics is proof against corruption, but only because there is no significant strategy involved. The ideal semantics is resistant to corruption, for each of the aims we studied. The resistance to collusion of the stage semantics varies according to the aim: It is resistant to collusion for the existential and uncontested aims, but not for the other aims. The stage semantics is also resistant to espionage for each of the aims. A summary of the results of this paper on resistance to collusion appears in Table 2.

There remain several argumentation semantics, including other extension-based semantics and ranking-based semantics [5], for which resistance to corruption has not been determined.

References

1. Baroni, P., Giacomin, M., Guida, G.: SCC-recursiveness: a general schema for argumentation semantics. Artif. Intell. **168**(1–2), 162–210 (2005)
2. Bartholdi, J.J., Tovey, C.A., Trick, M.A.: The computational difficulty of manipulating an election. Soc. Choice Welf. **6**(3), 227–241 (1989)
3. Baumann, R., Brewka, G.: Expanding argumentation frameworks: enforcing and monotonicity results. In: COMMA, pp. 75–86 (2010)
4. Bondarenko, A., Dung, P.M., Kowalski, R.A., Toni, F.: An abstract, argumentation-theoretic approach to default reasoning. Artif. Intell. **93**, 63–101 (1997)
5. Bonzon, E., Delobelle, J., Konieczny, S., Maudet, N.: A comparative study of ranking-based semantics for abstract argumentation. In: AAAI Conference on Artificial Intelligence, pp. 914–920 (2016)
6. Dung, P.M.: On the acceptability of arguments and its fundamental role in non-monotonic reasoning, logic programming and n-person games. Artif. Intell. **77**(2), 321–358 (1995)
7. Dung, P.M., Mancarella, P., Toni, F.: Computing ideal sceptical argumentation. Artif. Intell. **171**(10–15), 642–674 (2007). http://dx.doi.org/10.1016/j.artint.2007.05.003
8. Dunne, P.E.: The computational complexity of ideal semantics. Artif. Intell. **173**(18), 1559–1591 (2009). http://dx.doi.org/10.1016/j.artint.2009.09.001
9. Dunne, P.E., Dvořák, W., Woltran, S.: Parametric properties of ideal semantics. Artif. Intell. **202**, 1–28 (2013). http://dx.doi.org/10.1016/j.artint.2013.06.004
10. Dvořák, W., Woltran, S.: Complexity of semi-stable and stage semantics in argumentation frameworks. Inf. Process. Lett. **110**(11), 425–430 (2010). http://dx.doi.org/10.1016/j.ipl.2010.04.005
11. Governatori, G., Olivieri, F., Scannapieco, S., Rotolo, A., Cristani, M.: Strategic argumentation is NP-complete. In: Proceedings of European Conference on Artificial Intelligence, pp. 399–404 (2014)
12. Johnson, D.S.: A catalog of complexity classes. In: van Leeuwen, J. (ed.) Handbook of Theoretical Computer Science. Algorithms and Complexity, vol. A, pp. 67–161. Elsevier, Amsterdam (1990)
13. Maher, M.J.: Complexity of exploiting privacy violations in strategic argumentation. In: Proc. Pacific Rim International Conf. on Artificial Intelligence. pp. 523–535 (2014)
14. Maher, M.J.: Resistance to corruption of general strategic argumentation. In: Proceedings of International Conference Principles and Practice of Multi-Agent Systems, pp. 61–75 (2016)
15. Maher, M.J.: Resistance to corruption of strategic argumentation. In: AAAI Conference on Artificial Intelligence, pp. 1030–1036 (2016)

16. Rahwan, I., Larson, K., Tohmé, F.A.: A characterisation of strategy-proofness for grounded argumentation semantics. In: Boutilier, C. (ed.) IJCAI, pp. 251–256 (2009)
17. Verheij, B.: Two approaches to dialectical argumentation: admissible sets and argumentation stages. In: Proceedings of the 8th Dutch Conference on Artificial Intelligence, pp. 357–368 (1996)

Adaptive Multiagent Reinforcement Learning with Non-positive Regret

Duong D. Nguyen[1]([⊠]), Langford B. White[1], and Hung X. Nguyen[2]

[1] School of Electrical and Electronic Engineering, The University of Adelaide,
Adelaide, SA 5005, Australia
{duong.nguyen,lang.white}@adelaide.edu.au
[2] Teletraffic Research Centre, The University of Adelaide,
Adelaide, SA 5005, Australia
hung.nguyen@adelaide.edu.au

Abstract. We propose a novel adaptive reinforcement learning (RL) procedure for multi-agent non-cooperative repeated games. Most existing regret-based algorithms only use positive regrets in updating their learning rules. In this paper, we adopt both positive and negative regrets in reinforcement learning to improve its convergence behaviour. We prove theoretically that the empirical distribution of the joint play converges to the set of correlated equilibrium. Simulation results demonstrate that our proposed procedure outperforms the standard regret-based RL approach and a well-known state-of-the-art RL scheme in the literature in terms of both computational requirements and system fairness. Further experiments demonstrate that the performance of our solution is robust to variations in the total number of agents in the system; and that it can achieve markedly better fairness performance when compared to other relevant methods, especially in a large-scale multiagent system.

Keywords: Multiagent systems · Reinforcement Learning · Game theory · Correlated equilibrium · No regret

1 Introduction

Reinforcement learning (RL) is a popular adaptive procedure used in distributed system and has been widely studied in artificial intelligence (AI) research areas (for a survey on recent developed RL algorithms refer to [1]). A RL procedure [2–6] does not require the agents to know anything about the entire environment, except their local information. Each agent learns about the environment by observing its own payoffs. Overtime, using only this information, it can rationally choose the best course of actions to maximise its objective utility (payoff). Under mild conditions of finite payoffs and of stationary environment, an RL procedure is guaranteed to converge to a set of stable equilibria.

Despite this very attractive property, RL procedure applying in multiagent settings suffers from two well-known problems of slow convergence and of convergence to sub-optimal equilibrium points, especially in a distributed system with

© Springer International Publishing AG 2016
B.H. Kang and Q. Bai (Eds.): AI 2016, LNAI 9992, pp. 29–41, 2016.
DOI: 10.1007/978-3-319-50127-7_3

a very large number of agents [2]. Another challenge of RL-based algorithms is the inefficient of exploration. Since agents running RL procedure do not have a global knowledge of the whole system, they often require a high exploration times in order to converge to a stable equilibrium. In many application, these behaviours can result in undesirable outcomes [4,7].

This paper develops a new RL procedure that follows the regret-based principles [3,8] to overcome the disadvantage of slow speed and inefficient convergence of standard RL solutions. The notion of regret has been explored both in game theory and computer science [3,8–10]. Regret measures reflect how much worse in payoffs that an agent would experience if choosing other options instead of its current selection. In our problem formulation, we consider a multiagent non-cooperative repeated game with restricted information for the agents. Each agent only observe its own payoffs and know neither its payoff function nor the information on the other agents in the game. The goal of every agent is to guarantee no-regret in the long-term (average) payoffs.

Unlike most the existing regret-based algorithms that use only positive parts of regret measures to update the play probability and completely ignore negative regrets, we propose to use both positive and negative regrets to accelerate the convergence of the RL procedure. Our new approach is motivated by the observation that incorporation of negative regrets can help the agent to "explore" the environment more extensively as positive regrets decrease than the standard RL algorithm. The fact is that considering negative regrets can help agents make more "good" decisions by reducing unnecessary explorations on the actions that result in poor performances. Thus, more effective exploration has crucial impact on the convergence speed as well as the performance of the learning outcome.

However, since there is a negative impact on average performance by including more actions with negative regrets, our approach weights the impact of negative regrets on the probability distribution of actions in a manner that ensures (i) that actions with large (magnitude) negative regrets contribute less to the probability of choosing those actions than those with small (magnitude) negative regrets and (ii) that the contribution of negative regrets decreases to zero over time.

The main contribution of this paper are as follows:

1. *A Novel Adaptive Multiagent Reinforcement Learning Procedure:* We propose a novel fully distributed RL procedure that uses both positive and negative regret measures to improve convergence speed and fairness of the well-know regret-based RL procedure. We show that our solution is suitable for large-scale distributed multiagent systems.
2. *Our proof methodology:* We prove the convergence of our proposed procedure using differential inclusion (DI) technique. DI is a powerful theoretical framework that derived from the expected motion of a stochastic process. This paper demonstrates that the use of DI technique is particularly suitable to study the convergence behaviours of the regret based schemes and adaptive procedures in game theory, and provide a much more concise and extensible proof as compared to the classical approaches.

2 Background

This section reviews the background and notation used in this paper.

2.1 Game Model

We consider a game with A players denoted by the set $\{1, \ldots, A\}$ for some (finite) integer $A \geq 2$. Each player a has its set of actions (moves) $S_a = \{1, \ldots, m\}$, where m is the number of action of player a. The set of all possible moves is the Cartesian product $S = \Pi_{a=1}^{A} S_a$. We view the game from the point of view of player one. Let $\mathcal{I} = S_1$ denote the set moves of player one and $\mathcal{L} = S \setminus S_1$ the set of moves of all other players. Denote by X, the set of all probability mass functions (pmf) on $\mathcal{I}$ and Y the set of pmf on $\mathcal{L}$. Let Z denote the set of pmf on S, then $X \times Y$ is a subset of Z comprised of all pmf of the form $z = (x, y)$ where $x \in X$ and $y \in Y$, i.e. all pmf where the probability of the action of player one and the actions of all other players taken together, are statistically independent.

Let $U : S \rightarrow \mathbb{R}$ denote the payoff achieved by player one when the overall action taken by all players is $s \in S$. We represent a strategy in the form $s = (i, \ell)$ where i is the action of player one and ℓ is the action of all other players. We will consider the general formulation of game where users apply mixed strategies over the possible selection set S. Under randomised actions with overall probability (pmf) $z \in Z$, the payoff obtained by player one is defined by extending the domain of definition of U to Z according to

$$U(z) = \sum_{k \in S} z(k) U(k). \tag{1}$$

Notice that U is a linear function. The multiagent game model then can be denoted by $\mathcal{G} = (\mathcal{A}, (S_a)_{a \in \mathcal{A}}, (U_a)_{a \in \mathcal{A}})$.

2.2 Equilibrium States

In this paper, we are interested in a popular notion of rationality that generalises the Nash equilibrium called correlated equilibrium. It is an optimality concept introduced by Aumann [11]. It models possible correlation or co-ordination between players compared to the usual strategic equilibrium of Nash, where all players act independently. Correlated equilibrium is relevant to the probabilistic game, namely where strategies are determined probabilistically. Denote by ψ, a probability distribution defined in S, the ψ is said to be a correlated equilibrium for the game $\mathcal{G}$ if for every player $a \in \mathcal{A}$, and for every pair of action $j, k \in S^a$, it holds that

$$\sum_{s \in S : i = j} \psi(s)(U(k, \ell) - U(s)) \leq 0. \tag{2}$$

A correlated equilibrium results if each player does not benefit from choosing any other action, provided that all other players do likewise. When each player chooses their action independently of the other players, a correlated equilibrium is also a Nash equilibrium. We denote the set of correlated equilibria by CE.

2.3 Regret-Based Reinforcement Learning

A fully distributed procedure that can be used to reach the CE solution is the regret-based RL procedure [3]. The key idea of this method is to adjust the player's play probability proportional to the "regrets" for not having played other actions. Specifically, for any two actions $j \neq k \in \mathcal{I}$ at any time n, the regret of player one for not playing k is

$$[B_n]_{j,k} = \frac{1}{n} \sum_{t \leq n: i_t = j} U(k, \ell_t) - \frac{1}{n} \sum_{t \leq n: i_t = j} U(j, \ell_t). \tag{3}$$

This is the change in time average payoff that player one would have achieved if it substituted a given action j each time it was played in the past, with another action k. Since player one only knows his set of actions and his own payoffs, he cannot compute the first term. Thus, the regret in (3) needs to be replaced by an estimate that can be computed on the basic of the available information, as

$$[B_n]_{j,k} = \frac{1}{n} \sum_{t \leq n: i_t = k} \frac{p_t(j)}{p_t(k)} U(s_t) - \frac{1}{n} \sum_{t \leq n: i_t = j} U(s_t),$$

where, p_t denotes the play probabilities at time t, i.e., $p_t(k)$ is the probability of choosing k at time t and $U(s_t) = U(i_t, \ell_t)$ denotes the payoff at time t.

If $i_n = j$ is the action chosen by player one at time n, then the probability distribution that he chooses an action at time $n + 1$ is defined recursively as [3]

$$p_{n+1}(k) = \begin{cases} \left(1 - \dfrac{\delta}{n^\gamma}\right) \min\left\{ \dfrac{[B_n]_{j,k}^+}{\mu}, \dfrac{1}{m} \right\} + \dfrac{\delta}{n^\gamma} \dfrac{1}{m}, & k \neq j, \\ 1 - \sum_{k' \neq j} p_{n+1}(k'), & k = j, \end{cases} \tag{4}$$

with the initial play probabilities at $t = 1$ uniformly distributed over the set of possible actions; $\mu > 2mG$ is a constant, m is the cardinality of the set $\mathcal{I}$ and G is an upper bound on $|U(s)|$ for all $s \in \mathcal{S}$; $0 < \delta < 1$ and $0 < \gamma < 1/4$. We use the notation $[B_n]_{j,k}^+ := \max([B_n]_{j,k}, 0)$. By using $[B_n]_{j,k}^+$ in (4), the RL algorithm in [8] completely ignores negative regrets $[B_n]_{j,k} < 0$.

It is proven in [3] that if all players chooses their actions according to (4), the empirical distribution of all strategies played until time n, which is given by

$$z_n(s) = \frac{1}{n} \sum_{t=1}^{n} \mathbb{1}_{\{s_t = s\}},$$

converges almost surely as $t \to \infty$ to the CE set of the game $\mathcal{G}$. Note that this does not imply convergence to a specific point on CE set, but that the solution approaches the CE set.

The main drawback of this standard regret-based reinforcement learning procedure is that although guaranteeing convergence to the set of CE, it often requires long convergence time and sometime converges to an undesirable equilibrium (i.e. poor fairness). These issues motivate the reinforcement learning with non-positive regret in the next section.

3   Algorithm

In this section, we describe our proposed multiagent reinforcement procedure.

3.1   Reinforcement Learning with Non-positive Regret

The RL procedure in Sect. 2.3 does not use any negative regrets in determining the probability of plays. However, as discussed in Sect. 1, negative regrets contain information that could improve the performance of the learning procedure. We propose to complement the regret-based RL in [3] by taking into account additional negative regrets in updating the learning rule. To determine the probability distribution of its action at the next stage $n+1$, agent uses both its positive and negative parts of the time average regrets as follow

$$p_{n+1}(k) = \begin{cases} \delta_n \dfrac{1}{m}, & \text{if } k \neq j \text{ and } [B_n]_{j,k} = 0 \\[2ex] (1 - \delta_n) \, \dfrac{[B_n]_{j,k}^+}{\sum_k [B_n]_{j,k}^+} + \delta_n \dfrac{1}{m}, & \text{if } k \neq j \text{ and } [B_n]_{j,k} > 0 \\[2ex] (1 - \delta_n) \, \dfrac{1}{n^\alpha} \dfrac{\left([B_n]_{j,k}^-\right)^{-1}}{\sum_k \left([B_n]_{j,k}^-\right)^{-1}} + \delta_n \dfrac{1}{m}, & \text{if } k \neq j \text{ and } [B_n]_{j,k} < 0 \\[2ex] 1 - \sum_{k' \neq j} p_{n+1}(k'), & \text{if } k = j \end{cases} \tag{5}$$

where $\delta_n = \delta / n^\gamma$ for $0 < \delta \ll 1$ and $0 < \gamma < 1/2$; and $0 < \alpha \leq 1$. We use the notation $[B_n]_{j,k}^- := \min([B_n]_{j,k}, 0)$.

Our main insight here is that the negative regrets should be included in the update procedure to ensure that when n is small the algorithm keep exploring different solutions, including the solution that yields negative regret, to speed up the convergence. However, as the algorithm progresses, the negative regrets reduce to zero and the positive regrets become the dominant factors in determining the playing probabilities. We prove that our new RL algorithm converges almost surely to the CE set and show in simulations that this learning strategy provides very fast convergence toward equilibrium states.

3.2   Discussion

We discuss in detail here the major differences between our solution and the standard regret-based RL approach [3]. The main novelty in our approach is in the formula to update the play probability.

(a) Firstly, we do not use a constant proportional factor μ as in (4), but normalise the vector of regret to get a probability vector. The reason for doing this is to avoid being dependent on the appropriate choice of some arbitrarily large enough parameter μ. As discussed in [3], a higher value of μ results in a smaller probability of switching and thus leads to a slower speed of convergence.

(b) Secondly, in our solution, not only positive regrets but also negative values are contributing to the update procedure of the player. In particular, the play probability is proportional to the positive regret and is proportional to the inverse of the negative regret. This choice of play probability allows the action that yields larger positive regret to get a higher probability to be selected in the next state, while the action that yields larger negative regrets to receive a lower probability to be used in the future.

(c) Thirdly, in the standard approach, it is difficult to determine an appropriate $0 < \delta < 1$ in (4). A large δ will lead the convergence to a large distance from the CE set hence lead to lower total utility. However, small δ means to discourage the exploration processes, and agents tend to perform the same action and thus will cause slow convergence. In our proposed approach, the choice of δ is much simpler: we only need to set $0 < \delta \ll 1$. A much smaller value of δ not only improves the convergence rate but also reduces the instability properties caused by inaccurate estimates of regrets in the standard RL solution. The key point here is that δ can be taken smaller to still obtain a similar amount of "exploration" due to the inclusion of the negative regret terms.

(d) Lastly, the negative regrets vanish in the play probability as the time step goes to infinity due to the inclusion of $1/n^\alpha$ in the play probability for negative regrets in (5). This means that the agent no longer considers the selection that yields negative regret after sufficiently exploring all the potential options. Using negative regrets after the exploration phase would reduce the achievable payoffs.

3.3   Convergence Analysis

Theorem 1. *If an agent (i.e. player one) uses the proposed procedure, its time average conditional regret is guaranteed to approach the set of non-positive regrets in the payoff space almost surely, provided that other agents do likewise.*

We now provide a brief overview of the proof. We use the differential inclusion (DI) framework in [12] to prove our Theorem. DI is a generalisation of ordinary differential equation that is particularly suitable to study the asymptotic trajectory of the iterative process in game-theoretic learning, especially when the information available to a player is "restricted". Standard approach in game theory such as Blackwell's approachability theorem used in [3,8], however, cannot be trivially extended to prove the convergence of the proposed algorithm and will require a significant number of additional steps to handle the modifications of the play probabilities p_n. The use of DI technique yields a considerably simpler and shorter proof as compared to the classical approach in [3].

Proof. Let $C : Z \to \mathbb{R}^{m \times m}$ be defined by

$$[C(z)]_{j,k} = \sum_{\ell \in \mathcal{L}} z(j, \ell) \left(U(k, \ell) - U(j, \ell) \right),$$

which is the expected regret for player one when substituting action k for action j under the joint distribution z of actions. Suppose we consider player one playing some action i with probability one, then

$$[C(z^i)]_{j,k} = \sum_{\ell \in \mathcal{L}} \mathbb{1}_{\{i=j\}} \, y_\ell \left(U(k,\ell) - U(j,\ell) \right)$$

$$= \mathbb{1}_{\{i=j\}} \left(U(k,y) - U(j,y) \right).$$

Since player one cannot compute the first term as it only has access to the payoffs corresponding to actions it actually took, following [3], define an estimate of this term by

$$\tilde{U}(k,y) \, \mathbb{1}_{\{i=j\}} = \frac{p(j)}{p(k)} \, U(k,y) \, \mathbb{1}_{\{i=k\}}.$$

which is computed from the regrets associated with the alternative action k weighted proportional to the relative probabilities of player one choosing action j versus k when those actions were actually taken. The associated pseudo regret matrix at stage n is now

$$\tilde{C}_n(j,k) = \frac{p_n(j)}{p_n(k)} \, U(k,y_n) \, \mathbb{1}_{\{i_n=k\}} - U(j,y_n) \, \mathbb{1}_{\{i_n=j\}}.$$

Thus, we have

$$\mathbf{E}\left\{ \tilde{C}_n(j,k)|h_{n-1} \right\} = p_n(k)\frac{p_n(j)}{p_n(k)}U(k,y_n) - p_n(j)\,U(j,y_n)$$

$$= p_n(j)\left(U(k,y_n) - U(j,y_n) \right)$$

$$= \mathbf{E}\left\{ C_n(j,k)|h_{n-1} \right\},$$

where h_{n-1} is the action history of the game until stage $n-1$.

It can be seen that $C_n(j,k)$ and $\tilde{C}_n(j,k)$ are each bounded by $2mG/\delta_n$. The limit sets of the pair processes C_n and $\tilde{C}_n$ also coincide since they both have the same conditional expected values (see [3] for more details and discussions). Then Theorem 7.3 of [12] can be applied and thus the two processes exhibit the same asymptotic behaviour.

The average regret at stage n is thus a matrix B_n defined by

$$B_n(j,k) = \frac{1}{n} \sum_{t=1}^{n} \left[\frac{p_t(j)}{p_t(k)} U(k,y_t) \, \mathbb{1}_{\{i_t=k\}} - U(j,y_t) \, \mathbb{1}_{\{i_t=j\}} \right].$$

Hence, the discrete dynamics

$$\bar{B}_{n+1} - \bar{B}_n = \frac{1}{n+1} \left(B_{n+1} - \bar{B}_n \right)$$

is a discrete stochastic approximation of the DI

$$\dot{w} \in N(\mathbf{w}) - \mathbf{w} \quad (\text{with } w = B_n). \tag{6}$$

Now for $j \neq k$, define the matrix sequence

$$[M_n]_{j,k} = \begin{cases} 0, & \text{if } [B_n]_{j,k} = 0 \\ \dfrac{[B_n]_{j,k}^+}{\sum_k [B_n]_{j,k}^+}, & \text{if } [B_n]_{j,k} > 0 \\ \dfrac{1}{n^\alpha} \dfrac{\left([B_n]_{j,k}^-\right)^{-1}}{\sum_k \left([B_n]_{j,k}^-\right)^{-1}}, & \text{if } [B_n]_{j,k} < 0 \end{cases} \tag{7}$$

We set $[M_n]_{j,j} = 1 - \sum_{k \neq j} [M_n]_{j,k}$, which takes value in $[0,1]$ by virtue of (7). Thus M_n is a transition probability matrix on $\mathcal{S}$. So there is a probability vector μ_n such that $M_n^T \mu_n = \mu_n$.

The "non-positive regret set" $D^1 \subset \mathbb{R}^{m \times m}$ for player one is defined by

$$D^1 = \left\{ g \in \mathbb{C}^{m \times m} : g(j,k) \leq 0, \forall (j,k) \right\}.$$

Evidently, D^1 is a closed, convex subspace of $\mathbb{R}^{m \times m}$. Define the Lyapunov function $P(w) = \frac{1}{2}\|w\|^2$, with $\nabla P(w) = w$. Then P satisfies the following properties and thus is a potential function for D^1:

- P is continuously differentiable;
- $P(w) = 0 \Leftrightarrow w \in D^1$;
- $\langle \nabla P(w), w \rangle > 0$ for all $w \notin D^1$.

Let $\varphi : \mathbb{R}^{m \times m} \to 2^X$ given by

$$\varphi(w) = \begin{cases} (1 - \delta_n) \, \mu(w) + \dfrac{\delta_n}{m}, & w \notin D^1 \\ X, & w \in D^1 \end{cases} \tag{8}$$

where $\mu(w)$ denotes a probability vector computed from the matrix $w = B_n$ according to the process above. Define a correspondence N on $\mathbb{R}^{m \times m} \setminus D^1$ by

$$N(w) = C(\varphi(w) \times Y)$$

so that φ is N-adapted, which means $N(w)$ contains all resulting average regrets.

According to Lyapunov theory, to prove the approachability of w to D^1, we need then to show that for any $w \in \mathbb{R}^{m \times m} \setminus D^1$ and some positive constant β,

$$\frac{d}{dt} P(w) = \langle \nabla P(w), \dot{w} \rangle \in \langle \nabla P(w), N(w) - w \rangle \leq -\beta P(w),$$

meaning that we need the following result

$$\langle \nabla P(w), \theta - w \rangle \leq -\beta P(w)$$

for all $\theta \in N(w)$ and some constant $\beta > 0$ (see [12] for details).

Suppose $w \notin D^1$, let $\theta = \mathbf{E}\left\{\tilde{C}(\varphi(w), y)|h_{n-1}\right\}$, with $y \in Y$, which means

$$[\theta]_{j,k} = \varphi_j(w)\,(U(k, y) - U(j, y)).$$

Then consider

$$\langle \nabla P(w), \theta \rangle = \sum\nolimits_{j,k}^{m} \nabla P_{jk}(w)\,\varphi_j(w)\,(U(k, y) - U(j, y))$$

$$= (1 - \delta_n) \sum\nolimits_{j,k} \nabla P_{jk}(w)\,\mu_j(w)\,(U(k, y) - U(j, y))$$

$$+ \frac{\delta_n}{m} \sum\nolimits_{j,k} \nabla P_{jk}(w)\,(U(k, y) - U(j, y))$$

$$= (1 - \delta_n) \sum\nolimits_{j} U(j, y) \left(\sum\nolimits_{k} \mu_k(w)\,\nabla P_{kj}(w) - \mu_j(w) \sum\nolimits_{k} \nabla P_{jk}(w) \right)$$

$$+ \frac{\delta_n}{m} \sum\nolimits_{j,k} \nabla P_{jk}(w)\,(U(k, y) - U(j, y)). \tag{9}$$

In the second line we substituted for $\varphi_j(w)$ from (8), and in the last line we collected together all terms containing $U(j, y)$.

Let $\mu_j(w)$ be such an invariant measure. Suppose that for every $j = 1, \ldots, m$, it holds that

$$\mu_j(w) \sum\nolimits_{k} \nabla P_{jk}(w) = \sum\nolimits_{k} \mu_k(w)\,\nabla P_{kj}(w),$$

then the first term in the sum in (9) is equal to zero. Therefore, noting that the payoff function $|U(.)|$ is bounded by G, we obtain

$$\langle \nabla P(w), \theta \rangle = \frac{\delta_n}{m} \sum\nolimits_{j,k} \nabla P_{jk}(w)\,(U(k, y) - U(j, y))$$

$$\leq \|\nabla P(w)\| \frac{2G\delta_n}{m}. \tag{10}$$

Next, using $P(w) = \|w\|^2/2$ and $\nabla P(w) = w$, it can be show that

$$\langle \nabla P(w), w \rangle = \langle w, w \rangle = \|w\|^2 = 2P(w). \tag{11}$$

Therefore, it follows, using (10) and (11), that given $\epsilon > 0$, $\|w\| \geq \epsilon$, one can choose $\delta_n > 0$ small enough such that

$$\langle \nabla P(w), \theta - w \rangle = \langle \nabla P(w), \theta \rangle - \langle \nabla P(w), w \rangle$$

$$\leq \|\nabla P(w)\| \frac{2G\delta_n}{m} - 2P(w) \leq -P(w).$$

Consequently,

$$\frac{d}{dt} P(w(t)) \leq -P(w(t)),$$

so that

$$P(w(t)) \leq P(w(0))\,e^{-t}.$$

This implies that $P(w(t))$ goes to zero at exponential rate and the set D^1 is a global attractor for the DI (6). Hence, the time average regret B_n and its corresponding regret C_n will then approach D^1. This completes the proof.

Theorem 2. *If all agents follow the proposed procedure, the empirical distribution of joint play of all agents $z_n(s)$ converges almost surely as $t \to \infty$ to the set of correlated equilibria in the action space, for finite payoffs.*

Proof. The proof follows from how the "regret" measure is defined. Recall that

$$[C(z_n)]_{j,k} = \sum_{\ell \in \mathcal{L}} z_n(j, \ell_n) \left(U(k, \ell_n) - U(j, \ell_n) \right)$$
$$= \sum_{s_n \in S : i_n = j} z_n(s_n) \left(U(k, \ell_n) - U(s_n) \right),$$

where $s_n = (i_n, \ell_n)$ is the joint play made at stage n. On any convergent subsequence $\lim_{n \to \infty} z_n \to \Pi$, we get

$$\lim_{n \to \infty} [C(z_n)]_{j,k} = \sum_{s_n \in S : i_n = j} \Pi(s_n) \left(U(k, \ell_n) - U(s_n) \right) \le 0.$$

Next, comparing with the definition of CE as in (2) completes the proof.

4   Evaluation

In this section, we evaluate the performance of our proposed algorithm using a well-known multiagent Prisoner's Dilemma game (also known as the Tragedy of the Commons) [13]. Let's consider the game in which multiple agents ($A \ge 200$) compete for a limited common resource. Each agent has to make a binary decision – "yes" or "no" that models the agent decision of using the common resource or not, respectively. The agent that does not use the resource gets a fixed payoff. All the agents using the resource get the same payoff. Consequently, the more agents decided to use the resource, the smaller the obtainable payoff per agent; and when the number of agents sharing the resource is higher than a certain threshold, it is better for the others not to use the resource. A simple utility function reflecting this game can be expressed as follows:

$$U = \begin{cases} 1 & \text{if agent decision is "no"}, \\ 101 - \eta & \text{if agent decision is "yes"}. \end{cases}$$

with η being the number of agents making the same "yes" decision.

To evaluate the performance of our solution, we analyse the two metrics:

- Convergence speed (iterations): number of iterations to convergence. A fast convergence is preferable.
- System fairness index, which is derived as

$$J = \frac{\left(\sum_{a=1}^{A} x_a \right)^2}{A \times \sum_{a=1}^{A} x_a^2}, \tag{12}$$

where x_a is the average payoff of user a and A is the number of agents. Notes that $J = 1$ is the best fairness of the system, which guaranteeing the same payoff among the agents.

It can be seen that this game has two pure Nash equilibrium points when either 99 or 100 agents use the common resource. Any solutions that yield the average number of resource agents between 99 and 100 will be in the set of correlated equilibria. Among them, the equilibrium point when $\eta = 100$ provides the best system fairness since all agents will receive the same payoff of 1.

We compare our proposed algorithm with three other algorithms:

- CODIPAS-RL in [4]: Agents learn both the expected payoff and the strategies in order to make decisions. This is a popular state-of-the-art reinforcement learning algorithm and has been shown to be superior to the conventional RL scheme such as Q-learning.
- Regret-based RL in [3]: Agents update their play probability proportional only to the estimates of "positive regret" for not having played other options.
- Our proposed algorithm: Agents update their learning rules by considering both positive and negative regrets for not choosing other options.
- Exhaustive Search: A centralised controller with complete information of the game considers all possible associations involving all agents and assigns agents decisions in a way to maximise the system fairness. We use this algorithm as a benchmark since it leads to the highest performance in fairness.

Figures 1 and 2 show, respectively, the evolution of average number of agents using the resource (resource agents) and the system fairness index for the game with 200 agents. With the same initial probabilities, we observed that our proposed algorithm achieves the fastest convergence speed among all the reinforcement learning algorithms. Our algorithm converges to equilibrium states in a very small number of iterations (less than 150 iterations), where as it requires a longer time to converge for both CODIPAS-RL (up to 400 iterations) and

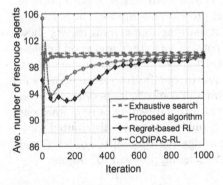

Fig. 1. Evolution of average number of resource agents by different algorithms.

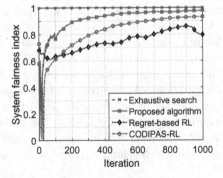

Fig. 2. Evolution of system fairness index by different algorithms.

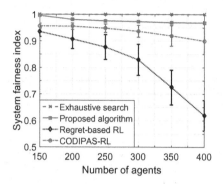

Fig. 3. Comparison of fairness between algorithms for the same number of iterations.

Regret-based RL (up to 900 iterations), especially the later. In fairness metric, our algorithm also leads to the highest system fairness index under the same number of iterations, as compared to the other RL schemes. The Regret-based RL scheme performs poorest due to its slow convergence speed.

To further study the impact of the total number of agents in the game on algorithms performance, we vary the agent number from 150 to 400 and measure the performances of all algorithms in fairness metric. The result is shown in Fig. 3. As we can see, proposed algorithm is quite robust in achieving system fairness to the change of the agent number. Increasing the total learning agents slightly reduces the system fairness index in our solution, but considerably bring down system fairness in other approaches, especially the Regret-based RL approach and when the total number of agents is very large.

5   Conclusion

We studied the problem of multiagent repeated games. We develop a fully distributed reinforcement learning procedure that takes advantage of both positive and negative regrets to speed up the learning process and improve the efficiency of the well-known regret-based reinforcement learning. Simulation results show that our solution is highly efficient with fast convergence speed and good fairness performance; and is more robust to the total number of agents in the system than other reinforcement learning algorithms. In our future research, we will study the rate of convergence of our algorithm and compare its performances on a broader set of benchmarks. As further work in this direction, a reinforcement learning framework for finding the global optimal solution in distributed multiagent system is still an open problem. Investigating the impact of irrational agents on the learning outcome is another challenging problem to consider.

Acknowledgment. This research is partially supported by the Australian Research Council Linkage Grant LP100200493.

References

1. Bhatnagar, S., Prasad, H., Prashanth, L.: Reinforcement learning. In: Bhatnagar, S., Prasad, H., Prashanth, L. (eds.) Stochastic Recursive Algorithms for Optimization, pp. 187–220. Springer, London (2013)
2. Sandholm, T.W., Crites, R.H.: Multiagent reinforcement learning in the iterated prisoner's dilemma. Biosystems **37**(1–2), 147–166 (1996)
3. Hart, S., Mas-Colell, A.: A reinforcement procedure leading to correlated equilibrium. In: Debreu, G., Neuefeind, W., Trockel, W. (eds.) Economics Essays, pp. 181–200. Springer, Berlin (2001). doi:10.1007/978-3-662-04623-4_12
4. Tembine, H.: Fully distributed learning for global optima. In: Distributed Strategic Learning for Wireless Engineers, pp. 317–359. CRC Press, UK (2012)
5. Kalathi, D., Borkar, V.S., Jain, R.: Blackwell's approachability in stackelberg stochastic games: a learning version. In: 53rd IEEE Conference on Decision and Control, pp. 4467–4472 (2014)
6. Bravo, M., Faure, M.: Reinforcement learning with restrictions on the action set. SIAM J. Control Optim. **53**(1), 287–312 (2015)
7. Borowski, H.P., Marden, J.R., Shamma, J.S.: Learning efficient correlated equilibria. In: 53rd IEEE Conference on Decision and Control, pp. 6836–6841 (2014)
8. Hart, S., Mas-Colell, A.: A simple adaptive procedure leading to correlated equilibrium. Econometrica **68**(5), 1127–1150 (2000)
9. Bowling, M.: Convergence and no-regret in multiagent learning. Adv. Neural Inf. Process. Syst. **17**, 209–216 (2005)
10. Cigler, L., Faltings, B.: Reaching correlated equilibria through multi-agent learning. In: The 10th International Conference on Autonomous Agents and Multiagent Systems, vol. 2, pp. 509–516 (2011)
11. Aumann, R.J.: Correlated equilibrium as an expression of Bayesian rationality. Econometrica **55**(1), 1.(1987)
12. Benam, M., Hofbauer, J., Sorin, S.: Stochastic approximations and differential inclusions, part II: applications. Math. OR **31**(4), 673–695 (2006)
13. Apt, K.R., Grädel, E.: A primer on strategic games. In: Apt, K.R., Grädel, E. (eds.) Lectures in Game Theory for Computer Scientists, pp. 1–37. Cambridge University Press (2011)

Composability in Cognitive Hierarchies

David Rajaratnam$^{(\boxtimes)}$, Bernhard Hengst, Maurice Pagnucco, Claude Sammut,
and Michael Thielscher

University of New South Wales, Sydney, Australia
{daver,bernhardh,morri,claude,mit}@cse.unsw.edu.au

Abstract. This paper develops a theory of node composition in a formal
framework for cognitive hierarchies. It builds on an existing model for
the integration of symbolic and sub-symbolic representations in a robot
architecture consisting of nodes in a hierarchy. A notion of behaviour
equivalence between cognitive hierarchies is introduced and node composition operators that preserve this equivalence are defined. This work
is significant in two respects. Firstly, it opens the way for a formal comparison between cognitive robotic systems. Secondly, composition, more
precisely *decomposition*, has been shown to be important to many fields,
and may therefore prove of practical benefit in the context of cognitive
systems.

1 Introduction

Building robots capable of interacting with humans and operating in unstructured environments is an open challenge. Such a robot needs to combine low-level
(sub-symbolic) sensor processing with high-level (symbolic) decision making.
Currently, there are two basic approaches to this challenge. The first is to combine both symbolic and sub-symbolic representations into a rich language, such
as Belle and Levesque's recent integration of probabilistic uncertainty into the
Situation Calculus for robot localisation [4]. The second approach is to separate
the required representations into interconnected sub-systems, for example using
an architecture such as the Robot Control System (RCS) [1].

The advantage of a combined representation is that it allows for formal systems analysis. On the other hand such languages can be difficult to master and
implement efficiently. For example, while GPU based processing is required for
scalable 3D Simultaneous Localisation and Mapping (SLAM) [15], incorporating
such specialised techniques into a rich expressive reasoner poses serious implementation challenges.

The alternative to the rich representation approach is to construct an architecture of loosely connected sub-systems. However, while having some practical
advantages, typical architecture based approaches either impose a particular representation language or are only described informally using text descriptions and
diagrams. The former lacks the flexibility required to combine disparate reasoning techniques, while the latter makes it difficult to formally establish properties
of such a system.

© Springer International Publishing AG 2016
B.H. Kang and Q. Bai (Eds.): AI 2016, LNAI 9992, pp. 42–55, 2016.
DOI: 10.1007/978-3-319-50127-7_4

The contribution of this paper is to take a step towards unifying these two approaches. We do this by extending a recently developed formal framework for integrating symbolic and sub-symbolic reasoning processes [9]. *In particular, we develop a formal notion of behaviour equivalence between two cognitive systems, and prove theoretical properties for node composability that preserves behaviour equivalence.*

The rest of this paper proceeds as follows. First, we introduce related work (Sect. 2) highlighting both the similarities and differences to existing research. We then provide a summary of the main features of the formal approach developed in [9]. In Sect. 4 we develop a notion of behaviour equivalence between cognitive hierarchies. Section 5 represents the main technical contribution of this paper, where node composition operators are introduced and formally shown to satisfy behaviour equivalence. It is further established that an arbitrary cognitive hierarchy can be reduced to a system consisting of a single node that nevertheless is formally behaviourally equivalent to the original. We conclude with a discussion and directions for future research.

2 Related Work

In this section we briefly highlight the different approaches to cognitive hierarchies, focusing on issues of formalisation and representation. We also briefly examine the varying notions of node composition that are prevalent in artificial intelligence (AI).

The use of hierarchies to build reasoning systems has a rich history in robotics and AI. While approaches vary, they can be broadly categorised into two groups: either they impose a particular representation language, or they are described informally.

The category of fixed language approaches include the popular cognitive architectures SOAR [14] and ACT-R [3], both of which employ symbolic based representations. Other examples include the logic-based Nilsson's triple-tower architecture [16], and the robot focused *dual dynamic* (DD) hierarchical behaviour system [12] formalised in terms of differential equations. The weakness of fixed language systems is that the representation language imposes limits on the type of problems that can be expressed. For example, the languages of SOAR and ACT-R are not easily applicable for representing the probabilistic uncertainty of robot localisation, while the differential equations used in the DD system cannot easily be adapted to perform traditional logical inference.

The alternative has been to describe architectures in an informal manner. The influential Robot Control System (RCS) model [1] consists only of textual descriptions and diagrams. While the seminal *subsumption architecture* [7] uses finite state machines for representing individual levels within the hierarchy [6], nevertheless the integration of these levels into an overall system is purely informal. An informal component integration is also true of the recent SOAR extension for dealing with sensor data [13]. Unfortunately, the informal approach has a number of weaknesses. In particular, it provides no basis on which to prove properties of the system as a whole, and there is no clear distinction between the architectural properties of an informal system and properties that arise from arbitrary implementation decisions.

Closely related to the construction of hierarchies is the notion of problem decomposition, where its benefits were evident in the context of hierarchical planning [17]. The more general notion of *factored planning* has been associated with the study of how to decompose a domain into sub-domains (know as *factors*) that can be solved independently and for which the solutions to each sub-domain can be combined [2,5]. This work is generalised in the field of general game playing, where a game is decomposed into sub-games that are solved independently, ensuring that the combined solution solves the original game [8]. As the dual to decomposition, *behaviour composition* [10] considers the problem of synthesising target behaviours from existing behaviours. However, the primitive behaviours in this case operate over a common transition system (i.e., the encoding of the environment) where as, composition in general-game playing considers composition over different transition systems. Our work is closer in intention to that of sub-game composition in general game playing than it is to behaviour composition.

3   Formal Architecture

In this section we summarise the formalisation developed in [9]. It avoids the weakness of existing formal models (discussed in Sect. 2) by adopting a meta-theoretic approach to cognitive hierarchies. In essence, it formalises the interaction between cognitive nodes in a hierarchy while making no commitments about the representation and reasoning mechanism within individual nodes. As such this framework complements, rather than competes with, existing hierarchical approaches.

3.1   Nodes

At the most basic level a cognitive hierarchy consists of a set of nodes. Nodes are tasked to achieve a goal or maximise future value. They have two primary functions: world-modelling and behaviour-generation. World-modelling is achieve through the maintenance of a *belief state*. A belief state is updated from lower-level nodes through *sensing*, which is the process of extracting *observations*. Behaviour-generation is achieved through a set of *policies*, where a policy maps a state to a set of actions and the current policy is determined by the actions of higher level nodes. A selected set of actions can also update a belief state, often referred to as an *expectation update*.

Definition 1. *A cognitive language is a tuple $\mathcal{L} = (\mathcal{S}, \mathcal{A}, \mathcal{T}, \mathcal{O})$, where $\mathcal{S}$ is a set of belief states, $\mathcal{A}$ is a set of actions, $\mathcal{T}$ is a set of task parameters, and $\mathcal{O}$ is a set of observations. A cognitive node is a tuple $N = (\mathcal{L}, \Pi, \lambda, \tau, \gamma, s^0, \pi^0)$ s.t:*

– $\mathcal{L}$ is the cognitive language for N, with initial belief state $s^0 \in \mathcal{S}$.

- Π a set of policies such that for all $\pi \in \Pi$, $\pi : S \rightarrow 2^A$, with initial policy $\pi^0 \in \Pi$.
- A policy selection function $\lambda : 2^T \rightarrow \Pi$, s.t. $\lambda(\{\}) = \pi^0$.
- A observation update operator $\tau : 2^O \times S \rightarrow S$.
- An action update operator $\gamma : 2^A \times S \rightarrow S$.

Note that Definition 1 provides a very abstract characterisation that captures only the interaction between nodes. No restrictions are made on internal representation and reasoning mechanisms, allowing, for example, a symbolic node to be created to encode a logical planner, or a stochastic node to encode a Kalman filter for robot localisation.

3.2   Cognitive Hierarchy

Nodes in the model are interlinked in a hierarchy, where the lowest level node corresponds to the interface to the real world, consisting of physical sensors and actuators (Fig. 1). Sensing data is passed up the *abstraction hierarchy*, while actions are sent down the hierarchy, eventually resulting in physical actions.

Definition 2. *A cognitive hierarchy is a tuple $H = (\mathcal{N}, N_0, F)$ s.t:*

- $\mathcal{N}$ is a set of cognitive nodes and $N_0 \in \mathcal{N}$ is a distinguished node corresponding to the external environment.
- F is a set of function pairs $\langle \phi_{i,j}, \psi_{j,i} \rangle \in F$ that connect nodes $N_i, N_j \in \mathcal{N}$ where:
  - $\phi_{i,j} : S_i \rightarrow 2^{O_j}$ is a sensing function, and
  - $\psi_{j,i} : 2^{A_j} \rightarrow 2^{T_i}$ is a task parameter function.
- Sensing graph: each $\phi_{i,j}$ represents an edge from node N_i to N_j and forms a directed acyclic graph (DAG) with N_0 as the unique source node of the graph.
- Action graph: the set of task parameter functions forms a converse to the sensing graph such that N_0 is the unique sink node of the graph.

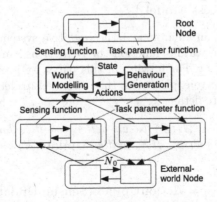

Fig. 1. An example cognitive hierarchy highlighting the sensing and action graphs.

Definition 2 establishes how the sensing and task parameter functions connect nodes in the hierarchy; forming sensing and action graphs. Sensing functions extract observations from lower-level nodes, while task parameter functions translate higher-level actions into task parameters for lower level nodes. N_0 models the external world.

3.3   Active Cognitive Hierarchy

The notions of a cognitive node and cognitive hierarchy capture only static components of a system and require additional details to model dynamic operational behaviour.

Definition 3. *An active cognitive node is a tuple $Q = (N, s, \pi, a)$ where: (1) N is a cognitive node with $\mathcal{S}$, Π, and $\mathcal{A}$ being the corresponding set of belief states, set of policies, set of actions respectively, (2) $s \in \mathcal{S}$ is the current belief state, $\pi \in \Pi$ is the current policy, and $a \in 2^{\mathcal{A}}$ is the current set of actions.*

Essentially an active cognitive node couples a (static) cognitive node with some dynamic information; in particular the current belief state, policy and set of actions.

Definition 4. *An active cognitive hierarchy is a tuple $\mathcal{X} = (H, \mathcal{Q})$ where H is a cognitive hierarchy with set of cognitive nodes $\mathcal{N}$ such that for each $N \in \mathcal{N}$ there is a corresponding active cognitive node $Q = (N, s, \pi, a) \in \mathcal{Q}$ and vice-versa.*

The active cognitive hierarchy captures the dynamic state of the entire hierarchy at some instance in time, where each node in the hierarchy has a current state, current policy and current set of actions. An *initial active cognitive hierarchy* specifies initial configurations for every node in the hierarchy initialised with each node's initial belief state and initial policy, as well as an empty set of actions.

3.4   Cognitive Process Model

In order to operate in an environment the cognitive system requires a *process model* that applies the various functions of Definitions 1 and 2 to update the active cognitive hierarchy. For brevity, in this paper we provide only the main definition and an intuitive explanation; the interested reader is instead referred to [9] for more complete details.

Definition 5. *Let $\mathcal{X} = (H, \mathcal{Q})$ be an active cognitive hierarchy with $H = (\mathcal{N}, N_0, F)$. The process update of $\mathcal{X}$, written $\mathbf{Update}(\mathcal{X})$, is an active cognitive hierarchy:*

$$\mathbf{Update}(\mathcal{X}) \stackrel{\mathrm{def}}{=} \mathbf{ActionUpdate}(\mathbf{SensingUpdate}(\mathcal{X}))$$

The functions **SensingUpdate** and **ActionUpdate** are intuitively easy to understand. **SensingUpdate** passes sensing information up the cognitive hierarchy, successively updating the belief states of the nodes in the hierarchy, while **ActionUpdate** passes actions down the hierarchy, causing the active policies and actions to be changed. These actions ultimately lead to task parameter changes for node N_0, which correspond to signals to the robot actuators. Crucially, both **SensingUpdate** and **ActionUpdate** update the nodes according to the partial ordering specified by the sensing and action graphs respectively. This guarantees that the functions are well-defined, since any sequence of node updates that satisfies the partial ordering will produce the same result.

The above formalism completes the technical background to this paper as developed in [9]. The remainder of this paper extends the formalism to allow for the specification of properties of behaviour equivalence and node composition.

4   Behaviour Equivalence

In this section we introduce the notion of behaviour equivalence between two cognitive hierarchies. This first involves establishing what it means for two cognitive systems to be behaviourally equivalent (with respect to the real-world) at some instant in time.

Definition 6. *Let $\mathcal{X}_i = (H_i, \mathcal{Q}_i)$ and $\mathcal{X}_j = (H_j, \mathcal{Q}_j)$ be two active cognitive hierarchies with $H_i = (\mathcal{N}_i, N_0, F_i)$ and $H_j = (\mathcal{N}_j, N_0, F_j)$. Then $\mathcal{X}_i$ and $\mathcal{X}_j$ are behaviour equivalent iff $T_i = T_j$, where:*

$$T_i = \bigcup \{\psi_{x,0}(s_x) \mid \langle \phi_{0,x}, \psi_{x,0} \rangle \in F_i \text{ for each } Q_x = (N_x, s_x, \pi_x, a_x) \in \mathcal{Q}_i\}$$
$$T_j = \bigcup \{\psi_{x,0}(s_x) \mid \langle \phi_{0,x}, \psi_{x,0} \rangle \in F_j \text{ for each } Q_x = (N_x, s_x, \pi_x, a_x) \in \mathcal{Q}_j\}$$

Definition 6 formalises the intuition that behaviour equivalence concerns how two cognitive systems behave with respect to the external world (i.e., N_0). The sets of task parameters T_i and T_j represent the actions that are carried out by the two cognitive systems. If the two sets are the same then the two systems perform the same actions.

Behaviour equivalence here is a theoretical notion, in effect saying that if we were able to observe the behaviour of the two systems operating at the same instant in time we would not be able to distinguish between them. This may not be possible to verify in practice, since we cannot actually replay time. Nevertheless the conceptual meaning is clear. We now consider how two active cognitive systems evolve *over* time.

Definition 7. *Let H_i and H_j be two cognitive hierarchies and $\mathcal{X}_i^0 = (H_i, \mathcal{Q}_i^0)$ and $\mathcal{X}_j^0 = (H_j, \mathcal{Q}_j^0)$ be their corresponding initial active cognitive hierarchies, such that N_0 is the distinguished node for both H_i and H_j. Then H_i and H_j are behaviour equivalent with respect to a cognitive process model Γ iff for every pair of sequences $[\mathcal{X}_i^0, \ldots, \mathcal{X}_i^n]$ and $[\mathcal{X}_j^0, \ldots, \mathcal{X}_j^n]$, where $\mathcal{X}_x^{k+1} = \Gamma(\mathcal{X}_x^k)$, each corresponding pair of active cognitive hierarchies $\mathcal{X}_i^k$ and $\mathcal{X}_j^k$ ($0 \leq k \leq n$) are behaviour equivalent.*

Intuitively, Definition 7 establishes that two behaviourally equivalent cognitive hierarchies will be interchangeable in their operations over time. It is worth noting that because this property deals with operational systems as they evolve, therefore it is necessary to reference the process model that is used for updating the active nodes.

5   Node Composition

The idea behind node composition is to replace a pair of nodes in a cognitive hierarchy with a single node that nevertheless encapsulates the behavioural properties of the original. Formally this takes the form of *composition operators*. We now introduce *parallel* and *sequential* composition operators and establish their foundational properties.

5.1   Parallel and Sequential Composition Operators

As a technical preliminary we first provide the following convenience functions that will be used in the definitions to follow. Given a pair $p = \langle x, y \rangle$, let $fst(p) = x$ and $snd(p) = y$. Furthermore, we overload these definitions in the obvious way when applied to a set of pairs. Namely, $fst(S) = \{x | \langle x, y \rangle \in S\}$ and $snd(S) = \{y | \langle x, y \rangle \in S\}$.

The parallel and sequential composition operators have a number of commonalities which we capture by introducing a *partial* composition operator. The definition of this operator is somewhat lengthy and notationally heavy, however it is intuitively straightforward. Essentially, a new hierarchy is created that is identical to the old except that the two nodes being composed are replaced with a single new node. The representational aspects of the new node (i.e., the set of states, actions, task parameters, and observations) needs to combine the representational aspects of the original. The most lengthy aspect of the definition involves updating the sensing and task parameter functions that connect nodes. Functions in the original hierarchy that connect to at least one of the original two nodes must be integrated into a new function for the composed node.

Definition 8 (Partial Composition).   *Let $H = (\mathcal{N}, N_0, F)$ be a cognitive hierarchy, and $N_i = (\mathcal{L}_i, \Pi_i, \lambda_i, \tau_i, \gamma_i, s_i^0, \pi_i^0)$ and $N_j = (\mathcal{L}_j, \Pi_j, \lambda_j, \tau_j, \gamma_j, s_j^0, \pi_j^0)$ be two cognitive nodes in H, where $\mathcal{L}_i = (\mathcal{S}_i, \mathcal{A}_i, \mathcal{T}_i, \mathcal{O}_i)$ and $\mathcal{L}_j = (\mathcal{S}_j, \mathcal{A}_j, \mathcal{T}_j, \mathcal{O}_j)$ are the cognitive languages for N_i and N_j respectively. The partial composition of N_i and N_j with respect to H is a cognitive hierarchy $H' = (\mathcal{N}', N_0, F')$ s.t:*

- *$\mathcal{N}' = \mathcal{N} \setminus \{N_i, N_j\} \cup \{N_x\}$ where $N_x = (\mathcal{L}_x, \Pi_x, \lambda_x, \tau_x, \gamma_x, s_x^0, \pi_x^0)$ where:*
  - *$\mathcal{L}_x = (\mathcal{S}_x, \mathcal{A}_x, \mathcal{T}_x, \mathcal{O}_x)$,*
  - *$\mathcal{S}_x = \mathcal{S}_i \times \mathcal{S}_j$, with initial state $s_x = \langle s_i, s_j \rangle$.*
  - *$\mathcal{A}_x = 2^{\mathcal{A}_i} \times 2^{\mathcal{A}_j}$, $\mathcal{T}_x = 2^{\mathcal{T}_i} \times 2^{\mathcal{T}_j}$, $\mathcal{O}_x = 2^{\mathcal{O}_i} \times 2^{\mathcal{O}_j}$*

  - *$\gamma_x(a_x, \langle s_i, s_j \rangle) \stackrel{\text{def}}{=} \begin{cases} \langle s_i, s_j \rangle & \text{if } |a_x| \neq 1, \\ \langle \gamma_i(a_i, s_i), \gamma_j(a_j, s_j) \rangle & \text{otherwise, for } a_x = \{\langle a_i, a_j \rangle\}. \end{cases}$*

- $F' = F \setminus \{\langle \phi_{k,l}, \psi_{l,k} \rangle \in F \mid k \in \{i,j\} \ \text{ or } \ l \in \{i,j\}\} \ \cup \bigcup_{k=1,\dots,6} F_k \quad \text{where:}$

- $F_1 = \{\langle \phi_{w,x}, \psi_{x,w} \rangle \mid w \text{ is a node}; \ \langle \phi_{w,i}, \psi_{i,w} \rangle \in F; \ \langle \phi_{w,j}, \psi_{j,w} \rangle \in F\} \ s.t:$

  $\phi_{w,x}(s_w) \overset{\text{def}}{=} \{\langle \phi_{w,i}(s_w), \phi_{w,j}(s_w) \rangle\}$

  $\psi_{x,w}(a_x) \overset{\text{def}}{=} \begin{cases} \{\} & \text{if } |a_x| \neq 1, \\ \psi_{i,w}(a_i) \cup \psi_{j,w}(a_j) & \text{otherwise, where } a_x = \{\langle a_i, a_j \rangle\}. \end{cases}$

- $F_2 = \{\langle \phi_{w,x}, \psi_{x,w} \rangle \mid w \text{ is a node}; \ \langle \phi_{w,i}, \psi_{i,w} \rangle \in F; \ \langle \phi_{w,j}, \psi_{j,w} \rangle \notin F\} \ s.t:$

  $\phi_{w,x}(s_w) \overset{\text{def}}{=} \{\langle \phi_{w,i}(s_w), \{\} \rangle\}$

  $\psi_{x,w}(a_s) \overset{\text{def}}{=} \begin{cases} \{\} & \text{if } |a_x| \neq 1, \\ \psi_{i,w}(a_i) & \text{otherwise, where } a_x = \{\langle a_i, a_j \rangle\}. \end{cases}$

- $F_3 = \{\langle \phi_{w,x}, \psi_{x,w} \rangle \mid w \text{ is a node}; \ \langle \phi_{w,i}, \psi_{i,w} \rangle \notin F; \ \langle \phi_{w,j}, \psi_{j,w} \rangle \in F\} \ s.t:$

  $\phi_{w,x}(s_w) \overset{\text{def}}{=} \{\langle \{\}, \phi_{w,j}(s_w) \rangle\}$

  $\psi_{x,w}(a_x) \overset{\text{def}}{=} \begin{cases} \{\} & \text{if } |a_x| \neq 1, \\ \psi_{j,w}(a_j) & \text{otherwise, where } a_x = \{\langle a_i, a_j \rangle\}. \end{cases}$

- $F_4 = \{\langle \phi_{x,w}, \psi_{w,x} \rangle \mid w \text{ is a node}; \ \langle \phi_{i,w}, \psi_{w,i} \rangle \in F; \ \langle \phi_{j,w}, \psi_{w,j} \rangle \in F\} \ s.t:$

  $\phi_{x,w}(\langle s_i, s_j \rangle) \overset{\text{def}}{=} \phi_{i,w}(s_i) \cup \phi_{j,w}(s_j)$

  $\psi_{w,x}(a_w) \overset{\text{def}}{=} \{\langle \psi_{w,i}(a_w), \psi_{w,j}(a_w) \rangle\}.$

- $F_5 = \{\langle \phi_{x,w}, \psi_{w,x} \rangle \mid w \text{ is a node}; \ \langle \phi_{i,w}, \psi_{w,i} \rangle \in F; \ \langle \phi_{j,w}, \psi_{w,j} \rangle \notin F\} \ s.t:$

  $\phi_{x,w}(\langle s_i, s_j \rangle) \overset{\text{def}}{=} \phi_{i,w}(s_i)$

  $\psi_{w,x}(a_w) \overset{\text{def}}{=} \{\langle \psi_{w,i}(a_w), \{\} \rangle\}.$

- $F_6 = \{\langle \phi_{x,w}, \psi_{w,x} \rangle \mid w \text{ is a node}; \ \langle \phi_{i,w}, \psi_{w,i} \rangle \notin F; \ \langle \phi_{j,w}, \psi_{w,j} \rangle \in F\} \ s.t:$

  $\phi_{x,w}(\langle s_i, s_j \rangle) \overset{\text{def}}{=} \phi_{j,w}(s_j)$

  $\psi_{w,x}(a_w) \overset{\text{def}}{=} \{\langle \{\}, \psi_{w,j}(a_w) \rangle\}$

It is worth highlighting some significant aspects of Definition 8. Firstly, the set of states for the composed node consists of the cross-product of the set of states for the original nodes. Hence exactly the same combination of states that can be represented by the two nodes in the original hierarchy can also be represented by the new node. Secondly, the set of actions (resp. task parameters and observations) of the new node consist of the cross-product of the *powerset* of the same property in the original two nodes. This is necessary to allow for the arbitrary combination of subsets of the original set of actions (resp. task parameters and observations). Finally, some of the connections in the old hierarchy need to be preserved while others need to be merged. $F_1, \dots, F_6$ in the definition cover the cases where connector functions need to be merged. The modified functions are defined in terms of the originals in order to preserve their behaviour, but are modified to deal with the merged representation of the new node.

We now turn to completing the definitions of the different composition operators. The more straightforward case is the parallel composition operator. This captures the situation where the two nodes being composed are not reachable from each other in the cognitive hierarchy (e.g., nodes N_4 and N_5 in Fig. 2(a)).

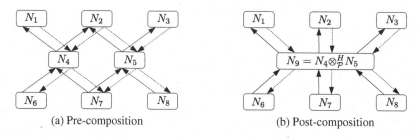

(a) Pre-composition   (b) Post-composition

Fig. 2. Parallel composition of nodes N_4 and N_5 in a cognitive hierarchy. Solid lines represent the sensing functions between nodes and dotted-lines represent the task parameter functions.

Definition 9 (Parallel Composition). *Let $H = (\mathcal{N}, N_0, F)$ be a cognitive hierarchy, and $N_i = (\mathcal{L}_i, \Pi_i, \lambda_i, \tau_i, \gamma_i, s_i^0, \pi_i^0)$ and $N_j = (\mathcal{L}_j, \Pi_j, \lambda_j, \tau_j, \gamma_j, s_j^0, \pi_j^0)$, where $\mathcal{L}_i = (\mathcal{S}_i, \mathcal{A}_i, \mathcal{T}_i, \mathcal{O}_i)$ and $\mathcal{L}_j = (\mathcal{S}_j, \mathcal{A}_j, \mathcal{T}_j, \mathcal{O}_j)$, be cognitive nodes in $\mathcal{N}$, that are distinct from each other and N_0, and furthermore that $N_i \not\leq N_j$ and $N_j \not\leq N_i$ for the partial order $\leq$ induced by the sensing graph of H. The parallel composition of N_i and N_j with respect to H (written $N_i \otimes_P^H N_j$) is a partial composition operator (Definition 8) with the additional requirements that for the composed node $N_x = (\mathcal{L}_x, \Pi_x, \lambda_x, \tau_x, \gamma_x, s_x^0, \pi_x^0)$:*

- *$\Pi_x = \{\pi_i \diamond \pi_j : \mathcal{S}_x \to 2^{\mathcal{A}_x} \mid \pi_i \in \Pi_i \text{ and } \pi_j \in \Pi_j\}$, and $\pi_x^0 = \pi_i^0 \diamond \pi_j^0$.*
- *$\lambda_x(t_x) \overset{\text{def}}{=} \lambda_i(T_i) \diamond \lambda_i(T_j)$, for any $t_x \subseteq \mathcal{T}_x$,*
  and where $T_i = \bigcup_{T \in fst(t_x)} T$ and $T_j = \bigcup_{T \in snd(t_x)} T$.
- *$\tau_x(o_x, \langle s_i, s_j \rangle) \overset{\text{def}}{=} \langle \tau_i(O_i, s_i), \tau_j(O_j, s_j) \rangle$, for any $o_x \subseteq \mathcal{O}_x, s_i \in \mathcal{S}_i, s_j \in \mathcal{S}_j$,*
  and where $O_i = \bigcup_{O \in fst(o_x)} O$ and $O_j = \bigcup_{O \in snd(o_x)} O$.

where the composition $\pi_x = \pi_i \diamond \pi_j$ is defined for $\pi_i \in \Pi_i$ and $\pi_j \in \Pi_j$ as:

$$\pi_x(\langle s_i, s_j \rangle) \overset{\text{def}}{=} \{\langle \pi_i(s_i), \pi_j(s_j) \rangle\}, \text{ for any } s_i \in \mathcal{S}_i, s_j \in \mathcal{S}_j.$$

There are a number of aspects to the parallel composition operator (Definition 9) that are worth highlighting. Firstly, the restriction that the two nodes being composed are not comparable under the sensing (or action) graph's partial ordering means that the two nodes are unrelated and the result (sensing or actions) of one node will not influence the results of the other node. This makes the definition of various node functions simpler. Secondly, the set of policies of the composed node are the cross-product of the policies of the two original nodes, and each composed policy simply applies the underlying policy to the appropriate component.

We now consider the more complicated case where the nodes to be composed are related by the partial ordering and therefore the results of one node effects the behaviour of the other node (e.g., nodes N_4 and N_5 in Fig. 3(a)).

Definition 10 (Sequential Composition). *Let $H = (\mathcal{N}, N_0, F)$ be a cognitive hierarchy, and $N_i = (\mathcal{L}_i, \Pi_i, \lambda_i, \tau_i, \gamma_i, s_i^0, \pi_i^0)$ and $N_j = (\mathcal{L}_j, \Pi_j, \lambda_j, \tau_j, \gamma_j,$*

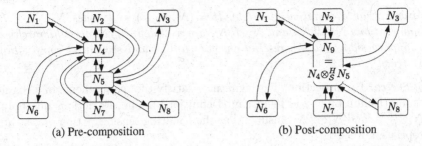

(a) Pre-composition   (b) Post-composition

Fig. 3. Sequential composition of nodes N_4 and N_5 in a cognitive hierarchy. Solid lines represent the sensing functions between nodes and dotted-lines represent the task parameter functions.

s_j^0, π_j^0), where $\mathcal{L}_i = (\mathcal{S}_i, \mathcal{A}_i, \mathcal{T}_i, \mathcal{O}_i)$ and $\mathcal{L}_j = (\mathcal{S}_j, \mathcal{A}_j, \mathcal{T}_j, \mathcal{O}_j)$, be cognitive nodes in $\mathcal{N}$, that are distinct from each other and N_0, and furthermore, $N_j \leq N_i$ for the partial order $\leq$ induced by the sensing graph of H and there does not exist a distinct node N_k such that $N_j \leq N_k$ and $N_k \leq N_i$. The sequential composition of N_i and N_j with respect to H (written $N_i \otimes_S^H N_j$) is a partial composition operator (Definition 8) with the additional requirements that for the composed node $N_x = (\mathcal{L}_x, \Pi_x, \lambda_x, \tau_x, \gamma_x, s_x^0, \pi_x^0)$:

- $\Pi_x = \{\pi_i \diamond t_j : \mathcal{S}_x \to 2^{\mathcal{A}_x} | t_j \subseteq \mathcal{T}_j \text{ and } \pi_i \in \Pi_i\}$, and $\pi_x^0 = \pi_i^0 \diamond \{\}$.
- $\lambda_x(t_x) \stackrel{\text{def}}{=} \lambda_i(T_i) \diamond T_j$, for any $t_x \subseteq \mathcal{T}_x$,
  and where $T_i = \bigcup_{T \in fst(t_x)} T$ and $T_j = \bigcup_{T \in snd(t_x)} T$.
- $\tau_x(o_x, \langle s_i, s_j \rangle) \stackrel{\text{def}}{=} \langle \tau_i(O_i \cup \phi_{j,i}(s_j'), s_i), s_j' \rangle$, for any $o_x \subseteq \mathcal{O}_x, s_i \in \mathcal{S}_i, s_j \in \mathcal{S}_j$,
  and where $O_i = \bigcup_{O \in fst(o_x)} O$,
  and $s_j' = \tau_j(O_j, s_j)$ such that $O_j = \bigcup_{O \in snd(o_x)} O$.

where the composition $\pi_x = \pi_i \diamond t_j$ is defined for $\pi_i \in \Pi_i$ and $t_j \subseteq \mathcal{T}_j$ as:

$$\pi_x(\langle s_i, s_j \rangle) \stackrel{\text{def}}{=} \{\langle \pi_i(s_i), \lambda_j(\psi_{i,j}(\pi_i(s_i)) \cup t_j)(s_j) \rangle\}, \quad \text{for any } s_i \in \mathcal{S}_i, s_j \in \mathcal{S}_j.$$

The definition of the sequential composition operator is more complicated than the parallel case because of the need to capture the interaction between the original two nodes. For example, consider the construction of the policies for the sequentially composed node in Fig. 3. Every action taken in N_4 will (potentially) change the policy for N_5. This behaviour needs to be preserved by the composition operator. So when the policy is applied for the composed node N_9, internally the behaviour that is applied to the N_5 component of the composed state depends on the result of the N_4 component of the composed state. A similar dependency exists for the sensing update operator.

5.2 Properties

We now consider the properties of the composition operators. Firstly, it is necessary to establish that they are in fact well-defined and generate valid cognitive hierarchies.

Lemma 1. *For a cognitive hierarchy $H = (\mathcal{N}, N_0, F)$ and nodes $N_i, N_j \in \mathcal{N}$, the compositions $N_i \otimes_{\mathcal{P}}^{H} N_j$ and $N_i \otimes_{\mathcal{S}}^{H} N_j$ are well-defined cognitive hierarchies, when applied subject to the restrictions for parallel and sequential composition respectively.*

Proof Sketch. By inspection. The signatures satisfy the requirements of a cognitive node (Definition 1) and hierarchy (Definition 2). The restrictions on applying $N_i \otimes_{\mathcal{P}}^{H} N_j$ and $N_i \otimes_{\mathcal{S}}^{H} N_j$ ensures that the resulting sensing/action graphs are appropriate DAGs.   □

Now we establish that the composition operators satisfy the property of behaviour equivalence. We do this separately for each composition operator.

Theorem 1. *Let $H = (\mathcal{N}, N_0, F)$ be a cognitive hierarchy and N_i and N_j be nodes in H that satisfy the requirements of the parallel composition operator (i.e., N_i and N_j are distinct and unrelated under the sensing graph partial ordering). Then H and $N_i \otimes_{\mathcal{P}}^{H} N_j$ are behaviour equivalent with respect to the cognitive process model* **Update.**

Proof Sketch. Let $H_c = N_i \otimes_{\mathcal{P}}^{H} N_j$ and let $\mathcal{X}^0 = (H, \mathcal{Q}^0)$ and $\mathcal{X}_c^0 = (H_c, \mathcal{Q}_c^0)$ be the initial active cognitive hierarchies for H and H_c respectively. Now let $\mathcal{X}^{i+1} = $ **Update**$(\mathcal{X}^i)$ and $\mathcal{X}_c^{i+1} = $ **Update**$(\mathcal{X}_c^i)$. We can show by induction that for all i that $\mathcal{X}^{i+1}$ is behaviour equivalent to $\mathcal{X}_c^{i+1}$. Proving the base case requires comparing $\mathcal{Q}^0$ and $\mathcal{Q}_c^0$ to ensure that the set of task parameters of $\mathcal{Q}^0$ and $\mathcal{Q}_c^0$ w.r.t. N_0 are the same (Definition 6). The induction step requires following every aspect of the application of the **Update** function to confirm that the two cognitive hierarchies are updated so as to preserve the beliefs and actions of the original hierarchy. This is lengthy but essentially straightforward.   □

Theorem 2. *Let $H = (\mathcal{N}, N_0, F)$ be a cognitive hierarchy and nodes N_i and N_j be nodes in H that satisfy the requirements of the sequential composition operator (i.e., that N_i and N_j are distinct, $N_j \leq N_i$ and there is no distinct node N_k where $N_j \leq N_k$ and $N_k \leq N_i$ under the sensing graph partial ordering). Then H and $N_i \otimes_{\mathcal{S}}^{H} N_j$ are behaviour equivalent with respect to the cognitive process model* **Update.**

Proof Sketch. The proof follows the same pattern as the parallel case (Theorem 1).   □

Theorems 1 and 2 establish that the application of the parallel and sequential composition operators does indeed preserve behaviour equivalence. But a composed cognitive hierarchy is just another cognitive hierarchy. Consequently, the composition process can potentially be repeated and it is not difficult to see that this process can be repeated successively until the hierarchy consists of only a single non-N_0 node.

Theorem 3. *Let $H = (\mathcal{N}, N_0, F)$ be an arbitrary cognitive hierarchy. Then there exists a cognitive hierarchy $H' = (\mathcal{N}', N_0, F')$ such that H and H' are behaviour equivalent with respect to cognitive process model* **Update** *and $|\mathcal{N}'| = 2$.*

Proof Sketch. Trivial for $|\mathcal{N}| = 2$. For other cases successively construct cognitive hierarchies by applying the parallel or sequential composition operators until a behaviour equivalent hierarchy of size 2 is reached. Showing that this is possible reduces to a graph property of a DAG that allows one or the other composition operator to be applied.   □

6 Discussion

The formal properties established in this paper have a number of interesting consequences. In the first place, the fact that any cognitive hierarchy is equivalent to a hierarchy with only two nodes (Theorem 3) establishes that the size of the cognitive hierarchy is unrelated to its behaviour. Rather the choice of the hierarchy is influenced by other factors such as the designers familiarity with a particular representational language, or theoretical and practical computational concerns, such as being able to apply pre-existing tools and techniques in components of a node.

Secondly, while our composition formalism does not in itself provide a method to automatically decompose existing systems it does provide some understanding about what sorts of features to look for in determining whether or not decomposition is possible. If the belief state of an existing node can be factored into two orthogonal components (of size N and M) and if the updating of these components can be performed either independently or sequentially then decomposition will likely follow. Importantly this could have computational benefits; for example instead of searching over an $N \times M$ state space the combined search for the two nodes would be over an $N + M$ state space. This is one of the key motivations that has lead to the study of decomposition in other AI fields (e.g., factored planning [5], general game playing [8]), and in hierarchical reinforcement learning (HRL) [11].

7 Conclusion and Future Work

In this paper we developed a theory of node composition for a formal model of cognitive hierarchies. Two node composition operators were defined and shown to satisfy a notion of behaviour equivalence. The property of behaviour equivalence was established with respect to changes manifested by a cognitive system on an external environment. Such a definition is very general because it makes no reference to the internal state or representation used by a particular cognitive system. Hence, for example, two cognitive systems could be implemented using very different algorithms and techniques but could nevertheless satisfy the property of behaviour equivalence.

A second development of this paper was to establish that for any cognitive hierarchy there is an equivalent one consisting of only a single node (excluding the node that represents the external world). Furthermore, the combination of the parallel and sequential node composition operators was shown to be sufficient to capture this property.

Finally, we argued that the results developed in this paper can serve as a tool to bridge between cognitive systems. This opens up interesting directions for future research to validate these ideas. In particular it is our intention to examine the extent to which a system modelled using a rich expressive language that can deal with both symbolic and sub-symbolic information, such as the language introduced in [4], can be functionally decomposed into a behaviourally equivalent system consisting of a hierarchy of nodes with simpler individual representations; potentially separating the symbolic and sub-symbolic components. This would allow a system modelled formally using an expressive language, from which formal properties can be established, to then be efficiently implemented as a provably equivalent system consisting of decomposed nodes that can take advantage of existing high-performance tools and techniques.

Disclaimer

Opinions, findings, and conclusions or recommendations expressed in this paper are those of the authors and do not necessarily reflect the views of the AOARD.

Acknowledgements. This material is based upon work supported by the Asian Office of Aerospace Research and Development (AOARD) under Award No: FA2386-15-1-0005. This research was also supported under Australian Research Council's (ARC) *Discovery Projects* funding scheme (project number DP 150103035). Michael Thielscher is also affiliated with the University of Western Sydney.

References

1. Albus, J.S.: Engineering of Mind: An Introduction to the Science of Intelligent Systems. Wiley, New York (2001)
2. Amir, E., Engelhardt, B.: Factored planning. In: Proceedings of IJCAI, pp. 929–935. Morgan Kaufmann (2003)
3. Anderson, J.R.: Rules of the Mind. Lawrence Erlbaum Associates Inc., New Jersey (1993)
4. Belle, V., Levesque, H.J.: Robot location estimation in the situation calculus. J. Appl. Logic **13**(4), 397–413 (2015)
5. Brafman, R.I., Domshlak, C.: Factored planning: how, when, and when not. In: Proceedings of AAAI, pp. 809–814 (2006)
6. Brooks, R.A.: A robust layered control system for a mobile robot. IEEE J. Robot. Autom. **2**(1), 14–23 (1986)
7. Brooks, R.A.: Elephants don't play chess. Robot. Auton. Syst. **6**, 3–15 (1990)
8. Cerexhe, T.J., Rajaratnam, D., Saffidine, A., Thielscher, M.: A systematic solution to the (de-)composition problem in general game playing. In: Proceedings of ECAI, pp. 195–200 (2014)
9. Clark, K., Hengst, B., Pagnucco, M., Rajaratnam, D., Robinson, P., Sammut, C., Thielscher, M.: A framework for integrating symbolic and sub-symbolic representations. In: Proceedings of IJCAI, pp. 2486–2492 (2016)
10. De Giacomo, G., Sardiña, S.: Automatic synthesis of new behaviors from a library of available behaviors. In: Proceedings of IJCAI, pp. 1866–1871 (2007)

11. Hengst, B.: Hierarchical approaches. In: Wiering, M., van Otterlo, M. (eds.) Reinforcement Learning: State of the Art. Adaptation, Learning, and Optimization, vol. 12, pp. 293–323. Springer, Heidelberg (2011). doi:10.1007/978-3-642-27645-3_9

12. Jaeger, H., Christaller, T.: Dual dynamics: designing behavior systems for autonomous robots. Artif. Life Robot. **2**(3), 108–112 (1998)

13. Laird, J.E., Kinkade, K.R., Mohan, S., Xu, J.Z.: Cognitive robotics using the soar cognitive architecture. Cognitive Robotics AAAI Technical report WS-12-06, pp. 46–54 (2012)

14. Laird, J.E., Newell, A., Rosenbloom, P.S.: SOAR: an architecture for general intelligence. Artif. Intell. **33**(1), 1–64 (1987)

15. Lee, D., Kim, H., Myung, H.: GPU-based real-time RGB-D 3D SLAM. In: Proceedings Ubiquitous Robots and Ambient Intelligence (URAI), pp. 46–48. IEEE (2012)

16. Nilsson, N.: Teleo-reactive programs and the triple-tower architecture. Electron. Trans. Artif. Intell. **5**, 99–110 (2001)

17. Sacerdoti, E.D.: Planning in a hierarchy of abstraction spaces. Artif. Intell. **5**(2), 115–135 (1974)

Enable Efficient Resource Deployment in Multiple Concurrent Emergency Events Through a Decentralised MAS

Jihang Zhang[1]([✉]), Minjie Zhang[1], Fenghui Ren[1], and Jiakun Liu[2]

[1] School of Computing and Information Technology,
University of Wollongong, Wollongong, NSW, Australia
jz718@uowmail.edu.au, {minjie,fren}@uow.edu.au
[2] School of Mathematics and Applied Statistics,
University of Wollongong, Wollongong, Australia
Jiakunl@uow.edu.au

Abstract. In metropolitan regions, emergency events could happen concurrently at different places with different severities, types, deadlines and resource requirements. Due to the complexity, unpredictability, dynamic natures and potential resource contention problems among these events, traditional resource allocation approaches may have difficulties to efficiently and effectively deploy resources to these emergency events concurrently, which may result in a considerable increase in fatalities. In this paper, an multi-agent based decentralised resource allocation approach is proposed to coordinate and allocate resources to multiple concurrent emergency events. Besides, an emergency resource deployment simulation system based on GoogleMaps is developed for testing the proposed approach in a virtual metropolitan environment.

1 Introduction

Resource deployment in emergency events usually involves the collaboration of different emergency departments for effectively conducting rescue missions in a strict time limit. These emergency events usually occur at various locations within a metropolitan region and each event may have different attributes, such as the severity, deadline, types and resource requirements. Nowadays, emergency departments of large cities in many countries still rely on human operators to allocate resources to emergency events, which may result in problems such as ineffective resource utilisation and allocation delay [5]. This is mainly because in a metropolitan region, large numbers of rescue resources with different functionalities and availabilities are distributed over an extensive area. Human operators usually have difficulties to efficiently find out the most suitable solution for multiple concurrent emergency events due to a large number of possibilities. Besides, potential resource contention problems among these events make them even harder to handle.

© Springer International Publishing AG 2016
B.H. Kang and Q. Bai (Eds.): AI 2016, LNAI 9992, pp. 56–68, 2016.
DOI: 10.1007/978-3-319-50127-7_5

Agent and Multi-Agent System (MAS) technologies can offer potential solutions for deploying resources for emergency events due to their abilities of complex environment modelling, autonomous reasoning, adaptive decision-making and collective group formulation. In recent years, many promising emergency resource deployment strategies have been proposed in MAS community, such as game theory-based resource allocation [6,10], predication-based resource allocation [1,2,7], auction-based resource allocation [4,8,9] and centralised resource allocation [3,12]. However, all of these approaches might have different problems or limitations when they are applied to allocate resources to multiple concurrent emergency events. More specifically, game theory-based approaches are mainly used to allocate security resources to reduce the risk of terrorist attacks and predication-based approaches are mainly used to pre-allocate rescue resources for large-scale disasters. Nevertheless, emergency events usually happen randomly in urban areas, and cannot be predicated and prevented in advance. Auction-based approaches are primarily used to coordinate distributed agents to generate resource allocation plans in decentralised manners. In auction-based approaches, the relationships between agents are usually competitive, which are not suitable for modelling the cooperative relationships between different departments in emergency resource deployment. Centralised approaches are good at finding the optimal resource allocation plan based on the comprehensive information about the events, the resources and the environments. However, acquiring the precise and updated global information could be impractical or extremely difficult due to the dynamics and uncertainties in real-world situations.

In this paper, an agent-based decentralised resource allocation approach for multiple concurrent emergency events is proposed to overcome the limitations of the current approaches in the state-of-the-art. In the proposed approach, the emergency events are handled by multiple agents in different roles. As each event may contain different requests of the resources, the domain transportation theory is adopted in the proposed approach to calculate the optimal resource allocation plan for each event. Besides, since in decentralised MASs, multiple agents could contend same resources due to the lack of global information, a resource coordination algorithm is also proposed to effectively coordinate the resources deployment between multiple concurrent events.

The major contributions of this paper are (1) the proposed approach is designed to efficiently allocate distributed resources in a decentralised manner, so it is more applicable and practical compared with centralised approaches in deploying distributed resources to concurrent emergency events, and (2) the proposed MAS framework can simultaneously handle the resource allocation for multiple concurrent emergency events with different severities in an optimal way by considering financial expenditure, time cost and potential resource contention problems among these events.

The rest of this paper is organised as follows. Section 2 formally defines the problem and domain notations. Section 3 introduces the theoretical foundation of the proposed resource allocation approach. Section 4 describes the proposed MAS as the implementation of the proposed approach. Section 5 demonstrates the experimental results and analysis. Section 6 gives the conclusion and outlines the future work.

2 Definitions and Problem Description

This section formally introduces the definitions that are used in the proposed approach and represents the resource allocation problem of multiple concurrent emergency events.

2.1 Definitions of Domain Knowledge

Definition 1 (Environment). An *environment* is defined as an undirected graph $\mathbb{G}$.

Definition 2 (Resource). A *resource* is defined as an six-tuple, $r = (typ, ser, fun, loc, vel, exp)$, where $typ \in \{facility, mobile\}$ represents the resource type; ser represents the emergency services that r can provide; fun represents r's functionality; loc represents r's location on $\mathbb{G}$; $vel \in (0, +\infty)$ represents r's average velocity and exp represents r's money expenditure in per hour when $typ = facility$, while exp represents r's money expenditure in per kilometre when $typ = mobile$.

Definition 3 (Task). A *task* is defined as a three-tuple, $t = (dl, ser, \mathbb{R})$, where dl represents the deadline; ser represents the emergency service that t requires to make the response and $\mathbb{R}$ represents a set of required resources.

Definition 4 (Event). An *event* is defined as a five-tuple, $e = (con, \mathbb{SER}, loc, sev, \mathbb{T})$, where con represents e's content; $\mathbb{SER}$ represents a set of emergency services required by e; $sev \in [1, 5]$ represents e's severity and $\mathbb{T}$ represents a set of tasks that need to be completed for e.

Definition 5 (Proposal). A *proposal* for the resource allocation of an event is defined as a two-tuple, $p = (e, \mathbb{R})$, where $\mathbb{R}$ represents a set of resources that be proposed for event e.

Definition 6 (Cost Function). A *cost function* for a single resource's allocation to task $e.t_y$ is defined by Eq. 1:

$$CR(e.t_y, r) = \begin{cases} E^m(r.exp), \text{ if } r.typ = facility \\ (w^t E^t(\frac{DS(r.loc, e.loc)}{r.vel}) + w^m E^m(r.exp)) \times \\ DL(e.t_y, r), \text{ if } r.typ = mobile, \end{cases} \quad (1)$$

where DS is a function that returns the distance of a passable road between resource location $r.loc$ and event location $e.loc$. E^m and E^t are two evaluation functions to convert the value of a resource's allocation time and money expenditure to a normalised value in-between 0 and 1. w^t and w^m represent the weighting (importance) of the resource's money expenditure and allocation time, respectively, which are calculated by following equations:

$$\begin{cases} w^t = \frac{e.ser}{5} \\ w^m = 1 - \frac{e.ser}{5} \end{cases} \tag{2}$$

Besides, DL is a function that is used to determine whether r can be allocated within $e.t_y$'s deadline, which is further defined by Eq. 3:

$$DL(e.t_y, r) = \begin{cases} 1, \text{ if } \frac{DS(r.loc,e.loc)}{r.vel} \leq e.t_y.dl \\ +\infty, \text{ if } \frac{DS(r.loc,e.loc)}{r.vel} > e.t_y.dl \end{cases} \tag{3}$$

Based on the above definitions, the cost function for allocating all required resources to task $e.t_y$ and event e are defined by Eqs. 4 and 5, respectively.

$$CT(e.t_y, p_y) = \sum_{r_n \in p_y.\mathbb{R}} CR(e.t_y, r_n) \tag{4}$$

$$CE(e, p) = \sum_{e.t_y \in e.\mathbb{T}} CT(e.t_y, p_y) \text{ subject to } p_y.\mathbb{R} \in p.\mathbb{R} \tag{5}$$

2.2 Problem Description

In a metropolitan region, multiple events $\mathbb{E}$ could happen concurrently. For a single event $e \in \mathbb{E}$, the objective of the resource allocation is to search for an optimal proposal p^* with the minimal allocation cost, which is defined by Eq. 6:

$$OE = \arg\min_{p^*} CE(e, p^*) \text{ subject to } p^*.\mathbb{R} \in \mathbb{R}^g \wedge e \in \mathbb{E}, \tag{6}$$

where $\mathbb{R}^g$ represents the available resources in the environment $\mathbb{G}$.

However, e might require resources from different emergency departments, which significantly complicated the searching for a complete p^*. Therefore, the proposed approach further creates a set of tasks $e.\mathbb{T}$ for e. For each task t_y in $e.\mathbb{T}$, it only requires resources that can provide the same emergency service (i.e. $r.ser$), but the resources' functionalities could be different (i.e. $r.fun$). By doing so, the searching of p^* for e is converted to the simultaneous searching of p_y^* for each task t_y in $e.\mathbb{T}$. The objective function OT for the resource allocation of t_y is defined by Eq. 7:

$$OT = \arg\min_{p_y^*} CT(e.t_y, p_y^*) \text{ subject to } p_y^*.\mathbb{R} \in \mathbb{R}^g \wedge e.t_y \in e.\mathbb{T} \tag{7}$$

In our previous work [13], we have shown that the domain transportation theory can be used to effectively generate the optimal proposal for each task in an event. Generally, domain transportation theory is a linear programming method to generate optimal solutions for multi-objectives problems in polynomial time, which is advisable to be used in the proposed approach to efficiently find the optimal resource allocation proposal in terms of resource allocation time and money expenditure. In domain transportation, the resource allocation problem of a task can be described as a resource mapping problem from the available resources on $\mathbb{G}$ to the required resources of the task (i.e. $p_y : \mathbb{R}^g \rightarrow t_y.\mathbb{R}$). A linear programming equation was used to find the optimal mapping solution that can fulfil OT. When it comes to the resource allocation for multiple concurrent events $\mathbb{E}$, there could also have a collection of proposals that can be used to deploy resources for $\mathbb{E}$. The objective function OA for all concurrent emergency events is defined by Eq. 8:

$$OA = \arg \min_{\mathbb{P}^* = \{p_1^*, \dots, p_x^*\}} \sum_{e_x \in \mathbb{E}} CE(e_x, p_x^*) \text{ subject to } p_x^*.\mathbb{R} \in \mathbb{R}^g, \tag{8}$$

For events that have no interrelationship, OA can be fulfilled by using decentralised agents to minimise the resource allocation cost of each individual event, respectively. Nevertheless, one problem that may occur between decentralised agents is resource contention, which will lead to the increase of system instability and the decline of system performance [11]. In the proposed approach, a resource contention problem refers to multiple tasks from different events require certain resources at same time, which results in resource allocation conflict. Under such a situation, simply minimising each events' resource allocation cost individually to fulfil OA becomes impossible, since each event' resource allocation may have interrelationship. In order to achieve better and more efficient resource allocation when a such problem occurs, an agent-based resource coordination mechanism is developed and integrated into the proposed approach. The detail of how to coordinate conflict resources between tasks in an optimal way is explained in next section.

3   Theoretical Foundation of the Optimal Resource Coordination

When a resource contention is found between tasks, the proposed approach considers that all involved tasks have interrelationships, regarding of their resource allocation. In order to effectively minimise the total allocation cost of these inter-related tasks, an optimal resource coordination algorithm (i.e. Algorithm 2 to be introduced later) is proposed to generate new resource allocation plans for these tasks.

More precisely, let $\mathbb{T}^e = \{e_1.t_1, \dots, e_x.t_y\}$ denote a set of tasks from different events. Let $\mathbb{R}_{xy}^c$ denote a set of candidate resources that can be allocated to $e_x.t_y \in \mathbb{T}^e$. Let $\mathbb{P}^e = \{p_{11}, \dots, p_{xy} | p_{xy}.\mathbb{R} \subset \mathbb{R}_{xy}^c\}$ denote a set of proposals that are sent out for executing by the tasks in $\mathbb{T}^e$. A resource contention problem between tasks in $\mathbb{T}^e$ is identified by Eq. 9:

$$RC(\mathbb{P}^e) = \begin{cases} 1, \text{ if } p_{11}.\mathbb{R} \cap \dots \cap p_{xy}.\mathbb{R} \neq \emptyset \\ 0, \text{ if } p_{11}.\mathbb{R} \cap \dots \cap p_{xy}.\mathbb{R} = \emptyset, \end{cases} \tag{9}$$

where $RC = 1$ denotes that the resource contention is identified, while $RC = 0$ denotes there is no resource contention. The objective function OC for the resource coordination of the resource contention in $\mathbb{T}^e$ is defined by Eq. 10:

$$OC = \arg \min_{\mathbb{P}^* = \{p_{11}^*, \ldots, p_{xy}^*\}} \sum_{e_x.t_y \in \mathbb{T}^e} CT(e_x.t_y, p_{xy}^*), \tag{10}$$

where the objective function indicates that for each task $e_x.t_y$ in $\mathbb{T}^e$, a new optimal proposal needs to be generated to minimise the total allocation cost of $\mathbb{T}^e$.

In order to fulfil OC, the proposed approach combines the candidate resource pool of each task $e_x.t_y$ in $\mathbb{T}^e$ to generate a common resource pool, and then selects resources from this pool to generate a new proposal for each task. In detail, let $\mathbb{R}^o = \{\mathbb{R}_{11}^c \cup \ldots \cup \mathbb{R}_{xy}^c\}$ represent the common resource pool. Based on the resources inside $\mathbb{R}^o$, a sort function is applied to sort the resource allocation cost for a task from the minimum value to the maximum value, which is defined as $SORT(e_x.t_y, \mathbb{R}^o)$. After the sorting, a new proposal (i.e. p_{xy}^*) with the minimal cost can be generated for task $e_x.t_y$ based on the sorting results.

Nevertheless, since each resource inside $\mathbb{R}^o$ is shared by all tasks in $\mathbb{T}^e$, which might cause new resource contention. When multiple tasks contend the same resource in $\mathbb{R}^o$, a resource reassignment algorithm needs not only to consider the cost of allocating this resource to an appropriate task, but also to minimise the cost of finding replacement resources for other tasks. The objective function OR for the reassignment of conflict resource r^c is formally defined by Eq. 11:

$$OR = \arg \min_{\begin{cases} e^*.t^* \in r^c.\mathbb{T}, \\ (e_1.t_1, r_{11}^r) \in (\mathbb{T}^u, \mathbb{R}^o), \\ \ldots, \\ (e_x.t_y, r_{xy}^r) \in (\mathbb{T}^u, \mathbb{R}^o) \end{cases}} CR(e^*.t^*, r^c) + \sum_{e_x.t_y \in \mathbb{T}^u} CR(e_x.t_y, r_{xy}^r), \tag{11}$$

where $r^c.\mathbb{T}$ represents a set of tasks that require $r^c.e^*.t^*$ represents the winning task that has gotten the conflict resource r^c. $\mathbb{T}^u = \{r^c.\mathbb{T} \setminus e^*.t^*\}$ represents a set of un-winning tasks that did not get r^c. The order pair $(e_x.t_y, r_{xy}^r)$ represents an un-winning task $e_x.t_y$ and a replacement resource r_{xy}^r that $e_x.t_y$ selects from the common resource pool $\mathbb{R}^o$ to substitute missing r^c.

In order to fulfil OR, the resource reassignment algorithm first needs to identify a replacement resources r_{xy}^r for each task $e_x.t_y$ in $r^c.\mathbb{T}$, which should has the minimal resource allocation cost for $e_x.t_y$ compared with other resources in $\mathbb{R}^o$. The algorithm will use recursion to handle any further resource contention, until all $e_x.t_y$ in $r^c.\mathbb{T}$ has found an unique replacement resources r_{xy}^r. Then, the resource reassignment algorithm needs to select the winning task $e^*.t^*$ for r^c, determining by the cost similarity comparison of allocating r^c and r_{xy}^r to each task $e_x.t_y$ in $r^c.\mathbb{T}$. The selection function of the winning task $e^*.t^*$ is calculated by Eq. 12:

$$e^*.t^* = \arg \min_{e_x.t_y \in r^c.\mathbb{T}} \frac{CR(e_x.t_y, r^c)}{CR(e_x.t_y, r_{xy}^r)} \tag{12}$$

The optimal resource reassignment process and resource coordination process are formally defined by Algorithms 1 and 2, respectively.

Algorithm 1. Optimal_Reassignment Process

Input: r^c, $\mathbb{R}^o$

1: **for all** $e_x.t_y \in r^c.\mathbb{T}$ **do**

2: $\quad r^r_{xy} = \arg\min_{r \in \mathbb{R}^o} CR(e_x.t_y, r)$ subject to
$$r \neq r^c \wedge r.fun = r^c.fun$$

3: $\quad r^r_{xy}.\mathbb{T} = r^r_{xy}.\mathbb{T} \cup \{e_x.t_y\}$

4: **end for**

5: **for all** $e_x.t_y \in r^c.\mathbb{T}$ **do**

6: $\quad$ **if** $|r^r_{xy}.\mathbb{T}| > 1$ **then**

7: $\quad\quad$ Optimal_Reassignment $(r^r_{xy}, \mathbb{R}^o)$

8: $\quad$ **end if**

9: **end for**

10: $e^*.t^* = \arg\min_{e_x.t_y \in r^c.\mathbb{T}} \frac{CR(e_x.t_y, r^c)}{CR(e_x.t_y, r^r_{xy})}$

11: assign r^c to $e^*.t^*$

12: $\mathbb{R}^o = \mathbb{R}^o \setminus \{r^c\}$

13: $\mathbb{T}^u = r^c.\mathbb{T} \setminus \{e^*.t^*\}$

14: **for all** $e_x.t_y \in \mathbb{T}^u$ **do**

15: $\quad$ assign r^r_{xy} to $e_x.t_y$

16: $\quad \mathbb{R}^o = \mathbb{R}^o \setminus \{r^r_{xy}\}$

17: **end for**

Algorithm 2. Optimal_Resource_Coordination Process

Input: $\mathbb{T}^e$, $\mathbb{P}^e$, $\mathbb{R}^c_{11}, ..., \mathbb{R}^c_{xy}$

1: **if** $RC(\mathbb{P}^e) = 1$ **then**

2: $\quad \mathbb{R}^o = \{\mathbb{R}^c_{11} \cup ... \cup \mathbb{R}^c_{xy}\}$

3: $\quad$ **for all** $e_x.t_y \in \mathbb{T}^e$ **do**

4: $\quad\quad$ create p_{xy}

5: $\quad\quad SORT(e_x.t_y, \mathbb{R}^o)$

6: $\quad\quad$ **for all** $r_n \in e_x.t_y.\mathbb{R}$ **do**

7: $\quad\quad\quad r^* = \arg\min_{r \in \mathbb{R}^o} CR(e_x.t_y, r)$ subject to
$$r.fun = r_n.fun \wedge r.ser = r_n.ser$$

8: $\quad\quad\quad r^*.\mathbb{T} = r.\mathbb{T} \cup \{e_x.t_y\}$

9: $\quad\quad\quad p_{xy}.\mathbb{R} = p_{xy}.\mathbb{R} \cup \{r^*\}$

10: $\quad\quad$ **end for**

11: $\quad$ **end for**

12: $\quad$ **for all** $e_x.t_y \in \mathbb{T}^e$ **do**

13: $\quad\quad$ create p^*_{xy}

14: $\quad\quad$ **for all** $r_n \in p_{xy}.\mathbb{R}$ **do**

15: $\quad\quad\quad$ **if** $|r_n.\mathbb{T}| > 1$ **then**

16: $\quad\quad\quad\quad$ Optimal_Reassignment $(r_n, \mathbb{R}^o)$

17: $\quad\quad\quad$ **else**

18: $\quad\quad\quad\quad p^*_{xy}.\mathbb{R} = p^*_{xy}.\mathbb{R} \cup \{r_n\}$

19: $\quad\quad\quad\quad \mathbb{R}^o = \mathbb{R}^o \setminus \{r_n\}$

20: $\quad\quad\quad$ **end if**

21: $\quad\quad$ **end for**

22: $\quad\quad$ send out p^*_{xy} for execution

23: $\quad$ **end for**

24: **end if**

4   Agent-Based Resource Allocation

This section introduces agents' definitions of five type of agents in the proposed approach, a MAS framework and the resource allocation progress in detail.

4.1   Definitions of Agents

Definition 7 (Response Agent). A *response agent* is represented by ra, which has the information of an event.

Definition 8 (Mobile Agent). A *mobile agent* is represented by ma, which has the information of a mobile resource $ma.r$.

Definition 9 (Facility Agent). A *facility agent* is represented by fa, which has the information of a facility resource $fa.r$ and a set of mobile agents, and $fa.\mathbb{R}$ represents all mobile and facility resources that under fa's management.

Definition 10 (Deployment Agent). A *deployment agent* is represented by da, which has the information of an event.

Definition 11 (Coordinator Agent). A *coordinator agent* is represented by ca, which is dynamically generated to address resource contention between multiple emergency events.

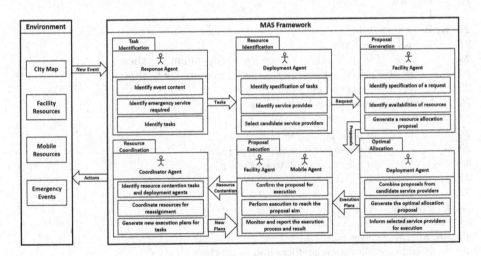

Fig. 1. The framework of the proposed resource Allocation system

4.2   A MAS-based Resource Allocation System

The proposed resource allocation approach is implemented by a decentralised MAS, which includes six modules (show in Fig. 1). During the resource allocation of concurrent events, each event is handled separately by a group of agents. The resource allocation process is formally described by Algorithm 3.

Algorithm 3. Resource Allocation Process

Input: e

1: assign e to ra
2: ra identifies $e.\mathbb{SER}$ based on $e.con$
3: ra identifies $e.\mathbb{T}$ based on $e.\mathbb{SER}$
4: ra sends $e.\mathbb{T}$ to da
5: da calculates circle communication area $da.com$
6: da locates $\mathbb{FA}$ in $da.com$
7: **for all** $t_y \in e.\mathbb{T}$ **do**
8:     da creates $\mathbb{P}_y$, $\mathbb{FA}_y$ and $\mathbb{R}_y^c$
9:     **for all** $fa_i \in \mathbb{FA}$ **do**
10:        **if** $fa_i.r.ser = t_y.ser$ **then**
11:            da updates $\mathbb{FA}_y = \{fa_i\} \cup \mathbb{FA}_y$
12:            da sends t_y to fa_i
13:            fa_i finds $p_{iy} : fa_i.\mathbb{R} \rightarrow t_y.\mathbb{R}$
14:            **if** $PDL(t_y.dl, e.sev)$ **then**
15:                fa_i submits p_{iy} to da
16:                da updates $\mathbb{R}_y^c = \{p_{iy}.\mathbb{R}\} \cup \mathbb{R}_y^c$
17:            **end if**
18:        **end if**
19:     **end for**
20: **end for**
21: **while** $|e.\mathbb{T}| > 0$ **do**
22:     **for all** $t_y \in e.\mathbb{T}$ **do**
23:        **if** $PDL(t_y.dl, e.sev) \vee \forall fa_i \in \mathbb{FA}_y : fa_i$ submit p_{iy} **then**
24:            **if** $\mathbb{P}_y$ does not contains enough resources for t_y **then**
25:                da expend $da.com$ by double
26:                process goes back to Line 7
27:            **else**
28:                da finds $p_y^* : \mathbb{R}_y^c \rightarrow t_y.\mathbb{R}$
29:                da updates $e.\mathbb{T} = e.\mathbb{T} \setminus \{t_y\}$
30:                da informs agents to execute p_y^*
31:            **end if**
32:        **end if**
33:     **end for**
34: **end while**
35: **for all** fa_c receives p_y^* **do**
36:     **if** resource allocation conflicts find in p_y^* **then**
37:        fa_c call coordinate agent ca_c
38:        ca_c identifies $\mathbb{P}^e$, $\mathbb{T}^e$, $\mathbb{R}_{11}^c, ..., \mathbb{R}_{xy}^c$
39:        ca_c executes Algorithm 2
40:     **else**
41:        fa_c executes p_y^*
42:     **end if**
43: **end for**

The resource allocation process shown in Algorithm 3 includes five steps, which are explained as follows.

Step 1: (Lines 1–4) When an emergency event e happenes, a response agent ra is assigned to e to analyse the event content $e.con$ and required emergency services $e.\mathbb{SER}$. Then, ra needs to identify the required resources of each service, which might be provided by human operators or other external agents. Finally, ra converts the information of emergency services to tasks and sends $e.\mathbb{T}$ to a deployment agent da.

Step 2: (Lines 5–12) After receiving $e.\mathbb{T}$, da first needs to calculate a communication area $da.com$, which is a circle centred at the event's location $e.loc$ and measured by square kilometres. After calculating $da.com$, da locates all facility agents inside $da.com$, which is represented by set $\mathbb{FA}$. Then, da sends each t_y to relevant facility agents.

Step 3: (Lines 13–16) After a facility agent fa_i receives t_y, fa_i uses the domain transportation theory to calculate an optimal proposal p_{iy} based on resources that under fa_i's management. Then, fa_i submits p_{iy} to da if current time has not exceeded task proposal deadline, which is calculated by function $PDL(t_y.dl, e.sev)$. Finally, da adds p_{iy} to $\mathbb{P}_y$.

Step 4: (Lines 22–31) After da receives t_y's proposals from all associated facility agents or t_y proposal deadline has been reached, da checks whether $\mathbb{P}_y$ has enough resources to generate a final plan for t_y. If the resources are not enough, da expands its original communication area $da.com$ by double to contact more facility agents, and then the process goes back to Step 2. Otherwise, da uses domain transportation theory to generate proposal p_y^* for t_y based on all resources in $\mathbb{P}_y$, which is represented by $\mathbb{R}_y^c$ (i.e. candidate resource pool). Finally, da sends p_y^* to facility agents to execute.

Step 5: (Lines 35–42) During the execution of proposals, if a facility agent finds that its resources is required by multiple proposals, a new coordination agent ca will be generated. Then, ca gathers the information of resource contention proposals (i.e. $\mathbb{P}^e$), resource contention tasks (i.e. $\mathbb{T}^e$) and a list of candidate resource pools for the tasks (i.e. $\mathbb{R}^c$). Finally, ca executes Algorithm 2 to solve the resource contention.

5   Experiments

5.1   Experimental Setting

In our experiments, an agent-based emergency resource allocation simulator based on GoogleMaps was implemented as the testbed. By using the GoogleMaps, real-world information regarding to emergency services, routes and live traffic can be easily accessed in the simulation. Therefore, the simulation results are more relevant to demonstrate the system performance in solving real-world problems. We conducted two experiments to evaluate our approach.

Experiment 1 is to evaluate the effectiveness of our resource coordination algorithm, which plays a key role in solving resource contention and optimising the solution in handing concurrent emergency events. Experiment 1 was conducted by using the coordinated allocation approach and the random allocation approach (i.e. randomly coordinate conflict resources) to deploy same resources to 10 concurrent events, respectively. In order to eliminate the difference and uncertainty in a real-world environment, the experiment was repeated by 1000 times, and the average allocation cost, the average number of resource contention events and the cost difference of the two approaches were finally reported.

In Experiment 2 is to evaluate the advantages and disadvantages of our approach, in which we compare the resource allocation performance between our approach and other two related work in agent-based emergency resource allocation. The first comparing approach is a decentralised ambulances allocation approach proposed by Lopez et al. [8], which uses an auction mechanism based on trust and time to assign ambulances to events. The second comparing work is a centralised resource allocation approach proposed by Gabdulkhakova et al. [3], implemented based on a service-oriented architecture. The resource allocation tests for each approach were also repeated for 1000 times and the average cost for 10 events was reported.

The results of Experiment 1 are shown in Fig. 2. As we can see from Fig. 2, when the event's average distance is 5000 m, the difference of the allocation cost between coordinated and random allocation is not significant (about 0.22). This is mainly because when the distance between events are large, resource contention problems have low possibility to happen, which results in less usage of our resource coordination algorithm. However, with the decrease of events' inter-distance from 5000 m to 1000 m, we can find that the total cost of random resource allocation has the obvious increment (from 5.84 to 6.87), while in the coordinated approach, the resource allocation cost decreases from 5.52 to 4.76. Furthermore, the cost difference between the two approachs has increased distinctly (from 0.22 to 1.96). These results indicate that when the events' inter-distance is close to each other, more events have involved in resource contention problems (from 2 to 8), the coordinated approach can effectively coordinate resources to minimise the total allocation cost for these events. Nevertheless, for random allocation approach, due to unable to reasonably coordinate

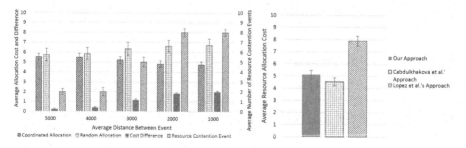

Fig. 2. The results of Experiment 1   **Fig. 3.** The results of Experiment 2

and reassign conflict resources between multiple tasks, the total allocation cost has increased gradually as more and more resource contention problems have occurred. Besides, the data variation lines (the black line on top of each bar) indicate that the coordinated allocation approach can produce more consistent cost compared with the cost generated by the random allocation approach.

The results of Experiment 2 are shown in Fig. 3, in which we can see that Gabdulkhakova et al.'s approach required the smallest cost to complete the resource allocation for 10 events, which performed slightly better than ours (4.56 verse 5.12). This is because Gabdulkhakova's approach was implemented with the centralised design, which could acquire the global information of all resources and allocate them through a centralised node. By doing so, Gabdulkhakova's approach can essentially avoid resource contention problems, thereby achieve better results. Nevertheless, acquiring all resources from different emergency departments or private companies and managing them by a centralised system are extremely difficult to be realised in the real-world environment. Our approach was implemented based on a decentralised MAS, so it is more practical compared with Gabdulkhakova's approach. Lopez's approach requires larger cost compared with our approach, because Lopez's approach fails to consider the factors of resource money expenditure and resource coordination. Besides, Lopez's approach is based on an auction mechanism and the relationships between contract agents are competitive. Therefore, the optimal resource allocation plan in Lopez's approach is the single solution provided by the winner agent of the auction. However, for most emergency situations, it is more reasonable for agents to act cooperatively rather than competitively. In our approach, deployment agents are used to combine proposals from different facility agents, thereby the optimal solution is a combined solution by integrating the advantages of each facility agent's proposal.

6   Conclusion

In this paper, an agent-based decentralised resource allocation approach was proposed to handle multiple concurrent emergency events. For each single event, the proposed approach breaks the event's requirements into different tasks and then the domain transportation theory is used to generate the optimal solution by considering time and money expenditure factors. Besides, when a resource contention problem has happened between tasks from different events, a resource coordination algorithm is used to effectively coordinate and reassign resources to appropriate tasks, thereby the total allocation cost of all events can be reduced indirectly. The experimental results indicate that our approach can effectively minimise the total cost of multi-event resource allocation by considering resource contention problems. Our future work will primarily focus on altering tasks goals when extra resources cannot be found for multiple events with resource conflict.

References

1. Chang, M.-S., Tseng, Y.-L., Chen, J.-W.: A scenario planning approach for the flood emergency logistics preparation problem under uncertainty. Transp. Res. Part E Logist. Transp. Rev. **43**(6), 737–754 (2007)
2. Dawson, R.J., Peppe, R., Wang, M.: An agent-based model for risk-based flood incident management. Nat. Hazards **59**(1), 167–189 (2011)
3. Gabdulkhakova, A., Konig-Ries, B., Rizvanov, D.A.: An agent-based solution to the resource allocation problem in emergency situations. In: 2011 Ninth IEEE European Conference on Web Services (ECOWS), pp. 151–157 (2011)
4. Gerding, E.H., Dash, R.K., Yuen, D.C.K., Jennings, N.R.: Bidding optimally in concurrent second-price auctions of perfectly substitutable goods. In: Proceedings of the 6th International Joint Conference on Autonomous Agents and Multiagent Systems, p. 53. ACM (2007)
5. Haddow, G., Bullock, J., Coppola, D.P.: Introduction to Emergency Management. Butterworth-Heinemann, Burlington (2013)
6. Kiekintveld, C., Jain, M., Tsai, J., Pita, J., Ordóñez, F., Tambe, M.: Computing optimal randomized resource allocations for massive security games. In: Proceedings of The 8th International Conference on Autonomous Agents and Multiagent Systems, pp. 689–696 (2009)
7. Li, L., Tang, S.: An artificial emergency-logistics-planning system for severe disasters. IEEE Intell. Syst. **23**(4), 86–88 (2008)
8. López, B., Innocenti, B., Busquets, D.: A multiagent system for coordinating ambulances for emergency medical services. IEEE Intell. Syst. **23**(5), 50–57 (2008)
9. Suárez, S., Collins, J., López, B.: Improving rescue operations in disasters: approaches about task allocation and re-scheduling. In 24rd Annual Workshop of the UK Planning and Scheduling Special Interest Group (PlanSIG), pp. 66–75 (2005)
10. Tsai, J., Yin, Z., Kwak, J., Kempe, D., Kiekintveld, C., Tambe, M.: How to protect a city: strategic security placement in graph-based domains. In: Proceedings of the 9th International Conference on Autonomous Agents and Multiagent Systems, pp. 1453–1454 (2010)
11. Youssefmir, M., Huberman, B.A.: Resource contention in multiagent systems. In: Proceedings of the First International Conference on Multiagent Systems (ICMAS), pp. 398–405 (1995)
12. Zayas-Cabán, G., Lewis, M.E., Olson, M., Schmitz, S.: Emergency medical service allocation in response to large-scale events. IIE Trans. Healthc. Syst. Eng. **3**(1), 57–68 (2013)
13. Zhang, J., Zhang, M., Ren, F., Liu, J.: A multiagent-based domain transportation approach for optimal resource allocation in emergency management. In: The Proceedings of the 2nd International Workshop on Smart Simulation and Modelling for Complex Systems, pp. 30–41 (2015)

AI Applications and Innovations

Forecasting Monthly Rainfall in the Western Australian Wheat-Belt up to 18-Months in Advance Using Artificial Neural Networks

John Abbot[1,2,3(✉)] and Jennifer Marohasy[1,3]

[1] Climate Modelling Laboratory, Noosaville, QLD, Australia
johnwabbot@gmail.com, jennifermarohasy@gmail.com
[2] Department of Engineering, University of Tasmania, Hobart, TAS, Australia
[3] Institute of Public Affairs, Melbourne, VIC, Australia

Abstract. Accurate medium-term rainfall forecasts are a significant constraint to dry land cropping. In Australia, official monthly forecasts for the Western Australian wheat-belt are currently based on output from the Bureau of Meteorology's general circulation model, the Predictive Ocean Atmosphere Model for Australia (POAMA). These forecasts are provided in a two-category format (above or below median rainfall) up to three months in advance for large grid areas, and are not considered reliable. An alternative approach is presented here for the three locations of Narrogin, Merredin and Southern Cross using artificial neural networks (ANNs) to forecast monthly rainfall up to 18 months in advance. Skilful monthly rainfall forecasts can be achieved at all lead times measured in terms of root mean square error (RMSE) and mean absolute error (MAE). This approach is of practical benefit to wheat growers in this region, with potential application to other locations with long historical temperature and rainfall records.

Keywords: Artificial neural network · Machine learning · Monthly rainfall forecast · Dry land cropping · Climate indices

1 Introduction

Wheat, *Triticum aestivum*, is a major crop in Australia, with 11–13 million hectares planted annually, approximately 40 percent of this in south-west Western Australia [1]. Aside from soil fertility and other agronomic considerations, the major constraints on production are meteorological and climatological [2–9]. Soil moisture profile at the time of planting influences crop success [10, 11], which in turn is influenced by rainfall during the summer-autumn fallow period [12]. Water availability during the growing period is the most critical factor affecting crop yields [10, 13, 14]. The timing of the arrival of adequate autumn rains for sowing, the so-called autumn 'break', is a key factor in the crop establishment phase [6]. In particular, the timing of rainfall events in relation to crop requirements is regarded as more important than the total rainfall received during the life cycle of the crop [6, 15]. Therefore, accurate and detailed forecasts of rainfall are important for wheat farmers worldwide [16], including Western Australia.

© Springer International Publishing AG 2016
B.H. Kang and Q. Bai (Eds.): AI 2016, LNAI 9992, pp. 71–87, 2016.
DOI: 10.1007/978-3-319-50127-7_6

Phenomena thought to affect rainfall, in particular the El Niño-Southern Oscillation (ENSO), have been extensively studied for Australia [17–21]. Relationships between ENSO and yields of wheat and maize have been studied for Argentina [22] and Brazil [23]. In the USA, yields of a variety of crops, including wheat, cotton and sugarcane have been found to have some dependence on ENSO [24, 25]. In Asia, relationships between ENSO and yields of wheat and rice have been studied for both China [26] and India [27].

Until May 2013, the official seasonal forecast for southwest Western Australia issued by the Australian Bureau of Meteorology (BOM) was based on an empirical statistical scheme using an ENSO index as the primary predictor in a relatively simple statistical model [28]. Reviewing the performance of these forecasts, Fawcett and Stone (2010) described the demonstrated skill level for seasonal rainfall forecasting as "only moderate, although better than climatology and randomly guessed forecasts". Since June 2013, the BOM has used output from a general circulation model (GCM), the Predictive Ocean Atmosphere Model for Australia (POAMA), to forecast for large grid areas (62,500 square kilometres) across the Australian continent. There has been no published detailed quantitative study enabling a direct comparison between the skills of the forecasts from POAMA versus the earlier statistical models, including for south-west Western Australia. Because the operational forecasts from POAMA are provided in a two-category format (above or below median rainfall), calculation of common measures of forecast skill including root mean square error (RMSE) are impossible. Furthermore, the BOM is unwilling to provide the output from POAMA used in operational forecasts in a form facilitating comparison with other methods, for example, as deterministic forecasts for point locations [29].

The BOM's seasonal forecasts are considered too unreliable as a basis for major cropping decisions [30]. This is consistent with earlier studies [31] that suggest accurate seasonal rainfall forecasts do not exist for Western Australia. More recent research findings [32] based on output from the latest version of POAMA (version 2.4), also indicate that for grid cells corresponding to the western Australian wheat-belt, monthly rainfall forecasts fall into categories corresponding to a zero, negative, or very low positive level of skill relative to climatology. This was even after extensive post-processing of POAMA 2.4 output with statistical bridging models [32]. Research [33] considering the reliability and skill of POAMA 2.4 relative to earlier versions for the entire Australian continent also indicates a forecast skill only comparable to climatology, with little or no skill beyond climatology after 3 months. An indication of the skill of POAMA after downscaling for locations in south eastern Australia again showing skill scores only comparable to climatology, with mean correlations of approximately 0.4 at one month lead, declining for longer leads [34].

POAMA, like other GCMs, attempts to simulate climate from an understanding of physical processes. An alternative approach to both POAMA, and the earlier statistical models [28, 35, 36] is machine learning using artificial neural networks (ANNs) [37–39]. In principle the forecasting of monthly rainfall in the wheat belt should be amenable to ANNs because there exist annual rainfall patterns modified by global ocean-atmospheric circulation patterns measurable though climatic indices including ENSO and the Indian Ocean Dipole. These climate indices, as well as historical rainfall and temperatures records for many locations in the wheat belt, potentially provide over 100 years of data for training and validation of ANN models.

Machine learning with ANNs focuses on prediction based on known properties learned from exposure to historical data sets during the training process. A core objective of the learning process is to be able to generalize from experience [40]. Machine leaning has been widely utilized in the hydrology area, for example in studying rainfall-runoff relationships [41]. Machine learning has also become important in the medical diagnostic field [42] where information needs to be combined from different tests, each carrying some relevant, but limited, diagnostic information. There may be no consistently useful method of combining the relevant information, with a traditional reliance on the skill and experience of the medical practitioner. A vast literature shows 13 difference climate indices can have some influence on seasonal rainfall across Australia, with their influences varying temporally and seasonally and forecasts improved when information is provided as a lag, typically from one to three months in length [43, 44].

In this study, arrays of data were constructed using 7 climate indices and 3 local climate attributes (rainfall, maximum and minimum temperature) with each of these input attributes lagged at monthly increments, for up to 12 months. The arrays were inputted into a state-of-the-art probabilistic ANN and models built that could be used for monthly rainfall forecasting from one to 18 months in advance for the three Western Australian wheat-belt towns of Narrogin, Merredin and Southern Cross, Fig. 1. Output, as monthly rainfall forecasts for the test period July 2004 to June 2014 from these models is presented as time-series charts, and the skill of the forecast measured using Pearson correlations, root mean square errors (RMSE), and mean

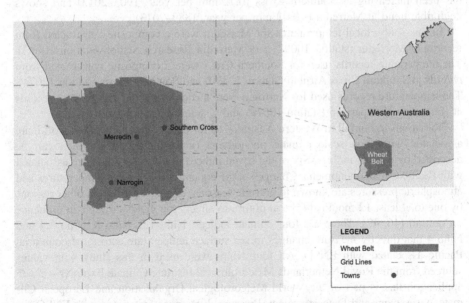

Fig. 1. Map of the Western Australian wheat belt showing the POAMA grid lines and the towns of Narrogin, Merredin and Southern Cross

absolute errors (MAE). Variations in forecast skill, between and within locations, and variability in the sensitivity of the input attributes was also analysed. This work builds on earlier research focused on the application of ANN models for forecasting monthly rainfall in north eastern Australia using a less sophisticated ANN platform [39].

2 Data and Method

The skill of a rainfall forecast from an ANN will depend on the quality and relevance of the data provided as input to the model. As with any statistical model, including ANNs, it is generally preferable to train and test with historical data sets extending as far back as possible. Narrogin, Merredin and Southern Cross have relatively long rainfall and temperature records, and fall within three of the four different BOM grid areas used in POAMA forecasts [32] covering the Western Australia wheat belt shown in Fig. 1.

All three towns experience predominantly winter rainfall, with Narrogin historically receiving significantly more rain. Narrogin (station number 10614, latitude 32.93°S, longitude 117.18°E., elevation: 338 m) began recording rainfall in 1891, and is still operating as a weather station today. Merredin (station number 10092, latitude 31.48°S, longitude 118.28°E, elevation: 315 m) opened in 1903 and is still operational. The original weather station at Southern Cross (station number 12074, latitude 31.23°S, longitude 119.33°E, elevation: 355 m) opened in 1889 and was closed December 2007. Another station was opened at Southern Cross Airfield in October 1996 (station number 12320, latitude 31.24°S, longitude 119.36°E, elevation: 347 m) and this remains operational. The historical data used in this study indicates long term decline in rainfall at Narrogin (−0.52 mm per year, 1891–2013). However, mean annual rainfall has been increasing at Southern Cross (0.72 mm per year, 1903–2013) and shows negligible trend at Merredin (−0.01 mm per year, 1889–2013).

In this study, local temperatures for Merredin were a composite constructed from records at Merredin (station 10092) and Merredin Research Station (station 10093). The temperature records used for Southern Cross were a composite constructed from records at Southern Cross Airfield (station 12370) and Southern Cross (station 12074). The temperature records used for Narrogin were a composite constructed from records at Katanning Comparison (station 10579) and Katanning (station 10916).

Variations in rainfall in Western Australia, as for many other parts of the world, are associated with large-scale climate phenomena (including ENSO), which can be described by climate indices. Six of the seven indices used in this study are associated with Pacific Ocean phenomena. Changes in the equatorial Pacific sea temperatures and atmospheric pressures are known to precede seasonal climate variation in other regions by one to at least 12 months [45] and in turn to affect crop yields through an influence on rainfall [46, 47]. There are four climate indices, Niño 1.2, Niño 3, Niño 3.4 and Niño 4 that directly measure changes in sea surface temperature across the equatorial Pacific associated with ENSO. All four Niños were used in this study with values sourced from the Royal Netherlands Meteorological Institute Climate Explorer – a web application that is part of the World Meteorological Organisation and European Climate Assessment and Dataset project. Pressure differences associated with ENSO are typically represented by the Southern Oscillation Index (SOI), calculated as the pressure difference between Tahiti and Darwin, with values for this study obtained from the BOM website. A sixth Climate Index, the Inter-decadal Pacific Oscillation (IPO), was also inputted with values sourced directly from Chris Folland at the UK Met Office. The IPO is considered a measure of temperature and pressure over the central North Pacific with its negative and positive phases thought to modulate the ENSO cycles described by the SOI and Niño values [48].

The Indian Ocean Dipole (IOD) is the only non-Pacific index used in this study. It is a coupled ocean-atmosphere phenomenon measured by the difference in sea surface temperatures in the western and eastern equatorial Indian Ocean [44]. The index is called the Dipole Mode Index (DMI) with values for this study sourced from the Royal Netherlands Meteorological Institute Climate Explorer.

These climate indices, and also local temperatures and rainfall, henceforth referred to as attributes, were input to Neurosolutions *Infinity* software and used to build ANN models to forecast monthly rainfall for Narrogin, Southern Cross and Merredin for lead times of 1, 3, 6, 9, 12 and 18 months, using a test period between July 2004 and June 2014. In some of our earlier reported investigations of applying neural networks to forecast rainfall in Australia, less sophisticated ANN software was available including Neurosolutions 6 for Excel [39] and Peltarion *Synapse* [49]. In those investigations, the approach to configuring the neural network was through manual trial and error, eventually selecting an Elman ANN. A major advantage of Neurosolutions *Infinity* is that it provides automation in testing many ANN configurations, so that an optimal network can be selected. In our more recent [50, 51] investigations using *Infinity*, we have studied monthly rainfall forecasting in several regions in eastern Australia, and found that, in each case studied so far, the configuration selected is a probabilistic ANN. The automated selection of a probabilistic ANN using *Infinity* after testing of various alternative configurations was also found with the present investigation.

All the forecasts were run with the full set of attributes, including local rainfall, each lagged at monthly increments, for up to 12 months. The concept of lag and lead are important in rainfall forecasting. Lag can be defined as the time period between the current value of a parameter, and the same parameter as some time in the past. Lead can be defined as the time period between the current value of a parameter and the same parameter as some time in the future. In this study, lags extended back 12 months, while leads (representing the forecast period) extended for 1, 3, 6, 9, 12 and 18 months forward relative to a particular current value.

The reliability of the probabilistic ANNs forecast was first tested by running seven identical experiments for Narrogin each time letting the *Infinity* software find a best model for Narrogin at a six-month lead. While the choice of attributes used by *Infinity* to arrive at the final solution varied somewhat between experiments, there was little variability in the skill of the monthly forecast for Narrogin for the test period July 2004 to June 2014 (r = 0.73 ± 0.01). This same technique was then used to predict monthly rainfall for Narrogin, Southern Cross and Merredin for 1, 3, 6, 9, 12 and 18 months in advance, for the 10 year test period July 2004 to June 2014.

For every run the total data was divided into training (70%), evaluation (20%) and test sets (10%). The test set was not used in network training, but was important in the choice of the final model. Pearson correlation coefficients (r), Mean Absolute Error (MAE) and Root Mean Square Error (RMSE) were used to compare the skill of the rainfall forecast from the best model for each ANN run against observed monthly rainfall for the test period, from July 2004 to June 2014.

Because total annual rainfall varies across locations, in order to compare RMSE and MAE across locations it is necessary to normalise for differences in rainfall. In this study we normalised only for RMSE by dividing the RMSE values by mean monthly rainfall for each location generating a normalised RMSE. The forecast skill was found to be lower, and more variable, at Merredin and Southern Cross as measured through normalised RMSE. Consequently, it was decided to also explore variability in the forecasts at the 6-month lead for Merredin and Southern Cross.

3   Results

Despite the lack of true seasonality at Narrogin during the test period from July 2004 to June 2014, the probabilistic ANN models generated surprisingly accurate and consistent forecasts of the peaks and troughs in monthly rainfall six months in advance, as shown in Table 1 and Fig. 2. The best and worst runs (in terms of RMSE), both forecast the unseasonal and heavy rainfall event in December 2011 of 143 mms with a high level of skill as illustrated in Fig. 2a and b. Other months with heavy rain were also reliably forecast, including May 2005, July 2007, June 2009 and September 2013, as shown in Fig. 2a and b.

Some of the runs overestimated, while others underestimated these events, with the most accurate forecast (r = 0.80, RMSE = 19.4, MAE = 14.7) generated by averaging the monthly output from the seven runs, referred to as the ensemble shown in Fig. 2c. This ensemble forecast had a better skill score than the mean of the 7 individual runs at the 6-month lead (r = 0.80 versus r = 0.73) as well as better than any of the individual forecast runs, Table 1, showing forecast skill from the ANN models and ensembles of models for Narrogin, Merredin and Southern Cross relative to climatology. Climatology was calculated by using the long term mean monthly rainfall listed at the BOM website for that location, and observed monthly rainfalls for the test period of July 2004 to June 2014. Ensemble means were calculated by averaging output for each month for the listed models for the test period, and also using observed monthly rainfall for the test period.

The skill of a monthly rainfall forecast from a GCM typically declines from one to three month-lead time [33, 34]. In contrast, the forecast 18 months in advance from the ANN was the most skilful measured in terms of highest Pearson correlation and lowest RMSE (r = 0.77, RMSE = 21.2), Table 1. This forecast was uniquely able to forecast the un-seasonally wet January months at Narrogin in 2006 and 2011. This counter-intuitive result would suggest that the predictors of monthly rainfall are in place up to 18 months in advance for Narrogin and that differences in the presentation of input data associated with the varying lead times (i.e. 6 months versus 18 months) are a more significant constraint on the skill of the ANN forecast.

Table 2 shows the relative contribution of the input attributes for the seven runs at Narrogin with a six-month lead and also for the six runs at the different lead times for Narrogin, Merredin and Southern Cross. 'Niño' includes Niño 1.2, Niño 3, Niño 3.4

Table 1. Forecast skills for Narrogin, Merredin and Southern Cross relative to climatology.

	Pearson correlation (r)	RMSE (mm)	MAE (mm)	Normalised RMSE
NARROGIN (mean annual rainfall 494 mm)				
Climatology	0.59	27.0	19.3	
Mean of 7 runs at 6-month lead	0.73 ± 0.01	23.9 ± 0.3	17.7 ± 0.3	
Single run at 1-month lead	0.73	22.7	17.4	0.55
Single run at 3-month lead	0.69	23.9	18.7	0.58
Single run at 6-month lead	0.74	23.1	17.4	0.56
Single run at 9-month lead	0.73	22.4	16.4	0.54
Single run at 12-month lead	0.73	22.2	16.6	0.54
Single run at 18-month lead	0.77	21.2	16.6	0.51
Ensemble mean for the 7 runs at 6-month lead	0.80	19.4	14.7	
MERREDIN (mean annual rainfall 326 mm)				
Climatology	0.32	21.3	15.9	
Mean of 7 runs at 6-month lead	0.58 ± 0.02	18.1 ± 0.5	13.8 ± 0.4	
Single run at 1-month lead	0.58	17.7	13.1	0.73
Single run at 3-month lead	0.61	21.3	15.2	0.87
Single run at 6-month lead	0.70	16.5	11.7	0.68
Single run at 9-month lead	0.60	18.8	13.0	0.77
Single run at 12-month lead	0.49	19.4	14.0	0.79
Single run at 18-month lead	0.65	16.2	11.3	0.66
Ensemble mean for 7 runs at 6-month lead	0.72	11.46	15.15	
Ensemble mean for the 6 runs each at leads of 1, 3, 6, 9, 12 and 18-months	0.75	14.36	10.63	
SOUTHERN CROSS (mean annual rainfall 293 mm)				
Climatology	0.15	22.3	17.3	
Mean of 7 runs at 6-month lead	0.56 ± 0.03	18.9 ± 0.4	14.2 ± 0.4	
Single run at 1-month lead	0.62	17.3	13.2	0.71
Single run at 3-month lead	0.59	18.5	14.6	0.76
Single run at 6-month lead	0.69	17.1	12.6	0.70
Single run at 9-month lead	0.64	17.3	12.6	0.71
Single run at 12-month lead	0.60	17.9	13.9	0.73
Single run at 18-month lead	0.59	18.1	13.7	0.74
Ensemble mean for 7 runs at 6-month lead	0.73	15.1	11.8	
Ensemble mean for 4 runs each at the leads of 6, 9, 12 and 18 months	0.77	14.0	10.8	
Ensemble mean for 3 runs at leads of 9 12 and 18 months and for the 7 runs at 6 months lead	0.76	14.5	11.5	

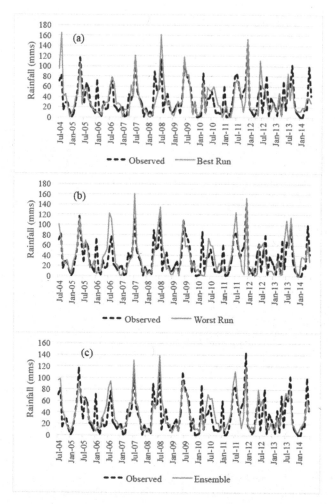

Fig. 2. Mean monthly rainfall for Narrogin, observed vs ANN output - test period July 2004 to June 2014

and Niño 4. 'Temps' includes maximum and minimum temperatures for each location. 'Complex' refers to situations where the ANN has combined attributes into a formula.

At Narrogin, there was some variability in the relative contribution of the different input attributes for the six-month forecasts that was also apparent comparing the forecasts for the different lead times (1, 3, 6, 9, 12 and 18 months), as shown in Table 2. Local atmospheric temperatures contributed on average 35% to the skill of the forecasts for both the 6-month lead forecast and other lead times. The Indian Ocean Dipole measured by the DMI, was the next most consistently important input attribute, contributing on average 14–15% of the skill, but with this contribution varying from 8–20% for individual runs. The contribution of the Niños and SOI was more variable, but on average they contributed 12.8 and 8.2% to the 7 forecast runs for the 6 month

Table 2. Relative contributions of the input attributes

	SOI	IPO	DMI	Niños	Temps.	Rainfall	Complex
Narrogin forecast for 6-month lead							
Run 1	14.1	4.7	13.6	7.4	35.1	7.0	18.1
2	6.8	0	19.1	0	52.0	15.7	6.3
3	14.7	0	20.3	11.2	30.3	8.1	15.4
4	20.1	0	7.8	7.9	39.3	5.9	19.0
5	14.7	0	18.9	7.3	27.8	3.0	28.3
6	9.1	0	9.6	0	28.2	12.6	31.0
7	10.1	5.7	14.2	23.3	33.9	8.2	4.5
Mean	**12.8**	**1.5**	**14.8**	**8.2**	**35.2**	**8.6**	**17.5**
Narrogin forecasts for different leads							
Lead 1	6.8	0	19.1	0	52	15.7	6.3
3	0	0	17.3	28	29.8	16.2	8.7
6	14.1	4.7	13.6	7.4	35.1	7	18.1
9	12.2	0	12	24.7	25	14.8	10.9
12	6.6	6	12.6	21.7	22.6	17.7	12.8
18	5.4	5	8.3	0	50.2	25.9	5.2
Mean	**7.5**	**2.6**	**13.8**	**13.6**	**35.8**	**16.2**	**10.3**
Merredin forecasts for different leads							
Lead 1	8.9	4.3	16.2	11.3	32.6	7.3	19.3
3	30.8	7.9	16.3	22	13.5	9.4	0
6	1.1	1.9	13	3	70.5	10.4	0.1
9	0	11.6	21.2	26.2	15.4	25.5	0
12	13.3	5.6	34.8	21.1	19.7	5.5	0
18	15	11.5	19.5	25.4	15.1	13.4	0
Mean	**11.5**	**7.1**	**20.2**	**18.2**	**27.8**	**11.9**	**3.2**
Southern Cross forecasts for different leads							
Lead 1	7.5	2.8	20.6	23.4	24.6	11.8	9.2
3	0	14.9	7.2	30.8	30.5	9.7	6.9
6	11.7	15.9	23.7	11.4	14.2	23	0
9	10.2	7.9	13.7	33	20.9	12.4	1.8
12	7.6	0	17.1	32.7	22.5	10.4	9.6
18	18.1	7.2	16.1	20	15.7	18.1	4.9
Mean	**9.2**	**8.1**	**16.4**	**25.2**	**21.4**	**14.2**	**5.4**

lead, and 7.5 and 13.6% to the single runs at the variable lead times, respectively. Table 2 also shows that the IPO contributed on average less than 3% to the skill of the forecasts.

While the ANN forecasts for Narrogin showed considerable skill in anticipating the peaks and troughs in monthly rainfall for the period of July 2004 to June 2014 including the heaviest rainfall events, as shown in Fig. 3, this was not the case for Merredin. At Merredin, each of the seven runs at the six-month leads failed to forecast

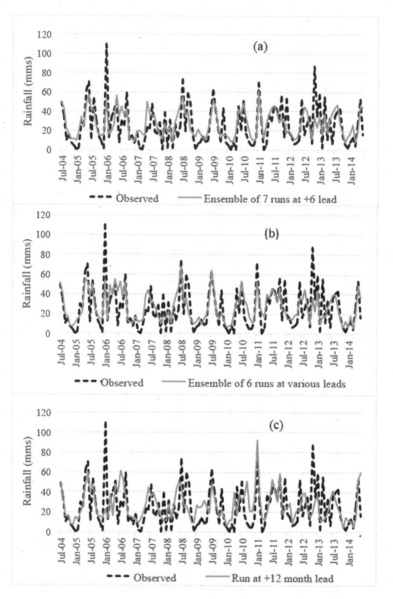

Fig. 3. Mean monthly rainfall for Merredin, observed vs ANN output - test period July 2004 to June 2014.

the two heaviest rainfall events, in January 2006 and November 2012. As for Narrogin, a better monthly rainfall forecasts at the 6-month lead was achieved by averaging the monthly output from the individual runs, and calculating a value for the ensemble, r = 0.72, Table 1. But even this ensemble showed no skill at forecasting the exceptionally wet months, as shown in Fig. 3a. The mean of the 7 runs gave a Pearson correlation of 0.58, as shown in Table 1.

The individual runs at forecast lead times of 1, 3, 6, 9, 12 and 18 months also failed to anticipate these periods of heavy rainfall, Fig. 4b. The worst individual run, Fig. 3c, which was at a lead-time of 12 months, with a Pearson correlation of 0.49, was nevertheless better than climatology at 0.32, as shown in Table 1.

The mean of the ensemble of the 6 runs at the leads of 3, 6, 9, 12 and 18 months gave a better score than the ensemble mean for the 7 runs at the 6-month lead, Table 1. As for the ensemble based on the 6-month lead, this forecast did not anticipate the exceptionally wet months, Fig. 4, but did have the best overall skill score, as shown in Table 1. While the two wettest months in the 10-year forecast period were not forecast, the ANN models consistently forecast the peaks in June 2007, and July 2008, 2009 and 2010. The ANN models also consistently forecast the un-seasonally heavy rain in January 2011 of 70.2 mms, shown in Fig. 3.

Consistent with Narrogin, local atmospheric temperatures were again the most important input followed by the Indian Ocean Dipole, Table 2. In the case of the worst forecast measured in terms of the Pearson correlation, which is the 12-month lead, the DMI contributed 34.8% to the skill of the forecast, as illustrated in Table 2.

At Southern Cross, as for Narrogin and Merredin, the best forecasts were achieved by creating ensembles, Table 1. Averaging the monthly output from the 6,9,12 and 18 month forecasts creates a forecast with the highest Pearson correlation, and lowest RMSE and MAE, as shown in Table 1. This ensemble is, in effect, a 6-month forecast, but with a better skill score than the ensemble created from the seven runs at the 6-month lead, or any of the individual run. Combining the seven runs at the 6-month lead with the individual 9, 12 and 18 month runs (not including the 6 lead because it is already counted in the 7 runs at 6 months) does not achieve a better skill score, as shown in Table 1.

While the Pearson correlations for the ensembles for Southern Cross are comparable to those achieved for Narrogin, Table 1, visual inspection of the output indicates that the forecasts are not as skilful, Fig. 3 versus Fig. 4. This is consistent with the higher normalised RMSE for Southern Cross and Merredin, relative to Narrogin, Table 1. While for Narrogin the ANN successfully forecast the wettest months during the 10 year period, at Southern Cross this was not the case. Also at Southern Cross, there was considerable variability in the individual runs. For example, the first run with the six-month lead time forecast only 9 mm for December 2011, Fig. 4b, while the third run forecast 118 mms, Fig. 4c. Results from the ANN were nevertheless better than climatology which suggests monthly rainfall at Southern Cross is historically erratic, as shown in Table 1. At Southern Cross the ANN did a surprisingly good job of forecasting the low winter rainfall of 2006 and 2007, as illustrated in Fig. 4.

At Southern Cross, the four Niños were consistently more important inputs contributing from 11 to 33% to the skill of the forecast, with a mean of 25%, as shown in Table 2. Atmospheric temperatures and DMI were still important contributing 21 and 16% respectively.

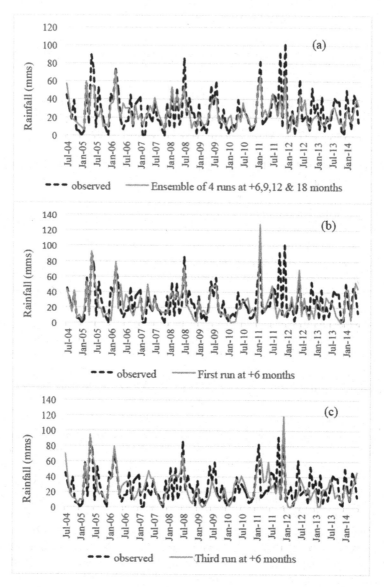

Fig. 4. Mean monthly rainfall for Southern Cross, observed vs ANN output - test period July 2004 to June 2014

4 Discussion and Conclusions

Western Australian wheat farmers would benefit from more skilful, site specific rainfall forecast information than currently provided in the operational forecasts from the BOM. In particular, they need more information on the likely quantity, duration and timing of rainfall [52–55]. Forecasts from ANNs are the type of deterministic, location-specific format that studies have identified as potentially most valuable for wheat farmers. Our results indicate skilful rainfall forecast can be achieved using neural networks for the three locations of Narrogin, Merredin and Southern Cross using ANNs, with skill scores above climatology consistently achieved up to 18-months in advance, as illustrated in Table 1.

It is well established that rainfall in the Western Australian wheat belt is affected by global atmospheric and oceanic circulation patterns that can be described by climate indices, also known as teleconnection signals. It is also well established that these climate indices exhibit non-linearity such that their influence varies in complex ways with season and geography [44]. Perhaps not surprisingly, the very nature of the problem of successfully forecasting monthly rainfall in the wheat belt is amenable to the use of machine learning using ANN as demonstrated by the relatively skilful rainfall forecasts achieved in this study as shown in Table 1 and Figs. 2, 3 and 4.

The skill of the monthly rainfall forecasts were calculated through direct comparisons of output from the best ANN models developed through the machine learning process with observed rainfall for the test period, in this study extending for ten years from June 2004 to July 2014. Table 1 shows that the most skilful forecasts were achieved through the use of ensembles whereby the forecast is calculated after averaging monthly output from multiple runs. Table 1 also shows that when the output from multiple runs was considered as an ensemble, Pearson correlation coefficients consistently exceeded 0.7 for all three locations. This result is better than any so-far reported for POAMA for any locations in Australia for any long lead-times [33, 34, 43].

The lack of deterioration in the forecast skill with lead-time is significant and suggests that some predictors of rainfall are in place more than one year in advance. Changes in ENSO are known to precede seasonal climate variation by up to 12 months [45]. The ANNs models used in this study appear to be utilizing and integrating such information in order to generate skilful forecasts up to 18 months in advance, particularly for Narrogin.

We have combined climate indices and other potentially predictive inputs (atmospheric temperature and local rainfall) into long arrays, each lagged incrementally for up to 12 months. We have then used this information to build location-specific models for monthly rainfall forecasting with analysis of the sensitivity of the input attributes, Table 2, in effect confirming the work of others [43, 44] that no-single climate index is likely to ever contributes more than 20% to forecast skill. Considering broad categories of affect, we found that the DMI could contribute 13.8 to 20.2% to the skill of a forecast for locations within the wheat-belt, as shown in Table 2. In the case of Southern Cross, a combination of the four Niños could contribute up to 25% to the skill of the forecast, and typically the contribution from individual climate indices was less than this.

The significantly more skilful forecasts for Narrogin relative to the forecasts for Merredin and Southern Cross may be a function of the lower mean rainfall for Merredin and Southern Cross of 325.8 mm and 292.8 mms respectively, relative to Narrogin at 494.4 mm. Generally the lower and more erratic the rainfall at a particular location, the more difficult it will be for an ANN to achieve a skilful forecast.

The forecasts for all three locations may be amenable to further improvement through the incorporation of additional climate indices. Schepen et al. [43] identified the

Indian Ocean East Pole Index (EPI), the ENSO Modoki Index (EMI) and Blocking as dominant influences for particular grid cells in the wheat belt. Other climate indices that should be considered in future research are the Madden Julian Oscillation (MJO) and the Southern Annual Mode (SAM), which are also measures of climate phenomena thought to impact Western Australia's climate. These indices were not considered here because they are only available from the 1950s. We chose to use longer datasets and were thus restricted to the climate indices available from the late 1800s.

The neural network approach illustrated here for three locations in the Western Australian wheat belt provides the potential for significant improvement in medium-term rainfall forecasts, particularly relative to forecasts currently reliant on general circulation models that have skill levels often little better than climatology particularly at long lead times [32, 34]. Several dozen other sites can be identified within the Western Australian wheat belt where long rainfall records are available and the neural network approach can also potentially be applied. These improved rainfall forecasts can be tailored to the specific requirements of farmers, and would likely translate into tangible economic benefits [3, 56].

The official rainfall forecasts issued by the BOM are produced using the general circulation model POAMA. Monthly rainfall forecasts with long lead times up to 8 months for south-west region of Western Australia have low skill, and are generally little better than climatology [32], a simple averaging of rainfall over an extended period. In contrast, skills significantly superior to climatology are achievable using the ANN approach. We have previously reported similar findings for studies of monthly rainfall forecasting for eastern Australia, including Queensland [50] and the Murray Darling Basin [51]. Low skill in rainfall forecasting with general circulation models has been reported for other regions of the world, for example in the United States [57]. There is significant scope to further improve the skill of the monthly rainfall forecasts presented here for sites in Western Australia. For example, our investigations have demonstrated that generating forecasts one month at a time with an ANN instead of for all months together produces an enhancement in forecast skill [50]. This approach has been particularly useful in forecasting extreme rainfall events, as for example during the flooding of Brisbane during summer of 2010/2011 [58].

Acknowledgements. This research was funded by the B. Macfie Family Foundation.

References

1. ABARE: Australian Crop Report No. 154. Technical report. Australian Bureau of Agricultural and Resource Economics, Canberra, ACT (2010)
2. Anderson, W.K., Hamza, M.A., Sharma, D.L., et al.: The role of management in yield improvement of the wheat crop—a review with special emphasis on Western Australia. Aust. J. Agric. Res. **56**, 1137–1141 (2005)
3. Anderson, W.K.: Closing the gap between actual and potential yield of rainfed wheat. The impacts of environment, management and cultivar. Field Crops Res. **116**, 14–22 (2010)

4. Del Cima, R., D'Antuono, M.F., Anderson, W.K.: The effects of soil type and seasonal rainfall on the optimum seed rate for wheat in Western Australia. Aust. J. Exp. Agr. **44**(6), 585–594 (2004)

5. Hunt, J.R., Kirkegaard, J.A.: Re-evaluating the contribution of summer fallow rain to wheat yield in southern Australia. Crop Pasture Sci. **62**, 915–929 (2011)

6. Pook, M., Lisson, S., Risbey, J., et al.: The autumn break for cropping in southeast Australia: trends, synoptic influences and impacts on wheat yield. Int. J. Climatol. **29**, 2012–2026 (2009)

7. Sharma, D.L., D'Antuono, M.F., Anderson, W.K., et al.: Variability of optimum sowing time for wheat yield in Western Australia. Aust. J. Agr. Res. **59**(10), 958–970 (2008)

8. Sprigg, H., Belford, R., Milroy, S., et al.: Adaptations for growing wheat in the drying climate of Western Australia. Crop Pasture Sci. **65**, 627–644 (2014)

9. Zhang, H., Turner, N.C., Simpson, N., et al.: Growing-season rainfall, ear number and the water-limited potential yield of wheat in south-western Australia. Crop Pasture Sci. **61**, 296–303 (2010)

10. French, R.J., Schultz, J.E.: Water use efficiency of wheat in a Mediterranean-type environment: I. The relation between yield, water use and climate. Aust. J. Agric. Res. **35**, 743–764 (1984)

11. Carberry, P.S., Hammer, G.L., Meinke, H., et al.: The potential value of seasonal climate forecasting in managing cropping systems. In: Hammer, G., Nicholls, N., Mitchell, C. (eds.) Application of Seasonal Climate Forecasting in Agricultural and Natural Ecosystems—The Australian experience, pp. 167–181. Kluwer Academic Publishers, Dordrecht (2000)

12. Dolling, P.J., Fillery, I.R.P., Ward, P.R., et al.: Consequences of rainfall during summer–autumn fallow on available soil water and subsequent drainage in annual-based cropping systems. Aust. J. Agric. Res. **57**, 281–296 (2006)

13. Stephens, D.J., Lyons, T.J.: Variability and trends in sowing dates across the Australian wheatbelt. Aust. J. Agric. Res. **49**, 1111–1118 (1998)

14. Stephens, D.J., Lyons, T.J.: Rainfall-yield relationships across the Australian wheatbelt. Aust. J. Agr. Res. **49**, 211–223 (1998)

15. Pook, M.J., Risbey, J.S., McIntosh, P.C.: The synoptic climatology of cool-season rainfall in the central wheatbelt of Western Australia. Mon. Weather Rev. **140**, 28–43 (2012)

16. Fox, G., Turner, J., Gillespie, T.: The value of precipitation forecast information in winter wheat production. Agr. Forest Meteorol. **95**, 99–111 (1999)

17. Anwar, M.R., Rodriguez, D., Liu, D.L., et al.: Quality and potential utility of ENSO-based forecasts of spring rainfall and wheat yield in south-eastern Australia. Aust. J. Agr. Res. **59** (2), 112–126 (2008)

18. Hammer, G.L., Hansen, J.W., Phillips, J.G., et al.: Advances in application of climate prediction in agriculture. Agr. Syst. **70**, 515–553 (2001)

19. Hammer, G.L., Holzworth, D.P., Stone, R.: The value of skill in seasonal climate forecasting to wheat crop management in a region with high climatic variability. Aust. J. Agr. Res. **47**, 717–737 (1996)

20. Marshall, G.R.: Risk attitude, planting conditions and the value of seasonal forecasts to a dryland wheat grower. Aust. J. Agr. Econ. **40**(3), 211 (1996)

21. Potgieter, A.B., Hammer, G.L., Meinke, H., et al.: Three putative types of El Nino revealed by spatial variability in impact on Australian wheat yield. J. Climate **18**(10), 1566–1574 (2005)

22. Podesta, G.P., Messina, C.D., Grondona, M.O., et al.: Associations between grain crop yields in central-eastern Argentina and El Nino-Southern Oscillation. J. Appl. Meteorol. **38** (10), 1488–1498 (1999)

23. Alberto, C.M., Streck, N.A., Heldwein, A.B., et al.: Soil water and wheat, soybean, and maize yields associated to El Nino Southern Oscilation. Pesquisa Agropecuaria Bras. **41**(7), 1067–1075 (2006)

24. Hansen, J.W., Jones, J.W., Irmak, A., et al.: El Nino-Southern Oscillation impacts on crop production in the southeast United States. In: Hatfield, J.L., Volenec, J.J. Dick, W.A. (eds.) Proceedings of Impacts of El Nino and Climate Variability on Agriculture, vol. 63, pp. 55–76. ASA Special Publication (2001)

25. Legler, D.M., Bryant, K.J., O'Brien, J.J.: Impact of ENSO-related climate anomalies on crop yields in the US. Clim. Change **42**(2), 351–375 (1999)

26. Shuai, J., Zhang, Z., Sun, D.Z., et al.: ENSO, climate variability and crop yields in China. Clim. Res. **58**(2), 133–148 (2013)

27. Selvaraju, R.: Impact of El Nino-southern oscillation on Indian food grain production. Int. J. Climatol. **23**(2), 187–206 (2003)

28. Fawcett, R.J.B., Stone, R.C.: A comparison of two seasonal rainfall forecasting systems for Australia. Aust. Meteorol. Oceanogr. J. **60**, 15–24 (2010)

29. National Climate Centre, BOM, pers. comm., July 2014

30. Grains Research and Development Corporation: Changing weather in the western wheatbelt, Australian government (2014). http://www.grdc.com.au/Media-Centre/Ground-Cover/Ground-Cover-Issue-101/Changing-weather-in-the-western-wheatbelt

31. Petersen, E.H., Fraser, R.W.: An assessment of the value of seasonal forecasting technology for Western Australian farmers. Agr. Syst. **70**, 259–274 (2001)

32. Hawthorne, S., Wang, Q.J., Schepen, A., et al.: Effective use of general circulation model outputs for forecasting monthly rainfalls to long lead times. Water Resour. Res. **49**, 5427–5436 (2013)

33. Langford, S., Hendon, H.H.: Improving reliability of coupled model forecasts of Australian seasonal rainfall. Mon. Weather Rev. **141**, 728–741 (2013)

34. Shao, Q., Li, M.: An improved statistical analogue downscaling procedure for seasonal precipitation forecast. Stoch. Env. Res. Risk A. **27**, 819–830 (2013)

35. Stone, R.C., Hammer, G.L., Marcussen, T.: Prediction of global rainfall probabilities using phases of the Southern Oscillation Index. Nature **384**, 252–255 (1996)

36. Vizard, A.L., Anderson, G.A., Buckley, D.J.: Verification and value of the Australian Bureau of Meteorology township seasonal rainfall forecasts in Australia, 1997–2005. Meteorol. Appl. **12**, 343–355 (2005)

37. Silverman, D., Dracup, J.A.: Artificial neural networks and long-range precipitation prediction in California. J. Appl. Meteorol. **39**(1), 57–66 (2000)

38. Goyal, M.K.: Monthly rainfall prediction using wavelet regression and neural network: an analysis of 1901–2002 data, Assam, India. Theor. Appl. Climatol. **118**(1–2), 25–34 (2014)

39. Abbot, J., Marohasy, J.: Input selection and optimisation for monthly rainfall forecasting in Queensland, Australia, using artificial neural networks. Atmos. Res. **138**, 166–178 (2014)

40. Bishop, C.M.: Pattern Recognition and Machine Learning. Springer, New York (2006)

41. Aytek, A., Asce, M., Alp, M.: An application of artificial intelligence for rainfall-runoff modelling. J. Earth Syst. Sci. **117**(2), 145–155 (2008)

42. Trambaiolli, L.R., Lorena, A.C., Fraga, F.J., et al.: Improving Alzheimer's disease diagnosis with machine learning techniques. Clin. EEG Neurosci. **42**(3), 160–165 (2011)

43. Schepen, A., Wang, Q.J., Robertson, D.: Improving rainfall forecasts for seasonal streamflow forecasts. In: Proceedings of the 34th Hydrology and Water Resources Symposium, pp. 1117–1124 (2012)

44. Risbey, J.S., Pook, M.J., Mcintosh, P.C., et al.: On the remote drivers of rainfall variability in Australia. Mon. Weather Rev. **137**, 3233–3253 (2009)

45. Montroy, D.L., Richman, M.B., Lamb, P.J.: Observed nonlinearities of monthly teleconnections between tropical Pacific sea surface temperature anomalies and central and eastern North American precipitation. J. Climate **11**, 1812–1835 (1998)
46. Chen, C.C., McCarl, B., Hill, H.: Agricultural value of ENSO information under alternative phase definition. Clim. Change **54**, 305–325 (2002)
47. Hill, H.S.J., Butler, D., Fuller, S.W., et al.: Effects of seasonal climate variability and the use of climate forecasts on wheat supply in the United States, Australia, and Canada. In: Proceedings of Impacts of El Nino and Climate Variability on Agriculture, vol. 63, pp. 101–123. ASA Special Publication (2001)
48. Power, S., Haylock, M., Colman, R., et al.: The predictability of interdecadal changes in ENSO activity and ENSO teleconnections. J. Climate **19**, 4755–4771 (2006)
49. Abbot, J., Marohasy, J.: Application of artificial neural networks to rainfall forecasting in Queensland, Australia. Adv. Atmos. Sci. **29**, 717–730 (2012)
50. Abbot, J. Marohasy, J.: Improving monthly rainfall forecasts using artificial neural networks and single-month optimisation: a case study of the Brisbane catchment, Queensland, Australia. In: Water Resources Management, vol. VIII, pp. 3–13. WIT Press (2015)
51. Abbot J., Marohasy, J.: Forecasting of monthly rainfall in the Murray Darling Basin, Australia. In: Miles as a case study. River Basin Management, vol. VIII, pp. 149–159. WIT Press (2015)
52. Hansen, J.W.: Realizing the potential benefits of climate prediction to agriculture: issues, approaches, challenges. Agr. Syst. **74**, 309–330 (2002)
53. Hansen, J.W.: Integrating seasonal climate prediction and agricultural models for insights into agricultural practice. Philos. Trans. R. Soc. London B. **360**, 2037–2047 (2005)
54. Mjelde, J.L., Dixon, B.L.: Valuing the lead time of periodic forecasts in dynamic production systems. Agr. Syst. **42**(1–2), 41–55 (1993)
55. Moeller, C.I., Smith, S., Asseng, F., et al.: The potential value of seasonal forecasts of rainfall categories—case studies from the wheatbelt in Western Australia's Mediterranean region. Agr. Forest Meteorol. **148**, 606–618 (2008)
56. Ash, A., McIntosh, P., Cullen, B., et al.: Constraints and opportunities in applying seasonal climate forecasts in agriculture. Aust. J. Agr. Res. **58**, 952–965 (2007)
57. Zimmerman, B.G., Vimont, D.J., Block, P.J.: Utilizing the state of ENSO as a means for season-ahead predictor selection. Water Res. Res. **52**(5), 3761–3774 (2016)
58. Abbot J., Marohasy, J.: Forecasting extreme monthly rainfall events in regions of Queensland, Australia using Artificial Neural Networks. Int. J. Sust. Dev. Plann (in press)

Forecasting Monthly Rainfall in the Bowen Basin of Queensland, Australia, Using Neural Networks with Niño Indices

John Abbot[1,2,3(✉)] and Jennifer Marohasy[1,3]

[1] Climate Modelling Laboratory, Noosaville, QLD, Australia
johnwabbot@gmail.com, jennifermarohasy@gmail.com
[2] Department of Engineering, University of Tasmania, Hobart, TAS, Australia
[3] Institute of Public Affairs, Melbourne, VIC, Australia

Abstract. For three decades there has been a significant global effort to improve El Niño-Southern Oscillation (ENSO) forecasts with the focus on using fully physical ocean-atmospheric coupled general circulation models (GCMs). Despite increasing sophistication of these models and the computational power of the computers that drive them, their predictive skill remains comparable with relatively simple statistical models. In this study, an artificial neural network (ANN) is used to forecast four indices that describe ENSO, namely Niño 1 + 2, 3, 3.4 and 4. The skill of the forecast for Niño 3.4 is compared with forecasts from GCMs and found to be more accurate particularly for forecasts with longer-lead times, and with no evidence of a Spring Predictability Barrier. The forecast values for Niño 1 + 2, 3, 3.4 and 4 were subsequently used as input to an ANN to forecast rainfall for Nebo, a locality in the Bowen Basin, a major coal-mining region of Queensland.

Keywords: ENSO · Niño · Sea surface temperature · Artificial neural network · General circulation model · Rainfall · Spring predictability barrier

1 Introduction

Australian rainfall is extremely variable with episodes of drought that often end with extreme flooding. During the austral summer of 2010–2011 flooding impacted Queensland with the capital city Brisbane inundated and 85% of Queensland coalmines either closed entirely or operating with restricted production [1, 2]. By May 2011, Queensland's coal mining sector had recovered to only 75% of its pre-flood output, with a loss of A\$5.7 billion, equivalent to 2.2% of Queensland's gross state product for the financial year ending June 2011. A report prepared for Australia's National Climate Change Adaptation Research Facility following that extreme event, concluded that available seasonal rainfall forecasts from the Australian Bureau of Meteorology (BOM) are not useful, lacking localised information, and other micro-details, to enable focused pro-active planning and risk management [1].

Intra-seasonal, inter-annual and decadal variability in Queensland rainfall has been linked to complex physical phenomena remote to the Australian land mass [3]. These phenomena, manifesting as recurrent patterns in sea surface temperature (SST) and air

© Springer International Publishing AG 2016
B.H. Kang and Q. Bai (Eds.): AI 2016, LNAI 9992, pp. 88–100, 2016.
DOI: 10.1007/978-3-319-50127-7_7

pressure, are described numerically by climate indices. The dominant phenomenon is El Niño-Southern Oscillation (ENSO), which can be described quantitatively by measuring the departure from the long-term average of sea surface temperature over specified Niño region of the Pacific Ocean. The terms La Niña and El Niño originally comes from the Spanish and represents opposites, with El Niño events mostly associated with below average rainfall in eastern Australia and warmer waters in the eastern Pacific. The extreme rainfall in the austral summer of 2010–11 was linked to an extraordinarily strong La Niña event [4].

There are five designated Niño regions spanning the tropical Pacific Ocean from which anomalies in SST are calculated. For widespread global climate variability, Niño 3.4 is commonly preferred, because the sea surface temperature variability in this region is thought to have the strongest effect on shifting rainfall in the western Pacific, which in turn modifies the location of the heating that drives other major global atmospheric circulation patterns. Relationships between weather patterns, particularly rainfall, and ENSO phenomena have been explored in many part of the world including Australia [5–7], north America [8, 9], the north Atlantic European region [10], China [11] and India and West Africa [12].

Over the last three decades, a whole suite of different models with varying degrees of complexity have been developed for ENSO prediction. They are generally categorised as (i) purely statistical models that depend on finding patterns in historical data, (ii) physical models that rely on simulating ocean-atmospheric interactions, and (iii) hybrids of the statistical and simulation models [13–15]. Most of the research effort has been into the fully physical coupled ocean-atmospheric models that run on supercomputers, however, their skill at forecasting remains comparable to simple statistical models [15–17].

Data on ENSO is available back to the mid-1800s, with the predictability of El Niño and La Niño varying across decades. For example, the predictability of these events, as measured by anomalous correlations and root mean square error (RMSE), is considered low for the period 1936–1955, while highest scores are achieved for the periods 1876–1895 and 1976–1995 [18]. Predictability also varies on a seasonal basis, with the austral autumn/boreal spring considered a period when there is a relatively small signal-to-noise ratio and is now known as the spring predictability barrier (SPB) [17–19]. The SPB has been extensively studied in the context of general circulation models (GCMs) [20–22].

Yan and Yu [23] found that although taking the mean of an ensemble of 10 GCMs models reduced the SPB, it still remained a feature. Duan and Wei [24] explain that the SPB model errors may come from many different sources, such as model parameter errors, the uncertainties of some physical processes, errors in external forces, and the uncertainties of the computation scheme, without concluding with a definitive determination of which type of model error plays the dominant role in producing prediction uncertainties.

This paper details an investigation into the use of artificial neural network (ANN) software to forecast Niño values for the period 1987 to 2013, and then the incorporation of these values into an ANN model to forecast monthly rainfall for Nebo, a locality in the Bowen Basin that was severely impacted by the strong La Niña event during the Austral summer of 2010–11. This represents an extension of previous investigations into the application of ANNs for monthly rainfall forecasting in

Queensland [25–28], in particular through the incorporation of both lagged and forecast values for the full complement of Niño regions. The forecast Niño 3.4 values are benchmarked against output from other statistical models and also general circulation models.

2 Materials and Method

ANNs are massive, parallel-distributed, information-processing systems with characteristics resembling the biological neural networks of the human brain. The mathematical fundamentals of neural networks and specific applications in hydrology, including rainfall, have been reviewed in the two-part series ASCE Task Committee on Application of Artificial Neural Networks in Hydrology [29]. The ANN models used in this study for forecasting Niño values, and also rainfall, were developed using NeuroSolutions 6 for Excel software (NeuroDimensions, Florida, USA). This software provides great versatility in the architecture of neural networks that can be deployed. For the purposes of this study, a limited number were tested in a preliminary investigations, without extensively attempting to optimise the ANN configuration.

For forecasts of the Niño indices, principal component analysis (PCA) was first deployed using an unsupervised neural network, followed by a supervised neural network comprising a multi-layer perceptron, with one hidden layer. The PCA component of the network consisted of a Sanger synapse which linearly projected the input onto a smaller dimensional space, while preserving maximum intensity of the original signal. This reduced dimension means fewer weights for the supervised network to follow, improving generalisation. Optimisation occurred over 6000 epochs, equally split between the unsupervised and supervised components.

For forecasts of rainfall, some preliminary exploratory testing was undertaken using neural networks with multilayer perceptron architectures with up to three hidden lawyers. However, it was found that superior results were obtained, with Jordan and Elman networks, as shown by lower RMSE in the training and test data sets. Jordan and Elman networks extend the MLP incorporating context units, processing elements that remember past activity. Context units enable the network to extract and utilise temporal information contained in the data. In the Elman network, the activity of the first hidden processing element is copied to the context unit, whereas for the Jordan network the output of the network is copied.

For forecasts of Nebo monthly rainfall reported in this study, a Jordan neural network was selected. For each input data set, the artificial neural network was optimised over 3000 epochs using a genetic optimisation algorithm for 10 or 20 generations. For both Niño forecasts and rainfall forecasts, for every run the total data was divided into training (70%), evaluation (20%) and test sets (10%). The test set was not used in network training, but was important in the choice of the final model. Pearson correlation coefficients (r), Mean Absolute Error (MAE) and Root Mean Square Error (RMSE) were used to compare the skill of the rainfall forecast from the best model for each ANN run against observed values for the test period.

The site of Nebo was chosen because of its proximity to a coal mine in the Bowen Basin and because it is a site with over 120 years of historical rainfall data available from the BOM (station 033054). Surface air temperature data for Nebo is not available, and so maxima and minima data from the Te Kowai Experimental Station in Mackay (station 033047), approximately 100 kms to the northeast of Nebo, were used with data available from 1908.

The first part of the study focused on forecasting ENSO by forecasting SST anomalies from four regions designated Niño 1 + 2 (0–10 S, 80–90 W), Niño 3 (5 S–5 N; 150 W–90 W), Niño 3.4 (5 S–5 N; 170 W–120 W) and Niño 4 (5 S–5 N: 160 E–150 W).

The other climate indices used for the rainfall forecasts were the Southern Oscillation Index (SOI), which is another measure of ENSO calculated using the pressure difference between Tahiti and Darwin [3]; the Inter-decadal Pacific Oscillation (IPO), which is thought to modulate the influence of ENSO on rainfall along the Australian east coast [30, 31]; and also the Dipole Mode Index (DMI), which is a coupled ocean and atmospheric phenomenon in the equatorial Indian Ocean [32].

Values for the Niño indices, SOI and DMI were sourced from the Royal Netherlands Meteorological Institute Climate Explorer, which is a web application that is part of the World Meteorological Organisation and European Climate Assessment and Dataset project. Values for the IPO were provided by Chris Folland from the United Kingdom's Met Office Hadley Centre.

In order to forecast the lead Niño values, fours unary data sets corresponding to Niño 4, Niño 3.4, Niño 3 and Niño 1 + 2 were initially constructed, each comprising the current monthly value, plus twelve lagged values for the previous twelve months. Niño data for the period January 1871 to 1987 was used as the training period, and then forecasts produced for the period August 1987 to August 2013.

The four Nino unary sets were used as input to forecast each of the four Niño SST values in turn, commencing with forecasts for lead month 0. An iterative process was then applied, so that forecast values were added to each unary set to provide the input data for a forecast corresponding to the following month. In this way, a one-month forecast for each of the four Niño-values was produced one-month to twelve months in advance.

A new unary dataset designated as "Ninos" was constructed comprising both lead and lagged values for each Niño index. Seven monthly values of each Niño index contributed to Ninos, comprising lag 3, lag 2, lag 1, current, lead 0, lead 1 and lead 2. Thus Ninos comprised 28 columns of data. The full complement of available lagged Ninos was not tested because input of very large datasets tends to degrade performance of ANN models.

In order to forecast rainfall for Nebo, the desired output, the observed rainfall, was assigned as the monthly rainfall with a lead-time of three months ahead of the current month (lead month 2). The test period, which is also the forecast period, was initially set for 137 months from August 2000 to December 2011 that included the exceptionally wet astral summer of 2010–2011. Five unary input data sets were constructed corresponding to monthly values of the DMI, SOI, IPO, maximum atmospheric temperature (MaxT), and minimum atmospheric temperature (MinT). Each unary data set comprised the current monthly value, plus twelve lagged values for the previous twelve months. The unary set Ninos has already been described.

Unary, binary and ternary combinations of these unary sets were used as inputs to forecast rainfall. A total of 15 combinations were tested.

3 Results and Discussion

3.1 Forecasting ENSO with an ANN

The skill of a model, whether a GCM or ANN, at predicting climate indices or rainfall, as a variable for a single locality or region can be measured by the RMSE. This is the difference between the observed and forecast values [33, 34]. The lower the RMSE the smaller the difference, and therefore the more accurate the forecast. There is an extensive literature evaluating RMSE relative to other statistical techniques [35, 36]. Acknowledging that this statistical measure gives higher proportionate weight to large errors, we consider it suitable for our type of data given the highly variable nature of Queensland rainfall and ENSO events and the importance of being able to accurately forecast extreme events.

Barnston et al. [17] used the RMSE between the forecast and observed succession of running three-month mean SST anomalies for Niño 3.4, as a measure of the skill of eight statistical and 12 dynamic general circulation models to forecast ENSO. In each case the forecast period commenced immediately after the latest available observed data value. The area between the lines designated upper and lower limit define the boundaries of the individual output from these 20 models, as illustrated in .Fig. 1. We have benchmarked output from our ANN against the envelope of values shown in Fig. 1. The ANN Niño 3.4 forecast was for the shorter interval of one month making it a more ambitious forecast. Nevertheless the ANN consistently produced a more skilful forecast than all of the models reviewed by Barnston et al. [17] except for our forecast

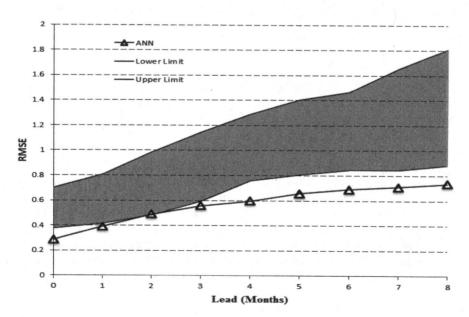

Fig. 1. A comparison of the RMSE for the ANN model versus the 20 models reviewed by Barnston et al. 2012 for Niño 3.4 forecasts

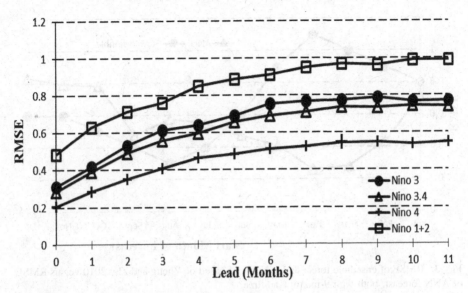

Fig. 2. RMSE as a function of lead-time for all seasons combined for 12 months. All Niño values generated from the ANN model.

with a lead of 2 months, which was equivalent to the best forecasts of the models reviewed by Barnston et al. [17], Fig. 1.

The specific years over which an ENSO forecast is made are considered important when measuring skill with the period 2002–10 and early to middle 1990s considered the more challenging years of the last thirty [18]. Using the ANN, we forecast for the 26 years from 1987 to 2013, which incorporates the more difficult periods.

Our ANN model is most skilful at forecasting Niño 4, and least skilful at forecasting Niño 1 + 2, as shown in Fig. 2. The RMSE for our ANN model increases as the lead time is increased from 0 to seven months, after which it is surprising constant as shown in Fig. 2, more so than for many other models [17, 19, 37].

3.2   The Spring Predictability Barrier

Studies using some GCMs and hybrid models show low skill at forecasting ENSO events during the boreal spring with high RMSE for the months of March to June, this is referred to as the Spring Predictability Barrier (SPB). It is an often-discussed characteristic of ENSO forecasts [38–40]. The SPB phenomenon is illustrated in Fig. 3 using results from the ensemble of forecasts reviewed in Zheng and Zhu [19]. The SPB corresponds to the months of March to May when the RMSE is in the vicinity of 1.0, compared to much lower values during other times of the year, for example July to September when the RMSE falls to values of about 0.3.

Output from our ANN model shows no such drop in skill for the March to May period, that is the SPB is not present, Figs. 3 and 4. The values in Fig. 4 were normalised by dividing the RMSE values by the corresponding absolute error for the corresponding

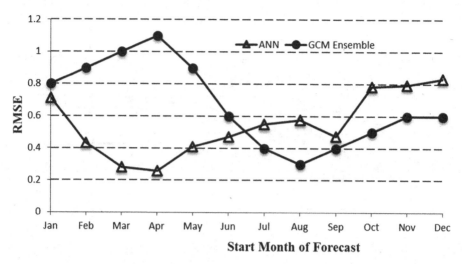

Fig. 3. RMSE of ensemble forecasts for Niño 3.4 based on Zheng and Zhu 2010 versus RMSE of ANN forecast, both with 9-month lead-time.

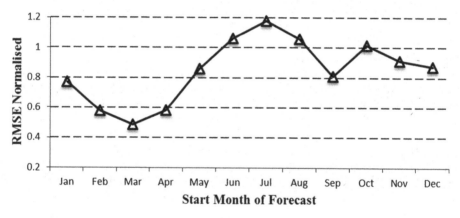

Fig. 4. Normalised RMSE of ANN forecast with 9-month lead-time

month. This produced some evidence for a barrier, but associated with the summer months around June, July and August rather than the spring March, April May.

The SPB is discussed most often in the literature with reference to values of Niño 3.4. The skill of the forecasts for Niño 4 from our ANN is highest (lowest RMSE) from April through to June, while the skill of the forecast for Niño 3 is highest (lowest RMSE) in February, as shown in Fig. 5.

In a review of ENSO predictability, Barnston et al. [17] have presented a set of contour plots of RMSE for Niño 3.4 showing variation with lead time and target season for 12 GCMs and 8 statistical models. Inspection of these figures suggests that forecast skill for each model does indeed vary with the time of year. However, the concept of a

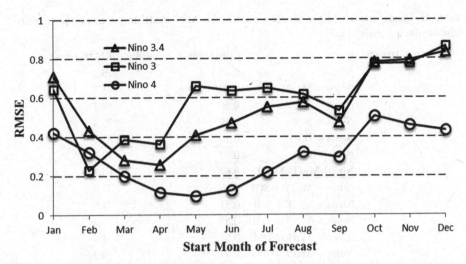

Fig. 5. Plot of RMSE against month for forecasts of Niño 3.4, 3 and 4 generated by the ANN

forecast barrier associated universally with the spring (March to May) is not as ubiquitous as suggested. Focussing on forecasts with 6 to 9 months lead-time, some models including the Scripps Hybrid Coupled Model do indeed exhibit higher RMSE in the boreal spring period. However, other GCMs including the University of Maryland Intermediate Coupled Model and the Australian POAMA GCM exhibit lower RMSE during the first part of the year compared to the latter part. Generally, the statistical models show a large variation in temporal designation of periods representing lower predictive skill for Niño 3.4 within the year. For example, the NOAA CPC-CA statistical model clearly shows higher predictive skill during the first half of the year.

Although the SPB has been extensively researched for many years, particularly in the context of GCMs, a satisfactory explanation remains unresolved [24]. Investigations have suggested that the SPB is probably a result of errors in the models themselves, with both initialisation errors and parameter errors being implicated [41]. Others have attempted to rationalise the SPB in terms of physical phenomena [20]. Results from the present study, together with an objective assessment of the results from a diversity of GCMs and statistical models would tend to suggest that different models have different profiles of skill over the annual cycle, and the emphasis on the SPB is a consequence of the prominence given to output from certain types of GCMs, rather than a real physical phenomenon.

3.3   Forecasting Rainfall for Nebo, Queensland

The unary dataset designated Ninos was used alone and in binary and ternary combinations to forecast rainfall for Nebo with a lead-time of three months ahead of the current month. Consistent with our previous studies using ANN to forecast rainfall in Queensland [25], combinations incorporating local maximum and minimum temperature gave superior rainfall forecasts, with the highest r value and lowest RMSE, as illustrated in Table 1.

Table 1. Combinations of input variables used in the ANN model to forecast Nebo monthly rainfall with a 3-month lead

Input data sets	Inputs	Correlation coefficient (r)	RMSE (mm)
Unary	Ninos	0.29	79.2
Binary	Ninos + MaxT	0.50	71.7
	Ninos + MinT	0.56	68.7
	Ninos + SOI	0.29	79.4
	Ninos + DMI	0.30	81.0
	Ninos + IPO	0.49	72.2
Ternary	**Ninos+MinT+MaxT**	**0.64**	**64.0**
	Ninos + MaxT + SOI	0.50	71.7
	Ninos + MinT + SOI	0.51	71.0
	Ninos + MaxT + DMI	0.51	71.3
	Ninos + MinT + DMI	0.51	71.6
	Ninos + MaxT + IPO	0.53	70.3
	Ninos + MinT + IPO	0.51	71.1
	Ninos + SOI + IPO	0.48	74.8
	Ninos + IPO + DMI	0.38	76.6
	Ninos + DMI + SOI	0.17	81.6

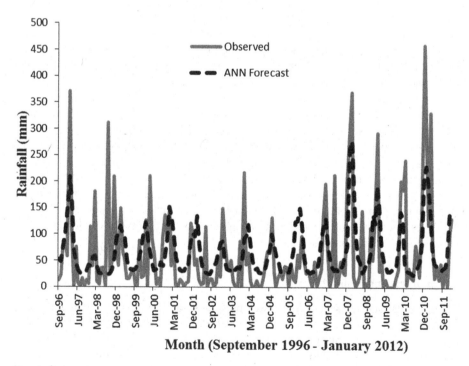

Fig. 6. Comparing observed rainfall (mm) with forecast rainfall (mm) for Nebo from September 1996 to January 2012

Figure 6 compares observed rainfall (mm) with forecast rainfall (mm) for Nebo from September 1996 to January 2012. The forecast was output from an ANN model after inputting current monthly values and 12 lagged values for minimum and maximum temperatures and the unary Nino combination of lagged and lead values.

A visual comparison of the observed versus forecast rainfall for the ternary combination, including the lagged and lead values for the Ninos with the current and lagged temperature inputs, indicates that the ANN consistently forecast too much rain for the drier months, as shown in Fig. 6. The ANN was able to give some indication that the astral summers of 1996–97, 2007–2008 and 2010–2011 were going to be wetter than average, but it did not forecast the exceptionally wet August of 1998 or December 2003 or adequately give an indication of the magnitude of the very wet months of February 1997 or December 2010.

Sensitivity analysis of the output indicated that the most influential inputs determining rainfall are Niño 3 (Lag 1), Niño 3.4 (Lag 1) and Niño 4 (Lead1).

4   Conclusions

Extreme rainfall during the austral summer of 2010–11 is linked with an extraordinarily strong La Niña [4]. This ENSO event was inadequately forecast by the BOM, and resulted in significant flooding across eastern Australia including major disruptions to coal mining operations in the Bowen Basin [1].

There has been a significant investment in GCMs over the last three decades and a major global research effort focused on improved ENSO forecasts. Barnston et al. [17] argue that GCMs can now outperform statistical models in their skill at forecasting ENSO, while Halide and Ridd [16] suggest that despite their complexity and the superior computational power of the super computers used to run them, GCMs are no better at forecasting ENSO than very simple models. Barnston et al. [17] and Halide and Ridd [16] both acknowledge problems associated with forecasting through the boreal SPB, with Barnston et al. [17] suggesting this will eventually be overcome as "science and engineering continues to advance" while Halid and Ridd [16] suggests the solution lies in better understanding the physical phenomena that result in anomalous warming and cooling of particular regions of the equatorial Pacific.

These, and other review papers [8, 19–21], fail to adequately consider the potential advantage of using the most advanced statistical models currently available, which are ANNs that have been developed largely independently of the mainstream climate science community. ANNs require a different skill set for implementation than GCMs but since at least 2006 has shown potential for forecasting ENSO with more skill than GCMs and simple statistical models [42]. Consistent with Wu et al. [42] the results from our study indicate that the lower RMSE for ANNs is at least in part a consequence of ANNs being able to forecast through the SPB.

Translating the improved ENSO forecast into an improved rainfall forecast provides an additional level of complexity. In previous studies we have shown that ANNs can produce a superior monthly rainfall forecast for localities in Queensland relative to output from the official GCMs [25–28]. The Bureau was unable to provide forecasts to enable benchmarking for the locality of Nebo in this study. We recognise that our best forecast measured in terms of lowest RMSE for the period from September 1996 to January 2012 has limitations, but it does demonstrate an improved skill through the inclusion of lead as well as lag values for Niño regions.

Acknowledgements. This work was funded by the B. Macfie Family Foundation.

References

1. Queensland Government Flood Commission of Inquiry, Chap. 13 Mining (2012). http://www.floodcommission.qld.gov.au/publications/final-report
2. Sharma, V., van de Graaff, S., Loechel, B., Franks, D.: Extractive resource development in a changing climate: learning the lessons from extreme weather events in Queensland, Australia. In: National Climate Change Adaptation Research Facility, Gold Coast, p. 110 (2013)
3. Risbey, J.S., Pook, M.J., Mcintosh, P.C.: On the remote drivers of rain variability in Australia. Mon. Weather Rev. **137**, 3233–3253 (2009)
4. Cai, W., van Rensch, P.: The 2011 southeast Queensland extreme summer Rain: a confirmation of a negative Pacific Decadal Oscillation phase? Geophys. Res. Lett. **39**, L08702 (2012)
5. Anwar, M.R., Rodriguez, D., Liu, D.L., et al.: Quality and potential utility of ENSO-based forecasts of spring rainfall and wheat yield in south-eastern Australia. Aust. J. Agri. Res. **59**, 112–126 (2008)
6. Clarke, A.J., Van Gorder, S., Everingham, Y.: Forecasting long-lead rainfall probability with application to Australia's Northeastern coast. J. Appl. Meteorol. Climatol. **49**, 1443–1453 (2010)
7. Hu, W., Clements, A., Williams, G., et al.: Dengue fever and El Nino/Southern Oscillation in Queensland, Australia: a time series predictive model. Occup. Environ. Med. **67**, 307–311 (2010)
8. Brigode, P., Mićović, Z., Bernardara, P., et al.: Linking ENSO and heavy rainfall events over coastal British Columbia through a weather pattern classification. Hydrol. Earth Syst. Sci. **17**, 1455–1473 (2013)
9. McCabe, G.J., Ault, T.R., Cook, B.I., et al.: Influences of the El Nino Southern Oscillation and the Pacific Decadal Oscillation on the timing of the North American spring. Int. J. Climatol. **32**, 2301–2310 (2012)
10. Bulic, I.H., Kucharski, F.: Delayed ENSO impact on spring precipitation over North Atlantic/European region. Clim Dynam. **38**, 2593–2612 (2012)
11. Xu, K., Zhu, C., He, J.: Two types of El Nino-related Southern Oscillation and their different impacts on global land precipitation. Adv. Atmos. Sci. **30**(6), 1743–1757 (2013)
12. Diatta, S., Fink, A.H.: Statistical relationship between remote climate indices and West African monsoon variability. Int. J. Climatol. **34**(2), 3348–3367 (2014)
13. Latif, M.T., Stockdale, J., Wolff, J., et al.: Climatology and variability in the ECHO coupled GCM. Tellus **46A**, 351–366 (1994)
14. Latif, M., Barnett, T.P., Cane, M.A., et al.: A review of ENSO prediction studies. Clim. Dynam. **9**, 167–179 (1994)
15. Tangang, F.T., Hsieh, W.W., Tang, B.: Forecasting the equatorial Pacific sea surface temperatures by neural network models. Clim. Dynam. **13**, 135–147 (1997)

16. Halide, H., Ridd, P.: Complicated ENSO models do not significantly outperform very simple ENSO models. Int. J. Climatol. **28**, 219–233 (2008)
17. Barnston, A.G., Tippett, M.K., L'Heureux, M.L., et al.: Skill of real-time seasonal ENSO model predictions during 2002–2011: is our capability increasing? B. Am. Meteorol. Soc. **93** (5), 631–651 (2012)
18. Chen, D., Cane, M.A.: El Nino prediction and predictability. J. Comput. Phys. **227**, 3625–3640 (2008)
19. Zheng, F., Zhu, J., Wang, H., et al.: Ensemble hindcasts of ENSO events over the past 120 years using a large number of ensembles. Adv. Atmos. Sci. **26**(2), 359–372 (2009)
20. Peng, Y., Duan, W., Xiang, J.: Can the uncertainties of Madden–Jullian Oscillation cause a significant "Spring Predictability Barrier" for ENSO events? Acta Meteorol. Sin. **26**(5), 566–578 (2012)
21. Duan, W., Zhang, R.: Is model parameter error related to a significant spring predictability barrier for El Nino events? results from a theoretical model. Adv. Atmos. Sci. **27**(5), 1003–1013 (2010)
22. Kramer, W., Dijkstra, H.A.: Optimal localized observations for advancing beyond the ENSO predictability barrier. Nonlin. Processes Geophys. **20**, 221–230 (2013)
23. Yan, L., Yu, Y.: The spring prediction barrier in ENSO hindcast experiments using the FGOALS-g model. Chinese J. Oceanology Limnol. **30**(6), 1093–1104 (2012)
24. Duan, W., Wei, C.: The 'spring predictability barrier' for ENSO predictions and its possible mechanism: results from a fully coupled model. Int. J. Climatol. **33**, 1280–1292 (2013)
25. Abbot, J., Marohasy, J.: Input selection and optimization for monthly rainfall forecasting in Queensland, Australia, using artificial neural networks. Atmos. Res. **128**(3), 166–178 (2014)
26. Abbot, J., Marohasy, J.: Application of artificial neural networks to rainfall forecasting in Queensland. Australia. Adv. Atmos. Sci. **29**(4), 717–730 (2012)
27. Abbot, J., Marohasy, J.: The application of artificial intelligence for monthly rainfall forecasting in the Brisbane Catchment, Queensland, Australia. WIT Trans. Ecol. Environ. **172**, 1743–3541 (2013)
28. Abbot, J., Marohasy, J.: The potential benefits of using artificial intelligence for monthly rainfall forecasting for the Bowen Basin, Queensland, Australia. WIT Trans. Ecol. Environ. **171**, 1743–3541 (2013)
29. ASCE Task Committee on Application of Artificial Neural Networks in Hydrology: Artificial neural networks in hydrology. I preliminary concepts. J. Hydrol. Eng. **5**, 115–123 (2000)
30. Verdon, D.C., Franks, S.W.: Long-term behavior of ENSO: Interactions with the PDO over the past 400 years inferred from paleoclimate records. Geophys. Res. Lett. **33**(6), L06712 (2006)
31. Power, S., Casey, T., Folland, C., et al.: Interdecadal modulation of the impact of ENSO on Australia. Clim. Dynam. **15**, 319–324 (1999)
32. Izumo, T., Vialard, J., Lengaigne, M., et al.: Influence of the state of the Indian Ocean Dipole on the following year's El Niño. Nat. Geosci. **3**, 168–172 (2010)
33. Singh, P., Borah, B.: Indian summer monsoon rainfall prediction using artificial neural network. Stoch. Environ. Res. Risk. Assess. **27**, 1585–1599 (2013)
34. Acharya, N., Chattopadhyay, S., Kulkarni, M.A., et al.: A neurocomputing approach to predict monsoon rainfall in monthly scale using SST anomaly as a predictor. Acta Geophys. **60**(1), 260–279 (2012)
35. Saigal, S., Mehrotra, D.: Performance comparison of time series data using predictive data mining techniques. Adv. Inf. Mining. **4**(1), 57–66 (2012)

36. Willmott, C.J., Matsuura, K.: Advantages of the mean absolute error (MAE) over the root mean square error (RMSE) in assessing average model performance. Clim. Res. **30**, 79–82 (2005)
37. Zhu, J., Zhou, G.Q., Zhang, R.H., et al.: Improving ENSO prediction in a hybrid coupled model with an embedded entrainment temperature parameterisation. Int. J. Climatol. **33**, 343–355 (2013)
38. Webster, P.J., Yang, S.: Monsoon and ENSO: selectively interactive systems. Quart. J. Roy. Meteor. Soc. **118**, 877–926 (1992)
39. Lau, K.M., Yang, S.: The Asian monsoon and predictability of the tropical ocean-atmosphere system. Quart. J. Roy. Meteor. Soc. **122**, 945–957 (1996)
40. McPhaden, M.J.: Tropical Pacific Ocean heat content variations and ENSO persistence barriers. Geophys. Res. Lett. **30**, 1480 (2003)
41. Zheng, F., Zhu, J.: Spring predictability barrier of ENSO events from the perspective of an ensemble prediction system. Global Planet. Change **72**, 108–117 (2010)
42. Wu, A., Hsieh, W.W., Tang, B.: Neural network forecasts of the tropical Pacific sea surface temperatures. Neural Netw. **19**, 145–154 (2006)

A Cluster Analysis of Stock Market Data Using Hierarchical SOMs

César A. Astudillo[1], Jorge Poblete[1], Marina Resta[2], and B. John Oommen[3(✉)]

[1] Department of Computer Science, Universidad de Talca,
Km 1. Camino a los Niches, Curicó, Chile
castudillo@utalca.cl
[2] DIEC, University of Genova, via Vivaldi 5, 16126 Genoa, Italy
resta@economia.unige.it
[3] School of Computer Science, Carleton University, Ottawa K1S 5B6, Canada
oommen@scs.carleton.ca

Abstract. The analysis of stock markets has become relevant mainly because of its financial implications. In this paper, we propose a novel methodology for performing a *structured* cluster analysis of stock market data. Our proposed method uses a tree-based neural network called the TTOSOM. The TTOSOM performs self-organization to construct tree-based clusters of vector data in the multi-dimensional space. The resultant tree possesses interesting mathematical properties such as a succinct representation of the original data distribution, and a preservation of the underlying *topology*. In order to demonstrate the capabilities of our method, we analyze 206 assets of the Italian stock market. We were able to establish topological relationships between various companies traded on the Italian stock market and visually inspect the resultant taxonomy. The results that we obtained, briefly reported here (but more elaborately in [10]), were amazingly accurate and reflected the real-life relationships between the stocks.

Keywords: TTOSOM · Stock market · Clustering · Hierarchical SOM · Tree-based SOM

1 Introduction

The focus of this paper is the cluster analysis of stock market data. There are several reasons motivating such an analysis. One of these reasons is that an *a priori* knowledge of the patterns that govern the movements of stocks provide a lucrative advantage to an intelligent investor. Also, the financial scenario

C.A. Astudillo—Assistant Professor. IEEE Member. The work of this author is partially supported by the FONDECYT grant 11121350, Chile.

B.J. Oommen—*Chancellor's Professor; Fellow: IEEE* and *Fellow: IAPR.* This author is also an *Adjunct Professor* with the University of Agder in Grimstad, Norway. The work of this author was partially supported by NSERC, the Natural Sciences and Engineering Research Council of Canada.

© Springer International Publishing AG 2016
B.H. Kang and Q. Bai (Eds.): AI 2016, LNAI 9992, pp. 101–112, 2016.
DOI: 10.1007/978-3-319-50127-7_8

encountered in the stock market is not too different from those encountered in other application domains in which data-driven decisions complement the views of human experts. Indeed, more recently, data mining has demonstrated to be a powerful tool for the prediction and forecasting of financial patterns.

A particular problem that arises in finance is to find the relationships between different stocks. The literature records solutions that simplistically assume an independent relationship between them except that the prices are modeled using various time series. From an opposing perspective, one can assume that all the stocks are dependent on each other, implying the creation of a completely connected graph that, in most cases, can be very complex to analyze.

The problem can be solved by using an alternate paradigm. One may opt to consider a stock as being a collection of features, and thus attempt to discover common patterns. From a Bayesian perspective, one could interpret those stocks as being pattern whose features are independent of each other [6]. On the other hand, if the features are considered to be dependent, the *forms* of the corresponding covariance matrices can render the problem to be intractable.

Our proposal is to apply a Neural Network (NN) processing paradigm, and to thus create a tree-based model from the data. The goal is to capture the correlation between the different commodities (*and not the features*)in the stock market. Unlike previous methods reported in the literature, our plan is to take advantage of the properties of Self-Organizing Maps (SOMs) to capture the stochastic and topological structure of the data. However, we achieve this goal without the necessity of generating the traditional SOM-grid. Instead, the method directly learns a Tree-Based SOM. The algorithm that we use is called the Tree-based Topology-Oriented Self-Organizing Map (TTOSOM) [2], explained below.

The algorithms, literature survey and results presented here are, by necessity, brief. More details of each of these aspects are found in [10].

2   Literature Review

The SOM [8] is a neural network that is typically trained using (un)supervised learning, so as to produce a neural representation in a space whose dimension is usually smaller than that in which the training samples lie. Further, the neurons attempt to preserve the topological properties of the input space.

The SOM concentrates all the information contained in a set of n input samples belonging to the d-dimensional space, say $\mathcal{X} = \{\mathbf{x}_1, \mathbf{x}_2, \ldots, \mathbf{x}_n\}, \mathbf{x} \in I\!\!R^d$, utilizing a much smaller set of neurons, $\mathcal{C} = \{\mathbf{c}_1, \mathbf{c}_2, \ldots, \mathbf{c}_m\}$, each of which is represented as a vector. Each of the m neurons contains a weight vector $\mathbf{w} = [w_1, w_2, \ldots, w_d]^t \in I\!\!R^d$ associated with it. These vectors are synonymously called "weights", "prototypes" or "codebook" vectors. The vector $\mathbf{w}_i$ may be perceived as the *position* of neuron $\mathbf{c}_i$ in the feature space. During the training phase, the values of these weights are adjusted simultaneously so as to represent the data's distribution and its structure. In each training step, an input $\mathbf{x}$ is presented to the network, and the neurons compete between themselves so as to identify which is the "winner", the Best Matching Unit (BMU), represented by:

$$s = s(\mathbf{x}) = \arg\min_{c \in \mathcal{C}} \| \mathbf{x} - \mathbf{w}_c \|, \tag{1}$$

where $\| . \|$ denotes the appropriate norm, which is here the Euclidean norm.

Once the BMU has been identified, it is migrated towards the input signal as per the so-called Update rule, given by the following equation:

$$\mathbf{w}(t+1) = \mathbf{w}_i(t) + h_{si}(t)[\mathbf{x}(t) - \mathbf{w}_i(t)], \tag{2}$$

where $h(t) = h(|c_i - c_j|; t)$ is a kernel function called the Bubble of Activity (BoA) and t is the time variable. $h(t)$ is a decreasing function that depends on the distance between two neurons.

In brief, the SOM algorithm can be summarized in the following steps:

1. Obtain a sample $\mathbf{x}$ from $\mathcal{X}$.
2. Find the Winner neuron, the BMU, i.e., the one which is most similar to $\mathbf{x}$.
3. Determine a subset of neurons close to the winner, the BoA.
4. Migrate the closest neuron and its neighbors in the BoA towards $\mathbf{x}$.
5. Modify the learning factor and radius as per the pre-defined schedule.

Although the SOM has demonstrated an ability to solve problems over a wide spectrum, it possesses some fundamental drawbacks. One of these drawbacks is that the user must specify the lattice *a priori*, which has the effect that he must run the NN a number of times to obtain a suitable configuration. Other handicaps involve the size of the maps, where a lesser number of neurons often represent the data inaccurately. The state-of-the-art approaches attempt to render the topology more flexible, so as to represent complicated data distributions in a better way and/or to make the process faster by, for instance, speeding up the task of determining the BMU. Other drawbacks concentrate on the quality of the resultant map [2] and the time necessary to achieve convergence [5]. Researchers have also focused on how to tackle the above mentioned problems, and as a result, different structured variations of the original SOM have been proposed[1]. A comprehensive comparison of selected variants can be found in [4].

The present paper applies the Tree-based Topology Oriented SOM (TTO-SOM) [2] to analyze the stock market. The TTOSOM is a tree-structured SOM which aims to discover the underlying distribution of the input data set $\mathcal{X}$, while also attempting to perceive the topology of $\mathcal{X}$ as viewed through the user's desired perspective. It works with an imposed tree-structured topology, where the codebook vectors are adjusted using a VQ-like strategy. Besides, by defining a user-preferred neighborhood concept, as a result of the learning process, it also learns the topology and preserves the prescribed relationships between the

---

[1] A paper, written by two of these present authors, which reported the preliminary results of a *dynamic* Tree SOM, won the Best Paper Award in a well-known international AI conference [1].

neurons as per this neighborhood. Thus, the primary consideration is that the concept of neurons being "near each other" is not prescribed by the metric in the space, but rather by the structure of the imposed tree. The BoA of the TTO-SOM defines a distinct scheme, differing from previous methods by two main aspects: The use of a tree structure and a semi-supervised learning paradigm.

The first task is to declare the user-defined tree, which, as explained in [2], is done in a recursive manner. The TTOSOM considers the concept of neural distance which is given by the minimum *number* of unweighted connections that separate them in the user-defined tree. Unlike other methods (c.f., [9]), this notion of distance is not dependent on whether the nodes are leaves or not. As in the case of the traditional SOM, the TTOSOM requires the identification of the BMU, i.e. the closest neuron to a given input signal, and involves a distance in the feature space (and not in the tree space) as per Eq. (1).

Intricately related to the notion of inter-node distance, is the concept of the BoA which is the subset of nodes "close" to the unit being currently examined. These nodes are essentially those which are to be moved toward the input signal presented to the network. This concept involves the consideration of a quantity, the so-called *radius*, which determines how big the BoA is. In particular non-leaf nodes can be part of the BoA. The question of whether or not a neuron should be part of the current bubble, depends on the number of connections that separate the nodes rather than the distance that separate the networks in the solution space (for instance, the Euclidean distance). This concept of neighborhood is distinct and different from the ones used in other approaches such as the ET [9] or the SOTM [7].

The training process of the TTOSOM involves positioning the neurons which describe the user-defined tree in the feature space so as to capture the distribution and topology of the data points. This process involves a loop of training steps which terminates when the convergence is acceptable to the user. The *Training step* involves requesting an input sample from the dataset, locating the BMU, computing the nodes within the *current* BoA, and migrating those neurons toward the input signal using a SOM-like philosophy.

3   Proposed Methodology

A fundamental question in our study is whether we should model the Stock Market's (SM's) topology using the SOM and then invoke a Maximum Spanning Tree (MST)-based post-processing phase, or whether we should rather utilize a neural strategy that inherently captures the structure of the underlying system. We propose that the neural scheme should be the TTOSOM over the SOM.

The first argument to support our assumption is the fact that unsupervised learning usually demands the deduction of the structure of the data. The tree topology used by the TTOSOM will "absorb" and display the properties of the input set. We shall now show how the representation obtained by the TTO-SOM is superior to both the ones obtained by using the SOM with a linear ordering of the neurons, and the one obtained by invoking a grid representation.

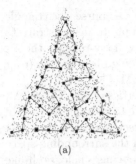 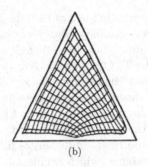 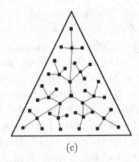

(a)   (b)   (c)

Fig. 1. How a triangle-shaped distribution is learned through unsupervised learning. (a) The positions of the SOM's neurons learned by a linear ordering, (b) The positions of the SOM's neurons learned by a grid-based ordering, and (c) The tree learned by the TTOSOM.

To demonstrate this, consider, for example, the following: A user may want to devise an algorithm that is capable of learning a triangle-shaped distribution as the one depicted in Fig. 1. In this case, we have displayed three different images of the same, explained below.

In Fig. 1a, we have not shown the boundary of the triangle, but rather the actual set of points from which the data points are sampled. The neurons in this case are arranged *linearly*, but not superimposed on any grid-like structure. Initially, the codebook vectors were randomly placed within the triangle and the so linear topology was completely lost due to the randomness of the data points. At the end of the training process, the list represents the triangle very effectively, as also reported in [8]. The SOM is capable of not only representing the whole distribution of the input samples, but also of preserving *this* linear topology. Indeed, on termination, the indices of the codebook vectors are arranged in an increasing order as seen in Fig. 1a. In this case, the one-dimensional list of neurons is evenly distributed over the triangle preserving the original properties of the 2-dimensional object, yielding a shape similar to the so-called Peano curve.

In the second case, in Fig. 1b, the SOM operates on an grid, and attempts to fit the grid within the overall shape (duplicated from [8]). In this case, we have shown only the boundary of the triangle. From our perspective, a grid does not naturally fit the triangular-shaped distribution, experiencing a deformation of the original lattice during training.

As opposed to the above two SOM solutions, Fig. 1c, shows the result of applying the TTOSOM [2] to the same dataset[2]. As one can observe from Fig. 1c, a 3-ary tree seems to be a far more superior choice for representing the particular shape in question. On closer inspection, Fig. 1c depicts how the complete tree fills in the triangle formed by the set of stimuli, and further, seems to do it *uniformly*. The final position of the nodes of the tree suggests that the underlying structure of the data distribution corresponds to the triangle. Additionally, the root of the

---

[2] Other examples of applying the TTOSOM are found in [2].

tree is placed roughly in the center of mass of the triangle. Of course, the triangle of Fig. 1c serves only as a very simple *prima facie* example to demonstrate to the reader, in an informal manner, how both techniques will try to learn the set of stimuli. We believe that the TTOSOM's topology-learning phenomenon (over the SOM's) makes it a superior choice in modeling the SM.

Recent studies have also shown other desirable properties of the TTOSOM. First of all, it possesses a holograph-like property. Figure 1c shows that each of the three main branches of the tree, cover the areas directed towards a vertex of the triangle respectively, and their sub-branches fill in the surrounding space around them in a recursive manner, which we identify as being a holograph-like behavior. This phenomenon is further investigated in [2], where more examples are available. The representation with a fewer number of neurons mimics the representation less accurately (with lesser resolution) than with a larger number.

The consequence of this to modeling the SM is immediate. It is, of course, obvious that we cannot model the SM using all the commodities and traded stocks. However, if we model it using a properly-sampled subset of the stocks and with fewer neurons, the model will be a lower-resolution model of the true model that would have been obtained by including all the stocks and if a larger NN was invoked. Such a conclusion cannot be drawn from a SOM-related strategy.

The TTOSOM has been shown to be useful for performing classification even in environments where the data contains missing values [3]. This, again, is crucial in modeling the SM because of the randomness of the task and the fact that all the transactions involving every single stock are not readily available.

Finally, the literature accepts the fact that the most time consuming task inherited from the SOM is the process of searching for the BMU. However, in the case of the TTOSOM, this task can be alleviated by incorporating tree-based data structures [5]. Such a scheme could be logarithmic, as opposed to linear, in the number of data points. This aspect could be crucial if the aim is to process even more complex economic data, such as intra-day time series involving numerous stocks, that can be in the order of Terabytes.

By virtue of all these arguments, we believe that the ideal neural strategy to model the SM is the TTOSOM.

4   Validation of the Model

To confirm the hypothesis, i.e., that the model obtained by the TTOSOM is truly a viable model, we performed experiments by examining data from the Italian market (the details of the companies, their prices etc. are found in [10]). This involved companies incorporated into the Mibtel index for the period between February 2013 and March 2014. This dataset included a total of 287 days and 206 stocks, which were classified into 37 productive sectors.

For the implementation of the proposed analysis, we selected all the available price levels available for the aforementioned period. For the generic asset i ($i = 1, 2, \ldots, N$, where $N = 206$ is the total number of companies available for the Mibtel index), we built the corresponding time-series of log-returns $S^{(i)} = \left\{ pv_k^{(i)} \right\}$ with length $T - 1$ (where $T = 287$ is the total number of days), being:

$$pv_k^{(i)} = \log \frac{pl_k^{(i)}}{pl_{k-1}^{(i)}}, \quad k = 2, ..., T, \tag{3}$$

where $pl_k^{(i)}$ is the price level at time k for the stock i.

By invoking such a transformation, we observed that it was possible to avoid trend effects on the data. The transformed data was then processed using the TTOSOM. In particular, our procedure consisted of three steps:

1. We defined a set of topologies for the TTOSOM tree. In this way it is possible to consider tree structures possessing different features.
2. We established a mechanism for optimizing the parameters, including the total number of iterations, the learning rate, the tree topology and the number of neurons. This stage allowed us to select the set of values for the parameters that produced the best results in the search space.
3. We utilized the best set of parameters to obtain a set of clusters for companies in the Italian SM.

The experiments considered all the combinations among the following values for the parameters: The total number of iterations which was either $50,000$ and $100,000$; The radius for the BoA equal to the maximum level of the depth in the tree (h) and also twice this level ($2h$); The learning rate of 0.5 or 0.9; Five different tree configurations that differed in the number of nodes, depth and branching factor, were also considered.

The validation of the experiments on our dataset was achieved using the TTOSOM and by measuring a modified version of the quantization error that considered the tree structure. Additionally, our method provided ways by which we could control the total number of clusters to be considered by the prior establishment of tree topologies.

For each configuration, we averaged 10 replications for each combination. Subsequently, we performed a comparison of the means, the aim of which was to detect significant differences between the results obtained. As a consequence of these tasks, we were able to obtain a hierarchical structure that relates the market sectors, which is equivalent to the skeleton framework presented in [11].

5   Experimental Results

In our experiments, each data point corresponded to the stocks of a company quoted in the Italian market. Also, in our model, each neuron of the tree corresponded to a single cluster. The best results were obtained by using a complete tree with 3 children per node and 3 levels of depth (resulting in a total of 13 nodes). This implied that we considered partitioning the stocks into 13 groups. Each of the resulting clusters described a representative time series which is

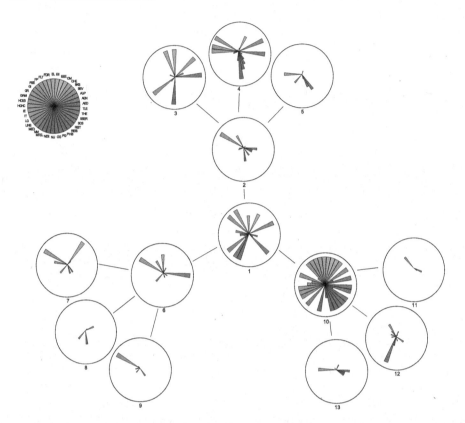

Fig. 2. The organization of the Italian market, as a result of using a TTOSOM trained with 50,000 iterations, a learning factor of 0.5 and an initial radius that equals 6.

equivalent to an average trend. Thus, essentially, we associated each stock to the BMU whose average trend was the closest. According to our observations, the best results were obtained when the TTOSOM was trained using an initial radius equal to twice the depth of the tree. We performed a total number of 50,000 iterations with an initial learning rate of 0.5. In a subsequent phase, we assigned each data point to its closest BMU.

Based on our algorithm[3], the 206 stocks were distributed in 13 clusters. Figure 2 shows a pictorial representation of the clusters. The pie chart within each cluster represents the component sectors by different coloured bars, and the cardinality of the sector by means of the length of the bars themselves. The number below each of the clusters, labels them in the correct order: The root of the tree corresponds to the Cluster identified by the index 1 and lies in the center of the figure; three children originated from it (namely: Clusters labelled

---

[3] The results presented here are brief in the interest of space. Additional results can be found in [10], and this paper can be sent to the Referees is required.

by indices 2, 6 and 10), and are symmetrically distributed in the figure. One should observe the holographic properties of the TTOSOM that emerged.

5.1   Discussion

Consider Fig. 2 which details the composition of the clusters. To demonstrate the salient aspects of our scheme, it would be helpful to record the following:

1. At a first glance, we observe that both the root of the tree and Cluster 10 (located at the second hierarchical level) are those with the largest numbers of companies and diversity of sectors.
2. Cluster 3, abbreviated as CL03, contains 12 assets belonging to the Banking Sector (BKS).
3. Cluster 1 (CL01) is strongly dominated by the Construction and Materials (CM), Automobile and Parts (AUP), and General Industries (GI).
4. CL02 has a homogeneous distribution in the Industrial Areas (IE and IT), the area of Technological Services (SCS and SSER), as well as in the Retail (GR) area.
5. CL03 and CL07 are composed mainly by the Bank Sector (BKS).
6. In CL05 and CL09, the prominent areas are Industrial Engineering (IE) and Household Goods and Home Construction (HGHC).
7. In CL04, we identify companies and services associated to the home, such as Personal Goods (PG), Household Goods and Home Construction (HGHC), Automobile and Parts (AUP) and health (HCES).
8. The clusters CL06 and CL09 constitute the more heterogeneous groups, including companies from the Electricity sector (EL), Media (MED), Financial Services (FSE) and Retail (GR).
9. CL08 relates to Media (MED), Oil (OG), Vehicles (AUP) and Banking (BKS).
10. We also record that CL10, CL11 and CL12 have an important percentage of companies related to Financial Services (FSE) and Technological Support (SSER).
11. Finally cluster CL13 groups a variety of companies, including Software (SCS), Hardware (THE) and Construction (CM and HGHC).
12. The analysis of the composition of the clusters reveals a clear dominance of the banking sector in CL03. We also identify a strong presence of the financial services in CL10, CL11 and CL12, while personal products are mostly grouped in CL04. Additionally, the construction area was present in CL01, the electricity and media in CL06 and media in CL08 and CL09.

The fact that the respective clusters collect stocks of similar types – without any prior knowledge of its constituent commodity type – is, in our opinion, truly amazing.

Additional insight can be gleaned by combining the analysis of the composition of the clusters to main statistics. The statistical analysis is detailed in Table 1, where we provided, for each cluster, the corresponding mean (mu),

Table 1. Statistics for the clusters of the Italian Stock Market between February 2013 and March 2014.

Cluster	mu	sd	sk	SR
1	−0,0007938	0,0039923	0,3311401	−19,883
2	−0,0006634	0,0050274	−0,4954843	−13,195
3	−0,0004778	0,0070769	0,4920971	−6,752
4	−0,0004604	0,0040726	0,5135600	−11,305
5	−0,0002265	0,0083026	−1,2349395	−2,728
6	−0,0007216	0,0052765	−0,1541048	−13,675
7	−0,0007557	0,0063180	−2,3810895	−11,961
8	−0,0001645	0,0083680	−0,2360082	−1,965
9	−0,0004073	0,0058753	−0,7328621	−6,932
10	−0,0004129	0,0019768	0,1761923	−20,885
11	−0,0008235	0,0045209	0,0844924	−18,216
12	−0,0002758	0,0033300	−0,3934497	−8,282
13	−0,0004909	0,0062031	−1,2187910	−7,914

standard deviation (sd), and skewness (sk). In addition, we included the Sharpe Ratio, SR, [12], which is a general indicator of the profitability for the group of companies included in each cluster: the higher the value of SR[4] the more profitable is the cluster.

As a results of such an analysis, we can report the following: All the groups exhibited negative average returns, and the variability was very low. *By combining these indications, we can obtain a simple snapshot of the Italian market in 2013, namely that it represented an economy that was deep into a stagnation phase.* The most alerting signals are those concerning the cluster with worst returns, i.e. mainly CL10 and CL11, predominantly containing companies dealing with financial services. Similar remarks can be made about companies in the construction area too, primarily grouped into CL01. It is pertinent to mention that these observations fit well with the actual situation that the financial services, the buildings and the industrial sectors in Italy were encountering – *all were going through very hard times.* The fact that we can accurately infer such economical facts from the TTOSOM is quite astonishing.

Finally, despite being negative, CL08 and CL09 characterized by the presence of companies in the media sector, were among the most attractive groups in the Italian market. It is worth mentioning that the banks (with their high presence in CL03) still maintained their (relative) attractiveness.

We hope that the reader appreciates the power of utilizing the TTOSOM in the clustering and analysis of such complex SM-based data.

---

[4] In this table, the SR values have been multiplied by 100 to make the results more readable.

6   Conclusions

In this paper, we have proposed the use of a self-organizing Neural Network (NN), based on a hierarchical structure, for analyzing the behavior of the Stock Market (SM). We successfully applied this NN for performing a static analysis of the Italian market for the period between February 2013 and March 2014, which was a critical economic period.

The main novelty of the paper was in suggesting the TTOSOM, a variant of the SOM that uses a tree-based topology, for the neurons. This NN was capable of defining clusters that are hierarchically connected. The main advantage of this tree-based SOM is that it produces a tree-based structure for the market topology. In doing this, it incorporates the tree structure into the self-organizing process, without any further processing, as one would have done if we would have worked with a standard SOM network.

With the proposed methodology we were able to visualize the relationships between the various companies traded on the Italian Stock Market. The automatic clustering of the companies into their relevant groups and sub-groups, achieved by identifying common winning neurons, was truly amazing. More specifically, the features of the clusters can lead the analyst to conclude that during the period under examination, the Italian market was moving towards a stagnation phase, with negative average returns and low dispersion. Without any external economic feedback, the method also highlighted that the clusters with higher concentration of companies from the sectors of both construction and financial services were the least attractive, while the media sector happened to be less negative.

To conclude, the technique that we have discussed is extremely promising, because it allows the analyst to infer the topology of the SM as the output of the learning procedure. The avenues for future research is vast. We recommend that future efforts be devoted to deepen the potential of tree-based SOMs to analyze markets topology, and those devoted to more risk-related stocks and derivatives. One could also apply our techniques to markets of different countries, and to the quotations of stocks sampled at higher time frequencies, i.e., moving from daily data to intra-day data.

References

1. Astudillo, C.A., Oommen, B.J. On using adaptive binary search trees to enhance self organizing maps. In: Nicholson, A., Li, X. (eds.) 22nd Australasian Joint Conference on Artificial Intelligence (AI 2009), pp. 199–209 (2009)
2. Astudillo, C.A., Oommen, B.J.: Imposing tree-based topologies onto self organizing maps. Inform. Sci. **181**(18), 3798–3815 (2011)
3. Astudillo, C.A., Oommen, B.J.: On achieving semi-supervised pattern recognition by utilizing tree-based SOMs. Pattern Recogn. **46**(1), 293–304 (2013)
4. Astudillo, C.A., Oommen, B.J.: Topology-oriented self-organizing maps: a survey. Pattern Anal. Appl. (2014). http://dx.doi.org/10.1007/s10044-014-0367-9

5. Astudillo, C.A., Oommen, B.J.: Fast BMU search in SOMs using random hyper-plane trees. In: Pham, D.-N., Park, S.-B. (eds.) PRICAI 2014. LNCS (LNAI), vol. 8862, pp. 39–51. Springer, Heidelberg (2014). doi:10.1007/978-3-319-13560-1_4

6. Gámez, J.A., Rumí, R., Salmerón, A.: Unsupervised Naive Bayes for data clustering with mixtures of truncated exponentials. In: Studený, M., Vomlel, J. (eds.) Third European Workshop on Probabilistic Graphical Models, 12-15 September 2006, Prague, Czech Republic, Electronic Proceedings, pp. 123–130 (2006)

7. Guan, L.: Self-organizing trees and forests: a powerful tool in pattern clustering and recognition. In: Campilho, A., Kamel, M.S. (eds.) ICIAR 2006. LNCS, vol. 4141, pp. 1–14. Springer, Heidelberg (2006). doi:10.1007/11867586_1

8. Kohonen, T.: Self-Organizing Maps. Springer-Verlag New York Inc., Secaucus (1995)

9. Pakkanen, J., Iivarinen, J., Oja, E.: The evolving tree – a novel self-organizing network for data analysis. Neural Process. Lett. **20**(3), 199–211 (2004)

10. Poblete, J.A., Resta, M., Oommen, B.J., Astudillo, C.A.: On representing the stock market topologically and inferring its economic indicators using tree-based SOMs (2016, to be submitted)

11. Resta, M.: The shape of crisis lessons from self organizing maps. In: Kahraman, C. (ed.) Computational Intelligence Systems in Industrial Engineering. Atlantis Computational Intelligence Systems, vol. 6, pp. 535–555. Atlantis Press, Amsterdam (2012)

12. Sharpe, W.F.: The sharpe ratio. J. Portfolio Manage. **21**(1), 4958 (1994)

A Generative Deep Learning for Generating Korean Abbreviations

Su Jeong Choi$^{(\boxtimes)}$, A-Yeong Kim, Seong-Bae Park, and Se-Young Park

School of Computer Science and Engineering, Kyungpook National University,
80 Daehakro, Buk-gu, Daegu 41566, South Korea
{sjchoi,aykim,sbpark,sypark}@sejong.knu.ac.kr

Abstract. An abbreviation is a short form of a sequence of words or
phrases. Abbreviations have been widely used as an efficient way of communicating within a human community, and nowadays they are used
more widely and more often because electronic communications such as
world wide web or twitter get available. One critical issue about abbreviations is that they are continuously generated whenever a new material
such as a new novel or a TV drama is made. Therefore, a method to
understand generation and detection of abbreviations is required for further processing of the abbreviations. The simple and well-known method
for abbreviation generation is to use the rules that are well designed by
human experts, but such rule-based methods are not appropriate for
Korean abbreviations. This is due to two major reasons. The first is that
Korean abbreviations are much irregularly generated compared to English ones, and thus the rules become too complex for managing all irregularities. The other is that many Korean abbreviations contain characters
or syllables that do not appear at the original sequence of words due to
a pronunciation issue. As a result, a great number of rules to generate
new characters or syllables should be made, which makes the rule-based
methods impractical. As a solution to this problem, this paper proposes a
generative deep learning architecture to generate Korean abbreviations.
The proposed architecture consists of two Long Short Term Memory
(LSTM) networks, in which one LSTM encodes a variable-length source
sequence into a fixed-length vector and the other LSTM decodes the
vector into a variable-length target shorter sequence. According to our
experiments on the Korean abbreviations set from National Institute
of Korean Language, the proposed method achieves 21.4% of accuracy,
which is 420% improved accuracy over a simple rule-based method. This
result proves that the proposed method is effective in generating Korean
abbreviations.

Keywords: Abbreviation · Korean irregularity · Recurrent Neural Network · Deep learning

1 Introduction

The abbreviations are widely used in human communications due to
their compact meaning and representation. In addition, they are more widely

© Springer International Publishing AG 2016
B.H. Kang and Q. Bai (Eds.): AI 2016, LNAI 9992, pp. 113–124, 2016.
DOI: 10.1007/978-3-319-50127-7_9

spread as electronic communications get common in everyday life. For instance, let us consider a Korean TV drama titled '너의 목소리가 들려/**Neo**-*ui* **Mok**sori-*ga* **Deul**-*yeo* (I can hear your voice)'. In many blogs and social network services such as Twitter and Facebook, its abbreviation '너목들/*Neo-Mok-Deul*' is much more commonly used rather than its full title.

Understanding and recognizing abbreviations helps many internet-based services including information retrieval, question answering, and chat bot operate interactively with their users by knowing that some entities and their abbreviations are same [6]. Thus, it is an important task to gather abbreviations for the services, but it is not straightforward to gather them manually. This is because new abbreviations are continuously generated whenever a new material such as a novel or a TV drama is made. Therefore, it is required to understand generation and recognition of abbreviations automatically.

There have been a number of studies on English abbreviations [8,11]. Especially in biological literatures, technical or area-specific terms are widely used [1,9,12,16]. Kuo et al. have proposed a novel set of rich features to learn rules for identifying abbreviations in biological literature [5], and their rules showed good performance in this task. Moses et al. suggested a system that generates abbreviations automatically by using a few simple rules [7], but the system has a limitation that it can generate only fixed-length abbreviations.

Note that it is difficult to apply the existing methods to Korean language. Korean abbreviations are generated much more irregularly than English ones. English abbreviations are generated mostly by combining the first characters of each word of a given word sequence. For instance, '*NY*' is an abbreviation of '*New York*', and '*as soon as possible*' is expressed as '*ASAP*'. There exist such abbreviations even in Korean like '너목들/*Neo-Mok-Deul*' as shown above. However, there are many cases that this rule can not cover. For instance, the abbreviation of a Korean drama '별에서 온 그대/**Byeol**-*eseo On* **Geudae** (My Love From the Star)' is '별그대/*Byeol-Geudae*'. Here, the word '온/*On* (from)' is omitted in the abbreviation. In addition, some abbreviations contain the words or syllables that do not appear at the original word sequence due to a pronunciation issue. As a result, it is difficult to generate Korean abbreviations automatically by hand-crafted rules [15]. This is because too many rules are required to cover all irregularities appearing in Korean abbreviations.

As a solution to this problem, this paper proposes a generative deep learning for generating Korean abbreviations. Unlike the previous rule-based methods, the proposed method regards abbreviation generation as a translation from a sequence of syllables to its shorter form. This method allows any syllables which are not observed in the original syllables to appear at the final short form in order to reflect the characteristics of Korean language. Various translation methods can be applied to this task, but a neural translation [2,13] is used as a generation model of Korean abbreviations in this paper. That is, a sequence-to-sequence

model is adopted for generating Korean abbreviations from a sequence of syllables. This model consists of two Long Short-Term Memory (LSTM) networks [3].
One LSTM encodes a long sequence of syllables into a fixed-length vector, and
the other LSTM decodes the vector into a short syllable sequence. Since LSTM
is known to consider long range temporal dependencies effectively [4], the model
adopts LSTM as its encoder and decoder. This sequence-to-sequence model has
another advantage. The number of syllables used in abbreviations is extremely
large. Thus, the task is apt to suffer from data sparseness. However, since the
proposed model represents an input syllable sequence as a limited dimensional
vector, it gets robust against data sparseness.

In the experiments on the Korean abbreviations set from National Institute
of Korean Language, the proposed method achieves 21.4% of accuracy. This
accuracy seems a little bit low, but this is 420% improved accuracy than that
of a simple rule-based method. In addition, it is much higher accuracy that
of the same method except that RNN is used instead of LSTM, which is just
16.5%. These results prove empirically that the proposed method is effective in
generating Korean abbreviations.

The rest of this paper is organized as follows. Section 2 reviews the previous
studies on abbreviations. Section 3 proposes a generative deep learning for generating Korean abbreviations, and then Sect. 4 presents the experimental results.
Finally, some conclusions and future work are drawn in Sect. 5.

2   Related Works

There have been a few studies for understanding and recognizing Korean abbreviations [10,15]. These studies, in general, aimed to build a Korean abbreviation
dictionary for named entities. For instance, Yoon et al. proposed a two-step approach to building a Korean abbreviation dictionary for companies, organizations
and universities [15]. Its first step generates all possible abbreviation candidates
for a given entity expression. For this, they handcrafted several rules for abbreviation generation. The basic principle of the rules is to drop a single noun or
to recompose the syllables of nouns within the entity expression. A few complex
rules are also designed by mixing the dropping and the recomposing. In the second step, the generated candidates are validated using a naive Bayes classifier.
Only one candidate is chosen as an abbreviation for the input entity. One drawback of this approach is that it requires large volume of labeled training data
to train the classifier, but such data are in general unavailable. On the other
hand, Park et al. utilized web search in choosing the final candidate instead of
training a classifier [10]. That is, the candidate which is frequently co-occurred
with a given named entity on the web is chosen as the abbreviation of the entity.
Though the rules used in the two studies above are easy to understand and reasonable, they have a tendency to overproduce fault abbreviations, which makes
the validation of candidates more difficult. Fine-grained rules can be used to
solve this problem. However, the finer the rules become, the more difficult it is
to maintain their consistency.

Most recent studies on English abbreviations are based on machine learning methods [8,9,11]. Especially in biology, a number of biological terms are used at an abbreviated form. Thus, many machine learning-based methods have been introduced to build an abbreviation dictionary for biological domain [1]. For instance, Kuo et al. proposed a set of novel and rich features for finding abbreviations of biological terms from literatures [5]. They proved the effectiveness of their proposed features by applying them to various machine learning models including support vector machine, naive Bayes classifier, and logistic regression. Such machine learning based approach needs a number of pairs of a term and its corresponding correct abbreviation as training data, but building the training data is more intuitive and easier than handcrafting rules for abbreviation generation. Unfortunately, to the best of our knowledge, there is no study that is based on machine learning to build a Korean abbreviation dictionary.

The proposed method in this paper is based on the sequence-to-sequence learning [2,13], a well-known deep learning for machine translation. In general, a deep neural network has a limitation that it is difficult to map a sequence to another sequence of various length. Thus, Sutskever et al. have proposed a sequence-to-sequence model to overcome the problem [13]. This model consists of two LSTMs, one LSTM for encoding an input sentence and the other LSTM for decoding the encoded sentence into a translated sentence. The key advantage of LSTM is that it can manage long range dependencies by remembering and forgetting words selectively. As a result, the method is reported to achieve similar to or higher than the state-of-the-art methods in English-to-French machine translation. On the other hand, Cho et al. proposed a modified version of the sequence-to-sequence model above [2]. The version has two new type of gates while removing three existing gates of LSTTM. As a result, it becomes simpler to compute and implement than LSTM. In addition, it is reported that the version achieves higher performance than the original sequence-to-sequence model. The sequence-to-sequence model is also utilized in conversational chat-bots, and has shown good performance compared to existing rule-based methods [14].

3   The Standard Sequence-to-Sequence Model

The sequence-to-sequence model is a general model to learn the conditional distribution over a variable-length sequence conditioned on another variable-length sequence. Figure 1 shows an illustration of the standard sequence-to-sequence model. The model consists of two components. The first component is an encoder depicted at the left side of this figure. It transforms an input sequence, $\mathbf{X} = (x_1, x_2, \dots, x_S)$, into a fixed-dimension vector $\mathbf{c}$. The second component is a decoder depicted at the right side. It generates an output sequence, $\mathbf{Y} = (y_1, y_2, \dots, y_T)$, from the vector $\mathbf{c}$. Here, the input and output sequence lengths S and T are not fixed and may differ with each other. The conditional probability of $\mathbf{Y}$ given $\mathbf{X}$ is then estimated as

$$p(y_1, y_2, \dots, y_T | x_1, x_2, \dots, x_S) = \prod_{t=1}^{T} p(y_t | \mathbf{c}, y_1, y_2, \dots, y_{t-1}). \qquad (1)$$

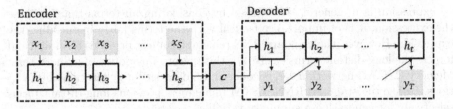

Fig. 1. The standard sequence-to-sequence model.

Note that $\mathbf{c}$ of Eq. 1 is an output of the encoder and the encoder is implemented with a recurrent neural network (RNN). At first, the neural network takes the first term x_1 as its input, and generates a hidden variable h_1 as its output. After that, the hidden variable h_t is given as an input of the network for time $t+1$ as well as the term x_{t+1}. For instance, h_1 and x_2 are used as input of the neural network at $t = 2$, and h_2 is generated as its output. When processing the last x_S, both h_{S-1} and x_S are the inputs of the network, and h_S is its output and becomes $\mathbf{c}$. Formally, a hidden variable h_t at each time t is updated by

$$\mathbf{h}_t = f(\mathbf{h}_{t-1}, x_t),$$

where f is a non-linear activation function, and $\mathbf{c} = \mathbf{h}_S$.

The probability at the decoder is also modeled with RNN. This RNN aims to generate the output sequence by predicting the next symbol y_t through the hidden variable $\mathbf{h}_t$. The input of the RNN for $\mathbf{h}_t$ is the previous symbol y_{t-1} and the previous hidden variable $\mathbf{h}_{t-1}$. Thus, the hidden variable $\mathbf{h}_t$ of the decoder is updated as the encoder does. That is, $\mathbf{h}_t$ at time t is given by

$$\mathbf{h}_t = f(\mathbf{h}_{t-1}, y_{t-1}). \tag{2}$$

After $\mathbf{h}_t$ is obtained, the conditional distribution of the next term y_t becomes

$$p(y_t | \mathbf{c}, y_1, y_2, ..., y_{t-1}) = g(\mathbf{h}_t, y_{t-1}), \tag{3}$$

where g is also a non-linear activation function. Note that the summary vector of the input sequence, $\mathbf{c}$, is the input of $\mathbf{h}_1$ of the decoder.

The two RNNs for the encoder and the decoder are jointly trained to maximize the conditional log-likelihood

$$\frac{1}{N} \sum_{n=1}^{N} \log\ p_\theta(\mathbf{Y}_n | \mathbf{X}_n),$$

where θ is a set of the model parameters and each $(\mathbf{Y}_n, \mathbf{X}_n)$ is a pair of input and output sequences from training data.

4 Sequence-to-Sequence Learning for Generating Abbreviations

Suppose that an expression is given as an input sequence for the encoder and its corresponding decoder aims to generate an abbreviation for the expression. Since an expression is in general abbreviated by considering all terms that compose the expression, it is of importance to deal with the terms fairly in encoding the expression. However, the standard RNN remembers a just previous term well and forgets the long-distant terms easily. For this reason, we adopt Long Short Term Memory (LSTM) network as an encoder and a decoder for generating abbreviations. Unlike the standard RNN, LSTM remembers previous long-distant terms selectively for generating a summary variable.

Figure 2 depicts the details of a LSTM cell for exploiting and remembering long-distant terms. In LSTM, the input gate and output gate control reading and writing information to cell. The forget gate decides what information to store in the memory and how long it is stored. The LSTM cell in this figure is implemented by the following composite functions,

$$i_t = \sigma(W_{xi}x_t + W_{hi}h_{t-1} + W_{ci}c_{t-1} + b_i) \tag{4}$$

$$f_t = \sigma(W_{xf}x_t + W_{hf}h_{t-1} + W_{cf}c_{t-1} + b_f) \tag{5}$$

$$z_t = \tanh(W_{xc}x_t + W_{hc}h_{t-1} + b_c) \tag{6}$$

$$c_t = f_t c_{t-1} + i_t z_t \tag{7}$$

$$o_t = \sigma(W_{xo}x_t + W_{ho}h_{t-1} + W_{co}c_t + b_o) \tag{8}$$

$$h_t = o_t \tanh(c_t) \tag{9}$$

where x_t is the input, i, f, o are input gate, forget gate, and output gate, respectively. c is the cell, z is a cell input activation, and h_t is the output of this LSTM. Finally, σ is the sigmoid function. The subscripts of the weight matrix W have their clear meaning. For example, W_{xi} is the weight matrix from input to input gate, while W_{cf} is from cell to forget gate. The weight matrices from cell to gate vectors are diagonal.

The decoder is also implemented with LSTM. However, it forgets the summary vector of an input sequence easily as the decoding steps progress. As a solution of this problem, we adopt the advanced sequence-to-sequence model

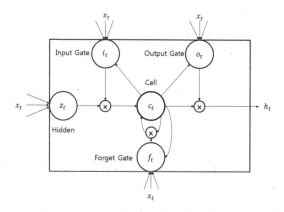

Fig. 2. The long short-term memory cell.

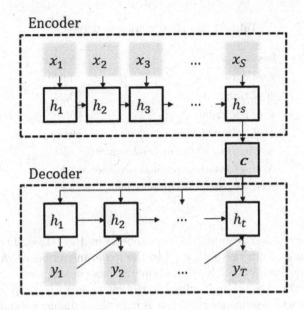

Fig. 3. A sequence-to-sequence model for remembering a summary vector bodily in each decoding step.

Table 1. The ratio of abbreviations which contain the syllable that are not observered at the original syllables.

Abbreviation type	Ratio
Some syllables are not observed at original syllables	36.61%
All syllables are observed at original syllables	63.39%

depicted in Fig. 3. Note that the key difference of Fig. 3 from Fig. 1 is that the summary vector **c** is given in each decoding step. That is, unlike the RNN in the standard sequence-to-sequence model, a hidden variable $\mathbf{h}_t$ of the decoder is inferred using the summary **c** as well as both $\mathbf{h}_{t-1}$ and y_{t-1}. Hence, the hidden variable at time t in Eq. (2) is modified as

$$\mathbf{h}_t = f(\mathbf{h}_t, y_{t-1}, \mathbf{c}).$$

Similarly, the conditional distribution of the next term in Eq. (3) becomes

$$p(y_t|\mathbf{c}, y_1, y_2, ..., y_{t-1}) = g(\mathbf{h}_t, y_{t-1}, \mathbf{c}).$$

In this paper, we assume that an abbreviation as a sequence of syllables. A syllable in Korean is a single unit of pronunciation. Thus, Korean abbreviations are usually a combination of several syllables from the original syllables. However, many Korean abbreviations contain some syllables that are not observed from the original syllables. For instance, Table 1 shows how large proportion

Table 2. Simple statistics on dataset.

Items	Values
# of data	4,930
# of train data	3,945
# of test data	985
# of unique syllable	948
Average of word in original sequence	1.93
Average length of original sequence	7.17
Average length of abbreviation	2.86

of Korean abbreviations contain such non-observered syllables. The ratios this table are obtained from the data used in the experiments below. According to this table, more than 36% of Korean abbreviations contain some syllables that are not observed at the original syllables. However, since the proposed model is a kind of sequence-to-sequence model, it is capable of managing generation and omission of some syllables.

5   Experiment

5.1   Experimental Settings

A Korean abbreviation set derived from National Institute of Korean Language is used for the experiments. The dataset consists of pairs of an original syllable sequence and its abbreviation syllable sequence Table 2 shows a simple statistics on the dataset, where '# of data'. '# of train data', '# of test data', and '# of unique syllables' denote the numbers of total data, train data, test data, and unique syllables on dataset, respectively. We split the dataset into training and test data with the ratio of 4:1. The average length of original sequence is 7.17, while average abbreviation length drops to one third of it.

For the experiments below, we use two LSTM networks with a single layer and 300 memory cells. They are trained with batch size of 100 and with 1,000 iteration time. The learning rate starts as 0.1 and gets halved at each epoch up to 0.0001. The activation function of decoder is the Log-Soft-Max function.

The superiority of the proposed method is demonstrated by comparing it with two baseline methods. The first baseline is a simple rule-based method named as `baseline` below. It generates abbreviations by combining the first character of each word in a given original syllable sequence. The other is same with the proposed method except that the standard RNN is used instead of LSTM. This is named below as `standard RNN`.

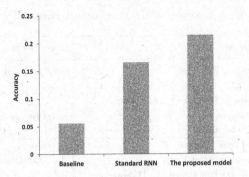

Fig. 4. Accuracy comparison for generating Korean abbreviations.

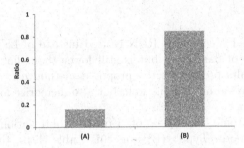

Fig. 5. A ratio of abbreviation types within correct abbreviations.

5.2 Experimental Results

Figure 4 presents the accuracies of generating abbreviations by the proposed method and the baseline methods. The accuracies of `baseline` and `standard RNN` are 0.05 and 0.165 respectively, while that of the proposed method is 0.214. The performance of the proposed method seems low overall. This is because abbreviation generation is an extremely difficult task. Nevertheless, the proposed method shows much higher accuracy than its competitors.

The `baseline` can not generate irregular abbreviations due to its rule-based approach. Thus, it shows the worst performance. On the other hand, the proposed method is capable of managing syllable addition or drop during abbreviation generation. Figure 5 proves it. We assumed that Korean abbreviations can be categorized into two types as shown in Table 1. This figure shows the ratio of each type within the correct abbreviations generated by the proposed method. Type '(A)' is the type in which some syllables are not observed at original syllables, and type '(B)' is the one in which all syllables are observed at original syllables. The type '(A)' takes about 16% of the abbreviations. That is, for the 16% of abbreviations, the proposed method performed correct syllable addition or omission. On the other hand, as explained above, RNN has a problem for memorizing the effect of long-distant words. Since `standard RNN` is same with

Table 3. Some examples of the generated Korean abbreviations.

Type	Original Sequence	Generated Sequence
(A)	해군·육군/**Hae-gun·Yuk**-gun	해륙/Hae-Ryuk
	전국 민주 학생 연맹 /**Jeon**-guk **Min**-ju **Hak**-saeng **Yeon**-maeng	전민학련/Jeon-Min-Hak-Lyeon
	덜거덕/**Deol-geo**-deonk	덜걱/Deol-Geok
	어긋물다/**Eo-geut-mul-da**	엇물다/Eon-Mul-Da
(B)	서력 기원/**Seo**-ryeok **Gi**-won	서기/Seo-Gi
	민법·상법/**Min**-beop·**Sang-beop**	민상법/Min-Sang-Beop
	대한 한의사 협회/Daehan **Han-ui**-sa **Hyeop**-hwoe	한의협/Han-Ui-Hyeop
	한국 여자 의사회/Han-guk **Yeo**-ja **Ui-sa**-hwoe	여의사회/Yeo-Ui-sa-hwoe
	고속 버스 터미널/**Go**-sok Bus **Ter**minal	고터/Go-Ter

the proposed method except that RNN is used instead of LSTM, its accuracy is higher than that of `baseline`, but is still lower than that of the proposed method. These results prove that the proposed method can not only manage irregularity of Korean abbreviations well, but also memorize long-distant words effectively.

Table 3 shows some examples of Korean abbreviations generated by the proposed method. Type '(B)' of this table lists the abbreviations are well generated by choosing some syllables from the original syllable sequence. For instance, '서기/Seo-Gi' is composed by the first syllables of '서력/Seo-ryeok (Anno Domini)' and '기원/Gi-won (beginning)'. Other abbreviations are generated by choosing some syllables of the original sequence. '한의협/Han-Ui-Hyeoop (The AssociationofKoreanMedicine)' is made of three syllables in the middle of '대한 한의사 협회/Daehan **Han-ui**-sa **Hyeop**-hwoe'. On the other hand, type '(A)' contains one or more syllables that do not appear the original sequence. The main reason that an abbreviation has new syllables is that the new syllables make it more comfortable to pronounce the abbreviation. For instance, '해륙/Hae-Ryuk (army and navy)' is generated from 해군·육군/**Hae-gun Yuk**-gun'. However, the syllable '륙/ryuk' does not appear at '해군·육군/**Hae-gun Yuk**-gun'. It is transformed from the syllable '육/yuk', and '해륙/Hae-Ryuk' is easier to pronounce than '해육/Hae-Yuk'.

Table 4 enumerates the example abbreviations which are generated wrong by the proposed method. The proposed method tends to commit a fault by making a given original sequence overly short. For example, '간접세/Gan-Jeop-Se' and '본방송/Bon-Bang-Song' are generated as '간/Gan' and '본/Bon', respectively. In addition, it misses the key syllable sometimes. The correct abbreviation of '신민주 연합당/**Sin-Min**-ju Yeon-hap-**Dang** (New Democratic Party)' is   '신민당/Sin-Min-Dang',   although   our   model

Table 4. Some examples of wrong Korean abbreviations generated by the proposed method.

Original Sequence	Correct Abbreviation	Generated Abbreviation
간접세/**Gan-Jeop-Se**	간세/Gan-Se	간/Gan
본방송/**Bon-Bang**-Song	본방/Bon-Bang	본/Bon
신민주 연합당 /**Sin-Min**-ju Yeon-hap-**Dang**	신민당/Sin-Min-Dang	신민련/Sin-Min-Lyeon
대한 결핵 협회 /Dae-han **Gyeol-haek Hyeop**-hwoe	결핵협/Gyeol-Haek-Hyeop	대한협/Dae-Han-Hyeop
금융결제원 /**Geum**-yung-**Gyeol-je-Won**	금결원/Geum-Gyeol-Won	결제원/Gyeol-je-Won

generates '신민련/Sin-Min-Lyeon' Since '당/Dang' means a political party, it is the key syllable for the abbreviation. However, the proposed method takes '연/Yeon' as a key syllable instead. As a result, a wrong abbreviation is generated. Similarly, '결핵/Gyeol-haek (tuberculosis)' is the most significant word in '대한 결핵 협회/Daehan Gyeolhaek Hyeopoe (The Korean National Tuberculosis Association)', but the proposed method misses it. Thus, a wrong abbreviation, '대한협/Dae-Han-Hyeop' is generated instead of '결핵협/Gyeol-Haek-Hyeop'. In making abbreviations, it is of importance to include the key syllables or words, but the proposed method fails generally in finding them.

6 Conclusion and Future Work

Korean abbreviations are far much irregular than English ones, and thus it is extremely difficult to generate Korean abbreviations by a rule-based approach. As a solution to this problem, we have proposed the generative deep learning for generating Korean abbreviations in this paper. The proposed method regards the abbreviation generation as a translation task, and adopts a sequence-to-sequence model for the task. As a result, the method is capable of generating irregular abbreviations such as addition or omission of some syllables that are not observed at original syllables. We also have proved empirically the proposed method is effective in Korean abbreviation generation. However, the method still has some room to improve. Many abbreviations include key syllables that deliver the meaning of an original sequence, but the proposed method can not handle them. Thus, it is our future work to suggest a model that considers key syllables in generating abbreviations.

Acknowledgements. This work was supported by Institute for Information &communications Technology Promotion(IITP) grant funded by the Korea government(MSIP) (No.R7117-16-0209,Smart Summary Report Generation from Big Data Related to a Topic) and ICT R&D program of MSIP/IITP[B0126-16-1002, Development of smart broadcast service platform based on semantic cluster to build an open-media ecosystem].

References

1. Adar, E.: SaRAD: a simple and robust abbreviation dictionary. Bioinformatics **20**(4), 527–533 (2004)
2. Cho, K., Merriënboer, V., Gulcehre, C., Bahdanau, D., Bougares, F., Schwenk, H., Bengio, Y.: Learning Phrase Representations using RNN Encoder-Decoder for Statistical Machine Translation. arXiv preprint arXiv:1406.1078 (2014)
3. Graves, A., Mohamed, A., Hinton, G.: Speech recognition with deep recurrent neural networks. In: Proceedings of 2013 IEEE International Conference on Acoustics, Speech and Signal Processing, pp. 6645–6649 (2013)
4. Hochreiter, S., Schmidhuber, J.: Long short-term memory. Neural Comput. **9**(8), 1735–1780 (1997)
5. Kuo, C., Ling, M., Lin, K., Hsu, C.: BIOADI: a machine learning approach to identifying abbreviations and definitions in biological literature. BMC Bioinform. **10**(15), 1–10 (2009)
6. Larkey, S., Ogilvie, P., Price, A., Tamilio, B.: Acrophile: an automated acronym extractor and server. In: Proceedings of the Fifth ACM Conference on Digital Libraries, pp. 205–214 (2000)
7. Moses, L., Ehrenreich, L.: Abbreviations for automated systems1. In: Proceedings of the Human Factors and Ergonomics Society Annual Meeting, vol. 25, pp. 132–135 (1981)
8. Okazaki, N., Ananiadou, S.: Building an abbreviation dictionary using a term recognition approach. Bioinformatics **22**(24), 3089–3095 (2006)
9. Pakhomov, S.: Semi-supervised maximum entropy based approach to acronym and abbreviation normalization in medical texts. In: Proceedings of the 40th Annual Meeting on Association for Computational Linguistics, pp. 160–167 (2002)
10. Park, Y., Kang, S., Yoo, B., Seo, J.: Title named entity recognition using wikipedia and making acronym. Korea Comput. Congr. **6**, 637–639 (2013)
11. Sánchez, D., Isern, D.: Automatic extraction of acronym definitions from the Web. Appl. Intell. **34**(2), 311–327 (2011)
12. Satx, P., Mogel, S.: An abbreviation of the WAIS for clinical use. J. Clin. Psychol. **18**, 77–79 (1962)
13. Sutskever, I., Vinyals, O., Le, V.: Sequence to sequence learning with neural networks. In: Advances in Neural Information Processing Systems, pp. 3104–3112 (2014)
14. Vinyals, O., Le, Q.: A Neural Conversational Model. arXiv preprint arXiv:1506.05869 (2015)
15. Yoon, Y., Song, Y., Lee, J., Lim, H.: Construction of Korean acronym dictionary by considering ways of making acronym from definition. In: Proceedings of the Korean Society for Cognitive Science Conference (2006)
16. Yoshida, M., Fukuda, K., Takagi, T.: PNAD-CSS: a workbench for constructing a protein name abbreviation dictionary. Bioinformatics **16**(2), 169–175 (2000)

Medical Prognosis Generation from General Blood Test Results Using Knowledge-Based and Machine-Learning-Based Approaches

Youjin Kim[✉], Jonghwan Hyeon, Kyo-Joong Oh, and Ho-Jin Choi

School of Computing, KAIST, 291 Daehak-ro, Yuseong-gu, Daejeon, Korea
{117kyjin,hyeon0145,aomaru,hojinc}@kaist.ac.kr

Abstract. In this paper, we present two approaches to generate prognosis from general blood test results. The first approach is a knowledge-based approach using ripple-down rules (RDR). The knowledge-based approach with RDR converts knowledge of pathologists into a knowledge base with the minimum intervention of knowledge engineers. The second approach is a machine-learning(ML)-based approach using decision tree, random forest and deep neural network (DNN). The ML-based approach learns patterns of attributes from various cases of general blood test. Our experimental results show that there are indeed some important patterns of the attributes in general blood test results, and they are adequately encoded by the both approaches.

Keywords: Clinical decision support system (CDSS) · Ripple-down rules (RDR) · Machine learning (ML) · Prognosis · General blood test

1 Introduction

In modern society, people suffer from stress, busy schedule, unhealthy diet and lack of exercise. In this situation, people are at risk of getting disease such as hypertension, hyperlipidemia, diabetes and obesity. There are no noticeable initial symptom in these illness so patients cannot recognize their bad health condition and do nothing for their own body. As time goes by, illness become worse leading complications and after-effects. Regular check and early detection increase the possibility of full recovery. Therefore, the medical examination which is typical form of preventive medicine has become more important. General blood test is one of the important test involved in medical examination. Through this, we can find various disease checking the number and the shape of cell in blood, hormone abnormality and metabolite value etc.

In this paper, we examine two approaches namely, knowledge-based and machine learning(ML)-based to generate prognoses from general blood test results. The knowledge-based approach is based on the ripple-down rules (RDR) and Induct RDR. The RDR is helpful to construct knowledge bases with the minimum intervention of knowledge engineers. The ML-based approach is based

© Springer International Publishing AG 2016
B.H. Kang and Q. Bai (Eds.): AI 2016, LNAI 9992, pp. 125–136, 2016.
DOI: 10.1007/978-3-319-50127-7_10

on decision tree, random forest and deep neural network (DNN). Machine learning methods can learn patterns of attributes automatically from various cases of general blood test. Clinical decision support system (CDSS) based on these approaches is an efficient way to deal with increasing general blood test demand, reducing the pathologist's burden and human error caused by repeating tasks.

There are two contributions in this paper. First, we construct a knowledge base for prognoses generation from real cases of the general blood test which has 685 types of subtests. Second, we train several machine learning models which can classify the prognoses from over 10,000 pathological multi-labeled cases.

The remainder of this paper is organized as follow. Section 2 explores the background for this research and Sect. 3 explains our methods. In Sect. 4, we illustrate experiments and in Sect. 5, Results and analysis are discussed. We conclude our study in Sect. 6.

2   Background

2.1   Clinical Decision Support System (CDSS)

Decision support system (DSS) involves various computer systems which help decision making process. It edits and reorganizes information by interacting with users and based on this, users could make reasonable decision [1,2].

Clinical decision support system (CDSS) refers to DSS in health care domain. It gets knowledge from expertise's experience and converts raw medical data to useful information. It helps physician in clinical decision process, reduces medical mistake and improves patient's safety. It is also useful in complex medical decision making process which has ambiguity and contradiction as well as general medical decision making process. It provides reasonable solution and strengthens the rationale of the solution [1,3–7]. CDSS is categorized by knowledge-based, ML-based and combination of knowledge-based and ML-based group [8] as shown in Fig. 1.

Knowledge-based CDSS are based on the logic and if-then statements, consisting of knowledge base, inference engine and communication mechanism. The knowledge base contains lots of clinical, medical information represented by a

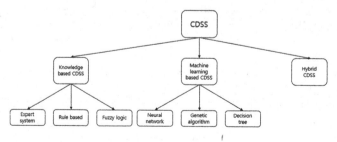

Fig. 1. Classification of CDSS. CDSS is categorized into knowledge-based, ML-based and hybrid system [8].

set of rules. The inference engine relate knowledge base rules to input data (new patient data) with if-then statements. Rule-based expert systems and fuzzy logic techniques are types of knowledge-based CDSS [2,7–9].

ML-based CDSS is a non-knowledge-based CDSS. It learns from clinical experiences finding patterns of attributes in a large amount of medical data [2,7]. Artificial neural networks, genetic algorithms and decision trees are involved in ML-based CDSS [8].

2.2 Knowledge-Based Approach

Ripple-Down Rules (RDR). Ripple-down rules (RDR) is a system which defines how to represent, infer and acquire rules. In this system, rules are represented as an n-ary tree, where a node corresponds to a rule which contains a set of conditions and a conclusion. The node can have children which are also rules.

The inference procedure of RDR is as follow:

1. Evaluate every rule in top level of rule tree.
2. Evaluate rules which is true in previous level.
3. Repeat procedure 2 until there are no more child rules or none of rules is evaluated as true.

Through this procedure, RDR obtains fired rule paths. When the process finished, a conclusion set of the inference consists of the conclusion of the last fired rule in each fired rule path.

RDR provides a systematic way to acquire knowledge so expert can easily modify the knowledge base. This is a crucial difference compared to other methods and makes RDR suitable for CDSS [10,11].

Induct RDR. A Induct RDR is an algorithm which produces a RDR knowledge base from a large dataset.

Domain experts easily access RDR form knowledge base. However, It requires lots of effort for domain expert to construct a knowledge base from A to Z. In this situation, Induct RDR can help domain expert to build an initial knowledge base. It extracts rules in the form of RDR applying statistical methods to data. After applying Induct RDR, domain experts modify and refine some errors [10–12].

2.3 ML Approach

Decision Tree. A decision tree illustrates decision making process as tree shape form. Logical thinking and comparison of various decision paths could be possible with decision tree.

Random Forest. A random forest consists of several decision trees. When it predicts new data, it merges several decision tree's prediction result. Generally, random forest shows good performance avoiding over fitting problem.

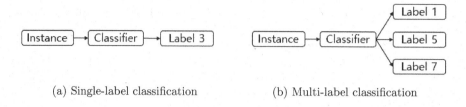

(a) Single-label classification   (b) Multi-label classification

Fig. 2. Single-label vs. multi-label classification

Deep Neural Network. A deep neural network imitates a process of human thinking and learns from examples. It consists of an input layer, hidden layers and an output layer and each layer has several neurons. Neurons are connected to neurons in the next layer by weighted connection. In this architecture, neural net analyses data, finds useful meaning and solves various problem [13].

2.4   Multi-label Classification

Single-label classification problem takes an instance and produces only one output label as shown in Fig. 2(a). In other words, the problem is to select a label $l \in L$ for each instance $d \in D$. In contrast, multi-label classification problem takes an instance and produces several output labels as shown in Fig. 2(b). In other words, let D be set of instances and L be set of labels. Then, the problem is to select a set of labels $S \subseteq L$ for each instance $d \in D$.

Pathological diagnosis is a kind of multi-label classification problem because one pathological test sample should produce multiple prognoses as output labels.

3   Methods

3.1   Knowledge-Based Approach

We use Induct RDR in constructing an knowledge base reducing expert's burden. We preprocess general blood test results rather than use it directly. First, we quantize numerical test results as low, normal, and high to reduce the complexity of statistical inferences. Also, we merge test codes which indicate the same test (e.g., test code 00011, 00530 indicate glucose test) into single test code.

Because Induct RDR handles only single-label classification problem, we cannot directly apply Induct RDR into the general blood test results which are multi-label classification problem. Therefore, we divide the data into 18 categories that is Anemia, Liver, Blood, Pancreas, Blood sugar, Rheumarthritis, Blood type, Stool, Electrolyte, Syphilis, Hepatitis virus, Thyroid, Infection, Tumor, Kidney, Urine, Lipid and Etc.

However, some categories are still multi-label classification. For example in case of the electrolyte category, a patient may have several abnormalities such as Na, K, Ca, etc. To handle this, we duplicate test value and split one data

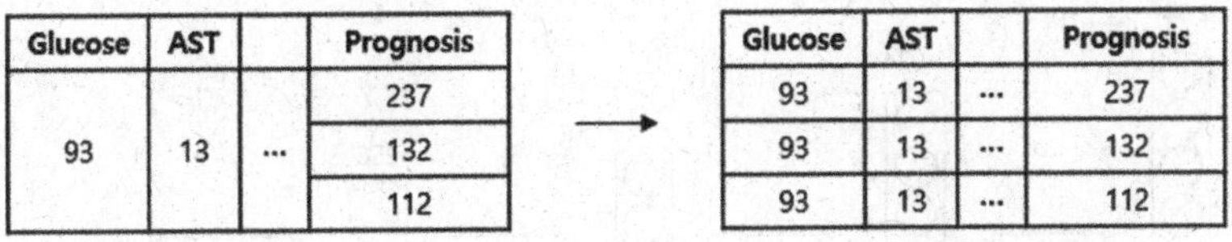

Glucose	AST		Prognosis
			237
93	13	...	132
			112

$\longrightarrow$

Glucose	AST		Prognosis
93	13	...	237
93	13	...	132
93	13	...	112

Fig. 3. Transformation from multi-label to single-label classification. We duplicate test value for each prognosis to make one data instance have one prognosis.

instance to several instances which have same test value but different prognosis as shown in Fig. 3.

After preprocessing the data as mentioned above, we apply Induct RDR for each category independently and then, we merge 18 knowledge bases into a single knowledge base.

3.2 ML-Based Approach

In case of ML-based approach, we preprocess data in other way. For this approach, test value needs to be transformed to vector form. We categorize test value by five categories and for each category, process data maintaining its original meaning. Table 1 shows conversion result of each class data.

Each patient has several prognoses based on the blood tests they take. Therefore, this could be considered as a multi-label classification, having patient's age and sex, blood test results as input and prognoses as output. It is important to choose appropriate machine learning methods to solve this problem because not all machine learning methods effectively solve multi-label classification. In this paper, we use decision tree, random forest and deep neural network.

Decision tree are constructed through Gini impurity and random forest consists of 200 decision trees. Deep neural network is comprised of four fully connected layers. Each layer has 1,041, 1,024, 512 and 256 neurons respectively and

Table 1. Conversion table for each test result category in the preprocessing step. *before* means raw result value and *after* indicates preprocessed value. Class a consists of nominal value and numeric value. Class b contains sign of inequality. Class c is comprised of two parts (value, class). Class d is nominal type data such as blood type. Class e has no value.

Category	Preprocessing	
	Before	After
a	Non-Reactive 1.0	Non-Reactive
b	<0.5	0.5
c	1.23 1.0	1.23
d	A, B, O, AB	1000, 0100, 0010, 0001
e	" "	empty

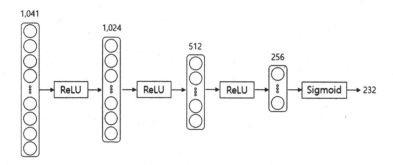

Fig. 4. Deep neural network comprised of four fully connected layers. Each layer has 1,041, 1,024, 512, 256 neurons respectively and output vector has 232 elements. ReLU is used as activation function and sigmoid is adopted for the last activation function.

output vector has 232 elements as shown in Fig. 4. Generally, softmax function is used as the last activation function in deep neural network to classify something. Instead however, we use a sigmoid function because this is a multi-label classification that each label needs probability distribution. Sigmoid gives n probability distributions and softmax gives one probability distribution when there are n labels. Therefore, we use sigmoid function as the last activation function. We assume that each element of output vector represents the probability of whether each prognosis is involved in the output or not to solve multi-label classification problem by deep neural network. If the probability is higher than 0.5, we include the prognosis in the output.

4   Experiments

4.1   General Blood Test Data

In this paper, we use anonymized dataset provided from Seegene Medical Foundation (http://www.seegene.co.kr) which consists of patient information, blood test results and prognoses.

General blood test has 685 types and consist of several subtest (blood sugar, kidney-gout-arthritis, liver function, electrolyte, lipid-cardiovascular system, hepatitis, venereal disease, iron, blood, blood type, pancreas, inflammation, urine, thyroid gland and tumor marker). Prognoses are made by experts and have 232 types.

Dataset consist of 14,479 data elements and one data element as shown in Fig. 5, consists of *Age, Sex, Test name, Test result* and *Prognosis* for each patient. *Test name* and *Test result* are came from tests each patient take. *Prognosis* is created considering all the test results one patient take.

In 14,479 number of data, we used 8610 data in RDR (training: 7,610, testing: 1,000) and we used all data when training and testing deep neural network.

Age			Sex		
47			M		
Test name	Test result	Test name	Test result	Test name	Test result
Creatinine	0.89	Triglyceride (TG)	77	Hbe Ab	Positive < 0.10
Glucose (FBS)	92	HDL-Cholesterol	60	AFP (α-fetoprotein)	2.89
AST (SGOT)	30	Hemoglobin (Hb)	15.8	HBV DNA정량(Realtime PCR)_01	< 116
ALT (SGPT)	23	HBs Ag(정밀)	Positive 4274.23	HBV DNA정량(Realtime PCR)_02	< 20
Γ-GTP (GGT)	37	HBs Ab (Anti-HBs) 정밀	Negative 1.01	HBV DNA정량(Realtime PCR)_03	< 0.0004
Cholesterol	222	HBeAg	Negative 0.29		
Prognosis					
① 공복 혈당이 정상입니다. 당뇨검사의 결과가 정상입니다.					
② 신장기능 검사 결과는 정상입니다.					
③ 간 기능 검사의 결과는 정상입니다.					
④ 지질검사의 결과 정상 입니다.					
⑤ B형간염항원 양성으로 B형간염 보균상태입니다.					
⑥ 빈혈이 없습니다.					
⑦ 종양표지자검사는 정상입니다. 종양 표지자는 종양 등에서 생성되어 혈액이나 체액으로 분비되는 여러 종류의 물질입니다.					

Fig. 5. Illustration of blood test results and prognoses by human experts. Prognoses are written in Korean. From the top, each comment sentence means that, ① Fasting blood sugar is normal. Result of diabetes mellitus test is normal. ② Result of renal function test is normal. ③ Result of liver function test is normal. ④ Result of lipid test is normal. ⑤ Hepatitis B antigen positive, hepatitis B carrier state. ⑥ Anemia is none. ⑦ Result of tumor marker test is normal. Tumor marker is created from tumor and is substance secreted to blood or body fluid.

4.2 Experiments for the Knowledge-Based Approach

After preprocessing the data as mentioned before, we apply Induct RDR for each category independently and then, we merge 18 knowledge bases into a single knowledge base. We use 7,610 general blood test data to create an knowledge base and evaluate the constructed knowledge base with unused 1,000 general blood test data.

4.3 Experiments for the ML-Based Approach

We construct decision tree through Gini impurity and we make random forest using 200 decision trees. Deep neural network are composed of four layers and we trained the neural network 100 times. We compare performance of decision tree, random forest and deep neural network and all evaluation was done using 5-fold cross validation.

5   Results and Analysis

5.1 Results for the Knowledge-Based Approach

We use an additional rate and a missing rate to evaluate the RDR-based expert system. The additional rate represents how many prognoses are over-generated compared to original prognoses. The missing rate represents how many prognoses are under-generated compared to original prognoses.

Table 2. Results of the knowledge-based approach

Missing rate	Additional rate
46.91%	96.01%

In conclusion, as shown Table 2, we can get 96.01% as the additional rate and 46.91% as the missing rate. We group the general blood test data into 18 categories so, 18 independent knowledge bases are generated. Therefore, the expert system should generate at minimum, 18 prognoses. We think this is the reason of the high additional rate.

5.2   Results for the ML-Based Approach

We use accuracy, precision, recall and F1-measure to evaluate the ML-based system. However, we cannot use general accuracy, precision, recall and F1-measure because multi-label classification problem partially correct concept [14]. Therefore, we use Godboles definition [15] to properly evaluate.

Let T, h and Z_i be defined as

T: dataset consisting of n data elements, (x_i, Y_i), $1 \leq i \leq n$
h: multi-label classifier
Z_i: classifier h's classification result, $h(x_i)$.

Then, accuracy, precision, recall and f1-measure are defined as

$$Accuracy = \frac{1}{n} \sum_{i=1}^{n} \frac{|Y_i \bigcap Z_i|}{|Y_i \bigcup Z_i|} \tag{1}$$

$$Precision = \frac{1}{n} \sum_{i=1}^{n} \frac{|Y_i \bigcap Z_i|}{|Z_i|} \tag{2}$$

$$Recall = \frac{1}{n} \sum_{i=1}^{n} \frac{|Y_i \bigcap Z_i|}{|Y_i|} \tag{3}$$

$$F1 - measure = \frac{1}{n} \sum_{i=1}^{n} \frac{2|Y_i \bigcap Z_i|}{|Y_i| + |Z_i|} \tag{4}$$

Furthermore, we use hamming-loss to take into account the prediction error (mistakenly predicting wrong label) and the missing error (missing out correct label).

Let I and k be defined as below

I: indicator function
k: the number of labels dataset T has.

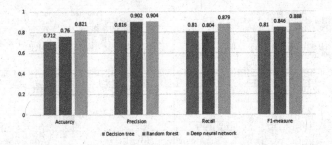

Fig. 6. Results of the ML-based methods. *Deep neural network* shows better performance than *decision tree* and *random forest* in every metric.

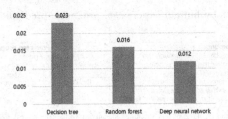

Fig. 7. Hamming-loss of the ML-based methods. *Deep neural network* has the lowest hamming-loss which means it is better than others.

Then, hamming-loss is defined as

$$Hamming-loss = \frac{1}{kn} \sum_{i=1}^{n} \sum_{l=1}^{k} [I(l \in Z_i \wedge l \notin Y_i) + I(l \notin Z_i \wedge l \in Y_i)] \quad (5)$$

We compare performance of decision tree, random forest and deep neural network. All evaluation was done using 5-fold cross validation and evaluation results can be checked in Figs. 6 and 7. We can find that deep neural network has better performance than decision tree and random forest in every metric. We think that if we have more data, then deep neural network would show the better performance.

In Fig. 8, we show prognosis by human experts and prognosis generated by deep neural network. The 5th, 8th and 15th prognosis from the top are in *real prognosis* but are not in *predicted prognosis*. This shows missing error. On the other hand, the last prognosis from the top is contained in *predicted prognosis* but not in *real prognosis* indicating prediction error. There are some errors but we can check deep neural net quite well predict.

Actual prognosis	Predicted prognosis
B형간염항체 양성으로 B형간염 면역상태입니다.	B형간염항체 양성으로 B형간염 면역상태입니다.
C형간염항체 음성입니다. C형간염이 없습니다.	C형간염항체 음성입니다. C형간염이 없습니다.
간 기능 검사의 결과는 정상입니다.	간 기능 검사의 결과는 정상입니다.
갑상선기능 검사 결과 정상입니다.	갑상선기능 검사 결과 정상입니다.
공복 혈당이 정상이고 당화혈색소(HbA1c)도 정상입니다. 당뇨검사의 결과가 정상입니다.	
류마티스인자(RF)가 정상입니다. 류마티스관절염의 30%에서 음성을 보이므로 임상증상과 영상의학적 소견 등을 참고하여야 합니다.	류마티스인자(RF)가 정상입니다. 류마티스관절염의 30%에서 음성을 보이므로 임상증상과 영상의학적 소견 등을 참고하여야 합니다.
빈혈이 없으며 철분도 부족하지 않습니다.	빈혈이 없으며 철분도 부족하지 않습니다.
소변에서 결정체가 발견되었습니다. amp.urate결정체는 정상에서도 발견될 수 있습니다.	
신장기능 검사 결과는 정상입니다.	신장기능 검사 결과는 정상입니다.
요산의 농도가 정상입니다. 요산은 통풍의 원인입니다.	요산의 농도가 정상입니다. 요산은 통풍의 원인입니다.
전해질 관련검사의 결과는 정상입니다.	전해질 관련검사의 결과는 정상입니다.
종양표지자검사는 정상입니다. 종양 표지자는 종양 등에서 생성되어 혈액이나 체액으로 분비되는 여러 종류의 물질입니다.	종양표지자검사는 정상입니다. 종양 표지자는 종양 등에서 생성되어 혈액이나 체액으로 분비되는 여러 종류의 물질입니다.
지질검사의 결과 정상 입니다.	지질검사의 결과 정상 입니다.
혈구질환 관련 검사의 결과 정상입니다.	혈구질환 관련 검사의 결과 정상입니다.
혈액형은 AB형 Rh 양성입니다.	
	소변 백혈구 양성입니다. 요로감염과 관련이 있습니다. 여성의 경우는 질분비물에 의한 오염이 흔합니다.

(a) Prognosis by human experts vs. prognosis predicted by DNN

Actual prognosis	Predicted prognosis
Hepatitis B antibody positive, hepatitis B immune state.	Hepatitis B antibody positive, hepatitis B immune state.
Hepatitis C antibody negative, hepatitis C is none.	Hepatitis C antibody negative, hepatitis C is none.
Result of liver function test is normal.	Result of liver function test is normal.
Result of thyroid gland function test is normal.	Result of thyroid gland function test is normal.
Fasting blood sugar is normal and glycosylated hemoglobins(HbA1c) is normal. Result of diabetes mellitus test is normal.	
Rheumatoid Factor(RF) is normal. 30% of rheumatoid arthritis is negative so clinical manifestation, radiological finding and etc. should be considered.	Rheumatoid Factor(RF) is normal. 30% of rheumatoid arthritis is negative so clinical manifestation, radiological finding and etc. should be considered.
Anemia is none and limatura ferri is not deficiency.	Anemia is none and limatura ferri is not deficiency.
Crystal is found in urine. amp.urate crystal can be found in normal.	
Result of renal function test is normal.	Result of renal function test is normal.
Concentration of uric acid is normal. Uric acid is cause of gout.	Concentration of uric acid is normal. Uric acid is cause of gout.
Result of electrolyte related test is normal.	Result of electrolyte related test is normal.
Result of tumor marker test is normal. Tumor marker is created from tumor and is several substance secreted to blood or body fluid.	Result of tumor marker test is normal. Tumor marker is created from tumor and is several substance secreted to blood or body fluid.
Result of lipid test is normal.	Result of lipid test is normal.
Result of hemocyte disease related test is normal.	Result of hemocyte disease related test is normal.
Blood type is AB Rh positive.	
·	Urine leukocyte positive. This is related to urinary tract infection. In cased of women, contamination caused by vaginal discharge is common.

(b) Translated version of (a)

Fig. 8. Illustration of predicted results for a patient

6   Conclusion

In this paper, we investigated two approaches for generating preliminary medical prognosis from patients' general blood test results.

The first approach was the RDR (and the induct RDR) framework to form a knowledge base for medical prognosis. By applying the RDR framework to CDSS, we can expect to build an environment where human experts (doctors) can freely modify the knowledge base with the reduced help of knowledge engineers when the need arises.

As the second approach, we utilized various machine learning algorithms including deep neural network to classify relevant prognoses using the pathological cases. Our experimental results showed that there are indeed some important patterns of the attributes in the blood test results, and they are adequately learned by this approach.

Acknowledgments. This work was supported by the Industrial Strategic Technology Development Program, 10052955, Experiential Knowledge Platform Development Research for the Acquisition and Utilization of Field Expert Knowledge, funded by the Ministry of Trade, Industry & Energy (MOTIE), Korea.

References

1. Foster, D., McGregor, C., El-Masri, S.: A survey of agent-based intelligent decision support systems to support clinical management and research. In: the 2nd International Workshop on Multi-Agent Systems for Medicine, Computational Biology, and Bioinformatics, pp. 16–34 (2005)
2. Abbasi, M.M., Kashiyarndi, S.: Clinical Decision Support Systems: A discussion on different methodologies used in Health Care. Marlaedalen University, Sweden (2006)
3. Smith, A.E., Nugent, C.D., McClean, S.I.: Implementation of intelligent decision support systems in health care. J. Manage. Med. **16**(2), 206–218 (2002)
4. Jao, C.S., Hier, D.B.: Clinical decision support systems: An effective pathway to reduce medical errors and improve patient safety. INTECH Open Access Publisher (2010)
5. Rokach, L., Maimon, O.: Data Mining with Decision Trees: Theory and Applications. World Scientific, River Edge (2014)
6. Harold, C.S., Michael, C., Higgins, D.K.: Owens: Medical Decision Making, 2nd edn. Wiley, Chichester (2013)
7. Moses, A.J., Lieberman, M., Kittay, I., Learreta, J.A.: Computer-aided diagnoses of chronic head pain: explanation, study data, implications, and challenges. J. Craniomandibular Sleep Pract. **24**(1), 60–66 (2006). Taylor & Francis
8. Aljaaf, A.J., Al-Jumeily, D., Hussain, A.J., Fergus, P., Al-Jumaily, M., Abdel-Aziz, K.: Toward an optimal use of artificial intelligence techniques within a clinical decision support system. In: Science and Information Conference(SAI), pp. 548–554. IEEE (2015)
9. Ye, Y., Diao, X., Jiang, Z., Du, G.: A knowledge-based variance management system for supporting the implementation of clinical pathways. In: International Conference on Management and Service Science, MASS 2009, pp. 1–4, IEEE (2009)
10. Compton, P., Edwards, G., Kang, B., Lazarus, L., Malor, R., Menzies, T., Sammut, C.: Ripple down rules: possibilities and limitations. In: the 6th AAAI Knowledge Acquisition for Knowledge-Based Systems Workshop, Calgary, Canada, University of Calgary, pp. 6–1. AAAI (1991)
11. Kang, B., Compton, P., Preston, P.: Multiple classification ripple down rules: evaluation and possibilities. In: the 9th Knowledge Acquisition for Knowledge Based Systems Workshop (1995)
12. Gaines, B.R., Compton, P.: Induction of ripple-down rules applied to modeling large databases. J. Intell. Inform. Syst. **5**(3), 211–228 (1995). Springer
13. Berner, E.S.: Clinical Decision Support Systems, pp. 3–22. Springer Science + Business Media, New York (2007)

14. Sorower, M.S.: A Literature Survey on Algorithms for Multi-label Learning. Oregon State University, Corvallis (2010)
15. Godbole, S., Sarawagi, S.: Discriminative methods for multi-labeled classification. In: Dai, H., Srikant, R., Zhang, C. (eds.) PAKDD 2004. LNCS (LNAI), vol. 3056, pp. 22–30. Springer, Heidelberg (2004). doi:10.1007/978-3-540-24775-3_5

Deep Learning for Classification of Malware System Call Sequences

Bojan Kolosnjaji[✉], Apostolis Zarras, George Webster, and Claudia Eckert

Technical University of Munich, Munich, Germany
{kolosnjaji,zarras,webstergd,eckert}@sec.in.tum.de

Abstract. The increase in number and variety of malware samples amplifies the need for improvement in automatic detection and classification of the malware variants. Machine learning is a natural choice to cope with this increase, because it addresses the need of discovering underlying patterns in large-scale datasets. Nowadays, neural network methodology has been grown to the state that can surpass limitations of previous machine learning methods, such as Hidden Markov Models and Support Vector Machines. As a consequence, neural networks can now offer superior classification accuracy in many domains, such as computer vision or natural language processing. This improvement comes from the possibility of constructing neural networks with a higher number of potentially diverse layers and is known as *Deep Learning*.

In this paper, we attempt to transfer these performance improvements to model the malware system call sequences for the purpose of malware classification. We construct a neural network based on convolutional and recurrent network layers in order to obtain the best features for classification. This way we get a hierarchical feature extraction architecture that combines convolution of n-grams with full sequential modeling. Our evaluation results demonstrate that our approach outperforms previously used methods in malware classification, being able to achieve an average of 85.6% on precision and 89.4% on recall using this combined neural network architecture.

1 Introduction

An increasing problem in large-scale malware detection and analysis is the high number of new malware samples. This number has exponentially increased throughout the years, which creates difficulty for malware analysts, as they need to extract information out of this large-scale data. Recent reports from McAfee reveal that tens of thousands of new distinct samples are being submitted on a daily basis [9]. Furthermore, statistics page of VirusTotal shows that, in just a single day, over a million newly retrieved samples had to be analyzed [29]. This surge of samples makes reverse engineering a challenging task. Although there exist efforts to automate the reverse engineering and malware analysis process, manual signature-based or heuristics-based detection and analysis procedures are still very prominent. Apart from the problem of sheer number of samples,

© Springer International Publishing AG 2016
B.H. Kang and Q. Bai (Eds.): AI 2016, LNAI 9992, pp. 137–149, 2016.
DOI: 10.1007/978-3-319-50127-7_11

these samples are of increased variety, which is usually caused by advances in malware development that utilize polymorphic and metamorphic algorithms to generate different versions of the same malware. This makes it extremely difficult for signature-based systems to correctly classify and analyze these samples.

To aid malware analysts in retrieving useful information from such a large amount of samples, we need to solve the problem of automatic classification under the existing statistical variance on a large scale. Existing signature-based malware detection systems cannot cope with this variance as they do not take statistical noise into account and thus this kind of systems can be easily evaded. Therefore, we need a robust alternative that can abstract away the noise and capture the essential information from static or behavioral malware properties.

Static analysis tools, such as PEInfo [1], offer extraction of different properties or metadata (e.g., entropy, histograms, section length) from malware code. This data can be very useful for characterizing malware samples. However, miscreants can easily obfuscate the malware code to the point where it is impossible to retrieve any useful information from static analysis. On the other hand, behavioral analysis tools are less sensitive to obfuscation, as they only record traces of activity retrieved from the execution of malware samples.

The most important traceable events for determining malware behavior are system calls. In order to execute malicious actions, malware needs to use the services from the operating system. For any meaningful action, such as opening a file, running a thread, writing to the registry, or opening a network connection, interaction with the operating system (OS) is necessary. This interaction is done through the system call API of the target OS. Therefore, in order to characterize the malware behavior, it is important to track the sequence of system call events during the execution of malware. Different malware families have different execution goals, which should be revealed by inspecting these traces.

Towards this direction, machine learning-based systems have been developed as a solution to the problems of large number and variety of malware samples. For instance, researchers utilized Hidden Markov Models in an attempt to model system call sequences [4]. Others used Support Vector Machines (SVM) with String Kernels to detect malware based on the executed system calls [21]. Apart from behavioral data, static code properties have also been used as data sources for statistical analysis [25]. Many other papers of similar content have tried to deal with this issue [5,6,22]. The problem with these approaches is that many times machine learning methods use simplifying assumptions. For example, some of these works utilize Markov assumption of memoryless processes in Hidden Markov Models or different kernel definitions that define similarity measures between samples. One exception is the work of Pascanu et al. [19], where they use recurrent networks for modeling system call sequences, in order to construct a "language model" for malware. They test Long Short-Term Memory and Gated Recurrent Units and report good classification performance. However, they do not test deep learning approaches. This may simplify the modeling, but on the other hand it can result in reduced classification accuracy.

In this paper, we focus on investigating the utilization of neural networks to improve the classification of newly retrieved malware samples into a predefined set of malware families. As a matter of fact, recent years have brought a significant development in the area of neural networks, mostly under a paradigm of *deep learning*. This paradigm encompasses a movement towards creating neural networks with a high number of layers to model complex functions of input data. Using modern hardware technology, such as General Purpose GPUs and novel algorithms developed in recent years, deep networks can be trained efficiently using high dimensional datasets on a large scale.

We construct and analyze two types of neural network layers for modeling system call sequences: *convolutional* and *recurrent* layers. These two types of layers use different types of approach in modeling sequential data. On the one hand, convolutional networks use sequences in a form of a set of n-grams, where we do not explicitly model the sequential position of system calls and only count presence and relation of n-grams in a behavioral trace. While this approach simplifies sequence modeling, it also potentially causes loss of information fidelity. On the other hand, recurrent networks train a stateful model by using full sequential information: the model contains dependency of certain system call appearance from the sequence of previous system calls. Since this model is more complex, it is more difficult to train. However, if trained properly, it could potentially offer better accuracy on sequential data, as it is able to capture more information about the training set. By combining those two layers in a hierarchical fashion, we can increase our malware detection capabilities. This is enabled by more robust automatic feature extraction, where we convolve n-grams of system calls and, furthermore, create sequential model out of convolution results. We stack layers in the neural network in accordance to the principles of *deep learning* [7]. In deep learning, one can construct deep neural networks in order to extract a hierarchy of features for classification. We use this technique to achieve improvement in capturing the relation between n-grams in system call sequences. In essence, our approach enables us to get average accuracy, precision, and recall of over 90% for most malware families, which brings significant improvement over previously used methods and thus can help analysts to classify malware more accurately.

In summary, we make the following main contributions:

- We construct deep neural networks and apply them to analyze system call sequences.
- We combine convolutional and recurrent approaches to deep learning for optimizing malware classification.
- We investigate neural unit activation patterns and explain the performance improvement of our models by illustrating the inner workings of our neural network.

2   Methodology

In this section, we describe the methodology we use to perform the task of malware classification. We first provide information regarding our set up environment and then proceed with our deployed techniques.

2.1   System Description

Our malware classification process is displayed in Fig. 1. It begins with a malware zoo, where the executable files are acquired and input data is retrieved by executing malware in a protected environment. The results of these executions are preprocessed in order to get numerical feature vectors. These vectors are then forwarded as inputs for neural networks, which in turn classify the malware into one of the predefined malware families.

We use the *Tensorflow* [3] and the *Theano* [8] frameworks to construct and train the neural networks. These frameworks enable us to design neural network architectures and precompile the training algorithms for execution on graphical processors. Since GPUs are designed for fast execution of linear algebra operations, such as matrix multiplications, we

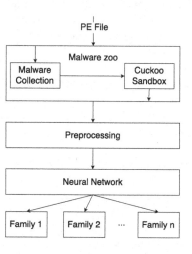

Fig. 1. Overview of our malware classification system.

utilize them to speed up our neural network training. This is a very popular approach in neural network applications, since training deep networks can be a very resource-intensive task. In our set up, we use the NVIDIA TITAN X GPU and the NVIDIA CUDA 7 software platform to accelerate the training algorithms.

2.2   Dataset

We collect malware samples and trace malware behavior using a *malware zoo* [32]. Our malware collection consists of samples gathered from three primary sources: Virus Share [23], Maltrieve [18] and private collections. We select these sources to provide a large and diverse volume of samples for evaluation.

Since malware authors can use code obfuscation and packers in order to subvert static analysis, we use dynamic malware analysis to gather data about malware behavior. Towards this direction, there exist multiple tools that enable tracing the execution of malware and gathering of logs of execution sequences [13,17]. We choose the Cuckoo sandbox which is open source, widely-used, and provides a controlled environment for executing malware. During the execution of malware samples we record calls to the kernel API that we later use to characterize these malware samples. For each malware sample we obtain a sequence of API calls (system calls) and employ this sequence for behavioral modeling.

We use kernel API call sequences (system calls) as features, but as we are doing supervised learning, we also need labels for training. These labels are obtained using services of VirusTotal [2]. In particular, for each malware sample we use, we extract antivirus labels from the VirusTotal web service. This service is used by uploading MD5 hashes of malware executables to VirusTotal and retrieving results from a large number of antivirus engines through the VirusTotal API. These engines compare the hash of the malware file to the data already contained in their own database. We leverage the VirusTotal services in order to access malware analysis results and particular signatures for the samples in our malware zoo. We are interested in the antivirus signatures, out of which we want to retrieve labels for supervised learning. Each antivirus that provides signatures through VirusTotal has its own methodology of labeling malware, which makes it difficult for machine learning classification tests, since we need to extract one numerical label per unique sample. Our approach was to execute clustering on the signatures from different antivirus programs and obtain ground truth classes from the resulting clusters. An ideal solution for labeling is to use reverse engineering and expert knowledge. However, on a larger scale that is not a realistic scenario. Therefore we postprocess the VirusTotal results in order to pull maximum information without manual inspection of malware samples.

2.3 Signature Clustering

We attempt to extract maximum amount of information from the VirusTotal signatures by performing a simplified version of signature clustering method introduced in VAMO [20]. To each malware sample we attach a boolean vector that contains information about presence or absence of different antivirus signatures. Using a variant of *cosine distance* and DBSCAN [12] algorithm we cluster signature vectors in order to detect regions of high similarity. We select the ten most populated clusters and use them to create classes for the evaluation of our methodology. These ten clusters contain 4753 malware samples in total and cover most of our sample set. The rest of the samples we consider as outliers. Usually a malware analyst is interested in extracting samples belonging to certain families and tries to differentiate them from the other executables in the dataset.

2.4 Feature Preprocessing

Before using the API call sequences as inputs to neural networks, we need to remove redundant data and convert the data to sequences of numerical feature vectors. First, we preprocess sequences by removing subsequences where one API call is repeated more than two times in a row. For example, this happens if a process tries to create a file repeatedly in a loop. We cut these subsequences by using maximum two consequent identical kernel API call instances in the resulting sequence. Furthermore, we use *one-hot encoding* to find a unique binary vector for every API call present in the dataset. This means that we create zero feature vectors of length equal to the number of distinct API calls and toggle one bit in a position unique for a particular kernel API call. This way, we have sequences of binary vectors instead of sequences of API call names provided by Cuckoo sandbox. Since our dictionary consists of only 60 distinct system calls, we do not face any challenge with the size of feature vectors.

2.5 Deep Neural Network

In order to maximize the utilization of the possibilities given by neural network methodology, we combine convolutional and recurrent layers in one neural network. Figure 2 depicts our neural network architecture. The convolutional part consists of convolution and a pooling layers. On the one hand, the convolutional layer serves for feature extraction out of raw one-hot vectors. Convolution captures the correlation between neighboring input vectors and produces new features. We use two convolution filters of size 3×60, which corresponds to 3-grams of instructions. As the results of convolution we take feature vectors of size 10 and 20 for the first and second convolution layer, for every input feature. After each convolutional layer we use max-pooling to reduce the dimensionality of data by a factor of two. Outputs of the convolutional part of our neural network are connected to the recurrent part. We forward each output of the convolutional filters as one vector. The resulting sequence is modeled using the LSTM cells. We use LSTM cells, as they are flexible in terms of training, even though the maximal sequence length was limited to 100 vectors. Using the recurrent layer we are able to explicitly model the sequential dependencies in the kernel API traces. Mean-pooling is used to extract features of highest importance from the LSTM output and reduce the complexity of further data processing. Furthermore, we use Dropout [26] in order to prevent overfitting and a softmax layer to output the label probabilities.

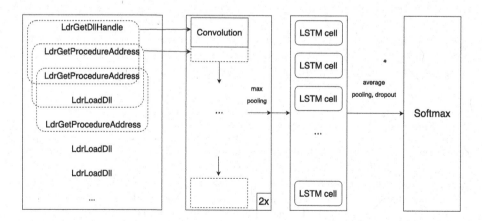

Fig. 2. Deep neural network architecture

3   Evaluation

In this section, we describe the outcomes of our experiments executed using the data from our malware zoo and the neural network architectures that we constructed. For our evaluation measurements we noted the performance for different families in terms of accuracy (ACC), precision (PR) and recall (RC).

In our evaluation we used a type of 3-fold crossvalidation. In each test, we chose two thirds of samples as training set, while one third of samples is assigned to the test set. As a matter of fact, to obtain a reliable performance estimate, we averaged the results of ten crossvalidation tests, executed each time with a new random dataset permutation.

In our first experiment, we investigated the performance of our neural network architecture with respect to the simpler, previously used architectures, such as feedforward neural network, or convolutional network. Table 1 displays the overall results in percentage. These results show that our configuration brings improvements in the classification performance. It also shows that for one family, *Mikey*, we do not get good accuracy. This family gets very often confused for the family *Zusy*, because of very similar behavioral traces.

Next, we wanted to measure the performance improvement of our methodology. Therefore, we compared the performance of deep hybrid architecture with Hidden Markov Models and Support Vector Machines (SVM), since these methods have been heavily used in previous works. Table 2 illustrates the results. This table shows high improvement in case of using neural networks as classifiers. The improvement is statistically significant, with a 5x2cv t-statistics value [11] in the 95% confidence level in comparison with the SVM.

Table 1. Evaluation of deep hybrid network.

Family	Results of various neural network architectures (%)								
	Feedforward network			Convolutional network			ConvNet + LSTM		
	ACC	PR	RC	ACC	PR	RC	ACC	PR	RC
Multiplug	99.6	100.0	99.3	98.9	98.9	98.9	98.9	99.8	99.0
Kazy	100.0	71.7	100.0	100.0	100.0	100.0	100.0	99.9	100.0
Morstar	0.0	0.0	0.0	100.0	100.0	100.0	100.0	99.9	100.0
Zusy	9.1	57.2	68.5	100.0	56.8	100.0	100.0	57.5	100.0
SoftPulse	100.0	100.0	98.2	100.0	99.6	100.0	100.0	99.1	100.0
Somoto	100.0	100.0	100.0	100.0	100.0	100.0	100.0	100.0	100.0
Mikey	89.1	37.1	28.8	0.0	0.0	0.0	0.0	0.0	0.0
Amonetize	100.0	98.4	100.0	99.1	100.0	99.1	99.1	100.0	99.6
Eldorado	100.0	100.0	96.4	99.2	100.0	99.2	99.4	100.0	99.5
Kryptik	100.0	100.0	100.0	95.0	100.0	95.1	96.6	100.0	96.2
Average	79.8	76.4	79.1	89.2	85.6	89.2	89.4	85.6	89.4

Table 2. Comparison with previously used methods.

Family	Deep neural network			Hidden Markov Model			Support Vector Machine		
	ACC	PR	RC	ACC	PR	RC	ACC	PR	RC
Multiplug	98.9	99.8	99.0	91.5	74.5	91.5	99.3	99.9	99.3
Kazy	100.0	99.9	100.0	73.1	95.1	73.1	96.6	93.1	96.6
Morstar	100.0	99.9	100.0	80.0	63.7	80.0	82.3	91.0	82.3
Zusy	100.0	57.5	100.0	65.4	45.1	65.4	100.0	58.4	100.0
SoftPulse	100.0	99.1	100.0	51.1	100.0	51.1	99.9	99.6	99.9
Somoto	100.0	100.0	100.0	50.0	37.6	50.0	99.8	100.0	99.8
Mikey	0.0	0.0	0.0	5.7	20.0	5.7	0.0	0.0	0.0
Amonetize	99.1	100.0	99.6	29.4	100.0	29.4	99.3	100.0	99.3
Eldorado	99.4	100.0	99.5	20.0	80.4	20.0	100.0	100.0	100.0
Kryptik	96.6	100.0	96.2	10.0	100.0	10.0	97.1	100.0	97.1
Average	89.4	85.6	89.4	47.5	71.6	47.6	87.4	84.2	87.4

Table 3. Comparison of neuron activation

API call array	Neuron activation value
LdrGetDllHandle, LdrGetProcedureAddress, LdrGetProcedureAddress, LdrGetDllHandle, LdrLoadDll	64.40814948
LdrGetDllHandle, LdrGetProcedureAddress, LdrGetProcedureAddress, LdrLoadDll, LdrGetProcedureAddress	72.43323427
LdrGetDllHandle, DeviceIoControl, DeviceIoControl, LdrGetDllHandle, LdrLoadDll	−59.49672516
LdrGetDllHandle, LdrGetProcedureAddress, LdrGetProcedureAddress, RegOpenKeyExW, RegOpenKeyExW	−60.89228849
ExitThread, LdrLoadDll, LdrGetProcedureAddress, LdrGetProcedureAddress, LdrGetDllHandle	−60.64636472
NtOpenFile, NtCreateSection, ZwMapViewOfSection, NtOpenFile, LdrLoadDll	−12.56477188

In our last experiment we attempted to illustrate the inner working of our neural network. Thus, we visualized intermediate results of the convolutional layer in the neural network trained on our dataset. Table 3 displays the correlation of 5-gram similarity and similarity of results from convolutional filters. We got the best classification results for using convolutional filters of width 3, but we use 5-grams here for visualization, as it is more illustrative to show longer n-grams. The activation values are normally vectors, but we reduced them to scalars using the t-SNE transformation [28], for better visualization. It is noticeable that the neuron activation values follow the similarity between n-grams of

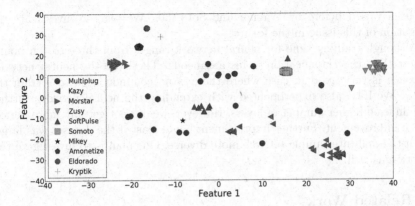

Fig. 3. t-SNE transform of neuron activations in the last layer of our deep hybrid network.

kernel API calls. The function that governs how this similarity is followed is optimized during the training of our neural network.

As the input values are propagated through the layers of the neural network, activations are trained in order to separate different malware classes. In the last layer the activation values should be able to totally separate between different classes, based on features learned in previous layers. Figure 3 shows the results of clustering neuron activation values before the softmax layer, when using malware data of different families as inputs. Again, the neuron activation vectors are reduced to two-dimensional representation using the t-SNE transformation.

4   Discussion

Our results show that deep learning indeed brings improvements in classification of malware system call traces. Although there exist classes where the classification accuracy is under 90%, on average we bring an improvement of performance with respect to Hidden Markov Models and SVM. We combine convolutional and recurrent layers to achieve this improvement. This combination helps us obtain slightly better results than with simpler architectures with only feedforward or only convolutional layers. Using only LSTM recurrent network also does not achieve accuracy as high as we get with our architecture, which can be explained with the relatively short length of the malware execution traces.

While dynamic analysis information is very important, in this work we only look at system call traces, which contain information of only the path taken by malware on execution. This path could also depend on input data, which we ignore in this work. By joining static and dynamic analysis, we could unify the two approaches in looking at malware features. Future work will be dedicated to this goal. One other limitation of our approach is that we do not consider evasion of malware detectors by inserting noise in the system call sequences. This issue has been raised before by Wagner and Soto [30]. We plan to investigate the mitigation of this issue in the future.

Although training time for neural network ranges from three to ten hours, on test time the classification is instantaneous. Therefore the neural network approach is very good in case where there is no common need to retrain the model. We have also experimented with extending the neural network further with an even higher number of layers. However, we did not get any more performance improvement. Further improvement could exist if the dataset was larger and if the malware sample set was more diverse. Our plan is to investigate this direction as well.

5 Related Work

This section contains the description of the research efforts that precede our paper. These efforts are mostly divided into research dedicated to (i) application of machine learning methods in malware analysis and (ii) using specifically neural networks for malware detection and classification. In the following we explain the evolution and current state of those two groups of methods separately.

5.1 Machine Learning Methods for Malware Detection

Machine learning-based malware detection and classification are topics of multiple research efforts. These efforts use various behavioral features of malware as input data for statistical models. The features are obtained by analyzing code or tracing events such as system calls [31], registry accesses [14], and network traffic [27]. This kind of event sequences are analyzed using unsupervised (e.g., clustering), semi-supervised, or supervised learning (classification) methods. Although most of papers investigate malware clustering [5,6], supervised learning is also a prominent topic.

There also papers in recent years that try to use advanced machine learning methods and extract more information from malware datasets. An example of this is given by recent application of statistical topic modeling approaches to the classification of system call sequences [34], which was further extended with a nonparametric methodology [16]. This approach was extended by taking system call arguments as additional information and including memory allocation patterns and other traceable operations [33]. Support vector machines with string kernels represent an another interesting sequence-aware approach [21].

5.2 Neural Networks for Malware Detection and Classification

There are also noticeable efforts to use neural networks for malware detection and analysis. For example, Dahl et al. [10] try to classify malware on a large scale using random projections and neural networks. However, they do not report advantages of increasing the number of neural network layers. Saxe et al. [24]

used feedforward neural networks to classify static analysis results. However, they do not consider dynamic analysis results in their research. In case of obfuscated binaries, the static analysis may not give satisfactory inputs for classification. Huang et al. [15], on the other hand, use up to four hidden layers of feedforward neural network, but focus on evaluating multi-task learning ideas. Pascanu et al. [19] use recurrent networks for modeling system call sequences, in order to construct a "language model" for malware. They test Long Short-Term Memory and Gated Recurrent Units and report good classification performance. However, they do not test deep learning approaches.

These papers give motivating conclusions that enhance the reputation of neural networks being applicable for malware datasets. However, these papers do not report any advantages from diversifying deep architectures in case of behavioral modeling of malware activity and they do not do extensive research in that direction.

6   Conclusion

In this paper, we construct deep neural networks to improve modeling and classification of system call sequences. By combining convolutional and recurrent layers in one neural network architecture we obtain optimal classification results. Using a hybrid neural network containing two convolutional layers and one recurrent layer we get a novel approach to malware classification. Our neural network outperforms not only other simpler neural architectures, but also previously widely-used Hidden Markov Models and Support Vector Machines. Overall, our approach exhibits better performance results when compared to previous malware classification approaches.

References

1. PEInfo Service. https://github.com/crits/crits_services/tree/master/peinfo_service
2. VirusTotal, May 2015. http://www.virustotal.com
3. Abadi, M., et al.: TensorFlow: large-scale machine learning on heterogeneous systems. arXiv preprint arXiv:1603.04467 (2015)
4. Attaluri, S., McGhee, S., Stamp, M.: Profile Hidden Markov Models and metamorphic virus detection. J. Comput. Virol. 5(2), 151–169 (2009)
5. Bailey, M., Oberheide, J., Andersen, J., Mao, Z.M., Jahanian, F., Nazario, J.: Automated classification and analysis of internet malware. In: Kruegel, C., Lippmann, R., Clark, A. (eds.) RAID 2007. LNCS, vol. 4637, pp. 178–197. Springer, Heidelberg (2007). doi:10.1007/978-3-540-74320-0_10
6. Bayer, U., Comparetti, P.M., Hlauschek, C., Kruegel, C., Kirda, E.: Scalable, behavior-based malware clustering. In: ISOC Network and Distributed System Security Symposium (NDSS) (2009)
7. Bengio, Y.: Learning deep architectures for AI. Found. Trends Mach. Learn. 2(1), 1–127 (2009)
8. Bergstra, J., Breuleux, O., Bastien, F., Lamblin, P., Pascanu, R., Desjardins, G., Turian, J., Warde-Farley, D., Bengio, Y.: Theano: a CPU and GPU math expression compiler. In: Python for Scientific Computing Conference (SciPy) (2010)

9. Bu, Z., et al.: McAfee Threats Report: Second Quarter 2012 (2012)
10. Dahl, G.E., Stokes, J.W., Deng, L., Yu, D.: Large-scale malware classification using random projections and neural networks. In: IEEE International Conference on Acoustics, Speech and Signal Processing (ICASSP) (2013)
11. Dietterich, T.G.: Approximate statistical tests for comparing supervised classification learning algorithms. Neural Comput. **10**(7), 1895–1923 (1998)
12. Ester, M., Kriegel, H.-P., Sander, J., Xu, X.: A density-based algorithm for discovering clusters in large spatial databases with noise. In: KDD (1996)
13. Guarnieri, C., Tanasi, A., Bremer, J., Schloesser, M.: The Cuckoo Sandbox (2012)
14. Heller, K., Svore, K., Keromytis, A.D., Stolfo, S.: One class support vector machines for detecting anomalous windows registry accesses. In: Workshop on Data Mining for Computer Security (DMSEC) (2003)
15. Huang, W., Stokes, J.W.: MtNet: a multi-task neural network for dynamic malware classification. In: Conference on Detection of Intrusions and Malware & Vulnerability Assessment (DIMVA) (2016)
16. Kolosnjaji, B., Zarras, A., Lengyel, T., Webster, G., Eckert, C.: Adaptive semantics-aware malware classification. In: Caballero, J., Zurutuza, U., Rodríguez, R.J. (eds.) DIMVA 2016. LNCS, vol. 9721, pp. 419–439. Springer, Heidelberg (2016). doi:10.1007/978-3-319-40667-1_21
17. Lengyel, T.K., Maresca, S., Payne, B.D., Webster, G.D., Vogl, S., Kiayias, A.: Scalability, fidelity and stealth in the DRAKVUF dynamic malware analysis system. In: Annual Computer Security Applications Conference (ACSAC) (2014)
18. Maxwell, K.: Maltrieve, April 2015. https://github.com/krmaxwell/maltrieve
19. Pascanu, R., Stokes, J.W., Sanossian, H., Marinescu, M., Thomas, A.: Malware classification with recurrent networks. In: IEEE International Conference on Acoustics, Speech and Signal Processing (ICASSP) (2015)
20. Perdisci, R., ManChon, U.: VAMO: towards a fully automated malware clustering validity analysis. In: Annual Computer Security Applications Conference (ACSAC) (2012)
21. Pfoh, J., Schneider, C., Eckert, C.: Leveraging string kernels for malware detection. In: Lopez, J., Huang, X., Sandhu, R. (eds.) NSS 2013. LNCS, vol. 7873, pp. 206–219. Springer, Heidelberg (2013). doi:10.1007/978-3-642-38631-2_16
22. Rieck, K., Holz, T., Willems, C., Düssel, P., Laskov, P.: Learning and classification of malware behavior. In: Zamboni, D. (ed.) DIMVA 2008. LNCS, vol. 5137, pp. 108–125. Springer, Heidelberg (2008). doi:10.1007/978-3-540-70542-0_6
23. Roberts, J.-M.: Virus Share, November 2015. https://virusshare.com/
24. Saxe, J., Berlin, K.: Features, deep neural network based malware detection using two dimensional binary program arXiv preprint arXiv:1508.03096 (2015)
25. Schultz, M.G., Eskin, E., Zadok, E., Stolfo, S.J.: Data mining methods for detection of new malicious executables. In: IEEE Symposium on Security and Privacy (2001)
26. Srivastava, N., Hinton, G., Krizhevsky, A., Sutskever, I., Salakhutdinov, R.: Dropout: a simple way to prevent neural networks from overfitting. J. Mach. Learn. Res. **15**(1), 1929–1958 (2014)
27. Tegeler, F., Fu, X., Vigna, G., Kruegel, C.: Botfinder: finding bots in network traffic without deep packet inspection. In International Conference on Emerging Networking Experiments and Technologies (CoNEXT) (2012)
28. Van der Maaten, L., Hinton, G.: Visualizing data using t-SNE. J. Mach. Learn. Res. **9**, 2579–2605 (2008). 85
29. VirusTotal. File Statistics, November 2015. https://www.virustotal.com/en/statistics/

30. Wagner, D., Soto, P.: Mimicry attacks on host-based intrusion detection systems. In: Conference on Computer and Communications Security (CCS) (2002)
31. Warrender, C., Forrest, S., Pearlmutter, B.: Detecting intrusions using system calls: alternative data models. In: IEEE Symposium on Security and Privacy (1999)
32. Webster, G.D., Hanif, Z.D., Ludwig, A.L.P., Lengyel, T.K., Zarras, A., Eckert, C.: SKALD: a scalable architecture for feature extraction, multi-user analysis, and real-time information sharing. In: Bishop, M., Nascimento, A.C.A. (eds.) ISC 2016. LNCS, vol. 9866, pp. 231–249. Springer, Heidelberg (2016). doi:10.1007/978-3-319-45871-7_15
33. Xiao, H., Eckert, C.: Efficient online sequence prediction with side information. In: IEEE International Conference on Data Mining (ICDM) (2013)
34. Xiao, H., Stibor, T.: A supervised topic transition model for detecting malicious system call sequences. In: Workshop on Knowledge Discovery, Modeling and Simulation (2011)

Similarity Matching of Computer Science Unit Outlines in Higher Education

Gaurav Langan, James Montgomery[✉], and Saurabh Garg

School of Engineering & ICT, Unversity of Tasmania, Hobart, Australia
{glangan,james.montgomery,saurabh.garg}@utas.edu.au

Abstract. With the globalisation of education, students may undertake higher education courses anywhere in the world. Yet there is variation between different universities' offerings. Even though web search engines can help one to locate potentially similar courses or subjects offered by different universities, judging the degree of similarity between each of them is currently a manual process in which a student or staff member has to go through subject/unit descriptions within a course to understand the different topics taught. In this paper, we study the application of text mining to evaluate the similarity or overlap between different units and propose a system that can help students and staff to make these judgements. The unit or course descriptions are generally short, containing 100–200 words, and exhibit very wide diversity in the ways they are written. Experimental results using data from Australian and international universities demonstrate the accuracy of the proposed system in calculating the similarity between different computing units.

1 Introduction

Higher education degrees are typically delivered via a collection of individual *units* of study (sometimes referred to as *subjects* or *papers*). A *unit outline* is a succinct summary of the learning and teaching aims and experiences that each unit offers, and serves as a mechanism for students to select which units they will study and for different institutions to judge how their own degrees compare with their competitors'. As higher education students are increasingly mobile, credit for prior learning may also be granted to students moving from one institution to another, and it is the unit outlines of a student's prior study that are used to determine which units they are credited.

At the core of this problem is the task of judging the similarity between unit outlines, as they serve as a proxy for the detailed content of the units themselves. However, determining the similarity between units of study based on such small documents is a challenging and difficult exercise. Unit outlines from different universities, even for the same topics, can vary vastly in terms of structure and content. Some unit outlines only specify the high level topics covered, while others may give detailed descriptions. Sometimes different units use synonyms for the same topics. For example, for computing units, some unit outlines use the term "programming languages", while others use specific programming language

© Springer International Publishing AG 2016
B.H. Kang and Q. Bai (Eds.): AI 2016, LNAI 9992, pp. 150–162, 2016.
DOI: 10.1007/978-3-319-50127-7_12

names, such as "Java". Comparing such diverse unit outlines from different universities is not trivial.

Educational data mining has emerged as an important research area, and various applications and tools have been developed to resolve educational research issues. These tools aim to make decision making processes efficient and thus enhance student learning and experience [11]. Romero and Ventura [9] have identified in details how different users and stakeholders can be benefited from educational data mining research. Data mining algorithms are used in planning and developing course curricula and creating meaningful learning outcome topologies [5]. Applications have been developed to give personalized recommendations on online learning content on the basis of learning style and web usage habits [14].

However, to best of our knowledge, there is no existing system that has been designed specifically for comparing unit content from different universities. Therefore, the goal of this paper is to investigate the application of text mining to measure the similarity between different unit outlines based on the degree of commonality between concepts. Finding the similarity between units is a step-wise process. The first step is to identify domain-relevant keywords from the units. The extracted keywords can also be considered as concepts. The second step is to obtain the degree of similarity between concepts found in different unit outlines. The similarity between units can then be calculated by combining the per-concept similarity scores. The unit-to-unit similarity score can then be used as an input to a recommendation system. Thus, the key contributions of this paper are following:

1. Finding the best techniques to accurately extract keywords and topics from a unit's description and learning outcomes.
2. A system to compute the similarity score between two unit outlines based on a Wikipedia-derived corpus.
3. Evaluation of the similarity between a subset of first year computing units offered by 31 Australian universities.

The next section describes the proposed system. Sections 3 and 4 discuss the mechanisms developed to extract relevant keywords and to assess the similarity between different units of study, respectively. We conclude the article with the future directions of this work.

2 Proposed System Architecture

The proposed system judges the similarity between unit outlines via the similarity of individual concepts that appear in each outline. The complete system workflow diagram is given in Fig. 1, and includes these major components:

1. **Keyword Extraction** is the initial step in the analysis of a unit outline's description or list of topics. Its role is to identify *concepts* likely to be characteristic of the unit's content and learning outcomes.

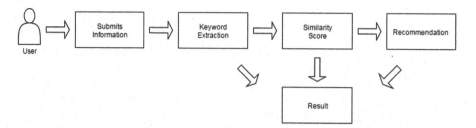

Fig. 1. System workflow diagram

Table 1. Introductory programming units. Some institution names have been replaced by commonly used abbreviations

Unit code	Unit title	Institution
COMP1100	Introduction to Programming	Australian National University
COIT11222	Programming Fundamentals	Central Queensland University
ITC106	Programming Principles	Charles Sturt University
SIT102	Introduction to Programming	Deakin University
COMP1102	Computer Programming	Flinders University
CAB201	Programming Principles	QUT
CSE100F	Java Programming	La Trobe University
COMP115	Introduction to Computer Science	Macquarie University
FIT1040	Programming Fundamentals	Monash University
ICT159	Foundations of Programming	Murdoch University
COSCI284	Programming Techniques	RMIT University
ISY00245	Principles of Programming	Southern Cross University
4478	Introduction to Information Technology	University of Canberra
COMP10001	Foundations of Computing	University of Melbourne
INFO1103	Introduction to Programming	University of Sydney
KIT101	Programming Principles	University of Tasmania
48023	Programming Fundamentals	UTS
300580.2	Programming Fundamentals	University of Western Sydney
CSCI103	Algorithms and Problem Solving	University of Wollongong
NIT1102	Introduction to Programming	Victoria University

2. **Similarity Between Units** is measured by pairwise comparisons between each of the concepts identified by keyword extraction in two unit outlines. Its role is to identify the most similar concepts between the two and their degree of similarity.

3. **Recommendation** is any use case-specific component that supports decisions based on the degree of similarity between units. Depending on the context, one can design different recommendation models. For example, the

system might recommend to not choose particular course after comparing all the units. One can even use the system for credit assessment after including other factors such as the level of units previously studied, a student's performance in those units, and the rank of the previous institution.

The present work deals with the implementation and evaluation of the first two components. For the evaluation, several alternative keyword extraction approaches were compared using 26 units from computer science degrees taught at Australian and international universities: 11 units (seven institutions) from Australia, 10 units (four institutions) from Malaysia, two units from an institution in Singapore and three units from an institution in Vietnam.[1] All unit outlines are in English. The data used to evaluate similarity matching approaches are from the unit outlines of 40 introductory and intermediate programming units taught by 31 Australian institutions: 20 introductory programming units (Table 1) and 20 intermediate level units that teach algorithms and data structures (Table 2).

Table 2. Data structure and algorithms units. Some institution names have been replaced by commonly used abbreviations

Unit code	Unit title	Institution
COMP3600	Algorithms	Australian National University
ITC322	Data Structures	Charles Sturt University
COMP1002	Data Structures & Algorithms	Curtin University
SIT221	Data Structures & Algorithms	Deakin University
CSP2348	Data Structures	Edith Cowan University
BIT208	Data Structures & Algorithms	KPM
CSE2ALG	Algorithms & Data Structures	La Trobe University
COMP225	Algorithms & Data Structures	Macquarie University
FIT2004	Algorithms & Data Structures	Monash University
ICT283	Data Structures & Algorithms	Murdoch University
COMPSCI1103	Algorithm Design & Data Structures	University of Adelaide
COMP10002	Foundations of Algorithms	University of Melbourne
COMP2230	Introduction to Algorithms	University of Newcastle
COMP582	Data Structures & Algorithms	University of New England
COMP9024	Data Structures & Algorithms	University of New South Wales
COMP3506	Algorithms & Data Structures	University of Queensland
CSC2401	Algorithms & Data Structures	USQ
INFO1105	Data Structures	University of Sydney
KIT205	Data Structures & Algorithms	University of Tasmania
300103.3	Data Structures & Algorithms	University of Western Sydney

---

[1] The full list is available upon request.

3   Extracting Relevant Keywords

While the structure of unit outlines varies between institutions, most include at least a *description* that presents topics in descriptive prose, and *content* section that is often presented as a list of individual topics. Either or both of these can be the source of keywords.

Keyword extraction involves the identification of important words from a piece of text. One of the most common measures of word importance is Term Frequency-Inverse Document Frequency (TF-IDF), which weights terms more highly if they appear frequently in a given document but infrequently in a wider corpus of documents [10]. As TF-IDF is most useful if the if the input text is large, and each unit outline is relatively small (and the importance of the topic is not determined by its frequency) such an approach cannot be used here. Instead we use a domain-specific dictionary, the *Oxford Dictionary of Computing* (which contains 6355 terms), to assess term relevance.

The text extraction process is as follows: text containing a brief description of all the topics covered in a unit is extracted manually; *tokenization* to identify individual words; *lemmatization* to standardise each token; grouping of adjacent lemmatized tokens into meaningful units; and finally filtering of terms using a dictionary. Two alternative approaches to recombining tokens into meaningful units are evaluated: *n-grams* and *noun phrase chunking*.

An n-gram is a list of n co-occurring words in a piece of text [3]. Using this approach, after the lemmatization step the list of candidate words is augmented by adding a combination of bi-grams and tri-grams of the same words. Noun phrase chunking is a process of extracting a combination of words from a text based on their part-of-speech. The heuristic filter used here is 'zero or more adjectives or nouns followed by another noun'. For example, from the sentence 'Advanced sorting algorithms and evaluation methods are covered', both 'advanced sorting algorithms' and 'evaluation methods' would be extracted.

3.1   Evaluation

To provide a gold standard for evaluating the quality of the keyword sets obtained from the two processes, genuinely relevant topics in each unit outline were first extracted manually. The quality of the automatically extracted keywords was then evaluated in terms of *recall* and *precision* [2]). Recall is the ratio of the number of relevant terms retrieved to the total number of relevant terms present (i.e., the number of manually identified relevant keywords). Precision is the ratio of the number of relevant terms retrieved to the total number of terms retrieved. The F-measure ($2 \times \frac{precision \times recall}{precision + recall}$) is also used to assess performance.

Figure 2 shows the plot of recall and precision of keywords extracted from each unit using the two methods. The marker for each unit is at its precision and recall score. The contours in the plot represent the F-measure of recall and precision. From these plots it is clear that the n-gram method of finding keywords gives better scores of recall and precision. The mean (and standard deviation) of the F-measures are 0.52 (0.16) for noun phrase chunking and 0.76 (0.11) for n-grams.

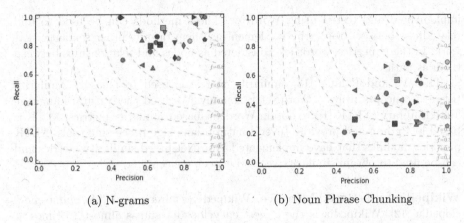

(a) N-grams   (b) Noun Phrase Chunking

Fig. 2. Recall and precision of extracted keywords

The two methods achieve similar precision but the n-gram method achieves higher recall. The n-gram method is able to extract greater number of relevant keywords than the noun phrase chunking method. A possible explanation of these better recall scores is that it increases the number of words to be matched within the dictionary by including bi-grams and tri-grams. The set of words obtained by noun phrase chunking has fewer words to match against the dictionary because it only extracts words that fit a predefined pattern. Consequently, the n-gram method is used to extract relevant keywords for the next stage in the comparison process.

4   Assessing Similarity of Keywords from Unit Outlines

Semantic similarity between concepts takes into account the meaning and context of the concepts. Semantic similarity is different from lexical similarity, which estimates similarity on the basis of syntactical representation. There are two approaches to find semantic similarity between concepts: knowledge-based and corpus-based. Knowledge-based similarity models use existing semantic networks to judge the similarity between concepts. Semantic networks based on a particular domain can also be used to determine the similarity between topics within that domain. Corpus-based methods, on the other hand, use an existing corpus of text to find semantic similarity between topics. Corpus-based methods have an advantage over knowledge-based methods that they can be used when a domain specific semantic network is not available. In the present study we compare the efficacy of WordNet [4] as a semantic network (knowledge-based approach) and Wikipedia as the corpus [6] for a corpus-based approach.

WordNet. WordNet is a freely available dictionary of nouns, verbs, adjectives and adverbs in English [4]. There are several metrics to measure semantic relatedness between concepts based on a taxonomy. Most of these metrics are implemented and tested using WordNet. These methods can be used to find relatedness between topics by implementing them on any other semantic network. In these metrics, words are considered as nodes and relationships between them as paths.

The Wu and Palmer [13] similarity metric is used to measure similarity between concepts using a WordNet taxonomy. We used the Natural Language Toolkit library (NLTK [1]) to obtain WordNet-based similarity metrics. NLTK is based in Python and provides several natural language processing tools. NLTK provides modules that have implemented WordNet based similarity metrics and can be used to obtain similarity scores between concepts.

Wikipedia Link-Based Measure. Wikipedia is a freely available, online encyclopedia [12]. Wikipedia is the largest encyclopedia and is almost 10 times as large as its nearest competitor, the *Encyclopedia Britannica*. Wikipedia can also be used as a corpus as it contains millions of articles in different languages.

Every month the Wikimedia Foundation releases Wikipedia in compressed XML format, also called the 'dumps'. Offline Wikipedia data can be used for various purposes such as backup, academic research and republishing. A dump of the English language Wikipedia dated 4 March 2015 is used in this study. This Wikipedia data set contains approximately 4.8M articles excluding the disambiguation and redirect pages.

The Wikipedia Link-based Measure (WLM) is a corpus-based method of determining the degree of similarity between concepts. The WLM uses the link structure of Wikipedia to estimate the similarity between concepts. Given any two keywords, WLM estimates the similarity between them in the range [0, 1]. A similarity score of 0 indicates that the terms are not similar at all and a similarity score of 1 means that the terms are identical. An open source implementation of WLM known as Wikipedia Miner toolkit [8] is used here. The toolkit provides a Java-based library that can be used with any version of Wikipedia. The toolkit extracts the link structure from the Wikipedia data set.

4.1 Determining Similarity Between Units

Given two short segments of text, we wish to calculate a score in the range [0, 1] that indicates the level of similarity between the two. First, relevant keywords are extracted from source unit outlines using the n-gram-based approach described in the previous section. Then, for a given pair of units, the maximum similarity score for all its keywords with the keywords from the other unit is obtained. The overall similarity between the two text segments T_1 and T_2 is given by the average of the directional similarity score [7]:

$$sim(T_1, T_2) = \frac{s(T_1, T_2) + s(T_2, T_1)}{2} \tag{1}$$

where $s(T, U)$ is the uni-directional similarity, defined as

$$s(T, U) = \frac{\sum_{w \in T} maxSim(w, U) \times idf(w)}{\sum_{w \in T} idf(w)} \tag{2}$$

Table 3. IDF weights of some of the keywords

Keyword	Count	IDF weight
Computer	28667	0.52104
System	23197	0.73277
Data	18377	0.96569
Set	18197	0.97553
Information	16575	1.06889
Software	15497	1.13614
Form	15469	1.13795
Process	13035	1.30915
Computer Science	8129	1.78135
Algorithm	8155	1.77816

where $maxSim$ is the maximum value of similarity for each keyword (i.e., its most similar term in the other document). These maximum similarity scores are summed and divided by the sum of inverse document frequency of the keywords $idf(w)$, giving their *specificity*. The IDF weights of the keywords are based on a domain specific corpus. IDF weights are used to decrease the significance of commonly occurring terms and increase the significance of rarely occurring terms. The IDF of a term w is given by:

$$idf(w) = log \left(\frac{Total\ number\ of\ documents\ in\ corpus}{Total\ number\ of\ documents\ containing\ w} \right) \tag{3}$$

The computer science domain corpus used to calculate IDF scores for keywords was derived from the 4 March 2015 Wikipedia dump by selecting articles that contain any of the following filter words: 'computer'; 'computer science'; 'data'; 'algorithm'. Articles containing lists and disambiguation are removed from the corpus. This produced a corpus of 48,269 documents (81,286,834 words).

Stop words are not removed from articles as it will not affect the calculation of IDF scores (as they are excluded by the initial keyword extraction process). IDF weights of all the dictionary keywords used in keyword extraction process are calculated. Table 3 shows the IDF weights of some of the keywords with high count and also for all four corpus selection words.

4.2   Evaluation

The evaluation of similarity metrics is on the basis of the similarity between segments of text taken from the unit descriptions. The units chosen for evaluation are from various Australian computer science curricula and can be divided into two groups. One group comprises units that are the first programming unit in a learning institute and can thus be called 'introductory programming' units

Table 4. Units that do not teach introductory or intermediate programming

Unit code	Unit title	Institution
JFA101	Aquatic Biology I	University of Tasmania
KPZ121	Ecology	University of Tasmania
CNA126	Health Care	University of Tasmania
JEE102	Mathematics II	University of Tasmania
KEA101	Understanding Earth Systems	University of Tasmania
FNW102	Fundamentals of Networking	AMC Malaysia
KIT108	Artificial Intelligence	University of Tasmania
KIT103	Computational Science	University of Tasmania
KIT102	Data Organization and Visualization	University of Tasmania
KIT109	Games Fundamentals	University of Tasmania

Table 5. Similarity score for introductory programming units compared with other such units (similar) or with the units in Table 4 (not similar)

	WordNet-based		Wikipedia-based	
	Similar	Not similar	Similar	Not similar
Samples	190	200	190	200
Minimum	0.34	0.0	0.61	0.0
Maximum	0.83	0.74	0.85	0.64
Mean	0.54	0.38	0.71	0.40
St Dev	0.09	0.21	0.04	0.18

(see Table 1). The second group are intermediate level units that teach basic algorithms and data structures (see Table 2). Both groups contain 20 units. To evaluate the discriminatory power of the approach, 10 additional units are included that should not be identified as similar: five are from computer science curricula (but not related to programming), and five are from other departments such as science or engineering (see Table 4).

Results for Introductory Programming Units. The similarity scores of all pairs from the 20 introductory programming units were calculated as well as each of those units compared with each of the 10 dissimilar units. Descriptive statistics are presented in Table 5 and the distributions of scores are shown in Fig. 3. Both approaches report higher mean similarity scores for units that are genuinely similar than those that are not. Mann-Whitney U tests comparing the similar and dissimilar distributions for both approaches report p-values < 0.01, indicating that both the WordNet- and Wikipedia-based measures are capable of discriminating similar units from dissimilar units.

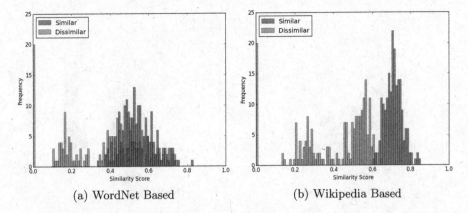

(a) WordNet Based   (b) Wikipedia Based

Fig. 3. Distribution of similarity scores for introductory programming units compared with other such units (similar) or with the units in Table 4 (not similar)

Table 6. Similarity scores for algorithms and data structures units compared with other such units (similar) or with the units in Table 4 (not similar)

	WordNet-based		Wikipedia-based	
	Similar	Not similar	Similar	Not similar
Samples	190	200	190	200
Minimum	0.41	0.0	0.61	0.0
Maximum	0.83	0.72	0.88	0.62
Mean	0.62	0.38	0.77	0.39
St Dev	0.08	0.19	0.05	0.18

However, the mean similarity score for similar units is higher using the Wikipedia-based measure, with a narrower spread, and exhibits less overlap between the distributions than with the WordNet-based similarity measure (Fig. 3a).

Algorithms and Data Structures. The similarity scores of all pairs from the 20 algorithms and data structures units (Table 2) were calculated as well as each of those units compared with each of the 10 dissimilar units. Table 6 presents descriptive statistics while Fig. 4 plots the distributions of the scores.

This group of units gives similar results to those for the introductory units. The mean similarity score is high among the similar units for both methods, and both are capable of differentiating between similar and dissimilar units. This is confirmed by performing Mann-Whitney U test between similarity scores of similar and dissimilar units for both the methods (p-value ~ 0).

For both introductory and data structures units, the Wikipedia based measure gives higher scores of similarity between the similar units (illustrated

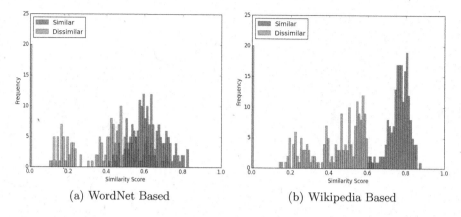

(a) WordNet Based   (b) Wikipedia Based

Fig. 4. Distribution of similarity scores for data structures and algorithms units with other such units (similar) or with the units in Table 4 (not similar)

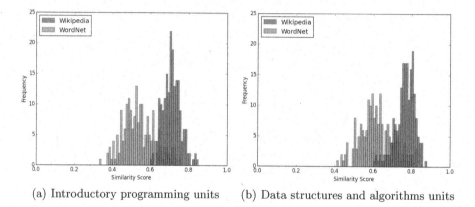

(a) Introductory programming units   (b) Data structures and algorithms units

Fig. 5. Distribution of similarity scores for Wikipedia-based and WordNet-based methods for genuinely similar units

in Fig. 5, which compares the two measures on similar units only). This hypothesis is confirmed by performing Welch's t-tests on both sets (p-value ~ 0). Also, there is a significant overlap in the distribution of similar and dissimilar similarity scores when using the WordNet-based measure (Figs. 3a and b). This implies that the Wikipedia-based method is comparatively better at discriminating between similar and dissimilar units, and should be used in preference to WordNet for comparing small domain-specific documents.

5   Conclusions and Future Directions

Knowing the similarity between different courses plays an important role in different educational processes such as assessing the coursework of different universities and in assigning credit for prior learning. To achieve this end we have proposed a novel text mining based system for assessing the similarity between different units based on their published outlines. The similarity scores computed by the system based on unit descriptions, augmented by other information such as the indicative level of each unit, can be used to provide different kinds of recommendation. We evaluated the accuracy of the system using several domestic (Australian) and international universities. The key findings of the evaluation of each component of the system are: the n-gram method of extracting dictionary keywords is an appropriate method for small segments of text; and the Wikipedia link-based measure, filtered by a small set of domain-specific keywords, can be used to obtain a valid similarity score for two units on the basis of their descriptions or lists of topics. It is plausible that the Wikipedia measure of similarity will be equally effective when applied in other curriculum domains.

In general, unit outlines are available in varying formats, including images, plain text or PDF. An obvious expansion of the present work is to include a module for extracting text via image processing or via text mining that can identify the most relevant section of a text-based document. Further, although the Wikipedia-based similarity metric achieved good separation between similar and dissimilar units, some manual calibration is required if the scores it produces are to be used to make the binary decision of whether two unknown units are indeed similar to not. Machine learning techniques such as clustering may be suitable for dealing with this problem, and will be investigated in the future.

References

1. Bird, S.: NLTK: the natural language toolkit. In: Proceedings of the COLING/ACL on interactive presentation sessions, pp. 69–72. Association for Computational Linguistics (2006)
2. Buckland, M.K., Gey, F.C.: The relationship between recall and precision. JASIS 45(1), 12–19 (1994)
3. Damashek, M., et al.: Gauging similarity with n-grams: language-independent categorization of text. Science 267(5199), 843–848 (1995)
4. Fellbaum, C.: WordNet. Wiley Online Library (1998)
5. Luan, J.: Data mining and knowledge management in higher education-potential applications (2002)
6. Medelyan, O., Milne, D., Legg, C., Witten, I.H.: Mining meaning from Wikipedia. Int. J. Hum. Comput. Stud. 67(9), 716–754 (2009)
7. Mihalcea, R., Corley, C., Strapparava, C.: Corpus-based and knowledge-based measures of text semantic similarity. In: AAAI, vol. 6, pp. 775–780 (2006)
8. Milne, D., Witten, I.: An open-source toolkit for mining Wikipedia. Artif. Intell. 194, 222–239 (2013)
9. Romero, C., Ventura, S.: Educational data mining: a review of the state of the art. IEEE Trans. Syst. Man Cybern. Part C Appl. Rev. 40(6), 601–618 (2010)
10. Salton, G., Wong, A., Yang, C.S.: A vector space model for automatic indexing. Commun. ACM 18(11), 613–620 (1975)

11. Tang, C., Lau, R.W., Li, Q., Yin, H., Li, T., Kilis, D.: Personalized courseware construction based on web data mining. In: Proceedings of the First International Conference on Web Information Systems Engineering, vol. 2, pp. 204–211. IEEE (2000)
12. Wikipedia: Main page (2015). https://en.wikipedia.org/
13. Wu, Z., Palmer, M.: Verbs semantics and lexical selection. In: Proceedings of the 32nd Annual Meeting on Association for Computational Linguistics, pp. 133–138. Association for Computational Linguistics (1994)
14. Zhang, L., Liu, X., Liu, X.: Personalized instructing recommendation system based on web mining. In: The 9th International Conference for Young Computer Scientists, 2008. ICYCS 2008, pp. 2517–2521. IEEE (2008)

Parallel Late Acceptance Hill-Climbing Algorithm for the Google Machine Reassignment Problem

Ayad Turky[1]([⊠]), Nasser R. Sabar[2], Abdul Sattar[3], and Andy Song[1]

[1] School of Computer Science and I.T., RMIT University, Melbourne, Australia
{ayad.turky,andy.song}@rmit.edu.au
[2] Queensland University of Technology, Brisbane, Australia
nasser.sabar@qut.edu.au
[3] Griffith University, Brisbane, Australia
a.sattar@griffith.edu.au

Abstract. Google Machine Reassignment Problem (GMRP) is an optimisation problem proposed at ROADEF/EURO challenge 2012. The task of GMRP is to allocate cloud computing resources by reassigning a set of services to a set of machines while not violating any constraints. We propose an evolutionary parallel late acceptance hill-climbing algorithm (P-LAHC) for GMRP in this study. The aim is to improve the efficiency of search by escaping local optima. Our P-LAHC method involves multiple search processes. It utilises a population of solutions instead of a single solution. Each solution corresponds to one LAHC process. These processes work in parallel to improve the overall search outcome. These LAHC processes start with different initial individuals and follow distinct search paths. That reduces the chance of falling into a same local optima. In addition, mutation operators will apply when the search becomes stagnated for a certain period of time. This further reduces the chance of being trapped by a local optima. Our results on GMRP instances show that the proposed P-LAHC performed better than single threaded LAHC. Furthermore P-LAHC can outperform or at least be comparable to the state-of-the-art methods from the literature, indicating that P-LAHC is an effective search algorithm.

Keywords: Google Machine Reassignment Problem · Optimisation · Resource allocation · Late Acceptance Hill-Climbing · Parallel search · Evolutionary algorithms

1 Introduction

The importance of optimising resource allocation for clouding computing is obvious as cloud computing is one of the fastest growing areas in ICT industry. Typical computational resources are CPUs and may also include storage, network bandwidth and peripheral services [2]. Service providers like Google and Amazon need to ensure high quality services for their users. Naturally optimisation of resources becomes an important part in cloud service management [6].

© Springer International Publishing AG 2016
B.H. Kang and Q. Bai (Eds.): AI 2016, LNAI 9992, pp. 163–174, 2016.
DOI: 10.1007/978-3-319-50127-7_13

Recently a benchmark problem on resource optimisation has been proposed by Google at ROADEF/EURO challenge 2012. That is called Google Machine Reassignment Problem (GMRP) [1]. The main aim of this challenge is to improve machine usage by reassigning a set of processes across a pool of servers while satisfying a set of constraints. A range of optimisation algorithms have been proposed for this GMRP challenge, including simulated annealing [18], variable neighbourhood search [11], constraint programming-based large neighbourhood search [16], large neighbourhood search [3], multi-start iterated local search [15], restricted iterated local search [13], evolutionary simulating annealing [22], grammatical evolution simulated annealing [19] and memetic algorithm [20].

In this work, we propose an evolutionary method which is named as Parallel Late Acceptance Hill-Climbing (P-LAHC) algorithm. The intention is to improve search efficiency for optimisation problems especially GMRP. This method is based on Late Acceptance Hill-Climbing algorithm (LAHC) and population based evolutionary search. LAHC itself is a recently proposed algorithm. The search of LAHC involves one heuristic and produces a single solution [4,5]. LAHC has been successfully used in various optimisation problems such as timetabling problems [9], travelling purchaser problem [12], exam scheduling problem [17], lock scheduling problem [23] and assembly line balancing problem [25].

LAHC can be considered as an improved hill climbing [4,21]. Different to standard hill climbing algorithm which is greedy, LAHC accepts worse solutions which may help the search process escape from local optima points. The main feature of LAHC is delaying the comparison between the newly generated solution with previous solutions. LAHC accepts a solution if it is not worse than a solution generated several steps before, instead of comparing the immediate previous solution as in standard hilling climbing.

The proposed Parallel LAHC (P-LAHC) extends the idea further by introducing a population of solutions instead of a single solution. Each solution is the result of its own LAHC search process. Multiple LAHC processes progress simultaneously in a parallel manner. Furthermore each LAHC process starts from a distinct initial solution. Hence these processes may follow different search paths. The chance of these processes converging to one local optima point is very small. In addition various mutation operators will be applied if a solution cannot be improved for a certain number of iterations. This mechanism further helps search avoid being trapped in local optima.

In this study the performance of the proposed P-LAHC algorithm is assessed on 20 GMRP instances from the original ROADED/EURO 2012 challenge. These instances are very diverse in size and features. They can be used to compare the proposed P-LAHC with standard LAHC in terms of optimisation performance. In addition the performance of P-LAHC will also be compared with the state-of-the-art algorithms from the literature. The comparison with the current best methods should be able to demonstrate the effectiveness of our proposed P-LAHC in solving GMRP.

The remainder of this paper is organised as follows: Sect. 2 presents the problem description of GMRP. Section 3 explains the proposed P-LHAC in details. Section 4 describes the experiment settings. The results are reported in Sect. 5. Finally the study is concluded in Sect. 6.

2   Problem Description

GMRP is a type of combinatorial optimisation problem originated from ROADEF EURO challenge 2012 [1]. The main elements of this problem are a set of machines M and a set of processes P. One service usually consists of multiple processes which can be distributed around machines. The goal is to find the optimal arrangement to allocate process $p \in P$ to machine $m \in M$ in order to improve the overall usage of a given set of machines. One machine is equipped with resources such as CPUs and RAM. One process can be moved from one machine to another when necessary. The allocation of processes must not violate the following hard constraints [1]:

- *Capacity constraints*: the sum of requirements of resource of all processes does not exceed the capacity of the allocated machine.
- *Conflict constraints*: processes of the same service must be allocated into different machines.
- *Transient usage constraints*: if a process is moved from one machine to another, it requires adequate amount of capacity on both machines.
- *Spread constraints*: the set of machines is partitioned into locations and processes of the same service should be allocated to machines in a number of distinct locations.
- *Dependency constraints*: the set of machines are partitioned into neighbourhoods. Then, if there is a service depends on another service, then the process of first one should be assigned to the neighbouring machine of second one or vice versa.

A feasible solution to GMRP is a process-machine assignment which satisfies all hard constraints mentioned above and minimises the weighted cost function shown below in Eq. (1):

$$f = \sum_{r \in R} weight_{loadCost}(r) \times loadCost(r) \tag{1}$$

$$+ \sum_{b \in B} weight_{balanceCost}(b) \times balanceCost(b)$$

$$+ weight_{processMoveCost} \times processMoveCost$$

$$+ weight_{serviceMoveCost} \times serviceMoveCost$$

$$+ weight_{machineMoveCost} \times machineMoveCost$$

In the equation R is a set of available resources; $loadCost$ represents the used capacity by resource r which exceeds the safety capacity; $balanceCost$ represents the use of available machine; $processMoveCost$ is the cost of moving a process from its current machine to a new machine; $serviceMoveCost$ represents the maximum number of moved processes over services; $machineMoveCost$ represents the sum of all moves weighted by relevant machine cost. Coefficients $weight_{loadCost}$, $weight_{balanceCost}$, $weight_{processMoveCost}$, $weight_{serviceMoveCost}$ and $weight_{machineMoveCost}$ indicate the importance of each individual cost.

Details about the constraints, the costs and their weights can be found on the challenge documentation [1]. Note that the quality of a solution is evaluated by the given solution checker, which returns the fitness measure of a solution generated the algorithm. Another important aspect of this challenge is the time limit. All methods have to finish within the 5-minute timeframe on prescribed models of CPUs to ensure the fairness of comparison.

3   Methodology

As mentioned early, our proposed method is based on Late Acceptance Hill Climbing (LAHC). It is a recently proposed single solution search mechanism [4]. LAHC starts from an initial solution as the basis, then generates a neighbourhood solution by changing one element of the solution. LAHC may accept worse solutions in attempt to escape from local optima points. For a newly generated

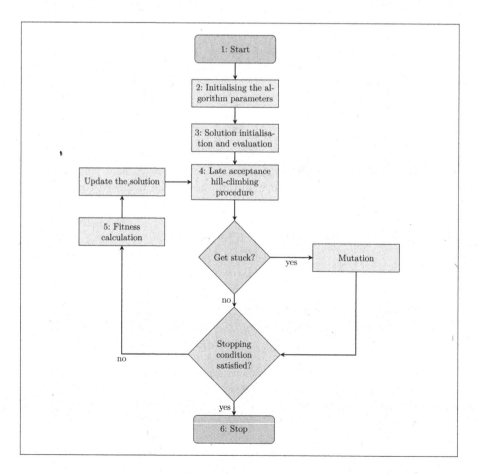

Fig. 1. Overview of a single LAHC process under P-LAHC

solution to be accepted, it only needs to be better than a solution generated several steps early. That means this new solution may be worse than the current best. Our P-LAHC extends this idea to multiple search processes.

Figure 1 shows the overview of a single LAHC process under P-LAHC. Firstly the parameter values are initialised (details in Sect. 3.1); then an initial solution is generated randomly and assigned with a fitness value by the evaluation function (see Sect. 3.2). For GMRP instances, Google provides an initial solution for every instance [1]. The initial solution in our P-LAHC is generated by randomly modifying this given seed solution for each problem instance. Next the standard LAHC procedure is called to improve this initial solution in an iterative manner, as shown under Block 4 of Fig. 1. If the solution cannot be improved for a certain number of iterations, mutation will applied on the solution (details in Sect. 3.4). Otherwise, the algorithm checks the termination condition. If it is satisfied, the search stops and returns the best solution. If not, the fitness of the newly generated solution will be evaluated and sent to be examined by a solution update strategy, which determines whether the current solution should be updated or not. Then the process goes back to Block 4 for LAHC procedure. This iterative process continues for a pre-defined number of times.

Details of the key components on Fig. 1 are discussed in the following subsections.

3.1 Initialising the Algorithm Parameters

LAHC parameters that are initialised in this step include the iteration counter I, which controls the number of steps in LAHC, L_{size} which determines the size of the solution list in LAHC. The latter sets how late the update is.

3.2 Solution Initialisation and Evaluation

For one GMRP instance the initial solution provided by Google is randomly modified by moving some processes from one machine to another [1]. Only feasible solutions will be used. Once a solution is initialised its fitness will be calculated by Eq. (1), which is the fitness measure for all solutions.

3.3 LAHC Procedure

During this procedure LAHC is applied to improve the solution. It always accepts a better local search move but also accept worse solution if it is not worse than the quality of a solution which was recorded at several steps earlier. Given an initial solution, S_0, a neighboured feasible solution, S_1, will be generated by randomly moving processes across machines. S_0 will be replaced by S_1 if the quality of S_1 is better than S_0 or better than f_v where f_v is the quality of v^{it} solution saved in list L. This list maintains the solutions from a number of previous steps to the current step. The value v is calculated as follows:

$$v = I \mod L_{size} \tag{2}$$

where L_{size} is the size of the list L and I is the iteration counter. The list is effectively a queue. At each step, LAHC will add the current solution to the beginning of L while removing the last one from the end.

3.4   Mutation Operators

Mutation plays a big role on the performance of evolutionary algorithms. One key contribution of mutation is helping escape from a local optima point by providing a new starting solution for the search. Therefore mutation is also introduced in our P-LAHC. Our approach uses four different mutation operators, each one being assigned to one LAHC process. They are:

1. **Single swap**: Randomly select two processes from two different machines and swap them.
2. **Double swap**: Randomly select four processes from two different machines and swap them.
3. **Single move**: Randomly select a process from a machine and move it to a different machine.
4. **Double move**: Randomly select two processes from a machine and move both to a different machine.

These mutation operators run independently from each other. The rationale behind assigning different operator to individual LAHC process is to minimise the impact of poorly chosen mutation operators. This is because these operators do perform differently for different problem instances and at different stages of search. Note there is no interaction between four LAHC processes.

4   Experiment Settings

In this section, we introduce the GMRP instances used in our experimental studies and the parameter settings of the proposed P-LAHC algorithm.

4.1   GMRP Instances

In this paper, 20 different instances provided by Google for ROADEF/EURO 2012 challenge were used for evaluation. These instances are divided into two groups which are named as a and b. These instances have various characteristics in terms of number of machines, the number of processes, neighbourhood, and so on. Table 1 shows the main characteristics of the these instances. In the table, R is the number of resources; TR is the number of resources that need transient usage; M is the number of machines; P is the number of processes; S is the number of services; L is the number of locations; N is the number of neighbourhoods; B is number of triples and SD is the number of service dependencies. In addition, the cost of initial solutions from Google are also listed.

Table 1. The characteristics of the 20 GMRP instances

Instance	R	TR	M	P	S	L	N	B	SD	Initial cost
a1_1	2	0	4	100	79	4	1	1	0	49, 528, 750
a1_2	4	1	100	1000	980	4	2	0	40	1, 061, 649, 570
a1_3	3	1	100	1000	216	25	5	0	342	583, 662, 270
a1_4	3	1	50	1000	142	50	50	1	297	632, 499, 600
a1_5	4	1	12	1000	981	4	2	1	32	782, 189, 690
a2_1	3	0	100	1000	1000	1	1	0	0	391, 189, 190
a2_2	12	4	100	1000	170	25	5	0	0	1, 876, 768, 120
a2_3	12	4	100	1000	129	25	5	0	577	2, 272, 487, 840
a2_4	12	0	50	1000	180	25	5	1	397	3, 223, 516, 130
a2_5	12	0	50	1000	153	25	5	0	506	787, 355, 300
b_1	12	4	100	5000	2512	10	5	0	4412	7, 644, 173, 180
b_2	12	0	100	5000	2462	10	5	1	3617	5, 181, 493, 830
b_3	6	2	100	20000	15025	10	5	0	16560	6, 336, 834, 660
b_4	6	0	500	20000	1732	50	5	1	40485	9, 209, 576, 380
b_5	6	2	100	40000	35082	10	5	0	14515	12, 426, 813, 010
b_6	6	0	200	40000	14680	50	5	1	42081	12, 749, 861, 240
b_7	6	0	4000	40000	15050	50	5	1	43873	37, 946, 901, 700
b_8	3	1	100	50000	45030	10	5	0	15145	14, 068, 207, 250
b_9	3	0	1000	50000	4609	100	5	1	43437	23, 234, 641, 520
b_10	3	0	5000	50000	4896	100	5	1	47260	42, 220, 868, 760

4.2 Parameters Settings

Our proposed P-LAHC has three parameters that need to be set by the user. These are: population size P, the size of the list L_{size} and the iteration counter I. In our study P was set as 4, L_{size} set as 20 and I defined as 10. The suggested values of these parameters are obtained based on a preliminary study. P-LAHC were tested with 31 independent runs using different parameters combinations.

5 Results and Comparisons

The effectiveness of P-LAHC is evaluated by comparing it with state of the art algorithms for GMRP. These leading algorithms used in our comparison include:

1. **MNLS**: Multi-neighborhood local search [24].
2. **VNS**: Variable neighbourhood search [11].
3. **CLNS**: CP-based large neighbourhood search [16].
4. **LNS**: Large neighbourhood search [3].
5. **MILS**: Multi-start iterated local search [15].

6. **SA**: Simulated annealing [18].
7. **RILS**: Restricted iterated local search [13].

Table 2 presents the results for comparison on instances from groups a and b. The Column "LAHC" shows the results from standard LAHC which has no mutation. Column "P-LAHC" shows the best results over 31 runs from the proposed P-LAHC on these instances. The other seven columns on the right show the results of the seven state-of-the-art algorithms. In addition the last three rows of the table presents the comparison summary. Row "better" (or "equal", "worse") means the number of instances on which P-LAHC performed better (or equal, worse) than the results listed on that column. The comparison is purely based on the cost of best solutions as all search had to finish within 5 min on single CPU. A result highlighted in bold is the best result on that particular instance, the best of the row. For GMRP the lower the cost the better the solution.

The comparison between standard LAHC and P-LAHC shows that the proposed parallel mechanism does have advantages and outperformed LAHC on all instances. Because multiple LAHC can lead the search to different local optima because of multiple paths, and mutation can help the search avoid being trapped in a local optima. This outcome is consistent with studies like [7,8,14].

Furthermore Table 2 shows that P-LAHC is highly competitive compared to the state-of-the-art algorithms. It achieved the best results on 11 instances. P-LAHC is better than MNLS [24] on 15 instances, VNS [11] on 14 instances, CLNS [16] on 18 instances, MILS [15] on 19 instances, better than LNS [3], SA [18] and RILS [13] on all instances. RILS has no group a results reported.

To verify this on a more formal basis, a multiple comparison using a Friedman statistical test with a significance level of 0.05 was conducted [10]. As a statistical analysis, Friedman's test [10] was first applied, followed by Holm and Hochberg tests as post hoc methods (if significant differences are detected) to obtain the adjusted p-values for each comparison between the control algorithm (the best-performing one) and the rest. Table 3 summarises the ranking obtained by the Friedman's test.

Table 3 shows that P-LAHC ranks first, followed by the VNS, LAHC, CLNS, SA, MILS and LNS. The p-values computed by the Friedman's test is 0.000, which is below the significance interval of 95% ($\alpha = 0.05$). This value shows that there is a significant difference among the observed results. Post hoc methods (Holm's and Hochberg's test) were also performed on the P-LAHC algorithm. Table 4 shows the adjusted p-values. We can see that Holm's and Hochberg's procedures reveal significant differences when using P-LAHC algorithm as a control algorithm, where P-LAHC algorithm is better than all algorithms, with $\alpha = 0.05$. Note that the RILS algorithm is not considered in the comparison due to the reason that this algorithm does not tested on set A instances.

Table 2. P-LAHC comparing with LAHC and the state of the art algorithms on 20 instances of GMRP

	LAHC	P-LAHC	MNLS	VNS	CLNS	LNS	MILS	SA	RILS
a1_1	44,306,501	44,306,501	44,306,501	44,306,501	44,306,501	44,306,575	44,306,501	44,306,935	–
a1_2	777,533,321	777,533,309	777,535,597	777,536,907	778,654,204	788,074,333	780,499,081	777,533,311	–
a1_3	583,006,901	583,005,810	583,005,717	583,005,818	583,005,829	583,006,204	583,006,015	583,009,439	–
a1_4	251,015,641	251,015,185	248,324,245	251,524,763	251,189,168	278,114,660	258,024,574	260,693,258	–
a1_5	727,579,557	727,578,310	727,578,309	727,578,310	727,578,311	727,578,362	727,578,412	727,578,311	–
a2_1	168	166	225	199	196	1,869,113	167	222	–
a2_2	720,685,171	720,671,543	793,641,799	720,671,548	803,092,387	858,367,123	970,536,821	877,905,951	–
a2_3	1,192,054,494	1,192,054,462	1,251,407,669	1,190,713,414	1,302,235,463	1,349,029,713	1,452,810,819	1,380,612,398	–
a2_4	1,680,587,618	1,680,587,596	1,680,744,868	1,680,615,425	1,683,530,845	1,689,370,535	1,695,897,404	1,680,587,608	–
a2_5	310,287,709	310,287,633	337,363,179	309,714,522	331,901,091	385,272,187	412,613,505	310,243,809	–
b_1	3,345,074,930	3,305,899,993	3,354,204,707	3,307,124,603	3,337,329,571	3,421,883,971	3,516,215,073	3,455,971,935	3,511,150,815
b_2	1,016,087,996	1,015,485,841	1,021,230,060	1,015,517,386	1,022,043,596	1,031,415,191	1,027,393,159	1,015,763,028	1,017,134,891
b_3	156,981,564	156,978,421	157,127,101	156,978,411	157,273,705	163,547,097	158,027,548	215,060,097	161,557,602
b_4	4,677,998,904	4,677,819,387	4,677,895,984	4,677,961,007	4,677,817,475	4,677,869,484	4,677,940,074	4,677,985,338	4,677,999,380
b_5	923,314,713	923,299,306	923,427,881	923,610,156	923,335,604	940,312,257	923,857,499	923,299,310	923,732,659
b_6	9,525,881,974	9,525,859,949	9,525,885,495	9,525,900,218	9,525,867,169	9,525,862,018	9,525,913,044	9,525,861,951	9,525,937,918
b_7	14,835,609,322	14,835,122,152	14,842,926,007	14,835,031,813	14,838,521,000	14,868,550,671	15,244,960,848	14,836,763,304	14,835,597,627
b_8	1,214,691,046	1,214,416,691	1,214,591,033	1,214,416,705	1,214,524,845	1,219,238,781	1,214,930,327	1,214,563,084	1,214,900,909
b_9	15,885,671,841	15,885,545,683	15,885,541,403	15,885,548,612	15,885,734,072	15,887,269,801	15,885,617,841	15,886,083,835	15,885,632,605
b_10	18,048,551,819	18,048,499,611	18,055,765,224	18,048,499,616	18,049,556,324	18,092,883,448	18,093,202,104	18,049,089,128	18,052,239,907
better	20/20	–	15/20	14/20	18/20	20/20	19/20	20/20	10/10
equal	0/20	–	1/20	2/20	1/20	0/20	1/20	0/20	0/10
worse	0/20	–	4/20	4/20	1/20	0/20	0/20	0/20	0/10

Table 3. Average ranking of Friedman test

#	Algorithm	Ranking
1	P-LAHC	1.425
2	VNS	2.775
3	LAHC	3.9
4	CLNS	3.925
5	SA	4.575
6	MILS	5.6
7	LNS	5.8

Table 4. The adjusted p-values of the compared algorithms

#	Algorithm	Unadjusted p	P Holm	P Hochberg
1	LNS	0	0	0
2	MILS	0	0	0 ·
3	SA	0.000004	0.000016	0.000016
4	CLNS	0.000253	0.000758	0.000582
5	LAHC	0.000291	0.000758	0.000582
6	VNS	0.048132	0.048132	0.048132

6   Conclusions

This paper proposed an evolutionary method based parallel late acceptance hill-climbing algorithm (P-LAHC). It performs search with parallel LAHC processes instead of only using one single LAHC process. Each LAHC process in P-LAHC starts with a distinct initial solution. Multiple mutation operators are introduced to redirect the search process when a solution does not improve for a certain period of time, a symptom of being trapped by local optima.

The comparison between P-LAHC and standard LAHC on 20 GMRP instances shows that the parallel mechanism introduced by us is effective. LAHC could not outperform P-LAHC on any instance. Comparisons were carried out between P-LAHC and the state of art algorithms. P-LAHC algorithm outperforms these algorithms. It produced the best solution more often than those algorithms from the literature. Note that all these methods have maximum five minutes to perform search on similar CPUs. The better solution and more frequent best solution are clear indication of a good search algorithm. Hence we conclude that P-LAHC is a highly competitive optimisation method, at least for solving GMRP.

In our future work the parallel mechanism and mutation will be studied to further improve the performance by tuning these two components. Additional evolutionary components may be added to enhance the search process.

Another important further work is applying P-LAHC on other resource optimisation problems as it is not just designed for GMRP but a general optimisation algorithm.

References

1. ROADEF/EURO challenge: Machine reassignment (2011). http://challenge. roadef.org/2012/en/
2. Armbrust, M., Fox, A., Griffith, R., Joseph, A.D., Katz, R., Konwinski, A., Lee, G., Patterson, D., Rabkin, A., Stoica, I., et al.: A view of cloud computing. Commun. ACM **53**(4), 50–58 (2010)
3. Brandt, F., Speck, J., Völker, M.: Constraint-based large neighborhood search for machine reassignment. Ann. Oper. Res. **242**, 1–29 (2012)
4. Burke, E.K., Bykov, Y.: A late acceptance strategy in hill-climbing for exam timetabling problems. In: PATAT 2008 Conference, Montreal, Canada (2008)
5. Burke, E.K., Bykov, Y.: The late acceptance hill-climbing heuristic. University of Stirling. Technical report (2012)
6. Calheiros, R.N., Ranjan, R., Beloglazov, A., De Rose, C.A., Buyya, R.: Cloudsim: a toolkit for modeling and simulation of cloud computing environments and evaluation of resource provisioning algorithms. Softw. Pract. Exp. **41**(1), 23–50 (2011)
7. Crainic, T.G., Toulouse, M.: Parallel meta-heuristics. In: Gendreau, M., Potvin, J.-Y. (eds.) Handbook of metaheuristics, vol. 146, pp. 497–541. Springer, US (2010). doi:10.1007/978-1-4419-1665-5_17
8. Domínguez, J., Alba, E.: Dealing with hardware heterogeneity: a new parallel search model. Natural Comput. **12**(2), 179–193 (2013)
9. Fonseca, G.H.G., Santos, H.G., Carrano, E.G.: Late acceptance hill-climbing for high school timetabling. J. Sched. **19**, 1–13 (2015)
10. García, S., Fernández, A., Luengo, J., Herrera, F.: Advanced nonparametric tests for multiple comparisons in the design of experiments in computational intelligence and data mining: experimental analysis of power. Inf. Sci. **180**(10), 2044–2064 (2010)
11. Gavranović, H., Buljubašić, M., Demirović, E.: Variable neighborhood search for Google Machine Reassignment Problem. Electron. Notes Discrete Math. **39**, 209–216 (2012)
12. Goerler, A., Schulte, F., Voß, S.: An application of late acceptance hill-climbing to the traveling purchaser problem. In: Pacino, D., Voß, S., Jensen, R.M. (eds.) ICCL 2013. LNCS, vol. 8197, pp. 173–183. Springer, Heidelberg (2013). doi:10.1007/978-3-642-41019-2_13
13. Lopes, R., Morais, V.W.C., Noronha, T.F., Souza, V.A.A.: Heuristics and matheuristics for a real-life machine reassignment problem. Int. Trans. Oper. Res. **22**(1), 77–95 (2015)
14. Manfrin, M., Birattari, M., Stützle, T., Dorigo, M.: Parallel ant colony optimization for the traveling salesman problem. In: Dorigo, M., Gambardella, L.M., Birattari, M., Martinoli, A., Poli, R., Stützle, T. (eds.) ANTS 2006. LNCS, vol. 4150, pp. 224–234. Springer, Heidelberg (2006). doi:10.1007/11839088_20
15. Masson, R., Vidal, T., Michallet, J., Penna, P.H.V., Petrucci, V., Subramanian, A., Dubedout, H.: An iterated local search heuristic for multi-capacity bin packing and machine reassignment problems. Expert Syst. Appl. **40**(13), 5266–5275 (2013)

16. Mehta, D., O'Sullivan, B., Simonis, H.: Comparing solution methods for the machine reassignment problem. In: Milano, M. (ed.) CP 2012. LNCS, vol. 7514, pp. 782–797. Springer, Heidelberg (2012). doi:10.1007/978-3-642-33558-7_56

17. Özcan, E., Bykov, Y., Birben, M., Burke, E.K.: Examination timetabling using late acceptance hyper-heuristics. In: IEEE Congress on Evolutionary Computation, 2009. CEC 2009, pp. 997–1004. IEEE (2009)

18. Ritt, M.R.P.: An algorithmic study of the machine reassignment problem. PhD thesis, Universidade Federal do Rio Grande do Sul (2012)

19. Sabar, N.R., Song, A.: Grammatical evolution enhancing simulated annealing for the load balancing problem in cloud computing. In: Proceedings of the 2016 on Genetic and Evolutionary Computation Conference, pp. 997–1003. ACM (2016)

20. Sabar, N.R., Song, A., Zhang, M.: A variable local search based memetic algorithm for the load balancing problem in cloud computing. In: Squillero, G., Burelli, P. (eds.) EvoApplications 2016, Part I. LNCS, vol. 9597, pp. 267–282. Springer, Heidelberg (2016). doi:10.1007/978-3-319-31204-0_18

21. Turky, A., Abdullah, S., McCollum, B., Sabar, N.R: An evolutionary hill climbing algorithm for dynamic optimization problems. In: The 6th Multidisciplinary International Conference on Scheduling: Theory and Applications (MISTA 2013), 27–30 August 2013

22. Turky, A., Sabar, N.R., Song, A.: An evolutionary simulating annealing algorithm for Google Machine Reassignment Problem. In: The 20th Asia-Pacific Symposium on Intelligent, Evolutionary Systems. Proceedings in Adaptation, Learning and Optimization. Springer Book Series (2016)

23. Verstichel, J., Berghe, G.V.: A late acceptance algorithm for the lock scheduling problem. In: Voß, S., Pahl, J., Schwarze, S. (eds.) Logistik Management, pp. 457–478. Physica, Heidelberg (2009). doi:10.1007/978-3-7908-2362-2_23

24. Wang, Z., Lü, Z., Ye, T.: Multi-neighborhood local search optimization for machine reassignment problem. Comput. Oper. Res. **68**, 16–29 (2016)

25. Yuan, B., Zhang, C., Shao, X.: A late acceptance hill-climbing algorithm for balancing two-sided assembly lines with multiple constraints. J. Intell. Manuf. **26**(1), 159–168 (2015)

Concept Drift Detection Using *Online* Histogram-Based Bayesian Classifiers

César A. Astudillo[1], Javier I. González[1], B. John Oommen[2(⊠)], and Anis Yazidi[3]

[1] Department of Computer Science, Universidad de Talca,
Km. 1 Camino a Los Niches, Curicó, Chile
castudillo@utalca.cl
[2] School of Computer Science, Carleton University, Ottawa, Canada
oommen@scs.carleton.ca
[3] Department of Computer Science,
Oslo and Akershus University College of Applied Sciences, Oslo, Norway
Anis.Yazidi@hioa.no

Abstract. In this paper, we present a novel algorithm that performs online *histogram*-based classification, i.e., specifically designed for the case when the data is dynamic and its distribution is non-stationary. Our method, called the Online *Histogram*-based Naïve Bayes Classifier (OHNBC) involves a statistical classifier based on the well-established Bayesian theory, but which makes some assumptions with respect to the independence of the attributes. Moreover, this classifier generates a prediction model using uni-dimensional *histograms*, whose segments or buckets are fixed in terms of their cardinalities but dynamic in terms of their widths. Additionally, our algorithm invokes the principles of information theory to automatically identify changes in the performance of the classifier, and consequently, forces the reconstruction of the classification model in run-time as and when it is needed. These properties have been confirmed experimentally over numerous data sets (In the interest of space and brevity, we present here only a subset of the available results. More detailed results are found in [2].) from different domains. As far as we know, our histogram-based Naïve Bayes classification paradigm for time-varying datasets is both novel and of a pioneering sort.

Keywords: Online Naïve Bayes Classifier · Online learning · Concept drift · Dynamic Histograms

C.A. Astudillo—This work was partially supported by the FONDECYT Grant No. 11121350, Chile.
B.J. Oommen—*Chancellor's Professor*; *Fellow: IEEE* and *Fellow: IAPR*. This author is also an *Adjunct Professor* with the University of Agder in Grimstad, Norway. The work of this author was partially supported by NSERC, the Natural Sciences and Engineering Research Council of Canada.

B.H. Kang and Q. Bai (Eds.): AI 2016, LNAI 9992, pp. 175–182, 2016.
DOI: 10.1007/978-3-319-50127-7_14

1   Introduction

In the fields of machine learning and statistical learning, supervised classification is a well-known problem that consists of identifying the category (or class) to which a new observation belongs, based on a sample data set that contains instances for which their respective categories are known. This task can be achieved using a range of methods, including linear classifiers, Support Vector Machines (SVMs), decision trees, neural networks, etc. Most of these families of supervised classifiers operate in an *offline* manner, i.e., the data set (training data) used for building the learning model is known in advance. Such algorithms operate in three phases: First, a learning model is constructed using the training data in which the instances are all labeled. In the second phase, the model that has been built is used to classify data instances from a test set, and a performance measure is obtained. Finally, in the third phase, the model is deployed and used to predict the category of an unlabeled data instance data [1].

It is possible to find a wide range of applications where the data arrives in the form of a stream, consisting of a (theoretically, infinite) sequence of instances that may become available at a very rapid rate. These applications include telecommunications data management, financial applications, sensory analysis, web history logs, etc. The analysis of dynamic and real-time data streams poses several challenges when compared to processing the data in an offline manner. In fact, *offline* classifiers assume that the training records can be examined and accessed several times. This is consistent with the three previously-mentioned phases of any supervised classification algorithm.

Unfortunately, in many real life situations, the rate of arrival of the data instances is so rapid that it is infeasible to store them for post-processing, forcing the algorithm to achieve the processing of a single instance at any given time instant. The work reported in [5] describes another difficulty with classifying data streams with respect to the dynamic nature of the data in the following terms: "Even if all the available examples can be handled by the system, the patterns discovered by an algorithm in the data from the past, may be hardly valid and useful for the new data obtained hours or even minutes later." *Online* algorithms [1] deal with such data streams, in which the labeled and unlabeled records are mixed. In this context, the phases of training, testing and deployment are interleaved. This is precisely the domain of this paper, and the results we contribute involve *histogram*-based Bayesian classifiers.

As we know, pattern classification is the discipline of building machines for classifying patterns based on prior knowledge or on statistical information extracted from the patterns [4]. However, as opposed to what we shall call traditional classification, we are interested in data that is generated in real time. Examples of sources of such streamed data are sensor networks on Mars, underwater sensors in the deep ocean, atmospheric measurements, etc. These potentially infinite sequences of information are usually known as "data streams" [1]. The data stream serves as an appropriate model when a large amount of data arrives for processing and where it is impractical to store it all, implying that a model that attempts to learn from it must achieve the task by processing each pattern at a time. Devising classifiers that work with data streams poses new challenges when compared to the standard classification algorithms, since the latter algorithms are able to examine the patterns repeatedly.

Algorithms for data streams are more complicated to design because they must be able to extract all the information needed with just a single examination of the patterns. According to the author of [1], algorithms capable of learning from streaming data must process the patterns in amortized $O(1)$ time. From this we can say that online learners are induction models that are trained – one instance at a time. The goal is to predict the classes of novel patterns as accurately as possible. The key feature that defines online learning is that shortly after the prediction is made, the actual class of the instance is discovered. This information can then be used to refine the algorithm's prediction hypothesis [6].

More formally, the classification in an online algorithm proceeds as follows. The sequence of operations can be decomposed into three phases. First of all, the algorithm receives an instance. Secondly, the algorithm predicts the class of the instance. Thirdly, the algorithm receives the true class of the instance [6]. The third phase is the most crucial one, since the algorithm can use this information to update the classification model.

Various algorithms that possess online learning properties have been reported in the literature. An example of one such algorithm, is the Online Random Forest [7]. In [7], the authors propose an online learning scheme based on the properties of Random Forests previously described in [3]. The main idea is to build a sequence of decision trees which learn from the data in an independent manner, and a final decision is made based on a function that depends on the output of all these trees. The Online Random Forest starts with a single node as the root of the tree, and systematically adds new nodes depending on a predefined criteria. When a new decision node is added to the tree, a series of functions of the form $g(x) < \theta$ divide the data. Both the functions and the constant θ are defined randomly. These functions produce a division of the feature space, and according to a performance measure, the best division is selected.

Due to space limitations, a more detailed survey of the field and the details of the background material is omitted. It is found in [2]. However, we shall concentrate on our contribution, namely, the formulation of our histogram-based OHNBC algorithm.

2   The OHNBC Algorithm

Based on the phenomena described above, we now detail the algorithmic characteristics and functionality of our Online *Histogram*-based Naïve Bayes Classifier (OHNBC). We first define the general structure of OHNBC. We then specify some of the algorithm's functionality in greater detail, and then describe how the components fit together. The general structure of the OHNBC algorithm is formalized in Algorithm 1, and also explained in [2] in a textual manner. It is omitted here in the interest of brevity and due to space limitations.

The two main interleaving phases of the algorithm's classification and training are obvious.

Algorithm 1. OHNBC $(\mathcal{X}, \lambda, \tau, \theta)$

Input:
  i) $\mathcal{X}$: Stream of instances for classification.
  ii) λ: Minimum number of training instances needed for classification.
  iii) τ: Size of the reference window.
  iv) θ: Threshold for the threshold.
Output:
  i) Confusion matrix with the classification results.
  ii) The histograms of the classes for the OHNBC.
Method:
1: **while** an instance $\mathbf{x} \in \mathcal{X}$ **do**
2:    **if** number of trained instances is less than λ **then**
3:        $\omega \leftarrow$ Original class of $\mathbf{x}$
4:        **for all** attribute $d \in \mathbf{x}$ **do**
5:            Get the histogram for the attribute d and the class ω.
6:            Add the value of the attribute d to the histogram.
7:        **end for**
8:        Update *prior* probabilities $\mathcal{P}$ associated with the class ω.
9:    **else**
10:       $P \leftarrow$ *prior* probabilities.
11:       **for all** attribute $d \in \mathbf{x}$ **do**
12:           **for all** ω_i class **do**
13:               $H(d, \omega_i) \leftarrow$ uni-dimensional histogram for the attribute d ω_i.
14:               **if** standard deviation of $(H(d, \omega_i))$ equals 0 **then**
15:                   Adjust all the $H(d, \omega_i)$.
16:               **end if**
17:               Compute the probability density function for $(H(d, \omega_i))$
18:               Multiply the the density functions over d with their *prior* values
19:           **end for**
20:       **end for**
21:       **if** original class of $\mathbf{x}$ equals the OHNBC's predicted class **then**
22:           Update the size of the windows.
23:           Update the classification model with the current correct prediction.
24:       **end if**
25:       Calculate the difference in entropy of the windows.
26:       **if** the If this difference is larger than θ **then**
27:           Reset the values of the windows.
28:           Discard the current classification model.
29:       **end if**
30:   **end if**
31: **end while**
End Algorithm

---

3   Experimental Design and Results

The solution that we propose here was tested for various artificial data sets[1] and three synthetic data which emulate conditions under which concept drift occurs [1]. These files were generated using the data generation tool DatGen proposed by Melli[2].

There are many situations that deal with the various scenarios involving concept drift [1]. The following describes each of the proposed scenarios:

- **Scenario 0:** This corresponds to that case when there is no concept drift. Here, instances are taken early in the stream, using which the learning model for the classifier is built, and it is then used to classify the instances.
- **Scenario 1:** In this scenario, we observe concept drift. Here, the data stream arrives in the form of data blocks. The data within each block possesses the same probability distribution. However, contiguous blocks have different data distributions. The cardinality of the blocks within this scenario is large enough to allow for the proper training of the classification model.
- **Scenario 2 and 3:** These scenarios inherit the essential properties of Scenario 1. The fundamental difference is that the sizes of the blocks are not large enough to allow the proper construction of the learning models. This corresponds to a complex environment where the instances available are not sufficient to correctly ensure the update of the classification model, especially at the boundaries between the blocks. In Scenario 2, the blocks have fixed sizes, while in Scenario 3, which is the more realistic one, the blocks themselves have random sizes and frequencies.

For the case of artificially generated data sets, we emulated scenarios 1, 2 and 3, which contained concept drift within their characteristics.

The first synthetic data set follows the characteristics of Scenario 1 and contains a total of three million instances, divided into three blocks of a million instances each, respectively. We consider three different categories ("C1", "C2" and "C3") in a five dimensional space (denoted by "A", "B", "C", "D" and "E").

Figure 1a is a graphical representation of the distribution of blocks for the synthetic data set '1'. The various distributions are represented using gray scales. The figure shows that for the first block of one million instances the distribution corresponds to a unique distribution, followed by a block of one million instances corresponding to the second distribution, and eventually the last block of one million instances correspond to the third distribution.

The second data set of 4 million instances, emulates the characteristics of Scenario 2. Figure 1b is a graphical representation of the data set, specifying the data distribution associated with a block at any given time. The set is divided into ten blocks of 400,000 instances each. The first block is composed of 90%

---

[1] As mentioned earlier, in the interest of space and brevity, we present here only a subset of the available results. More detailed results are found in [2].

[2] The data generation tool DatGen is publicly available at the following URL: http://www.datasetgenerator.com.

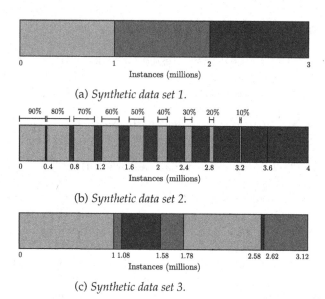

(a) *Synthetic data set 1.*

(b) *Synthetic data set 2.*

(c) *Synthetic data set 3.*

Fig. 1. Graphical representation of the synthetic data sets representing the three scenarios.

of instances belonging to the first distribution, while 10% corresponds to the second distribution. The next block has a ratio of 80% of instances of the first distribution and 20% of instances from the second distribution. This change in proportions is successively applied to the remaining blocks up to the tenth block, which contains 100% of instances of the second distribution.

Finally, the third set of data follows the characteristics of Scenario 3 and contains a total of 3 million and twelve hundred instances divided into several blocks of different sizes and frequencies. Figure 1c illustrates the distribution of the blocks for this synthetic data set. Analogous to the above sets, gray scales are used to differentiate the distributions.

3.1   Parameter Optimization

The OHNBC classifier requires three parameters: λ, which is the minimum number of instances are required to build the learning model, τ, which is the size of the reference window, and θ, which is the entropy threshold for identifying that a concept drift has occurred. The values chosen for each of these parameters are presented below:

- $\lambda \in \{2000, 5000, 10000, 20000, 50000, 100000\}$
- $\tau \in \{500, 1000, 2000, 5000, 10000\}$
- $\theta \in \{0.01, 0.05, 0.1, 0.2, 0.3, 0.4, 0.5, 0.6, 0.7, 0.8, 0.9\}$
- $\lambda \in \{500, 1000, 2000, 5000\}$
- $\tau \in \{50, 100, 200, 500, 1000\}$

Table 1. Final parameters for the OHNBC for each of the synthetic data sets.

Data sets	Parameters		
	OHNBC		
	λ	τ	θ
Synthetic-data-set1	2000	1000	0.10
Synthetic-data-set2	2000	500	0.10
Synthetic-data-set3	2000	2000	0.10

Table 1 lists the values found by applying a *Grid Search* to determine the best suitable parameters for the real and synthetic data sets. Each row indicates a specific data set, while in each column we list the values of the optimum parameters for each data set, respectively.

3.2 Results

This section presents an overview of the performance of the proposed algorithm, when compared to the Naïve Bayes algorithm (NB). Subsequently, we also performed a detailed analysis of the OHNBC algorithm relative to the data's concept drift. The results presented here are those obtained by averaging over ten runs of the classifier.

Table 2 presents the results for synthetic data sets. It summarizes the number of instances correctly classified ("Certainty") and the proportion of instances used for training the model ("Training") for the various synthetic data sets (rows). As one can see, in all these three scenarios, the accuracy of the OHNBC far surpasses that of the NB. We attribute the improved performance to the ability of the OHNBC to detect the changes in the corresponding distributions, which the NB clearly lacks. Given the artificial construction of these sets, we already know that there are concept changes in the stream, and the exact times when they occur in the data stream. Additionally, the way by which these schemes have been devised make them ideal for situations with a large number of instances, as in this case. Indeed, from the column titled "Training", we observe that it was not necessary to use a large percentage of training instances to build the learning models that achieved such high accuracy rates.

Table 2. Results obtained for the runs of the synthetic data sets.

Data sets	Accuracy		Fraction of training data set	
	NB	OHNBC	Total	Ratio
Synthetic-data-set1	0.824	0.991	3,000,000	0.002
Synthetic-data-set2	0.699	0.988	4,000,000	0.009
Synthetic-data-set3	0.746	0.984	3,120,000	0.005

Finally, in relation to the amount of concept drifts detected for each of the synthetic data sets, our algorithm is able to identify all the concept changes that were artificially embedded in the stream.

4 Conclusion

This article has tackled the problem of classification in environments where the instances are sequentially arriving in the form of a data stream, and whose statistical distribution potentially varies over time. We have proposed a novel method to identify these changes using Bayesian theory and the principles of Shannon's information theory. Our proposed classification algorithm, called the Online *Histogram*-based Naïve Bayes Classifier (OHNBC) follows the paradigm of online training outlined by three key phases: First, the algorithm receives an instance. Next, the algorithm predicts the class of the instance based on a histogram-based representation. Finally, the algorithm receives the true class of the instance and uses it to update the classification model. The results that we have obtained on synthetic and real data clearly demonstrate its power in classifying data streams characterized by non-stationary distributions.

References

1. Abdulsalam, H., Skillicorn, D., Martin, P.: Classification using streaming random forests. IEEE Trans. Knowl. Data Eng. **23**(1), 22–36 (2011)
2. Astudillo, C.A., Gonzalez, J., Oommen, B.J., Yazidi, A.: Concept drift detection using classifiers that are online, Bayesian and histogram-based. Unabridged Version of this paper. (In Preparation)
3. Breiman, L.: Random forests. Mach. Learn. **45**(1), 5–32 (2001)
4. García-Laencina, P., Sancho-Gómez, J., Figueiras-Vidal, A.: Pattern classification with missing data: a review. Neural Comput. Appl. **19**, 263–282 (2010). doi:10.1007/s00521-009-0295-6
5. Last, M.: Online classification of nonstationary data streams. Intell. Data Anal. **6**(2), 129–147 (2002)
6. Littlestone, N.: Learning quickly when irrelevant attributes abound: a new linear-threshold algorithm. Mach. Learn. **2**(4), 285–318 (1988)
7. Saffari, A., Leistner, C., Santner, J., Godec, M., Bischof, H.: On-line random forests. In: 2009 IEEE 12th International Conference on Computer Vision Workshops (ICCV Workshops), pp. 1393–1400, October 2009

Visual Odometry in Dynamic Environments with Geometric Multi-layer Optimisation

Haokun Geng[1]([✉]), Hsiang-Jen Chien[2], Radu Nicolescu[1], and Reinhard Klette[2]

[1] Department of Computer Science, The University of Auckland,
Auckland, New Zealand
hgen001@aucklanduni.ac.nz
[2] School of Engineering, Computing and Mathematical Sciences,
Auckland University of Technology, Auckland, New Zealand

Abstract. This paper presents a novel approach for optimising visual odometry results in a dynamic outdoor environment. Egomotion estimation is still considered to be one of the more difficult tasks in computer vision because of its continued computation pipeline: every phase of visual odometry can be a source of noise or errors, and influence future results. Also, tracking features in a dynamic environment is very challenging. Since feature tracking can only match two features in integer coordinates, there will be a data loss at sub-pixel level. In this paper we introduce a weighting scheme that measures the geometric relations between different layers: We divide tracked features into three groups based on geometric constrains; each group is recognised as being a "layer". Each layer has a weight which depends on the distribution of the grouped features on the 2D image and the actual position in 3D scene coordinates. This geometric multi-layer approach can effectively remove all the dynamic features in the scene, and provide more reliable feature tracking results. Moreover, we propose a 3-state Kalman filter optimisation approach. Our method follows the traditional process of visual odometry algorithms by focusing on motion estimation between pairs of two consecutive frames. Experiments and evaluations are carried out for trajectory estimation. We use the provided ground truth of the KITTI data-sets to analyse mean rotation and translation errors over distance.

1 Introduction

Visual odometry (VO) algorithms are widely used for solving "where-am-I" problems that require extensive and precise geometric calculations. For example, *driver-assistance systems* (DAS) in the automotive industry [1] require solutions in such problem domain. Computers are trained to listen, to see, and to sense the road geometry and the dynamic traffic environment, in order to provide comfort with safety, to assist drivers to follow traffic instructions, and to avoid road incidents. For instance, DAS should avoid that a sudden turn results in a pedestrian accident, or a braking manoeuvre in a collision. Demands for computer vision solutions, to be integrated in DAS, are immensely increasing these days towards holistic scene understanding.

© Springer International Publishing AG 2016
B.H. Kang and Q. Bai (Eds.): AI 2016, LNAI 9992, pp. 183–190, 2016.
DOI: 10.1007/978-3-319-50127-7_15

Motion data can be obtained by multiple types of sensors, including *inertial measurement units* (IMU), *global positioning system* (GPS) units, radar sensors, cameras, or laser range-finders. For the presented solution we choose to use only optical cameras (and not any additional sensor), and consider accuracy as being the first priority; optimizing run-time performance will need to be considered later. Accurately estimated motion among subsequent frames of a recorded video sequence can then be used, for example, for 3D roadside reconstruction.

Stereo cameras and stereo vision [2] offer in principle economy and robustness for understanding 3-dimensional (3D) scenes, but also require often more advances in computer vision techniques for solving tasks up to expectations. Camera-based egomotion estimation (also known as *visual odometry*) [3] is the method we use to determine the trajectory of the *ego-vehicle* (i.e. the vehicle where the cameras are operating in). This is typically a first step in a whole pipeline of processes for understanding the road environment.

2   Related Work

Nister et al. [3] firstly introduced visual odometry in 2004; the method estimates odometry information based on recorded stereo image pairs only, not using other data as, for example, the vehicles' yaw rate or speed. An advantage of visual odometry is that it avoids the influence of motion estimation errors in other sensors (e.g. the influence of mechanical issues such as wheel slips, or the still existing inaccuracy of cost-efficient GPS or IMU sensors).

Scaramuzza et al. [4] suggest that visual odometry methods usually lead to a smaller relative position error (in a range between 0.1% to 2% of actual motion) compared to traditional wheel odometry methods.

Maimone et al. [5] present a vision-based egomotion application used in Mars exploration rovers; they demonstrated the great capability of this technology on another planet for the first time. With these advantages and examples, vision-based egomotion analysis proves itself as being a valuable navigation technology, and a potential feature of mobile computer vision applications.

Existing vision-based egomotion algorithms take image data either from monocular cameras, stereo cameras, or omnidirectional cameras. A solid foundation of a stereo-vision method for visual odometry is provided by Matthies in [6,7]. The presented algorithm is for estimating the robot's positional data by tracking landmarks with a stereo vision system. It proved that a stereo vision system provides better estimation results than a monocular system.

A method for 3D rigid motion estimation from disparity images is also presented by Demirdjian and Darrell [8]. Following those pioneering contributions, various studies have been carried out, from 2D-to-2D matching to 2D-to-3D motion estimation, and finally to 3D-to-3D registration problems. A method for solving a 3D-to-3D point registration problem was presented by Demirdjian and Darrell in [8]. The given algorithm estimates the stereo cameras' 3D rigid transformation by directly taking the disparity images as the input.

Rabe et al. [9] defined concepts of 2-frame motion field estimation; the paper demonstrated a novel approach for calculating 3D scene flow data using a Kalman filter and parallel implementations.

Kalman filters are used for eliminating the noise that is introduced from every aspect of the VO production pipeline. The Kalman filter was first introduced by Rudolf Emil Kálmán [10], who is the primary developer of this theory, and this filter is named after him.

Schmidt [11] proposed a nonlinear extension of the original Kalman filter, which is known as the *extended Kalman filter* (EKF). The EKF has already been considered the *de facto* standard in the nonlinear filtering problem domain [12], such as for GPS.

3   Geometric Multi-layer Optimisation

The proposed *geometric multi-layer optimisation* (GMO) uses the information of the tracked features and their corresponding depth data in the scene for creating multiple layers, in order to optimise VO estimation for any two consecutive frames. Due to the properties of the disparity value's uncertainty, we assign different weight to the tracked features, in order to achieve more accurate results for translation and rotation measurements. For stereo matching algorithms it follows that features at larger distances have larger uncertainties in the disparity space, whereas closer features have relatively smaller uncertainties [13].

However, we found that features at further positions to the camera are usually better candidates for estimating the rotation, whereas features at closer positions are better for computing the translation. The weight of a feature for translation estimations:

$$\omega_t^i = \frac{d^i}{\sum\limits_{i=1}^{n} (d^i)} \tag{1}$$

and the weight for rotation estimations will be:

$$\omega_R^i = 1 - \omega_t^i \tag{2}$$

ω_t^i denotes the weight assigned to the i-th tracked feature in the phase of estimating the translation, and ω_R^i is the weight of the i-th tracked feature in the phase of estimating rotation. Symbol d denotes the distance from the camera to the relevant scene point.

The tracked feature set in the scene can be divided into three subsets based on the distance range of their 3D projections in the scene. Each subset is recognised as one layer, which should contain a similar number of features. We also pay attention on the tracked features' distribution in the input images. Each small region will have a pre-defined weight.

The combined weight of all the features from a subset layer defines the quality of the distribution. Based on the distribution quality factor, we can compute the

Fig. 1. Dynamic features on the truck can be accurately removed by our proposed GMO method. *Top:* Before applying the method. *Bottom:* After applying the method.

weight value of each layer, in order to bring the distribution quality into account. We use an adaptive threshold is to distinguish the dynamic and static features:

$$\tau = \sqrt{D^i_{drift} \times 2\%} + 0.5 \tag{3}$$

here, D_{drift} is the distance between the measured position of i-th feature in the scene and the transformed position of the same feature by using the optimal transformation matrix determined from the last frame.

In this way, each feature's threshold will be measured differently. If the distance difference is smaller than its measured threshold, the feature is considered to be static, otherwise it is a dynamic feature.

Figure 1 shows an example of the GMO result. The green dots indicate the tracked features from the previous frame to the current one. Figure 1 (*top*) shows that there are several image features on a moving truck recognised as being static features in the video sequence before using our GMO method. Figure 1 (*bottom*) shows that wrongly tracked features are accurately removed after applying our proposed method to the test sequence.

Our GMO method uses the previous motion data as one of the standard measures to distinguish static and dynamic features, in order to improve the egomotion estimations.

4   Three-State Kalman Filter

We consider cases when the ego-vehicle is moving slowly. Feature detectors can only detect pixel-level keypoints with integer coordinates. We can safely assume that there is a certain level of data loss when matching the features. The concept of a 3-state Kalman filter aims at using the results from linear Kalman filters of any three consecutive frames in order to minimise the data loss and to find optimal results for the subsequence.

We assume that the disparity values contain normal-distributed noise. Since disparity maps are the major input, we can use a set of linear Kalman filters to track the changes of the features' world coordinates, in order to build up the confidence of the selected features. Therefore, we can discard any features that appear at unexpected positions, for example a sudden change in depth. Those are considered to be outliers.

State Vector. The state vector is a 6×1 vector. It contains the 3D positional data and its velocity:

$$\mathbf{x}_k = [x \ y \ z \ x' \ y' \ z']^\top \tag{4}$$

Process Model. The process model relates to the state vector; it describes the state vector changes from the previous moment $k-1$ to the present moment k:

$$\mathbf{x}_k = \mathbf{A}_k \cdot \mathbf{x}_{k-1} + \mathbf{b}_k^\top + \mathbf{n}_k \tag{5}$$

where

$$\mathbf{A}_k = \begin{bmatrix} \mathbf{I}_3 & \Delta t \cdot \mathbf{I}_3 \\ 0_3 & \mathbf{I}_3 \end{bmatrix} \tag{6}$$

Here, $\mathbf{A}_k$ is the state transition matrix and $\mathbf{b}_k$ is the input-control vector.

Measure Model. The measurement is obtained by the mean of the two transformations $T_{k-1|k}$ (from frame $k-1$ to frame k) and $T_{k|k+1}$ (from frame k to frame $k+1$). The measurement model only observes the position data of the current state of the tracked image features:

$$\begin{bmatrix} x_k \ y_k \ z_k \end{bmatrix}^\top = \mathbf{H} \cdot \mathbf{x}_k + \mathbf{n}_k \tag{7}$$

where

$$\mathbf{H} = \begin{bmatrix} I_3 \ 0 \end{bmatrix} \tag{8}$$

Here, $\mathbf{H}$ is the observation model matrix, and $\mathbf{n}_k$ is the Gaussian noise. Because we aim at minimising the data loss of integer feature coordinates, we use the 3-state Kalman filter method to optimise this situation.

5   Evaluation

The experiments were carried out under the following assumptions: (1) the camera's intrinsic parameters are determined beforehand, and (2) the left and right images of the input stereo pairs are well rectified and time-synchronised.

We report about experiments with our proposed method on KITTI data sets [14]. Four sequences (6,548 stereo images in total) from the KITTI data sets are used in the evaluation. We compare the estimated trajectories of our VO method with the ground truth, and then show the measured translation and rotation errors for each sequence.

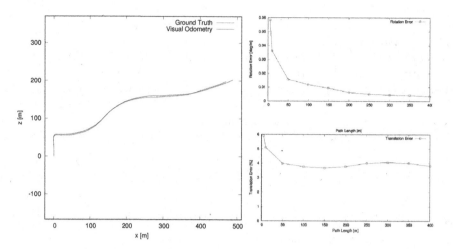

Fig. 2. Trajectory evaluation and computed VO errors for KITTI dataset 3.

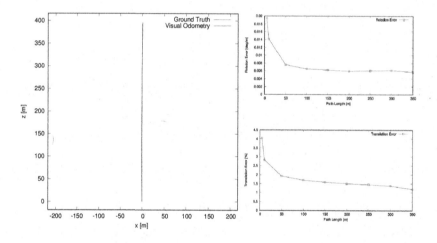

Fig. 3. Trajectory evaluation and computed VO errors for KITTI dataset 4.

Testing Sequence 3. The measurements of the rotation and translation errors are shown in Fig. 2. Because this sequence is recorded on the road that is surrounded by grass, bushes and trees, noise can be introduced by waving leaves in the feature tracking phase.

Testing Sequence 4. It appears that egomotion is accurately estimated, comparing with the ground truth of this data set. The translation error of this test sequence is almost as small as 1%; see Fig. 3.

Testing Sequence 6. Determining the correct rotation for sharp turns is considered as being another challenging task in the classic 2D-to-3D VO algorithms.

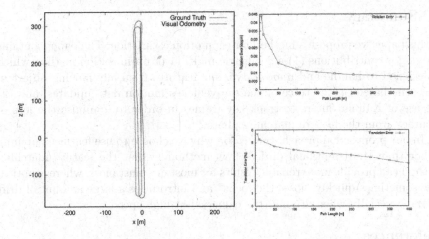

Fig. 4. Trajectory evaluation and computed VO errors for KITTI dataset 6.

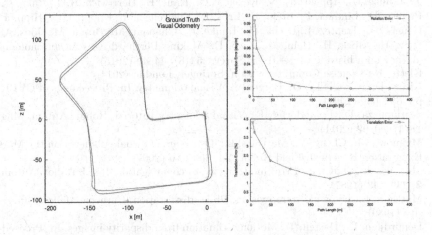

Fig. 5. Trajectory evaluation and computed VO errors for KITTI dataset 7.

This sequence contains two sharp turns; they may contribute more errors in the overall rotation estimation. The measured error in Fig. 4 also suggests that the rotation estimation brings more noise to the overall transformation.

Testing Sequence 7. Figure 5 presents an accurate trajectory estimation compared to the ground truth. The result of this test sequence demonstrates the performance of our method in a busy dynamic environment. The readings of the translation error reaches 1.5%, which tells that the proposed method can provide quality VO estimations for complex dynamic environments.

6 Summary

In this paper we propose a method for egomotion estimation. The paper contains two major contributions (1) a geometric multi-layer optimisation method which is designed to handle and remove dynamic features from any moving object in the scene based on the trace of the ego-vehicle's motion data updates, and (2) the use of Kalman filters to track key frames in order to minimise the loss of accuracy from the feature matching phase.

In our proposed approach, the reason why we choose to use feature matching rather than the traditional optical flow method is that the feature matching method can provide more reliable results for most circumstances, whereas optical flow sometimes quickly "loses the focus" and introduces a large amount of drift errors, especially when driving (the cameras) at high speed.

References

1. Ziegler, J., Bender, P., Schreiber, M., Lategahn, H., Strauss, T., Stiller, C., Dang, T., Franke, U., Appenrodt, N., Keller, C.G., Kaus, E., Herrtwich, R.G., Rabe, C., Pfeiffer, D., Lindner, F., Stein, F., Erbs, F., Enzweiler, M., Knoppel, C., Hipp, J., Haueis, M., Trepte, M., Brenk, C., Tamke, A., Ghanaat, M., Braun, M., Joos, A., Fritz, H., Mock, H., Hein, M., Zeeb, E.: Making Bertha drive - an autonomous journey on a historic route. IEEE Spectr. **51**(8), 44–49 (2015)
2. Klette, R.: Concise Computer Vision. Springer, London (2014)
3. Nister, D., Naroditsky, O., Bergen, J.: Visual odometry. In: Proceedings of CVPR, pp. 652–659 (2004)
4. Scaramuzza, D., Fraundorfer, F.: Visual odometry tutorial. Robot. Autom. Mag. **18**(4), 80–92 (2011)
5. Maimone, M., Cheng, Y., Matthies, L.: Two years of visual odometry on the Mars Exploration Rovers. J. Field Robot. **24**, 169–186 (2007)
6. Matthies, L., Shafer, S.: Error modeling in stereo navigation. Int. J. Robot. Autom. **3**, 239–248 (1987)
7. Matthies, L.: Dynamic stereo vision, Ph.D. dissertation, Carnegie Mellon University (1989)
8. Demirdjian, D., Darrell, T.: Motion estimation from disparity images. In: Proceedings of ICCV, vol. 1, pp. 213–218 (2001)
9. Rabe, C., Müller, T., Wedel, A., Franke, U.: Dense, robust, and accurate motion field estimation from stereo image sequences in real-time. In: Daniilidis, K., Maragos, P., Paragios, N. (eds.) ECCV 2010. LNCS, vol. 6314, pp. 582–595. Springer, Heidelberg (2010). doi:10.1007/978-3-642-15561-1_42
10. Kalman, R.E.: A new approach to linear filtering and prediction problem. J. Basic Eng. **82**, 35–45 (1960)
11. Schmidt, S.: Applications of state-space methods of navigation problems. J. Adv. Control Syst. **3**, 293–340 (1966)
12. Julier S.J., Uhlmann. J.K.: Unscented filtering and nonlinear estimation. In: Proceedings of IEEE, vol. 93, pp. 401–422 (2004)
13. Khan, W., Klette, R.: Stereo accuracy for collision avoidance for varying collision trajectories. In: Proceedings of IEEE Intelligent Vehicles Symposium, pp. 1259–1264 (2013)
14. Geiger, A., Lenz, P., Stiller, C., Urtasun, R.: Vision meets robotics: the KITTI dataset. Int. J. Robot. Res. **32**, 1231–1237 (2013)

High Resolution SOM Approach to Improving Anomaly Detection in Intrusion Detection Systems

Ayu Saraswati[✉], Markus Hagenbuchner, and Zhi Quan Zhou

School of Computing and Information Technology, University of Wollongong,
Wollongong, NSW 2522, Australia
sa783@uowmail.edu.au, {markus,zhiquan}@uow.edu.au
http://www.uow.edu.au

Abstract. Machine learning in general and artificial neural networks in particular are commonly used to address the problem of detecting anomalies in intrusion detection systems. Self-Organizing Maps (SOMs) have been shown to be a promising tool for this purpose, but the limitation of the cardinality of their display space has resulted in SOMs being a black box method and impeded the design of a simpler network architecture. High resolution SOMs are a very recent development that can overcome these problems. This paper explores how high resolution SOMs can help with anomaly detection in intrusion detection systems. Experiments on a large and well established benchmark problem show that high resolution SOMs improve results while allowing a simple network architecture. It is also shown that high resolution SOMs allow the development of better understanding of the results and the problem domain.

Keywords: Anomaly detection · High resolution neural network · Self organising map · Intrusion detection

1 Introduction and Problem Description

An Intrusion Detection System (IDS) is one of the security mechanisms to handle runtime system security. An IDS monitors and detects malicious intent from outside of the system. It is commonly implemented to detect intrusions in a computer network (network-based) and/or in a specific host. There are two general types of IDSs: misuse-based and anomaly-based. A misuse-based IDS checks the network traffic or system behaviour pattern against known attack signatures, whereas an anomaly-based IDS profiles the normal network traffic or system behaviour and identifies any deviation from the profile as a potential intrusion. The deployment of IDSs often combine these different strategies.

The Self-Organising Map (SOM), first proposed in 1982 by Kohonen [1], is an unsupervised neural network that produces a typically two dimensional projection of an input space that can be of any dimension. Due to their topology preserving properties, SOMs are commonly used as a tool to visualise and discover

© Springer International Publishing AG 2016
B.H. Kang and Q. Bai (Eds.): AI 2016, LNAI 9992, pp. 191–199, 2016.
DOI: 10.1007/978-3-319-50127-7_16

properties of the input data. SOMs have been applied to the anomaly detection problem in previous work [2–5]. These approaches produced promising results but were generally black box methods resulted from the low resolution of the SOMs. To alleviate the low resolution problem, some strategies were developed, such as an hierarchical approach where the result from one SOM is quantized to be the input to another SOM. But such a strategy increased the architectural complexity, and the black box problem remained [2–4].

This paper presents new insights obtained from deploying much larger SOMs to the anomaly detection problem, specifically in IDSs. The recent development in technology allows the training of SOMs up to two-orders of magnitude larger than was previously possible, resulting in SOMs with a finely granulated display space called the *high resolution SOMs* (HRSOMs) [6]. More specifically, the availability of massive parallel computing hardware such as found in graphics processing units (GPUs) and advances in algorithmic design expedite the training of HRSOMs [6], which was not available in previous literature on machine learning and anomaly detection. The possibility of obtaining new insights or better results motivated the present research.

This paper uses a well established benchmark problem viz. the KDD Cup 1999 datasets [7,8]. The dataset poses a challenging problem and is still widely used in recent literature for evaluating new methods. Related work in training relatively small SOMs produced competitive results with respect to the false positive and detection rates [2,3]. It is important to investigate how the training of much larger SOMs will affect the results.

The rest of this paper is organised as follows: Sect. 2 describes the learning problem. Section 3 presents selected features, the training condition and evaluation metric. Section 4 shows results and Sect. 5 concludes the paper.

2   KDD Cup 1999 Dataset

The MIT Lincoln Lab released a DARPA IDS evaluation dataset in 1998 [7–9]. They simulated a typical U.S. Air Force local-area network over seven weeks and recorded the TCP dump data [9]. Stolfo *et al.* derived the DARPA intrusion evaluation dataset for their research [8], and the derived dataset was then introduced to the third International Knowledge Discovery and Data Mining Tools Competition [7]. The competition task itself was to classify the data into the following five types of behaviour: Normal, Denial-of-service (DOS), Probe, User-to-Root (U2R), and Remote to Local (R2L). Some researchers reduced the classification types to only Normal and Attack behaviour [2–4].

The KDD 1999 dataset is available in three sets (1) whole dataset, (2) 10% subset and (3) Corrected dataset. The 10% subset is sampled from the whole dataset provided by the competition organiser. The 10% subset is commonly used as a training set [2,3,10,11]. A "corrected" dataset consists of 14 attacks that are neither in the whole dataset nor in the 10% subset.

Each entry has 41 features and is labeled as either *normal* or by the specific attack type, e.g., smurf, ftp_write, and ipsweep. These three sets do not share

Table 1. The first six features of the KDD Cup 1999 dataset consist of numeric and symbolic values that need to be encoded and standardized [8].

Feature name	Description	Type
duration	length (number of seconds) of the connection	numeric
protocol_type	type of the protocol, e.g. tcp, udp	symbolic
service	network service on the destination, e.g., http, telnet	symbolic
src_bytes	number of data bytes from source to destination	numeric
dst_bytes	number of data bytes from destination to source	numeric
flag	normal or error status of the connection	symbolic

the same probability distribution, and the "corrected" dataset includes unknown attacks. This is to simulate probable zero day attack in a real world environment.

Kayacik *et al.* [2,3] showed that the first six features (shown in Table 1) were sufficient for the detection of the attacks in the KDD dataset. They explored the capability of unsupervised methods, particularly SOMs, for anomaly detection.

3   Design of Experiments

Preprocessing: The first six features of the 10% subset and corrected dataset are used as the training and testing sets, respectively. The features consist of numeric and symbolic values as described in Table 1. The symbolic values need to be encoded into numeric values that SOMs can handle. Therefore, the symbolic features are encoded such that their Euclidean distance is equidistant, and the numeric values are standardized (using the z-score) to prevent any one dominating variable.

Training: A classic approach to training a SOM is described by Kohonen [1] whereas a training method for high resolution SOMs is described in [6]. We will refrain from repeating the details of the algorithms in this paper and instead focus on the use of the algorithms. A number of training parameters need to be set: The user can choose a map size $N \times M$ which means that there are $N \times M$ neurons on the map, a learning rate α which controls the magnitude by which the codebook vectors are updated, a neighbourhood function which defines the relationship (or neighbourhood) of the neurons in the grid, a neighbourhood radius σ which defines the size or extend of the neighbourhood, the topology which defines how the neurons are connected, and the number of training iterations.

We trained SOMs of four different sizes: 20×20 (M_{20}), 60×60 (M_{60}), 120×90 (M_{120}), and 250×188 (M_{250}). All maps were trained under similar conditions using the Gaussian neighbourhood function, hexagonal topology, 200 training iterations, $\alpha = 0.6$, and $\sigma = d$ (the diagonal of the map). The results will be presented in Sect. 4.

Evaluation Metrics: The KDD dataset comes with ground truth (class) labels which are not used during the training procedure. However, these labels are useful for evaluating the quality of results of a training session. A neuron location is labelled the positive (attack) cases or negative (normal) cases based on the majority of the labels mapped at each neuron location. The maps will be evaluated using a variety of indicators including precision (P), recall (R), accuracy (ACC) and the F_1 score. Previous work also used the false positive rate (FPR) and the detection rate (DR) as evaluation metrics [2–4].

4   Results of Experiments

The mappings of the data on a trained M_{20} do not show clear clusters that separate the five different classes, although the difference between the mappings of normal and attack cases can be roughly identified, as shown in Fig. 1: It can be seen that the majority of the mappings of the normal cases tend to be squashed towards the three corners. The granularity of the M_{20} is low, and the time cost required for training the SOM using the conventional approach is high. The consequence is that there are only a small number of available neurons to which data are mapped, and many different cases are mapped onto the location.

We then trained the M_{250}, and the results are shown in Fig. 2. There is a very significant improvement in the quality of the mappings. The figure shows that the normal cases form well defined clusters and that the M_{250} can differentiate many of the different attack cases, some of which form clusters themselves (such as the DOS cases in the center towards the left hand side of the map). It is

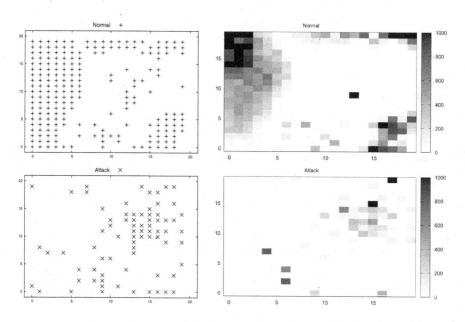

Fig. 1. Mappings of normal cases (top left) and attack cases (bottom left) on a trained M_{20}. The corresponding numbers of activations per neuron are shown right.

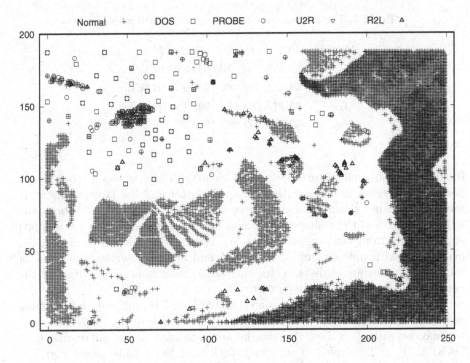

Fig. 2. Mappings of the five pattern classes in an M_{250}.

to be noted, however, that the majority of the attack cases are outliers, which suggests that an attack case does not necessarily follow any known attack pattern previously learnt from other attack cases.

There are three large well-formed normal case clusters on the map. One reason for the large sizes is that there are many more normal cases in the dataset than attack cases. Another reason is that the feature space of the normal cases is larger than that of the attack cases. One almost uniform cluster at the right hand side of the map stretches along the y-axis. Thus, cases that are mapped at one end of this cluster are very different from cases that are mapped at the other end of the cluster. Yet there are numerous other cases which are similar to both and are hence mapped in-between. There is also an S-shaped cluster and a fan-shaped cluster on the center of the map. The cases in one cluster are also very different from cases in the other cluster but because they are in separate clusters there are no cases which are similar to both. The results also show that there is not just one type of normal behaviour but rather there are cases of several kinds of normal behaviour.

The plot also shows a few normal cases that seem to be outliers (not in any cluster). There is a small pocket of normal cases on the left border of the map. These indicate rather unusual activities that are not marked as suspicious, or may refer to a type of normal cases which occur rather infrequent. The map shows similar occurrences on the right side where some attack cases are mapped

Table 2. The average performance scores of each map size.

Map size	P	R	ACC	F_1
M_{20}	94.28%	86.92%	93.03%	90.10%
M_{60}	98.68%	96.48%	98.09%	97.56%
M_{120}	99.39%	99.44%	99.54%	99.41%
M_{250}	99.51%	99.56%	99.63%	99.54%

to those neurons and distort the normal clusters. There is also a more sparse normal case region in between the S-shaped cluster and the largest cluster. It might be because the normal cases here share some similarity with both the largest cluster and the S-shaped cluster so that these normal cases are mapped slightly away from the largest cluster. We verified this result by running the experiment with different initial conditions and always observed that some of the normal cases are outliers. Thus, this observation is not an artifact of the network but a property of the dataset.

The attack cases are mainly sparsely distributed throughout the mapping space, except for one cluster in the upper left hand side of the map. This cluster may be caused by how the attacks are simulated in the environment by DARPA [9], or may refer to an attack path that has a wide attack corridor. Upon further inspection, the one well-formed cluster of attack cases are in fact DOS attacks, and it is known that DOS attacks can be conducted in many ways. In fact, the few small clusters of attacks tend to be either DOS or PROBE patterns. U2R and R2L cases are often mapped relatively close to normal cases and in some instances there is some overlap with normal cases. This shows that it is much harder to distinguish U2R and R2L attacks from normal cases than for the other types of attacks. For example, there is one R2L attack case mapped right in the middle of the largest normal cluster. While this could be a noise point, it could also indicate a breach of a given security policy in that there is an attack path that is virtually indistinguishable from normal behaviour. This observation is supported by Stolfo *et al.* [8] who highlighted that U2R and R2L attacks require content-based feature to be accurately detected whilst the features used to train the map is only network-based. The embedded R2L attack might become detectable if those extra features were included.

The average performance scores of different map sizes against the four evaluation metrics are shown in Table 2. To eliminate the influence of training errors, we repeated each experiment four times on the training set then report the average of the four results for each indicator and each map size. It is observed that the performance indicators generally improve with the map size. Table 2 shows a steep increase of all performance indicators from M_{20} to M_{60}. This means that the quality of results of the previous approach was affected negatively by the low resolution of the SOM. M_{120} performed at over 99% across all the performance metrics. This shows that a SOM of size 120 × 90 is sufficient for numeric analysis whereas for visual inspections we found that the M_{250} creates much clearer

Table 3. The relative improvement of SOM M_{250} and M'_{250} compared to relevant work

Related work	Approaches	Features used	metric metric	Relative improvement
Kayacik *et al.* [2]	Hierarchical-SOM	6	FPR, DR	$\delta(FPR) = -4.18\%$
				$\delta(DR) = 10.2\%$
Kayacik *et al.* [3]	Hierarchical-SOM	41	FPR, DR	$\delta(FPR) = -1.15\%$
				$\delta(DR) = 8.69\%$
Sarasamma *et al.* [4]	Hierarchical-SOM	20	FPR, DR	$\delta(FPR) = -1.77\%$
				$\delta(DR) = 5.83\%$
Fernando *et al.* [12]	SICLN	41	P, R, ACC	$\delta(P) = 0.59\%; \delta(R) = 0.04\%$
				$\delta(ACC) = -0.03\%$
de la Hoz *et al.* [13]	GHSOM	20	ACC, DR	$\delta(ACC) = -3.93\%$
				$\delta(DR) = -0.21\%$
Lin *et al.* [14]	CANN	19	FPR, DR,	$\delta(FPR) = -2.53\%$
				$\delta(DR) = 0.01\%$
				$\delta(ACC) = 0.17\%$

results. We also find that an increase beyond the size of M_{250} made no further gains of significance.

In order to compare our results with those obtained by other researchers [2,3] we also trained a 250×188 map on the corrected KDD 1999 dataset, which we will refer to as M'_{250}. We use *FPR* and *DR* to evaluate the system's ability as a unsupervised anomaly detection method. Kayacik *et al.* in [2] proposed an approach using hierarchical SOM using the first six feature (as before). Their method produced an $FPR = 4.6\%$ and $DR = 89.0\%$. Kayacik *et al.* expanded the work into all fourty-one features [3] and by using a single layer 20×20 SOM which reportedly yields $FPR = 1.53\%$ and $DR = 89.92\%$ and a dual layer SOM with yield $FPR=1.57\%$ and $DR=90.6\%$.

In comparison, the M'_{250} (a single layer map) yields $FPR = 0.42\%$ and $DR = 99.29\%$. Note that this very large improvement in results is obtained without increasing the complexity of the system architecture (i.e. hierarchical or multi-layered) and without using the full 41-dimensional features set, hence a clear demonstration of the advantages of the high resolution SOM.

The technology of high resolution SOMs was not available previously so researchers [2–4] attempted to circumvent the problem by training a second and third layer SOM on the most superimposed samples from the lower level SOM in order to further separate them. The approach is similar to the tree-SOM algorithm which uses a similar strategy but is based on the quantization error.

The more recent work in the same field used SOM in an ensemble method as reported in [12–14] and reviewed in [15]. The larger M_{250} map does not require such a strategy. The improvement of M_{250} map to relevant work is summarised in Table 3. The single layer large SOM generally improved the DR and reduced the FPR, except compared to the more recent work reported in [13].

5 Conclusion

This paper has shown that advances in computing technology and algorithmic design allowed the training of SOMs with a granularity increased by more than 150 times when compared with previous approaches. It has been demonstrated that high resolution SOMs greatly improves the mapping quality, so that it becomes possible to engage an unsupervised machine learning method for reliable detection of anomalies in intrusion detection systems. Moreover, it has been shown that the greatly improved level of details in the mappings of the data provided the means to explain the properties of the domain as well as the means to explain the quality of the results.

References

1. Kohonen, T.: Self-organized formation of topologically correct feature maps. Biol. Cybern. **43**(1), 59–69 (1982)
2. Kayacik, H.G., Zincir-Heywood, A.N., Heywood, M.I.: On the capability of an SOM based intrusion detection system. In: Proceedings of the International Joint Conference on Neural Networks, vol. 3, pp. 1808–1813, July 2003
3. Kayacik, H.G., Zincir-Heywood, A.N., Heywood, M.I.: A hierarchical SOM-based intrusion detection system. Eng. Appl. Artif. Intell. **20**, 439–451 (2007)
4. Sarasamma, S.T., Zhu, Q.A., Huff, J.: Hierarchical Kohonen net for anomaly detection in network security. IEEE Trans. Syst. Man Cybern. Part B (Cybern.) **35**(2), 302–312 (2005)
5. Kayacik, H.G., Zincir-Heywood, A.N., Heywood, M.I.: Using self-organizing maps to build an attack map for forensic analysis. In: Proceedings of the International Conference on Privacy, Security and Trust: Bridge the Gap Between PST Technologies and Business Services (PST), pp. 33:1–33:8, ACM (2006)
6. Nguyen, V.T, Hagenbuchner, M., Tsoi, A.C.: High resolution self-organising maps. In: The 29th Australasian Joint Conference on Artificial Intelligence (2016)
7. Hettich, S., Bay, S.D.: KDD Cup 1999 data. UCI KDD Archive, University of California, Irvine, Department of Information and Computer Science (1999)
8. Stolfo, S.J., Fan, W., Lee, W., Prodromidis, A., Chan, P.K.: Cost-based modeling for fraud and intrusion detection: results from the JAM project. In: Proceedings of DARPA Information Survivability Conference and Exposition (DISCEX 2000), vol. 2, pp. 130–144 (2000)
9. Lippmann, R.P., Cunningham, R., Fried, D., Garfinkel, S., Gorton, A., Graf, I., Kendall, K., McClung, D., Weber, D., Webster, S., et al.: MIT Lincoln laboratory offline component of DARPA 1998 intrusion detection evaluation. In: Presentation at MIT Lincoln Laboratory PI Meeting (1988) http://ideval.ll.mit.edu/intro-html-dir/. Accessed 14 Dec 1998
10. Tavallaee, M., Bagheri, E., Lu, W., Ghorbani, A.A.: A detailed analysis of the KDD CUP 99 data set. In: Proceedings of the IEEE Symposium on Computational Intelligence for Security and Defense Applications (CISDA) (2009)
11. Kayacik, H.G., Zincir-Heywood, A.N., Heywood, M.I.: Selecting features for intrusion detection: a feature relevance analysis on KDD 99 intrusion detection datasets. In: Proceedings of the 3rd Annual Conference on Privacy, Security and Trust (2005)

12. Fernando, Z.T., Thaseen, I.S., Kumar, C.A.: Network attacks identification using consistency based feature selection and self organizing maps. In: Proceedings 2014 First International Conference on Networks Soft Computing (ICNSC), pp. 162–166, August 2014
13. de la Hoz, E., de la Hoz, E., Ortiz, A., Ortega, J., Martnez-lvarez, A.: Feature selection by multi-objective optimisation: application to network anomaly detection by hierarchical self-organising maps. Knowl. Based Syst. **71**, 322–338 (2014)
14. Lin, W.-C., Ke, S.-W., Tsai, C.-F.: CANN: an intrusion detection system based on combining cluster centers and nearest neighbors. Knowl. Based Syst. **78**, 13–21 (2015)
15. Sharma, R.K., Kalita, H.K., Borah, P.: Analysis of machine learning techniques based intrusion detection systems. In: Proceedings of 3rd International Conference on Advanced Computing, Networking and Informatics (ICACNI), vol. 2, pp. 485–493 (2016)

Big Data

CPF: Concept Profiling Framework for Recurring Drifts in Data Streams

Robert Anderson$^{(\boxtimes)}$, Yun Sing Koh, and Gillian Dobbie

Department of Computer Science, University of Auckland, Auckland, New Zealand
rand079@aucklanduni.ac.nz, {ykoh,gill}@cs.auckland.ac.nz

Abstract. We propose the Concept Profiling Framework (CPF), a meta-learner that uses a concept drift detector and a collection of classification models to perform effective classification on data streams with recurrent concept drifts, through relating models by similarity of their classifying behaviour. We introduce a memory-efficient version of our framework and show that it can operate faster and with less memory than a naïve implementation while achieving similar accuracy. We compare this memory-efficient version of CPF to a state-of-the-art meta-learner made to handle recurrent drift and show that we can regularly achieve improved classification accuracy along with runtime and memory use. We provide results from testing on synthetic and real-world datasets to prove CPF's value in classifying data streams with recurrent concepts.

Keywords: Concept drift · Recurrent concepts · Data streams

1 Introduction

Classifying potentially infinite data streams requires a trade-off between accuracy of our predictions and speed and memory use of our approach. This trade-off is made more difficult by streams potentially changing over time *i.e.* the underlying distribution of data changing (*concept drift*). Classification of streams with drifting concepts benefits from drift handling techniques *e.g.* as described in [1]. For example, if we detect a significant change in the underlying distribution, we can build a new model to learn anew from it, avoiding bias learnt from previous distributions. However, we may lose all we have learnt up to that point.

Over time an underlying data-generating distribution may revert to a previously seen state. Through recognising and understanding these recurring concepts we can often perform better classification of incoming data in a stream by reverting to previously trained models [2]. Repeated patterns are regularly observed in the real world: seasons; boom-bust periods in financial markets; rush-hour in traffic flows; and battery states of sensors, for example. Where these exist, classification can be faster and more accurate by recognising and accounting for them. Identifying repeated patterns improves our understanding of our modelled problem. However, many state-of-the-art techniques use demanding statistical

© Springer International Publishing AG 2016
B.H. Kang and Q. Bai (Eds.): AI 2016, LNAI 9992, pp. 203–214, 2016.
DOI: 10.1007/978-3-319-50127-7_17

tests or ensemble approaches to identify and handle recurring concepts. These escalate memory use and increase runtime. In situations that rely on speed or low memory use, this overhead means these techniques cannot always be applied, and so recurring concepts will not be accounted for in the classification. Through this work, we address this problem by proposing a classification framework that utilises recurrent concepts while mininimising runtime and memory usage overhead compared to current state-of-the-art techniques.

We present the Concept Profiling Framework (CPF). This is a meta-learning approach that maintains a collection of classifiers and uses a drift detector. When our drift detector indicates a drift state *i.e.* that our current classifier is no longer suitable, we check our collection of classifiers for one better suited to the current stream. If one meets a set level of accuracy, we will select it as the current classifier; otherwise a new classifier is produced and trained on recent data. If this new classifier behaves similarly to a classifier in our collection (using a measure derived from *conceptual equivalence* [3]), we will choose that existing classifier as our current model (*i.e.* model *reuse*); otherwise we will add the new classifier to the collection and use that as our current classifier.

We introduce two techniques to allow efficient handling of recurrent concepts. First, we regularly compare behaviour of our classifiers, and over time, our certainty about their similarity will improve. If they behave similarly, we can use the older model to *represent* the newer one. Second, we implement a *fading* mechanism to constrain the number of models, a points-based system that retains models that are recent or frequently used. Through observing reuse patterns, we can understand how patterns recur in our stream.

Our contribution is a meta-learning framework that can: utilise observed model behaviour over time to accurately recognise recurring concepts, without relying on additional information about underlying concepts such as in [4]; and regularly outperform a state-of-the-art learning framework, the Recurring Concept Drifts framework (RCD [5]), in terms of accuracy, memory and runtime.

In the next section, we discuss related work that informed the creation of CPF. We then detail our proposed framework. We provide experimental results to: show that using our fading mechanism provides similar quality to a naïve implementation of our algorithm whilst improving memory and runtime; show our approach provides generally better accuracy than RCD, requiring less memory and runtime on eight synthetic datasets; and show CPF keeps this time and memory efficiency while matching RCD in accuracy on five commonly used real-world datasets. Next, we discuss the results and consider our technique in greater detail. Finally, we conclude by summarising our findings and suggesting sensible next steps for developing CPF.

2   Related Work

In this section, we discuss the previous work that has informed our development of CPF. Gama et al. [6] provide a fantastic overview of the problems faced while learning in drifting streams and solutions that have been proposed. Since data streams arrive incrementally, models such as Very Fast Decision Trees (VFDTs) [7] have been created to be built incrementally. They achieve a constant time and memory relationship with the number of instances seen, and are guaranteed to achieve performance similar to a conventional learner. This is through limiting the number of examples required to be seen at any node through utilising the Hoeffding bound [8], which describes the number of instances needed to be representative of an overall dataset within a set probability.

Drift-detection mechanisms try to detect underlying concept drift, so a learning framework can take a corrective action. DDM (Drift Detection Mechanism) [9] monitors the error rate of a classifier. When the mean error rate rises above thresholds based on the minimum error rate seen, it signals a warning or drift. EDDM (Early DDM) [10] introduces two changes to DDM. First, the mean distance between errors is measured instead of error rate. Second, it changes thresholds used for detecting warnings and drifts. Two user-set parameters, α and β decide its sensitivity. If the mean distance drops $\beta \times 2s.d.$ below the minimum mean distance seen, the detector signals a warning, and below $\alpha \times 2s.d.$, it signals a drift, where $1 > \beta > \alpha$. EDDM is made to detect drift more quickly than DDM, with less evidence required.

The approach in [11] handles recurring concepts by building a referee' model alongside every instance a classifier sees, and keeps a collection of these pairs. The referee model judges whether its classifier is likely to correctly classify an instance by learning whether it was correct for previous similar instances. When existing models aren't applicable, a new model is created. Two models are trained at once at any time, and no suggestions are made for constraining the total number of models built over time.

Gomes et al. [4] propose using an additional user-specified context stream alongside a data stream. Their approach relates the current context to the current classifier when it is performing well. Particular contexts become associated to classifiers that are useful in those contexts. After drift, a model is reused if the context learner feels it fits the current context. Otherwise, a new classifier is created. They use *conceptual equivalence* [3], which relates similar models through their classifying behaviour. This technique requires an additional context stream which is difficult to select for a problem that is not well understood.

Gonçalves and De Barros [5] propose RCD, a framework that maintains a classifier collection. For each classifier, a set of recent training instances are stored to represent its concept. When a drift is detected, a statistical test (k-NN) compares the instances in the warning buffer to the training instances for classifiers. If the instances are found to be significantly similar, it uses the corresponding model; otherwise, a new one is created. The statistical testing and buffer stores add significant runtime and memory requirements to their approach.

Ensemble approaches to recurrent learning use multiple classifiers examining a stream, allowing greater chance to find one that functions well at a given time. However, this gain in accuracy often comes with greater memory and runtime overhead compared to single classifier techniques as shown in [12].

3 Concept Profiling Framework (CPF)

In this section, we describe how our proposed technique functions. Our main goal is a learning framework that can recognise recurring concepts through model behaviour and use this to fit existing, pre-trained models rather than new models.

We use a meta-learning framework (Fig. 1) with a collection of one or more incremental classifiers. One is designated as our current classifier. A drift detector signals warning and drift states. On a warning state, the meta-learner will stop training the current classifier and store instances from the data stream in a buffer. If a drift state follows, the meta-learner looks for an existing model in the collection that classifies the warning buffer accurately to use as the current classifier. If it cannot find one, it will create a new model trained on even buffer instances. When an existing model behaves similarly to this new model (when tested on odd buffer instances) that model will be reused; otherwise the new model is trained on odd buffer instances and used. Every model in the collection is tested on the buffer, and the results will be compared and stored. Where it is found that models classify similarly to one another, the older model will *represent* the newer one.

Through regular reuse and representation of particular models (as described below), we hope for particular classifiers to model particular concepts very well over time. In addition, frequency of reuse and representation can show patterns of recurrence in the underlying data. CPF can pair with any classifier that is incremental and can perform in a streaming environment. We use Hoeffding Trees using Naïve Bayes - a version of a CVFDT [13] which creates a Naïve Bayes model at the leaves if it provides better accuracy than using the majority class. This is consistent with the implementation of RCD that the authors suggest in [5]. Our experiments use EDDM to be consistent with RCD. Technically, our technique can work using a drift detector that has no warning zone, but the buffer will be of a set minimum size rather than informed by the drift detector.

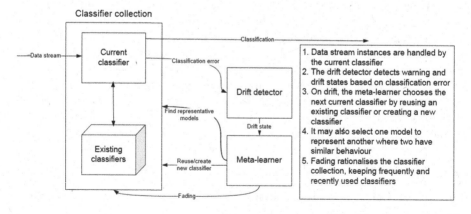

Fig. 1. The concept profiling framework

3.1   Model Similarity

Our approach uses a similarity measure based upon *conceptual equivalence* used in [4] for comparing models. We adapt their approach and do pairwise comparisons of models' respective errors when classifying given instances. When comparing classifier c_a and classifier c_b, we calculate a score per instance, where $c_a(x)$ and $c_b(x)$ is the classification error for c_a and c_b on a given instance x:

$$Score(x, c_a, c_b) = \begin{cases} 1 & \text{if } c_a(x) = c_b(x) \\ 0 & \text{if } c_a(x) \neq c_b(x) \end{cases}$$

We then calculate similarity for two models over a range of instances, using instances X seen during warning drift periods:

$$Sim(X_1...X_n, c_a, c_b) = \frac{\Sigma(Score(X, c_a, c_b))}{n}$$

If $Sim(X, c_a, c_b) \geq m$ ($m \leq 1$ and is a set similarity margin threshold), we describe the models as similar and likely to represent the same concept. We require a minimum of thirty common instances seen by two classifiers before we measure their similarity. Since new classifiers only train on even instances, we collect at least sixty instances. Our score function provides a Binomial distribution, and the Central Limit Theorem indicates that as we see thirty examples and beyond, this will approximate a Normal distribution. This gives assurance of finding a representative mean *Score* between two classifiers.

Every time a drift occurs, all existing classifiers will classify instances in the warning buffer. The results of each classifier will be compared as a bitwise comparison to see if they both had the same classification error. The similarity matrix stores pairwise comparisons between classifiers, through recording instances seen (or n) and total score (or $\Sigma(Score(X, c_a, c_b))$). We then check for similar models. Figure 2 shows an example of four classifiers being compared using a given warning buffer, with 0 representing a correct classification and 1 an incorrect classification by the classifiers. Of ten instances (X) in our buffer, we can see which our classifiers correctly or incorrectly classify, which is the behaviour we measure

Classifier (c_n)	Errors on warning buffer instances (X) x_1 x_2 x_3 x_4 x_5 x_6 x_7 x_8 x_9 x_{10}		Accuracy on buffer	$Sim(X, c_1, c_n)$	$Sim(X, c_{new}, c_n)$
Classifier 1 (c_1)	0 1 0 0 0 1 0 0 1 1		6 / 10	10 / 10	4 / 5
Classifier 2 (c_2)	1 1 0 0 1 1 0 1 1 1		3 / 10	7 / 10	2 / 5
Classifier 3 (c_3)	0 0 0 0 0 1 0 0 0 0		9 / 10	7 / 10	5 / 5
New classifier (c_{new}) *T = training instances	0 T 0 T 0 T 0 T 0 T		5 / 5	4 / 5	5 / 5

Fig. 2. Example of calculating model similarity for classifiers $c_1...c_{new}$ on warning buffer X

to find their Sim. If $m = 0.95$, we can see that $Sim(X, c_{new}, c_3) = 5/5 \geq m)$, so could treat our models as similar. In practice, we require instances in the buffer for that decision.

3.2 Reuse and Representation

When a drift occurs, CPF will check if any existing classifiers are suitable to reuse for the following data. This is by a two-step check. First, by checking from the oldest classifier, if any classifier achieves $\geq m$ accuracy (our similarity margin threshold), the model will be selected as the classifier to use. Otherwise, a new classifier will be trained on even instances in the buffer, then prequential test-and-trained on odd instances. If an old model has similarity $\geq m$ when tested on odd instances in the buffer, we will *reuse* the old model, as it is similar to the new model we would introduce. Our experiments suggest setting m as 0.95 for consistent performance across a range of datasets. This level should help prevent CPF from incorrectly assuming model similarity without fair evidence *i.e.* at least 95% similarity in behaviour as per classification errors. Lower values of m allow model reuse and representation with less evidence of similarity.

When two classifiers behave similarly, we assume that both classifiers describe the same concept. The older classifier is generally more valuable to keep, as it has often been trained on more instances and been compared to other classifiers in our collection more frequently. The older classifier *represents* the newer classifier as follows: if it was the current classifier, the older model becomes the current classifier; the older model receives the newer model's fading points (described below); and the newer model is removed from our collection. The older and newer model behave similarly, so we lose little by retaining only the older model.

3.3 Fading

Data streams can be of infinite length, and over time, the classifier collection in our framework may continually grow, risking an ever-expanding performance overhead. To avoid this, we use a fading mechanism to constrain the size of our classifier collection. Our fading mechanism prefers recent and frequently-used classifiers, and penalises others. At any stage, we maintain an array of fade points F. Fade points of a given classifier c_a, F_{c_a} can be expressed as follows:

$$
F_{c_a} = \begin{cases} 0 & \text{if represented by older model} \\ (r+1) \times f + \sum(F_{m_{c_a}}) - (d-r) & \text{otherwise} \end{cases}
$$

Here, r is drift points that a model is reused, f is a user-set parameter for points to gain on creation and reuse, d is the number of drifts the classifier has existed for (excluding the drift at which it is created) and $\sum(F_{m_{c_a}})$ is the sum of points for any models represented by c_a. Every drift in which a model is not reused, it loses a point, but when it is reused it gains f points. When newer models are represented by an older model, the older model inherits the newer model's points. When $F_{c_a} = 0$, the model is deleted. The user can control the size and number of models by selecting f. Removing models through fading gives our technique less opportunity to identify similarity with previous models, so the user can lose some information about recurrent behaviour through this step.

In our experiments, we set f to be 15. This is based on RCD's maximum of 15 models. This will constrain the total number of models to around that number unless models are regularly reused. Where there is a zero reuse and representation rate, we cannot have more than 15 models, and only have more models when we have reused models that may represent recurrent concepts.

3.4 Model Management in Practice

Figure 3 illustrates how our technique maintains its collection of classifiers. For the sake of simplicity, we have set $f = 3$ and $m = 0.95$.

1. After drift point T in a stream, in which our warning buffer had 100 instances, we have three models. Classifier c_1 was reused for concept A, and has just gained f points, while c_2 and c_3 lost a point each. Pairwise comparisons in the similarity matrix have been updated with the *Score* from 100 new instances seen in the warning buffer.

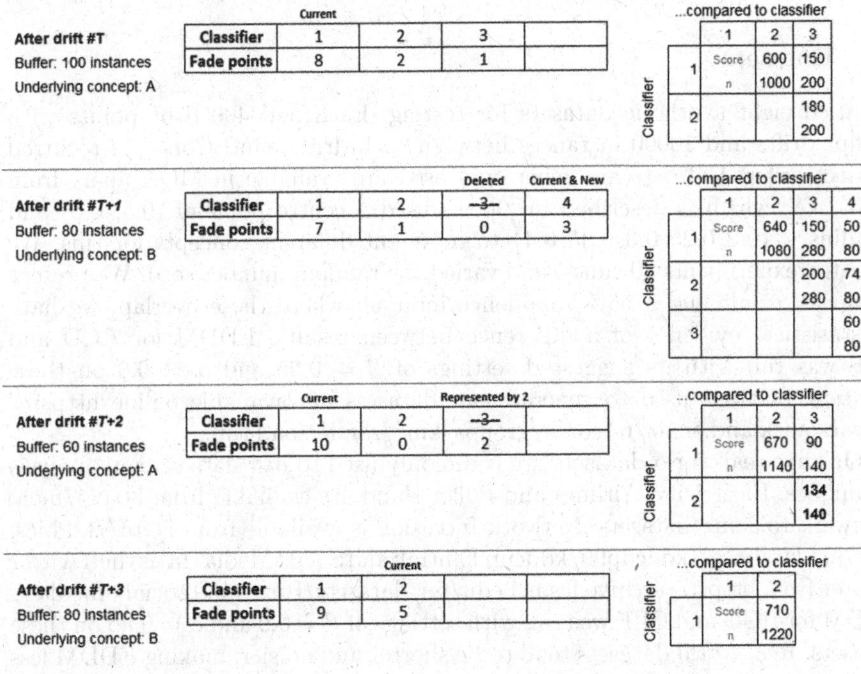

After drift #T — Buffer: 100 instances — Underlying concept: A

	Current		
Classifier	1	2	3
Fade points	8	2	1

...compared to classifier

Classifier		1	2	3
1	Score		600	150
	n		1000	200
2				180
				200

After drift #T+1 — Buffer: 80 instances — Underlying concept: B

			Deleted	Current & New
Classifier	1	2	~~3~~	4
Fade points	7	1	0	3

...compared to classifier

Classifier		1	2	3	4
1	Score		640	150	50
	n		1080	280	80
2				200	74
				280	80
3					60
					80

After drift #T+2 — Buffer: 60 instances — Underlying concept: A

	Current		Represented by 2
Classifier	1	2	~~3~~
Fade points	10	0	2

...compared to classifier

Classifier		1	2	3
1	Score		670	90
	n		1140	140
2				134
				140

After drift #T+3 — Buffer: 80 instances — Underlying concept: B

	Current	
Classifier	1	2
Fade points	9	5

...compared to classifier

Classifier		1	2
1	Score		710
	n		1220

Fig. 3. Example to illustrate reuse, representation and fading

2. After drift $T+1$, no existing classifiers match the buffer or the new classifier, so a new classifier becomes c_4 and gains f points, while the others lose a point each. Classifier c_3 now has zero points and is deleted, so $c_4 \rightarrow c_3$.
3. After drift $T+2$, concept A recurs, and c_1 matches the buffer through accuracy so gains f points and becomes the current classifier. The other two models lose a point each. Classifier c_2 would be deleted, but now $Sim(X, c_2, c_3) \geq m$ so c_3 is deleted and c_2 represents c_3, so gets its points.
4. After drift $T+3$, concept B recurs, and c_2 matches the buffer through accuracy so gains f points while c_1 loses a point.

4   Experimental Design and Results

To test CPF against RCD, we used the MOA API (available from http://moa. cms.waikato.ac.nz). We used the authors' version of RCD from https://sites. google.com/site/moaextensions, keeping default parameters (including limiting total models to a maximum of 15 at any given time). Experiments were run on an Intel Core i5-4670 3.40 GHz Windows 7 system with 8 GB of RAM. RCD and CPF maintain a buffer of instances if a detector is in a drift state. This can cause variations in memory requirements over a small space of instances so we excluded this buffer in our memory measurements for both. CPF was run with $f = 15$, $m = 0.95$ and a minimum buffer size of 60.

4.1   Datasets

We used eight synthetic datasets for testing. Each had 400 drift points with abrupt drifts and 10000 instances between each drift point. Concepts recurred in a set order. Data stream generators used are available in MOA apart from CIRCLES, which is described in [9]; we used a centre point of (0.5, 0.5) and a radius $\in \{0.2, 0.25, 0.3, 0.35, 0.4\}$ to represent different concepts for this. We repeated experiments 30 times, and varied the random number seed. We present the mean result and a 95% confidence interval; where these overlap, we have no statistical evidence of a difference between results. EDDM for RCD and CPF was run with its suggested settings of $\beta = 0.95$ and $\alpha = 0.9$ on these datasets. Settings used to generate our datasets are available online https:// www.cs.auckland.ac.nz/research/groups/kmg/randerson.html.

Our five real-world datasets are commonly used to test data stream analysis techniques. Electricity, Airlines and Poker Hand are available from http://moa. cms.waikato.ac.nz/datasets; Network Intrusion is available from http://kdd.ics. uci.edu/databases/kddcup99/kddcup99.html; and Social Media [14] is the Twitter dataset from https://archive.ics.uci.edu/ml/datasets/Buzz+in+social+media+. EDDM for RCD and CPF was run with settings of $\beta = 0.9$ and $\alpha = 0.85$ on these datasets. Real-world datasets tend to be shorter and noisier; making EDDM less sensitive avoids overly reactive drift detection.

4.2 CPF Fading Mechanism

For this experiment, we compared CPF with the fading mechanism and CPF without. We compared accuracy, memory use, runtime, mean models stored and mean maximum models stored in the classifier collection over the synthetic datasets. As per Table 1, our fading mechanism caused slight losses in accuracy for two datasets and a slight gain in another. Runtime and memory use were never significantly worse for the fading approach and were significantly better in most cases. Mean models and mean max models for the experiments show that the fading mechanism significantly constrains the number of models produced. This provides evidence that the fading mechanism works as an effective constraint for models considered by CPF and loses little accuracy.

Table 1. Effect of fading mechanism on CPF

Dataset	Approach	Mean accuracy	95% CI	Mean memory	95% CI	Mean time	95% CI	Mean models	95% CI	Max models	95% CI
Agrawal	CPF	78.18%	0.08%	9.91E+05	6.45E+04	2.60E+04	1.77E+02	16.94	0.12	21.83	0.23
	CPF - no fade	80.00%	0.77%	2.70E+08	1.67E+07	1.54E+06	1.14E+05	1183.04	51.30	2153.47	95.67
CIRCLES	CPF	96.39%	0.17%	9.42E+05	3.25E+04	5.15E+03	6.38E+01	9.75	0.45	18.93	0.82
	CPF - no fade	97.18%	0.12%	2.05E+06	1.35E+05	7.43E+03	3.10E+02	44.52	2.55	55.50	3.90
Hyperplane	CPF	93.61%	0.02%	5.15E+05	2.82E+04	2.13E+04	8.36E+01	17.04	0.14	24.37	0.35
	CPF - no fade	92.90%	0.03%	1.25E+08	1.47E+06	5.74E+05	1.10E+04	813.68	6.41	1500.77	9.16
LED	CPF	69.10%	0.31%	5.90E+05	1.01E+05	6.92E+04	2.05E+03	16.80	0.09	18.40	0.20
	CPF - no fade	68.95%	0.43%	6.19E+06	4.61E+06	1.15E+06	9.35E+04	491.48	28.94	971.43	57.29
RandomRBF	CPF	80.32%	0.90%	1.32E+06	8.01E+05	2.99E+04	1.32E+03	16.21	0.18	21.60	0.48
	CPF - no fade	81.13%	1.08%	1.84E+08	2.62E+07	9.43E+05	1.55E+05	797.00	106.02	1477.77	197.04
SEA	CPF	84.62%	0.14%	1.76E+06	2.66E+05	9.02E+03	5.87E+02	13.20	0.42	20.10	0.71
	CPF - no fade	84.46%	0.13%	5.06E+06	1.28E+06	1.31E+04	2.78E+03	39.87	5.78	60.87	8.85
STAGGER	CPF	99.33%	0.00%	4.30E+05	1.67E+03	3.05E+03	4.22E+01	4.05	0.07	4.83	0.31
	CPF - no fade	99.33%	0.00%	4.32E+05	3.62E+03	2.98E+03	2.55E+01	4.27	0.23	4.87	0.35
Waveform	CPF	78.15%	0.40%	1.15E+06	3.81E+05	1.00E+05	4.49E+03	16.47	0.41	26.47	0.61
	CPF - no fade	78.36%	0.07%	9.92E+07	4.48E+06	1.17E+06	4.39E+04	303.87	8.46	565.87	14.72

4.3 CPF Similarity Margin

We tested CPF with different values of m on synthetic and real datasets and ranked accuracy by m as per Table 2. For synthetic datasets, this was mean accuracy across 30 trials. We compared how often a setting for m did worse than most others *i.e.* was ranked 4th or 5th, as we wanted a setting that would perform consistently over a variety of datasets.

We chose $m = 0.95$ for our experiments, as it received few bad rankings for either synthetic or real datasets. Higher settings for m generally worked well on synthetic datasets except for Agrawal and RandomRBF. Choosing $m = 0.975$ would be justifiable based on real-world datasets. These had no guaranteed recurrence, so a higher setting for m could reduce spurious reuse of inappropriate models, avoiding resulting drops in accuracy.

We also tested CPF with differing minimum buffer sizes (30, 60, 120, 180, 240). For these approaches, we compared accuracy, memory use, runtime and drifts detected over the synthetic datasets. Our results showed some variation in accuracy, memory and runtime based upon buffer size but showed no clear discernable pattern across datasets.

Table 2. Comparison of CPF accuracy across different levels of m

Similarity margin (m)	Rank on synthetic datasets (lower is better)									Rank on real datasets					
	Agrawal	CIRCLES	Hyperplane	LED	RandomRBF	SEA	STAGGER	Waveform	Freq rank > 3	Electricity	Poker	Intrusion	Airlines	Social Media	Freq rank > 3
0.85	2	5	5	5	1	5	5	5	6	5	5	4	5	5	5
0.90	1	4	4	4	2	4	4	3	5	3	4	1	1	4	2
0.95	3	1	1	3	3	3	3	4	1	4	1	3	3	3	1
0.975	4	2	2	1.5	4	2	2	1	2	2	2	2	3	2	0
0.99	5	3	3	1.5	5	1	1	2	2	1	3	5	3	1	1

4.4   Comparison with RCD on Synthetic Datasets

For this experiment, we compared CPF's accuracy, memory and runtime against RCD's on eight synthetic datasets, as per Fig. 4. CPF consistently outperformed RCD in terms of memory usage and runtime. RCD stores a set of instances for each model, increasing memory usage, and runs k-NN to compare these instances to the warning buffer which increases runtime. CPF was always much faster, and reliably required less memory than RCD. CPF significantly outperformed RCD's accuracy on all datasets except for RandomRBF. RandomRBF creates complex problems through assigning classes to randomly placed overlapping centroids: k-NN is well-suited for identifying the current concept in this complex situation, while CPF relies on similarity between models' behaviour which may be too inexact to identify the current concept.

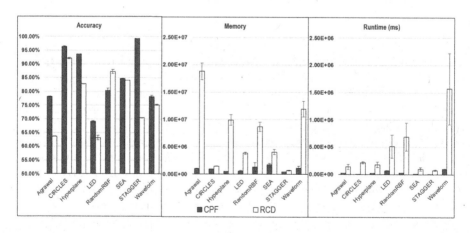

Fig. 4. Comparison of CPF and RCD on synthetic datasets

4.5   Comparison with RCD on Real-World Datasets

For this experiment, we compared CPF's performance against RCD's on our five real-world datasets in terms of accuracy, memory and runtime. As per Fig. 5, CPF consistently used less memory and was faster than RCD, taking less than 10% of the time of RCD for all datasets except for Airlines. In terms of accuracy, CPF performed similarly with Airlines and Intrusion, better with Poker and Social Media and worse with Electricity than RCD. CPF's accuracy is closer to RCD's than seen in our tests with synthetic data. These real datasets are shorter, often tend to be noisier and may not genuinely feature recurring concepts. This is likely to impact the techniques' relative performance.

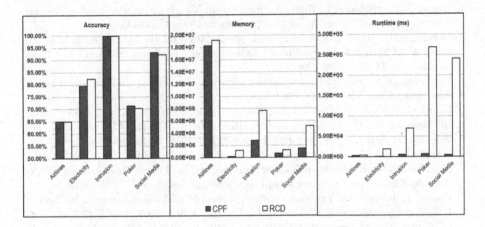

Fig. 5. Comparison of CPF and RCD on real-world datasets

5   Conclusion and Further Work

We have proposed CPF as a fast, memory-efficient, accurate framework for analysing streams with recurring concept drift. Through tracking similarity of model behaviour over the life of our stream, we can regularly reuse suitable models that have already been trained when a drift occurs. This avoids the need to train from scratch, and provides valuable insight on the recurrent patterns in drift for our data stream. Through our experiments, we have shown that our proposed approach is often more accurate than a state-of-the-art framework, RCD, while consistently requiring less memory and running much faster than this framework, on both synthetic and real-world datasets.

Future work to develop this technique could include: developing our similarity measure so that it considers classification and not just error in multi-class problems; finding ways to prioritise model-comparisons as part of our model similarity approach; investigating ways of combining or merging models where they are similar; and utilising our similarity measure/representation approach to simplify and speed up ensemble approaches to recurring drifts.

References

1. Tsymbal, A.: The problem of concept drift: definitions and related work. Technical report TCD-CS-2004-15, Trinity College Dublin (2004)
2. Widmer, G., Kubat, M.: Learning in the presence of concept drift and hidden contexts. Mach. Learn. **23**, 69–101 (1996)
3. Yang, Y., Wu, X., Zhu, X.: Mining in anticipation for concept change: proactive-reactive prediction in data streams. Data Mining Knowl. Disc. **13**, 261–289 (2006)
4. Gomes, J.B., Sousa, P.A., Menasalvas, E.: Tracking recurrent concepts using context. Intell. Data Anal. **16**, 803–825 (2012)
5. Gonçalves, P.M., De Barros, R.S.M.: RCD: a recurring concept drift framework. Pattern Recogn. Lett. **34**, 1018–1025 (2013)
6. Gama, J.A., Žliobaitė, I., Bifet, A., Pechenizkiy, M., Bouchachia, A.: A survey on concept drift adaptation. ACM Comput. Surv. **46**, 1–37 (2014)
7. Domingos, P., Hulten, G.: Mining high-speed data streams. In: Proceedings of the Sixth ACM SIGKDD International Conference on Knowledge Discovery and Data Mining, pp. 71–80. ACM (2000)
8. Hoeffding, W.: Probability inequalities for sums of bounded random variables. J. Am. Stat. Assoc. **58**, 13–30 (1963)
9. Gama, J., Medas, P., Castillo, G., Rodrigues, P.: Learning with drift detection. In: Bazzan, A.L.C., Labidi, S. (eds.) SBIA 2004. LNCS (LNAI), vol. 3171, pp. 286–295. Springer, Heidelberg (2004). doi:10.1007/978-3-540-28645-5_29
10. Baena-García, M., del Campo-Ávila, J., Fidalgo, R., Bifet, A., Gavalda, R., Morales-Bueno, R.: Early drift detection method. In: Fourth International Workshop on Knowledge Discovery from Data Streams, vol. 6, pp. 77–86 (2006)
11. Gama, J., Kosina, P.: Tracking recurring concepts with meta-learners. In: Lopes, L.S., Lau, N., Mariano, P., Rocha, L.M. (eds.) EPIA 2009. LNCS (LNAI), vol. 5816, pp. 423–434. Springer, Heidelberg (2009). doi:10.1007/978-3-642-04686-5_35
12. Brzezinski, D., Stefanowski, J.: Reacting to different types of concept drift: the accuracy updated ensemble algorithm. IEEE Trans. Neural Netw. Learn.Syst. **25**, 81–94 (2014)
13. Hulten, G., Spencer, L., Domingos, P.: Mining time-changing data streams. In: Proceedings of the Seventh ACM SIGKDD International Conference on Knowledge Discovery and Data Mining, pp. 97–106. ACM (2001)
14. Kawala, F., Douzal-Chouakria, A., Gaussier, E., Dimert, E.: Prédictions d'activité dans les réseaux sociaux en ligne. In: 4ième conférence sur les modèles et l'analyse des réseaux: Approches mathématiques et informatiques, pp. 16–28 (2013)

Meta-mining Evaluation Framework: A Large Scale Proof of Concept on Meta-learning

William Raynaut[✉], Chantal Soule-Dupuy, and Nathalie Valles-Parlangeau

IRIT UMR 5505, UT1, UT3, Universite de Toulouse, Toulouse, France
{william.raynaut,chantal.soule-dupuy,nathalie.valles-parlangeau}@irit.fr

Abstract. This paper aims to provide a unified framework for the evaluation and comparison of the many emergent meta-mining techniques. This framework is illustrated on the case study of the meta-learning problem in a large scale experiment. The results of this experiment are then explored through hypothesis testing in order to provide insight regarding the performance of the different meta-learning schemes, advertising the potential of our approach regarding meta-level knowledge discovery.

1 Introduction

Meta-mining designates the very general task of finding an efficient (or most efficient) way to solve a given data mining problem (Fig. 1). As such, it covers a very wide range of tasks, a good many of which have already been extensively studied. For instance, if we consider the very specific problem of Boolean Satisfiability (SAT), we can find different approaches, such as [27], based on the selection of a most efficient algorithm to solve a particular problem instance. Such approaches are designated as *portfolio* for the SAT problem, but have equivalents on many other problems. Their most common denomination would be *algorithm selection* methods, many of which have been studied for machine learning problems, such as classification [11], regression [7], or instance selection [12]. These many different problems have been well studied on their own, but the next step for *meta-mining* research is to start unifying some of them. In particular the problem of *data mining workflow recommendation* has received an increased interest over the last few years [20,22,28]. It consists in the elicitation of workflows (sequences of operators) solving a range of different data mining problems, but remains mostly focused on predictive modelling.

As new approaches emerges, we face a new challenge: *How to evaluate and compare those different meta-mining approaches?* Indeed, the criteria used by authors to evaluate their specific approaches will differ greatly, as they address very dissimilar and sometimes unrelated problems. In order to compare the existing and upcoming approaches able to cover a range of different problems, we will need a *unified* meta-mining evaluation framework. The development of such framework implies a number of new issues, which would be better illustrated on an example.

© Springer International Publishing AG 2016
B.H. Kang and Q. Bai (Eds.): AI 2016, LNAI 9992, pp. 215–228, 2016.
DOI: 10.1007/978-3-319-50127-7_18

Fig. 1. A general meta-mining experiment

2   Example of Meta-mining Experiment

In this example, we will perform algorithm selection for the well studied problem of classification (Fig. 2). Indeed, as it is one of the most prominent cases of the meta-learning framework, it is notably easier to find well described experiments. The whole set of data presented hereafter is extracted from OpenML [25], an important database of machine learning experiments.

For a given dataset, the objective is then to find a particular classifier maximising a specified criterion. For the sake of the example, let us simply use the traditional (albeit recognized insufficient [10]) criterion of predictive accuracy. To supply our classifier selection, we extract from OpenML two sets of data. The first one should describe the predictive accuracy of different machine learning algorithms over a number of datasets (Table 1), while the second should characterize those datasets according to a number of descriptors (Table 2).

The next step is to decide *how* to solve the meta-mining problem. In this example, we will make the naïve choice of identifying the meta-mining problem with a classification problem over the dataset illustrated in Table 3 (which will be referred as the *meta-dataset*). The *Class* label of a dataset instance of the metadataset identifies which algorithm performed best (i.e. had the highest predictive accuracy) on this dataset, according to the data in Table 1.

Fig. 2. An algorithm selection experiment for the classification problem

Table 1. Predictive accuracy of a set of classifiers over a range of datasets

	$classifier_1$	$classifier_2$	...	$classifier_{93}$
$dataset_1$	0.8	0.9	...	...
$dataset_2$	0.9	0.7	...	...
...	...	...	...	...
$dataset_{434}$	...	...	...	...

Table 2. Characterization of the datasets

	NumberOfInstances	NumberOfFeatures	...	MetaAttribute$_{105}$
dataset$_1$	100	62	...	...
dataset$_2$	5000	13	...	...
...	...	...	...	...
dataset$_{434}$	360	20	...	...

Table 3. Meta-dataset for a classification meta-problem

	NumberOfInstances	...	MetaAttribute$_{105}$	Class
dataset$_1$	100	...	4	classifier$_{18}$
dataset$_2$	5000	...	92	classifier$_7$
...	...	...	...	...
dataset$_{434}$	360	...	13	classifier$_{63}$

Next, we have to solve this classification problem. In this example, we will do so according to the following pseudocode:

foreach *dataset$_i$ (Dataset instance)* **do**
> Exclude *dataset$_i$* from the metadataset
> Apply ReliefF [19] attribute selection algorithm on the metadataset
> Learn a decision tree from the reduced metadataset using a C4.5 tree based classifier
> Use this decision tree to predict a class label *classifier$_j$* for *dataset$_i$*

For each of the datasets, we then have a predicted class label identifying which algorithm *should* perform best on it, according to a decision tree grown on every other dataset instances. We now want to evaluate the efficiency of this example experiment. For that purpose, we would require a criterion *as independent as possible* from all the particular choices made in the experiment. This can be achieved to some extent by the following:

Definition 1. *Let **x** be the actual value of the objective criterion (accuracy) achieved on **dataset$_i$** by the classifier **classifier$_j$** predicted by our experiment. Let **best** be the best value of the objective criterion achieved on **dataset$_i$** among the classifiers **classifier$_{1..m}$**. Let **def** be the actual value of the objective criterion achieved on **dataset$_i$** by the default classifier (majority class classifier). We define the performance of our example meta-mining experiment on **dataset$_i$**:*

$$perf(experiment,\ dataset_i) = 1 - \frac{|best - x|}{|best - def|}$$

This performance criterion is maximal at 1 when the predicted classifier achieves the best accuracy among the studied classifiers, and hits 0 when the predicted classifier achieves the same accuracy as the default classifier. Though simple, this criterion allows to compare the performance of meta-mining experiments solving different meta-problems, but needs to be supplied with a *default value* for the considered base criterion.

3   Dimensions of Study

The previous example details one single experiment, giving insight on the performance of one particular method of addressing a restricted area of the meta-mining problem. In order to gain a meaningful insight, one must explore a more significant domain of both problem and solution. This implies iterating the previous experiment over a number of dimensions illustrated in Fig. 3 by their particular values in the example.

Meta-problem. In the example, we chose to identify meta-mining with a classification problem. This was one of the first stances of meta-learning, and leads to a very simple experiment, but much more efficient formulations exist. We could for instance identify meta-mining with a set of regression problems, modelling the performances of the base classifiers, or a set of classification problems, modelling the applicability of the classifiers [1]. Meta-learning studies introduced many different definitions of the problem [4]. The approach followed in [7] consists in learning a model for each pair of base classifiers, predicting if one will significantly outperform the other on a given dataset. This pairwise vision is also adopted in [21], where particular sets of rules are used to compare the performance of the different base learners. [11] introduces active testing, a strategy minimizing the number of tests necessary to select a good classifier. Growing

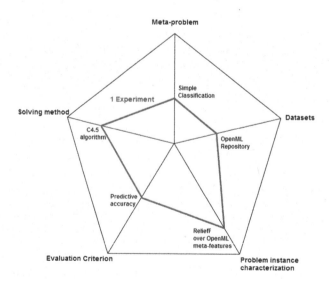

Fig. 3. Dimensions of the example experiment for the classification meta-problem

apart from the meta-learning framework, [14] identifies algorithm selection with a collaborative filtering problem, addressing in particular the problems of stochastic optimisation and boolean satisfiability. [22] tries to address the selection of different data mining operators, considering also the optimisation of their parameters. Finally, [6,15] use the DMOP Ontology to characterize data mining workflows, and learn models exploited in the construction of such workflow for new problem instances.

Repeating the experiment with those more complex definitions of the problem would allow to explore a greater area of the meta-mining problem. Iterating over other dimensions would then provide a sound comparison of those approaches.

Datasets. In order to allow some generality to the results, the datasets used in the experiment should reflect well what *"real world"* datasets are. This is a well know issue in machine learning and meta-learning, where the validation of new techniques requires a *good enough* population of test datasets. Yet, the inherent properties of *real world datasets* remains very unclear. In applications validated over relatively few datasets, it is common to find areas of absolute inefficiency when testing over new datasets. The common assumption is that a *large* sample of datasets provides enough guarantee of generalisability. But once again, *"large"* doesn't mean much, and seems to be often perceived as "larger than last year". In this context the meta-database of OpenML [25] provides a good number of datasets, coming from both the classic literature and from particular applications. To our knowledge, it could be considered one of the most accurate depiction of the set of *real world datasets* available to date, but such matters are difficult to assess and would deserve further studies.

Dataset Characterization. This problem has been addressed along two directions:

- In the first one, the dataset is described through a set of statistical or information theoretic measures. This approach, notably appearing in the STATLOG project [9,13], and in most studies afterwards [7,26], allows the use of many expressive measures. But its performance depends heavily on the adequateness of bias between the meta-level learner and the chosen measures. Experiments have been done with meta-level feature selection [8,23] in order to understand the importance of different measures. But the elicited optimal sets of meta-features to perform algorithm selection over two different pools of algorithms can be very different, revealing no significant tendencies among the measures themselves.
- The second approach to dataset characterization focuses, not on computed properties of the dataset, but on the performance of simple learners over the dataset. It was introduced as landmarking in [17], where the accuracies of a set of very simple learners are used as meta-features to feed a more complex meta-level learner. There again, the performance of the method relies heavily on the adequate choice of both the base and meta-level learner, with no absolute best combination. Further development introduced more complex measures than predictive accuracy over the models generated by the simple learners. For

instance, [16] claims that using as meta-features different structural properties of a decision tree induced over the dataset by simple decision-tree learners can also result in well performing algorithm selection.

OpenML features more than a hundred of such dataset characteristics, getting the most of both approaches. But as the few comparative experiments showed [8,23], the efficiency of a particular characterization mostly depends on its adequation with the problem at hand and the solving method. As some successful approaches focus on developing specific characterization adapted to restricted problems [12], we wish to experiment further by adapting the characterization to the particular set of datasets used in an experiment. In the example, the ReliefF algorithm [19] was used on the metadataset deprived from the ith dataset instance to select best suited dataset characteristics for these particular datasets. But other attribute selection methods exist, that would return potentially different subsets of dataset characteristics. Repeating the experiment with different attribute selection methods or different sets of potential attributes would allow to investigate the relation between the characterization and the other aspects of the problem, while also providing good comparison grounds between the diverse sets of dataset characteristics proposed in the literature.

Solving method. In the example, to predict which method would perform best on each dataset, we used a decision tree built by a particular C4.5 implementation, with a given (default) hyperparameter setting. As a model produced with a different method would possibly be very different, the method employed to solve the chosen meta-problem has to be considered as a dimension of the experiment. For the meta-problem of the example, any method capable of nominal classification could be used. If we also consider different possible hyperparameter settings of those methods, this dimension rapidly grows very large. However, exploring it as well as possible appears critical to the characterization of the different meta-problems. An ideal setup would be to use hyperparameter optimizations techniques on the different solving methods, making hyperparameter optimization a new separate dimension. However, for reasons of dimensional complexity (to be discussed later on), we will for now restrict ourselves to defined hyperparameter settings (hand picked or defaults).

Criterion. The chosen criterion is the measure we wish to enhance through the meta-mining process. In the example, we used predictive accuracy, which is a traditional comparison criterion for classification algorithms. But in another scenario, a different criterion could have made more sense. For instance, in a situation where false negatives are to be avoided in priority (such as in medical diagnosis), a more sensible criterion would have been recall. In practice, one should use a combination of measures to best describe the particular operating conditions of the data-mining experiment to be produced. However, for the sake of generality and simplicity, we will only consider a set of 11 simple measures, such as Cohen's *kappa*, or Kononenko's *Information score* [10].

Since different criteria will likely behave differently, optimal meta-mining processes will likely differ over them. This leads us to iterate experiments over this new dimension of criteria, in order to determine how the previous dimensions of study impact performance for each individual criterion.

Dimensional complexity. In order to get any insights, the space of meta-mining experiments, defined along those dimensions, has to be explored. This means actually realising as many as possible potential experiments we can build along those dimensions. Even with low estimations of the size of the different dimensions, it implies a number of individual experiments in the rough order of magnitude of the billion. As each individual experiment consists in one run of both meta-attributes-selection and a data-mining algorithm, for *each* considered dataset, the exploration of the full dimensional space could span over many years of machine time. However, as each experiment is completely independent from the others, the problem can be addressed through massively parallel computing, which makes exploration possible, even if still time consuming.

4   Experiment Setup

The metadataset is constructed from the OpenML database, which features more than 2700 datasets and 2500 base algorithms. As the construction of the metadataset requires a number of algorithms that were evaluated on the same datasets, we solved a maximal bi-clique problem with an efficient pattern enumerator [24], to find the largest sets of datasets and algorithms such as each element of both sets has been evaluated on every element of the other. This restricted us to 93 algorithms and 434 datasets from the OpenML database. We then extracted the evaluated values of 11 chosen criteria over those 40k runs, and the values of 105 dataset descriptors over each dataset (see *Ressources* section for listings).

In order to run the meta-level experiments, we needed to define a source for the candidate solution algorithms for the different meta-problems, and for the different feature selection algorithm. As it is one of the most widely used and features implementation of many state of the art algorithms, we decided to use the Weka [5] API framework. Spread over 4 classic meta-problems from the literature, we thus evaluate more than 2600 solving methods and 60 feature selection methods built from the Weka API (Fig. 4) (see *Ressources* section for listings).

The individual Weka experiments are then generated by a java program (see *Ressources* section for source code), that delegates their execution to a SLURM job scheduler system managing the *OSIRIM* 640 nodes cluster. The 800k resulting experiments sum up to more than three thousand billion individual executions of machine learning algorithms. Even with good computing power, these experiments are quite costly: 800k experiments of the magnitude order of the minute take almost 100 years of computer time, which reduces to 50 days of parallel execution over the 640 nodes.

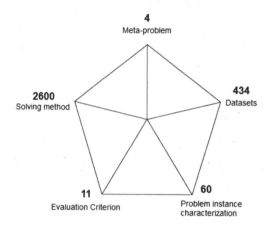

Fig. 4. Rough size of the dimensions

5   Results Interpretation

The setup described in the previous section yields a performance measure for each of those thousands of experiments. In practice, it takes the form of an important database, which can be seen as a population of individual experiment runs, characterized along the earlier described dimensions, and each associated with its performance. In this framework, every question we may aim to answer takes the form of a comparison of sub-populations of those experiment runs. For instance, comparing a new algorithm selection approach to other existing methods can be done by comparing the population of runs featuring the new technique to populations featuring the current state of the art for the problem.

A sheer comparison of mean performance already allows the discovery of tendencies. But to manage the risk of such a tendency not reflecting an actual difference in the sub-populations, one should conduct appropriate hypothesis testing over the results. Hypothesis testing has a reputation to be among the most misused tools in different research areas, and particular care has often been recommended in its application [2]. In this section, we will thus consider the appropriate tests for different situations, and demonstrate them over some cases. Since no assumption can be made regarding the underlying distribution of the performances, we will have to restrict ourselves to non-parametric tests. This implies a lower power than parametric tests could offer, but we will see that the scale of the experiment makes up for this loss.

The first situation we will review is the comparison of two matched sub-population of same size. For instance, let us say we want to compare the performance (in addressing our classifier selection task) of two variants of the Sequential Minimal Optimization algorithm [18] using different kernels. The sub-populations of the runs featuring those two variants have close means (difference

Table 4. Results of some Wilcoxon signed-rank tests

Population 1	Population 2	Difference of means over standard deviation	p-value	Effect size
SMO PolyKernel	SMO Puk	0,02	1,14E-09	0,4
Full set of meta-attributes	50 best from ReliefF	1,2	2,5E-89	0,48
Bagging of RandomForest	RotationForest of RandomForest	1,8	0,91	0,49
RotationForest of RandomForest	RotationForest of NaiveBayesTree	0,04	8,6E-137	0,32

of less than two percent of the population standard deviation), and we can legitimately wonder if there is an actual difference of performance. This situation is ideal for the Wilcoxon signed-rank test (for HO: all performances are identically distributed), which requires independent pairs of values that we can form with the runs of the two variants having all other dimensions equal. Wilcoxon's last assumption of ordinal measure is also met by our numeric performance criterion. The test results in a p-value of 10^{-9} with a 0.4 effect size. This implies that the observed difference between the two variants have a negligible chance not to represent an actual performance difference. Table 4 shows some results that can be obtained with such tests. To interpret these other examples, we could say that the selection of the 50 best meta-attributes with ReliefF will often result in a loss of performance on the whole studied meta-problems, relatively to keeping the full set. The difference of performance between Bagging and RotationForest as enhancer of RandomForest has little chance of betraying a general tendency on the whole studied meta-problems, while the way smaller gap between Random-Forest and NaiveBayesTree in a RotationForest has great chances of denoting an actual difference.

Another possibility would be to compare n matched populations of identical size. This is perhaps the most interesting setting, but also the most complex to study, and will feature the use of the Friedman test, with the same assumptions as the Wilcoxon signed-rank test. As an example, let us compare different kNN classifiers with varying k parameter and distance used, (as shown in Table 5), for use in ensembles of nested dichotomies (END) [3] addressing the classifier selection problem. Those approaches have very close mean performance, differing from one another by less than a percent of the global standard deviation. Yet the Friedman test concludes with a p-value of 10^{-90} that some are significantly different from the others.

Finding which one differ for the better will necessitate post-hoc tests, such as the Nemenyi test, in order to control the family-wise error rate (risk of making

Table 5. Mean performance of END built on different kNN classifiers

k	Distance	Mean performance
5	Manhattan	0,8828
5	Euclidean	0,8853
10	Euclidean	0,8899
20	Euclidean	0,8872
20	Manhattan	0,8832

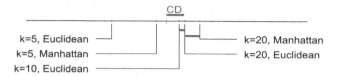

Fig. 5. Results of Nemenyi test, connected groups are not significantly different

at least one incorrect rejection among multiple hypotheses tests). This is often a real problem, as control of the family-wise error rate imposes much more conservative significance levels of the individual tests. But the scale of our experiment allows the extreme values it requires. Results are presented in Fig. 5, where the critical difference is adjusted following Nemenyi procedure to account for the 10 comparison being made (0.05 family-wise error rate). We can see that only the $k = 10$, Euclidian and $k = 20$, Manhattan variants cannot be considered different from the $k = 20$, Euclidian at this significance level, while all others differ by more than the critical difference. In particular, the $k = 10$, Euclidian and $k = 20$, Manhattan variants *are* significantly different from one another.

Running the Friedman test on a much larger number of populations also allows to draw interesting results. For instance, let us compare the different attribute selection methods used. We extracted from the results the performance of the runs featuring the attribute selection methods in identical setups, for over a million setups. This represents more than 50 million performance values to be ranked by Friedman's procedure. The test returns a p-value of zero (beyond machine precision), ascertaining the existence of differences among the methods performance. A Nemenyi test comparing every one of those attribute selection methods with all the others requires an important number of comparison setups in order to be able to find any significant differences. Figure 6 shows that even on a sample 100k setups, the test allows to build groups of methods of *equivalent* performance.

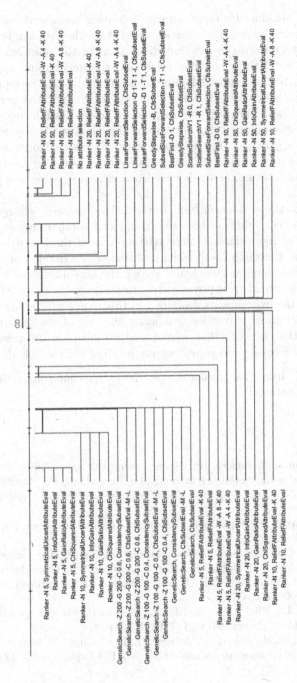

Fig. 6. Overview of a Nemenyi test over all the attribute selection methods

6   Conclusion

In this paper we introduced a meta-mining evaluation framework relying on a unified performance criterion, and demonstrated it on the problem of classifier selection. We characterized the different dimensions of the solutions, instantiated a large number of those classifier selection experiments, and applied statistical hypothesis testing methodologies to the results. These test procedures allowed to draw precise statistical results regarding the comparative performance of different approaches. They are able to produce general insight regarding the optimization potential of particular dimensions. This last result may reveal very interesting, as it can suggest that a second layer of (meta) algorithm selection could maximize the performance of the first. Such result meets the insights of [26] in the suggestion of a recursion of adaptive learners as a possible new paradigm of meta-learning.

Coming back to the evaluation framework, different aspects will require further work, such as the addition of a dimension of hyperparameter optimization, and the use of more dataset meta-attributes from the literature. To our best knowledge, no thorough comparison and review of existing dataset meta-attributes is available, and as they figure among the dimensions our framework allows to study, we intend to apply ourselves to such comparisons. The knowledge gained from the experiments described in this paper will be invaluable to that end, as it will allow to reduce drastically the size of the dimensions, by considering only the elements that were found significantly different. Similar approaches could be considered regarding any or all of the possible dimensions.

Ressources. All materials available at: https://github.com/WilliamR03/Meta-Mining-Evaluation.

References

1. Brazdil, P., Gama, J., Henery, B.: Characterizing the applicability of classification algorithms using meta-level learning. In: Bergadano, F., Raedt, L. (eds.) ECML 1994. LNCS, vol. 784, pp. 83–102. Springer, Heidelberg (1994). doi:10.1007/3-540-57868-4_52
2. Demšar, J.: Statistical comparisons of classifiers over multiple data sets. J. Mach. Learn. Res. **7**, 1–30 (2006)
3. Dong, L., Frank, E., Kramer, S.: Ensembles of balanced nested dichotomies for multi-class problems. In: Jorge, A.M., Torgo, L., Brazdil, P., Camacho, R., Gama, J. (eds.) PKDD 2005. LNCS (LNAI), vol. 3721, pp. 84–95. Springer, Heidelberg (2005). doi:10.1007/11564126_13
4. Giraud-Carrier, C., Vilalta, R., Brazdil, P.: Introduction to the special issue on meta-learning. Mach. Learn. **54**(3), 187–193 (2004)
5. Hall, M., Frank, E., Holmes, G., Pfahringer, B., Reutemann, P., Witten, I.H.: The weka data mining software: an update. ACM SIGKDD Explor. Newsl. **11**(1), 10–18 (2009)

6. Hilario, M., Nguyen, P., Do, H., Woznica, A., Kalousis, A.: Ontology-based meta-mining of knowledge discovery workflows. In: Jankowski, N., Duch, W., Grąbczewski, K. (eds.) Meta-Learning in Computational Intelligence. Studies in Computational Intelligence, vol. 358, pp. 273–315. Springer, Heidelberg (2011)

7. Kalousis, A., Hilario, M.: Model selection via meta-learning: a comparative study. Int. J. Artif. Intell. Tools 10(04), 525–554 (2001)

8. Kalousis, A., Hilario, M.: Feature selection for meta-learning. In: Cheung, D., Williams, G.J., Li, Q. (eds.) PAKDD 2001. LNCS (LNAI), vol. 2035, pp. 222–233. Springer, Heidelberg (2001). doi:10.1007/3-540-45357-1_26

9. King, R.D., Feng, C., Sutherland, A.: StatLog: comparison of classification algorithms on large real-world problems. Int. J. Appl. Artif. Intell. 9(3), 289–333 (1995)

10. Kononenko, I., Bratko, I.: Information-based evaluation criterion for classifier's performance. Mach. Learn. 6(1), 67–80 (1991)

11. Leite, R., Brazdil, P., Vanschoren, J.: Selecting classification algorithms with active testing. In: Perner, P. (ed.) MLDM 2012. LNCS (LNAI), vol. 7376, pp. 117–131. Springer, Heidelberg (2012). doi:10.1007/978-3-642-31537-4_10

12. Leyva, E., Gonzalez, A., Perez, R.: A set of complexity measures designed for applying meta-learning to instance selection. IEEE Trans. Knowl. Data Eng. 27(2), 354–367 (2015)

13. Michie, D., Spiegelhalter, D.J., Taylor, C.C.: Machine Learning, Neural and Statistical Classification. Ellis Horwood, Upper Saddle River (1994)

14. Misir, M., Sebag, M.: Algorithm selection as a collaborative filtering problem, p. 43 (2013). hal-00922840

15. Nguyen, P., Hilario, M., Kalousis, A.: Using meta-mining to support data mining workflow planning and optimization. J. Artif. Intell. Res. 51, 605–644 (2014)

16. Peng, Y., Flach, P.A., Brazdil, P., Soares, C.: Decision tree-based data characterization for meta-learning. In: IDDM-2002 p. 111 (2002)

17. Pfahringer, B., Bensusan, H., Giraud-Carrier, C.: Tell me who can learn you and i can tell you who you are: landmarking various learning algorithms. In: Proceedings of the 17th International Conference on Machine Learning, pp. 743–750 (2000)

18. Platt, J., et al.: Sequential minimal optimization: A fast algorithm for training support vector machines (1998)

19. Robnik-Šikonja, M., Kononenko, I.: Theoretical and empirical analysis of relieff and rrelieff. Mach. Learn. 53(1–2), 23–69 (2003)

20. Serban, F., Vanschoren, J., Kietz, J.U., Bernstein, A.: A survey of intelligent assistants for data analysis. ACM Comput. Surv. (CSUR) 45(3), 31 (2013)

21. Sun, Q., Pfahringer, B.: Pairwise meta-rules for better meta-learning-based algorithm ranking. Mach. Learn. 93(1), 141–161 (2013)

22. Sun, Q., Pfahringer, B., Mayo, M.: Full model selection in the space of data mining operators. In: Proceedings of the 14th Annual Conference Companion on Genetic and Evolutionary Computation, pp. 1503–1504. ACM (2012)

23. Todorovski, L., Brazdil, P., Soares, C.: Report on the experiments with feature selection in meta-level learning. In: Proceedings of the PKDD 2000 Workshop on Data Mining, Decision Support, Meta-learning and ILP: Forum For Practical Problem Presentation and Prospective Solutions, pp. 27–39. Citeseer (2000)

24. Uno, T., Asai, T., Uchida, Y., Arimura, H.: An efficient algorithm for enumerating closed patterns in transaction databases. In: Suzuki, E., Arikawa, S. (eds.) DS 2004. LNCS (LNAI), vol. 3245, pp. 16–31. Springer, Heidelberg (2004). doi:10.1007/978-3-540-30214-8_2

25. Vanschoren, J., van Rijn, J.N., Bischl, B., Torgo, L.: OpenML: net-worked science in machine learning. SIGKDD Explor. **15**(2), 49–60 (2013). http://doi.acm.org/10.1145/2641190.2641198
26. Vilalta, R., Drissi, Y.: A perspective view and survey of meta-learning. Artif. Intell. Rev. **18**(2), 77–95 (2002). http://dx.doi.org/10.1023/A:1019956318069
27. Xu, L., Hutter, F., Shen, J., Hoos, H.H., Leyton-Brown, K.: SATzilla2012: improved algorithm selection based on cost-sensitive classification models, pp. 57–58 (2012)
28. Zakova, M., Kremen, P., Zelezny, F., Lavrac, N.: Automating knowledge discovery workflow composition through ontology-based planning. IEEE Trans. Autom. Sci. Eng. **8**(2), 253–264 (2011)

Bayesian Grouped Horseshoe Regression with Application to Additive Models

Zemei Xu[1,2]([✉]), Daniel F. Schmidt[1], Enes Makalic[1], Guoqi Qian[2], and John L. Hopper[1]

[1] Centre for Epidemiology and Biostatistics, Melbourne School of Population and Global Health, The University of Melbourne, Parkville, VIC 3010, Australia
zemeix@student.unimelb.edu.au,
{dschmidt,emakalic,j.hopper}@unimelb.edu.au
[2] School of Mathematics and Statistics, The University of Melbourne, Parkville, VIC 3010, Australia
qguoqi@unimelb.edu.au

Abstract. The Bayesian horseshoe estimator is known for its robustness when handling noisy and sparse big data problems. This paper presents two extensions of the regular Bayesian horseshoe: (i) the grouped Bayesian horseshoe and (ii) the hierarchical Bayesian grouped horseshoe. The advantages of the proposed methods are their flexibility in handling grouped variables through extra shrinkage parameters at the group and within-group levels. We apply the proposed methods to the important class of additive models where group structures naturally exist, and we demonstrate that the grouped hierarchical Bayesian horseshoe has promising performance on both simulated and real data.

Keywords: Bayesian regression · Grouped variables · Horseshoe · Additive models

1 Introduction

Statistical variable selection, also known as feature selection, has become an indispensable tool in many research areas involving machine learning and data mining. The object is to select the best subset of predictors for fitting or predicting the response variable from a potentially large collection of candidate predictors. It is particularly important in high-dimensional problems, where there are potentially millions of predictors and only a few are associated with the outcome.

Consider the linear regression model:

$$y = X\beta + \epsilon, \tag{1}$$

where y is an n by 1 observation vector of the response variable, X is an n by p observation or design matrix of the regressors or predictors, $\beta = (\beta_1, \cdots, \beta_p)^T$ is a p by 1 vector of regression coefficients to be estimated, and ϵ is an n by 1 vector of i.i.d. $\mathcal{N}(0, \sigma^2)$ random errors with σ^2 unknown. Here, the vector β

© Springer International Publishing AG 2016
B.H. Kang and Q. Bai (Eds.): AI 2016, LNAI 9992, pp. 229–240, 2016.
DOI: 10.1007/978-3-319-50127-7_19

is assumed to be sparse in the sense that most of its components equal zero. Therefore, dimensionality reduction is necessary, especially for large-p problems.

Recent approaches for variable selection in ultra-high dimensions are based on penalised likelihood methods, which select a model by minimising a loss function that is usually proportional to the negative log likelihood plus a penalty term:

$$\hat{\beta} = \arg\min_{\beta \in \mathbb{R}^p} \left\{ (\mathbf{y} - \mathbf{X}\beta)^T (\mathbf{y} - \mathbf{X}\beta) + \lambda \cdot q(\beta) \right\}, \tag{2}$$

where $\lambda > 0$ is the regularisation parameter and $q(\cdot)$ is a penalty function. Many penalised regression approaches have been proposed in the literature, such as the non-negative garrote [3], the ridge regression [8], the adaptive lasso [14], the elastic net [15], and the smoothly clipped absolute deviation [7]. One of the most widely used penalised approaches is the lasso [11], which shrinks coefficients while setting other coefficients to zero, effectively producing a model that is simpler to interpret. However, penalised regression methods, such as the lasso, are designed for selecting individual explanatory variables.

1.1 Grouped Variables

Group structures naturally exist in predictor variables, and in this situation, groups of variables should be selected rather than individual variables. For example, a multi-level categorical predictor in a regression model can be represented by a group of dummy variables, a continuous predictor in an additive model can be expressed as a composition of basis functions, and prior knowledge such as genes in the same biological pathway can be used to form a natural group.

To find sparse solutions at the group level, several selection methods have been proposed. These include the group lasso [12], the group lasso for logistic regression [4], group selection with general composite absolute penalty [13], the group bridge method [9], and the bi-level selection in generalised linear models [2].

1.2 Bayesian Regression

An important class of alternative variable selection methods are the Bayesian penalised regression approaches. These are motivated by the fact that a good solution for β in (1) can be interpreted as the posterior mode of β in the Bayesian model when β follows a certain prior distribution. For example, the assumption of a Laplace prior distribution for β leads to the Bayesian interpretation of the lasso [11].

A natural way of estimating β in the Bayesian approaches is generating sparsity through a *spike and slab* prior for each element of β, with the spike concentrating near zero and the slab spreading away from zero:

$$\beta_j | \gamma_j \sim (1 - \gamma_j) \mathcal{N}(0, \tau_j^2) + \gamma_j \mathcal{N}(0, c_j \tau_j^2), \ j = 1, \cdots, p, \tag{3}$$

where γ_j are binary variables, τ_j^2 is small and $c_j > 0$ is large. A fully Bayesian approach with a spike and slab mixture for each β requires exploration of a model space of size 2^p, which becomes difficult when p is large.

Another way of estimating β in the Bayesian approaches is via the shrinkage priors, such as the Bayesian lasso. A competitive Bayesian alternative to the lasso is the Bayesian model resulting from the horseshoe prior, which is a one-component prior [5]. The horseshoe arises from the same class of multivariate scale mixtures of normals as the lasso does, but it is almost always superior to the double-exponential prior at handling sparsity [6]. Furthermore, the horseshoe prior is known for its robustness at handling large outlying signals and the estimator of the horseshoe model does not face the computational issues of the point mass mixture models. To date, grouped variable selection methods based on the Bayesian horseshoe (BHS) model have not been analysed in the literature.

In this paper, we extend the BHS model to handle grouped variables by introducing shrinkage parameters at the group level as well as within each group. The proposed methods enjoy the advantages of handling sparsity as well as strong signals, especially when there is the potential for grouped structures in the data. We apply the proposed methods to the important class of additive models where group structures naturally exist. Both simulated and real data experiments demonstrate the promising performance of the proposed methods.

2   Bayesian Grouped Horseshoe Models

In this section, the BHS model is briefly introduced and two extensions for grouped variables are proposed.

2.1   Bayesian Horseshoe Model

The horseshoe prior assumes that β_j are conditionally independent and each has a density function that can be represented as a scale mixture of normals. The horseshoe prior leaves strong signals unshrunk and penalises noise variables severely. Therefore, the horseshoe prior has the ability to adapt to different sparsity patterns, while simultaneously avoiding over-shrinkage of large coefficients.

Without loss of generality, $\mathbf{y}$ and $\mathbf{X}$ are assumed to be standardised for all models. The response $\mathbf{y}$ is centered, and the covariates $\mathbf{X}$ are column-standardised to have mean zero and unit length. The BHS estimator is defined as follows:

$$\mathbf{y}|\mathbf{X}, \boldsymbol{\beta}, \sigma^2 \sim \mathcal{N}(\mathbf{X}\boldsymbol{\beta}, \sigma^2\mathbf{I}_n)$$
$$\boldsymbol{\beta}|\sigma^2, \tau^2, \delta_1, \cdots, \delta_p \sim \mathcal{N}(\mathbf{0}, \sigma^2\tau^2\mathbf{D}_{\boldsymbol{\delta}}), \quad \mathbf{D}_{\boldsymbol{\delta}} = \mathrm{diag}(\delta_1^2, \cdots, \delta_p^2)$$
$$\delta_j \sim C^+(0, 1), \; j = 1, \cdots, p \tag{4}$$
$$\tau \sim C^+(0, 1)$$
$$\sigma^2 \sim \frac{1}{\sigma^2}d\sigma^2,$$

where $C^+(0,1)$ is a standard half-Cauchy distribution with the probability density function:

$$f(x) = \frac{2}{\pi(1+x^2)}, \ x > 0. \tag{5}$$

The scale parameters δ_j are local shrinkage parameters and τ is the global shrinkage parameter. A simple sampler proposed for the BHS hierarchy [10] enables straightforward sampling of the full conditional posterior distributions.

2.2 Bayesian Grouped Horseshoe Model

The original BHS model does not allow for grouped structures. Therefore, the Bayesian grouped horseshoe (BGHS) model is proposed. Suppose there are $G \in \{1, \cdots, p\}$ groups of predictors in the data and the gth group has size s_g, where $g = 1, \cdots, G$ (i.e. there are s_g variables in group g). The horseshoe hierarchical representation of the full model for grouped variables can be constructed as:

$$
\begin{aligned}
\mathbf{y}|\mathbf{X}, \boldsymbol{\beta}, \sigma^2 &\sim \mathcal{N}(\mathbf{X}\boldsymbol{\beta}, \sigma^2 \mathbf{I}_n) \\
\boldsymbol{\beta}|\sigma^2, \tau^2, \lambda_1, \cdots, \lambda_G &\sim \mathcal{N}(\mathbf{0}, \sigma^2 \tau^2 \mathbf{D}_{\boldsymbol{\lambda}}), \quad \mathbf{D}_{\boldsymbol{\lambda}} = \mathrm{diag}(\lambda_1^2 \mathbf{I}_{s_1}, \cdots, \lambda_G^2 \mathbf{I}_{s_G}) \\
\lambda_g &\sim C^+(0,1), \ g = 1, \cdots, G \\
\tau &\sim C^+(0,1) \\
\sigma^2 &\sim \frac{1}{\sigma^2} d\sigma^2,
\end{aligned}
\tag{6}
$$

where λ_g are the shrinkage parameters at group level. Instead of having shrinkage parameters for all predictors, we have shrinkage parameters λ_g for each group. Hence, the model either shrinks all variables in the same group towards zero or leaves them untouched.

2.3 Hierarchical Bayesian Grouped Horseshoe Model

By combining the original BHS model with the BGHS model, we develop a hierarchical Bayesian grouped horseshoe (HBGHS) model that does selection and shrinkage at the group level as well as within groups.

The total number of groups G is assumed to be greater than one because $G = 1$ implies that all predictors are in the same group and results in the regular BHS model. The full HBGHS model is:

$$
\begin{aligned}
\mathbf{y}|\mathbf{X}, \boldsymbol{\beta}, \sigma^2 &\sim \mathcal{N}(\mathbf{X}\boldsymbol{\beta}, \sigma^2 \mathbf{I}_n) \\
\boldsymbol{\beta}|\sigma^2, \tau^2, \lambda_1, \cdots, \lambda_G, \delta_1, \cdots, \delta_p &\sim \mathcal{N}(\mathbf{0}, \sigma^2 \tau^2 \mathbf{D}_{\boldsymbol{\lambda}} \mathbf{D}_{\boldsymbol{\delta}}) \\
\mathbf{D}_{\boldsymbol{\lambda}} = \mathrm{diag}(\lambda_1^2 \mathbf{I}_{s_1}, \cdots, \lambda_G^2 \mathbf{I}_{s_G}), \quad \mathbf{D}_{\boldsymbol{\delta}} &= \mathrm{diag}(\lambda_1^2 \mathbf{I}_{s_1}, \cdots, \lambda_G^2 \mathbf{I}_{s_G}) \\
\lambda_g &\sim C^+(0,1), \ g = 1, \cdots, G \\
\delta_j &\sim C^+(0,1), \ j = 1, \cdots, p \\
\tau &\sim C^+(0,1) \\
\sigma^2 &\sim \frac{1}{\sigma^2} d\sigma^2,
\end{aligned}
\tag{7}
$$

where $\delta_1, \cdots, \delta_p$ are the shrinkage parameters for each predictor variable and $\lambda_1, \cdots, \lambda_G$ are the shrinkage parameters for the group variables. Therefore, the model has a global shrinkage parameter τ, and local shrinkage parameters λ and δ, which control shrinkage at the group level and within group levels, respectively. The full conditional distributions required to sample from (7) for the HBGHS can be found in the Appendix.

3 Additive Models

Nonparametric regression methods, such as the additive model, are widely used in statistics. In an additive model, each predictor can be expressed as a set of basis functions that form a group structure. Given a data set $\{y_i, x_{i1}, \cdots, x_{ip}\}_{i=1}^n$, the additive model has the form:

$$y = \mu_0 + \sum_{j=1}^p f_j(X_j) + \epsilon, \tag{8}$$

where μ_0 is an intercept term and $f_j(\cdot)$ are unknown smooth functions. In the ideal situation, estimation of the selected smooth function is expected to be as close to the corresponding true underlying functions or target functions as possible.

There are various classes of basis functions that can be used to approximate the target functions. The basis functions include polynomials, spline functions, etc. Let $g_j(x), j = 1, \cdots, p$, be a set of basis functions. Each smooth function component in the additive model can be represented as:

$$f(x) = a_0 + a_1 g_1(x) + a_2 g_2(x) + \cdots + a_p g_p(x). \tag{9}$$

One limitation of the regular polynomial basis functions (i.e. $g_j(x) = x^{j-1}$) is that there is potentially a correlation between the data generated by the basis functions. As a result, orthogonal polynomials are widely used. In this paper, one of the orthogonal polynomial groups, the Legendre polynomials, are used for all polynomial expansions. The Legendre polynomials are defined on the interval $[-1, 1]$. Each Legendre polynomial $g_j(x)$ is a pth-degree polynomial and can be expressed as:

$$g_j(x) = \frac{1}{2^j j!} \frac{d^j}{dx^j} \left[(x^2 - 1)^j\right], \ p = 0, 1, \cdots. \tag{10}$$

The additive models allow for nonlinear effects and grouped structures, and therefore, form perfect test functions for our proposed methods.

4 Simulation Studies

In this section, we compared the prediction performance of four methods: (i) the regular BHS, (ii) the HBHS, (iii) lasso with regularisation parameter selected by the Bayesian information criterion (lasso-BIC) and (iv) the regular Bayesian horseshoe without expansion of the predictors (BHS-NE). The performances of the four methods were compared on three test functions (see Sect. 4.2).

To compare the performances of the four methods, we generated $\mathbf{X}$ with a large sample size n from a uniform distribution $U(-1,1)$ and computed the mean squared prediction error (MSPE) as the comparison metric:

$$\frac{1}{n}\sum_{i=1}^{n}[\mathrm{E}(y_i|\mathbf{x}_i)-\hat{y}_i]^2,\tag{11}$$

where y_i are the responses of the test functions and $\hat{y}_i$ are the fitted values based on the estimates from the four methods.

4.1   Simulation Procedures

For simulated data, we generated $t = 100$ data sets including $p = 10$ candidate predictors $\mathbf{X}_{n\times p}$ from the uniform distribution $U(-1,1)$. For each realisation, we calculated $\mathbf{y}$ based on the true test functions. Then, we expanded $\mathbf{X}$ using Legendre polynomials with degree K. Three methods, BHS, HBGHS, and lasso-BIC, were applied to 100 samples of expanded data $\tilde{\mathbf{X}}$. The benchmark method, BHS-NE, was applied to non-expanded covariates $\mathbf{X}$. We obtained t posterior samples $\boldsymbol{\beta}$ for each method and computed the posterior medians as the point estimates of each method.

4.2   Test Functions

Simulations were performed on data generated from the three true test functions shown below. The first true function is linear, the second is non-linear and the third consists of Legendre polynomials.

1. Function 1 (simple linear function):

$$y = X_1 + X_2 - X_3 - X_4.\tag{12}$$

2. Function 2 (nonlinear function):

$$y = \cos(8X_1) + X_2^2 + \mathrm{sign}(X_3) + |X_4| + X_5 + X_5^2 - X_5^3.\tag{13}$$

3. Function 3:

$$y = f_1(X_1) + f_2(X_2) + f_3(X_3),\tag{14}$$

where $f_j = \beta_{j1}P_1(X_j) + \beta_{j2}P_2(X_j) + \beta_{j3}P_3(X_j)$, $j = 1,2,3$ that consists of the Legendre polynomials of order up to three and the standardised true coefficients are: $\boldsymbol{\beta} = (2,1,1/2,1,1,1,-1,-4,1)'_{9\times 1}$.

Function 1 is a linear function where BHS-NE is expected to benefit the most. We expect all methods except BHS-NE to do well in Function 2, where there is a highly nonlinear structure in the model. In Function 3, we expect HBGHS to perform well because the true functions are built from Legendre basis functions.

To generate three test functions, we first generated $\mathbf{X}_{n \times m} = (\mathbf{X}_1, \cdots, \mathbf{X}_m)$ with a large sample size n from $U(-1, 1)$, where m is the number of non-zero components. We then scaled each component to ensure that all components have the same variance and contribute equally to the final function. The variance of the noise σ^2 can be computed to achieve a desired signal-to-noise ratio.

For all three test functions, we generated $p = 10$ predictors and varied the number of samples $n = \{100, 200\}$, the signal-to-noise ratio SNR= $\{1, 5, 10\}$ and the maximum degree of Legendre polynomial expansions $K = \{3, 6, 9, 12\}$.

4.3   Simulation Results

The BGHS results are not presented because they performed uniformly worse than the BHS and HBGHS in all experiments.

The average mean squared prediction error and the corresponding standard deviation for Function 1 are shown in Table 1. In the simple linear test, the BHS model without expansion of covariates unsurprisingly consistently produced the smallest MSPE values. Of the remaining three methods, HBGHS and BHS were competitive for the sample size $n = 100$. As the sample size grew, HBGHS showed significant improvement in terms of MSPE compared to BHS, and the MSPE of HBGHS was smaller in most cases when $n = 200$. The prediction performance of BHS, HBGHS and lasso-BIC dropped as the degrees of expansion increased.

For Function 2, where there is a highly nonlinear relationship in the model, the HBGHS improved performance in almost all scenarios, as shown in Table 2. The only scenario where the BHS slightly outperformed the HBGHS was when $n = 100$ and the signal-to-noise ratio was low (SNR $= 1$). The BHS-NE performed poorly in this case because it is unable to capture nonlinear effects.

For Function 3, the HBGHS was expected to achieve good performance because the true underlying additive model consists of a set of polynomial expansions. Indeed, referring to Table 3, the HBGHS gave the smallest mean MSPE for all scenarios. The performance of the BHS is better than the lasso-BIC, and the BHS-NE unsurprisingly performed poorly for these polynomial test functions.

In general, the HBGHS method outperforms the regular BHS. As an illustration, boxplots of component-wise mean squared prediction errors for the BHS and the HBGHS methods when $p = 10$, $n = 100$, SNR $= 5$, $K = 3$ of Function 3 are shown in Fig. 1. The first three components are associated with the response variable and the rest are noise variables. From the figure, we see that the HBGHS penalises noise variables more than the BHS does. When all the variables within a group have small effects, the HBGHS tends to apply extra shrinkage to the whole group and shrink them together.

Table 1. Mean and standard deviation (in parentheses) of squared prediction error for BHS, HBGHS, BHS-NE and lasso-BIC of Function 1, when sample size $n = \{100, 200\}$, signal-to-noise-ratio SNR $= \{1, 5, 10\}$, and the highest degree of Legendre polynomial expansions $K = \{3, 6, 9, 12\}$.

n	SNR	Degree	BHS	HBGHS	BHS-NE	lasso-BIC
100	1	3	0.1137(0.0786)	0.1107(0.0861)	0.0856(0.0513)	0.1941(0.1090)
		6	0.1258(0.0918)	0.1322(0.1034)	0.0855(0.0520)	0.2543(0.1312)
		9	0.1536(0.1274)	0.1635(0.1423)	0.0852(0.0511)	0.5293(2.3280)
		12	0.2247(0.3075)	0.2200(0.2793)	0.0857(0.0519)	17.810(18.670)
	5	3	0.0178(0.0125)	0.0173(0.0125)	0.0154(0.0086)	0.0375(0.0190)
		6	0.0173(0.0115)	0.0176(0.0123)	0.0153(0.0088)	0.0491(0.0225)
		9	0.0194(0.0144)	0.0196(0.0143)	0.0153(0.0086)	0.1029(0.4565)
		12	0.0377(0.0797)	0.0346(0.0759)	0.0154(0.0087)	3.1820(3.6550)
	10	3	0.0088(0.0062)	0.0086(0.0062)	0.0077(0.0043)	0.0185(0.0090)
		6	0.0085(0.0056)	0.0087(0.0060)	0.0077(0.0044)	0.0251(0.0113)
		9	0.0095(0.0070)	0.0096(0.0070)	0.0077(0.0043)	0.0516(0.2269)
		12	0.0186(0.0394)	0.0171(0.0380)	0.0077(0.0044)	1.4160(1.7350)
200	1	3	0.0442(0.0258)	0.0422(0.0256)	0.0378(0.0200)	0.0954(0.0461)
		6	0.0461(0.0293)	0.0468(0.0294)	0.0377(0.0198)	0.1224(0.0457)
		9	0.0473(0.0316)	0.0467(0.0318)	0.0378(0.0202)	0.1354(0.0516)
		12	0.0497(0.0352)	0.0479(0.0317)	0.0376(0.0202)	0.1487(0.0544)
	5	3	0.0084(0.0051)	0.0082(0.0052)	0.0076(0.0040)	0.0194(0.0100)
		6	0.0087(0.0058)	0.0089(0.0059)	0.0076(0.0039)	0.0242(0.0100)
		9	0.0089(0.0062)	0.0088(0.0062)	0.0076(0.0040)	0.0269(0.0100)
		12	0.0094(0.0071)	0.0091(0.0063)	0.0076(0.0040)	0.0297(0.0109)
	10	3	0.0042(0.0026)	0.0041(0.0026)	0.0038(0.0020)	0.0094(0.0047)
		6	0.0043(0.0030)	0.0044(0.0030)	0.0038(0.0020)	0.0120(0.0044)
		9	0.0044(0.0031)	0.0044(0.0031)	0.0039(0.0020)	0.0133(0.0053)
		12	0.0047(0.0037)	0.0045(0.0032)	0.0038(0.0020)	0.0148(0.0056)

5   Real Data

To evaluate the performance of the BGHS method and the HBGHS method, we applied them to the Electrical-Maintenance data set [1]. There are $n = 1056$ samples and $p = 4$ input variables in the Electrical-Maintenance data set. All variables are continuous.

To perform the real data analysis, we used hold-out validation methods by dividing data into two subsets, where the training data contains 75% of the samples and the testing data contains 25% of the samples.

All predictors were expanded using Legendre polynomials with degrees $K = \{2, 4, 6, 8, 10\}$ for each of the following methods: BHS, BGHS, HBGHS and lasso-

Table 2. Mean and standard deviation (in parentheses) of squared prediction error for BHS, HBGHS, BHS-NE and lasso-BIC of Function 2, when sample size $n = \{100, 200\}$, signal-to-noise-ratio SNR $= \{1, 5, 10\}$, and the highest degree of Legendre polynomial expansions $K = \{3, 6, 9, 12\}$.

n	SNR	Degree	BHS	HBGHS	BHS-NE	lasso-BIC
100	1	3	0.5008(0.1204)	0.5122(0.1357)	0.9182(0.0756)	0.6512(0.2203)
		6	0.4968(0.1518)	0.5279(0.1721)	0.9208(0.0771)	0.6850(0.2236)
		9	0.5524(0.1757)	0.5816(0.1877)	0.9192(0.0757)	0.9774(1.8780)
		12	0.6591(0.2845)	0.6874(0.3629)	0.9199(0.0760)	18.150(20.220)
	5	3	0.3055(0.0382)	0.2947(0.0360)	0.8579(0.0488)	0.3523(0.0682)
		6	0.1728(0.0354)	0.1578(0.0321)	0.8593(0.0505)	0.2345(0.0656)
		9	0.1357(0.0495)	0.1246(0.0669)	0.8582(0.0497)	0.5216(0.9430)
		12	0.1882(0.1893)	0.1819(0.2279)	0.8582(0.0492)	3.4360(4.0510)
	10	3	0.2823(0.0293)	0.2739(0.0271)	0.8490(0.0415)	0.3189(0.0561)
		6	0.1363(0.0234)	0.1245(0.0206)	0.8501(0.0431)	0.1781(0.0430)
		9	0.0859(0.0381)	0.0792(0.0475)	0.8491(0.0423)	0.2571(0.4638)
		12	0.1287(0.1521)	0.1353(0.2316)	0.8488(0.0414)	1.6830(2.0180)
200	1	3	0.3306(0.0491)	0.3271(0.0485)	0.8452(0.0379)	0.3985(0.0765)
		6	0.2330(0.0560)	0.2220(0.0574)	0.8450(0.0372)	0.3417(0.1006)
		9	0.2125(0.0646)	0.1973(0.0661)	0.8453(0.0372)	0.3517(0.1147)
		12	0.2329(0.0783)	0.2185(0.0740)	0.8449(0.0372)	0.3820(0.1175)
	5	3	0.2545(0.0188)	0.2500(0.0179)	0.8170(0.0023)	0.2795(0.0275)
		6	0.1098(0.0154)	0.1051(0.0150)	0.8168(0.0231)	0.1464(0.0269)
		9	0.0602(0.0128)	0.0551(0.0133)	0.8172(0.0233)	0.1034(0.0289)
		12	0.0639(0.0178)	0.0580(0.0152)	0.8168(0.0231)	0.1168(0.0313)
	10	3	0.2442(0.0135)	0.2404(0.0128)	0.8136(0.0216)	0.2623(0.0187)
		6	0.0937(0.0109)	0.0906(0.0104)	0.8134(0.0215)	0.1157(0.0167)
		9	0.0406(0.0076)	0.0374(0.0085)	0.8138(0.0217)	0.0667(0.0150)
		12	0.0413(0.0104)	0.0373(0.0093)	0.8134(0.0215)	0.0721(0.0182)

BIC. We also tested the BHS-NE with original non-expanded predictors as the benchmark.

The mean and standard deviation (in parentheses) of mean squared prediction errors are shown in Table 4. We see that there clearly exists a nonlinear pattern in the data because the mean squared prediction error decreases as the degree of expansions increases. On average, the HBGHS has the lowest MSPE, especially when the degree of polynomials grows. The BGHS has the smallest MSPE when each group has few variables ($K = 2$).

In conclusion, we have proposed the BGHS method and the HBGHS method for performing both group-wise and within group selection. We have shown the good performance of the HBGHS model in terms of the mean squared prediction

Table 3. Mean and standard deviation (in parentheses) of squared prediction error for BHS, HBGHS, BHS-NE and lasso-BIC of Function 3, when sample size $n = \{100, 200\}$, signal-to-noise-ratio SNR $= \{1, 5, 10\}$, and the highest degree of Legendre polynomial expansions $K = \{3, 6, 9, 12\}$.

n	SNR	Degree	BHS	HBGHS	BHS-NE	lasso-BIC
100	1	3	0.1564(0.0705)	0.1336(0.0699)	0.5514(0.0678)	0.2398(0.1146)
		6	0.1809(0.0789)	0.1645(0.0824)	0.5521(0.0677)	0.3128(0.1434)
		9	0.2142(0.0989)	0.2019(0.1079)	0.5514(0.0669)	0.6967(2.5440)
		12	0.2876(0.3553)	0.2808(0.4541)	0.5520(0.0670)	18.090(20.070)
	5	3	0.0329(0.0158)	0.0262(0.0141)	0.4951(0.0260)	0.0555(0.0247)
		6	0.0366(0.0172)	0.0318(0.0163)	0.4953(0.0261)	0.0732(0.0308)
		9	0.0421(0.0197)	0.0394(0.0194)	0.4951(0.0260)	0.1852(0.5761)
		12	0.0647(0.0969)	0.0618(0.1343)	0.4954(0.0256)	3.3320(4.3170)
	10	3	0.0179(0.0084)	0.0139(0.0073)	0.4896(0.0211)	0.0287(0.0127)
		6	0.0201(0.0086)	0.0173(0.0081)	0.4899(0.0212)	0.0386(0.0143)
		9	0.0230(0.0105)	0.0215(0.0105)	0.4896(0.0211)	0.0977(0.2842)
		12	0.0357(0.0521)	0.0339(0.0674)	0.4899(0.0209)	1.4920(1.9740)
200	1	3	0.0756(0.0318)	0.0629(0.0297)	0.4995(0.0259)	0.1235(0.0495)
		6	0.0905(0.0374)	0.0809(0.0355)	0.4994(0.0258)	0.1709(0.0702)
		9	0.0993(0.0413)	0.0894(0.0374)	0.4996(0.0262)	0.1918(0.0735)
		12	0.1056(0.0432)	0.0955(0.0396)	0.4993(0.0258)	0.2089(0.0763)
	5	3	0.0166(0.0069)	0.0133(0.0064)	0.4786(0.0110)	0.0289(0.0113)
		6	0.0192(0.0082)	0.0173(0.0080)	0.4787(0.0110)	0.0382(0.0125)
		9	0.0206(0.0087)	0.0189(0.0080)	0.4787(0.0110)	0.0456(0.0165)
		12	0.0218(0.0094)	0.0201(0.0082)	0.4785(0.0110)	0.0502(0.0178)
	10	3	0.0091(0.0038)	0.0071(0.0034)	0.4763(0.0093)	0.0150(0.0060)
		6	0.0105(0.0048)	0.0095(0.0045)	0.4763(0.0094)	0.0209(0.0073)
		9	0.0112(0.0050)	0.0104(0.0044)	0.4763(0.0093)	0.0242(0.0081)
		12	0.0118(0.0056)	0.0111(0.0047)	0.4762(0.0093)	0.0266(0.0084

Table 4. Mean and standard deviation of squared prediction errors for Electrical-Maintenance data set.

Degree	BHS	BGHS	HBGHS	BHS-NE	lasso-BIC
2	26866.9(2636)	26855.0(2637)	26879.8(2637)	27309.7(2489)	26858.6(2632)
4	13405.1(1285)	13437.1(1285)	13394.0(1281)	27314.3(2493)	13488.9(1291)
6	12939.2(1257)	13061.4(1255)	12939.9(1254)	27312.0(2498)	13038.6(1252)
8	11019.9(1141)	11054.0(1097)	10978.4(1136)	27307.9(2492)	11067.9(1100)
10	9970.9(1096)	10057.1(1075)	9958.4(1097)	27316.6(2492)	10022.4(1079)

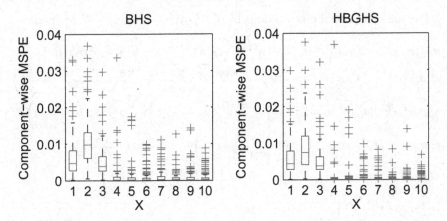

Fig. 1. Boxplots of component-wise mean squared prediction errors for the BHS and the HBGHS when there are $p = 10$ predictors, $n = 100$ samples, SNR $= 5$ and $K = 3$ degree of Legendre polynomial expansions of Function 3. The first three components are associated with the response variable.

error on both simulated data and real data. The proposed methods outperform the regular BHS method when applied to nonlinear functions and additive models. Even when there is no underlying group structure, the proposed methods are competitive with the regular BHS method. The results of real data analysis also support their promising performance.

Appendix: Full Conditional Distributions

The hierarchical specification of the complete model of the HBGHS is given in (7). By using the decomposition [10], the hierarchical representation becomes:

$$\mathbf{y}|\mathbf{X}, \boldsymbol{\beta}, \sigma^2 \sim \mathcal{N}(\mathbf{X}\boldsymbol{\beta}, \sigma^2 \mathbb{1}_n)$$

$$\boldsymbol{\beta}|\sigma^2, \tau^2, \lambda_1, \cdots, \lambda_G, \delta_1, \cdots, \delta_p \sim \mathcal{N}(\mathbf{0}, \sigma^2 \tau^2 \mathbf{D}_\lambda \mathbf{D}_\delta)$$

$$\mathbf{D}_\lambda = \text{diag}(\lambda_1^2 \mathbf{I}_{s_1}, \cdots, \lambda_G^2 \mathbf{I}_{s_G}), \quad \mathbf{D}_\delta = \text{diag}(\delta_1^2, \cdots, \delta_p^2)$$

$$\lambda_g^2|t_g \sim \mathcal{IG}\left(\frac{1}{2}, \frac{1}{t_g}\right), \; t_g \sim \mathcal{IG}\left(\frac{1}{2}, 1\right), \; g = 1, \cdots, G$$

$$\delta_j^2|c_j \sim \mathcal{IG}\left(\frac{1}{2}, \frac{1}{c_j}\right), \; c_j \sim \mathcal{IG}\left(\frac{1}{2}, 1\right), \; j = 1, \cdots, p$$

$$\tau^2|v \sim \mathcal{IG}\left(\frac{1}{2}, \frac{1}{v}\right), \; v \sim \mathcal{IG}\left(\frac{1}{2}, 1\right)$$

$$\sigma^2 \sim \frac{1}{\sigma^2} d\sigma^2.$$

The full conditional distributions of β, σ^2, $\lambda_1^2, \cdots, \lambda_G^2$, $\delta_1^2, \cdots, \delta_p^2$, τ are:

$$\beta|\sigma^2,\tau^2,\lambda_1^2,\cdots,\lambda_G^2,\delta_1^2,\cdots,\delta_p^2 \sim \mathcal{N}\left(\mathbf{A}^{-1}\mathbf{X}^T\mathbf{y}, \sigma^2\mathbf{A}^{-1}\right), \quad \mathbf{A} = \mathbf{X}^T\mathbf{X} + (\tau^2\mathbf{D}_\lambda\mathbf{D}_\delta)^{-1}$$

$$\sigma^2|\beta,\tau^2,\lambda_1^2,\cdots,\lambda_G^2,\delta_1^2,\cdots,\delta_p^2 \sim \mathcal{IG}\left(\frac{n-1+p}{2}, \frac{(\mathbf{y}-\mathbf{X}\beta)^T(\mathbf{y}-\mathbf{X}\beta) + \beta^T(\tau^2\mathbf{D}_\lambda\mathbf{D}_\delta)^{-1}\beta}{2}\right)$$

$$\lambda_g^2|\beta,\sigma^2,\tau^2,t_g,\delta_1^2,\cdots,\delta_p^2 \sim \mathcal{IG}\left(\frac{s_g+1}{2}, \frac{\beta_g^T(\mathbf{D}_{\delta_g})^{-1}\beta_g}{2\sigma^2\tau^2} + \frac{1}{t_g}\right), \quad t_g|\lambda_g^2 \sim \mathcal{IG}\left(1, \frac{1}{\lambda_g^2}+1\right)$$

$$\delta_j^2|\beta,\sigma^2,\tau^2,\lambda_1^2,\cdots,\lambda_G^2,c_j \sim \mathcal{IG}\left(1, \frac{\beta_j^2}{2\sigma^2\tau^2\lambda_{gj}^2} + \frac{1}{c_j}\right), \quad c_j|\delta_j^2 \sim \mathcal{IG}\left(1, \frac{1}{\delta_j^2}+1\right)$$

$$\tau^2|\beta,\sigma^2,\tau^2,\lambda_1^2,\cdots,\lambda_G^2,\delta_1^2,\cdots,\delta_p^2,v \sim \mathcal{IG}\left(\frac{p+1}{2}, \frac{\beta^T(\mathbf{D}_\lambda\mathbf{D}_\delta)^{-1}\beta}{2\sigma^2} + \frac{1}{v}\right)$$

$$v|\tau^2 \sim \mathcal{IG}\left(1, \frac{1}{\tau^2}+1\right).$$

References

1. Alcalá, J., Fernández, A., Luengo, J., Derrac, J., García, S., Sánchez, L., Herrera, F.: Keel data-mining software tool: data set repository, integration of algorithms and experimental analysis framework. J. Multiple-Valued Logic Soft Comput. **17**(2–3), 255–287 (2010)
2. Breheny, P., Huang, J.: Penalized methods for bi-level variable selection. Stat. Interface **2**(3), 369–380 (2009)
3. Breiman, L.: Better subset regression using the nonnegative garrote. Technometrics **37**(4), 373–384 (1995)
4. Bühlmann, P., Geer, S.V.D.: Statistics for High-Dimensional Data: Methods, Theory and Applications. Springer Science & Business Media, New York (2011)
5. Carvalho, C.M., Polson, N.G., Scott, J.G.: Handling sparsity via the horseshoe. JMLR **5**, 73–80 (2009)
6. Carvalho, C.M., Polson, N.G., Scott, J.G.: The horseshoe estimator for sparse signals. Biometrika **97**(2), 465–480 (2010)
7. Fan, J., Li, R.: Variable selection via nonconcave penalized likelihood and its oracle properties. J. Am. Stat. Assoc. **96**(456), 1348–1360 (2001)
8. Hoerl, A.E., Kennard, R.W.: Ridge regression: biased estimation for nonorthogonal problems. Technometrics **12**(1), 55–67 (1970)
9. Huang, J., Ma, S., Xie, H., Zhang, C.H.: A group bridge approach for variable selection. Biometrika **96**(2), 339–355 (2009)
10. Makalic, E., Schmidt, D.F.: A simple sampler for the horseshoe estimator. IEEE Signal Process. Lett. **23**(1), 179–182 (2016)
11. Park, T., Casella, G.: The Bayesian lasso. J. Am. Stat. Assoc. **103**(482), 681–686 (2008)
12. Yuan, M., Lin, Y.: Model selection and estimation in regression with grouped variables. J. R. Stat. Soc. Ser. B (Stat. Methodol.) **68**(1), 49–67 (2006)
13. Zhao, P., Rocha, G., Yu, B.: The composite absolute penalties family for grouped and hierarchical variable selection. Ann. Stat. **37**(6A), 3468–3497 (2009)
14. Zou, H.: The adaptive lasso and its oracle properties. J. Am. Stat. Assoc. **101**(476), 1418–1429 (2006)
15. Zou, H., Hastie, T.: Regularization and variable selection via the elastic net. J. R. Stat. Soc. Ser. B (Stat. Methodol.) **67**(2), 301–320 (2005)

Constraint Satisfaction, Search and Optimisation

Improving and Extending the HV4D Algorithm for Calculating Hypervolume Exactly

Wesley Cox and Lyndon While[✉]

Computer Science and Software Engineering,
The University of Western Australia, Perth, Australia
{wesley.cox,lyndon.while}@uwa.edu.au

Abstract. We describe extensions to the 4D hypervolume algorithm HV4D that greatly improve its performance in 4D, and that enable an extension of the algorithm to 5D. We add a facility to cope with dominated points, reducing the number of contribution calculations required; and a new representation of the front between slices, eliminating significant repeated work. The former also allows the algorithm to work efficiently with 5D data. The new algorithms can process sets containing 1,000 points in around 1 ms in 4D, and around 5–10 ms in 5D. They make a significant contribution to the state-of-the-art.

Keywords: Hypervolume · Multi-objective optimisation · Metrics

1 Introduction

Hypervolume [1] is a popular metric for comparing the performance of multi-objective optimisers. The hypervolume of a set of solutions is the size of the part of objective space that is dominated by those solutions collectively. Hypervolume captures in one scalar both the closeness of the solutions to the optimal set and their diversity. It also has nicer mathematical properties than other metrics [2,3]. However hypervolume is expensive to calculate; also it is sensitive to the relative scaling of the objectives, to the presence or absence of extremal points, and to the choice of the reference point against which solutions are compared.

The two fastest published algorithms for calculating hypervolume exactly in the general case are WFG [4] and QHV [5]. However a recent trend has seen low-complexity algorithms derived for specific small numbers of objectives: notably an optimal $O(n \log n)$ algorithm for 3D [6], and an $O(n^2)$ algorithm HV4D for 4D [7]. Both of these are dimension-sweep algorithms, where points are sorted in one objective and then processed first in this objective. Using intelligent representations and efficient operations, this approach makes for very fast algorithms.

This paper makes three principal contributions.

– An extension to HV4D which allows the contribution algorithm to handle dominated points directly. This reduces the number of contribution calculations required, thereby improving its efficiency without compromising its worst-case complexity. We call this algorithm HV4DR.

© Springer International Publishing AG 2016
B.H. Kang and Q. Bai (Eds.): AI 2016, LNAI 9992, pp. 243–254, 2016.
DOI: 10.1007/978-3-319-50127-7_20

– A reorganization of HV4D so that instead of storing only the non-dominated front between slices, it stores the decomposition of this front into the 3D boxes which are used in the main calculations. This eliminates a significant source of repeated work in the algorithm. We call this algorithm HV4DX.
– An extension of HV4DR to five objectives. Extension to 5D requires the algorithm to cope with dominated points in either four objectives or three objectives, which is facilitated by the first extension described above. We call this algorithm HV5DR.

We present a comprehensive performance evaluation that demonstrates the value of all of these changes: the new algorithms can process fronts with 1,000 points in around 1 ms in 4D, and around 5–10 ms in 5D.

The paper is structured as follows. Section 2 describes relevant background material. Section 3 describes HV4D. Section 4 describes our extensions to HV4D. Section 5 gives the experimental comparison. Section 6 concludes the paper.

2   Background Material

2.1   Multi-objective Optimisation

In a multi-objective optimisation problem, we aim to find the set of optimal trade-off solutions known as the Pareto optimal set. Pareto optimal solutions are characterised by the fact that improving any one objective means worsening at least one other objective. A key concept here is *domination*: we say that a solution $\overline{y}$ *dominates* a solution $\overline{z}$ iff $\overline{y}$ is at least as good as $\overline{z}$ in all objectives, and better in at least one. $\overline{y}$ is *non-dominated* wrt a set of solutions X iff no vector in X dominates $\overline{y}$. Precise definitions of these terms can be found in [8]. We will refer to the objectives as $\{x_1, x_2, ..., x_d\}$; and a d-objective solution p has objective values $\{p_1, p_2, ..., p_d\}$.

2.2   Hypervolume

Given a set of solutions S returned by a multi-objective optimiser, the *hypervolume* [1] of S is the size of the part of objective space that is dominated collectively by the solutions in S, relative to some reference point. If a set S has a greater hypervolume than a set S', S is taken to be a better set of solutions than S'.

The *exclusive hypervolume* (or *incremental hypervolume*) of a point p relative to an *underlying set* S is the size of the part of objective space that is dominated by p but is not dominated by any member of S. Exclusive hypervolume can be trivially defined in terms of hypervolume:

$$exc(p, S) = hyp(S \cup \{p\}) - hyp(S) \tag{1}$$

We also use the term *inclusive hypervolume* of a point p for the size of the part of objective space dominated by p alone:

$$inc(p) = hyp(\{p\}) \tag{2}$$

2.3   Previous Algorithms for Calculating Exact Hypervolumes

Several algorithms have been proposed for calculating hypervolumes and exclusive hypervolumes exactly. [9] gives a recent comprehensive list: we describe here only those algorithms which are relevant to HV4D and our extensions.

HSO (Hypervolume by Slicing Objectives) [10–12] slices the nD-volume into separate $n-1$D-volumes through the values in one of the objectives, then it sums the volumes of those slices. HSO's worst-case complexity is $O(m^{n-1})$ [12], but good heuristics for re-ordering objectives [13] deliver much better performance for typical data. Fonseca $et\ al.$ describe a highly-optimised version of HSO [14], including an optimal algorithm for the 3D case [6].

WFG (Walking Fish Group) [4] combines point-wise processing with a new bounding trick for calculating exclusive hypervolume quickly [15]. The exclusive hypervolume of p relative to S is the inclusive hypervolume of p minus the hypervolume of S after each point has been limited by p in every objective. Experiments show that this limiting operation produces a lot of points that are dominated within the new set, so subsequent calculations are very fast.

QHV (Quick Hypervolume) [5] uses a divide-and-conquer approach reminiscent of Quicksort, dividing up objective space using a pivot point, and analysing each of the individual sub-spaces recursively. It derives its performance from various clever implementation tricks, particularly in the data structures used.

An alternative to calculating hypervolume exactly is to use an approximation algorithm (e.g. [15–17]). This introduces a trade-off between precision and performance, and much improvement has been made on understanding this trade-off [3,18,19]. However, we do not compare with approximation algorithms.

3   The HV4D Algorithm

HV4D uses a combination of a HSO-style slicing step and a 3-objective contribution algorithm. Given n points in d objectives, when the points are sorted in one objective, then between adjacent points the cross-sectional hypervolume in the remaining $d-1$ objectives is constant. Thus the d-objective hypervolume can be split into n slices in $d-1$ objectives where the kth slice is determined by the non-dominating subset in $d-1$ objectives of the first k points sorted in the slicing objective. A 3-objective example of this process is illustrated in Fig. 1.

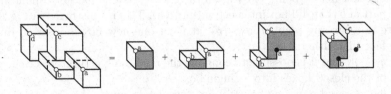

Fig. 1. HSO-style slicing in 3 objectives. The shape is split into four 2-objective slices. Solid circles represent points dominated in a particular slice in 2 objectives. The grey areas show the exclusive hypervolume of the added point in 2 objectives.

The total hypervolume is equal to the sum of the hypervolume of the slices each multiplied by its width, i.e. the difference in the slicing objective between adjacent points in the sorted list. For a front sorted in the slicing objective, if the $(d-1)$-objective hypervolume of the kth slice is known then the hypervolume of the $(k+1)$th slice is equal to that of the previous slice plus the exclusive hypervolume in $d-1$ objectives of the $(k+1)$th point relative to the previous k points. Thus a 4-objective hypervolume can be computed by performing a slicing step, then computing the size of each slice incrementally with a 3-objective contribution algorithm. The slicing objective here is chosen to be x_4.

The contribution algorithm works by decomposing $exc(p, S)$ into 3-dimensional boxes. A box b is a cuboid with edges parallel to the axes and defined by two opposing corners b_o and b_i. b_o, with objective values $\{b_{o1}, b_{o2}, b_{o3}\}$, denotes the corner further from the reference point r, and b_i, with objective values $\{b_{i1}, b_{i2}, b_{i3}\}$, denotes the corner closer to r. The algorithm maintains a set of partial boxes that represent part of the decomposition of $exc(p, S)$. When all six objective values of a box are known, it can be closed and its volume is added to the running total of $exc(p, S)$. The algorithm comprises three steps: splitting the set, initialising the boxes, and closing the boxes.

Splitting step. The splitting step divides S into two subsets $S_1(p)$, containing points greater in x_3 than p_3, and $S_2(p)$, containing points less in x_3 than p_3.

Initialisation step. As illustrated in Fig. 2, the points in $S_1(p)$ define the shape of the top "face" of $exc(p, S)$. Note that points in $S_1(p)$ that are dominated in the first two objectives do not contribute to this shape. To initialise the boxes, points in $S_1(p)$ are examined in increasing x_1 order, and each point dominated by p in the first two objectives defines a cut parallel to the x_2 axis. This divides the top face of $exc(p, S)$ into rectangles, each of which is the top face of one of the initial boxes. For each box created, $b_{o1}, b_{o2}, b_{i1}, b_{i2}$ are set to the corresponding values of the rectangle, and b_{o3} is set to p_3. b_{i3} is initially unknown and is computed in the closing step.

Closing step. To close the boxes, points in $S_2(p)$ are examined in decreasing x_3 order. Let q be the next point in $S_2(p)$. If $inc(q)$ intersects with any open box b in the first two objectives, then its b_{i3} value must be greater or equal to q_3 for the box to be contained within $exc(p, S)$. If b completely intersects with $inc(q)$ then it is closed with b_{i3} set to q_3. If b only partially intersects with $inc(q)$, then the outer corner values are reduced such that the box now represents the intersection before being closed, and a new box is created corresponding to the part of b that did not intersect with $inc(q)$. This is illustrated in Fig. 3(b). Note that for any point q in $S_2(p)$, at most one new box is inserted due to the fact that either:

1. $q_1 > p_1$ and it is possible to combine the remaining non-intersecting space of the closed boxes into a single box.
2. $q_2 > p_2$ and only one box will not completely intersect with $inc(q)$.
3. q dominates p in the first two objectives and all boxes are closed.

The first two cases are demonstrated in Fig. 3(c). Since at any point in the algorithm the current set of partial boxes will never overlap in x_1, they are stored in a linked-list sorted in b_{o1} (or b_{i1}), and new boxes will always be placed at either the front or end of the list and can be inserted in constant time. If all points in $S_2(p)$ have been examined, then any remaining boxes are closed with b_{i3} values set to r_3.

Note that this algorithm cannot handle points dominated by p in three objectives. When the point p to be inserted into S dominates some of the points in S, the current volume of the slice is subtracted by the contribution of any dominated point q relative to $S\backslash\{q\}$ and q is removed from S. Afterwards, the contribution of p is then computed relative to the set of points in S', which is the set of points in S not dominated by p.

HV4D is quadratic in the number of points [7]. The number of calls to the contribution function is linear, and each of the three steps of the contribution algorithm is linear. The splitting step examines each point in S, the initialisation step examines each point in S_1, and the closing step examines and closes every box, which is linear because the number of boxes is at most $|S|$.

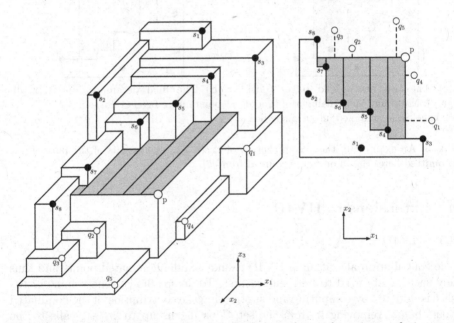

Fig. 2. A 3-objective example of the exclusive hypervolume of a point p relative to a set S. The right image shows certain features from a projection of the shape into the first two objectives. Points below p in x_3 are open circles, and points above are solid circles. The grey shape is the top "face" of $exc(p, S)$ and represents $exc(p, S_1(p))$ in the first 2 objectives. This face is decomposed into rectangles by examining points in $S_1(p)$ as per the initialisation step. Based on figure sourced from [7].

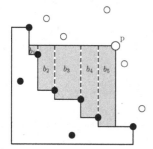

(a) The initial boxes (light grey).

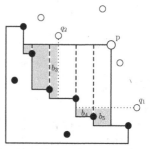

(b) The closed boxes (light grey) after examining two points in $S_2(p)$. Note that b_3, b_4, and b_5 are resized before being closed.

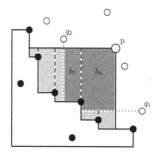

(c) The two new added boxes (dark grey). Note that the remainder of b_4 and b_5 have been combined into a single box.

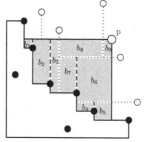

(d) The final decomposition after all points have been evaluated.

Fig. 3. An example of the decomposition of $exc(p, S)$. The heights of the boxes in x_3 is omitted here. Based on figure sourced from [7].

4   Extensions to HV4D

4.1   HV4DR

The contribution algorithm of HV4D cannot handle dominated points and thus any such points need to be removed first. To do so, the current volume of the slice is updated by computing and subtracting the contribution of the dominated point before removing it from the set. This means up to $2n - 1$ calls to the contribution function may be needed, with two calls for any point that is ever dominated. If the contribution function is modified to allow dominated points, this second call is no longer necessary and dominated points can simply be removed after computing the contribution of the point that dominated them.

HV4DR introduces a new case to the closing step of the contribution algorithm. Consider $exc(p, S)$, and suppose the next point $q \in S_2(p)$ being considered in the closing step is dominated by p in the first two objectives. The boxes that intersect with $inc(q)$ are found, as in Fig. 4(a). These will only partially intersect, and the outer corner values are updated to reduce the boxes to their intersection,

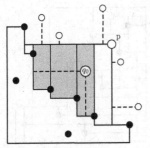 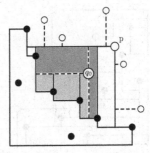

(a) The light grey rectangles represent boxes that are adjusted and closed.

(b) The dark grey rectangles represent two new created boxes.

Fig. 4. An example of the new case in HV4DR. The point q_0 is dominated by p in 3 objectives. Based on figure sourced from [7].

as with the previous cases. They are closed as normal with b_{i3} set to q_3. Unlike previous cases, two new boxes need to be created, as in Fig. 4(b).

The new case will be linear in the worst case due to the possibility of traversing the entire box list to find which boxes intersect with $inc(q)$, but the case can only occur once for each point that is dominated, after which the dominated point is removed from the set. Thus the new case adds a quadratic component to the (already quadratic) algorithm. The number of boxes for a single contribution call with the new case is now at most $2|S|$, which is still linear, thus the complexity of the rest of the algorithm is unchanged.

4.2 HV4DX

The contribution algorithm used in HV4D computes $exc(p, S)$ by decomposing it into 3D boxes and summing the volume of the boxes. The information required to create this decomposition is recreated for each contribution call. The proposed algorithm HV4DX is partly based on the main observation underlying the operation of WFG [4], i.e. that $exc(p, S)$ is equal to the inclusive hypervolume of p minus the hypervolume of the intersection of $inc(p)$ and $hyp(S)$. Thus $exc(p, S)$ can be computed by decomposing the intersection of $inc(p)$ and $hyp(S)$ into 3D boxes and subtracting the total volume from the inclusive hypervolume of p. HV4DX does not perform this operation explicitly, but uses the fact that $exc(p, S)$ can be computed by using the decomposition of $hyp(S)$. The key benefit comes from the fact that the decomposition of $hyp(S)$ can be stored and updated when inserting points. HV4D on the other hand, only stores S itself, and has to compute a new decomposition for each point.

HV4DX operates likes HV4D by building up slices incrementally in three objectives. The current non-dominating set S is not stored explicitly as a set of points but rather as a set of 3D boxes which form the decomposition of $hyp(S)$. The boxes are stored in a linked list that is sorted decreasing in the x_2 values of

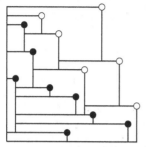

(a) A top down view of the hypervolume $hyp(S)$.

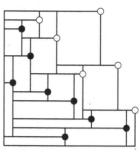

(b) A top down view of the box decomposition of $hyp(S)$.

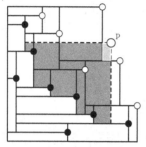

(c) The boxes (grey) involved in computing $exc(p, S)$. Note the light grey space not contained by any box.

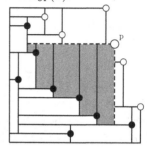

(d) The new boxes (grey) inserted after $exc(p, S)$ is computed.

Fig. 5. An example of HV4DX. Solid circles represent points above p in x_3, open circles are points below. Based on figure sourced from [7].

the outer corner, and then for equal x_2 values sorted decreasing in the x_1 values of the outer corner.

The contribution algorithm is illustrated in Fig. 5. Figure 5(a) shows a top-down view of $hyp(S)$. This can be seen as a set of 2D shapes, each representing the exclusive hypervolume in two objectives of the associated point relative to the set of points greater in x_3. These 2D shapes can be decomposed into rectangles through a set of cuts as seen in Fig. 5(b). This is a top-down view of the box decomposition of $hyp(S)$ where the b_{o3} value of each box b is the x_3 value of its associated point and the b_{i3} value is r_3. The contribution algorithm uses this set of boxes and compares it with the incoming point. Figure 5(c) shows the greyed area corresponding to the intersection of $inc(p)$ and $hyp(S_2(p))$. Note that the set of rectangles in the intersection are the same as the lower faces of the boxes that would be produced in the operation of HV4D, shown in Fig. 3(d).

Each box in the set is considered in order. Boxes whose outer corner is less in x_3 than p_3 are used to compute $exc(p, S)$. For any such box b that also intersects with $inc(p)$ in the first two objectives, the volume of $exc(p, S)$ is updated by adding the area of the intersection multiplied by the difference between b_{o3} and p_3. If b completely intersects with $inc(p)$ it is explicitly closed. Otherwise, b

is adjusted such that it no longer intersects with $inc(p)$, with a second box potentially added, implicitly closing the box represented by the intersection.

Boxes whose outer corner is greater in x_3 than p_3 determine the set of new boxes that need to be added with height p_3, as shown in Fig. 5(d). Note that the top faces match those created in the initialisation step of HV4D, shown in Fig. 2. Like in HV4D, only the points in $S_1(p)$ that are non-dominating in two objectives determine the shape of the new boxes. The order of the linked list guarantees that only these points are considered because any dominated box in this set will have an outer corner with a lower x_1 value than the last non-dominated box seen so far, and can thus be ignored. Given that the relevant boxes will thus be seen in decreasing x_2 order and increasing x_1 order, the new boxes are easily constructed.

The final step is to potentially compute any remainder of $exc(p, S)$. This corresponds to the portion of $inc(p)$ that does not intersect with any box, as observed in Fig. 5(c). Note that this space is not necessarily a simple shape, but can computed in constant time. The area of the top face of this remaining shape is simply the difference between the sum of the areas of the top faces of the new boxes, and the sum of the areas of the top faces of any closed boxes, implicit or explicit. The volume of the shape is thus this area multiplied by $(p_3 - r_3)$. The new boxes can be inserted in constant time by remembering the first open box seen that had an outer corner with lower x_2 than p_2, and inserting before that position. Once the contribution algorithm is completed the set of boxes will now represent the decomposition of $hyp(S \cup p)$.

Each call to the contribution function examines each box in the current list exactly once. For each box's top face, either its upper right corner coincides with a point or its lower right corner coincides with a point. Thus, the maximum number of boxes in the list at any one time is $2n$. Thus, the contribution algorithm is still linear, and the complete algorithm is still quadratic.

4.3   HV5DR

HV5DR is an extension of HV4DR to five objectives. HV4D (and its extensions) operate by reducing a 4-objective problem into a 3-objective problem with a combination of a slicing step and 3-objective contribution algorithm. HV5DR operates by slicing a 5-objective hypervolume and using a 4-objective contribution algorithm, with the slicing objective here chosen as x_5. The contribution algorithm reduces the exclusive hypervolume of a point p relative to a set S into a set of 4D hypercuboids, referred to here as boxoids. Like a 3D box, a boxoid has a top "face", in this case a 3D cuboid, and its "height" in the fourth objective needs to be determined.

The 4-objective contribution algorithm operates similarly to the 3-objective algorithm and features a splitting step, an initialisation step, and a closing step. The splitting step splits S into two subsets $S_1(p)$ and $S_2(p)$, here splitting by the x_4 value. Just as the initialisation step in HV4D involves decomposing the exclusive hypervolume in two objectives of p relative to $S_1(p)$ into rectangles, HV5DR decomposes the exclusive hypervolume in three objectives of p relative to $S_1(p)$ into cuboids. This can be done by using the 3-objective contribution algorithm for p relative to $S_1(p)$, but instead of deleting closed boxes, adding them to the boxoid list and setting their initial b_{o4} value to p_4. The closing step examines the points in $S_2(p)$ in decreasing x_4 order. For the next point q, any boxoids that intersect with $inc(q)$ are closed with adjusted corners and up to three new boxoids need be created for each boxoid that is closed but does not completely intersect with $inc(q)$.

Note that even if no points in S are dominated by p in four objectives, points in $S_1(p)$ may be dominated by p in three objectives, thus the 3-objective contribution algorithm used here requires the extra case introduced in HV4DR. This extra case is also used in the 4-objective contribution algorithm for improved performance, but there it is not strictly necessary.

HV5DR can easily be extended into higher numbers of objectives. In practise however, it performs poorly above five objectives as the number of boxoids created is exponential in the number of objectives.

5 Results

A series of experiments were run to investigate the performance of the new algorithms. The algorithms were compared to WFG and QHV on benchmark data introduced in [4]. The data comprised randomly generated fronts, and spherical and discontinuous fronts from the DTLZ test suite [20]; it is publicly available from www.wfg.csse.uwa.edu.au/hypervolume. The QHV implementation is the public release available at web.tecnico.ulisboa.pt/luis.russo/QHV/#down. All other algorithms are available at www.wfg.csse.uwa.edu.au/hypervolume. Timings were performed on a machine with an 8-core 3.20 GHz i7 processor, with 8 GB of RAM, running Red Hat Enterprise Linux 7.2.

All experiments were run as maximisation problems relative to 0*. Since QHV requires an upper bound to be specified in its operation, 10* was used for random and discontinuous data, and 1* was used for spherical data.

Results are shown in Fig. 6. All of the new algorithms are observed to outperform WFG and QHV, as well as HV4D in 4 objectives. HV4DR demonstrates consistent improvement over HV4D and HV4DX further demonstrates consistent improvement over HV4DR. Table 1 shows the reduction in runtime of HV4DR against HV4D, and HV4DX against HV4D and HV4DR.

Table 1. Average reduction in runtime of 4D algorithms for fronts with 1,000 points.

	Random	Spherical	Discontinuous
HV4DR vs. HV4D	36%	29%	34%
HV4DX vs. HV4DR	32%	38%	37%
HV4DX vs. HV4D	56%	56%	58%

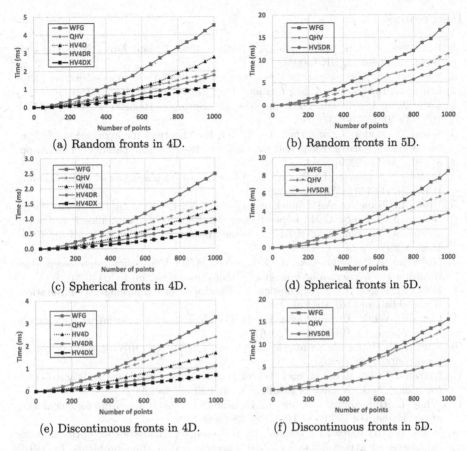

(a) Random fronts in 4D.

(b) Random fronts in 5D.

(c) Spherical fronts in 4D.

(d) Spherical fronts in 5D.

(e) Discontinuous fronts in 4D.

(f) Discontinuous fronts in 5D.

Fig. 6. Performance comparison of HV4D, HV4DR, HV4DX, and HV5DR with WFG and QHV. Each line plots the average processing time for twenty distinct fronts. Each experiment was run 200 times to ensure accurate timings.

6   Conclusions

We have described multiple extensions to the HV4D algorithm for calculating exact hypervolume. HV4DR includes a new case in the contribution algorithm to cope with dominated points, reducing the number of contribution calls. HV4DX uses a different structure to store slices, significantly reducing repeated work. HV5DR extends HV4DR to five objectives, and can be further generalised to any number of objectives. Experiments demonstrate HV4DX and HV5DR as the fastest known algorithms in four and five objectives respectively.

Future work includes extending HV4DX to higher numbers of objectives.

References

1. Purshouse, R.: On the Evolutionary Optimisation of Many Objectives. The University of Sheffield, UK (2003)
2. Fleischer, M.: The measure of Pareto optima applications to multi-objective metaheuristics. In: Fonseca, C.M., Fleming, P.J., Zitzler, E., Thiele, L., Deb, K. (eds.) EMO 2003. LNCS, vol. 2632, pp. 519–533. Springer, Heidelberg (2003). doi:10.1007/3-540-36970-8_37
3. Auger, A., Bader, J., Brockhoff, D., Zitzler, E.: Theory of the hypervolume indicator: optimal μ-distributions and the choice of the reference point. In: FOGA, pp. 87–102. ACM (2009)
4. While, L., Bradstreet, L., Barone, L.: A fast way of calculating exact hypervolumes. IEEE TEVC 16(1), 86–95 (2012)
5. Russo, L., Francisco, A.P.: Quick hypervolume. IEEE TEVC 18(4), 481–502 (2014)
6. Beume, N., Fonseca, C.M., López-Ibáñez, M., Paquete, L., Vahrenhold, J.: On the complexity of computing the hypervolume indicator. IEEE TEVC 13(5), 1075–1082 (2009)
7. Guerreiro, A.P., Fonseca, C.M., Emmerich, M.T.: A fast dimension-sweep algorithm for the hypervolume indicator in four dimensions. In: CCCG (2012)
8. Back, T., Fogel, D.B., Michalewicz, Z.: Handbook of Evolutionary Computation. IOP Publishing Ltd., Bristol (1997)
9. Cox, W., While, L.: Improving the IWFG algorithm for calculating incremental hypervolume. In: IEEE CEC (2016)
10. Zitzler, E.: Hypervolume metric calculation (2001). ftp://ftp.tik.ee.ethz.ch/pub/people/zitzler/hypervol.c
11. Knowles, J.: Local-Tearch and Hybrid Evolutionary Algorithms for Pareto Optimisation. The University of Reading, United Kingdom (2002)
12. While, L., Hingston, P., Barone, L., Huband, S.: A faster algorithm for calculating hypervolume. IEEE TEVC 10(1), 29–38 (2006)
13. While, L., Bradstreet, L., Barone, L., Hingston, P.: Heuristics for optimising the calculation of hypervolume for multi-objective optimisation problems. In: IEEE CEC, pp. 2225–2232 (2005)
14. Fonseca, C.M., Paquete, L., López-Ibáñez, M.: An improved dimension-sweep algorithm for the hypervolume indicator. In: IEEE CEC, pp. 3973–3979 (2006)
15. Bringmann, K., Friedrich, T.: Approximating the least hypervolume contributor: NP-hard in general, but fast in practice. In: Ehrgott, M., Fonseca, C.M., Gandibleux, X., Hao, J.-K., Sevaux, M. (eds.) EMO 2009. LNCS, vol. 5467, pp. 6–20. Springer, Heidelberg (2009). doi:10.1007/978-3-642-01020-0_6
16. Everson, R.M., Fieldsend, J.E., Singh, S.: Full elite sets for multi-objective optimisation. In: Parmee, I.C. (ed.) Adaptive Computing in Design and Manufacture V, pp. 343–354. Springer, London (2002)
17. Bader, J., Deb, K., Zitzler, E.: Faster hypervolume-based search using Monte Carlo sampling. In: Ehrgott, M., Naujoks, B., Stewart, T.J., Wallenius, J. (eds.) Multiple Criteria Decision Making for Sustainable Energy and Transportation Systems. Lecture Notes in Economics and Mathematical Systems, vol. 634, pp. 313–326. Springer, Heidelberg (2010). doi:10.1007/978-3-642-04045-0_27
18. Bringmann, K., Friedrich, T.: Don't be greedy when calculating hypervolume contributions. In: FOGA, pp. 103–112. ACM (2009)
19. Friedrich, T., Horoba, C., Neumann, F.: Multiplicative approximations and the hypervolume indicator. In: FOGA, pp. 103–112. ACM (2009)
20. Deb, K., Thiele, L., Laumanns, M., Zitzler, E.: Scalable multi-objective optimization test problems. In: IEEE CEC, pp. 825–830 (2002)

Local Search for Maximum Vertex Weight Clique on Large Sparse Graphs with Efficient Data Structures

Yi Fan[1]([⊠]), Chengqian Li[2], Zongjie Ma[1], Lian Wen[1], Abdul Sattar[1], and Kaile Su[1]

[1] Institute for Integrated and Intelligent Systems,
Griffith University, Brisbane, Australia
{yi.fan4,zongjie.ma}@griffithuni.edu.au,
{l.wen,a.sattar,k.su}@griffith.edu.au
[2] Department of Computer Science, Sun Yat-sen University, Guangzhou, China

Abstract. The Maximum Vertex Weight Clique (MVWC) problem is a generalization of the Maximum Clique problem, which exists in many real-world applications. However, it is NP-hard and also very difficult to approximate. In this paper we developed a local search MVWC solver to deal with large sparse instances. We first introduce random walk into the multi-neighborhood greedy search, and then implement the algorithm with efficient data structures. Experimental results showed that our solver significantly outperformed state-of-the-art local search MVWC solvers. It attained all the best-known solutions, and found new best-known solutions on some instances.

Keywords: Local search · Maximum vertex weight clique · Large sparse graphs · Data structures

1 Introduction

The Maximum Clique (MC) problem is a well-known NP-hard problem [17]. Given a simple undirected graph $G = (V, E)$, a *clique* C is a subset of V s.t. each vertex pair in C is mutually adjacent. The Maximum Clique problem is to find a clique of the maximum size. An important generalization of the MC problem is the Maximum Vertex Weight Clique (MVWC) problem in which each vertex is associated with a positive integer weight, and the goal is to find a clique with the greatest total vertex weight. This generalization is important in many real-world applications like [1,3,4,8,9,18,19,23]. In this paper we are concerned in finding a clique whose total vertex weight is as great as possible.

Both MC and MVWC are NP-hard, and the state-of-the-art approximation algorithm can only achieve an approximation ratio of $O(n(\log \log n)^2/(\log n)^3)$ [15]. Thus various heuristic methods have been developed to find a "good" clique within reasonable time. Up to now, there are two types of algorithms for the

© Springer International Publishing AG 2016
B.H. Kang and Q. Bai (Eds.): AI 2016, LNAI 9992, pp. 255–267, 2016.
DOI: 10.1007/978-3-319-50127-7_21

MVWC problems: complete algorithms and incomplete ones. Complete algorithms for MVWC include [2,14,21,30]. On the other hand, incomplete algorithms for MVWC include [7,10,22,27,28].

The aim of this work is to develop a local search MVWC solver named LMY-GRS[1] to deal with large crafted graphs. Firstly we incorporate random walk into the multi-neighborhood greedy search in [28]. Then we propose novel data structures to achieve greater efficiency.

We used the large crafted benchmark in [27][2] to test our solver. And we make two state-of-the-art solvers, MN/TS [28] and LSCC [27], as the competitors[3]. Experimental results show that LMY-GRS *attained all the best-known solutions* in this large crafted benchmark. Moreover, for a large proportion of the graphs LMY-GRS achieves better average quality. Among them there are four graphs where LMY-GRS reports new best-known solutions.

2   Preliminaries

2.1   Basic Notations

Given a graph $G = (V, E)$ where $V = \{v_1, \ldots, v_n\}$, an edge is a 2-element subset of V. Given an edge $e = \{u, v\}$, we say that u and v are adjacent to each other. Also we say that u and v are neighbors, and we use $N(v) = \{u | \{u, v\} \in E\}$ to denote the set of v's neighbors. The degree of a vertex v, denoted by $d(v)$, is defined as $|N(v)|$. We use $d_{max}(G)$ to denote the maximum degree of graph G, suppressing G if understood from the context. A clique C of G is a subset of V where vertices are pair-wise adjacent. An empty set is said to be an empty clique. A set that contains a single vertex is called a single-vertex clique.

Given a weighting function $w : V \rightarrow Z^+$, the weight of a clique C, denoted by $w(C)$, is defined to be $\sum_{v \in C} w(v)$. We use $age(v)$ to denote the number of steps since last time v changed its state (inside or outside the candidate clique).

2.2   The Large Crafted Benchmark

MC and MVWC solvers are often tested on the DIMACS [16] and BOSHLIB [29] benchmarks. Recently large real-world benchmarks have become very popular. Many of these graphs have millions of vertices and dozens of millions of edges. They were used in testing Graph Coloring and Minimum Vertex Cover algorithms [11,24,26], as well as the MC [25] and MVWC [27] algorithms.

In this paper we focus on the large benchmark. They were originally unweighted, and to obtain the corresponding MVWC instances, we use the same method as in [22,27,28]. For the i-th vertex $v_i, w(v_i) = (i \mod 200) + 1$.

---

[1] https://github.com/Fan-Yi/Local-Search-for-Maximum-Vertex-Weight-Clique-on-Large-Sparse-Graphs-with-Efficient-Data-Structures.

[2] http://www.graphrepository.com/networks.php.

[3] In [27], both solvers are incorporated with a heuristic named BMS to solve large instances. For simplicity, we write them as MN/TS and LSCC for short.

2.3   Multi-neighborhood Greedy Search

Usually the local search moves from one clique to another until the cutoff arrives, then it returns the best clique that has been found. There are three operators: add, swap and drop, which guide the local search. To ensure that these operators preserve the clique property, two sets are defined as below.

1. $AddS = \{v | v \notin C, v \in N(u)$ for all $u \in C\}$.
2. $SwapS = \{(u, v) | u \in C, v \notin C, \{u, v\} \notin E, v \in N(w)$ for all $w \in C \backslash \{u\}\}$.

So the add operator can only take an element from $AddS$ as its operand. Similarly the swap (resp. drop) operator can only take an element from $SwapS$ (resp. C).

Proposition 1. *If* $(u_1, v) \in SwapS$ *and* $(u_2, v) \in SwapS$, *then* $u_1 = u_2$[4].

That is, if a vertex v can go into C through a swap operation, then the vertex to be removed is unique. Then we have

Lemma 1. *For any* $P \subseteq SwapS$, $|P| = |\{v | (u, v) \in P\}|$.

So P can be projected to V by considering the second element in the swap-pairs.

We use Δ_{add}, Δ_{swap} and Δ_{drop} to denote the increase of $w(C)$ for the operations add, swap and drop respectively. Obviously, we have (1) for a vertex $v \in AddS$, $\Delta_{add}(v) = w(v)$; (2) for a vertex $u \in C$, $\Delta_{drop}(u) = -w(u)$; (3) for a vertex pair $(u, v) \in SwapS$, $\Delta_{swap}(u, v) = w(v) - w(u)$. Basically both MN/TS and LSCC obtain the best local move like Algorithm 1.

Algorithm 1. bestLocalMove

---

1   $v \leftarrow$ a vertex in $AddS$ with the biggest Δ_{add};
2   $(u, u') \leftarrow$ a pair in the $SwapS$ with the biggest Δ_{swap};
3   $x \leftarrow$ a vertex in C with the biggest Δ_{drop};
4   **if** $AddS \neq \emptyset$ **then**
5     $\lfloor$ $C \leftarrow (\Delta_{add} > \Delta_{swap})?(C \cup \{v\}) : (C \backslash \{u\}) \cup \{u'\}$;

6   **else**
7     $\lfloor$ $C \leftarrow (\Delta_{drop} > \Delta_{swap})?(C \backslash \{x\}) : (C \backslash \{u\}) \cup \{u'\}$;

---

[27] stated that *SwapS is usually very large when we solve large sparse graphs*. Yet we will show that this statement seems not to be the case.

2.4   The Strong Configuration Checking Strategy

[27] proposed a strategy named strong configuration checking (SCC): (1) in the beginning of the search, $confChange(v)$ is set to 1 for each vertex v; (2) when v is added, $confChange(n)$ is set to 1 for all $n \in N(v)$; (3) when v is dropped, $confChange(v)$ is set to 0; (4) When $u \in C$ and $v \notin C$ are swapped, $confChange(u)$ is set to 0.

---

[4] For any vertices u and v, we use $u = v$ to denote that u and v are the same vertex.

2.5   Best from Multiple Selections (BMS)

BMS is equivalent to *deterministic tournament selection* in genetic algorithms [20]. Given a set S and a positive integer k, it works as follows: *randomly select k elements from S with replacements and then return the best one.*

3   Local Move Yielded by Greedy and Random Selections

Based on Lemma 1, we have a theorem below which shows the bound of $|SwapS|$.

Theorem 1

1. If $C = \{w\}$ for some w, then $|SwapS| = |V| - |N(w)| - 1$;
2. If $|C| > 1$, then $|SwapS| \le 2d_{\max}$.

Proof. The first item is trivial, so we only prove the second one. Since $|C| > 1$, there must exist two different vertices u and w in C. Now we partition $SwapS$ to be: $P_1 = \{(x, y) \in SwapS | x = u\}$ and $P_2 = \{(x, y) \in SwapS | x \ne u\}$.

For any pair $(x, y) \in P_1$, since $x = u$, y must be a neighbor of w. Therefore, $\{y | (x, y) \in P_1\} \subseteq N(w)$. So $|\{y | (x, y) \in P_1\}| \le |N(w)|$. By the Lemma 1, we have $|P_1| = |\{y | (x, y) \in P_1\}| \le d(w)$, and thus $|P_1| \le d_{max}$.

For any pair $(x, y) \in P_2$, since $x \ne u$, y must be a neighbor of u. Therefore, $\{y | (x, y) \in P_2\} \subseteq N(u)$. So $|\{y | (x, y) \in P_2\}| \le |N(u)|$. By the Lemma 1, we have $|P_2| = |\{y | (x, y) \in P_2\}| \le d(u)$, and thus $|P_2| \le d_{max}$.

Therefore, $|SwapS| = |P_1| + |P_2| \le 2d_{max}$.

Since most large real-world graphs are sparse [5, 12, 13], we have $d_{max} \ll |V|$. Moreover, observing all the graph instances in the experiments, we find that $d_{max} < 3,000$. So for most of the time, $SwapS$ is a small set.

$SwapS$ becomes huge only when C contains a *single* vertex, namely u. In this situation, any vertex outside C but not adjacent to u, namely v, can go into C through a swap operation, and $\Delta_{swap}(u, v) = w(v) - w(u)$. Thus picking the best pair is somewhat equivalent to picking a vertex outside with the greatest weight. If we do so, we will obtain a single-vertex clique that contains a vertex with the greatest or near-greatest weight.

Usually there is only a tiny proportion of vertices whose weight is the greatest or close to the greatest. So when $|C| = 1$, if we pick the best swap-pair, we will obtain a clique from a tiny proportion of the single-vertex ones. Therefore, when $|C| = 1$ happens many times, the local search may revisit some areas.

We realize that when $|C| = 1$, [27] uses BMS, SCC and age to help diversify the local search. Yet it is unclear whether greedy search here is necessary, and we believe that random walk is feasible. In our solver, we will abandon BMS and use random walk instead, and the experimental performances are satisfactory.

4   The LMY-GRS Algorithm

LMY-GRS works with three operators add, swap and drop. With the current clique denoted by C, two sets S_{add} and S_{swap} are defined as below which are slightly different from $AddS$ and $SwapS$.

$$S_{add} = \begin{cases} \{v | v \notin C, v \in N(u) \text{ for all } u \in C\} & \text{if } |C| > 0; \\ \emptyset & \text{if } |C| = 0. \end{cases}$$

$$S_{swap} = \begin{cases} \{(u,v) | u \in C, v \notin C, \{u,v\} \notin E, v \in N(w) \text{ for all } w \in C\backslash\{u\}\} & \text{if } |C| > 1; \\ \emptyset & \text{if } |C| \leq 1. \end{cases}$$

Obviously we have

Proposition 2. $|S_{add}| \leq d_{\max}$, and $|S_{swap}| \leq 2d_{\max}$.

In our algorithm, the vertices of the operations are explicit from the context and thus omitted.

Basically LMY-GRS firstly initializes a clique via RandInitClique(G), and then uses local search to find better cliques. The initialization procedure is just the same as that in MN/TS and LSCC, which is shown in Algorithm 3.

Algorithm 2. LMY-GRS

---

input : A graph $G = (V, E, w)$ and the *cutoff*
output: A clique that was found with the greatest weight

1 *step* $\leftarrow$ 0; initialize the *confChange* array; $C \leftarrow$ RandInitClique(G);
2 **while** *elapsed time* < *cutoff* **do**
3   **if** $C = \emptyset$ **then** add a random vertex into C ;
4       $v \leftarrow$ a vertex in S_{add} s.t. *confChange*$(v) = 1$ with the biggest Δ_{add}, breaking ties in favor of the oldest one; otherwise $v \leftarrow$ NULL;
5       $(u, u') \leftarrow$ a pair in S_{swap} s.t. *confChange*$(u') = 1$ with the biggest Δ_{swap}, breaking ties in favor of the oldest u'; otherwise $(u, u') \leftarrow$ (NULL, NULL);
6   **if** $v \neq NULL$ **then**
7   **if** $(u, u') =$ *(NULL, NULL)* or $\Delta_{add} > \Delta_{swap}$ **then** $C \leftarrow C \cup \{v\}$;
8   **else** $C \leftarrow C\backslash\{u\} \cup \{u'\}$;
9   **else**
10          $x \leftarrow$ a vertex in C with the biggest Δ_{drop}, breaking ties in favor of the oldest one;
11   **if** $(u, u') =$ *(NULL, NULL)* or $\Delta_{drop} > \Delta_{swap}$ **then** $C \leftarrow C\backslash\{x\}$;
12   **else** $C \leftarrow C\backslash\{u\} \cup \{u'\}$;
13   *step* $\leftarrow$ *step* + 1; **if** $w(C) > w(C^*)$ **then** $C^* \leftarrow C$;
14   **if** *step* mod $L = 0$ **then**
15   drop all vertices in C; $C \leftarrow$ RandInitClique(G);
16   update *confChange* array according to SCC rules;
17 **return** C^*;

Algorithm 3. RandInitClique(G)

---

1 $C \leftarrow \emptyset$;
2 add a random vertex in C;
3 **while** $S_{add} \neq \emptyset$ **do** add a random vertex from S_{add} to C; $step \leftarrow step + 1$;
4 **if** $w(C) > w(C^*)$ **then** $C^* \leftarrow C$;
5 **return** C;

Like MN/TS and LSCC, we also exploit the multi-restart strategy. More specifically, every L steps we restart the local search. The difference from previous restarting methods is that we simply drop all vertices from C and regenerate a random maximal clique.

In LMY-GRS, C may sometimes become empty. In this situation, we add a random vertex in C and proceed to search for a better clique.

5 Data Structures

5.1 Connect Clique Degrees and Clique Neighbors

Definition 1. *Given a clique C and a vertex $v \notin C$, we define the connect clique degree of v to be $\kappa(v, C) = |\{u \in C | u$ and v are neighbors.$\}|$.*

So the connect clique degree of v is the number of vertices in C that are adjacent to v. Then we can maintain $\kappa(v, C)$ when a vertex is added into or dropped from C, based on the following proposition.

Proposition 3

1. $\kappa(v, C \backslash \{v\}) = |C| - 1$ *for any $v \in C$;*
2. $\kappa(v, C \backslash \{w\}) = \kappa(v, C) - 1$ *for all $v \in (N(w) \backslash C)$;*
3. $\kappa(v, C \cup \{w\}) = \kappa(v, C) + 1$ *for all $v \in (N(w) \backslash C)$.*

In [6], there is a notion named *the number of missing connections*. In fact the number of v's missing connections is $|C| - \kappa(v, C)$.

Definition 2. *Given a clique C, we define the clique neighbor set to be*

$$\mathcal{N}(C) = \begin{cases} \{v \notin C | v \in N(u) \text{ for some } u \in C\} & \text{if } |C| > 0; \\ \emptyset & \text{if } |C| = 0. \end{cases}$$

So clique neighbors are those vertices outside C which are adjacent to at least one vertex inside C.

When a vertex namely u is added into or dropped from C, $\mathcal{N}(C)$ is updated based on the proposition below.

Proposition 4

1. *For any vertex u, $u \notin \mathcal{N}(C \cup \{u\})$;*
2. *for any $u \in C$, $|C| > 1$ iff $u \in \mathcal{N}(C \backslash \{u\})$;*
3. *for any $v \notin C$, $\kappa(v, C) > 0$ iff $v \in \mathcal{N}(C)$.*

Lastly, we use the following proposition to maintain S_{add} and S_{swap}.

Proposition 5

1. *For any v, $v \in S_{add}$ iff $v \in \{w \in \mathcal{N}(C) | \kappa(w, C) = |C|\}$;*
2. *there exists $u \in C$ s.t. $(u, v) \in S_{swap}$, iff $v \in \{w \in \mathcal{N}(C) | \kappa(w, C) = |C| - 1\}$.*

This tells us that we can maintain S_{add} and S_{swap} simply by traversing $\mathcal{N}(C)$, so we do not have to traverse V. Previous local search solvers for MC or MVWC exclusively traverse all the vertices in V to maintain S_{add} and S_{swap}, so our implementation can sometimes be much more efficient. Notice that $|\mathcal{N}(C)| \ll |V|$ usually holds in huge sparse graphs.

5.2 A Hash Table for Determining Neighbor Relations

In graph algorithms there is a common procedure: *Given a graph G and two vertices u and v, determining whether u and v are neighbors.* In MN/TS, LSCC and LMY-GRS, this procedure is called *very frequently*, so we have to implement it efficiently. However, it is unsuitable to store the large sparse graphs by adjacency matrices. Therefore, we propose the following data structure which is both memory and time efficient.

We employ a one-to-one function $f : (Z^+, Z^+) \rightarrow Z^+$ and use a hash table to implement the procedure above. Given a graph $G = (V, E)$, for any $\{v_i, v_j\} \in E$ where $i < j$, we store $f(i, j)$ in a hash table $\mathcal{T}_h$. Then each time when we need to determine whether v_k and v_l ($k < l$) are neighbors, we simply check whether $f(k, l)$ is in $\mathcal{T}_h$. If so they are neighbors; otherwise, they are not.

In LMY-GRS, we adopt Cantor's pairing function as the function f above, *i.e.*, $f(x, y) = (x + y)(x + y + 1) \div 2 + y$. Then we can determine whether two vertices are neighbors in $O(1)$ complexity on average.

With the data structures above, our solver is able to perform steps faster than LSCC by orders of magnitude on huge sparse graphs.

6 Experimental Evaluation

In this section, we carry out extensive experiments to evaluate LMY-GRS on large crafted graphs, compared against the state-of-the-art local search MVWC algorithms MN/TS and LSCC.

6.1 Experiment Setup

All the solvers were compiled by g++ 4.6.3 with the '-O3' option. For the search depth L, MN/TS, LSCC and LMY-GRS set $L = 4,000$. Both MN/TS and LSCC employ the BMS heuristic, and the parameter k was set to 100, as in [27]. MN/TS employs a tabu heuristic and the tabu tenure TL was set to 7 as in [28]. The experiments were conducted on a cluster equipped with Intel(R) Xeon(R) CPUs X5650 @2.67 GHz with 8 GB RAM, running Red Hat Santiago OS.

Each solver was executed on each instance with a time limit of 1,000 s, with seeds from 1 to 100. For each algorithm on each instance, we report the maximum weight ($"w_{max}"$) and averaged weight ($"w_{avg}"$) of the cliques found by the algorithm. To make the comparisons clearer, we also report the difference ($"\delta_{max}"$) between the maximum weight of the cliques found by LMY-GRS and that found by LSCC. Similarly δ_{avg} represents the difference between the averaged weights. A positive δ_{avg} (resp. δ_{max}) indicates that LMY-GRS performed better, while a negative value indicates that LMY-GRS performed worse.

6.2 Main Results

We show the main experimental results in Tables 1 and 2. For the sake of space, we do not report the results on graphs with less than 1,000 vertices.

Quality Improvements. Table 1 shows the performances on the instances where LSCC and LMY-GRS returned different w_{max} or w_{avg} values. From the results in Table 1, we observe that:

1. LMY-GRS attained best-known solutions on all the graphs;
2. In a large proportion of the graphs, LMY-GRS returned solutions which had better average quality;
3. LMY-GRS found new best-known solutions in 4 graphs.

In fact these 4 graphs are the largest ones in the benchmark, and each of them has at least 10^6 vertices. Since LMY-GRS and LSCC present different solution qualities over these instances, it is inconvenient to evaluate the individual impacts of the heuristics and the data structures over these instances.

Time Improvements and Individual Impacts. Table 2 compares the performances on those instances where LSCC and LMY-GRS returned both the same w_{max} and w_{avg} values. We also present the averaged number of steps to locate a solution, and the number of steps executed in each millisecond.

Over these instances, we find that

1. The time columns show that LMY-GRS and LSCC performed closely well;
2. The step columns show that the heuristic in LMY-GRS is as good as the one in LSCC, since they needed roughly the same number of steps;
3. The last two columns show that LMY-GRS sometimes performed steps faster than LSCC, but sometimes slower.

To show the power of our data structures, we present the averaged number of steps per millisecond in some largest instances in Table 3. In this table we found that LMY-GRS is able to perform steps faster than LSCC by orders of magnitudes in some instances.

Based on the discussions above, we conclude that using random walk is roughly as good as using BMS, and our data structures are very powerful on huge sparse graphs.

Table 1. Results where LSCC and LMY-GRS returned different w_{max} or w_{avg} values

Graph	$|V|$	$|E|$	MN/TS w_{max} (w_{avg})	LSCC w_{max} (w_{avg})	LMY-GRS w_{max} (w_{avg})	$\delta_{max}(\delta_{avg})$
ca-coauthors-dblp	540486	15245729	37884 (27411.17)	37884 (34211.68)	37884 (**37884.00**)	0 (3672.32)
ca-dblp-2010	226413	716460	7575 (7256.55)	7575 (7470.61)	7575 (**7575.00**)	0 (104.39)
ca-dblp-2012	317080	1049866	14108 (11623.72)	14108 (13739.21)	14108 (**14108.00**)	0 (368.79)
ca-hollywood-2009	1069126	56306653	222720 (136456.22)	222720 (206446)	222720 (**219297.42**)	0 (12851.4)
ca-MathSciNet	332689	820644	2792 (2484.43)	2792 (2518.30)	2792 (**2792.00**)	0 (273.7)
inf-roadNet-CA	1957027	2760388	691 (598.72)	668 (622.60)	**752 (752.00)**	84 (129.4)
inf-roadNet-PA	1087562	1541514	637 (597.45)	599 (598.86)	**669 (669.00)**	70 (70.14)
sc-ldoor	952203	20770807	4074 (3874.34)	4081 (3966.68)	4081 (**4081.00**)	0 (114.32)
sc-msdoor	415863	9378650	4088 (3966.31)	4088 (4028.97)	4088 (**4088.00**)	0 (59.03)
sc-nasasrb	54870	1311227	4548 (4429.28)	4548 (4546.56)	4548 (**4548.00**)	0 (1.44)
sc-pkustk11	87804	2565054	5298 (4794.50)	5298 (5063.51)	5298 (**5298.00**)	0 (234.49)
sc-pkustk13	94893	3260967	6306 (5751.92)	6306 (5906.58)	6306 (**6287.10**)	0 (380.52)
sc-pwtk	217891	5653221	4596 (4384.80)	4620 (4606.56)	4620 (**4620.00**)	0 (13.44)
sc-shipsec1	140385	1707759	3540 (3084.78)	3540 (3294.51)	3540 (**3540.00**)	0 (245.49)
sc-shipsec5	179104	2200076	4524 (4320.50)	4524 (4407.96)	4524 (**4524.00**)	0 (116.04)
socfb-A-anon	3097165	23667394	2872 (2150.87)	2872 (2172.79)	2872 (**2872.00**)	0 (699.21)
socfb-B-anon	2937612	20959854	2537 (1911.92)	2662 (2008.74)	2662 (**2583.94**)	0 (575.2)
socfb-OR	63392	816886	3523 (3520.88)	3523 (3516.27)	3523 (**3523.00**)	0 (6.73)
soc-brightkite	56739	212945	3672 (3652.86)	3672 (3654.87)	3672 (**3661.98**)	0 (7.11)
soc-delicious	536108	1365961	1547 (1532.90)	1547 (1535.79)	1547 (**1546.93**)	0 (11.14)
soc-digg	770799	5907132	5287 (4742.21)	5303 (4712.43)	5303 (**4839.18**)	0 (126.75)
soc-flickr	513969	3190452	7083 (7032.71)	7083 (7069.15)	7083 (**7083.00**)	0 (13.85)
soc-flixster	2523386	7918801	3805 (3389.09)	3805 (3403.42)	3805 (**3805.00**)	0 (401.58)
soc-FourSquare	639014	3214986	3064 (**3057.10**)	3064 (3035.96)	3064 (2980.31)	0 (-55.65)
soc-gowalla	196591	950327	2335 (2249.70)	2335 (**2256.22**)	2335 (2253.51)	0 (-2.71)
soc-lastfm	1191805	4519330	1773 (1770.28)	1773 (1771.50)	1773 (**1773.00**)	0 (1.5)
soc-livejournal	4033137	27933062	7238 (2312.07)	15855 (2661.46)	**21368 (17783.19)**	5513 (15121.7)
soc-pokec	1632803	22301964	3191 (2132.10)	3191 (2008.17)	3191 (**3191.00**)	0 (1182.83)
soc-youtube-snap	1134890	2987624	1787 (1787.00)	1787 (1775.29)	1787 (**1787.00**)	0 (11.71)
tech-as-skitter	1694616	11094209	5703 (4894.34)	5703 (5076.34)	5703 (**5671.28**)	0 (594.94)
web-it-2004	509338	7178413	45477 (40088.88)	45477 (45380.95)	45477 (**45477.00**)	0 (96.05)
web-sk-2005	121422	334419	11925 (10583.83)	11925 (11892.72)	11925 (**11925.00**)	0 (32.28)
web-wikipedia2009	1864433	4507315	3823 (1752.62)	3823 (2100.06)	**3891 (3833.28)**	68 (1733.22)

Table 2. Performances on instances where they returned the same w_{max} and w_{avg} values

Graph	Time		#Step		#Step/ms	
	LSCC	LMY-GRS	LSCC	LMY-GRS	LSCC	LMY-GRS
bio-dmela	0.006	**0.005**	**140.9**	564.2	27.995	**113.956**
bio-yeast	0.002	**0.000**	**390.9**	489.16	94.419	**745.513**
ca-AstroPh	20.896	**11.057**	425754.9	**425117**	20.744	**40.648**
ca-citeseer	202.994	**0.647**	318659.9	**147513**	1.604	**231.164**
ca-CondMat	1.928	**0.168**	**32474.4**	36398	17.702	**228.675**
ca-CSphd	0.003	**0.000**	**523.5**	1085.49	57.501	**1069.836**
ca-Erdos992	0.003	**0.000**	**172.6**	183.19	**1099**	217.888
ca-GrQc	0.062	**0.009**	4882.4	**4496.11**	80.002	**297.030**
ca-HepPh	**0.086**	0.143	**2683.7**	5721.1	**33.919**	33.179
ia-email-EU	0.227	**0.079**	2129.3	**1053.53**	9.306	**15.749**
ia-email-univ	0.000	0.000	270.5	**232.85**	199.638	**348.546**
ia-enron-large	**6.280**	7.197	52895.0	**44609.6**	**8.575**	6.443
ia-fb-messages	0.000	0.000	21.8	**23.88**	**130.787**	111.793
ia-reality	0.048	**0.012**	**1325.7**	1550.71	26.814	**117.159**
ia-wiki-Talk	0.621	**0.433**	**1855.3**	2029.29	2.841	**3.666**
inf-power	**0.027**	0.051	**932.4**	1236.47	33.649	**1412.931**
rec-amazon	3.096	**0.000**	**4858.8**	66364.1	1.618	**1642.741**
socfb-Berkeley13	**49.960**	52.664	668700.0	**551266**	**13.601**	10.657
socfb-CMU	**3.312**	11.442	**127743.3**	145167	**39.716**	13.538
socfb-Duke14	**6.649**	19.876	**11332.0**	210134	**28.680**	11.280
socfb-Indiana	102.016	**90.841**	1073259.6	**1007020**	10.611	**12.112**
socfb-MIT	**3.246**	13.241	**134955.9**	172320	**42.597**	13.587
socfb-Penn94	**104.695**	135.512	782995.3	**397520**	7.638	**8.163**
socfb-Stanford3	**15.511**	36.116	**371036.8**	397520	**24.467**	11.514
socfb-Texas84	**79.057**	149.417	**673376.8**	868678	**8.510**	6.200
socfb-UCLA	**34.333**	45.327	**511804.3**	568163	**15.415**	13.202
socfb-UConn	**18.490**	19.268	**323312.8**	333033	**17.653**	17.254
socfb-UCSB37	**21.804**	30.156	426798.7	**410292**	**20.024**	14.249
socfb-UF	**70.877**	73.057	603137.5	**513143**	**8.575**	7.671
socfb-UIllinois	158.450	**140.481**	1539421.1	**1633500**	10.039	**12.798**
socfb-Wisconsin87	42.416	**34.065**	522928.6	**479857**	12.365	**15.266**
soc-BlogCatalog	**17.210**	122.788	45587.6	**31182.9**	**2.778**	0.267
soc-buzznet	92.703	**79.329**	211468.5	**17649.3**	**2.340**	0.233
soc-douban	1.770	**0.061**	**2789.6**	3718.46	1.801	**78.276**
soc-epinions	31.984	**17.557**	356916.5	**488393**	11.578	**28.758**
soc-LiveMocha	**7.809**	15.138	**21018.2**	31450.8	**2.719**	2.322
soc-slashdot	12.845	**1.557**	51360.6	**10392.7**	4.248	**6.993**
soc-twitter-follows	8.608	**0.511**	**3994.0**	6218.97	0.518	**12.417**
soc-youtube	**23.655**	33.585	11783.2	**17391**	**0.507**	0.487
tech-as-caida2007	0.094	**0.055**	951.9	**593.18**	**11.078**	3.257
tech-internet-as	**0.525**	0.803	**3871.9**	4119.28	**7.870**	3.503
tech-p2p-gnutella	0.968	**0.012**	3354.6	**3241.77**	3.582	**317.924**
tech-RL-caida	20.520	**0.741**	29238.2	**23989.1**	1.487	**27.937**
tech-routers-rf	**0.113**	0.116	**14551.5**	15513.4	127.910	**145.814**
tech-WHOIS	**0.426**	1.108	**16010.9**	21880.5	**38.349**	19.012
web-arabic-2005	30.543	**0.408**	57788.0	**51285**	1.897	**86.461**
web-Berkstan	0.077	**0.001**	1555.3	**1502.56**	22.143	**1439.152**
web-edu	2.885	**0.012**	191629.5	**2741.87**	67.589	**211.956**
web-google	0.117	**0.006**	25789.6	**15207**	220.953	**1644.486**
web-indochina-2004	0.041	**0.011**	**874.2**	1409.67	23.025	**124.223**
web-spam	**9.608**	26.431	417141.9	**392304**	**44.426**	14.681
web-uk-2005	33.803	**0.696**	**84260.7**	93855	2.567	**133.358**
web-webbase-2001	148.073	**2.234**	2393708.0	**57063.4**	16.748	**27.596**

Table 3. The number of steps per millisecond on huge sparse instances

Graph	LSCC	LMY-GRS	Graph	LSCC	LMY-GRS
ca-hollywood-2009	0.246	**0.365**	sc-ldoor	0.310	**356.573**
socfb-A-anon	0.070	**5.261**	soc-livejournal	0.057	**25.203**
socfb-B-anon	0.059	**5.751**	soc-pokec	0.125	**20.734**
inf-roadNet-CA	0.070	**1175.015**	tech-as-skitter	0.125	**0.785**
inf-roadNet-PA	0.126	**1085.205**	web-wikipedia2009	0.099	**11.713**

7   Conclusions and Future Work

In this paper, we developed a local search MVWC solver named LMY-GRS, which significantly outperforms state-of-the-art solvers on large sparse graphs. It attains best-known solutions on all the graphs in the experiments, and it finds new best-known solutions on some of them.

Our contributions are of three folds. Firstly we rigorously showed that the swap-set is usually small even when we are solving large sparse graphs. Secondly we incorporated random walk into the multi-neighborhood greedy search and showed that it is satisfactory. Thirdly we proposed efficient data structures that work well with huge sparse graphs.

In the future we will improve SCC to further avoid cycles. Also we will introduce more diversification strategies into LMY-GRS.

Acknowledgment. We thank all anonymous reviewers for their valuable comments. This work is supported by ARC Grant FT0991785, NSF Grant No.61463044, NSFC Grant No.61572234 and Grant No.[2014]7421 from the Joint Fund of the NSF of Guizhou province of China.

We gratefully acknowledge the support of the Griffith University eResearch Services Team and the use of the High Performance Computing Cluster "Gowonda" to complete this research.

References

1. Amgalan, B., Lee, H.: Wmaxc: a weighted maximum clique method for identifying condition-specific sub-network. PLoS ONE **9**(8), e104993 (2014)
2. Babel, L.: A fast algorithm for the maximum weight clique problem. Computing **52**(1), 31–38 (1994). http://dx.doi.org/10.1007/BF02243394
3. Balasundaram, B., Butenko, S.: Graph domination, coloring and cliques in telecommunications. In: Resende, M.G.C., Pardalos, P.M. (eds.) Handbook of Optimization in Telecommunications, pp. 865–890. Springer, Heidelberg (2006)
4. Ballard, D.H., Brown, C.M.: Computer Vision, 1st edn. Prentice Hall Professional Technical Reference, New York (1982)
5. Barabasi, A.L., Albert, R.: Emergence of scaling in random networks. Science **286**(5439), 509–512 (1999). http://www.sciencemag.org/cgi/content/abstract/286/5439/509

6. Battiti, R., Protasi, M.: Reactive local search for the maximum clique problem. Algorithmica **29**(4), 610–637 (2001)

7. Bomze, I.M., Pelillo, M., Stix, V.: Approximating the maximum weight clique using replicator dynamics. IEEE Trans. Neural Netw. Learn. Syst. **11**(6), 1228–1241 (2000). http://dx.doi.org/10.1109/72.883403

8. Brendel, W., Amer, M.R., Todorovic, S.: Multiobject tracking as maximum weight independent set. In: 24th IEEE Conference on Computer Vision and Pattern Recognition, CVPR 2011, Colorado Springs, CO, USA, 20–25 June 2011, pp. 1273–1280 (2011). http://dx.doi.org/10.1109/CVPR.2011.5995395

9. Brendel, W., Todorovic, S.: Segmentation as maximum-weight independent set. In: Advances in Neural Information Processing Systems 23: 24th Annual Conference on Neural Information Processing Systems 2010. Proceedings of a meeting held 6–9 December 2010, Vancouver, British Columbia, Canada, pp. 307–315 (2010). http://papers.nips.cc/paper/3909-segmentation-as-maximum-weight-independent-set

10. Busygin, S.: A new trust region technique for the maximum weight clique problem. Discrete Appl. Math. **154**(15), 2080–2096 (2006). http://dx.doi.org/10.1016/j.dam.2005.04.010

11. Cai, S.: Balance between complexity and quality: local search for minimum vertex cover in massive graphs. In: Proceedings of the Twenty-Fourth International Joint Conference on Artificial Intelligence, IJCAI 2015, Buenos Aires, Argentina, 25–31 July 2015, pp. 747–753 (2015). http://ijcai.org/papers15/Abstracts/IJCAI15-111.html

12. Chung, F., Lu, L.: Complex Graphs and Networks, vol. 107. American Mathematical Society (2006). https://books.google.com.au/books?id=BqqDsEKlAE4C

13. Eubank, S., Kumar, V.S.A., Marathe, M.V., Srinivasan, A., Wang, N.: Structural and algorithmic aspects of massive social networks. In: Proceedings of the Fifteenth Annual ACM-SIAM Symposium on Discrete Algorithms, SODA 2004, New Orleans, Louisiana, USA, 11–14 January 2004, pp. 718–727 (2004). http://dl.acm.org/citation.cfm?id=982792.982902

14. Fang, Z., Li, C., Qiao, K., Feng, X., Xu, K.: Solving maximum weight clique using maximum satisfiability reasoning. In: 21st European Conference on Artificial Intelligence, ECAI 2014, 18–22 August 2014, Prague, Czech Republic - Including Prestigious Applications of Intelligent Systems (PAIS 2014), pp. 303–308 (2014). http://dx.doi.org/10.3233/978-1-61499-419-0-303

15. Feige, U.: Approximating maximum clique by removing subgraphs. SIAM J. Discret. Math. **18**(2), 219–225 (2005). http://dx.doi.org/10.1137/S089548010240415X

16. Johnson, D.J., Trick, M.A. (eds.): Cliques, Coloring, and Satisfiability: Second DIMACS Implementation Challenge, Workshop, October 11–13, 1993. American Mathematical Society, Boston (1996)

17. Karp, R.M.: Reducibility among combinatorial problems. In: Proceedings of a Symposium on the Complexity of Computer Computations, 20–22 March 1972. At the IBM Thomas J. Watson Research Center, Yorktown Heights, New York, pp. 85–103 (1972). http://www.cs.berkeley.edu/luca/cs172/karp.pdf

18. Li, N., Latecki, L.J.: Clustering aggregation as maximum-weight independent set. In: Advances in Neural Information Processing Systems 25: 26th Annual Conference on Neural Information Processing Systems 2012. Proceedings of a meeting held 3–6 December 2012, Lake Tahoe, Nevada, United States, pp. 791–799 (2012). http://papers.nips.cc/paper/4731-clustering-aggregation-as-maximum-weight-independent-set

19. Ma, T., Latecki, L.J.: Maximum weight cliques with mutex constraints for video object segmentation. In: 2012 IEEE Conference on Computer Vision and Pattern Recognition, Providence, RI, USA, 16–21 June 2012, pp. 670–677 (2012). http:// dx.doi.org/10.1109/CVPR.2012.6247735

20. Miller, B.L., Goldberg, D.E.: Genetic algorithms, tournament selection, and the effects of noise. Complex Syst. **9**(3), 193–212 (1995)

21. Östergård, P.R.J.: A new algorithm for the maximum-weight clique problem. Nord. J. Comput. **8**(4), 424–436 (2001). http://www.cs.helsinki.fi/njc/References/ ostergard2001:424.html

22. Pullan, W.J.: Approximating the maximum vertex/edge weighted clique using local search. J. Heuristics **14**(2), 117–134 (2008). http://dx.doi.org/10.1007/s10732-007-9026-2

23. Ravetti, M.G., Moscato, P.: Identification of a 5-protein biomarker molecular signature for predicting alzheimer's disease. PLoS ONE **3**(9), e3111 (2008)

24. Rossi, R.A., Ahmed, N.K.: Coloring large complex networks. Soc. Netw. Analys. Min. **4**(1), 228 (2014). http://dx.doi.org/10.1007/s13278-014-0228-y

25. Rossi, R.A., Ahmed, N.K.: The network data repository with interactive graph analytics and visualization. In: Proceedings of the Twenty-Ninth AAAI Conference on Artificial Intelligence (2015)

26. Rossi, R.A., Gleich, D.F., Gebremedhin, A.H., Patwary, M.M.A.: Fast maximum clique algorithms for large graphs. In: 23rd International World Wide Web Conference, WWW 2014, Seoul, Republic of Korea, 7–11 April 2014, Companion Volume, pp. 365–366 (2014). http://doi.acm.org/10.1145/2567948.2577283

27. Wang, Y., Cai, S., Yin, M.: Two efficient local search algorithms for maximum weight clique problem. In: Proceedings of the Thirtieth AAAI Conference on Artificial Intelligence, 12–17 February 2016, Phoenix, Arizona, USA, pp. 805–811 (2016). http://www.aaai.org/ocs/index.php/AAAI/AAAI16/paper/view/11915

28. Wu, Q., Hao, J., Glover, F.: Multi-neighborhood tabu search for the maximum weight clique problem. Ann. OR **196**(1), 611–634 (2012). http://dx.doi.org/ 10.1007/s10479-012-1124-3

29. Xu, K., Boussemart, F., Hemery, F., Lecoutre, C.: A simple model to generate hard satisfiable instances. In: Proceedings of the 19th International Joint Conference on Artificial Intelligence, IJCAI 2005, pp. 337–342. Morgan Kaufmann Publishers Inc., San Francisco (2005). http://dl.acm.org/citation.cfm?id=1642293.1642347

30. Yamaguchi, K., Masuda, S.: A new exact algorithm for the maximum weight clique problem. In: ITC-CSCC: 2008, pp. 317–320 (2008)

Cascade Bayesian Optimization

Thanh Dai Nguyen[1]([✉]), Sunil Gupta[1], Santu Rana[1], Vu Nguyen[1],
Svetha Venkatesh[1], Kyle J. Deane[2], and Paul G. Sanders[2]

[1] Center for Pattern Recognition and Data Analytics,
Deakin University, Geelong 3216, Australia
{thanh,sunil.gupta,santu.rana,v.nguyen,svetha.venkatesh}@deakin.edu.au
[2] Materials Science and Engineering,
Michigan Technological University, Houghton, USA

Abstract. Multi-stage cascade processes are fairly common, especially in manufacturing industry. Precursors or raw materials are transformed at each stage before being used as the input to the next stage. Setting the right control parameters at each stage is important to achieve high quality products at low cost. Finding the right parameters via trial and error approach can be time consuming. Bayesian optimization is an efficient way to optimize costly black-box function. We extend the standard Bayesian optimization approach to the cascade process through formulating a series of optimization problems that are solved sequentially from the final stage to the first stage. Epistemic uncertainties are effectively utilized in the formulation. Further, cost of the parameters are also included to find cost-efficient solutions. Experiments performed on a simulated testbed of Al-Sc heat treatment through a three-stage process showed considerable efficiency gain over a naïve optimization approach.

1 Introduction

Cascade processes abound in real life. As example, consider the strength of materials produced by cascade heat treatment processes. From planes and cars to skyscrapers, there is a critical dependence on industrial alloys with specific properties - strength, creep resistance, weldability and so on. Heat treatments are applied to alloys to achieve such target properties and involve a cascade of steps, where the temperature is maintained for several hours in one stage before proceeding to a new temperature for the next step. This "cascade" matches the underlying physical processes, effectively allowing the earlier stages of lower temperature to determine the *nucleation* (the number of nuclei), and then *"growth"* or "coarsening" that dominates later stages and eventually controls final hardness. Temperature and duration of each stage influences the final alloy properties. In general, most of the industrial manufacturing processes are actually cascade of many different stages. Precursors or raw materials are transformed at each stage before being used as the input to the next stage. The parameters of each stage influence the final product quality, and cost. Searching for the right parameters that results in the highest quality product with the lowest cost can be a

© Springer International Publishing AG 2016
B.H. Kang and Q. Bai (Eds.): AI 2016, LNAI 9992, pp. 268–280, 2016.
DOI: 10.1007/978-3-319-50127-7_22

time-consuming process. Therefore, it is important to find them in as minimum number of trials as possible.

Bayesian optimization has recently emerged as a popular approach to efficiently optimize costly black-box functions, especially to tune hyperparameters of complex machine learning algorithms. Gaussian process is used as a prior for the unknown function, and the posterior distribution is maintained based on the current observation set. Next, a surrogate utility function named acquisition function is created by combining both the predictive mean and the variance of the posterior through a functional form. Acquisition function is optimized to find the parameters for the next trial. The acquisition function is crafted to exploit the areas of high predictive mean and explore high uncertain areas in a balanced fashion. Expected Improvement (EI) [1] is one such acquisition function that encodes the expected improvement over the current best for a given trial point. The maximizer of the EI is, therefore, expected to produce a high function value. Other choices of priors [2] and acquisition functions are possible [3].

The word "cascade" and "multistage" have been used in prior literature in many different contexts - it can mean "iteration" in a Gaussian process, it can mean "multiple stages" where the outputs of previous stages are coupled. Very early work in engineering design methodology refers to the term multistage Bayesian surrogates - the multistage here refers to consecutive iterations of a batch Gaussian process [4]. Wang et al. [5] propose multi-stage hyper-parameter optimization using Bayesian optimization that splits the data - each stage merely takes in increasing amounts of data. Multistage Gaussian process [6,7] cascades Gaussian processes, with a view for example to do multi-step ahead prediction - this is done by transmitting the mean and covariance of the previous GP to the next stage. This model differs from ours in a fundamental way: it does not allow new inputs at different cascade stages which is critical in experimental and industrial processes as each stage may be performing a different function. Our definition of cascade Bayesian optimization stems from the industrial cascade processes. Each intermediate stage must allow not only the output of the last stage to be coupled, but also allow the control parameters for this stage to be input. The target quality is the output of the final stage of the cascade. Intermediate stage measurements can be made in such processes. The idea that Bayesian optimization can be applied to such cascaded experimental processes is new. *To the best of our knowledge, no current solutions exist to incorporate the structure of cascade into Bayesian optimization frameworks.*

To solve this problem, we propose a cascade Bayesian optimization framework. Each of the cascade stage is modeled by an independent function through a separate Gaussian process. We assume that the output product quality of each stage is measurable. Since we need to maximize quality of the product from the final stage, we find the control parameters of the final stage and desired input material quality to the final stage by maximizing an acquisition function based on only the final stage Gaussian process posterior. The stage feeding to the final stage now thus only needs to supply the final stage with its desired input material quality. Clearly, this is also the case for all other intermediate stages which need to produce the desired material quality feeding the next stage by controlling its parameters and asking for a desired input material quality. We formulate a novel optimization problem to find the desired input quality and the control parameters of the intermediate stages through the inversion of the Gaussian process posterior, exploiting also the epistemic uncertainties in the model. A sequence of optimization problems is solved starting from the penultimate stage to the first stage. At the end of the cascade Bayesian optimization procedure we obtain the control parameter values of all the stages that need to be set for the next trial. Additionally, we can introduce costs associated with the control parameters into our optimization formulation to discover cost-efficient solutions.

We validate our algorithm on both synthetic and real world data. In synthetic data, we show that the algorithm is effective under diverse simulated conditions. In the first experiment, we fix the number of parameters per stage to 4 and vary the number of stages from 4 to 8. In the second experiment, the number of stages is kept fixed at 5, however, the number of parameters per stage is varied from 3 to 7. We compare our cascade Bayesian optimization algorithm with a naïve application of the standard Bayesian optimization approach which collapses all the control parameters into a joint optimization problem. We call that baseline method as joint Bayesian optimization. We show that the cascade Bayesian optimization is particularly effective when the number of stages is large or the number of parameters per stage is high. In our extreme example where the number of stages is 5 and the number of parameters per stage is 7, cascade Bayesian optimization is able to perform 11 times better than the joint Bayesian optimization.

For real world experiments, we consider industry standard simulation testbed of heat treatment for Al-Sc alloy. The testbed is based on the classical numerical precipitation model [8,9]. This model is built on molecular kinetic theory and can derive nucleation and growth of specific alloy compositions as the temperature is varied over time. Multiple processing steps are required as the radius of particles in the earlier stage affects the micro-structural evolution of the next stage. We design a three-stage heat treatment to achieve high hardness. We show that on average the cascade Bayesian optimization can achieve peak hardness in only 15 trials compared to around 20 trials required for joint Bayesian optimization. Since, a single heat-treatment trial can take anywhere from a day to 3 days, a saving of 5 trials can mean saving of at least 5 days. We further show that when cost is taken into account it is possible to save 30% of heat-treatment time while maintaining a similar level of hardness (<2% drop).

2   Background

Bayesian optimization [10] is a method to find the maximum of an unknown objective function f as

$$\mathbf{x}^* = \arg\max_{\mathbf{x} \in \mathcal{X}} f(\mathbf{x}) \tag{1}$$

where $\mathcal{X}$ is the domain of f.

Bayesian optimization has two key steps. The first step is to model the function so that we can predict the function value at any point in the function domain along with its uncertainty. Gaussian process [11] is a popular approach to achieve this step. The second step is to construct a surrogate function (or acquisition function) that has a tractable form that trades off between exploitation of high predicted value and exploration of highly uncertain regions. In the following, we provide a brief description of the Gaussian Process and Expected Improvement (EI) acquisition function.

2.1   Gaussian Process

Gaussian Process [11] is a generalization of a multivariate Gaussian distribution. While a multivariate Gaussian distribution is a distribution over vectors, we can think of Gaussian process as a stochastic process that describes the distributions over functions. Intuitively, if we can think of a function as an infinite dimensional vector then Gaussian process is a stochastic process defined over this vector. Gaussian process is specified by a mean function $m(\mathbf{x})$ and a covariance function $k(\mathbf{x}, \mathbf{x}')$. We denote a function $f(\mathbf{x})$ that is said to be drawn from Gaussian Process as follows:

$$f(\mathbf{x}) \sim \mathcal{GP}(m(\mathbf{x}), k(\mathbf{x}, \mathbf{x}'))$$

Assume that we have t data points $\{(\mathbf{x}_i, y_i)\}, i = 1, 2, ..., t$ where $y_i = f(\mathbf{x}_i)$ and $f(\mathbf{x})$ is drawn from a Gaussian distribution. Without loss of generality, we can consider mean function $m(\mathbf{x})$ is zero so the Gaussian Process depends only on the covariance function. We should choose a covariance function that makes a valid covariance matrix as follows:

$$\mathbf{K} = \begin{bmatrix} k(\mathbf{x}_1, \mathbf{x}_1) & \cdots & k(\mathbf{x}_1, \mathbf{x}_t) \\ \vdots & \ddots & \vdots \\ k(\mathbf{x}_t, \mathbf{x}_1) & \cdots & k(\mathbf{x}_t, \mathbf{x}_t) \end{bmatrix} \tag{2}$$

To be a covariance matrix, a matrix should be positive semi-definite. According to Mercer's theorem for kernels, a covariance function $k(\mathbf{x}, \mathbf{x}')$ should be a valid kernel function. One popular choice for kernel function is the squared exponential kernel:

$$k_{SE}(\mathbf{x}, \mathbf{x}') = \exp(-\frac{1}{2\tau^2} \|\mathbf{x} - \mathbf{x}'\|^2) \tag{3}$$

Both posterior and predictive distribution of a Gaussian process can be derived from the following construction. When a new data point $\mathbf{x}_{t+1}$ comes, y_{t+1} and $\mathbf{y}_{1:t}$ are jointly Gaussian by properties of Gaussian Process:

$$\begin{bmatrix} \mathbf{y}_{1:t} \\ y_{t+1} \end{bmatrix} \sim \mathcal{N}(0), \begin{bmatrix} \mathbf{K} & \mathbf{k} \\ \mathbf{k}^T & k(\mathbf{x}_{t+1}, \mathbf{x}_{t+1}) \end{bmatrix}$$

where $\mathbf{k} = [k(\mathbf{x}_1, \mathbf{x}_{t+1}), k(\mathbf{x}_2, \mathbf{x}_{t+1}), ..., k(\mathbf{x}_t, \mathbf{x}_{t+1})]$. Using the rules for conditioning Gaussian, we can predict the distribution of the function value at $\mathbf{x}_{t+1}$ as:

$$y_{t+1}|\mathbf{y}_{1:t}, \mathbf{x}_{1:t+1} \sim \mathcal{N}(\mu_t(\mathbf{x}_{t+1}), \sigma_t^2(\mathbf{x}_{t+1})) \tag{4}$$

$$\mu_t(\mathbf{x}_{t+1}) = \mathbf{k}^T \mathbf{K}^{-1} \mathbf{y}_{1:t} \tag{5}$$

$$\sigma_t^2(\mathbf{x}_{t+1}) = k(\mathbf{x}_{t+1}, \mathbf{x}_{t+1}) - \mathbf{k}^T \mathbf{K}^{-1} \mathbf{k} \tag{6}$$

2.2   Acquisition Functions for Bayesian Optimization

As the function to be optimized is unknown and expensive to evaluate, Bayesian optimization constructs an acquisition function that has a known form and can be optimized to efficiently determine the point where the next evaluation should be performed. Therefore, instead of maximizing the original function f in Eq. 1, we maximize the acquisition function

$$\mathbf{x}_{t+1}^* = \arg\max \alpha(\mathbf{x}; \mathcal{I}_t).$$

where $\mathcal{I}_t$ denotes the Gaussian process model estimated using t observations. We use $\mathcal{D}_t$ to denote the observations up to t, i.e. $\mathcal{D}_t = \{(\mathbf{x}_i, y_i)\}, i = 1, 2, ..., t$. The above maximization problem can be solved using standard numerical techniques such as local optimizers with multiple initialization, sequential quadratic programming or DIRECT [12]. In the following, we describe how to construct an acquisition function $\alpha(\mathbf{x}; \mathcal{I}_t)$ from the posterior distribution of Gaussian Process. Normally, acquisition functions are designed such that their high values yield high values of objective function $f(\mathbf{x})$, either because of high uncertainty or high function value or both. This is done to have a trade-off between exploiting promising area and exploring unknown space.

The expected improvement (EI): EI defines an improvement function as $I(\mathbf{x}) = \max\{0, f_{t+1}(\mathbf{x}) - f(\mathbf{x}^+)\}$. Using the posterior Gaussian process, its expectation is computed as

$$\alpha_{EI}(\mathbf{x}; \mathcal{D}) = (\mu(\mathbf{x}) - f(\mathbf{x}^+)) \Phi(z) + \sigma(\mathbf{x}) \phi(z) \tag{7}$$

where $z(\mathbf{x}) = \frac{\mu(\mathbf{x}) - f(\mathbf{x}^+)}{\sigma(\mathbf{x})}$, $\sigma(\mathbf{x}) > 0$ and be zero otherwise. Maximizing EI would suggest a trial point that has the highest average improvement over the current-best function value.

3   The Proposed Solution

Consider a process that consists of multiple stages (indexed as $s = 1, 2, ..., S$). In this process, each stage takes the output of the previous stage as input. Each stage also has control parameters that affect its output. Let us use $\mathbf{u}_t^s$, $\mathbf{v}_t^s$ and $\mathbf{w}_t^s$ to denote the input, control parameters and output of stage s for observation t, where $t = 1, 2, ..., T$. Further, let us use f_t^s to denote the underlying function at iteration t of stage s such that $\mathbf{w}_t^s = f_t^s(\mathbf{x}_t^s)$ where $\mathbf{x}_t^s = [\mathbf{u}_t^s, \mathbf{v}_t^s]$ is combination of input $\mathbf{u}_t^s$ and control parameters $\mathbf{v}_t^s$. This function is modeled by a Gaussian process $\mathcal{GP}_t^s$. Since the output of a stage s acts as the input to the next stage $s+1$, we have $\mathbf{w}^s = \mathbf{u}^{s+1}$. Let $y = \mathbf{w}^S$ be the output of the final stage. Given the input $\mathbf{u}^1$ at the first stage, the optimization problem becomes discovering the set of control parameters $\mathbf{V}^* = [\mathbf{v}^{1*}, \mathbf{v}^{2*}, ..., \mathbf{v}^{S*}]$ that yields the highest output y^* from the final stage in the minimum number of explorations. We call this problem as Cascade Bayesian Optimization.

We propose an algorithm that consists of two steps (see Algorithm 1).

Step-1: Tuning parameters at last stage. The problem of finding the maximum output value from the last stage can be represented as follows:

$$y^* = \max_{\mathbf{x}_t^S} f_t^S(\mathbf{x}_t^S) \tag{8}$$

Bayesian optimization is used to solve this optimization problem. At iteration t_0+1, the Gaussian process at last stage $\mathcal{GP}_{t_0}^S$ has t_0 observations: $\mathbf{x}_1^S, \mathbf{x}_2^S, ... , \mathbf{x}_{t_0}^S$. By using properties of Gaussian process, we have $\mathbf{w}_{t_0+1}^S$ following a Gaussian distribution whose mean and variance can be represented as functions of $x_{t_0+1}^S$:

$$\mu_{t_0}^S(\mathbf{x}_{t_0+1}^S) = \mathbf{k}^T\mathbf{K}^{-1}\mathbf{y}_{1:t_0} \tag{9}$$

$$\left[\sigma_{t_0}^S(\mathbf{x}_{t_0+1}^S)\right]^2 = k(\mathbf{x}_{t_0+1}^S, \mathbf{x}_{t_0+1}^S) - \mathbf{k}^T\mathbf{K}^{-1}\mathbf{k} \tag{10}$$

where $k(\mathbf{x}, \mathbf{x}')$ is the squared exponential kernel function (see Eq. 3), $\mathbf{K}$ is the covariance matrix (see Eq. 2) and $\mathbf{k} = [k(\mathbf{x}_1^S, \mathbf{x}_{t_0+1}^S), k(\mathbf{x}_2^S, \mathbf{x}_{t_0+1}^S), ..., k(\mathbf{x}_{t_0}^S, \mathbf{x}_{t_0+1}^S)]$.

We use Expected improvement (7) to suggest the next sample $\mathbf{x}_{t_0+1}^S$ for exploration. This is done by maximizing the acquisition function as

$$\mathbf{x}_{t_0+1}^S = \underset{\mathbf{x}}{\text{argmax}} \left(\mu_{t_0}^S(\mathbf{x}) - \gamma\right) \Phi(Z) + \sigma_{t_0}^S(\mathbf{x})\phi(Z) \tag{11}$$

where $Z = \frac{\mu_{t_0}^S(\mathbf{x}) - \gamma}{\sigma_{t_0}^S(\mathbf{x})}$ when $\sigma_{t_0}^S(\mathbf{x}) > 0$ and $Z = 0$ otherwise, and γ is the current best value.

Step-2: Tuning parameters at remaining stages. After finding $\mathbf{x}_{t_0+1}^S$ at the last stage, we search for the input and control parameters at previous stages $[\mathbf{u}_{t_0+1}^1, \mathbf{v}_{t_0+1}^1], [\mathbf{u}_{t_0+1}^2, \mathbf{v}_{t_0+1}^2], ..., [\mathbf{u}_{t_0+1}^{S-1}, \mathbf{v}_{t_0+1}^{S-1}]$ that produce $\mathbf{x}_{t_0+1}^S$. In other words, we have to compute $\mathbf{x}_{t_0+1}^1, \mathbf{x}_{t_0+1}^2, ..., \mathbf{x}_{t_0+1}^{S-1}$ given $\mathcal{GP}_{t_0}^1, \mathcal{GP}_{t_0}^2, ..., \mathcal{GP}_{t_0}^{S-1}$ and $\mathbf{u}_{t_0+1}^S$. Here we propose an approach to compute $\mathbf{x}_{t_0+1}^s$ given $\mathcal{GP}_{t_0}^s$ and $\mathbf{u}_{t_0+1}^s$ for $s = 1, 2, ..., S - 1$. Because a Gaussian process is built at each stage, $\mathbf{x}_{t_0+1}^s$ can be computed given the output $\mathbf{w}_{t_0+1}^s$. An intuitive approach to solve for $\mathbf{x}_{t_0+1}^s$ is to minimize the error between the output of Gaussian process $\mathcal{GP}_{t_0}^s$ and the desired output $\mathbf{w}_{t_0+1}^s$ (we note that $\mathbf{w}_{t_0+1}^s = \mathbf{u}_{t_0+1}^{s+1}$):

$$\min_{\mathbf{x}_{t_0+1}^s} \left\|\mu_{t_0}^s(\mathbf{x}_{t_0+1}^s) - \mathbf{u}_{t_0+1}^{s+1}\right\|$$

However, since Bayesian optimization deals with costly functions and there are only limited data points, the prediction of the model can not be overly trusted.

Algorithm 1. Cascade Bayesian Optimization Algorithm

Input: Input of the first stage $\mathbf{u}^1$

1: Initialize t_0 control parameters $\mathbf{v}_1^s, ..., \mathbf{v}_{t_0}^s$ and calculate $\mathbf{w}_1^s, \mathbf{w}_2^s, ..., \mathbf{w}_{t_0}^s$ for $s = 1, 2, ..., S$

2: Build $\mathcal{GP}_{t_0}^s, s = 1, 2, ..., S$

3: $t := t_0$

4: **repeat**

5:     $t := t + 1$

6:     Suggest a next sample to evaluate $\mathbf{x}_t^S$ at the last stage by maximizing Eq. 11

7:     **for** $s = S - 1$ down to 1

8:         $\mathbf{w}_t^s = \mathbf{u}_t^{s+1}$

9:         Compute $\mathbf{x}_t^s$ by minimizing Eq. 12

10:    **endfor**

11:    Add new data points $\mathbf{x}_t^1, \mathbf{x}_t^2, ..., \mathbf{x}_t^S$ to the Gaussian processes $\mathcal{GP}_t^1, \mathcal{GP}_t^2, ..., \mathcal{GP}_t^S$.

12:**until** y^* found or $t > T$

Output: $\mathbf{V}^*, y^*$

Thus, the uncertainty of Gaussian process should also be considered in the objective function. Therefore, we incorporate uncertainty in the optimization as

$$\min_{\mathbf{x}_{t_0+1}^s} \kappa_1 \frac{\left\| \mu_{t_0}^s(\mathbf{x}_{t_0+1}^s) - \mathbf{u}_{t_0+1}^{s+1} \right\|^2}{\left[\sigma_{t_0}^S(\mathbf{x}_{t_0+1}^S) \right]^2} + \kappa_2 \left\| \mu_{t_0}^s(\mathbf{x}_{t_0+1}^s) - \mathbf{u}_{t_0+1}^{s+1} \right\|^2 \left[\sigma_{t_0}^S(\mathbf{x}_{t_0+1}^S) \right]^2$$

subject to $\mathbb{E}\left[\mathbf{u}_{t_0+1}^{s+1} | \mathcal{GP}_{t_0}^s, \mathbf{x}_{t_0+1}^s \right] = \mu_{t_0}^s(\mathbf{x}_{t_0+1}^s)$

where κ_1 and κ_2 are model parameters balancing the requirement between exploitation (error term) and exploration (uncertainty) of the objective function. Since the function inversion problem may be ill-posed meaning there may be many different inputs that lead to the same output, we use a criteria for choosing one input among others. The final optimization problem is defined as follows:

$$\min_{\mathbf{x}_{t_0+1}^s} \kappa_1 \frac{\left\| \mu_{t_0}^s(\mathbf{x}_{t_0+1}^s) - \mathbf{u}_{t_0+1}^{s+1} \right\|^2}{\left[\sigma_{t_0}^S(\mathbf{x}_{t_0+1}^S) \right]^2} + \kappa_2 \left\| \mu_{t_0}^s(\mathbf{x}_{t_0+1}^s) - \mathbf{u}_{t_0+1}^{s+1} \right\|^2 \left[\sigma_{t_0}^S(\mathbf{x}_{t_0+1}^S) \right]^2 + c(\mathbf{x}_{t_0+1}^s)$$

(12)

subject to $\mathbb{E}\left[\mathbf{u}_{t_0+1}^{s+1} | \mathcal{GP}_{t_0}^s, \mathbf{x}_{t_0+1}^s \right] = \mu_{t_0}^s(\mathbf{x}_{t_0+1}^s)$

where $c(\mathbf{x}_{t_0+1}^s)$ is the cost function that is designed to choose among multiple input solutions. In our implementation, we use $c(\mathbf{x}_{t_0+1}^s) = \left\| \mathbf{x}_{t_0+1}^s \right\|^2$. At the first stage, since the input is fixed and only control parameters can vary, the inversion process will only estimate the control parameter $\mathbf{v}_1$ instead of $\mathbf{x}_1$.

The problem of finding maximum value of cascade process now becomes optimizing two functions (11) and (12). Here we derive the derivative of objective function of (12) which is useful if the objective function is optimized using local optimizers. Although it is possible to use global optimizers for this optimization, local optimizers are our only choice in high dimensional parameter space due to computational reasons. Let $\Psi(\mathbf{x})$ be the objective function of (12). Its derivative can be written as

$$\frac{d\Psi(\mathbf{x})}{d\mathbf{x}} = \kappa_1 \frac{d}{d\mathbf{x}}\left[\frac{(\mu(\mathbf{x}) - \mathbf{u})^2}{\sigma^2(\mathbf{x})}\right] + \kappa_2 \frac{d}{d\mathbf{x}}(\mu(\mathbf{x}) - \mathbf{u})^2\sigma^2(\mathbf{x}) + \frac{d}{d\mathbf{x}}c(\mathbf{x}) \quad (13)$$

$$= \kappa_1 \frac{A - B}{\left[\sigma^2(\mathbf{x})\right]^2} + \kappa_2(A + B) + \frac{d}{d\mathbf{x}}c(\mathbf{x}) \quad (14)$$

where $A = 2\sigma^2(\mathbf{x})(\mu(\mathbf{x}) - \mathbf{u})\frac{d\mu(\mathbf{x})}{d\mathbf{x}}$ and $B = (\mu(\mathbf{x}) - \mathbf{u})^2\frac{d\sigma^2(\mathbf{x})}{d\mathbf{x}}$. Finally, A and B can be rewritten as follows

$$A = -\frac{2}{\tau^2}\sigma^2(\mathbf{x})(\mu(\mathbf{x}) - \mathbf{u})\left[\mathbf{K}^{-1}\mathbf{y}\right]^T C$$

$$B = -\frac{2}{\tau^2}(\mu(\mathbf{x}) - \mathbf{u})^2\mathbf{k}^T\mathbf{K}^{-1}C$$

$$C = \begin{bmatrix} k(\mathbf{x}, \mathbf{x}_1)(\mathbf{x} - \mathbf{x}_1)^T \\ k(\mathbf{x}, \mathbf{x}_2)(\mathbf{x} - \mathbf{x}_2)^T \\ \vdots \\ k(\mathbf{x}, \mathbf{x}_{t_0})(\mathbf{x} - \mathbf{x}_{t_0})^T \end{bmatrix}$$

4   Experiments

In this section, we conduct a set of experiments using both synthetic and real data to demonstrate the efficiency of our cascade Bayesian optimization method. Experiments with synthetic data are performed to illustrate the behavior of our method in different scenarios e.g. how does the efficiency scale with growing number of stages in the cascade, how does the efficiency scale with growing number of parameters in a stage. Experiments with real data demonstrate the effectiveness of our method for alloy heat treatment process where we show the benefits of our method in two terms: the number of trial experiments to reach a desired alloy hardness and the total heating time required.

4.1   Baseline Method and Evaluation Measure

To evaluate the effectiveness of our proposed model, we compare its performance with a baseline named *joint Bayesian optimization*. When presented with a n-stage cascade process optimization, this algorithm does not take cascade process into account. Instead it combines all the control parameters of all stages into a single input vector and uses Bayesian optimization to optimize the overall underlying process in this combined space. The output of the last stage is used as target to be maximized. To the best of our knowledge, there are no other

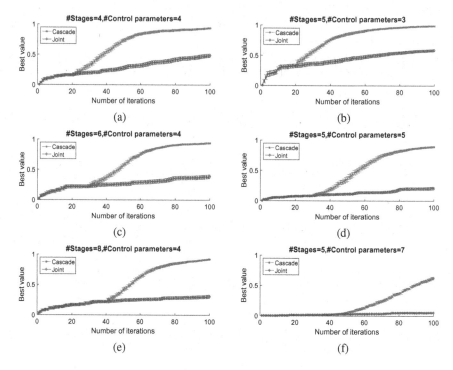

Fig. 1. Experimental results of Synthetic-I (left column) and Synthetic-II (right column): Best found value so far as a function of iteration. In the left column, number of control parameters per stage is fixed at 4 but the number of stages are varied. In the right column, number of stages is fixed at 5, but the number of control parameters per stage is varied.

methods that can perform optimization for cascade processes in sequential design setting.

To measure the effectiveness of our proposed and the baseline methods, we plot the best output value reached from the last stage as a function of number of iterations.

4.2 Experiments with Synthetic Data

We evaluate the proposed methods with two synthetic datasets. The first synthetic dataset is created to demonstrate the behavior of the proposed model and baseline method with increasing number of stages, whereas the second synthetic dataset is generated to illustrate performance of cascade Bayesian optimization and joint Bayesian optimization by varying number of control parameters for each stage.

Data Generation. Synthetic-I: This synthetic dataset is designed to illustrate the performance of the proposed method with varying number of stages.

We generate 4, 6 and 8 stage processes that have 4 control parameters for each stage. The underlying function at each stage is the probability density function of a multivariate Gaussian distribution. The input for s-th stage is the combination of control parameter of s-th stage and the output of the $s - 1$-th stage. Thus, our proposed method has to optimize in 5 dimensional space, whereas joint Bayesian optimization's input has 16, 24 and 32 dimensions respectively.

Synthetic-II: We set the underlying function of each stage as the probability density function of a multivariate Gaussian distribution. In this case, we keep the number of stages fixed to be 5 and vary the number of control parameters as 3, 5 and 7. This makes the number of dimensions for the input of cascade Bayesian optimization as 4, 6 and 8, whilst the input of baseline method has 15, 25 and 35 dimensions respectively.

Experimental Results. We generate 20 different initialization whose number of data points are proportion to number of stages and number of control parameters. The results reported below are averaged over these 20 initialization.

Synthetic-I: Figure 1a, c and e show the experimental results for the Synthetic-I dataset where number of control parameters is fixed and number of stages is varied. Figure 1a illustrates performance of cascade Bayesian optimization and baseline method for 4 stages and 4 control parameters for each stage. Starting with the same set of points, our proposed algorithm performs better than the baseline method in terms of best-found-value within a given number of iterations. Cascade Bayesian optimization reaches 0.5 after 38 iterations and reaches 0.7 after 48 iterations while joint Bayesian optimization can only reach 0.5 after 100 iterations. Figure 1c and e show similar results. Cascade Bayesian optimization gains 0.9 after 100 iterations whereas the joint Bayesian optimization reaches 0.4 and 0.3 when the number of stages is 6 and 8 respectively.

Synthetic-II: Figure 1b, d and f show the experimental results for Synthetic-II dataset where number of stages is fixed and number of control parameters at each stage is varied. When the number of control parameters is 3, joint Bayesian optimization has to solve the optimization problem with 15 parameters while cascade Bayesian optimization has to solve the optimization problem with 4 dimensional input. The maximum found value of cascade Bayesian optimization is almost doubled the one of the baseline in 100 iterations (Fig. 1b). As the number of control parameters increase, both algorithms perform worse. After 100 iterations, the best value found by cascade Bayesian optimization is 0.9 while the best value found by the joint Bayesian optimization is only 0.2 in the case the process has 5 control parameters for each stage (Fig. 1d). Figure 1f shows the results of the dataset with 7 control parameters for each stage. Joint Bayesian optimization has to find result in 35 dimensional space while cascade Bayesian optimization has to optimize in 8 dimensional space. As a result, joint Bayesian optimization suffered heavily and can not find any value bigger than 0.05 after 100 iterations. On the contrary, the best value found by the cascade Bayesian optimization is 0.65 after the same number of iterations.

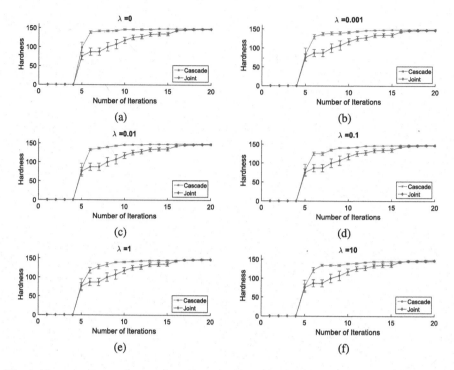

Fig. 2. Best hardness achieved as a function of number of iterations for both cascade BO (red) and joint Bayesian optimization (blue), for different cost parameter λ (Color figure online)

4.3   Experiments with Real Data

Alloy Heat Treatment. This is based on a simulation model of a real world heat treatment process of an Al-Sc alloy. The underlying physics of alloy strengthening is based on nucleation and growth. Nucleation is the process of either a new "phase" formation or clusters of atoms or precipitates through a self-organizing process. This process happens at lower temperatures over time. It is a stochastic process and thus difficult. The aim of the first step is to maximize nucleation, or the number of precipitates. The second step is growth. Through diffusion the initial precipitates grow and the requisite alloy property is achieved. We use the industrial standard precipitation model KWN [8,9] model for the kinetics of nucleation and growth. This tracks the precipitation nucleation, growth and coarsening over discrete time steps. It does so using Gibbs-Thomson relationship equations and nucleation theory. The model has several phases. For each heat treatment temperate, it iterates and calculates the precipitation for each time-step which is then adjusted using the Gibbs-Thomson equation. The outputs include hardness and precipitate.

We consider a three stage heat treatment process. The input to first stage is the alloy composition, the temperature and time. The nucleation output of this stage is input to the the second stage along with the temperature and time for the second stage. The input of final stage is the hardness of the alloy composition at second stage, temperature and time. The final output is hardness of the material. We seek to find the heat treatment that results in maximum hardness.

Experimental Results. Figure 2 shows the comparison of the cascade Bayesian optimization (red) and baseline method (blue) vs iteration, for varying cost parameter λ. Surprisingly, at four random initialization the hardness values remained very low. After that, the cascade Bayesian optimization outperformed the joint Bayesian optimization in terms of the speed at which it reaches higher hardness values. It takes only 6 iteration for cascade Bayesian optimization to reach the hardness of 120 whereas the baseline method needs 10 to reach the same level of hardness. After 10 iteration, our proposed method gets the hardness of 140 while the joint Bayesian optimization needs 15 iteration to get to a similar value.

4.4 Cost-Efficient Optimization

As we mentioned earlier, the optimization problem of finding the input given the output of a Gaussian process generally has multiple solutions, therefore, it is possible to choose a solution that meets a specific criteria. As energy costs are related to oven temperature and times, in the intermediate stages we encourage solution that minimizes the norm of time and temperature vector. Let us denote q_i as the time at i-stage of the process, then the *total time* taken for each process is computed as $\sum_{i=1}^{N} q_i$. Figure 3 shows this trade-off between hardness and time with respect to inverse Gaussian parameter λ. As λ increases from 0 to 0.1, the hardness drops slightly, however we could save almost 20% of time. When we set $\lambda = 10$, the hardness drops to 146 while the average time is 5 h which saves 30% of time compare with $\lambda = 0$.

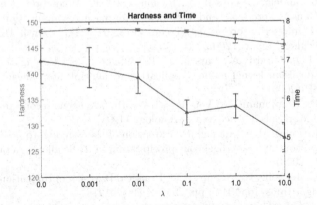

Fig. 3. Experimental result for Aluminum Scandium hardening: Hardness of alloy (red) and time (blue) vs cost parameter λ (Color figure online)

5   Conclusion

We proposed a novel cascade Bayesian optimization method to tune the parameters of a multi-stage cascaded system, more often found in industrial processes. Each of the stage of the process are separately modeled and appropriately optimized such that the final stage output is maximized. Novel optimization formulation is provided that also exploits the epistemic uncertainties of the underlying model. Additionally, the the formulation is also made cost-sensitive to find cost-efficient solutions in the small number of trials. On a simulated testbed of 3-stage heat-treatment for Al-Sc alloy the cascade Bayesian optimization showed superior performance over a naïve joint approach. When cost is taken into account, we were able to save 30% of the total time with only slight drop in hardness value.

Acknowledgement. This work is partially supported by the Telstra-Deakin Centre of Excellence in Big Data and Machine Learning.

References

1. Jones, D.R., Schonlau, M., Welch, W.J.: Efficient global optimization of expensive black-box functions. J. Global Optim. **13**(4), 455–492 (1998)
2. Lakshminarayanan, B., Roy, D.M., Teh, Y.W.: Mondrian forests for large-scale regression when uncertainty matters. arXiv preprint arXiv:1506.03805 (2015)
3. Srinivas, N., Krause, A., Seeger, M., Kakade, S.M.: Gaussian process optimization in the bandit setting: no regret and experimental design. In: ICML (2010)
4. Osio, I.G., Amon, C.H.: An engineering design methodology with multistage Bayesian surrogates and optimal sampling. Res. Eng. Design **8**, 189–206 (1996)
5. Wang, L., Feng, M., Zhou, B., Xiang, B., Mahadevan, S.: Efficient hyper-parameter optimization for NLP applications. In: Empirical Methods in Natural Language Processing (2015)
6. Quinonero-Candela, J., Girard, A., Rasmussen, C.E.: Prediction at an uncertain input for Gaussian processes and relevance vector machines-application to multiple-step ahead time-series forecasting. Technical report, IMM, Danish Technical University, Technical report (2002)
7. Candela, J.Q., Girard, A., Larsen, J., Rasmussen, C.E.: Propagation of uncertainty in Bayesian kernel models-application to multiple-step ahead forecasting. In: ICASSP (2003)
8. Wagner, R., Kampmann, R., Voorhees, P.W.: Homogeneous second-phase precipitation. In: Materials Science and Technology (1991)
9. Robson, J., Jones, M., Prangnell, P.: Extension of the N-model to predict competing homogeneous and heterogeneous precipitation in Al-Sc alloys. Acta Mater. **51**, 1453–1468 (2003)
10. Snoek, J., Larochelle, H., Adams, R.P.: Practical Bayesian optimization of machine learning algorithms. In: NIPS, pp. 2951–2959 (2012)
11. Rasmussen, C.E.: Gaussian processes for machine learning (2006)
12. Jones, D.R., Perttunen, C.D., Stuckman, B.E.: Lipschitzian optimization without the Lipschitz constant. J. Optim. Theory Appl. **79**, 157–181 (1993)

Assignment Precipitation in Fail First Search

Majid Namazi[1], Nina G. Ghooshchi[1,2], M.A. Hakim Newton[1(✉)],
and Abdul Sattar[1,2]

[1] IIIS, Griffith University, Nathan, Australia
mahakim.newton@griffith.edu.au
[2] Data61, CSIRO, Spring Hill, Australia

Abstract. Within a fail first search, we relocate assignments towards
the bottom of the upward-growing assignment stack. We do that when
the related variables, due to search dynamics, become more constrained
than were anticipated before. This fixes early variable selection mistakes.

Keywords: Constraints · Backtracking search · Variable ordering

1 Introduction

The fail first strategy within a constructive search selects more constrained variables first. However, due to the lack of search dynamics, variable selection mistakes might happen especially early in the search and even with a good dynamic variable ordering heuristic [1,3]. Such mistakes then can be recovered by using intelligent backtracking [4] and restart techniques [6]. However, these techniques suffer from the loss of the search effort already made. In this work, if a more constrained variable is identified later, we relocate the related assignment towards the bottom of the assignment stack, which presumably grows upward. The relocation process adjusts the effects of the constraint propagation and nogood learning as needed. Thus, the assignment reordering essentially fixes early variable selection mistakes and thereby saves significant effort that is wasted otherwise due to backtracking or restarting.

2 Constraint Preliminaries

A *constraint satisfaction problem (CSP)* $\mathcal{P}$ has a set of constraints $\mathcal{C}$ and a set of variables $\mathcal{V}$. An assignment $\alpha(v, k)$ assigns a value k to a variable $v = V(\alpha)$ from its *domain* $D(v)$. Each *constraint* $c \in \mathcal{C}$ restricts assignments involving a set of variables $V(c) \subseteq \mathcal{V}$. A *compound assignment* $\mathcal{A}$ is a set of assignments with no $\alpha, \alpha' \in \mathcal{A}$ such that $V(\alpha) = V(\alpha')$. We also take $\mathcal{A}$ as a sequence, if constructed over iterations. A *consistent* compound assignment satisfies all constraints that involve only the related variables. A consistent compound assignment $\mathcal{A}$ is a *(complete) solution* if $\mathcal{A}$ includes all variables in $\mathcal{V}$ and else is a *partial solution*. A compound assignment is a *nogood* if it is not a subset of *any* solutions.

© Springer International Publishing AG 2016
B.H. Kang and Q. Bai (Eds.): AI 2016, LNAI 9992, pp. 281–287, 2016.
DOI: 10.1007/978-3-319-50127-7_23

Backtracking search algorithms are complete and are used for solving *CSPs*. Backtracking search starts from an empty partial solution $\mathcal{A}$. In each step, an unassigned variable v is selected as the *current variable* and is assigned a value $k \in CD(v)$, if *current domain* $CD(v)$ of v is not empty and $\mathcal{A} \cup \{(v, k)\}$ is consistent. In this way, a solution to $\mathcal{P}$ is obtained when each variable in $\mathcal{V}$ is assigned a value. The *backtrack* method could revoke some previous assignment(s) when there is an unassigned variable v such that for any $k \in D(v)$, assignment (v, k) does not preserve consistency of the new partial solution. Below we discuss two relevant issues: the dom/wdeg fail first heuristic and elimination explanations.

The *dom/wdeg* fail first heuristic [3] is a very powerful dynamic variable ordering heuristic taking current domain sizes of the variables and the *weights* of the constraints into account. Each constraint c is initially given a weight of $W(c) = 1$ and the weight is incremented each time the constraint becomes the *latest cause* of a variable $v \in V(c)$ to have empty $CD(v)$. For an unassigned variable v, the *weighted degree* $WDeg(v)$ is the sum of the weights of the constraints that must involve v and also at least one other unassigned variable v'. According to the dom/wdeg heuristic, the next variable selected for assignment is an unassigned variable having the smallest $DomWDeg(v) = \frac{|CD(v)|}{WDeg(v)+1}$.

In *explanation based* backtracking algorithms, e.g. in *dynamic backtracking* [5], a value $k \in D(v)$ can be temporarily eliminated from $D(v)$ if k can not be assigned to v because of a compound assignment $\mathcal{A}' \subseteq \mathcal{A}$. Such an $\mathcal{A}'$ is considered as k's *elimination explanation*, denoted by $EE(v, k)$. Every elimination explanation $EE(v, k)$ remains valid as long as $EE(v, k) \subseteq \mathcal{A}$ holds and either v's assignment is located after all $\alpha \in EE(v, k)$ in $\mathcal{A}$ or v is yet unassigned. To preserve polynomiality of the memory needed, only one elimination explanation containing the lowest indexed assignment(s) in $\mathcal{A}$ is kept for every eliminated value of a variable [8]. After elimination of unassignable values, the reduced domain of v is called the *current domain* $CD(v)$ of v. For convenience, we define $EE(v) = \cup_{k \in D(v) \setminus CD(v)} EE(v, k)$ as the *elimination explanation* of a variable v.

3   Our Precipitation

In the backtracking search algorithms, in each step, the dom/wdeg heuristic in its own measure selects the most constrained variable. However, a more constrained variable with lower DomWDeg could be identified later for two reasons: increase in weights of some constraints and elimination of more values from a variable's domain. Whenever such a more constrained variable is identified, we relocate its assignment within $\mathcal{A}$ towards the lower indices. As such Procedure precipitate in Fig. 1, is called after every

```
1  procedure precipitate(assigned A)
2    v ← V(α|A|), here A = ⟨α1, . . . , α|A|⟩
3    if (|CD(v)| = 1 ∧ EE(v) = φ) freeze(A)
4    else condense(A)

5  procedure freeze(assigned A)
6    here A = ⟨α1, . . . , α|A|⟩
7    queue ← empty
8    queue.add(α|A|)
9    while ¬queue.isempty()
10      αi ← queue.remove()
11      FrozenCount = FrozenCount + 1
12      relocate(FrozenCount, i)
13      for each v ∈ V where αi ∈ EE(v)
14        EE(v) = EE(v) \ {αi}
15        if ∃αj ∈ A where v = vj and
16          |CD(vj)| = 1 and EE(vj) = φ
17          queue.add(αj)
```

Fig. 1. Precipitate and freeze

successful assignment to a variable. This procedure actually performs two types of precipitation: freezing and condensation.

Freezing Operation. Assume v is the current variable with $|CD(v)| = 1$ and $EE(v)$ empty. Each such variable v could essentially be treated like a constant. In every such a case (see Line 3 in Fig. 1), the freeze procedure is called which puts the assignment of the current variable v in a queue. As long as this queue is not empty, an assignment is removed from it and the assignment is moved to the end of the frozen part of the partial solution $\mathcal{A}$. This assignment is also eliminated from the elimination explanations of all other variables. Because of this, some other assigned variable(s) could appear as candidate(s) for assignment freezing and the same procedure is to be followed.

3.1   Condensing Operation

Procedure condense in Fig. 2 attempts to relocate the last assignment $\alpha_{|\mathcal{A}|}$ (Line 2). Condensation is a very costly operation and so we do not necessarily keep the whole partial solution $\mathcal{A}$ fully condensed. Because of this, we only consider those assignments in $\mathcal{A}$ that appear in $EE(v)$ (Line 3). However, we relocate at most two assignments that include $\alpha_{|\mathcal{A}|}$ (Lines 15,27). For relocation destination of $\alpha_{|\mathcal{A}|}$, each position i after FrozenCount (Line 4) is considered and $V(\alpha_i)$ is excluded from R (Line 5). For the other assignment, the destination is kept within an early threshold region of $\mathcal{A}$ (Line 6). This is motivated by the observation that most structured and real world constraint problems have minimal highly-constrained or hard to solve subproblems [9]. After some pilot experimentation, in this paper, we set the value of Threshold to $\text{FrozenCount} + \sqrt{|\mathcal{A}| - \text{FrozenCount}}$.

Let $\mathcal{A} = \langle \alpha_1, \ldots, \alpha_{|\mathcal{A}|} \rangle$ be the current partial solution and $\mathcal{A}_{i,j}$ be the subsequence $\langle \alpha_i, \ldots, \alpha_j \rangle$. For $i < j$, the *potential domain* $PD(i,j) = CD(V(\alpha_j)) \cup \{\hat{k} : (EE(V(\alpha_j), \hat{k}) \cap \mathcal{A}_{i,j-1}) \neq \phi\}$ of the assigned variable $V(\alpha_j)$, when α_j will be relocated from position j to position i, comprises the values in $CD(V(\alpha_j))$ and each of its eliminated values that has some assignments in $\mathcal{A}_{i,j-1}$ in its elimination explanation. The *potential weighted degree*

```
1 procedure condense(assigned 𝒜)
2   v ← V(α_{|𝒜|}), here 𝒜 = ⟨α₁,...,α_{|𝒜|}⟩
3   R ← {v} ∪ {V(α_j) : α_j ∈ EE(v)}
4   for i = FrozenCount + 1 to |𝒜| − 1
5     v_i ← V(α_i), R ← R \ {v_i}
6     if i > Threshold then R ← {v}
7     m ← min_{V(α_j)∈R} |PD(i,j)|
8     S ← {V(α_j) ∈ R : PD(i,j) = m}
9     if m < |CD(v_i)|
10      n ← max_{V(α_{j'})∈S} PWDeg(i,j')
11      T ← {V(α_{j'}) ∈ S :
12            PWDeg(i,j') = n}
13      v_j ← V(α_j), j = min_{V(α_{j'})∈T} j'
14      if PDWDeg(i,j) < DomWDeg(v_i)
15        relocate(i,j)
16        if v_j = v return else R = {v}
17      else if m = |CD(v_i)|
18        S = S \ {v}, if S = φ continue
19        T = {V(α_j) ∈ S :
20              PWDeg(i,j) > WDeg(v_i)}
21        if T = φ continue
22        p ← min_{V(α_{j'})∈T} |CPD(i,j',|𝒜|)|
23        U = {V(α_{j'}) ∈ T :
24              |CPD(i,j',|𝒜|)| = p}
25        v_j ← V(α_j), j = min_{V(α_{j'})∈U} j'
26        if |CPD(i,j,|𝒜|)| < |CD(V(α_{i+1}))|
27          relocate(i,j), relocate(i+1,|𝒜|)
28          return
```

Fig. 2. Pseudocode of condense

PWDeg(i,j) of $V(\alpha_j)$, when relocated from position j to position i, is defined as WDeg$(V(\alpha_j))$ when computed with the assumption of $\mathcal{A}_{1,i-1}$ being the current partial solution and so all variables in $\mathcal{A}_{i,|\mathcal{A}|}$ being treated as unassigned. Lastly, PDWDeg$(i,j) = \frac{|PD(i,j)|}{PWDeg(i,j)+1}$ denotes the potential dom/wdeg ratio. Based on these definitions, one can relocate an assignment from position j to an early position i in $\mathcal{A}$ if PDWDeg$(i,j) < $ DomWDeg$(V(\alpha_i))$ and also $|PD(i,j)| < |CD(V(\alpha_i))|$. The second condition's goal is to preserve the terminability of the backtracking search.

Next, we find the set of variables $S \subseteq R$ which have the minimum potential domain size m at position i (Lines 7,8). If $m < |CD(v_i)|$ we then find the set of variables $T \subseteq S$ that have the maximum PWDeg value n and also find variable $v_j \in T$ that has the smallest index in $\mathcal{A}$ (Lines 9–13). The assignment related to v_j is relocated to position i if PDWDeg$(i,j) < $ DomWDeg(v_i) (Lines 14,15). If this is the last assignment, then condensation ends, otherwise only the last assignment is considered for the next relocation (Line 16).

If $m = |CD(v_i)|$ (Line 9,17), we try to find a variable in EE(v) such that if the related assignment is relocated to position i and at the same time the last assignment is relocated to position $i+1$, the terminability of the backtracking search is preserved. To find such a variable, we define conditional potential domain. For $i < j' < j$, the *conditional potential domain* CPD$(i,j',j) = CD(V(\alpha_j)) \cup \{\hat{k} : (EE(V(\alpha_j),\hat{k}) \cap (\mathcal{A}_{i,j-1} \setminus \{\alpha_{j'}\})) \neq \phi\}$ of the assigned variable $V(\alpha_j)$, when assignment α_j is relocated to position $i+1$ while assignment $\alpha_{j'} \in EE(V(\alpha_j))$ to position i, comprises the values in CD$(V(\alpha_j))$ and its eliminated values whose elimination explanations include at least one assignment in $\mathcal{A}_{i,j-1} \setminus \{\alpha_{j'}\}$.

In Lines 18–21 in procedure condense in Fig. 2, we find the set of variables $T \subseteq S$ that are related to assignments α_j having PWDeg$(i,j) > $ WDeg(v_i). Then, we find the variable related to an α_j that has the smallest index in $\mathcal{A}$ and has the minimum CPD$(i,j,|\mathcal{A}|)$. The relocations of α_j to position i and $\alpha_{|\mathcal{A}|}$ to position $i+1$ are performed in Line 27 if $|CPD(i,j,|\mathcal{A}|)| < |CD(V(\alpha_{i+1}))|$, meaning preserving the terminability of the backtracking search.

3.2   Assignment Relocation

Procedure relocate(i,j) relocates assignment α_j from position j to position $i < j$ in the current partial solution $\mathcal{A}$ resulting into right shifting of the assignments $\alpha_i, \ldots, \alpha_{j-1}$ by one position. Because of the relocation, the current domains and the elimination explanations of the variables should be updated to preserve the soundness of the backtracking search as follows:

1. Each EE$(V(\alpha_j),k)$ that has an $\alpha \in \mathcal{A}_{i,j-1}$ will not be valid any more. Thus value k will be added to CD$(V(\alpha_j))$.
2. For each value $k \in CD(V(\alpha))$ where $\alpha \in \mathcal{A}_{i,j-1}$, if there is a constraint involving $V(\alpha_j)$ that does not allow value k to be assigned to variable $V(\alpha)$, then k is eliminated and the related elimination explanation becomes EE$(V(\alpha),k)$.
3. For each variable v and each eliminated value $k \in D(v) \setminus CD(v)$ where the last included assignment in EE(v,k) is in $\mathcal{A}_{i,j-1}$, if there is a constraint involving

$V(\alpha_j)$ that does not allow value k to be assigned to variable v, the corresponding elimination explanation is considered as new $EE(v, k)$. This ensures that the earliest elimination explanation is always maintained.

4 Related Work

Reordering assigned variables is a very rare behaviour in the polynomial space backtracking algorithms, especially in the centralised and forward ones. We refer to dynamic backtracking [5] and partial order backtracking [7] for some early work. The retroactive dynamic backtracking algorithm and its integration with the forward checking algorithm introduced in [10] are two centralized backtracking algorithms. In these algorithms, the assignment of the current variable v is relocated to a position after the last affecting variable. The retroactive asynchronous backtracking with dynamic ordering [12] is a distributed algorithm in which an unsuccessful variable or agent is relocated to a position after the second last affecting variable. The agile asynchronous backtracking [2], is another distributed backward backtracking algorithm which provides more flexibility for reoredring and selection of the backtracking variable such that the sequence of current domain sizes is reduced as much as possible. A new generalised version of this algorithm [8] considers reducing the sequence of dom/wdeg ratios.

5 Experimental Results

Our baseline backtracking algorithm, a variant of [11], uses forward checking and arc-consistency checking along with conflict directed back-jumping and nogood learning. On top of the baseline solver, we then separately incorporate our precipitation technique, the reordering heuristics in the centralised [10] and distributed [12] retroactive ordering algorithms, and all the different versions of the distributed agile backtracking algorithm [2,8]. We then compare our technique (called Precip) with the reordering heuristics of the centralised retroactive ordering (called CRetro) [10], distributed retroactive ordering (called DRetro) [12] and just the baseline solver (called Base). We denote these solvers by PrecipN, CRetroN, DRetroN, BaseN respectively. We also run each of these different techniques combined with a geometric restart approach and name these by PrecipR, CRetroR, DRetroR, BaseR respectively. The restart cutoff value is set initially to 10 backtracks, but on each restart, is multiplied by 1.5. Constraint weights are retained across restarts. We do not use agile-heuristics based implementations because they are very slow in most of our benchmark problems.

We ran all experiments on the same high performance computing cluster. Each node of the cluster is equipped with Intel Xeon CPU E5-2670 processors @2.60 GHz. We ran experiments with 4 GB memory limit and 3600 s timeout. We have run all algorithms on a number of standard benchmark problems from (http://www.cril.univartois.fr/~lecoutre/benchmarks.html). Note that we only use the instances that are solved by at least one of the eight solvers taking at most 1 h running time. Moreover, these instances exclude those that are solved

Table 1. Comparison of solvers on the numbers of problem instances solved and total time spent on all instances; each cell contains these two values.

Domain	#Ins	BaseN		CRetroN		DRetroN		PrecipN		BaseR		CRetroR		DRetroR		PrecipR	
Scens11	6	5	**3971**	5	4107	5	7246	5	5993	6	**2647**	6	3626	5	4185	5	4369
F-insert	10	10	3296	10	3271	10	3173	10	**3080**	10	3500	10	3923	10	3495	10	**3205**
Leigh15	7	7	4211	7	3921	7	4196	7	**3463**	6	5455	6	**4085**	6	4755	6	4105
SgbBook	11	11	1636	11	1656	11	**1419**	11	1506	11	**2174**	11	2361	11	2327	11	2572
HayStack	5	2	11999	3	8095	4	5545	5	**3593**	4	4034	4	4062	3	7247	5	**3475**
JShop	9	3	21850	4	21659	3	22225	7	**7388**	8	3608	9	38	8	3605	9	**33**
Tail5	26	19	33795	20	25457	22	26477	20	**23178**	24	11351	26	6748	26	7587	26	**2984**
QCP15	5	5	77	5	**66**	5	141	5	95	5	48	5	88	5	42	5	**19**
QCP20	13	11	19099	13	**12376**	11	21298	11	16360	11	14111	10	18252	10	17436	10	**13148**
QWH20	9	9	129	9	**115**	9	147	9	134	9	141	9	339	9	221	9	**83**
QWH25	9	5	24197	6	29353	6	24838	7	**20574**	1	30060	4	20823	2	26900	8	**16153**
SJShop	24	12	45703	14	41871	14	40016	17	**28704**	14	36236	18	25141	18	22943	24	**410**
STail4	8	8	663	8	723	8	1419	8	**539**	8	534	8	**209**	8	360	8	266
STail5	20	15	28031	14	26465	16	22909	17	**16657**	19	7795	18	8254	18	7989	19	**5387**
Queens	4	3	3682	3	3683	3	3682	4	**25**	4	135	4	138	4	139	4	**97**
Geom	17	17	274	17	274	17	282	17	**267**	17	436	17	424	17	408	17	487
Rand26	10	10	1772	10	1736	10	**1732**	10	1761	10	1987	10	2166	10	2082	10	**1943**
Rand27	10	10	3953	10	**3922**	10	3937	10	3976	10	5487	10	**5197**	10	5260	10	5323
MD0.9	71	71	2154	71	2062	71	**2029**	71	2093	71	1883	71	1749	71	1934	71	**1651**
Total	274	233		240		242		251		248		256		251		267	

by each of the solvers within 3 s. To compare the performance of the algorithms, we use CPU time (in seconds) spent by the solvers. The number of steps and especially constraint checks are mostly correlated with the times.

In Table 1, PrecipR comes the best by solving 267 of the total 274 instances over 19 domains. PrecipN is also the best among the solvers that do not use restarts. In terms of the time spent, PrecipR comes first in the highest number of 9 domains. Among the solvers that do not use restarts, PrecipN comes first in 11 domains and among those that use restarts, PrecipR comes first in the highest number of 13 domains. Overall, precipitation improves the number of

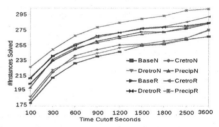

Fig. 3. Number of instances solved by each solver when different time cutoffs are used

instances solved both with or without restarts. Figure 3 shows when run with different time cutoffs, PrecipR using both precipitation and restarts comes the best performing solver maintaining a clear margin from CRetroR, which comes the second. Also, CRetro solvers are slightly better than DRetro solvers and PrecipN clearly better than the other no-restart solvers.

Acknowledgment. This work was partly supported by ARC Grant DP150101618.

References

1. Balafoutis, T., Stergiou, K.: Evaluating and improving modern variable and revision ordering strategies in CSPs. Fundam. Inform. **102**(3–4), 229–261 (2010)
2. Bessiere, C., Bouyakhf, E.H., Mechqrane, Y., Wahbi, M.: Agile asynchronous backtracking for distributed csps. In: ICTAI, pp. 777–784 (2011)
3. Boussemart, F., Hemery, F., Lecoutre, C., Sais, L.: Boosting systematic search by weighting constraints. In: ECAI, vol. 16, p. 146 (2004)
4. Chen, X., Van Beek, P.: Conflict-directed backjumping revisited. arXiv preprint arXiv:1106.0254 (2011)
5. Ginsberg, M.L.: Dynamic backtracking. JAIR **1**, 25–46 (1993)
6. Huang, J.: The effect of restarts on the efficiency of clause learning. IJCAI **7**, 2318–2323 (2007)
7. McAllester, D.A.: Partial order backtracking. JAIR **1** (1993)
8. Wahbi, M., Mechqrane, Y., Bessiere, C., Brown, K.N.: A general framework for reordering agents asynchronously in distributed CSP. In: Pesant, G. (ed.) CP 2015. LNCS, vol. 9255, pp. 463–479. Springer, Heidelberg (2015). doi:10.1007/978-3-319-23219-5_33
9. Williams, R., Gomes, C.P., Selman, B.: Backdoors to typical case complexity. In: IJCAI, vol. 3, pp. 1173–1178. Citeseer (2003)
10. Zivan, R., Shapen, U., Zazone, M., Meisels, A.: Retroactive ordering for dynamic backtracking. In: Benhamou, F. (ed.) CP 2006. LNCS, vol. 4204, pp. 766–771. Springer, Heidelberg (2006). doi:10.1007/11889205_67
11. Zivan, R., Shapen, U., Zazone, M., Meisels, A.: MAC-DBT revisited. In: Larrosa, J., O'Sullivan, B. (eds.) CSCLP 2009. LNCS (LNAI), vol. 6384, pp. 139–153. Springer, Heidelberg (2011). doi:10.1007/978-3-642-19486-3_9
12. Zivan, R., Zazone, M., Meisels, A.: Min-domain retroactive ordering for asynchronous backtracking. Constraints **14**(2), 177–198 (2009)

Knowledge Representation
and Reasoning

Update Policies

Abhijeet Mohapatra[✉], Sudhir Agarwal, and Michael Genesereth

Computer Science Department, Stanford University, Stanford, USA
{abhijeet,sudhir,genesereth}@cs.stanford.edu

Abstract. Underspecified transactions can be supported in databases by enabling administrators to specify update policies that complete underspecified transactions. We propose a language for expressing such update policies. We show that the problem of verifying whether or not a policy is sound and complete with respect to database constraints is undecidable in general. We identify decidable instances of this decision problem, and for such instances, present an algorithm that uses resolution to check whether or not a supplied policy is sound and complete with respect to database constraints.

1 Introduction

Many software systems use databases to model the state of the world. A database consist of base relations and derived relations (also called as *views*). Transactions on a database change the state of the database. A database may also contain *constraints* which characterize the set of allowable or *legal* database states and transactions.

For ease of use it is desirable to allow users to submit underspecified transactions. In such scenarios, database constraints may contain sufficient information for completing an incomplete transaction unambiguously. Otherwise, there are multiple ways to complete an underspecified transaction.

Underspecified transactions may either be legal or illegal. Database management systems typically reject illegal transactions. Prior works on repairing illegal transactions propose strategies for generating necessary and sufficient conditions for repairs. A survey of transaction repair strategies is presented in [4]. However, the proposed transaction repair techniques do not incorporate administrators' preferred strategies for completing underspecified transactions. A related approach is that of Courteous Logic Programs [3]. Courteous Logic Programs enable administrators to prioritize how different conflicts are resolved when evaluating queries on a database state. However, strategies for repairing or completing illegal transactions involve two database states.

In this paper, we investigate the problem of supporting underspecified transactions in database management systems. We present a formal framework for expressing *update policies* for updating databases. We identify a class of update policies called *inclusive* update policies which correspond to strategies for completing underspecified transactions. Transactions on databases with constraints and arbitrary update policies may potentially lead to illegal database states.

© Springer International Publishing AG 2016
B.H. Kang and Q. Bai (Eds.): AI 2016, LNAI 9992, pp. 291–302, 2016.
DOI: 10.1007/978-3-319-50127-7_24

To prevent such scenarios, we present a resolution-based technique to verify whether or not the specified update policies are sound and complete with respect to database constraints.

2   Problem Definition

A relational database consists of relations. An instance of a database is a subset of the set of all ground atoms that can be formed using the base relation constants and symbols from the domain. A database instance is updated through *transactions*.

Definition 1 (Transactions). *A transaction on a database instance D is a tuple $\langle T_i, T_d \rangle$, where T_i denotes the set of atoms that are inserted into the database instance, T_d denotes the set of atoms that are deleted from the instance. In addition, $T_i \cap T_d = \emptyset$, $T_i \cap D = \emptyset$, and $T_d \subseteq D$.*

A database may also contain *constraints* which characterize the set of legal database instances.

Definition 2 (Legal Transactions). *A transaction $\langle T_i, T_d \rangle$ on a database instance D is legal if the instance $(D \setminus T_d) \cup T_i$ is legal.*

Example 1. Consider a database that contains a single unary base relation p. Suppose the set of all ground atoms in the database is $\{p(a), p(b)\}$, and the database instance $\{p(b)\}$ illegal. As a result, $\langle \{p(b)\}, \{\} \rangle$, $\langle \{p(b)\}, \{p(a)\} \rangle$ and $\langle \{\}, \{p(a)\} \rangle$ are illegal transactions with respect to the database instances $\{\}$, $\{p(a)\}$, and $\{p(a), p(b)\}$ respectively.

The traditional approach for enforcing constraints is to reject illegal transactions. However, as shown in the following example, there are cases where it is desirable for users to specify only a subset of the intended transaction. For some of these cases, the specified input is a *complete specification* of the transaction. In other words, there exists a unique transaction which is a superset of the input, and is legal.

Example 2. Consider a modeling of the blocks world with two blocks a and b as a database with a binary relation *on*, and two unary relations *table* and *clear*. The relation *on* characterizes whether or not a block is placed directly on another block. The relations *table* and *clear* characterize whether or not a block is on the ground and whether or not a block has other blocks on it respectively.

Suppose blocks a and b are initially placed on the ground. The corresponding database instance contains the atoms *table(a)*, *table(b)*, *clear(a)*, and *clear(b)*.

Consider an action in the blocks world that places block a is on block b. This action corresponds to the insertion of the atom *on(a, b)* in the database. However, this correspondence does not explicitly specify whether or not the atoms *table(a)* and *clear(b)* are deleted from the database. However, these additional effects can be derived from the constraints on the blocks world.

In other scenarios, the input is an *incomplete specification* of the transaction i.e. multiple supersets of the input that are legal transactions. For example, suppose a third block c is added to the model of the blocks world in Example 2. Suppose that analogous to blocks a and b, c is initially placed on the ground. In this setting, the specification that a is placed on b in the next state does not clarify whether c is on the ground, or on a in the next state.

To deal with scenarios where users specify a subset of the intended transaction, we present a framework for administrators to author strategies for constructing *completed transactions*. In the following, we define the concepts of our framework.

Definition 3 (Transaction Requests). *A transaction request on a database instance is a tuple* $\langle R_i, R_d \rangle$, *where* R_i *denotes a set of atoms that are requested to be inserted into the database instance,* R_d *denotes the set of atoms that are requested to be deleted from the instance. In addition,* $R_i \cap R_d = \emptyset$, $R_i \cap D = \emptyset$, *and* $R_d \subseteq D$.

Definition 4 (Update Policies). *An update policy over a database is a function that takes as input a legal database instance* D *and a transaction request* $\langle R_i, R_d \rangle$ *on* D, *and outputs a legal transaction* $\langle T_i, T_d \rangle$ *on* D.

A policy is inclusive if for every input: D, $\langle R_i, R_d \rangle$, *the policy outputs* $\langle T_i, T_d \rangle$ *such that* $R_i \subseteq T_i$ *and* $R_d \subseteq T_d$.

An inclusive policy is minimal if for every input: D, $\langle R_i, R_d \rangle$, *the policy outputs* $\langle T_i, T_d \rangle$ *such that for all* T_i' *and* T_d', *where* $T_i' \subsetneq T_i$ *and* $T_d' \subseteq T_d$, *or* $T_i' \subseteq T_i$ *and* $T_d' \subsetneq T_d$, $\langle T_i', T_d' \rangle$ *is not a legal transaction on* D.

In our framework, an inclusive policy corresponds to an administrator's strategy for augmenting transaction requests, which may potentially be underspecified, to construct a *completed* transaction. It may also be desirable for an update policy to allow users to completely specify a transaction in their request. In other words, every legal state in the database should be reachable from every other legal state. However, an arbitrary inclusive policy may violate this condition as illustrated in the following example.

Example 3. Consider the setting in Example 1, and an update policy over the database with the following properties.

- The update policy results in the transaction $\langle \{p(a), p(b)\}, \{\} \rangle$ for the transaction requests $\langle \{p(a)\}, \{\} \rangle$ and $\langle \{p(b)\}, \{\} \rangle$ on the empty database instance.
- The update policy results in the transaction $\langle \{\}, \{p(a), p(b)\} \rangle$ for the transaction requests $\langle \{\}, \{p(a)\} \rangle$ and $\langle \{\}, \{p(b)\} \rangle$ on the database instance $\{p(a), p(b)\}$.

We note that an update policy that results in the above transactions satisfies Definition 4. However, as a result of such an update policy, the instance $\{p(a)\}$ is not reachable from the instances $\{\}$ or $\{p(a), p(b)\}$ through any transaction.

Proposition 1. *For every pair of legal database instances D and D', every minimal update policy maps a transaction request $\langle D' \setminus D, D \setminus D' \rangle$ on D to $\langle D' \setminus D, D \setminus D' \rangle$.*

In order to ensure that update policies do not prevent users from completely specifying a transaction in their request, it may be desirable that update policies be minimal. At first, the minimality requirement on inclusive policies may seem stringent. However, we show that there always is a minimal policy that ensures the simulation of a supplied legal transaction through a sequence of singleton transaction requests. Such a property is desirable when there are limits on the size of the transaction e.g. when submitting a batch API request to the Facebook API[1] or the Twitter API[2].

Theorem 1 (Unit Serializability). *For every database instance D, and a legal transaction $\langle T_i, T_d \rangle$ on D, there exists a minimal policy such that all of the following conditions are satisfied.*

- *S_T is the set of singleton transactions $\{\langle \{A\}, \{\} \rangle \mid A \in T_i\} \cup \{\langle \{\}, \{A\} \rangle \mid A \in T_d\}$.*
- *There exists an integer k such that $1 \le k \le |S_T|$, and a sequence of k distinct transactions from S_T such that applying these transactions in a sequence over D results in a database instance that is identical to the one obtained by applying $\langle T_i, T_d \rangle$ on D.*

Proof. We prove the above result using induction on size of $T_i \cup T_d$. The *base case* trivially holds i.e. when $|T_i| = 1$ or $|T_d| = 1$. In the *induction hypothesis*, we assume that for $|T_i \cup T_d| \le k$, the theorem holds.

For the *induction step*, we assume that $|T_i \cup T_d| = k + 1$. Consider a database instance D. There are two cases depending on whether or not there exists a pair of sets A and B such that $A \subseteq T_i$, $B \subseteq T_d$, $\langle A, B \rangle$ is a legal transaction on D, and $1 \le |A| + |B| \le k$. If there does not exist a pair of sets A and B that satisfy this property, then we construct a minimal policy by assigning $\langle T_i, T_d \rangle$ to some singleton sub-transaction on D. Otherwise, let D_1 be the database instance $(D \setminus B) \cup A$. By definition, D_1 is legal. Therefore, the transaction $\langle T_i \setminus A, T_d \setminus B \rangle$ is legal on D_1, and the theorem can be proved by applying the induction hypothesis on the transactions $\langle A, B \rangle$ and $\langle T_i \setminus A, T_d \setminus B \rangle$ on D and D_1 respectively.

3   Update Policy Language

In this section, we present a language that enables administrators to expressively specify update policies. Our update policy language is Datalog^u, which is an extension of standard Datalog with negation as failure, and *update* operators. First, we present a brief overview of Datalog. Then, we discuss the syntax of Datalog^u and discuss its expressive power.

---

[1] https://developers.facebook.com/docs/graph-api/making-multiple-requests.
[2] https://dev.twitter.com/tags/bulk-operations.

3.1   Overview of Datalog

A Datalog program is a finite set of *facts* and *rules*. Facts are ground atoms. An atom is a structure with the signature $r(\bar{t})$, where p is a relation constant with arity n, and $\bar{t}$ is a finite sequence of n *terms* $t_1, t_2, \ldots, t_n$. Datalog terms are either object constants or variables. Relation constants and object constants are denoted as strings which begin with a lowercase character e.g. *on*, *a*, and variables are denoted as strings which begin with an uppercase character e.g. X, Y. An atom is *ground* if it has no variables.

A Datalog rule is an expression of the form $H :\text{-} B_1, B_2, \ldots, B_k$, where k is finite and $k \geq 1$. Such a rule comprises of a distinguished atom H, which is the also called the *head* of the rule, and literals B_1, B_2, $\ldots$, and B_k, which comprise the *body* of the rule. A Datalog literal is an atom or the negation of an atom. In what follows, we represent negated atoms as atoms prefixed by the '$\neg$' symbol which represents *negation-as-failure*. For example, if $p(a, b)$ is an atom, then $\neg p(a, b)$ denotes the negation of this atom.

Semantically, a rule states that the conclusion of the rule (denoted as the head) is true whenever the conditions (denoted as the literals in the body) are true. Consider the following Datalog rule.

$$r(X, Y) :\text{-} p(X, Y), \neg q(Y).$$

Here, $r(X, Y)$ is the head of the rule. The body of the rule consists of the literals $p(X, Y)$ and $\neg q(Y)$. The rule above states that r is true of any object X and any object Y, if p is true of X and Y and q is not true of Y. For example, if we know $p(a, b)$ and we know that $q(b)$ is false, then, using this rule, we can conclude that $r(a, b)$ is true.

To ensure finite termination and unique minimal models of Datalog programs *safety* and *stratification* restrictions [7] are imposed on Datalog rules. A safe and stratified Datalog program can be evaluated in a top-down or in a bottom-up fashion [7] in time that is polynomial in the size of the program.

3.2   Constraints in Datalog

Constraints in a Datalog program are encoded by specifying the set of illegal database instances. A constraint is a rule of the form $\bot :\text{-} \phi$ where ϕ is conjunction of literals. A database instance D satisfies a constraint $\bot :\text{-} \phi$ if and only if $D \not\models \phi$. A database instance D satisfies a set of constraints C if and only if D satisfies every constraint in C.

3.3   Datalog^u: Extending Datalog with Update Operators

We propose to model update policies as Datalog^u programs. Datalog^u extends Datalog (with negation as failure) using four update operators δ^+, δ^-, Δ^+ and Δ^-. Each of these update operators takes an atom as an argument.

Update policies are encoded in Datalog^u as follows. The update operators δ^+ and δ^- are used to respectively encode insertions and deletions in a transaction request R. The operators Δ^+ and Δ^- are used to respectively encode the insertions and deleteinos in a transaction T. We require that δ^+ and δ^- must not appear in the head of a rule, and Δ^+ and Δ^- must not appear in the body of a rule.

An update policy (see Definition 4) is modeled as a safe stratified Datalog^u program which defines Δ^+ and Δ^- in terms of δ^+, δ^- and the current database instance.

Example 4. Consider Example 1. The constraint that the database instance $\{p(b)\}$ is illegal can be enforced by the following minimal update policy:

$$\Delta^+ p(a) \;:\text{-}\; \delta^+ p(b), \neg p(a).$$
$$\Delta^+ p(b) \;:\text{-}\; \delta^+ p(b).$$
$$\Delta^+ p(a) \;:\text{-}\; \delta^+ p(a).$$
$$\Delta^- p(a) \;:\text{-}\; \delta^- p(a).$$
$$\Delta^- p(b) \;:\text{-}\; \delta^- p(a), p(b).$$
$$\Delta^- p(b) \;:\text{-}\; \delta^- p(b).$$

The above policy ensures that if a transaction request requests the insertion of $p(b)$, then the next database instance contains both $p(a)$ and $p(b)$. This policy also ensures that if a transaction request requests the deletion of $p(a)$, then $p(b)$ is also deleted if $p(b)$ is present. We note that additional rules are not needed to handle transaction requests for insertion of $p(a)$ and deletion of $p(b)$. This is because the resulting database instance is legal.

3.4   Expressiveness of Datalog^u

As discussed in Sect. 2, an update policy defines a function from the set of database instances and the set of transaction requests to the set of transactions. If the set of database instances is finite then the set of transaction requests as well as the set of transactions are also finite. Therefore, there are only finitely many update policies. In this case, our update policy language, Datalog^u, can model all the functions since we can enumerate all the functions.

In case there are countably (infinitely) many database instances, there are also countably infinite transaction requests and countably infinite transactions. In this case, the number of possible update policies is uncountably infinite. Therefore, Datalog^u cannot model all possible update policies.

However, Datalog^u is Turing Complete, i.e., it can model all recursively enumerable functions. Functions which cannot be modeled using our Datalog^u also cannot be modeled by any other programming language.

Comparison with ECA-Rules: Database management systems use an approach that is similar to ours in enforcing constraints through *triggers*. Triggers are

event-condition-action (ECA) rules [9] that perform some actions such as applying additional updates or raising exceptions, when an update event occurs on
a database instance, and the requested update and the database instance satisfy the conditions of the trigger. ECA-rules can be modeled as Datalog^u rules
by characterizing the triggering event as a transaction request, the triggering
condition as a query, and the triggered action as a transaction.

Our update policy language, Datalog^u, is *strictly more expressive* than the
language used in triggers. In the following example, we present an update policy
which cannot be modeled as a trigger.

Example 5. Consider a database with three unary base relations p, q and r, and
the following update policy.

$$\Delta^+ q(X) \colonequals \delta^+ p(X), \, \delta^- r(X).$$

The above update policy inserts tuples of the form $q(X)$ in response to transaction requests which request the insertion of tuples of the form $p(X)$, and the
deletion of tuples of the form $r(X)$. Such an update policy *cannot* be implemented using triggers without changing the database's schema e.g. by introducing auxiliary tables. This is because triggers are executed in response to a *single*
type of event e.g. insert, update on *single table*.

Comparison with STRIPS Action Representation: The STRIPS [2] representation for an action consists of: (a) *preconditions*, a list of atoms that need to be
true for the action to occur, (b) a *delete list*, a list of those primitive relations no
longer true after the action, and (c) an *add list*, a list of the primitive relations
made true by the action.

A STRIPS action can be modeled as an update policy as follows. The update
policy contains one rule for each element in the add and delete lists. These rules
satisfy the following properties. The head of a rule is $\Delta^+ e$ or $\Delta^- e$ depending
on whether an element e is in the add list or the delete list. The body of a rule
is a conjunction of the atoms listed as the preconditions of the action, and $\delta^+ a$
where a is an atom representing the action. In the following example, we model
a STRIPS representation of an action in the blocks world as an update policy.

Example 6. Consider the modeling of the blocks world presented in Example 2.
Suppose we denote the act of placing a block X on a block Y as $place(X, Y)$.
The STRIPS representation of $place(X, Y)$ is as follows.

Action	Preconditions	Add list	Delete list
$place(X, Y)$	$clear(X), clear(Y)$	$on(X, Y)$	$clear(Y)$
$place(X, Y)$	$clear(X), clear(Y), on(X, U)$	$on(X, Y), clear(U)$	$clear(Y), on(X, Y)$

The STRIPS representation of $place(X, Y)$ can be equivalently modeled using the following update policy.

$$\Delta^+ on(X, Y) \text{ :- } clear(X),\ clear(Y),\ \delta^+ place(X, Y).$$
$$\Delta^- clear(Y) \text{ :- } clear(X),\ clear(Y),\ \delta^+ place(X, Y).$$
$$\Delta^+ clear(U) \text{ :- } clear(X),\ clear(Y),\ on(X, U),\ \delta^+ place(X, Y).$$
$$\Delta^- on(X, U) \text{ :- } clear(X),\ clear(Y),\ on(X, U),\ \delta^+ place(X, Y).$$

The presented update policy ensures the following behavior. First, requests for placing a block X on another block Y are completed only when both blocks are clear i.e. there is no other block on top of the blocks. Second, the block Y is no longer clear in the next state. Third, in the case that block X is on a block, say U, then X is no longer on U in the next state of the blocks world.

Comparison with a STRIPS Extension for Joint-Actions: Prior work presented in [1] presents a formalism for representing joint-actions and generating concurrent non-linear plans by extending STRIPS representations of actions. Joint-actions are represented by adding a possibly empty *concurrent list* to each action. A concurrent list specifies which actions may or may not co-occur with the given action in order to produce the described effect. Such joint-actions can be modeled as transaction requests in Datalog^u since multiple δ^+ and δ^- literals allowed in the body of Datalog^u rules.

In the formalism presented in [1], the concurrency lists and the effect of concurrent execution of actions a and b needs to described twice, once each for actions a and b. The concurrency list of a contains b and the concurrency list of b contains a, and analogously the effects. With Datalog^u we need to define the effect of concurrent execution of a and b only once.

In addition to and despite the simpler syntax of Datalog^u, Datalog^u is more expressive than the formalism presented in [1]. In Datalog^u, we can define effects that are recursively computed (see example below) from the current database instance and the requested transaction. Such effects cannot be modeled with the formalism presented in [1].

$$v(X, Y) \text{ :- } \delta^+ p(X, Y).$$
$$v(X, Z) \text{ :- } \delta^+ p(X, Y),\ v(Y, Z).$$
$$\Delta^+ p(X, Y) \text{ :- } v(X, Y).$$

Another interesting difference between the formalism presented in [1] and our update policy language, Datalog^u, concerns multiple occurrences of the same action. While the former does not allow multiple concurrent actions by the same agent, Datalog^u can deal with multiple occurrences of an action.

4 Verification of Update Policies

In this section, we consider the problem of verifying whether or not a given update policy enforces a given set of constraints. We formally define the verification problem as follows.

Definition 5 *(Decision Problem).* *Given an update policy P, and a set of static constraints Λ, the VERIFY-POLICY problem decides whether or not application of P on legal database instance D results in a legal transaction $\langle T_i, T_d \rangle$ for every transaction request $\langle R_i, R_d \rangle$.*

Case 1 (Finite Herbrand Base): First, we consider the case where the *Herbrand base* of the supplied database is finite, and is specified as a part of the input. In this case, a VERIFY-POLICY instance $\langle \Lambda, P \rangle$ can be *decided* as follows. Let B_D denote the Herbrand base of the database. Let L_D denote the set of legal database instances. In our case, a subset D of B_D is in L_D *iff* $D \cup \Lambda \nvdash \bot$. Let D_R be the set of transaction requests $\{\langle R_i, R_d \rangle \mid R_i, R_d \subseteq B_d \wedge R_i \cap R_d = \emptyset\}$. For every $D \in L_D$ and transaction request $\langle R_i, R_d \rangle$ in D_R, we compute the completion $\langle T_i, T_d \rangle$ using the supplied policy P, and the next state of the database $D' = D \setminus T_d \cup T_i$. If $D' \cup \Lambda \vdash \bot$ for some $D \in L_D$ and $\langle R_i, R_d \rangle$ in D_R, then we output *No* as the answer of the VERIFY-POLICY instance $\langle \Lambda, P \rangle$. Otherwise, we output *Yes*.

Case 2 (Non-recursive rules): Instead of exhaustively enumerating the set of all possible legal database instances and legal transactions, we can also decide VERIFY-POLICY instance $\langle \Lambda, P \rangle$ using *resolution* as follows. In this case, we lift the assumptions that the Herbrand base of the supplied database is finite, and s is supplied as input.

We initialize a set of clauses C to be empty.

1. **Assert Legality for Current Instance:** We add to C the clauses in $\neg(\exists \bar{X}_1 . \phi_1(\bar{X}_1) \wedge \exists \bar{X}_2 . \phi_2(\bar{X}_2) \ldots \wedge \exists \bar{X}_k . \phi_k(\bar{X}_k))$.

2. **Convert Delta and Policy Rules into Clausal Form:** For every constraint in Λ, we generate the *delta* rules [4], which serve as the necessary and sufficient conditions for determining the legality of a transaction on an instance of the supplied database. Let Δ_c denote the set of all delta rules where $\Delta^+ \bot$ appears in the head of the rule. For every delta rule in Δ_c of the form $\Delta^+ \bot :\!\!- \phi(\bar{X})$, we add the clause $\forall \bar{X} . \phi(\bar{X}) \implies \Delta^+ \bot$ to C. Suppose that that there are m delta rules of the form $\Delta^+ \bot :\!\!- \phi_i(\bar{X}_i)$ where $i \in [1, m]$. We add to C the clauses in $\Delta^+ \bot \implies \bigvee_i (\exists \bar{X}_i . \phi_i(\bar{X}_i))$. We convert the rules in P into clausal form in a similar manner, and add the generated clauses to C.

3. **Add Legality for Transaction and Transaction requests:** For every base relation r in the database we convert the sentences: (a) $\neg(\Delta^- r(\bar{X}) \wedge \Delta^+ r(\bar{X}))$, (b) $\Delta^- r(\bar{X}) \implies r(\bar{X})$, (c) $\delta^- r(\bar{X}) \implies r(\bar{X})$, (d) $\Delta^+ r(\bar{X}) \implies \neg r(\bar{X})$, and (e) $\delta^+ r(\bar{X}) \implies \neg r(\bar{X})$ into clausal form, and add the generated clauses to C.

4. **Assert Negation of Undefined Differentials:** For every differential relation $\Delta^a r$ where $a \in \{+, -\}$ and $\Delta^a r$ does not appear in the head of any rule in P, we add the clause $\{\neg \Delta^a r(\bar{X})\}$ to C.

5. **Resolution Step:** We add the goal clause $\{\Delta^+ \bot\}$ to C and perform resolution. If the empty clause is obtained, then the supplied update policy satisfies the constraints, and we output *Yes* as the answer to the VERIFY-POLICY instance $\langle \Lambda, P \rangle$. Otherwise, we output *No*.

The above resolution technique always terminates for Case 1. In addition, the above procedure can also be used to decide VERIFY-POLICY in cases where the view definitions in the supplied database are *non-recursive*. In such cases, the resolution is guaranteed to terminate. We illustrate our resolution technique in the following example.

Example 7. Consider a database that contains the following static constraint $\perp :- p(b)$. Suppose we want to verify whether or not the following update policy P satisfies the above constraint.

$$\Delta^- p(b) :- \Delta^+ p(X).$$

We initialize C to be the empty set of clauses, and add to C the following clauses.

$$
\begin{array}{lll}
c_1 : \{\neg p(b)\} & \text{(step 1)} \\
c_2 : \{\neg \Delta^+ p(b), \Delta^+ \perp\} & \text{(step 2)} \\
c_3 : \{\Delta^+ p(b), \neg \Delta^+ \perp\} & \text{(step 2)} \\
c_4 : \{\neg \Delta^- p(b), \Delta^+ p(k)\} & \text{(step 2)} \\
c_5 : \{\Delta^- p(b), \neg \Delta^+ p(X)\} & \text{(step 2)} \\
c_6 : \{\neg \Delta^+ p(X), \neg \Delta^- p(X)\} & \text{(step 3)} \\
c_7 : \{\neg \Delta^+ p(X), \neg p(X)\} & \text{(step 3)} \\
c_8 : \{\neg \delta^+ p(X), \neg p(X)\} & \text{(step 3)} \\
c_9 : \{\neg \Delta^- p(X), p(X)\} & \text{(step 3)} \\
c_{10} : \{\neg \delta^- p(X), p(X)\} & \text{(step 3)} \\
c_{11} : \{\neg \Delta^+ p(X)\} & \text{(step 4)} \\
c_{12} : \{\neg \Delta^+ \perp\} & \text{(step 5)}
\end{array}
$$

We can obtain the empty clause by resolving the clauses in C as follows.

$$
\begin{array}{lll}
d_1 : \{\neg \Delta^+ p(b)\} & \text{(resolve } c_2, c_{12}) \\
d_2 : \{\} & \text{(resolve } d_1, c_{11})
\end{array}
$$

Therefore, the supplied update policy satisfies the constraint on the database.

Complexity: The problem of deciding equivalence of two Datalog programs is reducible to VERIFY-POLICY. We present an outline of such a reduction in what follows. Consider two Datalog queries q_1 and q_2 that are defined using programs P_1 and P_2 respectively. We generate a VERIFY-POLICY instance with a single constraint $\perp :- b$, and the update rule $\Delta^+ a :- \delta^+ a, \neg b$ and the rules $\Delta^+ r(\bar{X}) :- \delta^+ r(\bar{X})$ and $\Delta^- r(\bar{X}) :- \delta^- r(\bar{X})$ for every base relation r in $P_1 \cup P_2$, where b is defined using the rules in P_1 and P_2, and the following.

$$
\begin{array}{l}
b :- q_1(\bar{X}), \neg q_2(\bar{X}). \\
b :- q_2(\bar{X}), \neg q_1(\bar{X}).
\end{array}
$$

In this case, answer to the generated instance of VERIFY-POLICY is *Yes* if and only q_1 and q_2 are equivalent. Since deciding equivalence of two Datalog programs is undecidable [6], we have the following intractability result.

Theorem 2. *In a general setting, VERIFY-POLICY is undecidable.*

Theorem 3. *If the policies and constraints are unions of conjunctive queries, then VERIFY-POLICY is NP-hard. If the policies and constraints contain negation or inequalities i.e. $\Delta^-, \leq, \geq$, then VERIFY-POLICY is Π_2^P-hard.*

Proof of the theorem follows directly from our reduction, and the complexity results presented in [5,8].

5   Conclusion and Outlook

In this paper, we have investigated the problem of supporting underspecified transactions in database management systems. We have presented a formal framework for expressing *update policies* for updating databases. We have identified a class of update policies called *inclusive* update policies which correspond to strategies for completing underspecified transactions. We have also presented a language, called, Datalog^u for expressing update policies.

Transactions on databases with constraints and arbitrary update policies may potentially lead to illegal database states. To prevent such scenarios, we have proposed a resolution-based technique that verifies whether or not specified update policies are sound and complete with respect to database constraints.

An alternative or complementary approach to the a-posteriori verification is to have administrators author update policies from the constraints in an interactive manner. Such an update policy authoring process can be facilitated by automatically generating all necessary and sufficient update policies, and letting the administrator pick one policy or combine multiple policies. We also note that the process of authoring minimal inclusive update policies can be simplified through a rule IDE that automatically adds rules of the form $\Delta^+ r(\bar{X}) :\text{-} \delta^+ r(\bar{X})$ and $\Delta^- r(\bar{X}) :\text{-} \delta^- r(\bar{X})$.

In this paper, we have considered only static database constraints. In such scenarios, every transaction between two legal database states is legal. In general, a database may also have contain *dynamic constraints* which place additional restrictions on legality of transactions. In the future, we plan to extend the presented approach to cover dynamic constraints.

References

1. Boutilier, C., Brafman, R.I.: Partial-order planning with concurrent interacting actions. J. Artif. Intell. Res. (JAIR) **14**, 105–136 (2001)
2. Fikes, R., Nilsson, N.J.: STRIPS: a new approach to the application of theorem proving to problem solving. Artif. Intell. **2**(3/4), 189–208 (1971)

3. Grosof, B.N., Labrou, Y., Chan, H.Y.: A declarative approach to business rules in contracts: courteous logic programs in XML. In: EC, pp. 68–77 (1999)
4. Orman, L.V.: Transaction repair for integrity enforcement. IEEE Trans. Knowl. Data Eng. **13**(6), 996–1009 (2001)
5. Sagiv, Y., Yannakakis, M.: Equivalences among relational expressions with the union and difference operators. J. ACM **27**(4), 633–655 (1980)
6. Shmueli, O.: Equivalence of datalog queries is undecidable. J. Logic Program. **15**(3), 231–241 (1993)
7. Ullman, J.D.: Principles of Database and Knowledge-Base Systems, vol. II. Computer Science Press, Rockville (1989)
8. Ullman, J.D.: Information integration using logical views. In: Afrati, F., Kolaitis, P. (eds.) ICDT 1997. LNCS, vol. 1186, pp. 19–40. Springer, Heidelberg (1997). doi:10. 1007/3-540-62222-5_34. http://dl.acm.org/citation.cfm?id=645502.656100
9. Widom, J., Ceri, S. (eds.): Active Database Systems: Triggers and Rules for Advanced Database Processing. Morgan Kaufmann Publishers Inc., San Francisco (1996)

Utilization of DBpedia Mapping in Cross Lingual Wikipedia Infobox Completion

Megawati, Saemi Jang, and Mun Yong Yi[✉]

Department of Industrial and Systems Engineering,
Graduate School of Knowledge Service Engineering,
KAIST, Daejeon, South Korea
{megawati, sammy1221, munyi}@kaist.ac.kr

Abstract. Wikipedia plays a central role in the web as one of the biggest knowledge source due to its large coverage of information that comes from various domains. However, due to the enormous number of pages and limited number of contributors to maintain all of the pages, the problem of missing information among Wikipedia articles has emerged, especially articles in multiple language versions. Several approaches have been studied to fix information gap in between cross- language Wikipedia articles. However, they can only be applied for languages that came from the same root. In this paper, we propose an approach to generate new information for Wikipedia infoboxes written in different languages with different roots by utilizing the existing DBpedia mappings. We combined mapping information from DBpedia with an instance-based method to align the existing Korean-English infobox attribute-value pairs as well as to generate new pairs from the Korean version to fill missing information in the English version. The results showed that we could expand up to 38% of the existing English Wikipedia attribute-value pairs from our datasets with 61% of accuracy.

Keywords: Infobox alignment · Infobox completion · DBpedia · Cross language Wikipedia

1 Introduction

Wikipedia is an online encyclopedia, which contains a massive amount of information from various domains provided collaboratively by its contributors. The information is easily accessible through the website[1] and is being continually updated by the contributors. Moreover, Wikipedia pages are also available in several languages (in May 2016, there are 282 different active Wikipedia language editions), creating an opportunity for people around the world to make contributions despite the language barrier. Thus, many practices have relied on Wikipedia as a knowledge source, such as Q&A systems, Linked Open Data (LOD), and intelligent agents.

Many Wikipedia pages usually contain an infobox, a small box located at the right side of the page, providing a summary of the page content in a structured manner. Due to its structure, the infoboxes are useful if we want to mine key information from a

---

[1] https://www.wikipedia.org/.

© Springer International Publishing AG 2016
B.H. Kang and Q. Bai (Eds.): AI 2016, LNAI 9992, pp. 303–316, 2016.
DOI: 10.1007/978-3-319-50127-7_25

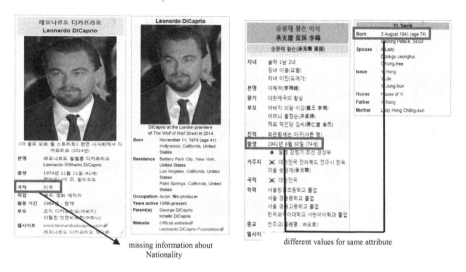

missing information about
Nationality

different values for same attribute

Fig. 1. Example of error Type II and Type III

particular page, which can take a lot more effort if we mine from free texts (the article itself). Generally, the infobox consists of three parts: (1) **template** represents type/category of the entity that is being discussed in the page (e.g. template Infobox Person is used in those pages related to a person, such as presidents, actors, soldiers), (2) **attributes** represents characteristics of the template (e.g. a person has a name, birth date, birth place, and parents), and (3) **values** are the instances of the attribute.

Although Wikipedia is considered as a reliable knowledge source, its information is not flawless, given the fact that it is entered by people. Problems such as the use of different names for an entity are common to be found across the pages. Moreover, the number of Wikipedia pages is growing rapidly, making it hard to maintain all the existing pages, including their corresponding pages in different languages. Consequently, information incompleteness and inconsistencies have emerged as serious issues and must be tackled to maintain information quality in Wikipedia. This paper focused on maintaining completeness and consistencies between infoboxes in multilingual Wikipedia pages. In terms of infoboxes, we did an observation with some pairs of random Wikipedia pages in different language and found three types of errors regarding the information in the infoboxes.

1. Type I

   This error is related to the missing infobox in one of the pages. For example, articles about Takeo Takagi has an infobox in English version of Wikipedia while the infobox does not exist in the Korean version.

2. Type II

   This error happens when both pages have infoboxes but one of them have missing attributes that exist in the other version. Figure 1 (left) shows comparison between infoboxes Leonardo DiCaprio from Korean and English version. In the Korean version there is an attribute that describes his Nationality which should also be presents in the English version.

3. Type II

Similar with error Type II, error Type III can be found when both pages have infoboxes. This error happens when the same attribute have different values, such as values for attribute 출생 and Born that are being shown in Fig. 1 (right).

As a knowledge source, Wikipedia should maintain the quality of the available information at a high level to ensure its reliability. However, huge efforts will be needed to go through all the existing pages and check for error one by one. Therefore, we conducted a study whose aim is to enhance information quality of Wikipedia infoboxes by correcting the Type I and II errors. We developed a model that is able to automatically generate new infoboxes for pages that do not have any infobox or adding more information to the existing infoboxes with the help from DBpedia mappings. We found the possibility that the DBpedia mappings might be useful as a translation tool, eliminating the need to involve any bilingual dictionary. Instead, the mappings can map an attribute from one language to another since DBpedia facts are also available in multilingual environment. If the two attributes are mapped to the same property of the same entity, then they are very likely matching words.

The purpose of this study is to evaluate the use of DBpedia mappings to align infobox attributes and templates in multilingual environments. First, we evaluated the capability of the existing mappings to translate infobox templates and attributes from Korean to English and we measured how many new pairs could be generated. Then, we tried to search other possible translations that are not covered in the mapping by using instance-based method that were used in [2, 3] to expand the number of generated pairs.

The organization of this paper is as follows. Section 2 will explain about the related studies about various approaches for cross-lingual schema matching and DBpedia enrichment that have already been done by other researchers. In Sect. 3, we will explain in detail about our proposed model and the techniques that we used in aligning infoboxes. Section 4 will describe the detail of our experiment and the results. Section 5 presents the conclusion that we could draw from the experiment and discusses some possibilities for future research.

2 Related Work

Schema Matching. The infobox alignment problem can be considered as a schema matching problem. Studies related to this area have already been done by many researchers [6]. Several studies have been conducted to develop approaches for aligning multilingual schemas as well as ontologies. Wang et al. [12] tried to identify correspondences between Chinese and English attributes from multilingual schemas by transliterating Chinese characters into alphabets and took the first letter of each syllable to replace the original attribute name. A domain ontology built by human was used to determine the mapping between translated attribute and English attribute. [13, 14] proposed models that can align multilingual ontologies by translating the source ontology into the target language and match them by using the existing monolingual ontology matching approaches. Unfortunately, the proposed approaches mentioned above are hardly applicable to matching infobox data. Schemas and ontologies have a well-defined structure and metadata while infoboxes are much looser on their data type constraints. Thus, for infobox matching, comparing only metadata and structures might be insufficient to solve the problem. Moreover, the approaches to align multilingual schemas were fully based on translation tools, which have clear limitations when applied to aligning infoboxes, as we discussed in the previous chapter.

Cross-Language Infobox Alignment. Studies about aligning Wikipedia infoboxes in different language have been done by several researches. [15] proposed an approach to align Dutch and English Wikipedia templates and attributes by utilizing multilingual nature of Wikipedia as well as cross-language links between pages with precision 65%. Moreover, the approach can be used to generate new attribute-value pairs in Dutch Wikipedia by 50%. [16] developed WikiMatch, a tool that can align two infoboxes in different language without using dictionary or translator. They combined three similarity measures to determine the similarity between two attributes; value similarity, link similarity, and cosine similarity from attribute co-occurrence vectors, which were decomposed by using Latent Semantic Indexing (LSI). [2] used a three stages approach to align multilingual infoboxes from six languages; entity matching, template matching, and attribute matching. In our work, we adapted a similar template matching from the paper to determine the template for the generated infoboxes. We also adapted attribute matching techniques that had been used in [2] to determine potential new attribute mappings. Another approach was used by [4] who exploited machine learning approach to align Wikipedia infoboxes while [5] has developed an information extraction tool, Kylin, to extract information from Wikipedia text and predict possible attribute-value pairs from the sentences by using CRF classifier. However, Kylin was tested by using only English Wikipedia articles and, as for our knowledge, there has not been a study yet that try to measure Kylin's capability of generating new infoboxes from different language.

Cross-Language DBpedia Enrichment. DBpedia, which is essentially structured information extracted from Wikipedia, is also dealing with inconsistencies and incompleteness issues due its multilingual nature. These problems are being solved by crowd sourcing effort from the community members. Various approaches have been studied to develop an automatic system that can better align multilingual DBpedia. [17] proposed an approach that can automatically extend the existing alignment of multilingual DBpedia chapters by using mapping frequencies of two properties and then integrate the results to a question answering system over linked data. A study in [18] has been conducted to find semantically corresponding properties from Korean and English DBpedia datasets by using the triple-conceptualization technique. The enrichment can also be done by using the existing Wikipedia data and map them to DBpedia ontology, like what have been done by [3, 20].

3   Cross Language Infobox Completion

We developed a model that compares two infoboxes from the Korean Wikipedia and English Wikipedia to find which information should be added from the Source to the Target infobox. Later in this paper, we refer the Korean Wikipedia infoboxes as the Source infoboxes and English Wikipedia infoboxes as the Target infoboxes. We used Korean infoboxes as source because the localized version might still cover more information about the topics related to the local culture though English version has the largest information coverage. Therefore, we could introduced such information to the people outside the culture as well as contribute to expanding the information coverage in the English Wikipedia. The overview of the model is shown in Fig. 2. It basically consists of 4 main parts: Mapping Table, Infobox Alignment, Infobox Generator, and Infobox Populator. Details of each part will be elaborated in the following subsections.

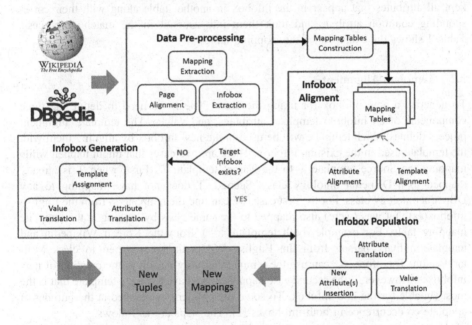

Fig. 2. Overview of cross language infobox completion model

3.1   Mapping Tables

The mapping tables contain mapping information extracted from DBpedia. DBpedia is a knowledge base built based on structured information from Wikipedia, i.e. infoboxes. Up to this day, DBpedia community members manually map Wikipedia infobox templates to DBpedia ontology classes as well as Wikipedia infobox attributes to DBpedia ontology properties. The results are available on the Web[2] and can also be downloaded as xml files.

---

[2] http://wiki.dbpedia.org/Downloads2015-10.

Table 1. Examples of attribute mapping tables

Attribute_ko	Attribute_en	DBpedia_property
이름	name	foaf:name
출생지	birth_place	dbo:birthPlace
사망지	death_place	dbo:deathPlace
개교	established	dbo:established
학생수	students	dbo:numberOfStudents

For each language, we built two kinds of mapping tables; the template mapping table and the attribute mapping table. From these tables, we could find pairs of attributes/templates that are semantically similar. We tackled one-to-many mappings by only picking one common attribute to be included in the mapping table. However, we kept all attributes that appear in the infobox in another table along with their corresponding common attribute and took them into account in the matching process. Table 1 shows the examples of the mapping tables.

3.2 Template Alignment

To generate a new infobox for fixing the error Type I, we need to define the three components of an infobox: template, attributes, and values. The template alignment process defines which template will be used in the new infobox by aligning them with the template used in the existing infobox. There are two cases that might happen while mapping the Source template T to the Target template T'. First, when T is already mapped to a DBpedia ontology class. Second, T does not have mapping to any DBpedia ontology class. For the first case, we can find the template(s) from the English infobox(es) that was (were) also mapped to the same class by looking at the template mapping table. For example, both template 군인 from the Korean Wikipedia and template military person from the English Wikipedia are mapped to class MilitaryPerson. Therefore, we can use the template military person in the creation of new infoboxes. However, if the second case happened, we have to pick a template that is the most suitable for the new infobox. To solve the problem, we looked at the number of template co-occurrence in both infoboxes [2]. The steps are as follows.

1. Let P_S be a set of articles in Source languages and $P_{S'}$ be a set of articles in Target language that are connected to element in P_S through interlanguage links. Let T_S be the Source template and $T_{S'}$ be the Target template that we are trying to define
2. Calculate the total occurrence number of each template that is being used by the members in $P_{S'}$
3. Template with the highest occurrence number will become $T_{S'}$

3.3 Attribute Alignment

The purpose of the attribute alignment process was to find pairs of cross language attributes that are semantically similar. Similar to the template alignment, two cases might happen while mapping Source attribute a to Target attribute a'. The first case is when both a and a' are connected via their mapping to the same DBpedia ontology property. The second case happens when either a or a' does not have mapping information to any property in DBpedia ontology so they are not connected to each other. While in the first case we could easily look up to the mapping tables we had already constructed to obtain the mapping information, we need another way to find new mappings from potential attribute pairs that do not have any connection yet. Therefore, we decided to use instance-based method introduced in [2] to find new alignments between such attributes. The steps are as follows.

1. Let S be a set of article pairs P_l-$P_{l'}$ where l is the Source language and l' is the Target language and each P_l contains an infobox with template T
2. Let A be the set of attributes from all P_l and A' be a set of attributes from all $P_{l'}$ where each element in A does not exist in the mapping table. For each attribute pair $(a_l, a_{l'})$, we compute sim_a

$$sim_a(a_l, a_{l'}) = \frac{\sum_{s \in S} sim_{instance}(a_l, a_{l'})}{|S|}.$$ (1)

The algorithm that we used to calculate the similarity between the attributes is further explained in detail in Sect. 3.5
3. All $(a_l, a_{l'})$ whose $sim_a < \alpha$ will be discarded
4. For each a_l, find $(a_l, a_{l'})$ with the maximum value and add to matching set M_a
5. Add M_a to the mapping table

We decided to use an instance-based method due to the format-loose nature of infobox values. Wikipedia does not provide a specific convention that must be followed to define infobox value (e.g. birth_date must be in YYYYMMDD or DDMMYYYY). Instead, it let authors use their own style. We found it difficult to use other similarity measures to compare two different strings that are semantically similar but are written in different forms. Moreover, infobox values are often composed of several elements other than texts. Therefore we use an instance-based method as a heuristic to measure the similarity of two strings of infobox value.

3.4   Infobox Generation and Population

We could use all information about templates and attributes alignments to align and complete the cross language infoboxes. First of all, we had to determine whether an infobox exists in the Target article. If the Target infobox I' does not exist, it means that we have to create a new infobox in the Target article by translating all information available from the Source infobox into the Target language. The process is called the infobox generation process. Otherwise, we had to compare both infoboxes from a pair of article, which talks about a same topic (later we refer to it as article pair) to determine whether new attribute-value pairs should be added to the Target infobox.

The infobox population process adds new potential attribute-value pairs which are not yet available to the Target infobox. Basically, both processes consists of three steps.

1. Template assignment
   As mentioned before, to create new infobox we need to define all the components. This step finds a mapping of T by looking at the mapping table and assigns it the corresponding template of T as the template of the new infobox. This step is particularly important in the generation process while in the population process we could skip it because the Target infobox usually already has its own template.
2. Attribute translation and insertion
   This step maps a set of attributes A from the Source infobox to its corresponding mapping in the Target language by looking at the attribute mapping table and assigns them as attributes for the new infobox. In the generation process, we inserted all translated attributes to the new infobox while in the population process we omit attributes that already exist in the Target infobox and insert the new ones.
3. Value translation
   For each attribute, we translated its value to the Target language as the new values by using translator API[3]. For the values that contain links, we substitute them with the corresponding link in the Target language by utilizing Wikipedia interlanguage links. For example, 1990년 03월 09일 <미국> will be translated as 09-03-1990 <United States> where the brackets denote a link.

3.5   Similarity Measure

To get $sim_{instance}$ of an attribute pair we break down the value into four parts: text, number, date, and links, and then calculate a similarity score for each part. We then aggregate the results to find the final similarity score. We adapted the methods in [2] to calculate the similarity score for each part and aggregate them in Table 2.

To determine the overall instance similarity between a pair of attribute values, [2] took into account the portion of respective components in the original attribute value string. f_{s1} and f_{s2} are the fraction of string values of both attribute values, while f_n represents fraction of number, and f_d is fraction of date. len_1 and len_2 are the length of the original attribute values. Then, f_s, f_n, and f_d can be defined as follow.

$$f_s = \frac{len_1 \cdot f_{s1} + len_2 f_{s2}}{len_1 + len_2} \quad f_n = \frac{len_1 \cdot f_{n1} + len_2 \cdot f_{n2}}{len_1 + len_2}$$

$$f_d = \frac{len_1 \cdot f_{d1} + len_2 \cdot f_{d2}}{len_1 + len_2}$$

After the similarity scores from each value components were obtained, similarities of texts, numbers, and dates were calculated and then weighted together with similarities of links to produce the overall instance similarity value.

---

[3] https://www.microsoft.com/en-us/translator/translatorapi.aspx.

Table 2. Similarity measures for number, date, link, and text extracted from attribute values

Number similarity	$sim_{num}(n_1, n_2) = \begin{cases} 1, & if\ n_1 = n_2 \\ 0.5 \cdot \dfrac{\min\{|n_1|, |n_2|\}}{\max\{|n_1|, |n_2|\}}, & otherwise \end{cases}$
	$sim_{numset}(N_1, N_2) = \dfrac{\sum_{<n_1, n_2> \in M_n} sim_{num}(n_1, n_2)}{\max\{|N_1|, |N_2|\}}$
Date similarity	$sim_{date}(d_1, d_2) = 1 - \dfrac{|d_1 - d_2|}{maxDate - minDate}$
	$sim_{dateset}(D_1, D_2) = \dfrac{\sum_{<d_1, d_2> \in M_n} sim_{date}(d_1, d_2)}{\max\{|D_1|, |D_2|\}}$
Link similarity	$sim_{wikilinks}(w_{l1}, w_{l2}) = \dfrac{2 \cdot |w_{l1} \cap w_{l2}|}{|w_{l1}| + |w_{l2}|}$
	$sim_{exlinks}(e_{l1}, e_{l2}) = \dfrac{2 \cdot |e_{l1} \cap e_{l2}|}{|e_{l1}| + |e_{l2}|}$
Text similarity	$sim_{str}(s_1, s_2) = \dfrac{|T_1 \cap T_2|}{|T_1 \cup T_2|}$

$$sim_{val}(a_1, a_2) = \frac{w_s \cdot f_s \cdot sim_{str} + w_n \cdot f_n \cdot sim_{numset} + w_s \cdot f_s \cdot sim_{dateset}}{w_s \cdot f_s + w_n \cdot f_n + w_d \cdot f_d} \quad (2)$$

$$sim_{instance}(a_1, a_2) = \frac{w_v \cdot sim_{val}(a1, a2) + w_w \cdot sim_{wikilinks} + w_e \cdot sim_{exlinks}}{w_v + w_w + w_e} \quad (3)$$

According to [2], the weights for each data type portion and links were largely determined empirically. The weight used in the experiment are $w_s = 0.11$, $w_n = 0.44$, $w_d = 0.44$, $w_v = 0.3$, $w_w = 0.6$, and $w_e = 0.1$. Note that the links similarity could be calculated if both set have at least one member. Otherwise, the weight, either w_w or w_e will become 0.

4 Experiment

In our experiment, we used Korean Wikipedia and English Wikipedia articles dump[4] and DBpedia mappings[5]. In the data pre-processing step, we extracted infobox data by using infobox2rdf [25][6], which robustly extracts the infoboxes from xml files, cleanses them, and transforms them into RDF triples. To test our model, we picked five different infobox templates from Korean Wikipedia that became the Source infoboxes; 군인 (Military person), 학교 (School), 왕 (Monarch), 회사 (Company), and 대학 (University). We then extracted Korean-English article pairs from each template as our dataset in the experiment by using interlanguage links.

---

[4] https://dumps.wikimedia.org/.

[5] DBpedia mapping (http://mappings.dbpedia.org/) version 5 March 2016.

[6] https://github.com/thomlee/infobox2rdf.

4.1   Infobox Alignment by Instance-Based Method

As the number of Wikipedia attributes is huge, it is almost impossible to map all existing Wikipedia attributes into the DBpedia properties. We wanted to find other attributes in Source language that have potential to be mapped to the Target language to expand our mapping tables. Therefore, we used instance-based approach to find new alignments.

First, we did the data pre-processing step to filter and cleanse the attributes. It is important to note that we only focused on the attributes whose values are text, number, date, or links, so we omitted any attributes whose value is related to pictures, logos, captions, or signatures. Then, we matched each attribute against the candidate attributes, which are the attributes that already exist in the mapping table. We set a threshold 0.6 to filter the similarity scores of each attribute pair. The score was ranged from 0 to 1. The higher score means the higher probability of two attributes being similar. Each attribute pairs with the score lower than the threshold would be ignored. We only considered attributes that occur ≥ 10 times in the whole articles for the same template. We added another constraint to only accept the alignments whose number of the matching article pairs ≥ 5. For the pairs that did not pass, we depended on human judgment to determine whether they are acceptable. There are two criteria we used to determine whether the matching is acceptable or not; the types of their values (e.g. currency, location, organization, etc.) and the values themselves. We compared the values of each pair with other values from the same attribute. For example, for 관할관청–district, we looked at all possible values of 관할관청 attribute in the Korean Wikipedia and all possible values of district in the English Wikipedia. If both attributes share the same values from the same type for at least five different articles, we defined it as acceptable.

After we aligned all possible attribute pairs, we could generate total 41 new mappings for attributes in our dataset. The discovery of the new mappings might also help in the generation process to add more attributes that exist in the Source article but do not exist in the Target article.

4.2   Infobox Attribute-Value Pairs Generation

After the Source and Target infoboxes were aligned, we applied our generation technique to generate new infobox tuples to solve error Type I and II in the infoboxes. Figure 3 shows an example of the original attribute-pairs for infobox and the new attributes that have been generated by our approach.

We tested our approach to generate new attribute-value pairs for all articles from five templates that we had chosen. Table 3 and Fig. 4 show the comparison between the number of the existing tuples before alignment and after alignment.

4.3   Evaluation

We evaluated the accuracy of our method in two ways. First, we compared similarity between the newly generated values in English with their original tuples in Korean to see the number of attributes that had been correctly translated. Second, we evaluated

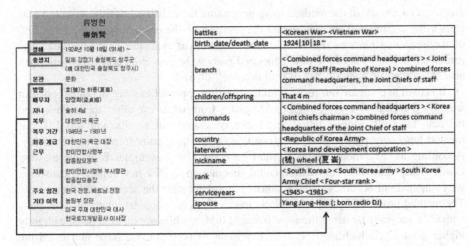

battles	<Korean War> <Vietnam War>
birth_date/death_date	1924	10	18 ~
branch	< Combined forces command headquarters > < Joint Chiefs of Staff (Republic of Korea) > combined forces command headquarters, the Joint Chiefs of staff
children/offspring	That 4 m
commands	< Combined forces command headquarters > < Korea joint chiefs chairman > combined forces command headquarters of the Joint Chief of staff
country	<Republic of Korea Army>
laterwork	< Korea land development corporation >
nickname	(號) wheel (夏 齋)
rank	< South Korea > < South Korea army > South Korea Army Chief < Four-star rank >
serviceyears	<1945> <1981>
spouse	Yang Jung-Hee (; born radio DJ)

Fig. 3. Attribute-value pairs for article Lew Byung Hyun (류병현) from the original infobox and new attribute-value pairs generated by our method

Table 3. Statistics about the number of attribute-value pairs before and after generation process

Template	Total article pairs	Existing tuples		Tuples after alignment		Expanded (%)
		Ko	En	DBpedia	DBpedia + IB	
군인/Military person	457	5249	6669	7478	8444	21.02
학교/School	219	3000	2426	3846	3940	38.43
왕/Monarch	584	5640	6654	7333	8273	19.57
회사/Company	1568	21016	18831	27466	27827	32.33
대학/University	879	9523	14991	18694	18788	20.21

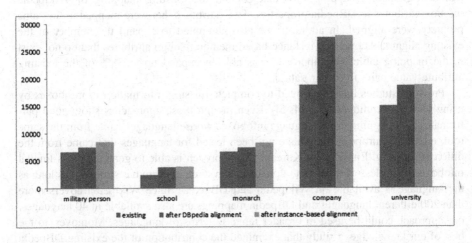

Fig. 4. Comparison of the number of attribute-value pairs before and after generation process

the overall accuracy of the method by re-generating the existing English attribute-value pairs using our method and comparing the results with the original ones. We took 20% of the generated pairs as our sample and used human evaluator to do the task. The results show that 73% of the new generated pairs were translated correctly while they also show that our method has overall accuracy of 61%.

It is hard to compare our result with the existing approaches since in our experiment we only used data from five infobox templates from while the other approaches used the data from all infobox templates. If we compared our results with [4, 15], our method has better performance in expanding the existing pairs (tuples) and accuracy. According to [15], their approach could generate 27% new tuples from the existing Dutch Wikipedia while our method could generate up to 38% new tuples. Meanwhile, we compared our accuracy with [4] since [15] did not state the accuracy of their result. The method proposed by [4] was able to match cross language template-attribute pairs with 60% accuracy by using the most frequent tuples, while our method has the slightly higher accuracy, which is 61%. It is important to note that we have not tested the performance against the whole infobox data. Therefore, the number might still be changed. We will leave the evaluation for the future works.

While performing evaluation, we also found common errors that occurred in the results. They happened due to the translation errors, API errors, link errors, and inconsistencies. We found that this approach might be useful to detect value inconsistencies for the same attribute. However, we did not do any validation to resolve the problem and left it for future work.

5   Conclusion and Future Work

The purpose of the present study is to fix information gap between cross language Wikipedia articles. We have proposed an approach that takes advantages from the existing DBpedia mappings to align two attributes in different languages that are semantically similar by constructing the mapping tables, which were derived from the extracted DBpedia mapping files that contain the existing mapping of Wikipedia infobox attributes to DBpedia properties. Two attributes that were mapped to the same property were aligned. In addition, we also attempted to expand the number of the existing alignments by using instance-based method to align attributes that do not exist in the mapping table. Our approach was able to expand up to 38% of the existing attribute-value pairs from our dataset.

Previous studies have attempted to complete missing information in infoboxes by using various techniques (e.g. [4, 5]). Even though those approaches show good performances on aligning cross language infoboxes whose languages came from the same root, e.g. Indo-European, they have not been tested for languages that came from the different root, e.g. English and Korean. Our approach is able to generate cross-lingual infoboxes regardless of their root, alphabetical system, or grammar structure, as long as the language is available on Wikipedia and DBpedia. Since Wikipedia covers more than 200 different languages and DBpedia mappings are also available in 40 languages, our approach could be used in broader range in terms of languages. Moreover, to the best of our knowledge, a study that examined the contribution of the existing DBpedia mapping to the infobox completion process among Wikipedia pages has not been conducted yet. Thus, our study represents the first step into this new direction, thereby making a new contribution to this research field.

Our approach needs to be further refined. Given that our approach relies on human intervention, we would like to do some improvements to reduce the human effort, such as using a robust XML parser to construct the mapping tables. We also would like add a validation component to resolve inconsistencies in the aligned attribute values. Finally, we are planning to expand our dataset to the whole Korean Wikipedia and to other language versions and evaluate our approach to get a holistic view about its performance.

Acknowledgments. This work was supported by the Industrial Strategic Technology Development Program, 10052955, Experiential Knowledge Platform Development Research for the Acquisition and Utilization of Field Expert Knowledge, funded by the Ministry of Trade, Industry & Energy (MI, Korea).

References

1. Bizer, C., Lehmann, J., Kobilarov, G., Auer, S., Becker, C., Cyaniak, R., Hellmann, S.: DBpedia - a crystallization point for the web of data. Web Sem. Sci. Serv. Agents World Wide Web 7(3), 154–165 (2009)
2. Rinser, D., Lange, D., Naumann, F.: Cross-lingual entity matching and infobox alignment in Wikipedia. Inf. Syst. 38(6), 887–907 (2013)
3. Palmero Aprosio, A., Giuliano, C., Lavelli, A.: Towards an automatic creation of localized versions of DBpedia. In: Alani, H., et al. (eds.) ISWC 2013, Part I. LNCS, vol. 8218, pp. 494–509. Springer, Heidelberg (2013)
4. Adar, E., Skinner, M., Weld, D.S.: Information arbitrage across multi-lingual Wikipedia. In: Proceedings of the Second ACM International Conference on Web Search and Data Mining. ACM (2009)
5. Wu, F., Weld, D.S.: Autonomously semantifying Wikipedia. In: Proceedings of the Sixteenth ACM Conference on Conference on Information and Knowledge Management. ACM (2007)
6. Rahm, E., Bernstein, P.A.: A survey of approaches to automatic schema matching. VLDB J. 10(4), 334–350 (2001)
7. Melnik, S., Garcia-Molina, H., Rahm, E.: Similarity flooding: a versatile graph matching algorithm and its application to schema matching. In: Proceedings of 18th International Conference on Data Engineering. IEEE (2002)
8. Li, W.-S., Clifton, C.: SEMINT: a tool for identifying attribute correspondences in heterogeneous databases using neural networks. Data Knowl. Eng. 33(1), 49–84 (2000)
9. Nottelmann, H., Straccia, U.: Information retrieval and machine learning for probabilistic schema matching. Inf. Process. Manag. 43(3), 552–576 (2007)
10. Kohonen, T.: Adaptive, associative, and self-organizing functions in neural computing. Appl. Opt. 26(23), 4910–4918 (1987)
11. Fuhr, N.: Probabilistic datalog—a logic for powerful retrieval methods. In: Proceedings of the 18th Annual International ACM SIGIR Conference on Research and Development in Information Retrieval. ACM (1995)

12. Wang, H., et al.: Identifying indirect attribute correspondences in multilingual schemas. In: 17th International Workshop on Database and Expert Systems Applications, 2006. DEXA 2006. IEEE (2006)

13. Fu, B., Brennan, R., O'Sullivan, D.: Cross-lingual ontology mapping – an investigation of the impact of machine translation. In: Gómez-Pérez, A., Yu, Y., Ding, Y. (eds.) ASWC 2009. LNCS, vol. 5926, pp. 1–15. Springer, Heidelberg (2009)

14. Dos Santos, C.T., Quaresma, P., Vieira, R.: An API for multilingual ontology matching. In: Proceedings of 7th Conference on Language Resources and Evaluation Conference (LREC). No commercial editor (2010)

15. Bouma, G., Duarte, S., Islam, Z.: Cross-lingual alignment and completion of Wikipedia templates. In: Proceedings of the Third International Workshop on Cross Lingual Information Access: Addressing the Information Need of Multilingual Societies. Association for Computational Linguistics (2009)

16. Nguyen, T., et al.: Multilingual schema matching for Wikipedia infoboxes. Proc. VLDB Endow. 5(2), 133–144 (2011)

17. Cojan, J., Cabrio, E., Gandon, F.: Filling the gaps among DBpedia multilingual chapters for question answering. In: Proceedings of the 5th Annual ACM Web Science Conference. ACM (2013)

18. Kim, E.-K., Choi, K.-S.: Cross-lingual property alignment for DBpedia ontology using triple conceptualization (2014)

19. Palmero Aprosio, A., Giuliano, C., Lavelli, A.: Automatic expansion of DBpedia exploiting Wikipedia cross-language information. In: Cimiano, P., Corcho, O., Presutti, V., Hollink, L., Rudolph, S. (eds.) ESWC 2013. LNCS, vol. 7882, pp. 397–411. Springer, Heidelberg (2013). doi:10.1007/978-3-642-38288-8_27

20. Kim, E.K., et al.: An approach for supplementing the Korean Wikipedia based on DBpedia. Liliana Cabral (Open University, UK) Tania Tudorache (Stanford University, USA), p. 7 (2010)

21. Mahdisoltani, F., Biega, J., Suchanek, F.: Yago3: a knowledge base from multilingual Wikipedias. In: 7th Biennial Conference on Innovative Data Systems Research. CIDR Conference (2014)

22. Tacchini, E., Schultz, A., Bizer, C.: Experiments with Wikipedia cross-language data fusion. In: Workshop on Scripting and Development (2009)

23. Spohr, D., Hollink, L., Cimiano, P.: A machine learning approach to multilingual and cross-lingual ontology matching. In: Aroyo, L., et al. (eds.) ISWC 2011, Part I. LNCS, vol. 7031, pp. 665–680. Springer, Heidelberg (2011)

24. Salhi, A., Camacho, H.: A string metric based on a one-to-one greedy matching algorithm. Res. Comput. Sci. 19, 171–182 (2006)

25. Lee, T.Y., et al.: Automating relational database schema design for very large semantic datasets. Technical report, Department of Computer Science, University of Hong Kong (2013)

26. Lehmann, J., et al.: DBpedia–a large-scale, multilingual knowledge base extracted from Wikipedia. Semant. Web 6(2), 167–195 (2015)

A Multi-linguistic-Valued Modal Logic

Jinsheng Chen and Xudong Luo[✉]

Institute of Logic and Cognition, Department of Philosophy,
Sun Yat-sen University, Guangzhou 510275, China
luoxd3@mail.sysu.edu.cn

Abstract. This paper develops a multi-valued modal logic, in which a logic formula takes a value of truth in linguistic terms. In other words, *truth* in our logic is regarded as a linguistic variable and its values are linguistic terms, which can be modelled as fuzzy sets. In particular, we define negation and implication on linguistic truth values such that their truth tables accord with those in conventional three-valued logic. Moreover, we also prove the soundness and completeness of our logic.

Keywords: Fuzzy logic · Linguistic variable · Multi-valued logic · Modal logic

1 Introduction

In our daily life, when talking about the age of somebody, we do not always give a statement with a precise number. Instead, often we simply say 'pretty young', which, unlike a precise number (*e.g.*, 23), is vague in its meaning. The prevalent phenomenon that we use words with fuzzy meaning imposes a challenge on characterising human reasoning. To take this challenge, Zadeh [17] introduces the concept of linguistic variable, which takes values in the forms of words or sentences in a natural language. For example, *age* could be a linguistic variable, taking values of *young, very young, not young, quite young, very old, not very old*, and so on, rather than 20, 21, 22, and so on. A linguistic value of *age* like *young* can be modelled as a fuzzy set with the numerical age as domain.

The reason why *truth* can be seen as linguistic variable is from the observation on our daily discourse that we frequently use expressions such as *very true, quite true, essentially true* to characterise a degree to which a statement is true. The phrases like *very true, quite true*, and *essentially true* are the linguistic values of *truth*. Thus, with *truth* as linguistic variable, it is required to define a fuzzy linguistic logic in which the truth value of every logic formula is a linguistic value. In fact, there are some efforts in this direction. However, few work is about fuzzy multi-valued modal logic (see Sect. 5 for detailed discussion).

So, in this paper, we will propose a modal logic with five linguistic truth values. In particular, we define the implication on linguistic truth values in a way that its truth table is consistent with that in conventional three-valued logic. Moreover, we prove the soundness and completeness of our logic system.

© Springer International Publishing AG 2016
B.H. Kang and Q. Bai (Eds.): AI 2016, LNAI 9992, pp. 317–323, 2016.
DOI: 10.1007/978-3-319-50127-7_26

This paper has the following structure: Sects. 2 and 3 present the syntax and semantics of our logic. Section 4 proves its soundness and completeness. Section 5 discusses the related work. Finally, Sect. 6 concludes our work.

2 Syntax

This section presents the syntax of our fuzzy multi-valued logic (denoted as F_5).

Definition 1. *The well-formed formulas of F_5 are given by:*

$$\phi ::= p \mid \neg\phi \mid \varphi \to \phi \mid \Diamond\phi, \tag{1}$$

where p ranges over elements of the proposition letter set Φ.

The possible interpretations of $\Diamond$ are *'is possible that'*, *'is permitted that'*, *'believe that'* and so on [3]. In our logic, we also need the following abbreviations:

$$\overline{\phi} := (\phi \to \neg\phi), \tag{2}$$

$$\Box\phi := \neg\Diamond\neg\phi. \tag{3}$$

Definition 2. *Suppose that φ_1, φ_2 and φ_3 are formulas in F_5. The axioms of F_5 are as follows:*

A1: $\varphi_1 \to (\varphi_2 \to \varphi_1)$,
A2: $(\varphi_1 \to \varphi_2) \to ((\varphi_2 \to \varphi_3) \to (\varphi_1 \to \varphi_3))$,
A3: $(\overline{\varphi_1} \to \varphi_1) \to \varphi_1$, .
A4: $(\neg\varphi_1 \to \neg\varphi_2) \to (\varphi_2 \to \varphi_1)$,
K: $\Box(\varphi_1 \to \varphi_2) \to (\Box\varphi_1 \to \Box\varphi_2)$,
D: $\Diamond\varphi_1 \leftrightarrow \neg\Box\neg\varphi_1$.

Among the above axioms, A1 to A4 are those use in [4], which with the rules of proof below ensures that we can make use of results in [4].

1. Modus ponens: if ϕ and $\phi \to \varphi$, then φ.
2. Uniform substitution: if ϕ, then φ, if φ is obtained from ϕ by uniformly replacing proposition letters in ϕ by arbitrary formulas.
3. generalisation: if ϕ, then $\Box\phi$.
4. Diamond generalisation: if $\Diamond\phi$, then $\Box\phi$.

3 Semantics

In this section, we present the semantics of our logic.

Definition 3 (Truth as a linguistic variable). *Truth is a linguistic variable characterized by a triple $(truth, T(truth), [0, 1])$, where*

$$T(truth) = \{very\text{-}true, fairly\text{-}true, fairly\text{-}false, very\text{-}false, undecided\}$$

and $[0, 1]$ is a numerical interval of truth values.

Table 1. The truth table of $\phi \to \varphi$ and $\neg\varphi$ on linguistic truth values

ϕ \ φ	F_1	F	N	T	T_1
F_1	T_1	T_1	T_1	T_1	T_1
F	T	T	T	T	T_1
N	N	N	T	T	T_1
T	F	F	N	T	T_1
T_1	F_1	F	N	T	T_1

φ	$\neg\varphi$
F_1	T_1
F	T
N	N
T	F
T_1	F_1

For convenience, we use some abbreviation such that T_1 stands for *very true*, T for *fairly true*, F for *fairly false*, F_1 for *very false*, and N for *undecided*. Thus, $T(truth) = \{T_1, T, F, F_1, N\}$. It is noted that it accords with our daily use that the values of *truth* are fuzzy sets with numerical truth values as domain.

Definition 4 (Membership functions of linguistic truth values).

$$\mu_{T_1}(x) = x^6, \ \mu_T(x) = \sqrt{x}, \ \mu_F(x) = \sqrt{1-x}, \ \mu_{F_1}(x) = (1-x)^6,$$
$$\mu_N(x) = e^{-26(x-0.5)^4}.$$

What we are going to do next is to use the extension principle (denoted as operator $\otimes$) [16] and the linguistic approximation method (denoted as operator $\odot$) [14] in fuzzy set theory to define negation and implication operating on linguistic truth values based on the operation on intervals.

Definition 5 (Negation on linguistic truth values). *Suppose* $\tau \in T(truth)$, *the negation of* τ, *denoted as* $\neg\tau$, *is a fuzzy set defined as*

$$\mu_{\neg\tau}(x) = \odot(\otimes(\tau, 1-x)). \tag{4}$$

Next, we extend Lucasiewicz implication to linguistic truth values.

Definition 6 (Implication on linguistic truth values). *Suppose* $\tau_1, \tau_2 \in T(truth)$, *the value of* τ_1 *implying* τ_2 *is a fuzzy set defined as*

$$\mu_{\tau_1 \to \tau_2}(x) = \odot(\otimes(\tau_1(x_1), \tau_2(x_2), \min\{1, 1 - x_1 + x_2\})). \tag{5}$$

By (4) and (5), we can calculate the negation and implication on linguistic truth values through MATLAB and then the result is showed in Table 1.

With the definition of truth as linguistic variable and operation on it, we can now introduce the concept of model satisfaction in our logic F_5 as follows:

Definition 7 (Model satisfaction). *Suppose* w *is a state in a model* $\mathcal{M} = (W, R, V)$. *Then we recursively define the notion of a formula being satisfied in* $\mathcal{M}$ *at a possible world* w *as follows:*

1. $\mathcal{M}, w \models p$ *iff* $V(p, w) \in \{T_1, T\}$, *where* $p \in \Phi$.

2. $\mathcal{M}, w \models \neg\phi$ iff $\overline{V}(\phi, w) \in \{F_1, F\}$.
3. $\mathcal{M}, w \models \phi \rightarrow \varphi$ iff $\overline{V}(\phi, w) \in \{F_1, F\}$ or $\overline{V}(\varphi, w) \in \{T_1, T\}$ or $\overline{V}(\phi, w) = \overline{V}(\varphi, w) = N$.
4. $\mathcal{M}, w \models \Diamond\phi$ iff $\exists v(wRv \wedge \mathcal{M}, v \models \phi)$.
5. $\mathcal{M}, w \models \Box\phi$ iff $\forall v(wRv \rightarrow \mathcal{M}, v \models \phi)$.

Table 1 shows the result of implication and negation on linguistic truth values, which corresponds to our definition of truth assignment for formulas and model satisfaction.

4   Soundness and Completeness

In this section, we present the soundness and completeness of our logic F_5.

Intuitively, the soundness of a logic system means that for a formula of a logic, if it is correct in the sense of syntax, then it is correct in the sense of semantics. The completeness of a logic system means that for a logic formula, if it is correct in the sense of semantics, then it is correct in the sense of syntax[1].

Theorem 1 (Soundness of F_5). F_5 *is sound with respect to the class of frames with the property that* $\forall y(Rxy \rightarrow \forall z(Rxz \rightarrow y = z))$*, i.e.,* $\vdash \phi$ *implies* $(W, R, V) \models \phi$*, where* (W, R) *is a frame with the required property and V is arbitrary.*

Definition 8 (Canonical model). *The canonical model $\mathcal{M}_{F_5}^*$ of F_5 is the triple (W^*, R^*, V^*), where:*

1. W^* *is the set of all maximally consistent sets of F_5;*
2. R^* *is the binary relation on W^* such that wR^*u if for all formulas ψ, $u \vdash \psi$ implies $w \vdash \Diamond\psi$ or $w \not\vdash \neg\Diamond\psi$; and*
3. *the valuation V is defined as follow:*
   (a) $V^*(p, w) = T$ *if $w \vdash p$,*
   (b) $V^*(p, w) = F$ *if $w \vdash \neg p$,*
   (c) $V^*(p, w) = N$ *if $w \not\vdash p$ and $w \not\vdash \neg p$.*

The canonical model we propose here is different from the classic one in two aspects: the definition of canonical relation and valuation. The change in canonical relation can help us prove the completeness of our logic. The classic canonical valuation is defined as follows: $V^-(p, w) = T$ if $p \in w$, and $V^-(p, w) = F$ if $\neg p \in w$. Since we have three categories of truth values in our logic, the classic canonical valuation does not work here.

Theorem 2 (Completeness). *Suppose that ϕ is a formula in F_5 and Γ is a set of formulas in F_5. If $\Gamma \models \varphi$, then $\Gamma \vdash \varphi$.*

---

[1] For the sake of page limit, we cannot present the details of our theorems' proofs in this paper, but we will present them in the extended version of this paper.

5   Related Work

Recently, there have been many studies about modal logic. Kontinen *et al.* [9] introduce a logic called modal independence logic that can explicitly talk about independence among propositional variables. Formulas of their logic are evaluated in sets of worlds, rather than a single world. Herzig and Lorini [6] present a logic, which can reason about the relationship between an agent's belief and the information that the agent obtains. Bozzelli *et al.* [1] present a multi-agent refinement modal logic, which contains an operator ∀, and standard box modalities for each agent. A refinement is like a bisimulation where only the 'atoms' and 'back' requirements need to be satisfied. Operator ∀ is a quantifier over the set of all refinements of a given model. Sack and Hoek [13] present a modal logic for games that allow mixed strategies and demonstrate its soundness and strong completeness. However, our logic can reflect the fuzziness of modal formulas, but all of them above cannot. In addition, Jung *et al.* [8] set up a Kripke semantics for a modal expansion of bilattice logic, which is based on a four-valued logic. They prove the soundness and completeness of their logic with respect to four-valued Kripke frames. In our work, we set up Kripke frames of five truth values, which can be divided into three categories. Besides, we also prove the soundness and completeness of our logic.

There are also some studies on fuzzy modal logic, because this kind of logic is a powerful tool for dealing with fuzzy information and has been applied in many areas. Hájek [5] studies a fuzzy variant of classical modal logic $S5$. He proposes a recursively axiomatised logic. He also proposes three kinds of Kripke models and the corresponding systems. His work mainly focuses on the fuzzy extension of the classical modal logic, while our work focuses on the multi-valued extension of modal logic, the truth values of which are fuzzy linguistic terms. Rodrigues and Godo [12] define modal uncertainty logics with fuzzy neighborhood semantics. Different from them, the semantics we use is Kripke semantics. Pan and Xu [11] deal with a propositional fuzzy modal logic with evaluated syntax based on MV-algebras and they show its application to fuzzy decision implications. However, the syntax of our logic is different from theirs. Cintula *et al.* [2] explore a more general semantics of fuzzy modal logics, namely a fuzzified version of the classical neighborhood semantics. While they use neighborhood semantics, we use the relational semantics. Vidal *et al.* [15] study the modal extension of product fuzzy logic with both algebraic semantics and relational semantics based on Kripke structures with crisp accessibility relations. They prove the completeness for both kinds of semantics. While they study the modal extension of product fuzzy logic, we study fuzzy extension of the classic three-value logic. Jing *et al.* [7] propose a fuzzy modal logic on the basis of the work of Luo *et al.* [10]. However, our logic is different from theirs. Ours is a five-valued modal logic, but theirs is a nine-valued one; and our modal operator does not have a fixed meaning, while theirs mean 'know' and 'believe'.

6   Conclusion

Linguistic variable is a powerful tool to characterize human reasoning. In this paper, to reflect the fact that in real life people regard *truth* as a linguistic variable with its values being fuzzy sets, we define the implication on linguistic truth values, and thus develop a linguistic multi-valued modal logic. Moreover, after presenting the syntax and semantics of our logic, we prove its soundness and completeness. Since modal logic has been used as the base for epistemic logic, deontic logic, temporal logic and so on [3], we believe that the idea behind our logic can be applied to these fields and produces more interesting results in the future.

Acknowledgments. This research is supported by the Bairen Plan of Sun Yat-sen University, the Natural Science Foundation of Guangdong Province, China (No. 2016A030313231) and China National Foundation of Social Science (No. 13BZX066).

References

1. Bozzelli, L., van Ditmarsch, H., French, T., Hales, J., Pinchinat, S.: Refinement modal logic. Inf. Comput. **239**, 303–339 (2014)
2. Cintula, P., Noguera, C., Rogger, J.: From Kripke to neighborhood semantics for modal fuzzy logics. In: Carvalho, J.P., Lesot, M.-J., Kaymak, U., Vieira, S., Bouchon-Meunier, B., Yager, R.R. (eds.) IPMU 2016. Communications in Computer and Information Science, vol. 611, pp. 95–107. Springer, Switzerland (2016). doi:10.1007/978-3-319-40581-0_9
3. Goble, L.: The Blackwell Guide to Philosophical Logic. Blackwell, Malden (2001)
4. Goldberg, H., Leblanc, H., Weaver, G.: A strong completeness theorem for 3-valued logic. Notre Dame J. Form. Logic **15**(2), 325–330 (1974)
5. Hájek, P.: On fuzzy modal logics S5($\mathscr{C}$). Fuzzy Sets Syst. **161**(18), 2389–2396 (2010)
6. Herzig, A., Lorini, E.: A modal logic of perceptual belief. In: Lihoreau, F., Rebuschi, M. (eds.) Epistemology, Context, and Formalism. Synthese Library, vol. 369, pp. 197–211. Springer, Switzerland (2014). doi:10.1007/978-3-319-02943-6_12
7. Jing, X., Luo, X., Zhang, Y.: A fuzzy dynamic belief logic system. Int. J. Intell. Syst. **29**(7), 687–711 (2014)
8. Jung, A., Rivieccio, U.: Kripke semantics for modal bilattice logic. In: 28th Annual IEEE/ACM Symposium on Logic in Computer Science, pp. 438–447 (2013)
9. Kontinen, J., Müller, J.-S., Schnoor, H., Vollmer, H.: Modal independence logic. arXiv preprint arXiv:1404.0144 (2014)
10. Luo, X., Zhang, C., Jennings, N.R.: A hybrid model for sharing information between fuzzy, uncertain and default reasoning models in multi-agent systems. Int. J. Uncertain. Fuzziness Knowl. Based Syst. **10**(04), 401–450 (2002)
11. Pan, X., Xu, Y.: Semantics of propositional fuzzy modal logic with evaluated syntax and its application to fuzzy decision implications. Int. J. Comput. Intell. Syst. **8**(sup1), 85–93 (2015)
12. Rodriguez, R.O., Godo, L.: Modal uncertainty logics with fuzzy neighborhood semantics. In: IJCAI-13 Workshop on Weighted Logics for Artificial Intelligence, pp. 79–86 (2013)

13. Sack, J., van der Hoek, W.: A modal logic for mixed strategies. Studia Logica **102**(2), 339–360 (2014)
14. Schmucke, K.J.: Fuzzy Sets: Natural Language Computations, and Risk Analysis. Computer Science Press, Incorporated (1984)
15. Vidal, A., Esteva, F., Godo, L.: On modal extensions of product fuzzy logic. J. Log. Comput. (2015). doi:10.1093/logcom/exv046
16. Zadeh, L.A.: Fuzzy sets. Inf. Control **8**(3), 338–353 (1965)
17. Zadeh, L.A.: The concept of a linguistic variable, its application to approximate reasoning. Inf. Sci. Part I **8**(4), 199–249 (1975). Part II **8**(4), 301–357 (1975). Part III **9**(4). 43–80 (1975)

An Empirical Study of a Simple Naive Bayes Classifier Based on Ranking Functions

Kinzang Chhogyal[(✉)] and Abhaya Nayak

Macquarie University, Sydney, Australia
{kin.chhogyal,abhaya.nayak}@mq.edu.au

Abstract. Ranking functions provide an alternative way of modelling uncertainty. Much of the research in this area focuses on its theoretical and philosophical aspects. Approaches to solving practical problems involving uncertainty have been, by and large, dominated by probabilistic models of uncertainty. In this paper we investigate if ranking functions can be used to solve practical problems in an uncertain domain. In particular, we look at the problem of identifying spam e-mails, one of the earliest success stories of probabilistic machine learning techniques. We show how the probabilistic naive Bayes classifier can easily be translated to one based on ranking functions, and present some experimental results that demonstrate its efficacy in correctly identifying spam e-mails.

1 Introduction

Probabilities represent degrees of belief where as ranking functions (Spohn 1988) represent degrees of disbelief. Both are formalisms for representing uncertainty, and the former has received much more attention compared to the latter that was conceived relatively recently. We provide a quick overview of ranking functions in Sect. 2. Most Machine Learning (ML) applications are based on probabilistic methods whose success is mainly due to Bayes' rule whereas research on ranking functions has been confined to studying its theoretical and philosophical aspects, and so if they are to be practically useful, it is important to develop and explore a ranking analogue of the probabilistic Bayes' rule. Indeed such an analogue exists. In Sect. 2, we work through a simple example using both the probabilistic as well as the ranking function versions of Bayes' rule. It is quite common in ML to make simplifying assumptions to make problems computationally more amenable. A common assumption made in probabilistic methods is the conditional independence or the *naive Bayes* assumption that has proven to be highly successful in classification tasks. It would be interesting to see if ranking functions also lend themselves to such analogous assumptions. As we will see, they do so relatively easily. We show how this may be done by focussing on the binary classification problem in Sect. 3, and report on its performance as a spam classifier in Sect. 4. Experimental results suggest that the performance of

This research has been partially supported by the Australian Research Council (ARC), Discovery Project: DP150104133.

© Springer International Publishing AG 2016
B.H. Kang and Q. Bai (Eds.): AI 2016, LNAI 9992, pp. 324–331, 2016.
DOI: 10.1007/978-3-319-50127-7_27

the ranking function based naive Bayes classifier is comparable to the probabilistic naive Bayes classifier used in the Natural Language Toolkit (NLTK) (Bird et al. 2009). Finally, after a brief discussion pertaining to our work in Sect. 5 we conclude the paper.

2 Ranking Functions $\kappa(\cdot)$

Ranking functions were first introduced in Spohn (1988) as *ordinal conditional functions* to model an agent's epistemic state and its dynamics.

Definition 1. *Let $\mathcal{A}$ be an algebra over a non-empty set of possible worlds, W. Then, κ is a* negative ranking function *for $\mathcal{A}$ iff κ is a function from $\mathcal{A}$ into $\mathbb{N} \cup \{\infty\}$ such that for all $A, B \in \mathcal{A}$: (1) $\kappa(W) = 0$ and $\kappa(\emptyset) = \infty$, (2) $\kappa(A \cup B) = min\{\kappa(A), \kappa(B)\}$, and (3) either $\kappa(A) = 0$ or $\kappa(\bar{A}) = 0$ or both.*

$\kappa(A)$ is called the *negative rank* or simply the *rank* of A. Intuitively, κ represents an agent's degree of disbelief in some proposition (*event*). If $\kappa(A) = 0$, A is not disbelieved at all where as if $\kappa(A) > 0$, A is disbelieved to degree $\kappa(A)$. A is believed iff $\kappa(\bar{A}) > 0$, where $\bar{A} = W \backslash A$. Also, note that it is perfectly fine for $\kappa(A) = \kappa(\bar{A}) = 0$, i.e. neither proposition is disbelieved, which is a representation of ignorance. For all propositions $A, B \in \mathcal{A}$, the *conditional ranking function* $\kappa(\cdot|\cdot)$ is defined as: $\kappa(A|B) = \kappa(A \cap B) - \kappa(B)$ if $A \neq \emptyset$ and ∞ otherwise. It follows that $\kappa(A|B) = 0$ or $\kappa(\bar{A}|B) = 0$ or both (Spohn 2009). To show how ranking functions work, we take a simple example and analyse it first using Bayes' rule from probability theory followed by its analogue for ranking functions.

Example 1. Consider that patients are being screened for cancer. Let c represent the belief that the patient has cancer and t the belief that the result of a test for cancer is positive. Assume we have the *prior* probability of cancer, $P(c) = 0.01$, the probability that the test comes out as positive given the person has cancer, $P(t|c) = 0.8$, and the probability that the test comes out as positive in absence of cancer, $P(t|\bar{c}) = 0.096$. It follows from the standard rules of probability that: $P(\bar{c}) = 0.99$, $P(\bar{t}|c) = 0.2$ and $P(\bar{t}|\bar{c}) = 0.904$. Using Bayes' Rule, $P(c|t) \approx 0.078$. So, the chance of having cancer has increased from 1% to about 8% if the test is positive. Similarly, we can compute the probability of having cancer given that the test is negative, $P(c|\bar{t}) = 0.002$, which is less than 1%.

It is not very clear how one can go about assigning ranks $\kappa(\cdot)$ to propositions. However, one way is to view ranking functions as order-of-magnitude approximations of probabilities (Spohn 1988; Pearl 1990) or in other words, ranks may be viewed as logarithms of probabilities with some infinitesimally small base ϵ (Spohn 2009).

Definition 2. *Given ϵ and a probability value p, the rank of p denoted as $\kappa(p)$ is equal to $\arg\max_i$ such that $\frac{p}{\epsilon^i} \leq 1$, $0 < \epsilon < 1$.*

Based on Definition 2, Darwiche and Goldszmidt (1994) proposed the algorithm below, which we call ϵ-Rank, that generates the rank of a proposition whose probability is p. In practice, if ϵ is infinitesimally small, we get a *flat* structure where all propositions have a rank of 0.

1. if $p = 0$, then print ∞
2. $k \leftarrow 0$
3. $p \leftarrow p / \epsilon$
4. if $p > 1$, print k, otherwise $k \leftarrow k + 1$
5. go to 3

Algorithm ϵ-Rank: Generating ranks from probabilities

Table 1. Joint distribution table along with frequency estimate from the probabilities, and the rank of each world.

c	t	P(.)	Frequency	Frequency-Rank	ϵ-Rank
c	t	0.008	8	887	41
c	$\bar{t}$	0.002	2	893	53
$\bar{c}$	t	0.095	95	800	20
$\bar{c}$	$\bar{t}$	0.895	895	0	0

The analogue of Bayes' Rule for ranking functions (Halpern 2005) is given below:

$$\kappa(A|B) = \kappa(B|A) + \kappa(A) - \kappa(B) \quad \text{(Bayes' Rule for ranking functions)} \quad (1)$$

Its similarity with Bayes' Rule, $P(A|B) = P(B|A) \cdot P(A)/P(B)$, is striking. The translation from probabilities to ranks is straightforward: sums (respectively products, quotients) of probabilities turn into the minimum (sums, difference) of ranks, respectively (Spohn 2009; Pearl 1991). In Table 1, the third column, $P(\cdot)$, shows the probabilities of the atomic events as computed from the probabilities in Example 1. The fourth column, *Frequency*, shows the frequencies of these atomic events that we would expect from the experiment in question repeated 1000 times. We consider two ways to generate the ranks and they are shown in the last two columns:

1. *Frequency-Rank*: The computation of the Frequency Rank is done by taking the highest amongst the frequency of all outcomes, and then subtracting from it each of the other frequencies including itself. For instance, the highest frequency is 895. To get the rank of ct, we subtract 8 from 895 to get 887. The intuition is that the more frequently obtained outcomes are less disbelieved.
2. *ϵ-Rank*: The ranks here are calculated based on the ϵ-Rank algorithm. As far as we know, the value of ϵ used for generating ranks seems to be rather ad hoc. In Table 1 above, ϵ was set to 0.89, slightly below the highest probability assigned to any outcome. This satisfies condition (1) of a ranking function that $\kappa(W) = 0$.

Let us continue by using Eq. 1 and the ϵ-Ranks from Table 1 to get $\kappa(c|t)$, the rank of cancer given that the test has come out positive. First, we compute $\kappa(c)$, which is equal to $min\{\kappa(ct), \kappa(c\bar{t})\} = min\{41, 53\} = 41$. Similarly, we get $\kappa(t) = 20$ and $\kappa(t|c) = 0.$[1] Thus, $\kappa(c|t) = \kappa(t|c) + \kappa(c) - \kappa(t) = 0 + 41 - 20 = 21$. The prior rank, $\kappa(c)$, was 41 and the new rank of c after learning that the test is positive is 21. So, it is now less disbelieved that the patient has cancer. Similarly, the new rank of cancer given that the test came out negative, $\kappa(c|\bar{t})$, can be computed to be 53. Not surprisingly, if the test comes out negative, the disbelief in cancer gets enhanced. It can be easily verified that the change in the degree of disbelief using the Frequency-Rank produces similar results albeit the magnitude of the changes are greater.

In computing the ranks above, we used a fully specified joint probability distribution table. As it is computationally intensive to generate such tables, it is of interest to see if this can be avoided along the lines of how we computed $P(c|t)$ in Example 1. Specifically, we generate the ranks of $\kappa(c)$, $\kappa(t|c)$, and $\kappa(t|\bar{c})$ directly from $P(c)$, $P(t|c)$ and $P(t|\bar{c})$ in Example 1, and then utilize those values to obtain $\kappa(c|t)$. Using algorithm ϵ-Rank with $\epsilon = 0.89$, we get the following values: (1) $\kappa(c) = 39$, (2) $\kappa(t|c) = 1$, and (3) $\kappa(t|\bar{c}) = 20$. As $\kappa(c|t) = \kappa(t|c) + \kappa(c) - \kappa(t)$, we need to obtain $\kappa(t)$ first. We know $\kappa(t) = min\{\kappa(ct), \kappa(\bar{c}t)\}$ which in turn is equal to $min\{\kappa(t|c) + \kappa(c), \quad \kappa(t|\bar{c}) + \kappa(\bar{c})\}$. Plugging in the values, we get $\kappa(t) = 20$. Finally, $\kappa(c|t) = 1 + 39 - 20 = 20$. This result agrees with the previous result in the sense that if the test comes out positive, it is less disbelieved that the patient has cancer. However, the ranks obtained do differ. For instance, the prior rank $\kappa(c)$ was 41 previously but is now 39. We suspect that this maybe due to the value of ϵ and we will defer to the discussion section for more on this. It is easy to verify that $\kappa(\bar{c}t)$ is 0 as in the previous case and thereby satisfies the condition that $\kappa(W) = 0$.

3   The Naive Bayes Classifier for Ranking Functions

In the classification problem, given a set C of classes, $\{c_1, c_2, \ldots\}$, and an input vector X consisting of features $x_1, x_2, \ldots$, the classifier must predict the class to which X belongs. Though there is a substantial amount of work on this problem, it is still an active area of research. We begin by presenting the chain rule for ranking functions which is easily derived from the conditional ranking function. Given n events, $E_1, E_2, \ldots, E_n$:

$$\kappa(E_1, E_2, \ldots E_n) = \kappa(E_1) + \kappa(E_2|E_1) + \kappa(E_3|E_1, E_2) + \ldots$$
$$\ldots + \kappa(E_n|E_1, E_2, \ldots, E_{n-1}). \qquad (2)$$

We focus on the *binary classification* problem, where the set of class labels, $C = \{c, \bar{c}\}$, has only two elements. We assume that we are dealing with n binary features, i.e. each feature is either observed or not. Given input/observation X_i, the classification task boils down to determining the new rank of c and $\bar{c}$, and then

---

[1] $\kappa(ct) - \kappa(c) = 41 - 41 = 0$.

using some decision rule to predict the class label. Our presentation is restricted[*] to computing $\kappa(c|X_i)$ since the procedure is similar is $\kappa(\bar{c}|X_i)$. We resort to Eq. 1, the ranking function analogue of Bayes' rule. Given $X_i = \{x_1, x_2, \ldots, x_n\}$, $\kappa(c|X_i)$ is defined as:

$$\kappa(c|x_1, x_2, \ldots x_n) = \kappa(x_1, x_2, \ldots x_n|c) + \kappa(c) - \kappa(x_1, x_2, \ldots x_n) \qquad (3)$$

Since $\kappa(x_1, x_2, \ldots x_n|c) + \kappa(c)$ is $\kappa(x_1, x_2, \ldots x_n, c)$, we can rewrite Eq. 3 using the product rule as follows:

$$\kappa(c|x_1, x_2, \ldots x_n) = \kappa(x_1|x_2 \ldots x_n, c) + \kappa(x_2|x_3 \ldots x_n, c) + \ldots$$
$$\ldots + \kappa(x_n|c) + \kappa(c) - \kappa(x_1, x_2, \ldots x_n) \qquad (4)$$

We now make the *naive* conditional assumption (Norvig and Russell 2010). In ranking terms, once we know that the class is c, our degree of disbelief in feature x_i is independent of whether feature x_j is true or not, where $i \neq j$. We use the subscript NB to indicate that we are making the conditional independence assumption.

$$\kappa_{NB}(c|x_1, x_2, \ldots x_n) = \kappa(x_1|c) + \kappa(x_2|c) + \ldots + \kappa(x_n|c) + \kappa(c) - \kappa(x_1, x_2, \ldots x_n)$$

(Naive Bayes Classifier for ranking functions)

Proposition 1. *Given the set of binary class labels, $C = \{c, \bar{c}\}$, and an input feature vector $X_i = \{x_1, x_2, \ldots x_n\}$, $\kappa_{NB}(c|x_1, x_2, \ldots x_n)$ is either 0 or greater than 0.*

Proof. We know that $\kappa(x_1, x_2, \ldots x_n) = min\{\kappa(x_1, x_2, \ldots x_n, c), \kappa(x_1, x_2, \ldots, x_n, \bar{c})\}$. Assume $\kappa(x_1, x_2, \ldots x_n)$ is equal to $\kappa(x_1, x_2, \ldots x_n, c)$. Now $\kappa(x_1, x_2, \ldots x_n, c)$ under the naive conditional assumption is simply the minuend $\kappa(x_1|c) + \kappa(x_2|c) + \ldots \ldots \kappa(x_n|c) + \kappa(c)$. Thus, $\kappa_{NB}(c|x_1, x_2, \ldots x_n) = 0$. On the other hand, if $\kappa(x_1, x_2, \ldots, x_n)$ is equal to $\kappa(x_1, x_2, \ldots x_n, \bar{c})$, then it must be smaller than $\kappa(x_1, x_2, \ldots x_n, c)$, and hence $\kappa_{NB}(c|x_1, x_2, \ldots x_n)$ will be greater than 0. $\square$

Proposition 1 assures that $\kappa_{NB}(c|x_1, x_2, \ldots x_n)$ will never be negative as required by ranking functions. In a probabilistic naive Bayes classifier, the counterpart of the subtrahend $\kappa(x_1, x_2, \ldots x_n)$ is the divisor $P(x_1, x_2, \ldots x_n)$. It is a constant divisor and safely omitted in the calculations since it does not influence the prediction of the classifier. This is especially useful if we have many class labels as it can save computational costs. Ranking functions offer the same advantage. For the binary classification task, we must compute $\kappa_{NB}(c|x_1, x_2, \ldots x_n)$ and $\kappa_{NB}(\bar{c}|x_1, x_2, \ldots x_n)$. Let X and Y respectively denote their minuends. As we saw above, the subtrahend $\kappa(x_1, x_2, \ldots x_n)$ is then $min\{X, Y\}$. Thus,

$$\kappa_{NB}(c|x_1, x_2, \ldots x_n) = X - min\{X, Y\}$$
$$\kappa_{NB}(\bar{c}|x_1, x_2, \ldots x_n) = Y - min\{X, Y\} \qquad (5)$$

Since $min\{X, Y\}$ is subtracted in both cases, we can tell which term $\kappa_{NB}(c|x_1, x_2, \ldots x_n)$ or $\kappa_{NB}(\bar{c}|x_1, x_2, \ldots x_n)$ is ranked higher solely based on X and Y as shown below:

1. If $X > Y$ then $\kappa_{NB}(c|x_1, x_2, \ldots x_n) > 0$ and $\kappa_{NB}(\bar{c}|x_1, x_2, \ldots x_n) = 0$ since $min\{X, Y\} = Y$.
2. If $X < Y$ then $\kappa_{NB}(c|x_1, x_2, \ldots x_n) = 0$ and $\kappa_{NB}(\bar{c}|x_1, x_2, \ldots x_n) > 0$ since $min\{X, Y\} = X$.
3. If $X = Y$ then both $\kappa_{NB}(c|x_1, x_2, \ldots x_n)$ and $\kappa_{NB}(\bar{c}|x_1, x_2, \ldots x_n)$ are 0.

The observations above lead to the following decision rule:

Definition 3. *Given the set of binary class labels, $C = \{c, \bar{c}\}$, and an input feature vector $X_i = \{x_1, x_2, \ldots x_n\}$, we say X_i belongs to class $\bar{c}$ iff $\kappa_{NB}(c|x_1, x_2, \ldots x_n) > 0$. Otherwise, we say X_i belongs to class c.*

The intuition behind this definition is as follows:

1. If $\kappa_{NB}(c|x_1, x_2, \ldots x_n) > 0$, then $\kappa_{NB}(\bar{c}|x_1, x_2, \ldots x_n) = 0$. The degree of disbelief in $\bar{c}$ being the correct class is less than that of c, so we predict label $\bar{c}$.
2. Else, that is, if $\kappa_{NB}(c|x_1, x_2, \ldots x_n) = 0$, there are two cases:
   (a) Either $\kappa_{NB}(\bar{c}|x_1, x_2, \ldots x_n) > 0$ and $\kappa_{NB}(c|x_1, x_2, \ldots x_n) = 0$ in which case we should predict c.
   (b) Or $\kappa_{NB}(\bar{c}|x_1, x_2, \ldots x_n) = \kappa_{NB}(c|x_1, x_2, \ldots x_n) = 0$, in which case we should neither predict c or $\bar{c}$, but we give the benefit of the doubt to class c. In practice however, we suspect such examples are rare.

Table 2. Experimental results for the Enron1 test dataset averaged over 5 runs.

Enron1 test dataset		
Classifier	Prediction accuracy	Avg. classification time (secs)
$\epsilon = 0.001$	0.911	3.36
$\epsilon = 0.01$	0.932	3.52
$\epsilon = 0.1$	0.944	3.60
$\epsilon = 0.5$	0.943	4.6
NLTK Naive Bayes	0.947	4.04

4   Experimental Evaluation

To test the feasibility of the ranking functions based classifier, denoted *R-NB*, we chose the binary spam classification task where probabilistic naive Bayes classifiers are known to do well. The set of class labels is $C = \{spam, \overline{spam}\}$. The input to the classifier is a feature vector of words obtained after preprocessing and represents the email to be classified. For comparison, we used the Natural Language Toolkit (NLTK) naive Bayes classifier (Bird et al. 2009), which we will simply refer to as the NLTK classifier.

Dataset and Implementation. We used two datasets Enron1 and Enron2 from the Enron corpus to train and test both the R-NB and NLTK classifiers. The Enron1 dataset consists of 5172 e-mails (3672 ham, 1500 spam) and the Enron2 dataset consists of 5857 e-mails (4361 ham,1496 spam). For each dataset, 80% of the e-mails were reserved for training and the rest for testing. The first step consisted of training the NLTK classifier. As it stores the prior probabilities of the two class labels and the conditional probability of each word encountered in the training set given each class label, it does part of the work for training the R-NB classifier. Test inputs were then classified using the NLTK classifier and the number of correct predictions was stored. We then completed the training of the R-NB classifier by using the stored probabilities from the NLTK classifier to generate the ranks using the algorithm ϵ-Rank. The test input was then classified using Definition 3. Smoothing is not employed and instead, both classifiers ignore features that were not seen during the training phase but are present in test inputs.

Results. Tables 2 and 3 show the results on the test set for the Enron1 and Enron2 datasets, respectively. The first four rows show the values of ϵ used to compute the ranks for the R-NB classifier. The adjacent columns show the *prediction accuracy*, i.e. how many emails were correctly identified as spam, and the *classification time*, which measures how long it took to classify e-mails in the test data. The prediction accuracy and classification time were averaged over 5 runs. The last row shows the prediction accuracy and the classification time for the NLTK classifier which was also averaged over 5 runs. For both test datasets, it is clear that as ϵ increases, the prediction accuracy of the R-NB classifier gets closer to that of the NLTK classifier. Once ϵ reaches 0.1, any increase in ϵ does not significantly improve the accuracy. Note that the average classification time steadily increases as ϵ increases for both datasets. It is less for the Enron1 test dataset compared to Enron2 test dataset as there are more test e-mails to be classified in the latter case (1172 vs. 1035). The classification time proportionally increases with ϵ and this is due to algorithm ϵ-Rank requiring more iterations to satisfy the condition $p > 1$ as ϵ increases. We can conclude that the R-NB classifier performs as well as the NLTK classifier provided ϵ is chosen appropriately.

Table 3. Experimental results for the Enron2 test dataset averaged over 5 runs.

Enron2 test dataset		
Classifier	Prediction accuracy	Avg. classification time (secs)
$\epsilon = 0.001$	0.960	4.87
$\epsilon = 0.01$	0.981	5.02
$\epsilon = 0.1$	0.992	5.02
$\epsilon = 0.5$	0.989	6.04
NLTK Naive Bayes	0.993	5.91

5   Discussion and Conclusion

We presented a simple naive Bayes classifier based on ranking functions and we experimentally demonstrated such classifiers are feasible. Our experimental results show that the performance of the R-NB classifier is comparable to the NLTK classifier. However, since ranking functions can be viewed as order-of-magnitude probabilities, it is tempting to wonder whether they are really different. We suspect they are not. The problem is, many machine learning applications are based on the frequency of the datum and thus the data automatically attains a probabilistic interpretation.[2] We could circumvent this reliance on probabilities by using frequencies directly to calculate ranks as in Sect. 2. This will be an interesting area to explore in future. It will be interesting to investigate further an intuitive explanation behind the values for ϵ that give us good performance. This may also help account for the discrepancy of the ranks obtained in Example 1 when using ϵ-Rank with atomic probabilities and conditional probabilities.

References

Bird, S., Klein, E., Loper, E.: Natural Language Processing with Python. O'Reilly Media, Inc., Sebastopol (2009)

Darwiche, A., Goldszmidt, M.: On the relation between kappa calculus and probabilistic reasoning. In: Proceedings of the Tenth International Conference on Uncertainty in Artificial Intelligence, pp. 145–153. Morgan Kaufmann Publishers Inc. (1994)

Halpern, J.Y.: Reasoning About Uncertainty. MIT Press, Cambridge (2005)

Norvig, P., Russell, S.J.: Artificial intelligence (a modern approach) (2010)

Pearl, J.: Probabilistic semantics for nonmonotonic reasoning: a survey. In: Readings in Uncertain Reasoning, pp. 699–710. Morgan Kaufmann Publishers Inc. (1990)

Pearl, J.: Epsilon-semantics, in Computer Science Department. Technical report, UCLA (1991)

Spohn, W.: A dynamic theory of epistemic states. In: Harper, W., Skryms, B. (eds.) Causation in Decision, Belief Change, and Statistics, II, pp. 105–134. Kluwer (1988)

Spohn, W.: A survey of ranking theory. In: Huber, F., Schmidt-Petri, C. (eds.) Degrees of Belief. Synthese Library, vol. 342, pp. 185–228. Springer, The Netherlands (2009). doi:10.1007/978-1-4020-9198-8_8

---

[2] This is the frequentist view of probability.

Cognitive-Task-Based Information Aid Design for Clinical Diagnosis

Dong-Gyun Ko, Youkyoung Park, Yoochan Kim, Juyoun Kim,
and Wan Chul Yoon[✉]

Knowledge Service Engineering, KAIST, Daejeon, Korea
{kodonggyun,park60,yoochana,juyounkim,
wcyoon}@kaist.ac.kr

Abstract. Clinical diagnosis systems are increasingly supported by advanced information technology but enhancing human diagnosis performance with intelligent aiding has not been found guaranteed. Diagnosis is an intense cognitive task for which the supporting intelligent features should be designed and integrated to match human decision process. This study conducted cognitive work analysis to identify the user strategies and needs of information, and designed an effective interface based on a set of design principles for diagnosis aiding. The target system was a medical service that provided comprehensive blood test reviews and diagnosis to local hospitals. An interim evaluation by experts showed highly promising service value scores.

Keywords: Clinical decision support system · Intelligent interface design · Cognitive work analysis

1 Background

Clinical diagnosis systems are increasingly supported by advanced information technology for integration of sub-processes and tasks from initial registration of patients through various tests to diagnostic conclusions. Designing intelligent interface to enhance the human expert's diagnosis is essential in developing an advanced CDSS (Clinical decision support system). If rightly designed, such decision aiding approach may achieve more accurate diagnosis with fewer errors. Brian and Boren [1] reported that 13 studies (76%) out of 17 studies in clinical decision support systems (CDSS) yielded improved outcomes. Garg et al. [2] conducted a broader survey to report that, out of 97 cases of CDSS, the human performance was enhanced in 62 cases (64%). However, it should be noted that only 4 of 10 diagnostic systems they surveyed were more effective than other types of clinical task support such as reminder systems (76%), disease management systems (62%), and prescribing systems (66%). This result suggests that diagnosis, being cognitively more complex than managerial tasks, does not easily benefit from a computer aid.

The reduction of diagnostic errors is the most important purpose of diagnosis aiding. Many researchers have listed cognitive biases in clinical diagnosis that increased the probability of errors [3–5]. Conservatism biases such as anchoring, confirmation bias, diagnosis momentum, and triage cueing lead the decision maker to

© Springer International Publishing AG 2016
B.H. Kang and Q. Bai (Eds.): AI 2016, LNAI 9992, pp. 332–337, 2016.
DOI: 10.1007/978-3-319-50127-7_28

stick to the initial hypothesis while suppress suggestions for other hypotheses. Satisficing heuristic, premature closure of search, availability heuristic, and representative bias may also contribute to incorrect diagnosis. There are also simpler errors such as slips and lapses. Weakened attention due to mental or visual fatigue tends to increase errors, too.

In this research, we are developing a computer aiding system with intelligent features including partially automated diagnosis, selecting similar cases to compare with, and providing relevant texts in literature that are text-mined with recognized keywords. This paper introduces the development of an intelligent interface that integrates these machine intelligence ingredients and the human expert's decision capability. A cognitive work analysis is conducted to identify the user needs of information and the opportunity of intelligent support. The interface is design based on a set of interaction design principles for diagnosis aiding by a computer.

2 The Task and Cognitive Task Analysis

2.1 The System and Task

SG Company in Seoul provides a laboratory service receiving patient blood samples from hospitals and conducting a comprehensive set of blood tests. Most of the blood samples are collected as a part of routine medical examinations and analyzed in SG laboratories. The medical experts in SG review the test results and conclude comments on the reports. The number of indicator readings is somewhere around 30 ~ 50 usually. The task takes less than 5 min for most patients, but may require more than 20 min when the conclusion is not immediately obvious. Hospitals may apply different range for a same indicator so that the report should accordingly be adjusted.

2.2 Cognitive Work Analysis

Cognitive Work Analysis (CWA) is a comprehensive framework of system analysis developed by Rasmussen [6] to understand complex sociotechnical systems that embraces human decision-making as the essential ingredient. Work domain analysis (WDA), the first stage of CWA, tries to identify necessary functionalities and the relationships among them across functional means-ends layers: system purpose, abstract functions, generalized functions, physical functions, and physical forms.

Figure 1 shows a brief overview of the WDA result of SG Company's blood test review. The abstract function of *hypothesis management* encompasses all required functions for decision-making to produce diagnostic conclusions based on the measured indicator values. The concluded diagnosis is then dealt with by the other abstract function of *diagnosis management*, which comprises of the report producing tasks considering client requests and other necessary adjustment due to factors outside the indicator readings themselves.

The next step in cognitive task analysis is cognitive task analysis. The actual paths of cognitive actions by the human decision maker are described. Two SG experts were interviewed for a variety of cases that demanded different diagnosis strategies.

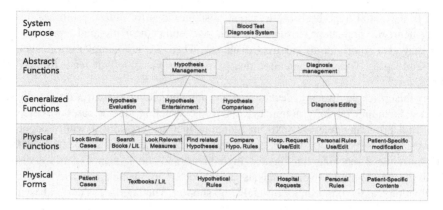

Fig. 1. The work domain analysis of blood test review

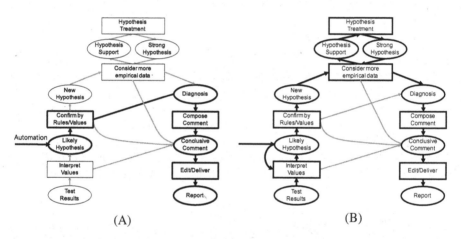

Fig. 2. Cognitive task analysis: (A) Overall framework (B) Pattern recognition (C) Decision table search (D) Hypothesis generation and test

Figure 2 depicts the SG expert's possible paths of cognitive process identified through the interview in an input-process-output framework. Human experts usually try to follow more economic and reliable paths using shortcuts for obvious choices. Indeed, the most frequent strategy is one shown as bold lines in Fig. 2(A). In contrast, Fig. 2(B) models the most complicated strategy, where the expert needs to evaluate and select between competing hypotheses. There exist other strategies between the two extremes.

2.3 Information Requirements and Design Principles

The strategy analysis enables us to determine the information needs for each strategy. Aiding features can then be devised to help reduce mental effort and hence error

Table 1. Information and representation requirements for pattern recognition

Task	Information Req.	Representation Req.
Confirm suggested Hypothesis	Suggested Hypothesis	Central and clear
	Supportive readings	Grouped, Hypo-related
	Unsupportive readings	Positivized, Hypo-related
	Unrelated non-normal readings	Marked, Hypo-indicating
Compose Comments	Hospital requests/rules	Automated modification
	Personal rules	Automated/referenced
	Patient peculiarity	Highlighted, if any
	Importance of results	Prioritized
	Certainty of results	Visual scan of strength

possibility. Information and representation requirements for the pattern recognition strategy shown in Fig. 2(B), for example, are summarized in Table 1.

We set interaction design principles for effective diagnosis aiding based on the identified strategies:

- Alternative hypotheses for current data should be made easy to access and evaluate.
- Switching hypotheses back and forth should be supported saving working memory.
- Data that are treated together should be visually grouped by position or color.
- Relations between data and hypothesis should be visualized by links or associative highlighting.
- Navigation or operations should be able to exploit visual association (e.g., links).
- Information should be displayed to match the human strategies that are expected to utilize it. Examples are sorting for visual scan, trimming irrelevant data, and highlighting words in text.

3 The Design of User Interface

Figure 3 shows one of the main user interfaces that was designed upon the above-derived requirements of information and aiding features. In the upper-left part, the interface provides important indicators and their values. To its right hand side is the corresponding diagnosis part and comment part that delivers the results to the client and the patient, respectively. Each part of the interface operates as follows.

1. The values are highlighted when they fall outside of the normal ranges. The degree of deviation is intuitively shown for the expert assessment.
2. The computer suggests automated diagnosis based on the readings. The expert can adjust the diagnosis to one's judgment. Manually choosing a diagnosis item will highlight the related readings in part 1. The ground and the conclusion are clearly shown together.
3. The machine generates comments on the basis of the diagnosis. The expert can edit them as necessary.

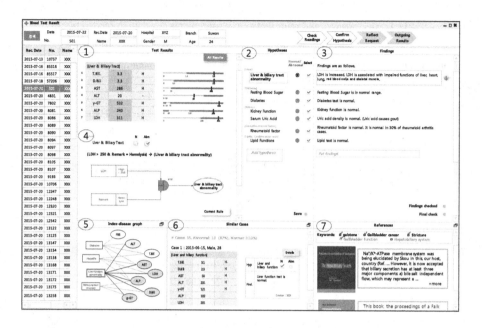

Fig. 3. The aiding interface of SG blood test review

4. For each hypothesis, the necessary conditions in terms of expected readings are shown in a causal diagram. This visual aid is based on the diagnosis rule base. The expert can easily see what conditions are met and what are not.

5. The indicator-hypothesis network for knowledge navigation. The expert can traverse according to his interest to find out new related hypotheses or the indicators to check.

6. Similar cases recommended based on similarity to the current case. It is very helpful when the conclusion is somewhat ambiguous.

7. Textbook or literature that are automatically searched using the current prominent words in highlighted readings and hypotheses.

As an overall principle, the interface responds to the movement of user attention; when a diagnosis, comment, or indicator is selected by mouse click, all the logically related information panes are changed, selected, highlighted, detailed, or expanded so that the screen remains coherent and informative as a whole.

4   A Brief Evaluation and Conclusion

The system completed its prototype and is currently under a full implementation. A preliminary evaluation, employing SERVPERF [7] was conducted for the service value of the aiding, beyond mere interface usability. Three SG medical experts participated in the evaluation. Since the medical diagnosis task is very professional and specialized task, a non-expert evaluation would not have much value. We performed the evaluation in five facets of the service quality that were paraphrased for the given service. They were (1) Tangibles: does the interface look good and react well, (2) Reliability: is the interaction believable, predictable, and understandable, (3) Responsiveness: does the system provide desired information in right time, (4) Assurance: does the system give confidence in its proficiency (5) Empathy: is the system adaptive for individual needs.

The evaluation was done for six information units (windows) in the display, and for six frequent task elements. On a 7-point Likert scale, the information units scored at 5.85 overall compared with 2.5 in the current system. The task elements, which were helped by the aiding features, were rated at 6.4 in contrast to the current system's 1.7 showing higher acceptance for the actual aiding performance.

It is emphasized that the functions of machine intelligence should be designed to fit human cognitive tasks from the beginning. The features should also be integrated in an interface based on the study of human strategies in particular task domains. After the ongoing full implementation, the system will be installed for actual service by SG Company; then we expect to acquire more evidences for the effectiveness of designed interface.

Acknowledgement. This work was supported by the Industrial Strategic Technology Development Program, 10052955, Experiential Knowledge Platform Development Research for the Acquisition and Utilization of Field Expert Knowledge, funded by the Ministry of Trade, Industry & Energy (MI, Korea).

References

1. Bryan, C., Boren, S.A.: The use and effectiveness of electronic clinical decision support tools in the ambulatory/primary care setting: a systematic review of the literature. Inform. Prim. Care **16**(2), 79–91 (2008)
2. Garg, A.X., Adhikari, N.J., McDonald, H.; et al.: Effects of computerized clinical decision support systems on practitioner performance and patient outcomes: a systematic review. JAMA **293**(10), 1223–1238 (2005)
3. Croskerry, P.: The importance of cognitive errors in diagnosis and strategies to minimize them. Acad. Med. **78**, 775–780 (2003)
4. Wickens, C., Hollands, J.: Engineering Psychology and Human Performance. Prentice Hall, Upper Saddle River (1999)
5. Lee, C.S., Nagy, P.G., Weaver, S.J., et al.: Cognitive and system factors contributing to diagnostic errors in radiology. AJR Am. J. Roentgenol. **201**, 611–617 (2013)
6. Vicente, K.J.: Cognitive Work Analysis: Toward Safe, Productive, and Healthy Computer-Based Work. CRC Press, Boca Raton (1999)
7. Cronin, J.J., Taylor, S.A.: Measuring service quality: a re-examination and extension. J. Mark. **56**(3), 55–68 (1992)

Domain Ontology Construction
Using Web Usage Data

Thi Thanh Sang Nguyen[1(✉)] and Haiyan Lu[2]

[1] School of Computer Science and Engineering, International University
– Vietnam National University, Hochiminh City, Vietnam
`nttsang@hcmiu.edu.vn`
[2] Decision Systems and e-Service Intelligence (DeSI) Lab,
Faculty of Engineering and Information Technology, School of Software,
Centre for Quantum Computation and Intelligent Systems (QCIS),
University of Technology, Sydney, P.O. Box 123 Broadway,
Ultimo, NSW 2007, Australia
`haiyan.lu@uts.edu.au`

Abstract. Ontologies play an important role in conceptual model design and the development of machine-readable knowledge bases. They can be used to represent various knowledge not only about content concepts, but also explicit and implicit relations. While ontologies exist for many application domains of websites, the implicit relations between domain and accessed Web-pages might be less concerned and unclear. These relations are crucial for Web-page recommendation in recommender systems. This paper presents a novel method developing an ontology of Web-pages mapped to domain knowledge. It will focus on solutions of semi-automating ontology construction using Web usage data. An experiment of Microsoft Web data is implemented and evaluated.

Keywords: Domain ontology construction · Knowledge representation · Web usage data · Software products

1 Introduction

Ontology is a knowledge representation technology in new generation Web like Semantic Web [1]. It defines the links of things that exist in an application domain. Because of its powerful expression strength, ontology is popular in many fields, e.g. education and industry. Ontology can be used to represent not only domain knowledge, but also other extended knowledge. By discovering the relations between Web access activities, and the temporal and event attributes in Web logs, a Web usage ontology is generated to represent personalized usage knowledge [2]. Alternatively, available contextual information from the recommended items and the recommendation process in a recommender system are also able to be described by ontology in a proposed semantics-based approach [3]. This approach allows the system to make up-to-date recommendation at run-time. In other words, the quality of recommender system is improved considerably thanks to ontology technology [4].

© Springer International Publishing AG 2016
B.H. Kang and Q. Bai (Eds.): AI 2016, LNAI 9992, pp. 338–344, 2016.
DOI: 10.1007/978-3-319-50127-7_29

In the context of Web-page (page) recommendation, it is necessary to recognize the meaning of Web-pages for studying Web browsing behaviors and discovering interesting topics. We motivate that knowledge representation containing domain concepts and accessed Web-pages is necessary to facilitate the process of semantic page recommendation. However, a domain ontology is able to be constructed manually, but integrating Web-pages is a slow process when the number of Web-pages is numerous. In order to remedy the shortcomings, we aim to semi-automatically constructing the ontology of accessed Web-pages mapped to domain knowledge.

The paper is structured as follows: Sect. 2 presents a domain ontology model of a website. Subsequently, Sect. 3 proposes a new method for domain ontology construction of a website, and Sect. 4 presents some issues of ontology development using a case study of the Microsoft website. The obtained ontology will be evaluated and discussed in Sect. 5. Finally, we give conclusions in Sect. 6.

2 Domain Ontology Model of a Website

It is assumed that a Web-page title contains important information in the page content, in other words, it contains the keywords embracing the semantics of the page. The rationale behind this assumption can be seen from two aspects: (1) Well-designed Web-pages have the TITLE tags containing the meaningful keywords which are relatively short and attractive to support Web search or crawling; (2) the fact that the terms in page titles are usually given higher weights by search engines, e.g. Google [5], because of their importance. Therefore, this study makes use of the page titles in order to discover domain knowledge for modelling the domain ontology of a website.

2.1 Domain Ontology Model of a Website

According to the definitions of an ontology in [1], this study defines a domain ontology model of a website as follows:

Definition 1 (**Domain ontology model of Web-pages - DomainOntoWP**). A domain ontology structure of a website is defined as a four-tuples: $O_{man} :=$ $<C, D, P_{MAN}, A>$, where C represents terms extracted from the Web-page titles within the given website, D represents the Web-pages, P_{MAN} represents properties defined in the ontology, and A represents axioms, such as, an instantiation axiom assigning an instance to a class, an assertion axiom assigning two instances by means of a property, a domain axiom for a property and a class, and a range axiom for a property and a class. In details, C, D, and P_{MAN} are further divided into sets:

$C = C \cup T_{man}$ comprises a set of general domain terms (concepts) C, and a set of specific domain terms (instances of the concepts) T_{man},
$D = SemPage \cup D$ comprises class $SemPage$ which represents Web-page instances, and a set of Web-pages D,
$P_{MAN} = R_{man} \cup A_{man}$ comprises a set R_{man} of the relations between terms (R_c) and the relations between terms and Web-pages (R_p), and a set of attributes A_{man} defined

in the ontology. In particular, R_c will be specified depending on the application domain. $R_p = hasPage \sqcup isAbout$, where the 'hasPage' relation states that a domain term may have some Web-pages, and the 'isAbout' relation is the inverse of the 'hasPage' relation. That means each domain concept class has the 'hasPage' object property referring to class *SemPage*, and class *SemPage* has the 'isAbout' object property referring to the domain concept classes.

3   New Method for Domain Ontology Construction of a Website

This section proposes a new method that would reduce human effort in constructing a domain ontology of a website. According to Grimm et al. [6], a generic methodology of ontology engineering comprises three main steps: requirement analysis, conceptualization, and implementation. The new method consists of three parts corresponding to the main steps in the generic methodology.

3.1   Requirements Analysis

The goal of a website domain ontology is to represent the semantics of Web-pages within the given website. In this study, we use Web-page titles to represent the Web-page contents. In order to model the domain knowledge related to user interests for supporting Web-page recommendation, and to reduce the processing load, this study focuses on the Web-pages that have been visited by users (referred to as accessed or visited Web-pages), which can be obtained from Web server logs [7]. The URLs of visited Web-pages can be extracted from the Web logs, and the titles of these Web-pages can be retrieved using the Web crawling technique [8].

3.2   Conceptualization

Based on the page titles, domain experts can identify the domain terms for the website. From general to specific or vice versa, the domain experts need to classify the terms to be general or specific and the *belongs* relations between the general and specific terms. The general terms will become the domain concepts (C), the specific terms become the instances of the general terms (T_{man}) and the *belongs* relations become the instantiation axioms between the instances and the domain concepts. Domain experts also need to build the ontology schema at the general level by identifying the relations between the domain concepts, and the associations between them and *SemPage*, i.e. R_{man}, and the attributes of the domain concepts and *SemPage*, i.e. A_{man}. A special attribute, namely keyword string, is required for each domain term. With C, *SemPage*, R_{man} and A_{man}, the design of ontology schema is accomplished.

3.3   Implementation

The domain ontology can be easily implemented using a popular ontology language, i.e. OWL. Protégé (http://protege.stanford.edu/) as a commonly used OWL-based ontology editor can be used to implement such an ontology.

A special effort has been devoted to define the keyword strings for each specific term. Table 1 shows the BNT-like grammar, which are used as the composition rules of keyword expressions. It is noticed that OR operator is not included in the syntax because keyword strings can be input separately as a set of attribute values in the ontology editor tool, i.e. Protégé, and that set is handled as the union of keyword strings.

Table 1. Keyword expressions

Keyword Expressions	
K : = \<kw> | [\<kw>] I : = K | %syn%K A : = I | !I Exp : = A(&&A)*	where, - \<kw> : a keyword - [\<kw>]: a keyword in capitals - %syn%: the synonyms of keyword K - !: not contain keywords specified in I - &&: AND operator - *: REPEAT operator for the expression in the two brackets () - Exp: a keyword expression

Based on this set of grammar rules, the domain experts can easily define keyword strings for specific terms in the ontology. *By using these keyword strings, the page instances will be mapped and added automatically to the corresponding specific terms.* The details about how the page mapping is carried out is presented in Algorithm 1. In which, the input data includes the domain ontology and the page titles. The keyword expression (Exp) is used for parsing the keyword strings of each domain term. WordNet is used to query for the synonyms of keywords during the mapping process. A Web-page whose keywords meeting a keyword string are mapped to the respective term. If no match is found, new specific terms will be generated based on the keywords in the Web-page title to be added into the concept which has most relevant terms, and then the Web-page instance is created and mapped to these terms.

Algorithm 1. Mapping Web-pages to domain terms using the keyword expressions

```
For each Web-page{
   For each concept in the domain ontology{
      For each instance of the concept{
         Check If the title of the Web-page meets the keywords of the domain concept instance:
               Using the rules of keyword expressions to parse the keyword strings
               Using WordNet to query synonyms if necessary
            Then create a new SemPage instance of the Web-page, and map it to the domain concept in-
            stance via the 'isAbout' and 'hasPage' object properties.
} } }
```

4   Case Study: Develop a Domain Ontology of the MS Website

The Microsoft anonymous Web data which is the Web usage data of the MS website created in 1998 is used as a case study of domain ontology construction in this study. The dataset is found from http://kdd.ics.uci.edu/databases/msweb/msweb.html. In this dataset, the Web-page titles and paths of the MS website are available, so we do not need to crawl the website. Based on the method of domain ontology construction presented in Sect. 3, the domain ontology of the MS website is developed as Fig. 1, which focuses on the application scenario of MS software products.

The considered domain concepts are *Manufacturer*, *Application*, *Product*, *Category*, *Solution*, *Support*, *News*, *Misc*, and *SemPage*. The 'consistsOf', 'includes' and 'belongsTo' relations are taxonomic relationships. The 'provides', 'has' and 'hasPage' relations are non-taxonomic relationships. Their inverse relations are respectively the 'isProvided', 'isAppliedFor' and 'isAbout' relations. Based on this built domain ontology, terms and Web-pages are populated.

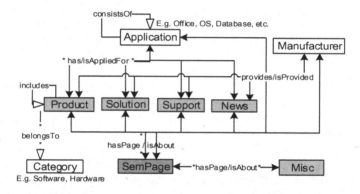

Fig. 1. Domain ontology schema of the MS website

5   Evaluation and Discussion

According to Tartir et al. [9], the schema metrics and instance metrics, i.e., Relationship Richness (RR), Class Richness (CR), Class Instance Distribution (CID), Class Connectivity, Class Importance, and Relationship Utilization (RU), are used to evaluate the domain ontology model. Table 2 summarizes the ontology metrics of the domain ontology constructed in the case study. The values of RU, which equal 1, reflect the all defined relationships are used.

The *class importance* shows that class *SemPage* is most important because more than 50% of instances in the ontology are Web-pages. About *class connectivity*, class *SemPage* is the most connected class with 216 connections to other concept instances out of a total of 715 connections. The second and third most connected classes are *Product* and *Support* with 148 and 151 connections, respectively. This indicates that

Table 2. Evaluation of the domain ontology of the MS website

Class	# Instances	Connectivity	Importance	RU		
Application	17	85	0.0675	1		
Category	4	0	0.0159	1		
Manufacturer	1	86	0.0040	1		
Misc	3	3	0.0119	1		
News	5	21	0.0198	1		
Product	19	148	0.0754	1	**SUM of Instances**	252
SemPage	173	216	0.6865	1	**RR**	1
Solution	1	5	0.0040	1	**CR**	1
Support	29	151	0.1151	1	**CID**	55.2

Product and *Support* are more focal classes than the remaining concept classes. The CID of the domain ontology which indicates how instances are spread across the classes of the schema, the standard deviation in the number of instances per class is 55.2.

Overall, the above results emphasizes the profit of the automated page mapping process when a huge amount of instances are Web-pages. Keyword strings are added into domain terms by the designer, so it is biased. However, its advantage is that the correctness can be optimized by changing the keyword strings of domain terms to have more Web-pages properly mapped. As a result, the built ontology has 97% of the mapped Web-pages accepted. Compared with most of the existing domain ontologies, which take much time and labor to manually construct, the proposed method alleviates these problems by semi-automatically constructing the domain ontology.

6 Conclusions

Although many domain ontologies have been built for knowledge representation in Web recommender systems, mapping pages to domain ontologies may not been presented so far. We have therefore addressed how pages should appear with regard to the domain ontology, and how to develop the ontology for a website. The ontology structure includes two parts: domain concepts and pages, allowing more flexibility in utilizing the ontological model. Manually constructing domain ontology allows the ontology being confident, rich and detailed by basing on an expert's experience. Moreover, the used relationship types are useful for extension. We can expand not only the domain part, but also update pages in later development. Furthermore, the proposed domain ontology model is efficient for semantically reasoning pages or terms within the website and making effective page recommendations.

References

1. Antoniou, G., van Harmelen, F.: A Semantic Web Primer. MIT Press, Cambridge (2008)
2. Zhou, B., Hui, S.C., Fong, A.C.M.: Web usage mining for semantic web personalization. In: PerSWeb 2005 Workshop on Personalization on the Semantic Web, Edinburgh, UK (2005)

3. Loizou, A., Dasmahapatra, S.: Recommender systems for the semantic web. In: ECAI 2006 Recommender Systems Workshop, Trento, Italy (2006)
4. Chen, L.-C., Kuo, P.-J., Liao, I.-E.: Ontology-based library recommender system using MapReduce. Cluster Comput. **18**(1), 113–121 (2015)
5. Liu, B.: Information retrieval and web search. In: Liu, B. (ed.) Web Data Mining: Exploring Hyperlinks, Contents, and Usage Data, pp. 183–236. Springer, Heidelberg (2011)
6. Grimm, S., et al.: Ontologies and the semantic web. In: Domingue, J., Fensel, D., Hendler, J.A. (eds.) Handbook of Semantic Web Technologies, pp. 507–580. Springer, Heidelberg (2011)
7. Liu, B., Mobasher, B., Nasraoui, O.: Web usage mining. In: Liu, B. (ed.) Web Data Mining: Exploring Hyperlinks, Contents, and Usage Data, pp. 527–603. Springer, Heidelberg (2011)
8. Markov, Z., Larose, D.T.: Data Mining the Web: Uncovering Patterns in Web Content, Structure, and Usage. Wiley, New Britain (2007)
9. Tartir, S., et al.: OntoQA: metric-based ontology quality analysis. In: IEEE Workshop on Knowledge Acquisition from Distributed, Autonomous, Semantically Heterogeneous Data and Knowledge Sources (2005)

Learning Functional Argument Mappings for Hierarchical Tasks from Situation Specific Explanations

Gavin Suddrey[✉], Markus Eich, Frederic Maire[✉], and Jonathan Roberts

Science and Engineering Faculty, School of Electrical Engineering
and Computer Science, Queensland University of Technology,
Gardens Point, Brisbane, QLD 4000, Australia
{g.suddrey,f.maire}@qut.edu.au

Abstract. Hierarchical tasks learnt from situation specific explanations are typically limited in how well they generalise to situations beyond the explanation provided. To address this we present an approach to learning functional argument mappings for enabling task generalisation regardless of explanation specificity. These functional argument mappings allow subtasks within a hierarchical task to utilise both arguments provided to the parent task, as well as domain knowledge, to generalise to novel situations. We validate this approach with a number of scenarios in which the agent learns generalised tasks from situation specific explanations, and show that these tasks provide equal performance when compared to tasks learnt from generalisable explanations.

1 Introduction

In using natural language to explain complex tasks, it is intuitive to describe such tasks as collections of smaller, more manageable tasks, called subtasks. For example, the task of hosting a dinner party might include subtasks such as setting the table, and preparing the meal. Each subtasks can then itself decomposed into its own collection of subtasks. Describing tasks as collections of subtask forms the basis of *Hierarchical tasks*.

A key challenge in learning *hierarchical tasks* from natural language involves determining how parameters for any given subtask should be mapped. Prior approaches have constrained mapping these parameters to either constants [7], or parameters of the parent task [9]. The inability to exploit domain knowledge however limits the generalisability of these tasks. Other work has sought to address this limitation by introducing the ability to map parameters to functions of parameters of the parent task, using what we call *functional argument mappings* [10]. These *functional argument mappings* enable tasks to generalise to a broader range of situations by taking advantage of domain knowledge.

While *functional argument mappings* provide increased generalisability, they currently depend on the user providing generalisable explanations, containing predicate relations, to generate the correct mappings. For instance, the generalisable explanation for clearing a table would be "put away everything on the

© Springer International Publishing AG 2016
B.H. Kang and Q. Bai (Eds.): AI 2016, LNAI 9992, pp. 345–352, 2016.
DOI: 10.1007/978-3-319-50127-7_30

table", where the objects being put away are a function of the table. Conversely, in the situation specific explanation, "put away the three cups", it is unclear how the cups relate to the table, in which case the agent will forever link clearing the table with putting away the three cups, regardless of where they are located.

To remove the dependence on users giving generalisable task explanations, this paper presents an approach to learning *functional argument mappings* directly from situation specific explanations. Importantly, we show that our approach allows agents to learn *functional argument mappings* to enable task generalisation with only (a) a small number of examples, and (b) domain knowledge.

2 Related Work

Previous work has demonstrated the use of natural language as means for commanding robots to complete complex real-world tasks including both manipulation tasks [12], and navigation tasks [11]. Research has also shown that agents can make use of natural language in learning everyday tasks from sources including the World Wide Web [13], and situated interactions with users [9].

There exists many different approaches to representing tasks learnt from natural language. This includes representing tasks as sequences of behaviours, in which repeated demonstrations and conditional branching are used to provide generalisability [3,6,7]. Other approaches involves representing each task, not as a sequence of behaviours, but rather as a goal state for a planning problem, and generating a new plan each time the task is executed [8].

Task hierarchies provides a means for combining tasks, creating new and more complex tasks [7,9]. However, the degree to which these approaches are able to generalise depends on how parameters of subtasks are mapped. This may involve limiting the mapping of parameters within these hierarchies to constant values [7]. These tasks however are then restricted to acting only on the set of objects used during training. Less restrictive approaches add the ability to map parameters of subtasks to the parameters of their respective parent tasks [9]. However, any parameter that cannot be mapped to a parameter of the parent task must still be mapped to a constant, limiting overall generalisability.

More recent work has sought to address this limitation by adding the ability to explain tasks using binary relations, which allow parameters of a subtask to map to functions of parameters of the parent [10]. While this approach allows tasks to generalised to a broader variety of situations however, it depends on the user articulating any relational predicates necessary to ensuring generalisation of the task while avoiding the use of situation specific examples.

3 Hierarchical Tasks

Hierarchical tasks provide a mechanism for decomposing complex everyday tasks, such as serving dinner, into sequences of smaller, more manageable tasks, called subtasks. Formally, a task hierarchy is composed of a collection of *tasks*, *methods* and *operators* (Fig. 1). The following definitions define *tasks* and *methods*.

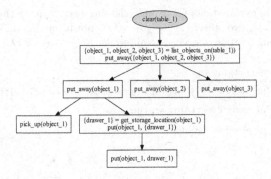

Fig. 1. A decomposition tree for a simple clearing task. This task is initiated with the instruction *"clear the table"* and demonstrates iteration over a set of objects.

Definition 1. *A **task** is any activity that can be undertaken by the agent, and is comprised of a name and a parameter list* [5]. *A task is either **compound**, and can be decomposed into subtasks through the use of associated methods; or **primitive**, and maps to operators that enables the agent to act within its world.*

By definition, a task does not specify how it should be accomplished, rather, this information is deferred to its associated *methods* or *operators*. The following formal definition for *methods* is based on previous work in learning generalisable tasks from natural language [10].

Definition 2. *A **method** details how to implement a given compound task. We define a method as a 5-tuple, $m = (N, P, E, \Pi, \Sigma)$, where N is the name of the task T; P is the set of preconditions for m; E is the set of positive and negative effects of m; Π is a sequence of subtasks providing a partial plan for completing T; and Σ contains an ordered list of **argument mappings** for each $\pi_i \in \Pi$.*

Each $\sigma_i \in \Sigma$ corresponds to a subtask $\pi_i \in \Pi$ of T, and specifies to what value each parameter of π_i will be mapped. For any argument to π_i, its corresponding element of σ_i dictates if it will map to a *term*, where a term is a parameter or function of a parameter of T, or a *constant* identifying an object in the domain.

4   Problem Definition

In order to teach an agent a new compound task, the user provides an explanation that describes the steps in completing the task. Table 1 provides three alternative explanations for a table clearing task T.

For each general explanations, binary relations relate the table to the objects that should be put away. This allows us to derive argument mappings that maps from the parameters of the *put away* subtask to terms of T.

For the specific explanation, the set of objects to be put away is provided explicitly. This has the effect that argument mappings learnt for the *put away*

Table 1. Alternate explanations for a table clearing task. Among the objects on the table $O = \{o_1, ..., o_n\}$, two objects $E^+ = \{o_1, o_2\}$ do not belong on the table. Note that OG indicates the instruction is over generalised, entirely covering O.

Quality	Instruction
General	Put away everything on the table that does not belong on the table
General (OG)	Put away everything on the table
Specific	Put away o_1 and o_2

subtask will map to constants identifying these objects. Future invocations of T will result in the agent not clearing T as intended, but rather tracking down and putting away the objects described in the initial explanation of T.

Given a specific explanation of T, and the *put away* subtask π_0, we therefore wish to induce a function f, that for a parameter β of T, produces $f(\beta) = \Lambda$, where Λ is the set of objects provided for π_0. That is, we wish to find a *functional argument mapping* for the parameter of π_0 describing the objects to be put away.

In order to induce the function f, we reformulate our problem as an Induction Logic Programming (ILP) problem [4], where the set of positive examples E^+ describes the set Λ, while the set of negative examples E^- is initially empty.

5   Approach

The following section describes our approach to learning functional argument mappings from situation-specific explanations.

5.1   Background Knowledge

Knowledge is expressed using OWL, the Web Ontology Language, and provides a complete description of the set of concepts (unary predicates) (Fig. 2), roles (binary predicates) (Table 2), and individuals (objects) within the domain. To Reason about this knowledge, we use *Racer Knowledge Representation System* and the *new Racer Query Language* (nQRL) [1].

5.2   Finding Pre-existing Roles

We now describe our induction algorithm (See Algorithm 1), that for a given argument β of the task T, and a set of objects Λ supplied for some parameter of the subtask π_i, will attempt to learn a function f such that $f(\beta) = \Lambda$.

Lines 4 to 11 of Algorithm 1 *search* the knowledge base for any role $P(x, y)$ such that $\forall \lambda \in \Lambda$, $P(\beta, \lambda)$. For each role $P(x, y)$, we take the functional form $f(x) = \{y | P(x, y)\}$. If $f(\beta) = \Lambda$, execution halts and f is returned. Otherwise, we update the set of negative examples $E^- = E^- \cup f(\beta) \backslash \Lambda$, and move to the next role. If no function f is found such that $f(\beta) = \Lambda$, we move to rule induction.

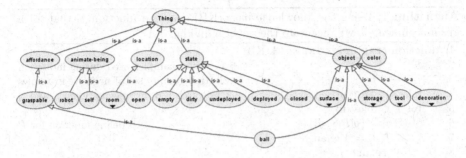

Fig. 2. A partial view of the taxonomy of concepts used during the evaluation described in Sect. 6. Nodes containing triangles can be expanded further.

5.3 Rule Induction

Lines 12 to 22 of Algorithm 1 begin by inducing a rule $R(x, y)$ from the set of positive E^+ and negative E^- examples. We induce this rule using the *DL-Reasoner* [2] ILP system with the standard CELOE algorithm and a closed-world reasoner. To ensure generalisation of the induced rule, we bias the search such that predicates used in generating the rule are limited to only the concepts associated with β, as well as any concepts for the elements in Λ that encode state (e.g. Dirty). The generated rule $R(x, y)$ with the highest training accuracy, defined as the proportion of true results (both positive and negative) of the total number of examples $|E^+ \cup E^-|$ is then returned. If two or more rules have an equally high training accuracy, the shortest rule is returned.

If DL-Learner is unable to find a rule with a training accuracy of 1, or the generated rule has been induced previously, further iterations will fail to find a valid result. Execution is therefore halted, and null is returned to indicate failure.

If a rule $R(x, y)$ is generated with a training accuracy of 1, we take the functional form $f(x) = \{y | R(x, y)\}$. We then test if f meets the requirement $f(\beta) = \Lambda$. If $f(\beta) = \Lambda$, the function f is returned and execution halted. Otherwise, any new false positives are added to E^-, and the process is repeated.

Table 2. A description of the set of roles that comprise the domain in which our agent reasons. For simplicity, we express each role in predicate logic.

Role	Inverse	Example
is_on(x,y)	has_on(y,x)	is_on(cup, table)
is_in(x,y)	has_in(y,x)	is_in(cup, cupboard)
is_located(x,y)	has_located(y,x)	is_located(table, kitchen)
is_owner(x,y)	has_owner(y,x)	is_owner(agent, gripper)
belongs(x,y)	not used	belongs(vase, table)
is_colored(x,y)	not used	is_colored(ball, red)

Algorithm 1. Using the knowledge base (KB), learns a function f, that for a given argument β, will generate the set of individuals Λ

```
 1: function LEARN_FUNCTION(β, Λ, KB)
 2:     E⁺ ← Λ                                    ▷ Positive examples used during induction
 3:     E⁻ ← ∅                                    ▷ Initially empty set of negative examples
 4:     roles ← SEARCH(β, Λ, KB)                  ▷ Search KB for roles between β and Λ
 5:     for P ∈ roles do
 6:         f(x) ≝ {y|P(x,y)}                     ▷ define function for P
 7:         if f(β) = Λ then
 8:             return f
 9:         end if
10:         E⁻ ← E⁻ ∪ f(β)\Λ                      ▷ Append false positives to negative examples
11:     end for
12:     while true do
13:         accuracy, R ← INDUCE(β, E⁺, E⁻, KB)
14:         if accuracy < 1 or R generated previously then
15:             return null                        ▷ No solution found
16:         end if
17:         f(x) ≝ {y|R(x,y)}                      ▷ define function associated with R
18:         if f(β) = Λ then
19:             return f
20:         end if
21:         E⁻ ← E⁻ ∪ f(β)\Λ                      ▷ Append false positives to negative examples
22:     end while
23: end function
```

6 Experiments

The following section details our evaluation using a series of task learning scenarios; and provides a comparison of tasks learnt with and without our approach.

6.1 Scenarios

Each scenarios involves learning a task from a situation specific explanation using the task learning approach described in [10], extended with Algorithm 1.

Scenario 1 - Putting Away Objects: This scenario involved learning a *put away* task T, initiated with the command "put away the knife". T contained only a single subtask π_0 explained with: "Move the knife to the kitchen drawer."

As the first argument provided for π_0, the *knife*, matched the argument provided for T, the associated parameter of π_0 was mapped directly to the parameter of T. However, as the argument provided for the second parameter of π_0 did not match any argument to T, the agent initiated a learning problem to explain this parameter, in which $\beta = knife$ and $\Lambda = \{kitchen\ drawer\}$.

Given β and Λ, the Algorithm 1 learnt the function $f(x) = \{y|belongs(x,y)\}$. We then apply $f(knife)$ which correctly generates $\{kitchen\ drawer\}$.

Table 3. Completion rates and number of primitive tasks for each planning problem.

Task type	Put away		Clear 1		Clear 2	
	Complete	Primitives	Complete	Primitives	Complete	Primitives
Generalised	Yes	104	Yes	128	Yes	104
Specific (Learning)	Yes	104	Yes	128	Yes	104
Specific (No Learn)	No	104	No	120	No	80

Scenario 2 - Clearing the Table I: For the second scenario the agent learnt a *clear* task T. The explanation provided for T, initiated with the command "clear the table", contained only a single subtask π_0, explained as follows: "Put away the red cup, the blue cup and the green cup."

In the explanation provided, the subtask π_0, was provided with a set of *cups* O. As O was not provided for T, the agent initiated a learning problem in which $\beta = table$, and $\Lambda = O$. For this scenario, O contained all objects on the table.

From these inputs, we learnt the function $f(x) = \{y|has\_on(x,y)\}$. Calling $f(table)$ then correctly generates the set O. This function will however include any objects that belong on the table, if present in future.

Scenario 3 - Clearing the Table II: The third scenario involved the agent learning a *clear* task from the explanation provided previously in Scenario 2. However, in this scenario, a *vase* was added to the table prior to learning. This *vase* belongs on the *table* and should therefore not be moved.

For this scenario, O now contains the set of cups and a vase. Given $\beta = table$ and $\Lambda = O \backslash \{vase\}$, the learning algorithm generates the function $f(x) = \{y|has\_on(x,y) \wedge \neg belongs(y,x)\}$. Calling $f(table)$ correctly generated Λ.

6.2   Quantitative Evaluation

Table 3 compares the problem solving ability of tasks learnt in Sect. 6.1 (Specific (Learning)) to those without the ability to learn functional mappings (Specific (No Learn)); and tasks learnt with generalised explanations (Generalised).

The generalised *put away* task was explained using the instruction "move the knife to where it belongs". The first generalised *clear* task was explained with: "put away everything on the table". The final clear task was explained with: "put away everything on the table that does not belong on the table".

Each problem required the agent to affect a number of object. This included putting away 13 items in the first problem, and clearing 5 tables in the remaining two. The number of *primitive* tasks was recorded for each problem.

7   Conclusion

In this paper we have presented for approach for learning functional argument mappings from situation specific task explanations. We show that an agent was able to learn generalisable tasks from a situation specific task explanations in a number of scenarios. We also provide a comparison of these tasks to tasks learnt from both generalisable explanations, and situation specific explanations in which the agent was unable to learn functional mappings. We show that tasks learnt using our approach were capable of equalling the performance of tasks learnt from generalisable explanations. At present our approach relies on two assumptions. First, that the agent has complete knowledge of its domain, and second, that generated functions will generalise in a way that matches the intention of the user. We plan to address these assumptions in future work.

References

1. Haarslev, V., Möller, R.: Description of the RACER system and its applications. In: International Workshop on Description Logics, vol. 1, pp. 132–142 (2001)
2. Lehmann, J.: DL-learner: learning concepts in description logics. J. Mach. Learn. Res. **10**, 2639–2642 (2009)
3. Meriçli, C., Klee, S., Paparian, J., Veloso, M.: An interactive approach for situated task teaching through verbal instructions. In: AAAI (2013)
4. Muggleton, S.: Inductive logic programming. New Gener. Comput. **8**(4), 295–318 (1991)
5. Nau, D., Au, T.C., Ilghami, O., Kuter, U., Murdock, J.W., Wu, D., Yaman, F.: SHOP2: an HTN planning system. J. Artif. Intell. Res. **20**(1), 379–404 (2003)
6. Nicolescu, M.N., Mataric, M.J.: Natural methods for robot task learning: instructive demonstrations, generalization and practice. In: Proceedings of the Second International Joint Conference on Autonomous Agents and Multiagent Systems (AAMAS), p. 241 (2003)
7. Rybski, P., Yoon, K., Stolarz, J., Veloso, M.: Interactive robot task training through dialog and demonstration. In: 2nd ACM/IEEE International Conference on Human-Robot Interaction (HRI), pp. 49–56. IEEE (2007)
8. She, L., Yang, S., Cheng, Y., Jia, Y., Chai, J.Y., Xi, N.: Back to the blocks world: learning new actions through situated human-robot dialogue. In: 15th Annual Meeting of the Special Interest Group on Discourse and Dialogue, vol. 89 (2014)
9. Shiwali, M., Laird, J.: Learning goal-oriented hierarchical tasks from situated interactive instruction. In: Proceedings of the Twenty Eighth AAAI Conference on Artificial Intelligence, Québec (2014)
10. Suddrey, G., Lehnert, C., Eich, M., Maire, F., Roberts, J.: Teaching robots generalisable hierarchical tasks through natural language instruction. IEEE Robot. Autom. Lett. **2**, 1 (2016)
11. Talbot, B., Schulz, R., Upcroft, B., Wyeth, G.: Reasoning about natural language phrases for semantic goal driven exploration. In: Proceedings of the Australasian Conference on Robotics and Automation (2015)
12. Tellex, S., Kollar, T., Dickerson, S., Walter, M.R., Banerjee, A.G., Teller, S.J., Roy, N.: Understanding natural language commands for robotic navigation and mobile manipulation. In: Proceedings of the Twenty-Fifth AAAI Conference on Artificial Intelligence, San Francisco, pp. 1507–1514 (2011)
13. Tenorth, M., Nyga, D., Beetz, M.: Understanding and executing instructions for everyday manipulation tasks from the World Wide Web. In: 2010 IEEE International Conference on Robotics and Automation, pp. 1486–1491. IEEE, May 2010

Ontology Based Data Access with Referring Expressions for Logics with the Tree Model Property
(Extended Abstract)

David Toman[(✉)] and Grant Weddell

Cheriton School of Computer Science, University of Waterloo, Waterloo, Canada
{david,gweddell}@uwaterloo.ca

Abstract. Certain answer computation for a query has usually entailed the search for constants as substitutions for the query variables that make the query logically entailed by a knowledge base. Such constants are simple examples of *referring expressions*, that is, syntactic artifacts that identity objects in an underlying domain. In earlier work, we have begun to explore how more general referring expressions can be used to allow more descriptive and useful object identification. In this paper, we present a novel approach to ontology based data access in this more general setting for logics with the so-called tree model property. The proposed solution remedies a problem with our earlier work in certain answer computation with referring expressions in which a process of extending the knowledge base with new constants and assertions that depended on a particular query is required.

1 Introduction

In [1], we introduce a *referring expression type language*, Rt, for specifying the format of answers to *conjunctive queries* (CQs) over a first order *knowledge base* (KB) $\mathcal{K}$, showing in particular how such answers can be computed with the aid of an oracle for traditional query answering. Our main result in this earlier work was to exhibit a procedure for computing so-called *certain R-answers* for a CQ ψ annotated with referring expression types, in which answer substitutions for the answer variables of ψ were formulas free in one variable that conformed to such types. This enabled, for example, answers to MEETING-ROOM(x) to return substitutions of the form

$$x \mapsto (RoomNumber(x) = 1234) \wedge (Building \circ Location(x) = \text{``MATH''}).$$

In contrast, the substitutions in traditional query answering produced by the oracle assumed by the procedure have the form "$x \mapsto a$" where a corresponds to constants occurring in $\mathcal{K}$.

A limitation of the procedure is that it requires additional constant symbols and assertions to be added to $\mathcal{K}$ that are query dependent, a property that is

© Springer International Publishing AG 2016
B.H. Kang and Q. Bai (Eds.): AI 2016, LNAI 9992, pp. 353–361, 2016.
DOI: 10.1007/978-3-319-50127-7_31

at odds with efficient implementation techniques for *ontology based data access* (OBDA). Indeed, in most current approaches to OBDA, $\mathcal{K}$ is formulated in terms of a *description logic* (DL), and is comprised of a TBox $\mathcal{T}$ defining an ontology and an ABox $\mathcal{A}$ consisting of facts or observations. In this setting, efficient implementations of OBDA entail that query answering can usually be realized by a pair of functions REF and CLOSE for respective rewriting of CQs and ABoxes, according to a given TBox, in such a way that evaluating $\text{REF}_{\mathcal{T}}(\psi)$ over $\text{CLOSE}_{\mathcal{T}}(\mathcal{A})$ yields the traditional certain answers to ψ over $\mathcal{K}$. However, these techniques rely crucially on the underlying DL having the so-called *tree model property*, that is, on the property that all consistent knowledge bases over the DL have a model that is tree-like outside of $\mathcal{A}$.

Observe that the CLOSE function in OBDA computes the necessary closure of a given ABox *independently* of queries, unlike the procedure for computing certain R-answers presented in [1]. In this paper, we address this deficiency for any logic with the tree-model property together with additional properties fundamental to OBDA and to the variety of referring expressions that are desirable:

1. A KB $\mathcal{K}$ over the logic can be expressed as a combination of a TBox $\mathcal{T}$ and ABox $\mathcal{A}$;
2. Query answering for CQs with respect to a knowledge base $\mathcal{K} = (\mathcal{T}, \mathcal{A})$ can be reduced to evaluating query $\text{REF}_{\mathcal{T}}(\psi)$ over $\text{CLOSE}_{\mathcal{T}}(\mathcal{A})$, as described above; and
3. The logic can express *path functions*, that is, expressions that are a composition of total functions (or of binary predicates that are functional).

One example of a logic with these properties is the DL dialect $\mathcal{CFDI}_{nc}^{\forall-}$ [3]. In [2], it is shown how the logic *instantiates* this general framework, in particular by exhibiting REF and CLOSE functions for computing certain answers to conjunctive queries over $\mathcal{CFDI}_{nc}^{\forall-}$ KBs.

Our results are given in Sect. 3: we show how the results for plain OBDA can be lifted to more general functions that replace the procedure in [1] for computing certain R-answers for such queries, thereby further establishing a practical basis for OBDA with referring expressions. We also extend the Rt language in [1] with an additional construct for expressing more general referring terms. The development culminates in the following theorem:

Theorem 1. Let ψ be a typed conjunctive query with answer variables $x_1, \ldots, x_k$ and with normalized head $\mathsf{Hd}(\psi)$, and $\mathcal{K} = (\mathcal{T}, \mathcal{A})$ a KB. Then

$$(b_1, \ldots, b_k) \in \text{REF}_{\mathcal{T}}^{\mathsf{Hd}(\psi)}(\text{CLOSE}_{\mathcal{T}}(\mathcal{A}))$$

if and only if the k tuple of referring expressions $(b_1, \ldots, b_k)$ is an R-answer to ψ over $\mathcal{K}$ conforming to $\mathsf{Hd}(\psi)$.   □

2   Preliminaries

We begin by introducing a space of referring expressions and by formally defining what constitutes certain R-answers for such expressions. The expressions conform to a variable-free syntax for formulas with one answer variable common in DLs, and in particular to the syntax for so-called *concept descriptions* in the logic $\mathcal{CFDI}_{nc}^{\forall-}$ given in [2].

Definition 2 (Referring Expressions and Certain R-Answers). A *referring expression* is a concept description E given by the following grammar:

$$ E \quad ::= \quad \{a\} \mid E_1 \sqcap E_2 \mid A \mid \forall \mathsf{Pf}.E \mid E.\mathsf{Pf} $$

with the standard semantics [2], for example

$$ (E.\mathsf{Pf})^{\mathcal{I}} = \{(\mathsf{Pf})^{\mathcal{I}}(x) \mid x \in (E)^{\mathcal{I}}\}. $$

An *R-substitution* generalizes a substitution by allowing variables to be replaced by referring expressions. Let $\mathcal{K}$ be a knowledge base and $\psi = \{\bar{x} \mid \varphi\}$ a CQ. A *certain R-answer* to ψ over $\mathcal{K}$ is a vector of referring expressions $\bar{E}$ for $\bar{x}$ for which: (a) $\mathcal{K} \models \exists \bar{x} : (\psi \wedge \bigwedge_i E_i(x_i))$; and (b) $\mathcal{K} \models \forall \bar{x} : \bigwedge_i E_i(x_i) \rightarrow \psi$. □

To illustrate, consider our introductory example that presents a hypothetical certain R-answer to the query $\psi = \text{MEETING-ROOM}(x)$. The answer would be expressed as the R-substitution

$$ x \mapsto (\forall RoomNumber.\{1234\}) \sqcap (\forall Location.Building.\{\text{"MATH"}\}). $$

We now introduce a type language for referring expressions, largely from [1], to control the space of possible R-substitutions when computing certain R-answers by attaching such types to the free variables of conjunctive queries. (We refer the reader to this paper for an more in-depth discussion of the merits of the various parts of this language.)

Definition 3 (Referring Expression Types). A *referring expression type* Rt is given by the following grammar:

$$ Rt \quad ::= \quad T?.\mathsf{Pf} \mid \mathsf{Pf} = ? \mid Rt_1 \sqcap Rt_2 \mid T \rightarrow Rt \mid Rt_1; Rt_2 $$

where T is a conjunction of primitive concepts. We write $\text{RE}(Rt)$ to denote a set of referring concepts *generated* by Rt that is inductively defined as follows (with respect to a $\mathcal{CFDI}_{nc}^{\forall-}$ knowledge base $\mathcal{K}$ determined by context):

$$ \text{RE}(T?.\mathsf{Pf}) = \{(T \sqcap \{a\}).\mathsf{Pf} \mid a \text{ a constant}\} $$
$$ \text{RE}(\mathsf{Pf} = ?) = \{\forall \mathsf{Pf}.\{a\} \mid a \text{ a constant}\} $$
$$ \text{RE}(Rt_1 \sqcap Rt_2) = \{E_1 \sqcap E_2 \mid E_i \in \text{RE}(Rt_i)\} $$
$$ \text{RE}(T \rightarrow Rt) = \{T \sqcap E \mid E \in \text{RE}(Rt)\} $$
$$ \text{RE}(Rt_1; Rt_2) = \text{RE}(Rt_1) \cup \{E_2 \in \text{RE}(Rt_2) \mid \neg \exists E_1 \in \text{RE}(Rt_1) : \mathcal{K} \models E_1 \equiv E_2\} $$

A referring expression type Rt is *homogeneous* if it is free of any occurrence of the constructor ";" (which expresses, among other things, *preference* among referring expressions). □

Definition 4 (Typed CQs and Certain R-Answers). Let $\psi = \{\bar{x} \mid \varphi\}$ be a CQ over a KB $\mathcal{K}$. A *head* for ψ, written $\mathsf{Hd}(\psi)$, associates a referring expression Rt_i with each $x_i \in \bar{x}$. We say that a query is *typed* if it has a head. A *certain R-answer* for a typed CQ ψ is a certain R-answer $\bar{E}$ to ψ for which each E_i occurs in $\mathrm{RE}(Rt_i)$, where $(x_i : Rt_i) \in \mathsf{Hd}(\psi)$. $\qquad\square$

The final pair of definitions, again from [1] and needed by our development in the next section, introduce a normal form for referring expression types together with the notion of a *induced homogeneous head* for typed CQs.

Proposition 5 (Normal Form Typed CQs). For every referring expression type Rt, there is an *equivalent normal form* $Rt_1; \cdots; Rt_n$, denoted $\mathrm{NORM}(Rt)$, that consists of homogeneous referring expression types given by

$$T_i \to (Rt_{i,1} \sqcap \cdots \sqcap Rt_{i,m_i}),$$

where each $Rt_{i,j}$ is in turn of the form "$\mathsf{Pf} = ?$" or "$T ? . \mathsf{Pf}$". Here, the ";" referring expression type constructor is assumed to be left associative, and *equivalent* means that, for any KB $\mathcal{K}$, $\mathrm{RE}(Rt)$ coincides with $\mathrm{RE}(\mathrm{NORM}(Rt))$.

A typed query ψ is *normalized* whenever $(x : Rt) \in \mathsf{Hd}(\psi)$ implies Rt is in normal form. $\qquad\square$

Definition 6 (Induced Homogeneous Heads). Let ψ be a normalized typed query, where $\mathsf{Hd}(\psi) = \{x_1 : Rt_1, \ldots, x_k : Rt_k\}$, in which each Rt_i is in normal form and, for each $0 < i \leq k$, is given by the form $Rt_{i,1}; \cdots; Rt_{i,n_i}$. Given a k-tuple $\langle j_1, \ldots, j_k \rangle$ for which $0 < j_i \leq n_i$, we write $\mathsf{H}_{j_1,\ldots,j_k}$ to denote a *homogeneous head* $\{x_1 : Rt_{1,j_1}, \ldots, x_k : Rt_{k,j_k}\}$ of ψ, written $\mathsf{H}_{j_1,\ldots,j_k} \in \mathsf{Hd}(\psi)$ where ψ will be clear from context.

We write $\mathsf{H}_{j'_1,\ldots,j'_k} < \mathsf{H}_{j_1,\ldots,j_k}$ for a pair of *distinct* homogeneous heads whenever $j'_i \leq j_i$, for $0 < i \leq k$. $\qquad\square$

3 Query Reformulation with Referring Expressions

In this section, we focus on the main contribution of the paper: we show how the query reformulation $\mathrm{REF}_{\mathcal{T}}$ can be extended to account for typed CQs.

The main intuition is that one can appeal to the tree model property outside of an explicit (completion of an) ABox when evaluating CQs. Intuitively, in such a setting, whenever a path in a referring expression leads to a certain answer *that lies outside* of the explicit ABox, there will always be a *unique "last"* object in the explicit ABox before the path *exits* the ABox and starts traversing the implicit part of the universal model.

With the above observation in mind, consider two special cases of referring expression types (a formal definition of the first case follows):

1. Referring expression types that have a *common suffix* in all path expressions participating in the type; and
2. Referring expression types that have a *common prefix* in all path expressions participating in the type.

In both cases, the common suffix or prefix can be *mapped* to the anonymous part of the KB models, and thus care needs to be taken to retain the ability to resolve preferences among alternative referring expressions specified in the head of the given typed query.

This is the main deviation from the technique developed in [1]: we completely avoid the need for query-specific completion of the KB, albeit at the price of requiring the underlying logic to have the tree-model property.

Our main focus for the remainder of this section is on the common suffix case. We also assume, without loss of generality, that the query heads are represented as sets of *induced homogeneous heads* with an appropriate preference ordering (see Definition 6). At the end of this section, we briefly comment on the prefix case, and argue that all other cases reduce to either the suffix or prefix cases.

The case of the common suffix: Consider where referring expressions in the head of a query have the form

$$T \to (T^1\,?.\,Pf^1.\,Pf \sqcap \cdots \sqcap T^m\,?.\,Pf^m.\,Pf),$$

where Pf is the *maximal common suffix* of all path expressions occurring in the form.

Example 7. Consider a typed query $\psi = \{x \mid A(x)\}$ with head

$$\mathsf{Hd}(\psi) = \{x : C?.f.g.h; D?.f.g.h\}.$$

The R-answers will be referring expressions of the form "$(C \sqcap \{a\}).f.g.h$" and (less preferably) "$(D \sqcap \{b\}).f.g.h$".

The three principal cases for how R-answers to this query that refer to an A-object can be realized in an (completed) ABox are illustrated in Fig. 1:

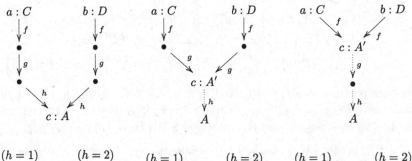

$$(h = 1) \qquad (h = 2)$$
$$(p = id)$$

$$(h = 1) \qquad (h = 2)$$
$$(p = h)$$

$$(h = 1) \qquad (h = 2)$$
$$(p = g.h)$$

Fig. 1. Possible ABox Matches for $\{x : A(x)\}$ and $x : C?.f.g.h; D?.f.g.h$.

1. (the left-most case) The A-object is denoted by a constant c.
2. (the middle case) The $c : A'$-object *does not* have an explicit h successor in the ABox, but the TBox implies $A' \sqsubseteq \forall h.A$.[1]
3. (the right-most case) The $c : A'$-object *does not* have an explicit g successor in the ABox, but the TBox implies $A' \sqsubseteq \forall g.h.A$.

In all cases, one can arbitrate preferences among R-answers based on an explicit retrieval of c as an additional auxiliary component of the answer. Hence, in our example, for each of the above three cases and for each of the heads, we define a binary query that retrieves the pairs (c, a) (resp. (c, b)) in our sample ABoxes in Fig. 1. The c component is then used to disambiguate preferences between the R-answers "$(C \sqcap \{a\}).f.g.h$" and "$(D \sqcap \{b\}).f.g.h$" (constructed from the retrieved constants a and b, respectively).

Note that there is one more pair of cases: when $C(a)$ (or $D(b)$) exists in an ABox, when neither has an explicit outgoing f, and when the TBox implies $C \sqsubseteq \forall f.g.h.A$ ($D \sqsubseteq \forall f.g.h.A$). In such cases, however, disambiguation is no longer necessary since $c = a$ ($c = b$), respectively.   □

The idea in the above example generalizes to k-ary queries using the following definition (in which the queries generated in the first part — with the help of query rewriting $\text{REF}_{\mathcal{T}}$ for $\mathcal{CFDI}_{nc}^{\forall-}$ — effectively *double* the number of answer variables: intuitively, the first k stand for the *preference-disambiguating* objects and the later k-tuple for the referring expressions that describe the actual answers[2]):

Definition 8. Let ψ be a query with answer variables $x_1, \ldots, x_k$ and *homogeneous heads* $\mathsf{Hd}(\psi)$, each of which attaches referring expressions of the form

$$T_i \rightarrow (T_i^1 ? . Pf_i^1 . Pf_i \sqcap \cdots \sqcap T_i^{m_i} ? . Pf_i^{m_i} . Pf_i)$$

to variables x_i. In addition, let Pf_i be a *maximal* common suffix of all the path expressions in the particular homogeneous head and $\mathsf{S}(H)$ the set of all tuples $\langle Pf_1^R, \ldots, Pf_k^R \rangle$ with Pf_i^R a suffix of Pf_i.

We define an $\mathsf{Hd}(\psi)$-reformulation as a union of queries in two steps as follows:

1. For every head $\mathsf{H} \in \mathsf{Hd}(\psi)$, we define

$$\psi_{\mathsf{H}}^{\mathsf{S}} = \big\{ (x_1', \ldots, x_k', Rt_1[y_1^1, \ldots, y_1^{m_1}], \ldots, Rt_k[y_k^1, \ldots, y_k^{m_k}]) :$$
$$\text{REF}_{\mathcal{T}} \Big(\exists x_1, \ldots, x_k.\psi \wedge \bigwedge_{i=1}^{k} (T_i(x_i) \wedge x_i'.Pf_i^R = x_i \wedge$$
$$\bigwedge_{j=1}^{m_i} (T_i^j(y_i^j) \wedge y_i^j.Pf_i^j.Pf_i^L = x_i')) \Big) \wedge \bigwedge_{i=1}^{k} (\neg \exists z(x_i'.f_i = z)) \big\}$$

for each partition of Pf_i to Pf_i^L and Pf_i^R, where f_i is the first feature of Pf_i^R and where $Rt_i[y_i^1, \ldots, y_i^{m_i}]$ constructs a *term* representation of the referring expression assigned to x_i in H, and where S is a tuple $\langle Pf_1^R, \ldots, Pf_k^R \rangle$.

---

[1] Using "$\sqsubseteq$" in this way is standard DL notation for expressing set containment of concepts. E.g., for any pair of referring expressions E_1 and E_2, we write $E_1 \sqsubseteq E_2$ as shorthand for the boolean formula $\forall x.(E_1(x) \rightarrow E_2(x))$.

[2] To simplify the notation we allow ground terms representing the referring expressions to be bound to query variables.

2. The final rewriting of the query ψ then is

$$\bigvee_{H \in \mathsf{Hd}(\psi)} \bigvee_{S \in \mathsf{S}(H)} \left(\exists x_1, \ldots, x_k . \psi_H^\mathsf{S}(x_1, \ldots, x_k, y_1, \ldots, y_k) \wedge \right.$$
$$\left. \bigwedge_{H' < H} \left(\neg \exists y'_1, \ldots, y'_k : \psi_{H'}^\mathsf{S}(x_1, \ldots, x_k, y'_1, \ldots, y'_k) \right) \right).$$

We denote the result of the reformulation by $\mathrm{REF}_\mathcal{T}^{\mathsf{Hd}(\psi)}(\psi)$. □

Note that distributing the subformula $\neg \exists z(x'_i.f_i = z)$ into the disjunction produced by $\mathrm{REF}_\mathcal{T}$ will prune trivially unsatisfiable disjuncts (and this way significantly reduce the size of the reformulation).

Example 9. Consider again the typed query $\psi = \{x \mid A(x)\}$ with the head

$$\mathsf{Hd}(\psi) = \{x : C?.f.g.h; D?.f.g.h\}.$$

This assignment yields two homogeneous query heads $\{x : C?.f.g.h\}$ and $\{x : D?.f.g.h\}$ (below we denote the referring types in these heads Rt_1 and Rt_2, respectively). Assume, in addition, that the type assignment is *weakly identifying*.[3] The component rewritings of ψ are then the following:

1. $\psi_{x:Rt_1}^{\langle id \rangle} = \{x, Rt_1[y] \mid A(x), C(y), y.f.g.h = x\}$;
2. $\psi_{x:Rt_2}^{\langle id \rangle} = \{x, Rt_2[y] \mid A(x), D(y), y.f.g.h = x\}$;
3. $\psi_{x:Rt_1}^{\langle h \rangle} = \bigvee_{A' : \mathcal{T} \models A' \sqsubseteq \forall h..A} \{x, Rt_1[y] \mid A'(x), C(y), y.f.g = x, \neg \exists z(x.h = z)\}$
4. $\psi_{x:Rt_2}^{\langle h \rangle} = \bigvee_{A' : \mathcal{T} \models A' \sqsubseteq \forall h..A} \{x, Rt_2[y] \mid A'(x), D(y), y.f.g = x, \neg \exists z(x.h = z)\}$
5. $\psi_{x:Rt_1}^{\langle gh \rangle} = \bigvee_{A' : \mathcal{T} \models A' \sqsubseteq \forall g.h.A} \{x, Rt_1[y], \mid A'(x), C(y), y.f = x, \neg \exists z(x.g = z)\}$
6. $\psi_{x:Rt_2}^{\langle gh \rangle} = \bigvee_{A' : \mathcal{T} \models A' \sqsubseteq \forall g.h.A} \{x, Rt_2[y], \mid A'(x), D(y), y.f = x, \neg \exists z(x.g = z)\}$
7. $\psi_{x:Rt_1}^{\langle fgh \rangle} = \bigvee_{A' : \mathcal{T} \models A' \sqsubseteq \forall f.g.h.A} \{x, Rt_1[y] \mid A'(x), C(y), y = x, \neg \exists z(x.f = z)\}$
8. $\psi_{x:Rt_2}^{\langle fgh \rangle} = \bigvee_{A' : \mathcal{T} \models A' \sqsubseteq \forall f.g.h.A} \{x, Rt_2[y] \mid A'(x), D(y), y = x, \neg \exists z(x.f = z)\}$

where valuations map y to the constant value needed to construct the actual referring expression and x to a constant that uniquely determines the actual retrieved object. Cases (1) and (2) address cases in which the entire referring expression can be found in an ABox, cases (3) and (4) address the case in which the last feature of the referring expression, h, is implied by $\mathcal{T}$, and so on.

The overall reformulation is then

$$\exists x(\psi_{x:Rt_1}^{\langle id \rangle}(x, y) \vee (\psi_{x:Rt_2}^{\langle id \rangle}(x, y) \wedge \neg \exists y . \psi_{x:Rt_1}^{\langle id \rangle}(x, y))) \vee$$
$$\exists x(\psi_{x:Rt_1}^{\langle h \rangle}(x, y) \vee (\psi_{x:Rt_2}^{\langle h \rangle}(x, y) \wedge \neg \exists y . \psi_{x:Rt_1}^{\langle h \rangle}(x, y))) \vee$$
$$\exists x(\psi_{x:Rt_1}^{\langle gh \rangle}(x, y) \vee (\psi_{x:Rt_2}^{\langle gh \rangle}(x, y) \wedge \neg \exists y . \psi_{x:Rt_1}^{\langle gh \rangle}(x, y))) \vee$$
$$\exists x(\psi_{x:Rt_1}^{\langle fgh \rangle}(x, y) \vee (\psi_{x:Rt_2}^{\langle fgh \rangle}(x, y) \wedge \neg \exists y . \psi_{x:Rt_1}^{\langle fgh \rangle}(x, y))).$$

Note that, in the second part, the reformulation selects the most preferred referring expression for each answer to the query. □

---

[3] A type assignment is weakly identifying when an ontology $\mathcal{T}$ ensures all referring expressions E occurring in certain R-answers are interpreted as singleton sets; see [1] for the full development.

To establish correctness of query reformulation in Definition 8, we first show that each of the components generated in the first phase of the reformulation produces certain R-answers.

Lemma 10. A k tuple of referring expressions $a_1, \ldots, a_k$ is a certain R-answer to ψ over $\mathcal{K} = (\mathcal{T}, \mathcal{A})$ that conforms to an induced homogeneous head H if and only if

$$(c_1, \ldots, c_k, a_1, \ldots, a_k) \in \psi_H^S(\text{CLOSE}_T(\mathcal{A}))$$

for some constants $c_1, \ldots, c_k$ and a tuple of suffixes S in $\mathsf{S}(H)$.

To *disambiguate* among the R-answers that refer to the *same* certain answer based on our preferences expressed in the head of the query, we use the following lemma to justify the second part of the reformulation.

Lemma 11

1. If $(c_1, \ldots, c_k, a_1, \ldots, a_k) \in \psi_H^S(\text{CLOSE}_T(\mathcal{A}))$ and $(c_1, \ldots, c_k, b_1, \ldots, b_k) \in \psi_{H'}^S(\text{CLOSE}_T(\mathcal{A}))$, then the referring expressions $(a_1, \ldots, a_k)$ and $(b_1, \ldots, b_k)$ refer to the same certain answer to ψ.
2. If $(c_1, \ldots, c_k, a_1, \ldots, a_k) \in \psi_H^S(\text{CLOSE}_T(\mathcal{A}))$ and $(d_1, \ldots, d_k, b_1, \ldots, b_k) \in \psi_{H'}^{S'}(\text{CLOSE}_T(\mathcal{A}))$ and $S \neq S'$ or $c_i \neq d_i$ for some $0 < i \leq k$, then the referring expressions $(a_1, \ldots, a_k)$ and $(b_1, \ldots, b_k)$ refer to distinct certain answers to ψ.

Analogous results can be shown for *prefix heads* (i.e., cases where the heads are of the form $T \rightarrow (Pf . Pf^1 = ? \sqcap \cdots \sqcap Pf . Pf^m = ?)$. Finally, for homogeneous heads that are neither prefix nor suffix (for a particular query variable), it is easy to see that such a variable must be instantiated by an $\mathcal{A}$ object. Altogether the Definitions and Lemmas yield Theorem 1.

4   Summary

We have shown how the computation of certain answers for conjunctive queries over logics with the tree model property, and for which REF and CLOSE functions to implement OBDA have been developed, can be lifted to the computation of certain R-answers for typed conjunctive queries. This yields an important alternative to the computation of such answers presented in [1] in which the ABox completion is inherently query dependent. In doing so, we have extended the definition of the referring expression type language in [1] with an additional construct for expressing more general terms as referring expressions.

There are a number of avenues for further work. To begin, further extensions to our Rt referring expression type language is also desirable. Indeed, the richer the language, the greater the opportunities for finding additional query answers. Also, extensions to logics that do not enjoy the tree model property but could otherwise be limited in some fashion to maintain decidability and even low complexity for certain R-answer computation also merit consideration.

References

1. Borgida, A., Toman, D., Weddell, G.: On referring expressions in query answering over first order knowledge bases. In: Principles of Knowledge Representation and Reasoning, pp. 319–328 (2016)
2. St. Jacques, J., Toman, D., Weddell, G.E.: Object-relational queries over $\mathcal{CFDI}_{nc}$ knowledge bases: OBDA for the SQL-Literate. In: Proceedings of International Joint Conference on Artificial Intelligence, IJCAI, pp. 1258–1264 (2016)
3. Toman, D., Weddell, G.: On adding inverse features to the description logic $\mathcal{CFD}_{nc}^{\forall}$. In: Pham, D.-N., Park, S.-B. (eds.) PRICAI 2014. LNCS (LNAI), vol. 8862, pp. 587–599. Springer, Heidelberg (2014). doi:10.1007/978-3-319-13560-1_47

Machine Learning and Data Mining

Artificial Prediction Markets for Clustering

Sina Famouri[✉], Sattar Hashemi, and Mohammad Taheri

School of Electrical and Computer Engineering, Shiraz University, Shiraz, Iran
{sina.famouri,mtaheri}@cse.shirazu.ac.ir, s_hashemi@shirazu.ac.ir

Abstract. There exist a lot of clustering algorithms for different purposes. But there is no general algorithm that can work without considering the context. This means clustering is not an application independent
problem. So there is a need for more flexible frameworks to engineer
new clustering algorithms for the problems at hand. One way to do this
is by combining clustering algorithms. This is also called consensus or
ensemble clustering in the literature. This paper presents a framework
based on prediction markets mechanism for online clustering by combining different clustering algorithms. In real world, prediction markets
are used to aggregate wisdom of the crowd for predicting outcome of
events such as presidential election. By using the prediction markets
mechanism and considering clustering algorithms as agents or market
participants, an artificial prediction market is designed. Here clustering
is viewed as a prediction problem. Beside working online, the proposed
method provides flexibility in combining algorithms and also helps in
tracking their performance in the market. Based on this framework an
algorithm for center-based clustering algorithms (like k-means) is proposed. The first set of experiments show the flexibility of the algorithm
on synthetic datasets. The results from the second set of experiments
show that the algorithm also works well on real-world datasets.

1 Introduction

In many fields such as data mining, image segmentation, genetics, marketing and
etc. the structure that lies beneath datasets is very important. Cluster analysis
is used as a technique to figure out this structure. A lot of different clustering
algorithms have been proposed (k-means, Spectral Clustering, DBSCAN, etc.).
However, as far as the authors know, there is no clustering method to extract
the structure of all types of datasets. Moreover, in the real world, datasets come
from different sources with various structures that makes the problem even more
challenging.

An alternative way to tackle machine learning problems is to combine models
that already exist in order to model more complex problems. These methods
have proven to be useful in supervised classification, which provoked the idea
that it can also be applied to clustering algorithms. This approach is called
consensus clustering or ensemble clustering. This method can be challenging for
the following reasons:

© Springer International Publishing AG 2016
B.H. Kang and Q. Bai (Eds.): AI 2016, LNAI 9992, pp. 365–377, 2016.
DOI: 10.1007/978-3-319-50127-7_32

1. Aggregating clusters is not that easy since different clustering algorithms produce different number of clusters with different labels.
2. Tracking the best clustering algorithm is one of the main reasons for consensus clustering, and since there are no labels it should be done based on some objective function.
3. Distribution of the data changes over time and many clustering algorithms can't adapt to these changes.

Prediction markets is a mechanism that can be used to aggregate wisdom of the crowd in the real world. This mechanism can also be used in machine learning (also called artificial prediction markets, machine learning markets). Prediction markets also have the potential to work online. Therefore, designing a prediction mechanism for consensus clustering seems to solve challenges 1 and 3. Since there are no labels in unsupervised learning algorithms, more intelligent market maker and agents are needed to overcome challenge 2.

One should also take into account that clustering is a domain specific task [10]. In other words, there is no application-independent clustering algorithm that solves all kinds of problems. In this paper, instead of crafting a new algorithm from scratch, a flexible framework is purposed to engineer new clustering methods by combining different clustering algorithms. Since the framework is based on prediction markets, despite its modularity, it works in an online setting.

2   Related Work

Several methods have been proposed to combine different clustering methods. In general, these methods are made up of two steps: Generation and Consensus Function [3]. In the first step, some clusterings are generated. The final partitions[1] which is made by aggregating clusters are made by the consensus function.

In [4] some traditional cluster ensembles like CSPA (Cluster-based Similarity Partitioning Algorithm), BCE (Bayesian Clustering Ensemble), MCLA (Meta-CLustering Algorithm) are compared. CSPA is based on a co-association matrix. After building the matrix, a graph partitioning algorithm is used to produce the consensus partition. In MCLA each base partition is considered as a vertex. Then by linking clusters with different partitions a meta-graph is generated. The edges are weighted based on the similarity of the clusters. Then this meta-graph is partitioned, and the consensus is produced by assigning each object to a meta-cluster. BCE assumes that each data point has a mixed membership of consensus clusters, and tries to learn a distribution over consensus clusters [5,6].

Online Clustering with Experts (OCE) is proposed in [7]. In OCE, a modified version of k-means objective is proposed as an analog to a loss function. The main idea is to extend algorithms used in supervised learning to the unsupervised learning setting in order to track the best algorithms in its expert set. Another feature of OCE is that it gives approximation guarantees with respect to k-means objective.

---

[1] The term cluster and partition is used interchangeably.

Another example of consensus clustering algorithm is Evidence Accumulation Clustering (EAC) [9]. It transforms the cluster ensemble into a pairwise co-association matrix. The algorithm proposed in [8] is based on EAC. It determines the probability that a data point is assigned to a cluster by minimizing an objective function based on Bergman divergence.

3   Prediction Markets

Prediction markets have three main components:

- Contracts
- Market Participants/Agents
- Market Maker

The only tradable goods in this kind of markets are contracts. Contracts are based on the outcome of events. Agents or market participants will buy and sell these contracts in the market. It is shown that prices of these contracts are good estimators for predicting the outcome of the events [11,12]. Unlike stock markets, prediction markets are controlled by a market maker. In this context, the market maker has the responsibility to define contracts, compute prices and reward the agents when the outcome is known.

The reader may have already noticed, replacing the market participants with machine learning agents that have the ability to do cluster analysis, can result in a mechanism for aggregating clusters.

There are two main formulations that use prediction market mechanism in the field of machine learning. One of them is proposed by A. Storkey which is called Machine Learning Markets and the other one is proposed by A. Barbu and N. Lay which is called Artificial Prediction Markets.

In *Machine learning markets* Storkey modeled the prediction market as an optimization problem. Each agent tries to maximize it's expected utility. It is also shown that combinatorial models that already exist can be derived by using various utility functions [1].

Artificial prediction markets also introduce a mathematical model for prediction markets. Both models aim to find the equilibrium prices. These prices can be interpreted as conditional probabilities given some evidence $\mathbf{x}$ [2].

In this context *Artificial Prediction Markets for Classification* is modified to create an ensemble of clustering algorithms. Since it is an unsupervised task the labels are not available. Hence, market maker doesn't have any idea about the ground truth. This may raise some questions. How should the market maker define the contracts if it doesn't have any idea about the outcomes? How should the market maker reward the agents when the ground truth may never be revealed to the market maker? This section focuses on a general framework based on the idea behind prediction markets that answers these questions.

4   General Framework

4.1   Definitions and Notation

As mentioned, there are three main components in designing a prediction market mechanism: Contracts, Agents and Market Maker. Here each component is separately defined for the task of cluster analysis. The mathematical model for prediction market is inspired by Artificial Prediction Markets [2].

Suppose we have a set of data-points X. Each data-point $x^t \in X$ belongs to a cluster $c_k \in C$. Where $t = \{1, 2, ..., n\}$ and $k = \{1, 2, ..., K\}$ are indices.

Contracts are defined based on clusters in C. So if the data is partitioned into K clusters then K contracts are defined accordingly. For each x^t each contract has a price p_k. What the price actually estimates is:

$$p_k = Pr(c_k|x^t) \tag{1}$$

which is the probability that x^t belongs to the cluster associated with c_k. Note that the above equation results in:

$$\sum_{k=1}^{K} p_k = 1 \tag{2}$$

Agents or Market Participants buy and sell contracts based on a buying function $\phi(.)$, observation x^t and budget β_i. So each agent A_i is defined by a tuple $(\beta_i, \phi_i(x^t, p))$. The buying function specifies what percentage of β_i should be invested in each contract. For instance a buying function can have the form:

$$\phi_i^k(x, p) = \phi_i^k(x) = \gamma e^{-L_i^k(x)} \tag{3}$$

A buying function that doesn't depend on the price is called a static buying function. $L_i^k(x)$ can be considered as a loss function that measures how much x diverges from contract k based on what A_i learned from the data so far. If $L_i^k(x) = 0$ then A_i will risk all of its budget on one instance. So $\gamma \in (0, 1)$ is added to prevent bankruptcy or in other words A_i won't risk all of it's budget on one instance.

Market Maker has the responsibility to create contracts, update prices and evaluate agents. As mentioned before, in an unsupervised task like clustering, ground truth doesn't exist. So the market maker has to analyze the data and the results provided by agents. This means it has to be more intelligent.

The market maker can be fully automatic which means the machine does everything itself and there is no need for human interaction. But when it comes to clustering not all problems are purely unsupervised. Usually there exists some level of insight over the problem. So one can design and modify a market maker, that works better for a specific problem by using the knowledge available about that problem.

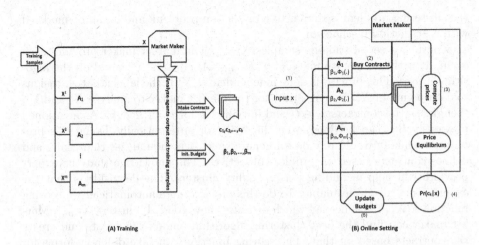

Fig. 1. (A) Training - The market maker provides some training samples for agents. Then the samples and clustering results of each agent is analyzed in order to define contracts and initialize the agents budget. **(B)** The Online Setting - This figure depicts how the agents and the market maker interact in the market: (1) The input is given to agents. (2) agents buy contracts based on their beliefs. (3) market maker computes the equilibrium price. (4) Prices estimate the probability that x belongs contract c_k. (5) The market maker updates the agents budget based a similarity or distance measure.

In both scenarios, some samples are needed to analyze the problem and agents clustering quality. These samples can be considered as the training data. The market maker can have scoring function $S_i(.)$ that evaluates clustering quality of A_i (Note that $S_i(.)$ can be any intrinsic or extrinsic scoring function like Silhouettes, purity and etc.). For example, a fully automatic market maker can select A_j with the highest score. Then define the contracts based on the output of A_j. Now consider another situation, one knows that there exist k categories in a specific problem. So the market maker can be designed in a way to split the data into k partitions by defining k contracts accordingly. An advantage of the first example is that, there is no need for an expert or further knowledge about the data. The machine does everything itself. But one should also consider that with more knowledge the algorithm can be modified to better fit the problem.

4.2   The Framework

Regardless of the problem being online or not, the dataset should be divided into two parts. The first part is used for training, which will be an initialization for the next part. The rest of the data-points are analyzed in an online setting. In other words, first the agents and market maker will get a grasp on how the dataset looks like and how many partitions it should be divided to. Then for every new point their task is to predict what category it belongs to and based on that prediction agents will invest on the related contract. This way, the clustering problem is similar to a prediction problem, and it can be modeled with a prediction mechanism.

Consider a set of training samples X_{train}. The market maker provides each agent A_i with a set of samples $X^i \subseteq X_{train}$. A_i clusters X^i based on the algorithm it uses. This process is done independently. Y^i is the labels that A_i assigns to data-points in X^i. Now it's time for the market maker to analyze X^i and Y^i, then define the contracts $c_k \in C$ and initialize β_i for $i = 1, 2, .., m$. As mentioned previously this can be done either manually or automatically. Doing this procedure manually would rely on a human expert, which analyzes the results and defines contracts based on problem objectives. The knowledge that the expert provides will help in finding better quality clusters. The downside is that the expert is not always available. To do this process more automatically, a scoring function $S(X^i, Y^i)$ is needed which measures how good A_i clusters X_{train}. Market maker can find the best clustering algorithm among the agents and create the contracts based on that. The scoring function can take different forms for different kinds of agents. Here for simplification only one scoring function is used for all of them, so one could easily initialize the budget this way:

$$\beta_i = \frac{S(X^i, Y^i)}{\sum_{j=1}^{m} S(X^j, Y^j)} \times B \tag{4}$$

where B is the total budget in the market and $S(X^i, Y^i) > 0$. For example, one can use Silhouette score incremented by 1 to find the best agent and best number of clusters. Now the market is established and ready to use in an online environment.

As you can see in Fig. 2 the framework for online clustering generally looks like the one used in artificial prediction markets for classification. The budget update formula and the market equation also looks the same. So:

$$\beta_i^{t+1} = \beta_i^t - \sum_{k=1}^{K} \beta_i^t \phi_i^k(x^t, p) + \frac{\beta_i^t \phi_i^y(x^t, p)}{p_y} \tag{5}$$

Where β_i^t shows the budget for agent i after investing on t new samples. Since $\sum_{i=1}^{m} \beta_i^t = \sum_{i=1}^{m} \beta_i^{t+1} = B$:

$$\sum_{i=1}^{m} \sum_{k=1}^{K} \beta_i^t \phi_i^k(x^t, p) = \sum_{i=1}^{m} \frac{\beta_i^t \phi_i^k(x^t, p)}{p_k} \tag{6}$$

Solving the above equation results in finding the equilibrium price. By considering a static buying function like (2) for $p_k > 0$:

$$p_k = \frac{\sum_{i=1}^{m} \beta_i^t \phi_i^k(x^t)}{\sum_{i=1}^{m} \sum_{k=1}^{K} \beta_i^t \phi_i^k(x^t)} \tag{7}$$

The only thing left, is that how should the market maker reward the agents without knowing the outcome. In clustering, data points are partitioned into a number of groups based on their similarity. This means the data points should be assigned to the cluster with minimum distance. The distance or similarity measure may change based on the nature of the problem. For simplicity from now on consider all distances are computed in Euclidean space. Suppose that we can compute the distance between clusters of A_i and a data point x with the following function:

$$d_i^k = \min f(x, b_i^k) \tag{8}$$

Where for each A_i, b_i^k is the subset of clusters related to contract c_k. How they are related depends on the Market Maker design. So d_i^k is the minimum distance between x and clusters in b_i^k. The winner contract is the one with minimum average distance:

$$l(k) = \frac{1}{m} \sum_{i=1}^{m} d_i^k \tag{9}$$

To find the winner contract c_{k^*}:

$$k^* = \operatorname*{argmin}_{k} l(k) \tag{10}$$

This is a general framework that one can use to create a combinatorial model for clustering. The focus of the next section is mainly on implementing an algorithm based on this framework.

5   Algorithm Design

In this section a new ensemble clustering algorithm is designed by combining centroid-based algorithms like k-means++ [15]. The training algorithm is given in Algorithm 1. Consider we have a set of agents $A = \{A_i\}$, $i = 1, ..., m$ and each A_i has a centroid-based clustering algorithm associated with it. The number of clusters and associated parameters can be different among agents. The agents will be trained independently by the training data provided for them then each agent will be associated with a set of cluster centers according to the number of clusters it has found. Now the market maker can initialize their budget by using Eq. (4). The scoring function can be considered Silhouette score incremented by 1 to avoid negative values [16].

The market maker also has a clustering algorithm which doesn't have to be centroid based like the agents. It gathers all the centers and partitions them into K clusters. K contracts are defined respectively. After this step, a mapping between centers and contracts is made. To fine tune the centers and the budgets the market will re-sample from training data I times and predicts the clusters for each x in the re-sampled dataset. This final step also makes the algorithm more robust since Algorithm 2 is online and may be affected by the order of the instances.

Algorithm 1. Market Train

1 $X_{train} = \{x_1, x_2, ..., x_n\}$ is the set of data-points in an Euclidean space in $\mathbb{R}^d$
   given as input
2 $A = \{A_1, A_2, ..., A_m\}$ is a set of clustering agents
3 Re-sample $X^i \subseteq X^{train}, i = 1, 2, ..., m$ for training A_i
4 $centers = \{\}$
5 Initialize I
   // Agents cluster data-points associated with them independently
6 **forall** A_i and X^i **do**
7   Run clustering algorithm associated with A_i
8      $Y^i \leftarrow$ set of labels associated with X^i
9   Append the centers b_i to $centers$
10 **end**
   // The following tasks are done by market maker
11 **forall** A_i and X^i and Y^i **do** Perform budget initialization
12 Make contracts by clustering $centers$
13 $b_i^k \leftarrow$ centers in b_i which is assigned to cluster k
14 **for** j in range 1 to I **do**
15   Re-sample X^j
16   **forall** x in X^j **do** Run Market
17 **end**

As a data-point x gets closer to j^{th} center of A_i agents belief on assigning that point to cluster j becomes stronger. In other words A_i is more likely to invest on contract associated with cluster j. So consider $L_i^k(x)$ in (3):

$$L_i^k(x) = \min \frac{\|x - b_i^k\|^2}{R} \qquad (11)$$

where R is a constant scalar. Since each cluster is represented by its center d_i^k is computed with the following formula:

$$d_i^k = \min \|x - b_i^k\| \qquad (12)$$

And the winner contract is selected with (10).

After the winner contract is known its time to update agents. Suppose $b_{ij}^{k^*}$ denotes the jth center of A_i which is involved in investing on the winner contract. Agent update is done by updating $b_{ij}^{k^*}$:

$$b_{ij}^{k^*} \leftarrow \eta x + (1 - \eta)b_{ij}^{k^*} \qquad (13)$$

Budget update is done by formula (5) and prices are computed using (7).

Algorithm 2. Run Market

    input : A data point $x \in \mathbb{R}^d$

    output: Prices of contracts $\{p_1, p_2, ..., p_K\}$

1 Agents buy contracts based on their buying functions
2 Market maker aggregates investments
3 Market maker computes equilibrium price
4 Update agents
5 Update budgets

6 Experimental Results

Two sets of experiments are devised. In the first set which is done on random synthetic datasets, the goal is to show how the algorithm can be modified to fit a specific problem. The second set of experiments is to see how the algorithm works on real world datasets. It should be noted that base clustering algorithms and random data generation is done by using Scikit-Learn package [17]. As the reader may have noticed, our view of the problem differs from similar approaches in the literature, so the algorithms are tested against true labels.

6.1 Case Study I - Synthetic Data

Three kinds of random datasets are generated. They are shown in Fig. 2. The settings for the algorithms are depicted in Table 1. A, B and C represent datasets shown in the first row of Fig. 2. MM is the market maker and A_is are agents. The numbers in the table show the number of partitions for MM and each A_i.

Table 1. Number of clusters for market maker and agents in case study I

Data	MM	A_1	A_2	A_3	A_4	A_5
A	3	3	6	9	12	15
B, C	2	40	45	50	55	60

A, B and C each have 600 instances. 200 samples are randomly selected for training and the rest is for test. Second row of Fig. 2 shows the partitions of the test-set when MM and all A_is use k-means++. As you can see A is perfectly partitioned but B and C are not. By changing the MM's algorithm to Spectral clustering instead of k-means++, B and C will also be perfectly partitioned as shown in Fig. 3.

For interpreting the results, consider B and C. Increasing the number of clusters in k-means results in extracting more detailed structures. But the problem is that the actual number of clusters are far less than that. The trick here is to somehow relate the centers provided by the agents and create the contracts

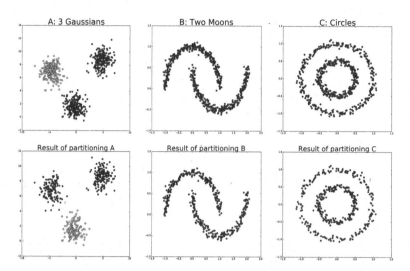

Fig. 2. Illustration of three synthetic data sets used for case study 1. The first row shows the whole dataset with correct partitions. The second row is the result of applying algorithm (with k-means++ for MM) to test-sets of A, B and C.

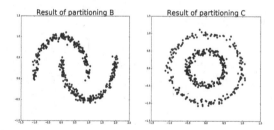

Fig. 3. Using spectral clustering instead of k-means++ for MM and running the algorithm again will result in perfect clustering for test-sets of B and C.

based on that. MM does this by clustering those centers itself. In situations like B and C a center-based clustering algorithm like k-means is not suitable for the task. On the other hand Spectral clustering uses the spectrum of Laplacian matrix to partition the similarity graph, constructed from data [19]. So clearly Spectral clustering works better than k-means for B and C.

As you can see the framework provides flexibility to configure the algorithm in order to fit the problem. Besides flexibility, it works online. The assignment of each point to each cluster is done by the prices of contracts. The price of contract c_k for input x can be interpreted as the probability that point x is assigned to cluster c_k. The probabilistic nature of this assignment will help in specifying overlap of clusters.

6.2   Case Study II - Experiments on Real-World Datasets

In this section the algorithm is tested on five different datasets. For each dataset the algorithm is tuned. The characteristics of datasets and the algorithm used to cluster them are shown in Table 2.

The first column of Table 2 shows the experiment numbers. The MM column shows the algorithm for market maker. K stands for k-means++ and S stands for Spectral clustering. The number of clusters is also written near the algorithm. In experiments 1 and 5, 1/3 of instances are used for training and the rest is used for test. Experiment 2, 3 and 4 had very small number of instances so 2/3 of the instances are used for training and the rest is used for test. Test and train split is done randomly. Each experiment is repeated 20 times.

After running the algorithm the results were compared with true labels. Table 3 contains average, maximum and minimum homogeneity score of experiments [18]. To satisfy homogeneity, datapoints in a single cluster should be members of a single class. The score falls in range [0, 1], where 1 means the clusters are perfectly homogeneous. The last column shows the homogeneity score of only using one k-means++ with number of clusters equal to the number of classes.

The small number of instances in experiments 2, 3 and 4 causes the score of the algorithm to be close to a single agent with k-means++ and k equal to number of classes. But as the dataset grows larger like experiment 1 and 5, the improvement is more observable (2.4% for 1 and 13.9% for 5).

Table 2. Datasets, number of instances and market setting for case study II

Exp.#	Data	# Instances	MM	Agents
1	Breast cancer wisconsin	699	K, 2	$[3, 4, 5, 6, 7]$
2	Wine	178	K, 2	$[3, 4, 5, 6, 7]$
3	Iris	150	K, 3	$[3, 6, 9, 12]$
4	Glass	214	S, 2	$[20, 30, 40]$
5	Digits	1797	S, 10	$[40, 50, 60]$

Table 3. Homogeneity score of the market and single k-means tested on random splits of data

Exp.#	Avg	Max	Min	Var	k-means avg
1	0.765	0.813	0.701	0.8×10^{-3}	0.741
2	0.854	1.0	0.735	0.6×10^{-2}	0.849
3	0.798	0.926	0.698	0.3×10^{-2}	0.780
4	0.400	0.536	0.280	0.3×10^{-2}	0.396
5	0.699	0.738	0.640	0.8×10^{-3}	0.560

The experiments show that the aggregation of the clusters in an online setting provides a better average score than one single k-means++.

7   Conclusion and Future Work

The purpose of this research was to show how to use prediction market mechanism in the literature of clustering. The mechanism provides the flexibility to combine different clustering algorithms in order to reach higher quality clusters based on the objectives of the problem. A new and simple algorithm based on the general framework, discussed in Sect. 4, is proposed to show how the framework works.

The experimental results in Sect. 6.2, show aggregating clusters, provided by agents in the market, work better than single k-means. The improvement is more obvious in bigger datasets because the algorithm works online and adapts it self as more samples are given to the market.

For future work one can design a market with agents capable of using different kinds of clustering algorithms, not only center based ones. This will lead to more variety in the structure of clusters extracted by agents. The agents can exchange information instead of working independently. For more illustration, consider an agent that relies on the opinion of other agents to invest in a specific contract. Designing a more intelligent market maker can also help in defining contracts and higher quality clusters. The buying function can also be modified for better results based on application. As discussed in Sect. 6.1, the algorithm can be easily modified to extract more complicated structures. So one can put the framework into use for different applications and create new algorithms.

References

1. Storkey, A.: Machine learning markets. In: International Conference on Artificial Intelligence and Statistics (2011)
2. Barbu, A., Lay, N.: An introduction to artificial prediction markets for classification. J. Mach. Learn. Res. **13**(1), 2177–2204 (2012)
3. Vega-Pons, S., Ruiz-Shulcloper, J.: A survey of clustering ensemble algorithms. Int. J. Pattern Recogn. Artif. Intell. **25**(03), 337–372 (2011)
4. Piantoni, J., Faceli, K., Sakata, T.C., Pereira, J.C., Souto, M.C.P.: Impact of base partitions on multi-objective and traditional ensemble clustering algorithms. In: Arik, S., Huang, T., Lai, W.K., Liu, Q. (eds.) ICONIP 2015. LNCS, vol. 9489, pp. 696–704. Springer, Heidelberg (2015). doi:10.1007/978-3-319-26532-2_77
5. Wang, H., Shan, H., Banerjee, A.: Bayesian cluster ensembles. Stat. Anal. Data Min. **4**(1), 54–70 (2011)
6. Strehl, A., Ghosh, J.: Cluster ensembles–a knowledge reuse framework for combining multiple partitions. J. Mach. Learn. Res. **3**, 583–617 (2003)
7. Choromanska, A., Monteleoni, C.: Online clustering with experts. In: International Conference on Artificial Intelligence and Statistics, pp. 227–235 (2012)
8. Louren, A., et al.: Probabilistic consensus clustering using evidence accumulation. Mach. Learn. **98**(1–2), 331–357 (2015)

9. Fred, A.L.N., Jain, A.K.: Data clustering using evidence accumulation. In: Proceedings of 16th International Conference on Pattern Recognition, vol. 4. IEEE (2002)
10. Von Luxburg, U., Williamson, R.C., Guyon, I.: Clustering: science or art? In: ICML Unsupervised and Transfer Learning, pp. 65–80 (2012)
11. Gjerstad, S., Hall, M.: Risk aversion, beliefs, and prediction market equilibrium. Economic Science Laboratory, University of Arizona (2005)
12. Manski, C.F.: Interpreting the predictions of prediction markets. Econ. Lett. **91**(3), 425–429 (2006)
13. Chen, Y., Pennock, D.M.: Designing markets for prediction. AI Mag. **31**(4), 42–52 (2010)
14. Hanson, R.: Logarithmic markets coring rules for modular combinatorial information aggregation. J. Predict. Mark. **1**(1), 3–15 (2012)
15. Arthur, D., Vassilvitskii, S.: k-means++: the advantages of careful seeding. In: Proceedings of the Eighteenth Annual ACM-SIAM Symposium on Discrete Algorithms. Society for Industrial and Applied Mathematics (2007)
16. Rousseeuw, P.J.: Silhouettes: a graphical aid to the interpretation and validation of cluster analysis. J. Comput. Appl. Math. **20**, 53–65 (1987)
17. Pedregosa, F., et al.: Scikit-learn: machine learning in Python. J. Mach. Learn. Res. **12**, 2825–2830 (2011)
18. Rosenberg, A., Hirschberg, J.: V-measure: a conditional entropy-based external cluster evaluation measure. In: EMNLP-CoNLL, vol. 7 (2007)
19. Von Luxburg, U.: A tutorial on spectral clustering. Stat. Comput. **17**(4), 395–416 (2007)

Transfer Learning in Probabilistic Logic Models

Pouya Ghiasnezhad Omran$^{(\boxtimes)}$, Kewen Wang, and Zhe Wang

Giriffith University, Brisbane, Australia
pouya.ghiasnezhadomran@griffithuni.edu.au,
{k.wang,zhe.wang}@griffith.edu.au

Abstract. Several approaches to learning probabilistic logic programs have been proposed in the literature. However, most learning systems based on these approaches are not efficient for handling large practical problems (especially, in the case of structure learning). It has been a challenging issue to reduce the search space of candidate (probabilistic) logic programs. There is no exception for SLIPCOVER, a latest system for both parameter and structure learning of Logic Programs with Annotated Disjunction (LPADs). This paper presents a new algorithm T-LPAD for structure learning of LPADs by employing transfer learning. The new algorithm has been implemented and our experimental results show that T-LPAD outperforms SLIPCOVER (and SLIPCASE) for most benchmarks used in related systems.

Keywords: Transfer learning · Probabilistic logic programs

1 Introduction

In logic programming, researchers have realized the importance of modeling uncertainty for a long time and thus various types of probabilistic logic programs have been proposed and studied. Consequently, significant attention has been paid to the issue of learning probabilistic logic programs. Recently, several approaches have been proposed for parameter learning (i.e., for a candidate logic program, the task is to determine probability degrees for certain objects such as rules and/or atoms). For instance, the PRISM [13] is well known for its distribution semantics. LeProbLog [5] is based the technique of gradient descent while LFI-ProbLog [6] and EMBLEM [2] adopt an Expectation Maximization approach in which the expectations are computed directly using binary decision diagrams (BDDs). In many realistic applications, we need also to learn candidate logic programs as well as learning parameters. This paradigm of probabilistic logic program learning is referred to as structure learning. Obviously, this task is much more difficult than parameter learning only. ProbLog is proposed by De Raedt et al. [4] for learning the structure of (probabilistic) logic programs while SEM-CP-logic is developed for learning ground LPAD programs. More recently, based on a new beam search, SLIPCASE [10] is developed for learning LPAD programs. SLIPCASE can learn general LPADs including non-ground programs. An improved version of SLIPCASE, named SLIPCOVER, is described in [1].

© Springer International Publishing AG 2016
B.H. Kang and Q. Bai (Eds.): AI 2016, LNAI 9992, pp. 378–389, 2016.
DOI: 10.1007/978-3-319-50127-7_33

In learning algorithms for both SLIPCASE and SLIPCOVER, beam search is performed in the space of LPADs using the log likelihood of training data as the guiding heuristics and theory refinements are achieved using EMBLEM. However, there is still significant room for improving the efficiency of these algorithms. Especially, in these algorithms, the search space of LPADs is still very large and thus it would be useful to reduce the search space for structure learning algorithms using information from another application domain that shares a certain similarity with the domain of interest.

Human beings can make analogy across different domains by determining the structural similarities even in seemingly irrelevant domains, for example, we can easily understand the analogy between the domain of movie information and the domain of academic information. Even though the movie domain has nothing in common with the academic domain, we can still make the analogy based on certain similarity. For example, the predicate "movie(title, person)" with two arguments of types "title" and "person" is similar to the predicate "publication(person, title)" in the academic domain; the predicate "director(person)" is similar to "professor(person)"; the predicate "actor(person)" is similar to "student(person)".

In fact, reusing knowledge across different domains for learning has been actively pursued in machine learning communities [3,12] and is usually referred to as *transfer learning*. It aims to learn a more accurate set of rules by using additional data from a source domain with less training data from target domain and less running time.

In this paper, we apply transfer learning in learning LPADs and describe a new algorithm T-LPAD for structure learning of LPADs (here 'T' in T-LPAD is for 'Transfer learning'). Specifically, suppose that our task is to produce a set of LPAD rules for a domain of interest (i.e., target domain) based on a given training (relational) dataset. At the same time, we are given a set of LPAD rules in another domain (i.e., source domain) that is independent of the target domain but shares a kind of structural similarity with the target domain. By employing a technique from ontology matching [7], we are able to figure out a similarity degree for a pair of predicates in the source domain and the target domain. Using these similarity degrees, we construct rules in the target domain based on those rules in the source domain. This will allow us to narrow down the search space of constructing candidate rules in structure learning of LPADs. Note that these rules are standard (disjunctive) rules but not LPAD rules since no probability degrees are assigned to head atoms. For example, from the academic domain, we can produce a rule $director(person) \vee starring(person) \leftarrow movie(title, person), bigname(person)$. In the next stage, we figure out the best probability for each predicate in the rule head for each disjunctive rule using existing parameter learning algorithms such as EMBLEM. For instance, we can come up with an LPAD rule like $(director(person) : 0.4) \vee (starring(person), 0.6) \leftarrow movie(title, person), bigname(person)$.

We have developed a prototype implementation for T-LPAD algorithm and conducted experiments on four benchmarks **IMDB**, **UW-CSE**, **WebKB**, and

Twitter, which are widely used and publicly available [1,14]. The experimental results show that T-LPAD outperforms SLIPCOVER (and SLIPCASE), two latest algorithms for structure learning of LPADs.

In the rest of the paper, we briefly recall basics of probabilistic logic programs and structure learning of such logic programs in Sect. 2, our new learning algorithm T-LPAD is described through a running example in Sect. 3, some experimental results are reported in Sect. 4, the related works are reviewed briefly in Sect. 5, and finally we conclude the work in Sect. 6.

2   Learning with Probabilistic Logic Programs

In this section, we review some basics of learning probabilistic logic programs, especially, structure learning of logic programs with annotated disjunctions (LPADs).

LPADs allow disjunction in the heads of program rules to express probabilistic multiple choices [15]. Similar to other classes of (probabilistic) logic programs, the fragment of LPAD logic is based on three types of symbols: constants, variables, and predicates. Constants are names for specific objects in the domain, variables range over objects in the domain, and predicates represent relations or features among objects. We use c (possibly with subscripts) or words starting with a lower case symbol for constants; x, y and z or with subscripts for variables; P or words starting with a capital symbol for predicates. Variables and constants are typed. An atom is of the form $P(t_1, \ldots, t_n)$ where P is a predicate and each s_i is a constant or variable $(1 \leq i \leq n)$. $P(t_1, \ldots, t_n)$ is a ground atom if every term t_i is a constant. A probabilistic atom is a pair (a, p) where a is an atom and p is a probability degree $(0 \leq p \leq 1)$. (a, p) is to represent that the atom a is true with the probability p. A literal is an atom or its negation.

Formally, an *LPAD rule* is of the form

$$(a_1 : p_1); \cdots ; (a_m : p_m) \leftarrow b_1, \ldots, b_n$$

where $(a_1 : p_1), \ldots, (a_m : p_m)$ are probabilistic atoms, $b_1, \ldots, b_n$ are literals. ';' is for disjunction. Informally, the above rule reads that if $b_1, \ldots, b_n$ are true, then at least one of $(a_1 : p_1), \ldots, (a_m : p_m)$ is true.

The semantics of LPADs is defined by the distribution semantics, which is first introduced for PRISM [13]. In the distribution semantics for LPADs, each world is established by choosing one atom from the head of grounding form of an LPAD rule, and the probability p_i of head atom (a_i, p_i) is computed by accumulating the probability of worlds whose model evaluates a_i as true. We refer the reader to [11] for further technical details.

Parameter learning for LPAD is carried out by the EMBLEM [2], which performs the Expectation Maximization (EM) over BDDs. A typical input for EMBLEM will be a set of ground atoms, a set of LPAD rules, and a set of so-called goal predicates. For each goal predicate, EMBLEM creates a BDD encoding its explanations and starts the EM cycle, in which the steps of Expectation and Maximization are continued until the log likelihood of the examples achieves a local maximum.

For structure learning of probabilistic logic programs, the learner needs to construct a logic program as well as determining relevant parameters. For instance, SLIPCOVER learns an LPAD program by first searching for promising ones from the space of disjunctive rules, looking for good refinements in terms of LL (likelihood) of the data, and finally performing EMBLEM on the best target disjunctive program. SLIPCOVER is able to learn general LPADs including non-grounded programs.

3   T-LPAD

In this section, we present a new approach to learning LPAD rules through transferring knowledge across domains. We will first provide a sketch of our learning algorithm T-LPAD and then explain further technical details in the subsections through a running example.

Suppose that we have two application domains in hand. One is the source domain, which is relatively well understood, and the other is the target domain in which we want to learn new knowledge in the form of LPAD rules. More formally, the source domain has a set of relational data and a set of LPAD rules that have previously obtained. The target domain has only a set of relational data but does not have any rules. The transfer learning task for the given domains is that, given a predicate in the target domain, we construct a set of LPAD rules for the target domain that are compatible with the (training) data in the target domain.

In our T-LPAD algorithm, for each domain we first construct a language bias called *predicate description (PD)*, which contains information about predicate arity, predicate arguments, and their types. A PD can also be represented as a matrix (PD matrix). Based on two PD matrices for the source domain and the target domain, for each target predicate P, we are able to determine a set $S(P)$ of source predicates that are structurally similar to the given target predicate. In this way, we can construct a set of LPAD rules for the target domain from an LPAD rules in the source domain, by replacing each source predicate with a target predicate that is structurally similar to it.

In the next two subsections, we will explain our method using the following example.

Example 1 (Running Example). The source domain and the target domain are Academic and Movie, respectively. The academic covers knowledge about people in an academy department (i.e. students and professors) and their relationships. The movie domain contains information about movies, their directors, and the actors of the movies. The predicate descriptions for a Academic domain and the Movie domain are specified as follows.

Academic PD: $advisedby(person, person)$, $professor(person)$, $student(person)$, $publication(person, title)$, $position(person, +pos)$

Movie PD: $workunder(person, person)$, $actor(person)$, $movie(title, person)$, $director(person)$

Academic domain contains one rule $professor(x) : 0.2 \leftarrow publication(x,y)$.

Based on the above information, we want to learn rules for Movie about *director* (i.e., rules with *director* in the head).

3.1   PD Graph and Similarity Matrix

To extract similarity between the predicates in the source domain and those in the target domain, we make use of the PDs, which describes the predicates in the domain, their arities, the types of their arguments, and how each argument appears in rules (i.e., as variables or constants).

The predicate description for a domain can be conveniently represented as a graph called the *PD graph* for the domain.

Definition 1. *Given a domain D, its PD graph, denoted G_D, is defined as follows:*

1. *The vertex of G_D is either a predicate P/m with m denoting its arity, or a type T or $+T$ with $+$ denoting it is a constant type.*
2. *If the k-th argument of a predicate P is the type T, then there is a directed edge labelled $\#k$ from vertex P/m to vertex T.*

The PD graphs for those two domains in Example 1 are shown in Fig. 1.

Fig. 1. PD graphs for two domains UWC-SE and IMDB.

We want to construct a similarity matrix using PD graphs, in which each pair of source and target predicates is assigned a similarity degree as a real number. To achieve this, we adapt existing graph matching techniques which can extract similarity degree between the vertices of two graphs based on their structural similarity. Our PD graphs can be conveniently represented as RDF triples [7], where an RDF triple represents one edge or a labelled vertex of a PD graph.

Our RDF encoding contains three kinds of triples:

1. The predicate-type triples $(P, \#k, T)$ expressing that the kth argument of predicate P has the type T. Each triple of this type encodes one edge in the PD graph.
2. The predicate triples $(P, n, predicate)$, encoding that vertex P represents a predicate with arity n.
3. The type triples which has form $(T, cons/var, type)$, encoding that vertex T represents a constant or variable type.

Encoding PD graphs as RDF triples allows us to adapt the RDF matcher GMO [7] to extract a graph matching matrix, denoted GM, which can be seen as a function that maps each pair of predicates to a real number. We use GM as a basis to construct the similarity matrix between source and target predicates.

The graph matching matrix GM is completely based on structural similarity, which does not necessarily reflect the different labels in graphs, for example, whether a vertex is a binary predicate or a constant type. Hence, we need to refine GM to reflect the different reasons for two predicates to be considered similar, i.e., due to the same arity or similar argument types. We compute a label matrix LM to capture the similarity due to different labels, by comparing labels using an adapted string matching method [7]. We refine GM by compositing GM and LM linearly with parameter β into a refined matrix RM as the following.

$$RM = \beta * LM + (1 - \beta)GM \tag{1}$$

The refined matrix RM can be divided into two matrices, RM_t and RM_p which present the similarity between types and predicates respectively. To obtain the final similarity matrix SM, we further refine RM_p through RM_t according to the argument types of predicates. For this refinement step, we use the linear combination (2). Again, each matrix can be seen as a function mapping each pair of predicates (or types) to a real number. For a predicate P, T^P is the set of all the types of the arguments of P, and T_i^P is the type of ith argument of P. To define SM, for a pair of predicates P and Q,

$$SM(P,Q) = (1 - \alpha) * RM_p(P,Q) + \alpha * \frac{1}{m} * \sum_n \max_{i,j}(RM_t(T_i^P, T_j^Q)) \tag{2}$$

where $m = \max(|T^P|, |T^Q|)$ and $n = \min(|T^P|, |T^Q|)$. The parameter α determines the degree how much the similarity degrees in RM_p are refine by the similarity between types in RM_t. In particular, the second part of Eq. (2) aggregates the similarities degrees of types for the two predicates P and Q. For example, to compute the similarity degree between two predicates $student(person)$ and $movie(title, person)$, we have $m = 2$ and $n = 1$. Taking $\alpha = 0.5$, suppose $RM_p(student, movie) = 0.3$, $RM_t(person, title) = 0.3$, and $RM_t(person, person) = 1$, we can compute $SM(student, movie) = 0.5 \times 0.3 + 0.5 \times 0.5 \times \sum_1 max(0.3, 1) = 0.4$.

We summarize the computation of the similarity matrix in Algorithm 1.

Example 2. (Cont'd Example 1). Taking as input the PDs of Academy and Movie domains, Algorithm 1 outputs the similarity matrix, $SM_{Academy,Movie}$ as follows:

$$SM_{Academy,Movie} = \begin{array}{r} \\ advised\_by \\ student \\ professor \\ position \\ publication \end{array} \begin{array}{cccc} worked\_under & actor & director & movie \\ \left[\begin{array}{cccc} 0.85 & 0.5 & 0.5 & 0.7 \\ 0.5 & 0.8 & 0.8 & 0.4 \\ 0.5 & 0.8 & 0.8 & 0.4 \\ 0.7 & 0.4 & 0.4 & 0.8 \\ 0.7 & 0.4 & 0.4 & 0.8 \end{array}\right] \end{array}$$

Algorithm 1. Compute the similarity matrix

Input: source PD S and target PD T
Output: similarity matrix SM
1: Encode S and T into RDF documents
2: $GM \leftarrow GMO^*(S,T)$ ▷ GMO^* is our adapted GMO method
3: $LM \leftarrow LabelStringMatch(S,T)$
4: $RM \leftarrow Refine(GM,LM)$ ▷ $Refine$ is defined in Eq. (1)
5: $SM \leftarrow FurtherRefine(RM_p, RM_t)$ ▷ $FurtherRefine$ is defined in Eq. (2)
     return SM

3.2   Rule Construction

In what follows, we show how rules can be constructed for the target domain based on the similarity matrix. In contrast to existing approaches, which transfer second order template, we use rule templates of the form $P_1; \cdots; P_m \leftarrow Q_1, \ldots, Q_n$ where P_i and Q_j are predicates (without arguments). We first show how our method works for rules with a single-atom head, and then extend the approach to rules with multiple-atom heads. For the convenience of discussion, we omit the weight of LPAD rules in our examples.

Let SM be the similarity matrix obtained previously, for each predicate P in the target (or source) domain, let $S(P)$ (resp., $T(P)$) be the set of source (resp., target) predicates that have the highest similarity degree to P in SM. To construct target rules with goal predicate G, our rule construction method consists three steps.

In the first step, for each source predicate $P \in S(G)$ and each rule r of the form $P(t) \leftarrow Q_1(t_1), \ldots, Q_n(t_n)$, in the source domain, we obtain a set $R_0(r)$ of rule templates in the target of the form $G \leftarrow Q'_1, \ldots, Q'_n$, where $Q'_i \in T(Q_i)$. For example, suppose r is $A(x,y) \leftarrow B(y), C(x,z)$, $T(B) = \{B', B''\}$, and $T(C) = \{C'\}$. Then $R_0(r)$ contains two rule templates: $G \leftarrow B', C'$ and $G \leftarrow B'', C'$.

In the second step, we assign variables and constants as arguments to the predicates in $R_0(r)$. If a predicate has an argument type that is a constant type, we construct rules by assigning all possible constants of that type. Assignment of variable is less straightforward. We first assign a distinct variable to each argument which is not a constant, and then unify certain variables based on the variable sharing information in r, which we capture using *variable sharing constraints*. A variable sharing constraint in r is of the form $x(P_{m_1}, \ldots, P_{m_k}, Q_{n_1}, \ldots, Q_{n_l})$, where P_{m_i}'s and Q_{n_i}'s are the head and body predicates that have x as an argument. In the previous example, the variable sharing constraints in r are $x(A,C)$ and $y(A,B)$. Based on variable sharing constraints, we unify the assigned variable according to the following three conditions: (1) Two variable can unify only if they have the same type. (2) If two predicates share a variable in r, the corresponding predicates in the rule template will have one variable unified whenever possible. (3) Unification is performed only when it is required by condition (1). In this way, we obtain a set of rules $R_1(r)$ from $R_0(r)$ through assignment. In the above example, suppose arities of G, B',

B'' and C' are respectively 2, 1, 2 and 1, $R_1(r)$ may contain some of the following rules (regardless of types): $G(x,y) \leftarrow B'(x), C'(x)$, $G(x,y) \leftarrow B'(y), C'(y)$, $G(x,y) \leftarrow B''(x,z), C'(y)$, $G(x,y) \leftarrow B''(y,z), C'(x)$, $G(x,y) \leftarrow B''(z,x), C'(y)$, and $G(x,y) \leftarrow B''(z,y), C'(x)$.

Example 3. (Cont'd Example 2). To learn a rule about *director* in the Movie domain, we obtain from our similarity matrix $S(director) = \{professor, student\}$ and source rule $r : professor(x) : 0.2 \leftarrow publication(x,y)$. Also, we have $T(publication) = \{movie\}$. In this case we get one rule template in $R_0(r)$, that is $director \leftarrow movie$. After assigning arguments based on the single variable sharing constraint in r, that is $x(professor, publication)$, we obtain one rule in $R_1(r)$, that is $director(x) \leftarrow movie(y,x)$.

T-LPAD handles rules with multiple-atom head in a similar way. We illustrate this using an example. Suppose the source rule r is $A(x,y); B(x,y) \leftarrow C(x), D(y)$ and from the similarity matrix we obtain $T(A) = \{A', B'\}$, $T(B) = \{C'\}$, $T(C) = \{D'\}$, and $T(D) = \{E'\}$, then $R_0(r)$ consists of the following two rule templates: $A'; C' \leftarrow D', E'$ and $B'; C' \leftarrow D', E'$. The assignment of argument works exactly as for rules with single-atom heads, and hence is not repeated here.

The rules constructed from the above two steps do not have weights in their head. Hence, in the third step, we generate weights for them. We initialise the weights of head atoms evenly, through dividing 1 by the number of head atoms. For example, in the cases with one head atom, noting that there is an implicit null atom, we assign 0.5 to each of these two. Then, we feed these rules to the parameter learner EMBLEM [2] to induce weights. Finally, rules with probability degrees below a threshold are eliminated from the candidates.

4 Experiments

In our experiments, we compare our T-LPAD with the state-of-the-art LPAD learner SLIPCOVER [1]. T-LPAD learns rules through transferring knowledge from an unrelated domain, whereas SLIPCOVER learns rules directly from facts. We try to address the following three questions:

- Can T-LPAD learn rules with high accuracy comparable to what SLIPCOVER can learn?
- Can T-LPAD learn such rules with (much) smaller amount of data than SLIP-COVER?
- Does T-LPAD learn such rules (much) faster than SLIPCOVER?

We used four domains **IMDB**, **UW-CSE**, **WebKB**, and **Twitter**, which are widely used and publicly available [1,14]. Part of the **IMDB** domain about movies and the **UW-CSE** domain about academics have been used in the running example. The **WebKB** domain describes web pages from the computer science departments of four universities, and the **Twitter** domain contains tweets

Table 1. Comparison between T-LPAD and SLIPCOVER.

Number of folds	LL						AUROC					
	SLIPCOVER			T-LPAD			SLIPCOVER			T-LPAD		
	1	4	6	1	4	6	1	4	6	1	4	6
UW-CSE→IMDB	−891	−590	−3	−606	−589	−39	0.90	0.88	1.00	0.90	0.88	0.99
IMDB→UW-CSE	−8301	−199	−155	−601	−193	−153	0.90	0.91	0.90	0.90	0.93	0.93
Twitter→WebKB	−1429	−533	−152	−727	−531	−220	0.82	0.87	0.95	0.85	0.85	0.89
WebKB→Twitter	−1316	−506	−308	−907	−483	−301	0.40	0.48	0.46	0.37	0.56	0.50

about Belgian soccer matches. As with existing approach, the facts in each domain are divided into smaller sets called folds [1,14]. To evaluate the performance of learners on small amount of data, we further divide the existing folds into smaller ones. In our experiment, **IMDB** has 10 folds of facts, **UW-CSE** has 10 folds, **WebKB** has 8 folds, and **Twitter** has 8 folds.

To perform transfer learning, we paired up the domains into two pairs: **IMDB** with **UW-CSE** and **WebKB** with **Twitter**; The two domains in each pair serve as source and target respectively in one round of evaluation, and swap roles in another round. The goal predicate (i.e., that occurring in the head of rules to learn) for **IMDB** is *workunder*, for **UW-CSE** is *advisedby*, for **WebKB** is *coursepage*, and for **Twitter** is *accountfan*. These predicates were picked following the existing literature. Rules in each source domain either came with the domain or learnt from the facts.

For all learning tasks, we evaluated T-LPAD under the same configuration: graph matching steps ≤ 30 and convergence threshold $10E - 9$, $\alpha = 0.5$, and $\beta = 0.3$. We used EMBLEM [2] to learn weights for the candidate rules. For SLIPCOVER, we adopted the parameters recommended by [1] for each domain. For a fair comparison, we disabled domain-dependent heuristics like "Lookahead" in SLIPCOVER. All the evaluation was conducted on a PC with 8 G RAM and corei5 CPU.

To evaluate and compare T-LPAD and SLIPCOVER, we adopted the standard measurement test set Log Likelihood (LL) and Area Under the Receiver Operating Characteristic Curve (AUROC) for measuring accuracy of the learnt rules [1]. LL directly measures the quality of the probability estimates produced, and the advantage of the AUROC is that it is insensitive to the large number of true negatives and it used both sensitivity and specificity for all possible thresholds.

Table 1 shows the LL and AUCROC for SLIPCOVER and T-LPAD with varying amounts of training data. In particular, "Number of folds" refer to the number of folds used for training the learners (T-LPAD uses training data only as inputs of EMBLEM), whereas the remaining data will be used for evaluating the learnt rules. The results show that regarding to the accuracy of learnt rules, T-LPAD was comparable to SLIPCOVER, and in several cases outperformed SLIPCOVER. The result was surprising as T-LPAD uses limited knowledge from

Table 2. Training times with 6 folds training data.

	SLIPCOVER	T-LPAD
UW-CSE→IMDB	1.1	0.1
IMDB→UW-CSE	37.63	0.1
Twitter→WebKB	5.25	3.15
WebKB→Twitter	2.1	0.4

an unrelated domain; yet on the other hand, it suggests the tight connection between the PD and rule structure of a domain. Through graph matching and the similarity matrix, T-LPAD is able to exploit such connection to a large extent and extract critical information for rule construction. The advantage of T-LPAD become obvious when the training data is reduced. In particular, with 1 fold of training data, SLIPCOVER suffered from insufficient training data whereas the performance of T-LPAD was reasonably stable. This is due to the fact that T-LPAD learns by transferring knowledge from another domain and hence is not data hungry.

We also measured the times needed for training with 6 folds, as shown in Table 2. All times are in minutes. From the results, it is clear that T-LPAD learns faster than SLIPCOVER when more training data are involved.

To conclude, with sufficient training data, the performance of T-LPAD is comparable to SLIPCOVER, while T-LPAD learns faster. In the case where only a small amount of training data is available, T-LPAD outperforms SLIPCOVER regarding to the quality of rules learnt.

5   Related Works

T-LPAD belongs to the class of deep transfer learning methods which are capable of generalizing knowledge cross distinct domains. Conceptually, the closest transfer learning approaches to T-LPAD are TAMAR [9], DTM [3], TODTLER [14], and transfer learning with type matching [8] which perform deep transfer in the context of Markov Logic Networks. One difference between our approach and these existing ones is about what knowledge is transferred. In particular, T-LPAD produces a similarity matrix, which associates the predicates of two domains based on their similarity in the predicate descriptions.

In [3,14], the knowledge from source domain is transferred through an intermediate knowledge language called second-order templates. Recall our running example, from the source rule $professor(x) : 0.2 \leftarrow publication(x, y)$, a second order template $X(x) \leftarrow Y(x, y)$ can be obtained. Assume we want to learn a rule about *director* in the Movie domain, by initialising the second template, a desired rule $director(x) \leftarrow movie(y, x)$ cannot be obtained from the second order template. This is due to the fact that second order templates cannot capture the similarity between predicates beyond their arity and the order of their arguments. Yet from the PDs and especially the graphs in Fig. 1, it is clear that predicate *movie* in the Movie domain is most similar to the *publication* in the Academy domain.

Existing approach [8] learns by transferring, instead of second order rule templates, type sharing knowledge among predicates. In particular, *professor* and *publication* share one type *person* and one variable x in the rule $professor(x)$: $0.2 \leftarrow publication(x, y)$ of Academy. Rules $director(x) \leftarrow workunder(x, y)$ and $director(x) \leftarrow actor(x)$ can be constructed for Movie in [8] as candidates, as *director* shares one type *person* and one variable x with *workunder* in the former rule and *actor* in the latter one. Yet again, these candidates are counterintuitive. Matching types allows more flexibility than second order template (e.g., in the above example, *actor* can be in the place of *publication*), but it does not take into account the similarity between predicates.

Another distinguishing feature of our approach is that our system does not require fact level data for structure learning. All the other approach require a refinement procedure to eliminate inaccurate rules, which is often data hungry. Also, the computational cost of such refinement is often high compared to the process of similarity matrix and rule construction in our case.

6   Conclusion

We have proposed an algorithm T-LPAD for LPAD structure learning using transfer learning, based on the similarity between two independent problem domains. Our algorithm identifies such similarity by matrix matching, and uses it to guide the candidate rule crafting for target domain in the presence of rules in the source domain. We have implemented the T-LPAD algorithm and conducted experiments for Web and social network domains. Our experimental results show that T-LPAD outperforms SLIPCOVER, a major structure learning algorithm for LPADs.

Acknowledgement. This work was supported by Australian Research Council (ARC) under grant DP130102302. We would like to thank Fabrizio Riguzzi for sharing his LPAD implementation cplint, Jan Van Haaren and Jesse Davis for providing data.

References

1. Bellodi, E., Riguzzi, F.: Structure learning of probabilistic logic programs by searching the clause space. In: Theory and Practice of Logic Programming, pp. 169–212. Cambridge University Press (2015)
2. Bellodi, E., Riguzzi, F.: Expectation maximization over binary decision diagrams for probabilistic logic programs. Intell. Data Anal. **17**(2), 343–363 (2013)
3. Davis, J., Domingos, P.: Deep transfer via second-order Markov logic. In: Proceedings of the 26th Annual International Conference on Machine Learning. ACM (2009)
4. De Raedt, L., Kimmig, A., Toivonen, H.: ProbLog: a probabilistic prolog and its application in link discovery. IJCAI **7**, 2462–2467 (2007)

5. Gutmann, B., Kimmig, A., Kersting, K., De Raedt, L.: Estimating the parameters of probabilistic databases from probabilistically weighted queries and proofs [extended abstract]. In: Inductive Logic Programming, Late Breaking Papers, pp. 38–43 (2008)

6. Gutmann, B., Thon, I., Raedt, L.: Learning the parameters of probabilistic logic programs from interpretations. In: Gunopulos, D., Hofmann, T., Malerba, D., Vazirgiannis, M. (eds.) ECML PKDD 2011. LNCS (LNAI), vol. 6911, pp. 581–596. Springer, Heidelberg (2011). doi:10.1007/978-3-642-23780-5_47

7. Hu, W., Jian, N.S., Qu, Y.Z., Wang, Y.B.: A graph matching for ontologies. In: K-Cap 2005 Workshop on Integrating Ontologies, pp. 43–50 (2005)

8. Kumaraswamy, R., Odom, P., Kersting, K., Leake, D., Natarajan, S.: Transfer learning via relational type matching. In: 2015 IEEE International Conference on Data Mining (ICDM), pp. 811–816 (2015)

9. Mihalkova, L., Huynh, T., Mooney, R.R.J.: Mapping and revising Markov logic networks for transfer learning. In: AAAI, vol. 7, pp. 608–614 (2007)

10. Bellodi, E., Riguzzi, F.: Learning the structure of probabilistic logic programs. In: Muggleton, S.H., Tamaddoni-Nezhad, A., Lisi, F.A. (eds.) ILP 2011. LNCS (LNAI), vol. 7207, pp. 61–75. Springer, Heidelberg (2012). doi:10.1007/978-3-642-31951-8_10

11. Riguzzi, F., Swift, T.: Probabilistic logic programming under the distribution semantics, Festschrift in honor of David S. Warren, 5 (2013). http://coherentknowledge.com/wp-content/uploads

12. Pan, S.J., Yang, Q.: A survey on transfer learning. IEEE Trans. Knowl. Data Eng. 22, 1345–1359 (2010)

13. Sato, T., Kameya, Y.: PRISM: a language for symbolic-statistical modeling. IJCAI 97, 1330–1339 (1997)

14. Van Haaren, J., Kolobov, A., Davis, J.: TODTLER: two-order-deep transfer learning. In: Proceedings of the Twenty-Ninth AAAI Conference on Artifcial Intelligence (2015)

15. Vennekens, J., Verbaeten, S., Bruynooghe, M.: Logic programs with annotated disjunctions. In: Demoen, B., Lifschitz, V. (eds.) ICLP 2004. LNCS, vol. 3132, pp. 431–445. Springer, Heidelberg (2004). doi:10.1007/978-3-540-27775-0_30

RETRACTED CHAPTER: Co-clustering for Dual Topic Models

Santosh Kumar[(✉)], Xiaoying Gao, and Ian Welch

School of Engineering and Computer Science, Victoria University of Wellington,
Wellington, New Zealand
santosh.kumar@ecs.vuw.ac.nz

Abstract. Biclustering is a data mining method that allows simultaneous clustering of two variables row and columns of a matrix. A bicluster typically corresponds to a sub-matrix that presents some coherent tendency. A traditional biclustering task for categorical variables is to determine heavy sub-graphs correspond to significant biclusters, i.e., biclusters with high co-occurrence values. Though algorithms have been proposed to extract sub-graphs biclusters, they present limited knowledge about the relevant importance of individual bicluster, as well as an importance of the variables for each bicluster. To address above problems, there have been several attempts to employ Bayesian method or mixture models using information theory. Although they can rank the biclusters and the variables for specific bicluster; they do not aim at extracting heavy sub-graphs biclusters. Moreover, these models force the search for biclusters in such a way that each cell in the matrix must engage in some bicluster. We attempt to mitigate these constraints employing dual topic models. In particular first, we propose a generalised Latent Dirichlet Allocation (LDA) topic model that obtains dual topics, i.e., topics in opposite directions: row and column topics. To achieve better topics, it applies joint reinforcement, i.e., considering column-topics while creating row-topics, and vice versa. Heavy sub-graphs biclusters, the high co-occurred association, are extracted using thresholds. We demonstrate that our proposed model Co-clustering for Dual Topic is useful for obtaining heavy sub-graphs biclusters by testing over a simulated data, a text corpus and a microarray gene expression data. The experimental results show that biclusters extracted by Co-clustering for Dual Topic model are better than traditional biclustering models.

1 Introduction

Biclustering is a data mining method that provides concurrent clustering of two variables, i.e., objects or rows, and features or columns of a matrix. Unlike traditional clusterings, such as K-means and hierarchical clustering, it does not significantly partition the rows into a number of clusters. Arguably, biclusters are closer to substance than conventional clusters. It is also known as co-clustering [1] or block clustering [11].

The original version of this chapter was retracted: The retraction note to this chapter is available at https://doi.org/10.1007/978-3-319-50127-7_67

B.H. Kang and Q. Bai (Eds.): AI 2016, LNAI 9992, pp. 390–402, 2020.
DOI: 10.1007/978-3-319-50127-7_34

This paper focuses on a particular type of biclusters where the matrix cells are co-occurrences between the similar row and column. For an instance, mobility data is where the rows are users and columns are locations that the users visit. A bicluster represents strong interrelationships between a subset of users and locations, i.e., which group of users visit which group of locations. We label them heavy sub-graph biclusters, which are described quasi-bicliques when the matrix is described by a weighted bipartite graph [21]. This is distinct from other bicluster models where the numerical relationship between values is a means for evaluation; heavy sub-graph biclusters are those where co-occurrence values are high. Low co-occurrences are not important. In Table 1, rows correspond to users and columns to locations. Individually cell value corresponds to a number of times a user visited a place. Apart from the mobility data, the heavy sub-graph biclusters are also applied to other data types. These other data types are: (a) author document data where rows of the matrix are authors and columns are words and cells contain the number of times an author wrote a word, (b) user song matrix where rows represent the users and columns represent the songs and each cell contains the number of times a user listened a song. Table 2 shows the adverse effect of such a constraint over biclustering structure. User U3, who visits all the places, is expected to be assigned to both clusters. But because of the restriction, U3 is assigned to only one bicluster. Existing methods are incapable of assigning relative ranks to the biclusters. Furthermore, they cannot determine the relative importance of the two variables for each bicluster. These disadvantages are slightly mitigated by replacing deterministic models with probabilistic models. For example, mixture models are employed in methods [7,22]. Even though they can assign comparative ranks to the biclusters and define the corresponding value of the two variables for each bicluster, they furthermore require each cell of the matrix to be assigned to a bicluster. Table 3 shows, that to satisfy this constraint, users U1-U3 are grouped together with locations L4-L7 where clearly U3 is very different from U1 and U2. Similarly, users U4-U6 are arranged together with locations L4-L7 where apparently U6 is the odd man out. Considering the constraints of existing methods, this paper propose a method to extract heavy sub-graph biclusters with the following characteristics:

- technique to determine corresponding importance of the two variables for each bicluster,
- technique to allow similar ranks to the biclusters,
- flexibility, or independence from any structure limitation, in finding biclusters.

Table 1. Sample Mobility dataset for 6 users and 7 locations The expected heavy sub-graph biclusters are presented in blue and red.

	L_1	L_2	L_3	L_4	L_5	L_6	L_7
U_1	18	20	21	0	0	1	0
U_2	15	18	16	0	0	3	0
U_3	19	17	17	8	10	6	0
U_4	0	0	0	10	7	5	12
U_5	0	1	0	9	9	4	0
U_6	0	0	1	0	2	0	0

Table 2. Extraction of heavy biclusters assuming an exclusive row and column structure

	L_1	L_2	L_3	L_4	L_5	L_6	L_7
U_1	18	20	21	0	0	1	0
U_2	15	18	16	0	0	3	0
U_3	19	17	17	8	10	6	0
U_4	0	0	0	10	7	5	12
U_5	0	1	0	9	9	4	0
U_6	0	0	1	0	2	0	0

Table 3. Extraction of heavy biclusters assuming a chequerboard structure

	L_1	L_2	L_3	L_4	L_5	L_6	L_7
U_1	18	20	21	0	0	1	0
U_2	15	18	16	0	0	3	0
U_3	19	17	17	8	10	6	0
U_4	0	0	0	10	7	5	12
U_5	0	1	0	9	9	4	0
U_6	0	0	1	0	2		

The biclusters showed in Table 1 demonstrates these characteristics. We observe that users U1-U3 manage to visit locations L1-L3 often. Consequently, they are allocated to one bicluster. This is the identical case for users U3-U5 and locations L4-L6. Therefore, they are allotted to a second bicluster. We observe that user U3 visits all the locations in the dataset and thus he is supposed to be allotted to both biclusters. On the contrast, there are not sufficient visiting records for user U6, and therefore he should not be allotted to any bicluster. A comparable case befalls in location L7. Though, a robust model should allow partial association for L7 to the red bicluster. We propose a novel method in an attempt to provide above mentioned objectives. The proposed method is based on dual topic models. In the user location domain, this dual-nature is observed in user- topics that are distributions over users, and location topics that are distributions over locations. We introduce a generalised LDA topic model that extracts these dual topics. A method of mutual reinforcement is employed to benefit from their dual nature. To extract biclusters, we examine their inter-relationships. Thresholds are used to extract these biclusters.

The rest of this paper is structured as follows: In Sect. 2, we propose our model based on the limitations of existing methods. Then, we present our proposed model, its relationship with other topic models and how we can use the model to extract biclusters. In Sect. 3, we present the preliminary results and finally, in Sect. 4, we conclude our work and propose future research directions.

2   Dual Topics for Bicluster Model

We propose Dual Topics for Bicluster model by providing the three characteristics mentioned in the introduction section. In this section, we present an overview of our proposed model accompanied by a detailed description.

2.1   Proposed Dual Topics for Biscluter Model

For a given matrix $|R| \times |C|$, where R is set of rows or, as in user location domain, set of users, and C is set of columns or locations. r and c represents for a single row and column, respectively. Each cell (r, c) in the matrix contains a real number conferring frequency of user c visiting location r. Regard that, without any loss of generalisation; we use rowcolumn and userlocation reciprocally.

Practising the Latent Dirichlet Allocation (LDA) model [5], if we analyse the documents to be the rows and the words to be the columns of the matrix, then we can extract column topics z^c. Where the topics are distributions over columns, parameterized by ϕ^c that follows a Dirichlet distribution with parameter β^c. Furthermore, if use an LDA considering the documents to be the columns and the words to be the rows of the matrix, we could extract row-topics z^r, where the topics are distributions over rows, parameterized by ϕ^r that follows a Dirichlet distribution with parameter β^r. We label the second LDA as the Dual LDA, since we use the same input to obtain complementary information. The two sets of topics parameterized by ϕ^c and ϕ^r are named as dual topics in this paper. ϕ^c values signify the importance of columns in a column- topic, and similarly, ϕ^r values imply importance of rows in a row-topic, thus leading to the fulfillment of the first characteristic. We use dual topics to extract heavy sub-graph biclusters, i.e., biclusters with high co-occurrence values. These biclusters are formed by inter-relationships between column- and row-topics. In other words, we need to model the mutual dependency between the column topics and the row-topics of an LDA and its dual LDA, respectively.

Topic models and so have not studied the mutual dependency between two types of topics, we address this necessity by defining a probability distribution. To model the mutual dependency between the cluster descriptors z^c and z^r, we assume that the two kinds of topics are independent only if we know the parameters that describe their dependency. That is, we assume a common cause dependency between the latent variables [14].

Because, we require a probability distribution to describe the simultaneous generation of two variables, we define a Bivariate Probability distribution. Additionally, it would be ideal to use a categorical distribution because categorical distributions are commonly used to capture the mixing components of the topics. In our case, with a categorical distribution, we can use the parameter $\theta_{i,j}$ to obtain the interrelation between the ith column topic and the jth row topic. Parameters with high values will represent strong dependencies. Using these specifications, we can define the support, parameters and probability mass function of the distribution. The help of a probability distribution represents the set of results that may appear when sampling from the distribution; the parameters correspond to the set of variables that represent the probability mass function, and the probability mass function maps every element in the help to a probability value. Assuming K_c column topics and K_r row topics, we describe the Bivariate Categorical Distribution in Definition 1.

Definition 1. *A Bivariate Categorical Distribution represented by in terms of the Categorical distribution: Given a random variable $x \sim Categorical^2(\theta, K_c, K_r)$ is a discrete probability distribution parameterized $\theta \in \mathbb{R}_{[0,1]}^{K_c.K_r}$ and $K_c, K_r \in \mathbb{N}_{>0}$. The dependence of the distribution is $X \in \mathbb{N}_{[1:K_c]} \times \mathbb{N}_{[1:K_r]}$. The probability mass function is given by $f_\theta(\vec{x}) = \sum_i^{K_c} \sum_j^{K_r} \mathbb{I}[x_0 = i \wedge x_1 = j] \theta_{i,j}$.*

Lemma 1. *The Bivariate Categorical Distribution is a substantial distribution because it can be represented regarding the categorical distribution: Given a random variable $x \sim Categorical^2(\theta, K_c, K_r)$. We need to use a mapping g so that $y = g(x) \sim Categorical(\theta)$. This mapping is presented by $g : \mathbb{N}_{[1:K]} \rightarrow \mathbb{N}_{[1:K_c]} \times \mathbb{N}_{[1:K_r]}$ where $g(x) = (x \div K_c, x(mod)K_r)$.*

Because g is a bijection, we are confirmed that we can always represent a Bivariate Categorical Distribution in terms of the Categorical Distribution and viceversa. The last characteristic is achieved by modeling heavy biclusters as the interdependent row topics and column topics. In this way, every cell in the matrix may be part of none, one or many biclusters. This is different from other probabilistic approaches where every row/column pair is assigned to one bicluster generating a partition of the input matrix [20,22]. Applying the above, we propose our model Dual Topics for Biclusters.

2.2 Generative Method

We use the generative model to sample pairs of values corresponding to the pair of random variables c, r. First, we utilise a Dirichlet distribution parameterized by α to generate the low dimensionality representation of the joint distribution of the two variables; so, we obtain $K_c \times K_r$ parameters representing the partition of the data into $K_c \times K_r$ sections. Second, we apply the bivariate categorical distribution to sample one section of the partition which will point to one column topic and one row topic. Third, we use the column topic to sample a column following the same procedure as in LDA, and finally, we use the row topic to sample one row. Algorithm 1 presents a description of the generative process, given the number of column and row topics.

2.3 Bayesian Inference Process

For the Bayesian inference process, we use collapsed Gibbs sampling. Accordingly, we switch from a matrix to a tabular representation of the data. Every instance of a cell in the matrix become a row in the table so that cells with higher values will incur greater evidence for the topic assignments. This is the same concept that applies to LDA when using collapsed Gibbs sampling.

Algorithm 1. Dual Topics for Biclusters Generative Model

1: **procedure** DUAL TOPICS FOR BICLUSTERS MODEL
2: Draw a distribution over topics $\theta \sim Dirichlet\left(\alpha\right)$
3:     **for** $i \in 1 \ldots K_c$ **do**
4:         Draw a distribution over the columns $\phi_i^c \sim Dirichlet\left(\beta^c\right)$
5:     **end for**
6:     **for** $j \in 1 \ldots K_r$ **do**
7:         Draw a distribution over the columns $\phi_j^r \sim Dirichlet\left(\beta^r\right)$
8:     **end for**
9:     **for do** $i \in 1 \ldots N$
10:        Draw simultaneously a row-topic and a column-topic $z_{i}^{c}, z_i^r \sim$ $Categorical^2\left(g\left(\theta\right)\right)$
11:        Draw a row $c \sim Categorical\left(\phi_{z_i^c}^c\right)$
12:        Draw a column $r \sim Categorical\left(\phi_{z_i^r}^r\right)$
13:     **end for**
14: **end procedure**

Equation 1 presents the following distribution that we use to update the column topic z_i^c and the row topic z_i^r assigned to the ith record in the dataset. For each record $(m,n)\,\epsilon N$, we sample from this distribution to update its topic assignments; the posterior distribution must be recomputed after each assignment. After the model updates all the records, we compute the log-likelihood of the model using Eq. 2. Then, we iterate the same procedure until the log-likelihood converges.

$$P\left(z_i^c = j, z_i^r = k \mid c_i = m, r_i = n, z_{-i}^c, z_{-i}^r\right) \propto \phi_{mj}^c \cdot \phi_{nk}^r \cdot \theta_{jk} \tag{1}$$

where

$$\phi_{mj}^c = \frac{C^{|C|K_c} + \beta^c}{\sum_{m'} C_{m'j}^{|C|K_c} + |C|\,\beta^c}, \quad \phi_{nk}^r = \frac{C_{nk}^{|R|K_r} + \beta^r}{\sum_{n'} C_{n'k}^{|R|K_r} + |R|\,\beta^r}, \quad \theta_{jk} \propto C_{jk}^{K_c K_r} + \alpha$$

Following the notation in [19], C_{xy}^{XY} denotes the counting of the xth instance of a variable with size X and the yth instance of a variable with size Y. The inference equation has an intuitive interpretation. First, the parameter ϕ^c expresses the probability of the mth column to be assigned to the jth column-topic z^c. Second, the parameter ϕ^r represents the probability of the nth row to be assigned to the kth row topic z^r. Finally, θ maintains the counting over the current relationship between row and column topics capturing the mutual dependence between the topics. The log-likelihood of the model given the topic distributions is given by:

$$\mathcal{L}\left(c, r \mid \theta, \phi^c, \phi^r\right) = \sum_{m,n}^{R} \sum_{j}^{K_c} \sum_{k}^{K_r} \theta_{jk} \phi_{mj}^c \phi_{nk}^r \tag{2}$$

2.4   Bicluster Extraction Using Dual Topic for Bicluster Model

We obtain the biclusters from the parameters θ, ϕ_c, ϕ_r obtained by the inference process. First, we get the groups of columns from the column topics or distributions over the set of columns ϕ_c. Accordingly, we require an automatic mechanism to select the elements of a bicluster for each row topic. Bicego et al. have proposed different mechanisms for selecting the most relevant variables in a probability distribution. They assume that a threshold rule results in the best performance [3]. As an outcome, we set a threshold γ_r to select which rows are more appropriate to each distribution.

Second, we obtain the groups of rows from the row-topics or distributions over rows ϕ_r. Furthermore, we set a threshold γ_r to select which rows are more relevant to each distribution. Finally, we extract which groups of rows have a substantial connection with which groups of columns. In other words, we extract the heavy biclusters. This information is explained by the parameter θ because θ encodes the relationship between every distribution in ϕ_c and every distribution in ϕ_r. We set a threshold γ to select the connections are more appropriate in the dataset.

2.5   Mutual Reinforcement in Dual Topic for Bicluster Model

Mutual reinforcement of both inter and intra-document statistics has explained improvement over a single approach in non-probabilistic clustering in the mobility domain [2] and the text domain [6]. Let rows r correspond to the set of documents d with cardinality D and the columns c correspond to the vocabulary of words w with cardinality V. Theorem 1 states that if we assign the document-topics using a one-to-one correspondence between the documents d and the document topics z^r. Then, the inference equation of Dual Topic for Bicluster Model is reduced to the inference equation of LDA.

Theorem 1. *If $z_i^r = d_i \forall i \epsilon N$, then Dual Topic for Bicluster Model is reduced to the LDA model.*

Proof: Because of the document-topic assignments z^r are in- dependent among documents, the inference of Dual Topic for Bicluster Model only depends on z^c. Recall that the parameters ϕ^r in Eq. 1 do not depend on z^c, so we can replace ϕ^r with a constant value. Then, we analyse the parameters θ. Since $z_i^r = d_i \forall i \epsilon R$, we can replace z_r with d_i, and then we end up with the inference equation of LDA [12]:

$$P\left(z_i^c = j \mid w_i = m, d_i = n, z_{-i}^c\right) \propto \frac{C_{mj}^{WK_c} + \beta}{\sum_{m'} C_{m'j}^{WK_c} + V\beta} C_{nj}^{DK_c} + \alpha \qquad (3)$$

In LDA, the hyper-parameter of the Dirichlet prior on the per-column topic distributions α is independent of the given column to avoid overfitting. We can break the dependence of α and the given column by setting α to the same value for all column-topics. Similar relation can be inferred between Dual Topic for Bicluster Model and Dual LDA. This observation strengthens our intuition that Dual Topic for Bicluster Model reinforces both models to bi-clustering.

3 Experimental Evaluation

Datasets: We use a synthetic dataset and two real-life data sets. (1) Synthetic data (2) NIPS: Text data obtained from papers accepted by NIPS from 1988 until 2003 [10]. The data set contains 40,552,970 records generated by 2865 authors who wrote 2483 documents using a vocabulary of 14036 words. (3) Microarray: Eight different microarray gene expression sets from the Gene Expression Omnibus (GEO). Microarray Data contains the information about the gene expression profile from tumour cells and normal cells from a particular part of the body. Each dataset can be represented by a matrix, where the rows represent the genes and the columns represent different biological samples. Discovering which set of samples express which set of genes is the common objective of biclustering algorithms.

Parameter Settings: For all runs of our model, we set the hyperparameters α with a value of $50/(K_c K_r)$ and β_c, β_r with a value of 0.01. Similar to previous research, we find that these values work well with most datasets [12,19]. The first two experiments require the extraction of biclusters. In these experiments, we set $\gamma = E[\theta]$, $\gamma_c = E[\phi_c]$, and $\gamma_r = E[\phi_r]$.

3.1 Biclustering Using Synthetic Data

In this section, we evaluate the adaptability of our model to extract heavy biclusters assuming different bicluster structures in a synthetic dataset. We study three biclustering structures: (i) Every column and every row must be assigned to exactly one bicluster, (ii) Every column must be assigned to one bicluster, but the rows may, Nonetheless, to more than one bicluster, (iii) Every column/row may or may not be allocated to any bicluster. We use normal distributions to generate the data. The cells corresponding to a bicluster naturally have a higher mean when compared with those cells that do not belong to any bicluster. The variance is used to control the amount of noise in the data. The specification of each dataset is shown in Fig. 1. We initialize Dual Topic for Bicluster Model by randomly sampling the topic assignments z^c and z^r for each entry in the dataset. Also, we initialized our model using an LDA and its dual LDA. LDA used to initialize z^c, and the Dual LDA used to initialize z^r.

For the evaluation metric, we use recovery $S(\varepsilon, \beta)$ and relevance $S(\beta, \varepsilon)$ as an established metric used to validate the biclustering extraction [8]. ε as the list of expected biclusters, β is the list of extracted biclusters and the function S is defined by:

$$S(\beta_1, \beta_2) = \frac{1}{|\beta_1|} \sum_{b_1 \epsilon \beta_1} max_{b_2 \epsilon \beta_2} s(b_1, b_2)$$

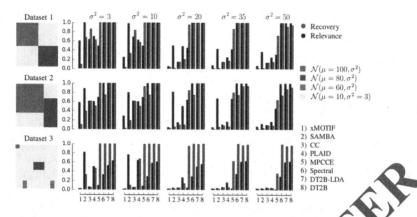

Fig. 1. Recovery and Relevance results for bicluster classification. The datasets are generated from four normal distributions. The distributions colored with blue, green and red correspond to the expected biclusters (Color figure online)

Here $s(b_1, b_2)$ is the Jaccard similarity applied for the sub-matrix elements defined by each bicluster. Intuitively, these metrics are an analogy to Recall and Precision.

We compared results with the bipartite spectral graph (Spectral) [6] and SAMBA [21] since these algorithms extract heavy biclusters. We also compared with the most common baseline in the biclustering task [13] and with the PLAID model [16] because of its good performance on Gene Expression data. Finally, we compared with a probabilistic approach called xMOTIF [17] and a non-parametric model (MPCCE) [22]. We repeated the experiment ten times, and we report the average results in Fig. 1. In the first dataset, we observe an improvement of Spectral over other models; this is because Spectral assumes that all columns and all rows must be allocated to exactly one bicluster and thus is suitable for the first dataset. Nonetheless the performance of Dual Topics for Bicluster Model is also competitive in this dataset. When we relax the assumption of the first dataset in the other datasets, the results of Spectral degrade compared to our model. MPCCE models the matrix as a partition, and thus its recovery has a good performance, but its relevance is bounded by those regions where there are no expected biclusters. SAMBA can recognise the biclusters, but since it does not use the values in the matrix, it is not robust to noise, and it does not often discover all the rows and columns involved in the bicluster. Overall Dual Topics for Bicluster Model complies the best results.

3.2   Biclustering in Microarray Data

Biclustering is important usually applied to microarray data. These datasets have been examined extensively. Thus, the association between groups of genes with disease phenotypes is available through a method called Gene Ontology Enrichment [9]. Besides, many types of biclusters have been found to coexist in microarray datasets [18]. Therefore, we evaluate the proposed method Dual Topics for Bicluster Model on microarray datasets. In this experiment, we show that Dual Topics for Bicluster Model can mine biclusters from DNA microarray data by following the evaluation procedure proposed by Eren et al. [8]. That is, we extract biclusters from microarray data, and then, we perform gene ontology enhancement analysis of the extracted biclusters. A bicluster is considered to be enriched if at least one term from the Biological Process Gene Ontology was enriched at the $P = 0.05$ level after Benjamini and Hochberg multiple test correction [15]. Evaluation is carried out by extracting the relationships of the improved biclusters to the extracted biclusters aggregated for the eight microarray datasets. The proportion of improved biclusters is a broadly used method to distinguish different bicluster methods [8].

For Dual Topics for Bicluster Model, we set the number of both row-topics (gene-topics) and column topics (sample-topics) to be 15. The data must be preprocessed since microarray data contains negative values and our model is intended for discrete data. We propose to analyse the positive values and the negative values separately. Figure 2(a) shows the proportions of the improved biclusters to the extracted biclusters aggregated for the eight datasets. We also demonstrate the effect of increasing the significance level α. Regard that in this section; α does not belong to a hyperparameter. Conclusions recommend that the heavy bicluster is a bicluster type worth studying in microarray data. The proportion of improved biclusters of Dual Topics for Bicluster Model is higher than those of other methods. This is particularly significant when the α level is selected to be greater than 0.1%.

3.3   Reviewer Recommendation

On the text data, to evaluate our method quantitatively, we use the information extracted from the model instead of evaluating the generated biclusters directly. In this experiment we use text data to evaluate Dual Topics for Bicluster Models ability of modelling the joint distribution of authors and words. As suggested by Rosen-Zvi et al. Automated Reviewer Recommendations is an application that requires modelling the relationship between authors and words [19]. Therefore, we evaluate the usefulness of Dual Topics for Bicluster Model in the reviewer recommendation application. We will also qualitatively evaluate the results of Dual Topics for Bicluster Model on text data by visualising the generated biclusters in next subsection. We use Dual Topics for Bicluster Model to model the rows as authors, the columns as words and the values in each cell as the number of times an author wrote a word. With this configuration, we obtain the similarity between authors with author topics z_a, the similarity between words with word-topics z_w and the relationships between the two types of topics.

Given a document D represented by a set of words w, we can use Dual Topics for Bicluster Model to compute the probability of an author a to write a word w by:

$$P(a, w) = \sum_{z_a} \sum_{z_w} P\left(a \mid \phi_{z_w}^w, \beta_w\right) P\left(a \mid \phi_{z_a}^a, \beta_a\right) P\left(z_a, z_w \mid \theta\right) \qquad (4)$$

Then, we use $P(D \mid a) = \prod_{w \in D} P(w \mid a)$ to obtain the probability of the document being written by an author a. For the recommendation, we obtain the top n authors based on $P(D \mid a)$. The current biclustering methods cannot be used for the reviewer recommendation task. For illustration, we consider other topic models, including LDA [5], the correlated topic model (CTM) [4] and the author-topic model (ATM) [19]. Note that our model Dual Topics for Bicluster Model groups authors into author topics and it groups words into word topics. Oppositely, other topic models group the words using topics, but they do not explicitly group the authors. We compute $P(w \mid a)$ using an LDA by expressing the authors at the document level and leaving the word level unchanged. In this model, the probability of an author a to write a word w is calculated from $P(w \mid a) = \sum_z P(a \mid \phi_z) P(z \mid \theta)$. We use a similar formulation to estimate $P(w \mid a)$ using the correlated topic model (CTM) [4] and the author-topic model (ATM) [19]. For this experiment, we equally divided the papers into the training set and testing set. After deleting all papers in the test set with one author and those papers where the authors are not in the training set, we obtain as the test set 516 papers with two authors, 453 with three authors and 131 with four authors. We evaluate the precision and recall by selecting the top n authors based on $P(D \mid a)$. In all the models we set the number of topics at fifty. In the case of Dual Topics for Bicluster Model both author-topics and word, topics are set to fifty as well. We do not compare with other biclustering algorithms given that they do not provide a probabilistic framework for obtaining $P(D \mid a)$. We present the results in Fig. 2(b). We mention improvement of $P(D \mid a)$ over

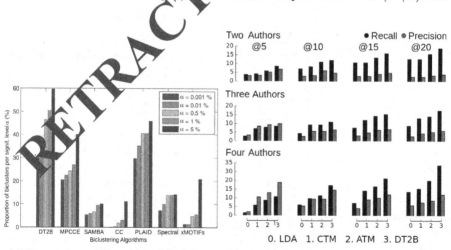

(a) Proportion of the enriched biclusters   (b) Recall and precision results for the author classification task

Fig. 2. Experimental results

ATM and LDA. This is assumed as Dual Topics for Bicluster Model generalizes both models. Depending on the grouping of the documents carried out by the document topics, Dual Topics for Bicluster Model can be decreased to ATM or LDA. Theorem 1 shows that grouping each document with its document topic reduces Dual Topics for Bicluster Model inference equation to LDAs inference equation. In the case of ATM, Dual Topics for Bicluster Model decreases to ATM if we assign the document topics for document d_i by uniformly sampling the authors that have written d_i. Besides, ATM and CTM improve LDA which is expected since both the models generalise LDA. Overall, we observe how papers with the larger number of authors have a better performance when trying to identify the authors of the paper. This is because the number of authors increases the chances for the algorithms to predict the author that has written the document.

4 Conclusion and Future Work

We have shown how combining an LDA and its dual into a model called Dual Topics for Bicluster generates a representation of the co-occurrence relationship between two types of variables. Dual Topics for Bicluster Model generalises the modelling of LDA by capturing similarity between rows as represented by row-topics. In text data, this translates to a model where, in addition to word topics, we capture the similarity between documents with document-topics. We also modelled the interrelation between the row-topics and column topics with an adhoc probability distribution. The inter-relation between topics can be used to extract heavy biclusters in count data. We have shown the quality of the extracted biclusters using experiments on synthetic data, Microarray data and text data.

Several extensions may be useful for the analysis of bivariate data. For instance, we may consider relaxing the assumptions about knowing in advance the number of row topics and the number of column topics. However, since this case includes bivariate analysis of data, some alternative solution must be considered. Also, considering a temporary extension of the model or considering multidimensional datasets may be helpful in several domains.

References

1. Aggarwal, C.C., Reddy, C.K.: Data Clustering: Algorithms and Applications. CRC Press, Boca Raton (2013)
2. Bao, J., Zheng, Y., Mokbel, M.F.: Location-based and preference-aware recommendation using sparse geo-social networking data. In: Proceedings of the 20th International Conference on Advances in Geographic Information Systems, pp. 199–208. ACM (2012)
3. Bicego, M., Lovato, P., Ferrarini, A., Delledonne, M.: Biclustering of expression microarray data with topic models. In: 2010 20th International Conference on Pattern Recognition (ICPR), pp. 2728–2731. IEEE (2010)

4. Blei, D., Lafferty, J.: Correlated topic models. In: Advances in Neural Information Processing Systems, vol. 18, pp. 147 (2006)

5. Blei, D.M., Ng, A.Y., Jordan, M.I.: Latent Dirichlet allocation. J. Mach. Learn. Res. **3**, 993–1022 (2003)

6. Dhillon, I.S.: Co-clustering documents and words using bipartite spectral graph partitioning. In: Proceedings of the Seventh ACM SIGKDD International Conference on Knowledge Discovery and Data Mining, pp. 269–274. ACM (2001)

7. Dhillon, I.S., Mallela, S., Modha, D.S.: Information-theoretic co-clustering. In: Proceedings of the Ninth ACM SIGKDD International Conference on Knowledge Discovery and Data Mining, pp. 89–98. ACM (2003)

8. Eren, K., Deveci, M., Küçüktunç, O., Çatalyürek, Ü.V.: A comparative analysis of biclustering algorithms for gene expression data. Briefings Bioinform. 14(3), 279–292 (2013)

9. Falcon, S., Gentleman, R.: Using gostats to test gene lists for go term association. Bioinformatics **23**(2), 257–258 (2007)

10. Globerson, A., Chechik, G., Pereira, F., Tishby, N.: Euclidean embedding of co-occurrence data. J. Mach. Learn. Res. **8**, 2265–2295 (2007)

11. Govaert, G., Nadif, M.: Block clustering with bernoulli mixture models: comparison of different approaches. Bioinformatics **52**(6), 3233–3245 (2008)

12. Griffiths, T.L., Steyvers, M.: Finding scientific topics. Bioinformatics **101**(suppl 1), 5228–5235 (2004)

13. Hartigan, J.A.: Direct clustering of a data matrix. Bioinformatics **67**(337), 123–129 (1972)

14. Hitchcock, C.: Probabilistic causation. In: Stanford Encyclopedia of Philosophy (2010)

15. Hochberg, Y., Benjamini, Y.: More powerful procedures for multiple significance testing. Bioinformatics **9**(7), 811–818 (1990)

16. Lazzeroni, L., Owen, A.: Plaid models for gene expression data. Statistica sinica **12**, 61–86 (2002)

17. Murali, T., Kasif, S.: Extracting conserved gene expression motifs from gene expression data. Bioinformatics 8, 77–88 (2003)

18. Pontes, B., Giráldez, R., Aguilar-Ruiz, J.S.: Biclustering on expression data: a review. Bioinformatics 57, 163–180 (2015)

19. Rosen-Zvi, M., Griffiths, T., Steyvers, M., Smyth, P.: The author-topic model for authors and documents. In: Proceedings of the 20th Conference on Uncertainty in Artificial Intelligence, pp. 487–494. AUAI Press (2004)

20. Shan, H., Banerjee, A.: Bayesian co-clustering. In: 2008 Eighth IEEE International Conference on Data Mining, pp. 530–539. IEEE (2008)

21. Tanay, A., Sharan, R., Shamir, R.: Discovering statistically significant biclusters in gene expression data. Bioinformatics **18**(suppl 1), S136–S144 (2002)

22. Wang, P., Laskey, K.B., Domeniconi, C., Jordan, M.I.: Nonparametric Bayesian co-clustering ensembles. In: SDM, pp. 331–342. SIAM (2011)

Optimization of Traffic Signals Using Deep Learning Neural Networks

Saman Lawe[1(✉)] and Ruili Wang[2]

[1] Auckland Transport, Auckland, New Zealand
saman.lawe@at.govt.nz
[2] Massey University, Palmerston North, New Zealand
ruili.wang@massey.ac.nz

Abstract. Reducing traffic delay at signalized intersections is a key objective of intelligent transport systems. Many existing applications do not have the intelligence embedded to learn about the environmental parameters (such weather, incident etc.) that influence traffic flow; therefore, they are passive to the dynamic nature of vehicle traffic. This report proposes a deep learning neural networks method to optimise traffic flow and reduce congestion at key intersections, which will enhance the ability of signalized intersections to respond to changing traffic and environmental conditions. The input features of the proposed methods are composed of historical data of all the movements of an intended intersection, time series and environmental variables such as weather conditions etc. The method can learn about the region and predict traffic volumes at any point in time. The output (i.e. predicted traffic volume) is fed into the delay equation that generates best green times to manage traffic delay. The performance of our method is measured by root mean squared error (RMSE), against other models: Radial Basic Function, Random Walk, Support Vector Machine and BP Neural Network. Experiments conducted on real datasets show that our deep neural network method outperforms other methods and can be deployed to optimize the operations of traffic signals.

Keywords: Deep learning · Intelligent Transport Systems (ITS) · Machine learning · Multi-layer neural networks · Neural networks

1 Introduction

Over the past years, improving the vehicle traffic flow has been the focal point of many researchers [1]. A large portion of the research and innovations aim to minimize, manage, and mitigate disruptions to traffic arising from traffic congestions [3]. To deal with the inevitable effect of traffic flow on road safety, the economy, the environment and social life, it is prudent to design traffic systems such that they can cater for the ever-growing number of vehicles and all the potential influencers of traffic flow [2]. Recent studies that address the issue of optimizing traffic lights timing and reducing delays can be classified into the following three categories [2]:

© Springer International Publishing AG 2016
B.H. Kang and Q. Bai (Eds.): AI 2016, LNAI 9992, pp. 403–415, 2016.
DOI: 10.1007/978-3-319-50127-7_35

1. Time-series approach: This approach attempts to predict traffic light phases by determining patterns of the temporal variation of traffic flow [2].
2. Stochastic approach: The stochastic approach utilizes probabilistic models to forecast traffic flow.
3. Nonparametric approaches: The algorithms of nonparametric approaches have no (or very little) prior knowledge about the form of the true function that is being modelled [2].

The work of this paper falls under nonparametric approaches with deep neural networks being utilized to address the factors that influence traffic delay at signalized intersections. Furthermore, those influencing factors are injected into the deep neural network as input features.

The contributions of this paper can be summarised as follows:

1. To the best of authors knowledge, this paper is the first attempt to utilize DL in optimizing traffic signals through accurate prediction of vehicle flow at signalized intersections.
2. The experiments conducted in researches for this paper take into account all the past and future states of input features that are known to be impacting traffic flow of arterial networks. Some of the input features are: time, weather, events, car population, congestion etc.
3. Contrary to the conventional techniques of traffic prediction that are used in setting traffic signal phases, the method proposed in this paper has two interdependent regression tasks; one for vehicle flow, and the other one for phases of intersection movements.

The rest of this paper is organized as follows: Sect. 2 reviews the studies on traffic light phase prediction. Section 3 sets the scene on the general architecture of the network while Sect. 4 discusses the experimental results. Concluding remarks are described in Sect. 5.

2   Literature Review

To date, research on urban traffic optimization has been relatively scarce due to the complex and random nature of intra-city traffic in comparison to highways. However, the desire to model this complexity and the dire need for improving it, and traffic flow, recently attracted serious interest from both academia and commercial entities. The following sections review some recent works from the field of traffic signal optimization and traffic flow, focusing on approaches that are categorised under time-series, stochastic and nonparametric.

2.1   Time-Series Approach

One of the predominant time-series models that transportation researchers have developed is to do with prediction of short-term freeway traffic flow [4], called Autoregressive Integrated Moving Average (ARIMA) [7]. Tselentis et al.

used ARIMA and its variations to compare the performance of statistical and Bayesian combination models with single time series models for short-term traffic forecasting [8]. Their results show that linear regression combination techniques may provide more accurate prediction than Bayesian combination models [8]. However, the results are based on short-term prediction and only cater for specific applications. Moayedi et al. applied an ARIMA model for traffic volume prediction in intercity streets [9]. Xu et al. propose a step-up model from ARIMA called ST-BMARS that proves to have lower Root Mean Square Error (RMSE) than ARIMA and its variations [10]. The findings in [10] are limited to applications found in freeway interstates.

2.2 Stochastic Approach

Advocates of this approach believe that traffic flow prediction and optimization should be treated stochastically due to the nature of traffic demand, driver behaviour, unprecedented events, etc. [1]. Ge et al. focused on the energy-sustainable traffic signal timings in which they modelled the issue of traffic network equilibrium stochastically, taking into account vehicle delays at signalized intersections and travel demand [12]. The findings of Yun and Park propose a novel algorithm that proves the usefulness of a fuel surcharge policy in China [11]. Barimani et al. use the Kalman filter method to minimise the prediction error of traffic flow by applying adaptive time variant transformation from primary space to reproduce kernel Hilbert space [10].

2.3 Nonparametric Approaches

Despite its popularity and promising capabilities, deep learning has not been utilized to its full potential to address congestion problems stemming from traffic signals. On the other hand, conventional Artificial Neural Networking has been applied extensively in all fields of vehicle traffic optimizations and prediction [3,6,13,14]. In their novel study, De Oliveira and Neto used simulation software and presented a novel proposal about the optimization of traffic lights using Multiple Neural Network technique [14]. The authors suggested a pre-timed and actuated phase time in their work [15]. However, the authors do not suggest when to use either option. This would be necessary mainly because traffic can be quite sporadic and does not follow fixed patterns. One of the disadvantages of not following what is evident when obtaining signal timing.

So far the above studies have not proven that Artificial Neural Network techniques, such as deep learning, can provide a better alternative to the existing technologies that aim to optimize traffic light operations. The work in this paper extends and improves the implementation of deep learning in analysing the historical timing of traffic lights, traffic congestion and future events. Ultimately, the objective is to train the traffic lights to function proactively in response to traffic conditions. Thus, to train the deep neural networks, this paper will use Microsoft Azure Machine Learning tool that promises high processing power and virtually unlimited capacity with scripting and programming options for tuning the neural network and manipulation of mathematical formulae. Some of the options are Python, R Programming and Microsofts propriety Net#.

3   Background

3.1   Deep Learning

Deep learning, as a branch of artificial neural networks, was instantiated by the works of Hinton et al. as a type of Machine Learning technique branching out from Artificial Neural Networks (ANN) in 2006 [3]. Soon after, Yann LeCun from Facebook AI Research, and Yoshua Bengio from Universit de Montral joined to further improve the technique [3].

Historically, DL implementations have been limited to teaching machines how to classify and recognize patterns in images and sounds, motion modelling and dimensionality reduction. However, DLs promising capabilities have attracted many researchers to extend its application into the other domains that are extremely complex in nature. Inter-city traffic flow prediction and analysis is one of those complex systems that has a large number of influencing factors and is subjected to many variations that are highly stochastic and provide no prior knowledge to rely on when studied. The solution discussed here attempts to minimize delay time at an intersection through finding the best effective green time for the movements of the intersection approaches.

The proposal discussed in this paper models the number of vehicles that pass through signalized intersections based on regression techniques derived from deep learning neural networks. To make the regression model more effective, this paper suggests using Multitask Learning (MTL) [2] such that tasks (traffic lights phase times and traffic congestion) are trained separately. However, due to control over the tasks and the resulted congestion, the tasks could influence each other. The gist of this study is an attempt to analyse historical behaviours and future activities (e.g. growing number of vehicles or road works) of the roads using DL to optimize the future state of traffic signals so congestions are controlled in the best way possible.

3.2   Problem Description

Because a poorly designed traffic light can lead to increased vehicle delay and road incidents, causing traffic congestion, and may encourage drivers to take alternative routes that may not be suitable for traffic (such as routes through residential neighbourhoods) [5]. Therefore, congestion needs to be studied under a generalized form of delay modelling.

3.3   Mathematics Formulation

The formula that describes average delay per vehicle derived by the Highway Capacity Manual (HCM2010) is found to be [16]:

$$d = d_1 PF + d_2 + d_3 \tag{1}$$

where d is the average signal delay per vehicle in seconds, d1 is the average delay per vehicle due to uniform arrivals in seconds, PF is the progression adjustment factor used to account for the effect of signal progression on traffic flow, d2 is the average delay per vehicle due to random arrivals in seconds, and d3 is the average delay per vehicle due to initial queue at start of analysis time period, in seconds.

$$d_1 = \frac{0.5C(1 - \frac{g}{C})}{(1 - [min(1, X)\frac{g}{C}])} \tag{2}$$

where g is effective green time in seconds, C is the cycle length in seconds, and X is ratio of lane group.

$$X = \frac{v}{c} \tag{3}$$

where v is the Traffic Volume in veh/h and c is capacity (the maximum hourly volume that can pass through an intersection from a lane or group of lanes under prevailing roadway, traffic and control conditions) in veh/h.

$$d_2 = 900T[(X - 1) + \sqrt{(X - 1)^2 + \frac{8kIX}{cT}}] \tag{4}$$

where T is the duration of analysis in hours, k is the delay adjustment factor that is dependent on the signal controller mode and I is the upstream filtering/metering adjustment factor.

$$I = \frac{\mu}{\sigma^2} \tag{5}$$

where μ is the mean arrivals per cycle, and 2 is the variance of arrivals per cycle.

$$s = \frac{3600}{h} \tag{6}$$

where h is the saturation headway in seconds, and s is saturation flow rate in veh/h, defined as the maximum number of vehicles per hour per lane, which can pass through the intersection.

$$c = s \times \frac{g}{C} \tag{7}$$

3.4   Computing Best Green Time

To minimize delay at any intersection, traffic engineers need to select the best green time for every single signalized movement. To achieve that, we will need to calculate the total vehicle delay at an intersection using the formulae aforementioned. Assuming analysing an isolated intersection (PF = 1) and $d_3 = 0$, the delay equation yields:

$$d_i = d_1 i + d_2 i \tag{8}$$

Total delay of all vehicles at an intersection is given during T time by:

$$\sum_{i=1}^{R}\sum_{t=0}^{T} D_i t = \sum_{i=1}^{R}\sum_{t=0}^{T} d_1 it + \sum_{i=1}^{R}\sum_{t=0}^{T} d_2 it \qquad (9)$$

where i identifies the movement 1, 2, .. R, and t identifies the delay calculation time 0, 1, 2, T in seconds. To set the optimal signal time, the effective green time should be determined such that dDt/dgijk = 0; the derivative of total delay with respect to effective delay of movements. Combining Eqs. (2) and (4) gives us:

$$D_t = \frac{0.5C(1 - \frac{g}{C})}{(1 - [min(1, X)\frac{g}{C}])} + 900T\left[(X - 1) + \sqrt{(X - 1)^2 + \frac{8kIX}{cT}}\right] \qquad (10)$$

Therefore, when X > 1

$$D_t = 0.5C + 900T\left[(\frac{vC}{sg} - 1) + \sqrt{(\frac{vC}{sg} - 1)^2 + (\frac{8kIvC^2}{(sg)^2T})}\right] \qquad (11)$$

Considering D_t as a function of g and taking partial derivatives will yield:

$$\frac{\delta D}{\delta g} = 900T\left(\frac{-\frac{2Cv(\frac{Cv}{sg} - 1)}{sg^2} - \frac{16IC^2kv}{g^2Tg^3}}{2 \times \sqrt{\frac{8IC^2kv}{s^2Tg^2} + \frac{Cv}{sg} - 1)^2}} - \frac{Cv}{sg^2}\right) \qquad (12)$$

However, when $X \le 1$

$$D_t = \frac{0.5C(1 - \frac{g}{C})}{1 - \frac{sg^2}{C^2}} + 900T\left[(\frac{vC}{sg} - 1) + \sqrt{(\frac{vC}{sg} - 1)^2 + \frac{8kIvC^2}{T(sg)^2}}\right] \qquad (13)$$

This time we get $\delta D/\delta g$ as

$$\frac{\partial D}{\partial g} = -\frac{1}{2\left(1 - \frac{sg^2}{C^2}\right)} + \frac{sg\left(1 - \frac{g}{C}\right)}{C\left(1 - \frac{sg^2}{C^2}\right)^2} + 900T\left(\frac{-\frac{2Cv\left(\frac{Cv}{sg} - 1\right)}{sg^2} - \frac{16IC^2kv}{s^2Tg^3}}{2\sqrt{\left(\frac{8IC^2kv}{s^2Tg^2} + \left(\frac{Cv}{sg} - 1\right)^2\right)}} - \frac{Cv}{sg^2}\right) \qquad (14)$$

To determine the g value of a particular movement in relation to the other movements during time T (and assuming all other values are predetermined), the sum derivatives of delays across all other movements needs to be substituted into either Eqs. (11) or (13) depending on X value such that:

$$\frac{\partial D}{\partial g} \sum_{i=1}^{R} \sum_{t=0}^{T} D_i t = 0 \tag{15}$$

To calculate the ideal green time, real-time traffic volume is practised using the deep learning network, and the value is plugged into the aforementioned delay formulae. Other elements of the equation are either pre-set (constant) or calculated using relevant formulae.

Since the determining factor for the optimization of traffic signals is congestion reduction in essence traffic volume and the analysis is specific for an isolated intersection, a few assumptions are made:

- $h = 1.9\,s$, based on researches mentioned in [?],
- $s = 1750$ v/h; a theoretical value assumed by most researchers [5],
- $PF = 1.0$,
- $k = 0.5$,
- $I = 1.0$,
- $T = 5/60\,h$; the captured traffic volume is in 5 min intervals.
- $C = 85\,s$, based on the average cycle lengths obtained from Auckland Transport authority. This value will be recalibrated using Eq. (7) once effective green time is determined, anyways.
- Plugging-in the above assumptions into Eqs. (3), (11) to (14) can be simplified to: $c = s \times g/C$

To determine X (v/c) both values need to be measured in hours, but since the dataset captures traffic volume v in intervals of five minutes and capacity c is calculated in hours, the arithmetic below assumes measurements of five minutes instead of sixty.

Through basic Calculus and with the help of Matlab, all the values of effective green time g were determined and added to the table for every corresponding movement.

Effectively, when $X(v/c) > 1$,

$$D_t = 42.5 + 900T \left[\left(\frac{v}{22g} - 1 \right) + \sqrt{\left(\frac{v}{22g} - 1 \right)^2 + \frac{v}{121.35g^2 T}} \right] \tag{16}$$

The partial derivative of D with respect to g would be equal to:

$$\frac{\partial D}{\partial g} = 900T \left(\frac{-\frac{v\left(\frac{v}{22g}-1\right)}{11g^2} - \frac{40v}{2427Tg^3}}{2\sqrt{\left(\frac{20v}{2427Tg^2} + \left(\frac{v}{22g} - 1\right)^2\right)}} - \frac{v}{22g^2} \right) \tag{17}$$

To solve the equation for g, we will need $\partial D/\partial g = 0$. However, this does not yield real numbers, i.e. imaginary values are not an option here. That proves the logical concept of having the volume greater than the capacity of the intersection. This is because mathematical manipulation will not provide a realistic green time value if other mechanical and geometric engineering of the intersection are not fixed. However, the study of this phenomenon is beyond the scope of this research.

4   The Experiment

4.1   Scenario Set-Up

During the initial steps of the experiments traffic data was the only input fed into the network. However, to attain more knowledge about the environment being analysed, the neural network needs to be fed with as many environment related parameters as possible. Some of those parameters are traffic volume of other movements and intersections, weather condition and incidents and events nearby. Due to lack of data, this experiment was conducted using time series, weather condition and traffic volume of the same intersection as inputs of the ANN.

For road network and traffic demand, a data set for the intersection Redoubt Road and Great South Road of South of Auckland has been provided by Auckland Transport a public transport branch of Auckland Council. To experience the impact of missing values in a large dataset, the month of May was selected as an ideal period that virtually contained no missing values as opposed to the other months. Hence, two sets of experiments were attempted, each with many trials using different permutations; one for Oct 2014–Oct 2015, and the other for May 2015.

The information presented in the dataset shows the traffic volume of every signalized movement time stamped by date and hours of the day. To make it computationally easier and provide meaningful time stamps, the date and hours are merged into a separate cell called Date_Time which corresponds to the traffic volume of all the movements in a period of five minutes (300 s). It is worth noting that the missing values in the data set were substituted with the median of the entire column containing the missing value. Even though, those substituted data were not realistic values, it was determined that they are better than no data.

Figure 1 shows a representation of one of the experiments conducted during this study. The figure does not show all the neurons and layers due to lack of space. The input layer is composed of timestamp, raining rate and traffic volume of all the other movements except the movement being forecasted. The output layer is used for predicting the interested movement and the predicted vehicle volume, which is not included in the input layer.

The relationship between metrological conditions is a well know phenomenon noticed by road users, and backed by many researchers including [16] in which it proves the reduction of traffic flow due to weather conditions (rain, fog, mist, haze, or snow); however, this experiment does not attempt to establish this association. It rather, in addition to the time factor, adds more realistic parameters to the network. Out of that dataset, the hourly ratio (in millimetres) of rain was

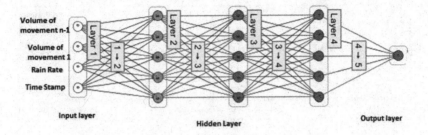

Fig. 1. Representation of the ANN used in the experiment

selected for Oct 2014 to Oct 2015 the same period as the traffic volume dataset. However, since our traffic volume data was structured at a 5 min frequency, the hourly rain data needed to be divided by five so it could be fed into the neural network along with traffic volume in parallel.

To take advantage of the big data and fast computation capability of the Cloud, the dataset was uploaded to an experiment created on Azure Machine Learning platform. From there, a Softmax based activation function was created to train deep learning neural networks.

The rest of the experiment was configured with the following permutations:

- Number of input layers was selected to be 25; composed of raining rate in mm, timestamp and the 23 movements. The total number of movements was 24, but every time the traffic volume rate of one movement is predicted. Hence, total 1.
- Number of hidden layers is described in Table 1.
- Number of neurons in the hidden layers is selected to 26 per hidden layer, i.e. N+1 [18] where N is the number of input layers.
- Learning rates: 0.01, 0.02, 0.04, 0.05, 0.06 consecutively
- Number of iterations: 20, 40, 80, 160, 320, 640
- The type of normalizer: Min-Max
- The momentum: 0
- Activation functions: Sigmoid

To reduce the experiment time and choose the best parameters, the above settings were applied in one go.

During the experiment it was determined that best traffic volume forecast for a particular move was obtained when participating all the other signalized movements of the intersection. In other words, when constructing the neural network architecture, timestamp and all the movements (minus the movement in question) should serve as inputs of the neural network. This process should be repeated until the network learned about all the movements.

4.2   Resuts

The initial experiment which was conducted on a subset of data which belonged to only the month of May 2015 and had no missing values, gave a substantial

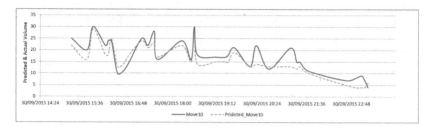

Fig. 2. Movement 10 actual traffic volume vs. predicted volume with five hidden layers

approximation between real traffic volumes and predicted ones. This experiment only took around 5 min to complete, and subsequent runs less than eight seconds as the network had already learnt about the network. However, the network took more than three hours to be trained when a whole years data was fed into the network, and subsequently was again fast less than eight seconds. In this scenario the impact of missing values was apparent on the disparity between real and predicted values, even though much more data was present which should have made the approximation even closer.

A bigger data set (Oct 2014–Oct 2015) and including rain data for movement 10. The set-up was later conducted using various permutations of number of hidden layers, and monitored by observing monitoring Mean Absolute Error (MAE), Root Mean Squared (RMS), Relative Absolute Error (RAE), Relative Squared Error (RSE) and Coefficient of Determination (CoD). Table 1 shows different Activation Functions and respective performance with Number of Nodes (Fig. 2).

Since mean square error is the mechanism in back propagation for training the networks, RMSE was picked for the comparison between different combinations of numbers of hidden layers. Table 1 shows that the best performance is gained when the number of hidden layers is five. Three hidden layers gave the worst performance.

As can be seen in the results, the proposed model does not seem to be performing well in low traffic flow conditions, and that confirms the results found in [2]. This is due to the fact that the lower difference between predicted and actual values can result in larger relative error when the traffic flow rate is

Table 1. Activation function and number of nodes performance

No. of layers hidden layers	MAE	RMSE	RAE	RSE	CoE
5	2.4919238	3.349391	0.315202	0.130396	0.869604
4	2.51361645	3.359593	0.317946	0.131191	0.868809
7	2.54743383	3.398744	0.322224	0.134267	0.865733
6	2.62368841	3.502543	0.331869	0.142593	0.857407
3	2.67769041	3.567632	0.3387	0.147942	0.852058

Table 2. Activation function and number of nodes performance

Model	RMSE	MAE
Deep Learning	3.4507	2.5283
Radial Basic Function	7.4766	6.1887
Random Walk	7.5299	6.2209
Support Vector Machine	7.8930	6.1887
BP Neural Network	9.7005	7.7974

small. However, this outcome was not a huge concern in this situation as the concern here was heavy traffic prediction for the purpose of reducing delay and congestion.

When compared RMSE and MAE results generated by experiments in this paper against their counterparts of different prediction models, it was determined that deep learning presents more accurate approach for traffic prediction than other models. Table 2 lists RMSE and MAE for BP Neural Network, Random Walk (RW), Support Vector Machine (SVM) and Radial Basic Function (RBF) [2] as well as deep learning neural networks.

In this experiment, traffic volume data, time and weather condition (raining) were the only inputs for the analysis, even though other factors such as incidents and events could have been used as well. However, incidents are quite random and unknown. The author did not have access to the database that included a schedule of events.

To determine the lowest accessible delay, traffic engineers can plug in predicted traffic volume and series of green times into Eq. (18). As an example, this experiment calculated the delay of 2 min per vehicle when the effective green time was set to 5 seconds and there were 12 vehicles waiting at the movement.

5   Conclusions and Future Works

The deep learning technique of Neural Networking is very effective in establishing numerical regression. This paper shows that the deep learning approach had greater performance advantage in forecasting the traffic volume at isolated intersections over other approaches discussed in this paper. The experiments have also shown that the system used here works under normal conditions and can be applied across all traffic lights. Whenever the traffic condition violates the normal pattern, the algorithm learns about it and starts implementing it when similar patterns are expected. This behaviour would make traffic signals more intelligent such that rather than being reactive to changes, they would become proactive to a large degree.

When performing the experiment, it was determined that feeding traffic data of all the movements of the intersection to the network provided significant results. The difference was due to the extra knowledge concluded from the all the approaches connecting to the neighbouring arterial, in which they all together could explain the nature of the traffic at the time of the experiment.

Challenges were not absent in the works of this experiment. It was represented in implementation of the theories and results in the real world. The other challenges included access to commercial simulation packages, as well as traffic control systems (such as SCATS) to try out the new set up. Furthermore, the datasets used in this experiment had missing information which were substituted by the median value of the entire column. This led to repetition of many values in the same column, and ultimately unrealistic data to reflect on the prediction model.

Future work - multiple steps can be completed to expand the work of this research and effectively boost the algorithm so it covers an entire arterial network. Researchers can extend the work to include traffic behaviour of neighbouring intersections by observing volume and delay of the egress of a neighbouring movement that connects to the ingress of the movement of the next intersection. Furthermore, delay can be measured using real time volume and allocated green times. Consequently, delay can become another input to the neural network to make the forecasting more accurate and at the same time predict the delay component as well to get effective green time using Eq. (19).

References

1. Huang, Y., Weng, Y., Zhou, M.: Modular design of urban traffic-light control systems based on synchronized timed petri nets. IEEE Trans. Intell. Transp. Syst. **15**(2), 530–539 (2014)

2. Huang, W., Song, G., Hong, H., Xie, K.: Deep architecture for traffic flow prediction: deep belief networks with multitask learning. IEEE Trans. Intell. Transp. Syst. **15**(5), 1–11 (2014). doi:10.1109/TITS.2014.2311123

3. Hinton, G.E., Salakhutdinov, R.R.: Reducing the dimensionality of data with neural networks. Science **313**(5786), 504–507 (2006)

4. Lv, Y., Duan, Y., Kang, W., Li, Z., Wang, F.-Y.: Traffic flow prediction with big data: a deep learning approach. IEEE Trans. Intell. Transp. Syst. **16**(2), 865–873 (2015). doi:10.1109/TITS.2014.2345663

5. Mannering, F.L., Washburn, S.S.: Principles of Highway Engineering and Traffic Analysis, p. 227 (2012). ISBN-10:1118120140

6. Blythe, P., Ji, Y., Guo, W., Wang, W., Tang, D.: Short-term forecasting of available parking space using wavelet neural network model. Intelt. Transp. Syst. IET **9**, 202–209 (2015). doi:10.1049/iet-its.2013.0184

7. Williams, B.M., Hoel, L.A.: Modeling and forecasting vehicular traffic flow as a seasonal ARIMA process: theoretical basis and empirical results. J. Transp. Eng. **129**(6), 664–672 (2003)

8. Tselentis, D.I., Vlahogianni, E.I., Karlaftis, M.G.: Improving short-term traffic forecasts: to combine models or not to combine? Intell. Transp. Syst. IET **9**(2), 193–201 (2015). doi:10.1049/iet-its.2013.0191

9. Zare Moayedi, H., Masnadi-Shirazi, M.A.: Arima model for network traffic prediction, and anomaly detection. In: International Symposium on Information Technology, ITSim 2008, 26–28 August 2008, vol. 4, pp. 1–6 (2008). doi:10.1109/ITSIM.2008.4631947

10. Barimani, N., Kian, A.R., Moshiri, B.: Real time adaptive non-linear estimator/predictor design for traffic systems with inadequate detectors. Intell. Transp. Syst. IET **8**(3), 308–321 (2014)
11. Yun, I., Park, B.: Stochastic optimization for coordinated actuated traffic signal systems. J. Transp. Eng. **138**(7), 819–829 (2012)
12. Ge, X.-Y., Li, Z.-C., Lam, W.H.K., Choi, K.: Energy-sustainable traffic signal timings for a congested road network with heterogeneous users. IEEE Trans. Intell. Transp. Syst. **15**(3), 1016–1025 (2014). doi:10.1109/TITS.2013.2291612
13. Blythe, P., Ji, Y., Guo, W., Wang, W., Tang, D.: Short-term forecasting of available parking space using wavelet neural network model. IET Intel. Transp. Syst. **9**, 202–209 (2015). doi:10.1049/iet-its.2013.0184
14. Chen, X., Gao, Y., Wang, R.: Online selective kernel-based temporal difference learning. IEEE Trans. Neural Netw. Learn. Syst. **24**(12), 1944–1956 (2013). doi:10.1109/TNNLS.2013.2270561
15. De Oliveira, M.B.W., De Almeida Neto, A.: Optimization of traffic lights timing based on multiple neural networks. In: Proceedings of the International Conference on Tools with Artificial Intelligence, ICTAI, pp. 825–832 (2013). doi:10.1109/ICTAI.2013.126
16. Koonce, P., Rodergerdts, L., Lee, K.: Signal Timing Manual. Federal Highway Administration (2008)
17. LeCun, Y., Yoshua Bengio, G.H.: Deep learning. Nature **521**(7553), 436–444 (2015)
18. Hunter, D., Yu, H., Pukish, M.S., Kolbusz, J., Wilamowski, B.M.: Selection of proper neural network sizes and architectures: a comparative study. IEEE Trans. Ind. Inf. **8**(2), 228–240 (2012)

An Online Competence-Based Concept Drift Detection Algorithm

Anjin Liu[1]([✉]), Guangquan Zhang[1], Jie Lu[1], Ning Lu[2], and Chin-Teng Lin[1]

[1] QCIS, University of Technology Sydney, Ultimo, NSW 2007, Australia
anjin.liu@student.uts.edu.au, {guangquan.zhang,jie.lu}@uts.edu.au
[2] SAS Institute Inc., Lane Cove, NSW 2066, Australia
ning.lu@sas.com, chintenglin@gmail.com

Abstract. The ability to adapt to new learning environments is a vital feature of contemporary case-based reasoning system. It is imperative that decision makers know when and how to discard outdated cases and apply new cases to perform smart maintenance operations. Competence-based empirical distance has been recently proposed as a measurement that can estimate the difference between case sample sets without knowing the actual case distributions. It is reportedly one of the most accurate drift detection algorithms in both synthetic and real-world data sets. However, as the construction of competence models have to retain every case in memory, it is not suitable for online drift detection. In addition, the high computational complexity $O(n^2)$ also limits its practical application, especially when dealing with large scale data sets with time constrains. In this paper, therefore, we propose a space-based online case grouping strategy, and a new case group enhanced competence distance (CGCD), to address these issues. The experiment results show that the proposed strategy and related algorithms significantly improve the efficiency of the current leading competence-based drift detection algorithm.

Keywords: Case base reasoning · Concept drift · Online clustering

1 Introduction

In machine learning, concept drift is a barely predictable change in the learning environment that can cause the induced pattern to become incompatible with new data [7]. Such changes can be caused by the fast pace of personal preference changing, new emerging technologies or changes in national policies. Real-world examples include: personal assistance applications that deal with information filtering, macroeconomic forecasts, bankruptcy prediction and individual credit scoring [25]. Without considering the existence of concept drift, the accuracy of a well-trained learning model will deteriorate over time [7]. To minimize the damage caused by the drifts, a fast and accurate drift detection algorithms is desired.

There are two basic categories of concept drift: (1) virtual concept drift (drift in data distribution) and (2) real concept drift (drift in decision concepts) [6].

© Springer International Publishing AG 2016
B.H. Kang and Q. Bai (Eds.): AI 2016, LNAI 9992, pp. 416–428, 2016.
DOI: 10.1007/978-3-319-50127-7_36

These two basic drift categories can be sub-divided into many different types according to the properties of the drift, such as sudden drift which has a fast drift speed or seasonal drift which occurs once per season. [17]. Since Schlimmer and Granger Jr. [18] first raised the problem of concept drift, numerous algorithms have been proposed for addressing this problem. These include rule-based learning [16,23], decision trees [8,10], info-fuzzy networks [11] and ensemble learners with different base models [5,17,22]. A case-based reasoning drift detection algorithm [14] was recently proposed based on competence models. It has the ability to highlight conflicting cases and has been reported as one of the best case-based reasoning drift detection algorithms [13]. It had been tested with different types of concept drifts and achieved a significant success. However, as a lazy learner, this algorithm is criticized for its high maintenance cost. It has to store every case in memory to analysis the discrepancy. Its applications for online machine learning tasks have been greatly limited by these shortcomings, especially for data stream mining problems that require a one pass process with very short time frame and minimum storage.

Motivated by these issues, we propose an online competence-based drift detection algorithm. We also give the formal definition of case groups and the definition of case group enhanced competence-based empirical distance (or in shorter form, case group competence-based distance CGCD). By adopting a space-based case grouping strategy and applying the proposed CGCD, our algorithm can significantly reduced the computing time, and storage requirements. Moreover, the computational complexity of the proposed algorithm can be controlled by an input parameter in the case grouping process. It is more flexible for handling data sets with different size. Experiments shows that the computation time and storage costs have been significantly reduced without accuracy drop, which makes our algorithm more suitable for online tasks.

This paper is organized as follows: Sect. 2 briefly reviews previous work on competence-based concept drift detection. Section 3 presents the proposed CGCD and CGCD-based drift detection. Section 4 contains experiment descriptions, results and analysis. The paper concludes with Sect. 5.

2   Related Works and Preliminaries

2.1   Concept Drift and Concept Drift Detection Methods

Concept drift is defined as a phenomenon in which the statistical properties of a target domain change over time in an arbitrary way [7]. These changes might be caused by hidden variables or by features that cannot be measured directly. Formally, concept drift is defined as: at time t_i a set of observation cases is given, denoted as $S_i = \{c_0, ..., c_m\}$, for each case $c_i = (x_i, y_i)$, where $x_i = (x_{i_1}, ..., x_{i_n}) \in X$ is the feature vector, $y_i \in Y$ is the true label; at time t_i, the S_i follows a distribution $F_i(X, Y)$. Concept drift is identified whenever there is a statistically significant difference between any two observed case sample sets S_i, S_j that $F_i(X, Y) \neq F_j(X, Y)$. The state-of-the-art concept drift detection algorithms are often implicitly related to sudden drift detection [14].

These algorithms usually adopt a statistical test that tracks the change of: (1) raw data distribution [3,9]; (2) the output (error rates) of learners [6,12]; (3) the parameters of selected learners [21]. More details about the pros and cons of each drift detection category can be found in a recent survey [7]. Generally, category (1) algorithms provide a statistical significance level that ensures detected drifts are not caused by sampling errors, but they cannot easily explained the detected drift regions. Category (2) algorithms have much less computational complexity but they are not sensitive to local and gradual drift. To the best of our knowledge, only one attempt [21] has been made in respect of category (3). The authors' framework for modeling concept drift is creative and can be applied to many parameterized learning models. However, it is not suitable for knowledge-based learners and cannot interpret detected drifts well.

Competence-based drift detection is a special case of category (1) given above. It monitors the difference between raw data distribution through an estimation of the case competence distribution. As a result, the detected drift regions can be explained in competence-related terms. However, as it is highly dependent on the competence model, which incurs a high computational cost to build and to maintain, its practical application is limited. Therefore, an online version of competence-based drift detection is desirable.

2.2   Drift Detection via Competence Models

Competence-Based Empirical Distance. Given a case base CB, and two case sample sets S_1, $S_2 \subseteq CB$, the competence-based empirical distance is a rough estimation of the competence discrepancy between S_1, S_2 in CB. The intuitive explanation of the competence discrepancy is the difference between how well S_1 supports the case base and how well S_2 supports the case base. Competence-based empirical distance can reflect the discrepancy between two case sample sets without prior knowledge of their actual distribution [14]. The formal definition is presented below:

Definition 1 (see [19]). *For a case base $CB = \{c_1, \ldots, c_n\}$, a Coverage Set of a case denoted as $CoverageSet(c) = \{c^{'} \in CB : Solves(c, c^{'})\}$, where $Solves(c, c^{'})$ means that c can be retrieved and be used as a support case to solve $c^{'}$.*

Definition 2 (see [19]). *$ReachabilitySet(c) = \{c^{'} \in CB : Solves(c^{'}, c)\}$ is a set of cases that can be retrieved and applied to solve case c.*

Definition 3 (see [20]). *For a case c, the union set of its CoverageSet and ReachabilitySet is defined as $RelatedSet(c) = CoverageSet(c) \bigcup Reachability Set(c)$, denoted as $R^{CB}(c)$.*

Definition 4 (see [15]). *For $c \in CB$, a RelatedClosure of c with regard to CB is defined as $\Re^{CB}(c) = \{R^{CB}(c_i) : \forall c_i \in CB, \exists R^{CB}(c_i) st.c \in R^{CB}(c_i)\}$. For a group of cases $S \subseteq CB$, the RelatedClosure of S with regard to CB is defined as $\Re^{CB}(S) = \cup_{c \in S} \Re^{CB}(c)$.*

The *RelatedClosure* is a set of *RelatedSet*. It is able to uniquely model the entire case base.

Definition 5 (see [14]). *For a case base CB, let $\Re^{CB}(S) = \{\Re_1^{CB}(S), \ldots, \Re_n^{CB}(S)\}$, the density of $\Re_i^{CB}(S)$ is defined as:*

$$w^*(\Re_i^{CB}(S)) = \frac{1}{|S|} \times \sum_{c_i \in S} \frac{|\Re_i^{CB}(S) \cap \Re^{CB}(c_j)|}{|\Re^{CB}(c_j)|} \tag{1}$$

Definition 6 (see [14]). *Given a case base CB, and a case sample set $S \subseteq CB$, denote the power set of $\Re^{CB}(CB)$ as $\mathfrak{p}(\Re^{CB}(CB))$. Considering $\mathfrak{p}(\Re^{CB}(CB))$ as the measurable space $\mathfrak{A}$, for $A \in \mathfrak{A}$, the competence-based empirical weight of S with regard to A over CB is defined as:*

$$S^{CB}(A) = \sum_{\Re_i^{CB}(S) \in A \cap \Re^{CB}(S)}^{|A \cap \Re^{CB}(S)|} w^*(\Re_i^{CB}(S)) \tag{2}$$

For any case sample set S, the competence-based empirical weight presents how well a case set S supports a competence area represented by A. The higher the weight is, the better is the support that competence area A can obtain from case set S. Consequently, the difference between the weights of two sample sets in the same competence area reflects the difference between the importance of two sample sets related to this competence area. The cumulated difference between the weights of any two case sample sets S_1, S_2 over the entire case base CB can therefore be used as a metric to detect the global difference between S_1 and S_2.

Definition 7 (see [14]). *For any two case sample sets S_1, $S_2 \subseteq CB$, the competence-based empirical distance between S_1 and S_2 is defined as:*

$$d^{CB}(S_1, S_2) = 2 \times \sup_{A \in \mathfrak{A}} |S_1^{CB}(A) - S_2^{CB}(A)| \tag{3}$$

The competence-based empirical distance provides a rough estimation of the difference between the competence distribution of two case sets. This distance is a relaxation of the total variation distance. This method has three reported advantages: (1) achievement of a high detection rate, (2) robustness on small sample size, and (3) ability to quantify and describe the changes it detects, which makes it highly suitable for handling local concept drift problems [14]. However, concept drift detection for large scale data streams is still a challenging topic.

Statistical Guarantee. The measurement of the difference between two case sample sets is only the first step of detecting drifts. The second step is to proof the changes are statistical significant. To fairly compare with Lu's algorithm, we applied the same statistic test, which is the *two-sample non-parametric permutation test* [4]. In this case, null hypothesis H_0 is that there is a concept drift. The smaller the *p-value* is, the stronger the evidence against H_0.

Let us denote the *test statistic* of the permutation test as $\hat{\theta}$. In this case, the *test statistic* is the *competence-based empirical distance* between two given case sample sets S_1 and S_2. The *achieved significance level (ASL)*, which is also known as *p-value*, can be attained by counting the random variable $\hat{\theta}^*$ greater or equal to the observed $\hat{\theta}$ and dividing by the total number of tests:

$$ASL_{perm} \approx A\hat{S}L_{perm} = \frac{\#\{\hat{\theta}^* \geq \hat{\theta}\}}{N} \tag{4}$$

As discussed in Lu's paper, if we do not want the Monte Carlo Error to affect our estimation by more than 30%, we need to run at least 100 tests. This precision can be improved with larger N.

3 Case Group Enhanced Competence-Based Empirical Distance

3.1 Case Group Competence-Based Empirical Distance

The fundamental idea of our algorithm is to gather the same or similar cases into one weighted case group. Instead of calculating the competence-based empirical distance for each case, it is more efficient to detect the changes in terms of weighted cases.

Definition 8. *Let* $cg = \{c_1, \ldots, c_n\} \subseteq CB$, $cg \neq \phi$, *we say* cg *is a Case Group if for any* c_i, $c_j \in cg$, $c_i \neq c_j$ *and* $R^{CB}(c_i) = R^{CB}(c_j)$.

In other words, a case group is a set of the same or very similar cases that can solve exactly the same case set and càn be solved by exactly the same case set. As a result, two cases from the same case group have the same related set and the same related closure. This can be formally presented in Proposition 1.

Proposition 1. *For any cases* c_i *and* c_j, *if* c_i, $C_j \in cg$, *then* $c_i \in R^{CB}(c_j)$, $c_j \in R^{CB}(c_i)$ *and* $\Re^{CB}(c_i) = \Re^{CB}(c_j)$.

Proof. From Definition 3, we have $c_i \in R^{CB}(c_i)$, $c_j \in R^{CB}(c_j)$ and $\Re^{CB}(c_i) = \Re^{CB}(c_j)$, therefore, $c_i, c_j \in R^{CB}(c_i) = R^{CB}(c_j)$; from the Definition 4, which is $\Re(c) = \{R^{CB}(c_0, \ldots, R^{CB}(c_n))\}$ where $c_i \in R^{CB}(c)$. As any cases in a *Case Group* has the same $R^{CB}(c)$, therefore $\Re^{CB}(c_i) = \Re^{CB}(c_j)$ if $c_i, c_j \in cg$.

Definition 9. *The density of* $\Re_i^{CB}(S)$ *with regard to* S, *presented by Case Group is:*

$$w_{cg}^*(\Re_i^{CB}(S)) = \frac{1}{\sum_{cg_j \in S} |cg_j|} \times \sum_{cg_j \in S} |cg_j| \times \frac{|\Re_i^{CB}(S)| \cap |\Re^{CB}(c^{cg_j})|}{\Re^{CB}(c^{cg_j})} \tag{5}$$

where c^{cg_j} *is the core case of a case group* cg_j

Example 1. Let $S = \{c_3, c_4\}$ be a case sample set taken from the case base $CB = \{c_1, c_2, c_3, c_4\}$, $R^{CB}(c_1) = \{c_1, c_2\}$, $R^{CB}(c_2) = \{c_1, c_2, c_3, c_4\}$, $R^{CB}(c_3) = R^{CB}(c_4) = \{c_2, c_3, c_4\}$. So we have a *Case Group* $cg_1 = \{c_3, c_4\}$ and $c^{cg_1} = c_3 = c_4$. The *Related Closure* of S is $\mathfrak{R}^{CB}(S) = \{\{c_1, c_2, c_3, c_4\}, \{c_2, c_3, c_4\}\}$ and $\mathfrak{R}_1^{CB}(S) = \{c_1, c_2, c_3, c_4\}$. Substitute into Eq. 5, we have:

Theorem 1. *The density of $\mathfrak{R}_i^{CB}(S)$ presented by Case Group is equal to the density of $\mathfrak{R}_i^{CB}$ presented by single cases, namely $w_{cg}^*(\mathfrak{R}_i^{CB}(S)) = w^*(\mathfrak{R}_i^{CB}(S))$.*

Proof. If c_1, $c_2 \in cg_i$ and c_1 is a core case in cg_i, then $\mathfrak{R}(c_1) = \mathfrak{R}(c_2)$ (refer to Proposition 1). As a result, we obtain $\frac{|\mathfrak{R}_i^{CB}(S) \cap \mathfrak{R}^{CB}(c_1)|}{\mathfrak{R}^{CB}(c_1)} = \frac{|\mathfrak{R}_i^{CB}(S) \cap \mathfrak{R}^{CB}(c_2)|}{\mathfrak{R}^{CB}(c_2)}$. Therefore, $\sum_{c_i \in cg_i} \frac{|\mathfrak{R}_i^{CB}(S) \cap \mathfrak{R}^{CB}(c_i)|}{\mathfrak{R}^{CB}(c_i)} = |cg_i| \times \frac{|\mathfrak{R}_i^{CB}(S) \cap \mathfrak{R}^{CB}(c_1)|}{\mathfrak{R}^{CB}(c_1)}$ where $cg_i \subseteq S$.

As $w_{cg}^*(\mathfrak{R}_i^{CB}(S)) = w^*(\mathfrak{R}_i^{CB}(S))$, the *competence-based empirical distance* calculated based on single cases is equal to the distance calculated on the basis of case groups.

3.2 Space-Based Case Grouping Approximation

In the real world, the construction of ideal case groups, which have to maintain a $n \times n$ distance matrix to retrieve the *k nearest neighbour*, is a complicated process. Accordingly, we adapt micro-cluster clustering algorithms as an alternative process for constructing the case groups. We exploit the cluster-features of micro-clusters as a case grouping approximation. The cluster-features store the statistical summary of the grouped data points. As a result, a significant amount memory can be released. Moreover, the *additive* and the *subtractive* properties of cluster-feature vectors make it easy to maintain the entire case base in an online manner. Algorithm 1 is the case grouping procedure. Algorithm 2 is the CGCD-based drift detection algorithm.

In this grouping process, we only take the advantages of cluster-features. All other unnecessary information, such as core micro-cluster analysis and time features [1,2,24], is omitted. The center of a micro-cluster is an approximation of the core case in a case group. The weight of a micro-cluster is the size of a case group. As a result, the number of cases that need to be considered to construct the competence model can be significantly reduced. The computational complexity is now determined by the spacial granularity $\frac{\max(distance(c_i, c_j))}{\epsilon}$ instead of the number of cases n.

The first step of Algorithm 2 is to group similar cases. For each cluster-feature vector, we can acquire the group center and its weight. Secondly, a weighted competence model can be constructed. If the distance between two core cases from two case groups is smaller than a given threshold d_ϵ, then these two case groups are mutually solvable. Thirdly, with the new competence model, a case group enhanced competence-based empirical distance can be established. Lastly, we shuffle the two sample sets S_i, S_j and repeat for N times to run the permutation test. Another advantage of using cluster-features is that the number of

---

Algorithm 1. Case Grouping

---

    input : Case sample sets S_i
                Case similarity threshold ϵ
    output: Case Group Set S_i^{cg}

1   **for** c_i *in* S_i **do**
2   **if** $S_i^{cg} = \phi$ **then**
3   create new case group cg_1;
4   set $c^{cg_1} = c_i$;
5        $S_i^{cg} = S_i^{cg} \cup \{cg_1\}$
6   **else**
7   find c_i nearest case group cg_i in S_i^{cg};
8   **if** $dist(c_i, c^{cg_i}) \leq \epsilon$ **then**
9   merge c_i into cg_i;
10   update $c^{cg_i} \leftarrow c = \frac{\overline{CF1}}{w}$;
11   **else**
12   create new case group cg_k;
13   set $c^{cg_k} = c_i$;
14           $S_i^{cg} = S_i^{cg} \cup \{cg_k\}$
15   **end**
16   **end**
17 **end**
18 **return** S_i^{cg}

---

cluster-feature vectors will not increase along with the increasing of the cases. The number of cluster-feature vectors will only increase when there is an expansion in the feature space $\max(distance(c_i, c_j))$. The reason is that the enhanced competence model is build on space granularity, but not on the cases.

3.3   Discussion About Case Similarity Threshold

The selection of the case similarity threshold ϵ will directly affect the sensitivity of the drift detection. On the one hand, if ϵ is too big, dissimilar cases may be grouped together and the number of cases will be over-reduced. As a result, minor drift will not be detected. On the other hand, if ϵ is too small, no cases will be grouped together and the storage requirement cannot be satisfied. Therefore, the best choice is to select the ϵ to meet computational complexity requirements first, and then to set the ϵ as small as possible so that CGCD can be in its most sensitive state to detect concept drift. In addition, ϵ can be treated as a parameter to control the preferred drift detection level. As all cases with similarity less than ϵ will be grouped together, it is guaranteed that no concept drift can be detected in that case group region. Lastly, if there are no computational limits or other unfavorable conditions, $\epsilon = [0.05d_\epsilon, 0.2d_\epsilon]$ is recommended.

Algorithm 2. Case Group Competence Distance

---

 input : Case sample sets S_i, S_j
     Case similarity threshold ϵ
     Case solve threshold d_ϵ
     Number of permutation tests N
 output: Statistical test p-$value$

1 **for** $perm \leftarrow 0$ to N **do**
2 $S_i^{cg} = CaseGrouping(S_i, \epsilon)$;
3 $S_j^{cg} = CaseGrouping(S_j, \epsilon)$;
4 **for** $cg_i \in S_i^{cg} \cup S_j^{cg}$ **do**
5   **for** cg_j in $S_i^{cg} \cup S_j^{cg}$ **do**
6   **if** $dist(c^{cg_i}, c^{cg_j} \leq d)$ **then**
7     $R^{CB}(c^{cg_i}) = R^{CB}(c^{cg_i}) \cup \{cg_j\}$;
8   **end**
9   **end**
10   $\Re^{CB} = \Re^{CB} \cup \{R^{CB}(c^{cg_i})\}$;
11 **end**
12 $d_{cg}^{CB}(S_i^{cg}, S_j^{cg}) = \frac{1}{2} \sum_{R_i^{CB} \in \Re^{CB}}^{|\Re^{CB}|} |w_{cg}^*(\Re_i^{CB}(S_i^{cg})) - w_{cg}^*(\Re_i^{CB}(S_j^{cg}))|$;
13 **if** $perm = 0$ **then**
14   $\hat{\theta} = d_{cg}^{CB}(S_i^{cg}, S_j^{cg})$;
15 **else**
16   $\hat{\theta}^* = d_{cg}^{CB}(S_i^{cg}, S_j^{cg})$;
17 **end**
18 $shuffle(S_i, S_j)$;
19 **end**
20 **return** $1 - \frac{\#\{\hat{\theta}^* \geq \hat{\theta}\}}{N}$;

---

4 Evaluation

In this section, we conduct three experiments to compare the CGCD-based drift detection algorithm with the current leading competence-based drift detection algorithm [14]. Experiment 1 compares the execution time and the sensitivity to distribution change of these two algorithms. Experiment 2 compares the drift detection accuracy of these two algorithms on seven synthetic data sets. All experiments were conducted on a 3.1 GHz 8 core CPU and 128 GB RAM cluster node with unique access.

Experiment 1: Evaluating the Competence-Based Empirical Distance.
In this experiment, series cases are coming from a synthetic data stream with 22 time steps from 11 consecutive 1-D normal distributed data, with a batch size of 5000. The batch size in the original work is 50, but to demonstrate the efficiency of our method, we increased the data size to 5000. Each normal distribution has a

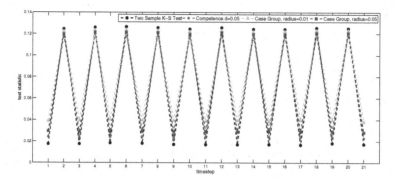

Fig. 1. Test statistic between normal distributed data with varying μ

Table 1. Running time and number of cases in memory

Parameter settings	Execution time (min)	Cases in memory
$d_\epsilon = 0.05$, $\epsilon = N/A$	426.88	10,000
$d_\epsilon = 0.05$, $\epsilon = 0.01$	2.61	181
$d_\epsilon = 0.05$, $\epsilon = 0.005$	4.56	337

fixed standard deviation of $\sigma = 0.2$ and a moving mean of $\mu = 0.2 + 0.06 \times (i-1)$, where i represents the ith distribution. When $t = 2 \times i - 1$ we compare two samples that are both drawn from the ith distribution; when $t = 2 \times i$ we compare two samples drawn from the ith and $(i+1)^{th}$ distribution. We run this test on 100 synthetic data sets and plot the average distance in Fig. 1. The case solving threshold is set as $d_\epsilon = 0.05$ for all tests in this experiment. The max radii of micro-cluster is set as $\epsilon = 0.005$ $(0.01d_\epsilon)$ and $\epsilon = 0.001$ $(0.02d_\epsilon)$. The efficiency of the original work and CGCD is shown in Table 1, and how the empirical distance changes as the distributions vary is shown in Fig. 1. From Table 1 we can find that the execution time of CGCD is much shorter than Lu's work [14]. The reason is that CGCD has much fewer cases to calculate after applying case group approximation process. This process can effectively reduce the number of cases for computing. Table 1 also shows that larger micro-cluster radius ($\epsilon = 0.01$) can reduce more cases than a smaller radius ($\epsilon = 0.05$), as expected. Regarding to the distance changes, we added *two samples Kolmogorov − Smirnov test* (two-sample K-S test) as a base line. Two-samples K-S test is a nonparametric test of the one-dimensional probability distributions that can be used to test whether two underlying one-dimensional probability distributions are differ. In Fig. 1, as the differences between samples depend only on the mean values, all peaks have nearly the same height. The two-sample K-S test is the most sensitive test statistics to the change, as the margin between the peaks and valleys is large. The second one is competence-based empirical distance. The third and firth one are CGCD with different micro-cluster radius. It can be seen that the case

Table 2. Results of case group enhanced competence-based drift detection

Stream	Window size(n)	Detected	Late	False	Missed	Cases in memory
M(0.05)	10,000	99(99)	0(0)	5(2)	0(0)	2846(20000)
M(0.02)	10,000	91(87)	7(5)	13(12)	1(7)	2847(20000)
C(0.01)	10,000	31(33)	18(19)	3(2)	50(47)	1277(20000)
C(0.15)	10,000	83(81)	7(10)	7(7)	9(8)	2906(20000)
C(0.15)	5,000	57(55)	19(24)	9(14)	23(20)	2113(10000)
C(0.2)	5,000	84(82)	9(9)	8(8)	6(8)	1997(10000)
P(0.1)	10,000	80(76)	9(10)	5(5)	10(13)	1229(20000)
P(0.2)	10,000	99(99)	0(0)	5(5)	0(0)	1277(20000)

group approximation process only has very little impact on the competence-based empirical distance. It means that our method retain the power of the discrepancy detection.

Experiment 2: Evaluating the Change Detection Method. In this experiment, we test the CGCD-based drift detection algorithm on seven synthetic data streams. Each data stream has 5,000,000 data instances with two features following a specific data distribution. For every 50,000 instances, the parameters of the corresponding distribution will vary randomly within a certain range. These variations produce 99 controllable concept drifts. They are the same as in [3,14]. The parameters for the permutation test have been configured as: significance level $\alpha = 0.01$ and permutation size $N = 500$, for both algorithms.

In $M(\triangle)$ streams, feature x_1 and x_2 follow an independent normal distribution with mean μ_1 and μ_2 moving in $[0.2, 0.8]$. For every 50,000 instances, the mean μ_1 and μ_2 are randomly updated with a step size $[-\triangle, -\triangle/2] \cup [\triangle/2, \triangle]$. In $C(\triangle)$ streams, the two features follow a two-dimensional normal distribution with a moving ρ, which starts at 0 and then randomly walks within $[-1, 1]$, with step size chosen randomly in $[-\triangle, -\triangle/2] \cup [\triangle/2, \triangle]$. The parameters for the $M(\triangle)$ and $C(\triangle)$ streams are $d_\epsilon = 0.05, \epsilon = 0.2d_\epsilon$, as in Lu's work.

In $P(\triangle)$ streams, the two features follow a Poisson distribution with $Poisson$ $(500(1 - \rho), 500(1 - \rho), 500\rho)$, where ρ starts at 0.5 and then performs a random walk between 0 and 1 with step size $\triangle = 0.2$ or $\triangle = 0.2$. The parameters for $P(\triangle)$ streams are $d_\epsilon = 10, \epsilon = 0.2d_\epsilon$, as in Lu's work.

We run both the CGCD-based drift detection and the original competence drift detection [14] on these data sets. The results are shown in Table 2. The numbers in brackets are the results of the original competence drift detection. In Table 2, the figure under *Detected* indicates the number of times the drift detection algorithm detected a significant difference right after the parameter update; *Late* indicates the number of times a significant difference was detected within the next two time steps later than the update; *False* indicates the number of false alarm with no parameter update; *Missed* indicates the number of times no significant difference was found between two updates. The results demonstrate that CGCD-based drift detection detects concept drifts as accurately as the original work while it takes fewer storage requirements. Since the permutation test can be conducted with multi threads, and we have already compared the execution time of both algorithms in Experiment 1, we intentionally leave out the running time of each data set for this experiment.

5 Conclusion

In this paper, we have proposed a case group enhanced competence drift detection algorithm and a case group approximation method. The improved algorithm can dramatically reduce the computational complexity of the original algorithm and it has offered a solution for calculating the competence-based empirical distance on very large scale data sets. Moreover, the computational complexity of the improved algorithm will not increase along with the growth of the case base size, unless there is an expansion in the case feature space. The reason is that the case group approximation method we adopted in this work can keep absorbing the incoming data instances based on spacial granularity. The experiments showed that the proposed algorithm can perform well on most types of concept drift.

The case group enhanced competence model provides a solution for calculating competence-based empirical distance for large volumes of data. However, the permutation test of the statistical guarantee is still time consuming. In the future, replacing the *two-sample non-parametric permutation test* by a more appropriate statistical guarantee method for CGCD is desired.

Acknowledgment. This work is supported by the Australian Research Council (ARC) under discovery grant DP150101645. Also, the authors would like to thank the anonymous reviewers for their valuable feedback and all members of the Decision Systems and e-Service Intelligence laboratory of University of Technology Sydney for discussion.

References

1. Aggarwal, C.C., Han, J., Wang, J., Yu, P.S.: A framework for clustering evolving data streams. In: Proceedings of the Twenty-Ninth International Conference on Very Large Data Bases, vol. 29, pp. 81–92. VLDB Endowment (2003)
2. Cao, F., Ester, M., Qian, W., Zhou, A.: Density-based clustering over an evolving data stream with noise. In: Proceedings of the Sixth SIAM International Conference on Data Mining, vol. 6, pp. 328–339. SIAM (2006)
3. Dasu, T., Krishnan, S., Venkatasubramanian, S., Yi, K.: An information-theoretic approach to detecting changes in multi-dimensional data streams. In: Proceedings of the Symposium on the Interface of Statistics, Computing Science, and Applications, 24-27 May 2006, pp. 1–24. Citeseer (2006)
4. Efron, B., Tibshirani, R.J.: An Introduction to the Bootstrap. Chapman and Hall, New York (1994)

5. Elwell, R., Polikar, R.: Incremental learning of concept drift in nonstationary environments. IEEE Trans. Neural Netw. **22**(10), 1517–1531 (2011)
6. Gama, J., Medas, P., Castillo, G., Rodrigues, P.: Learning with drift detection. In: Bazzan, A.L.C., Labidi, S. (eds.) SBIA 2004. LNCS (LNAI), vol. 3171, pp. 286–295. Springer, Heidelberg (2004). doi:10.1007/978-3-540-28645-5_29
7. Gama, J., Zliobait, I., Bifet, A., Pechenizkiy, M., Bouchachia, A.: A survey on concept drift adaptation. ACM Comput. Surv. **46**(4), 1–37 (2014)
8. Hulten, G., Spencer, L., Domingos, P.: Mining time-changing data streams. In: Proceedings of the Seventh ACM SIGKDD International Conference on Knowledge Discovery and Data Mining, pp. 97–106. ACM (2001)
9. Kifer, D., Ben-David, S., Gehrk, J.: Detecting change in data streams. In: Proceedings of the Thirtieth International Conference on Very Large Data Bases, vol. 30, pp. 180–191. VLDB Endowment (2004)
10. Kolter, J.Z., Maloof, M.A.: Dynamic weighted majority: an ensemble method for drifting concepts. J. Mach. Learn. Res. **8**, 2755–2790 (2007)
11. Last, M.: Online classification of nonstationary data streams. Intell. Data Anal. **6**(2), 129–147 (2002)
12. Li, P., Hu, X., Liang, Q., Gao, Y.: Concept drifting detection on noisy streaming data in random ensemble decision trees. In: Perner, P. (ed.) MLDM 2009. LNCS (LNAI), vol. 5632, pp. 236–250. Springer, Heidelberg (2009). doi:10.1007/978-3-642-03070-3_18
13. Lu, N., Lu, J., Zhang, G., de Mantaras, R.L.: A concept drift-tolerant case-base editing technique. Artif. Intell. **230**, 108–133 (2016)
14. Lu, N., Zhang, G., Jie, L.: Concept drift detection via competence models. Artif. Intell. **209**, 11–28 (2014)
15. Lu, N., Zhang, G., Lu, J.: Detecting change via competence model. In: Bichindaritz, I., Montani, S. (eds.) ICCBR 2010. LNCS (LNAI), vol. 6176, pp. 201–212. Springer, Heidelberg (2010). doi:10.1007/978-3-642-14274-1_16
16. Maloof, M.A., Michalski, R.S.: Incremental learning with partial instance memory. Artif. Intell. **154**(1), 95–126 (2004)
17. Minku, L.L., White, A.P., Xin, Y.: The impact of diversity on online ensemble learning in the presence of concept drift. IEEE Trans. Knowl. Data Eng. **22**(5), 730–742 (2010)
18. Schlimmer, J.C., Granger Jr., R.H.: Incremental learning from noisy data. Mach. Learn. **1**(3), 317–354 (1986)
19. Smyth B., Keane, M.T.: Remembering to forget: a competence-preserving case deletion policy for case-based reasoning systems. In Proceedings of the Fourteenth International Joint Conference on Artificial Intelligence, 20-25 August 1995, pp. 377–382. Morgan Kaufmann(1995)
20. Smyth, B., McKenna, E.: Modelling the competence of case-bases. In: Smyth, B., Cunningham, P. (eds.) EWCBR 1998. LNCS, vol. 1488, pp. 208–220. Springer, Heidelberg (1998). doi:10.1007/BFb0056334
21. Su, B., Shen, Y.-D., Xu, W.: Modeling concept drift from the perspective of classifiers. In: IEEE Conference on Cybernetics and Intelligent Systems, 21–24 September 2008, pp. 1055–1060. IEEE (2008)
22. Wang, H., Fan, W., Yu, P.S., Han, J.: Mining concept-drifting data streams using ensemble classifiers. In: Proceedings of the Ninth ACM SIGKDD International Conference on Knowledge Discovery and Data Mining, pp. 226–235. ACM (2003). doi:10.1145/956750.956778
23. Widmer, G., Kubat, M.: Learning in the presence of concept drift and hidden contexts. Mach. Learn. **23**(1), 69–101 (1996)

24. Zhang, T., Ramakrishnan, R., Livny, M.: BIRCH: an efficient data clustering method for very large databases. In: Proceedings of the Sixteenth International Conference on Management of Data, pp. 103–114. ACM (1996). doi:10.1145/233269.233324
25. Zliobaite, I.: Learning under concept drift: an overview. Report, Faculty of Mathematics and Informatics, Vilnius University (2009)

Bayesian Robust Regression
with the Horseshoe+ Estimator

Enes Makalic$^{(\boxtimes)}$, Daniel F. Schmidt, and John L. Hopper

Centre for Epidemiology and Biostatistics, The University of Melbourne,
Carlton, VIC 3053, Australia
{emakalic,dschmidt,j.hopper}@unimelb.edu.au

Abstract. The horseshoe+ estimator for Gaussian linear regression models is a novel extension of the horseshoe estimator that enjoys many favourable theoretical properties. We develop the first efficient Gibbs sampling algorithm for the horseshoe+ estimator for linear and logistic regression models. Importantly, our sampling algorithm incorporates robust data models that naturally handle non-Gaussian data and are less sensitive to outliers. The resulting software implementation provides a powerful, flexible and robust tool for building prediction and classification models from potentially high-dimensional data and represents the state-of-the-art in Bayesian machine learning techniques.

1 Introduction

Bayesian regression models are becoming increasingly common in the big data domain. Consider the following Bayesian regression hierarchy for data $\mathbf{y} = (y_1, \ldots, y_n)^{\mathrm{T}} \in \mathbb{R}^n$:

$$y_i | \mathbf{X}, \boldsymbol{\beta}, \beta_0, \omega_i^2, \sigma^2 \sim \mathcal{N}(\mathbf{x}_i^T \boldsymbol{\beta}, \omega_i^2 \sigma^2), \tag{1}$$

$$\omega_i^2 \sim \pi_\omega(\omega_i^2) d\omega_i^2, \tag{2}$$

$$\sigma^2 \sim \sigma^{-2} d\sigma^2, \tag{3}$$

$$\beta_j | \lambda_j^2, \tau^2, \sigma^2 \sim \mathcal{N}(0, \lambda_j^2 \tau^2 \sigma^2), \tag{4}$$

$$\beta_0 \sim d\beta_0, \tag{5}$$

$$\lambda_j \sim \mathcal{C}^+(0, \eta_j), \tag{6}$$

$$\eta_j \sim \mathcal{C}^+(0, 1), \tag{7}$$

$$\tau \sim \mathcal{C}^+(0, 1) \tag{8}$$

where $i = (1, \ldots, n)$, $j = (1, \ldots, p)$, $\mathbf{X} \in \mathbb{R}^{n \times p}$ is a matrix of predictor variables (not necessarily full rank), $\beta_0 \in \mathbb{R}$ and $\boldsymbol{\beta} \in \mathbb{R}^p$ are the unknown regression coefficients, $\mathcal{N}(\cdot, \cdot)$ denotes the Gaussian distribution and $\mathcal{C}^+(0, 1)$ is the standard half-Cauchy distribution with probability density function

$$p(z) = \frac{2}{\pi(1 + z^2)}, \quad z > 0.$$

© Springer International Publishing AG 2016
B.H. Kang and Q. Bai (Eds.): AI 2016, LNAI 9992, pp. 429–440, 2016.
DOI: 10.1007/978-3-319-50127-7_37

The hierarchy introduced in (1)–(8) comprises two groups: (i) the sampling distribution of the data $\mathbf{y} \in \mathbb{R}^n$ given by (1)–(3) and (ii) the prior distributions of the regression coefficients β_0 and $\boldsymbol{\beta}$ given by (4)–(8). The data model is built from a scale mixture of normal distributions, which is a standard technique for representing a wide range of statistical distributions [1]. For example, by appropriate choice of the prior distribution $\pi_\omega(\cdot)$, one may model the data $\mathbf{y}$ as arising from a heavy-tailed distribution, such as the Cauchy or the Student-t distribution, or even a distribution for categorical data, such as the logistic model.

In most regression problems, the main task is to estimate the unknown regression parameters as well as determine which predictors should be included in the model. It is becoming more common that the dimensionality of the predictors p is large, often much larger than the sample size n (i.e., $p \gg n$), and the performance of the aforementioned Bayesian estimator depends crucially on the particular choice of prior distributions (6)–(8). In big data problems, it is common to assume that the number of predictor variables associated with the outcome $\mathbf{y}$ is small relative to the overall dimensionality p. In other words, the majority of the regression coefficients $\boldsymbol{\beta}$ are assumed to be exactly equal to zero (i.e., the sparse model assumption).

Bhadra et al. [2] recently introduced the Bayesian horseshoe+ hierarchy for ultra-sparse regression problems, which enjoys many favourable theoretical properties. The usual implementation of Bayesian hierarchical regression models is by using standard Markov Chain Monte Carlo (MCMC) approaches, such as the Gibbs sampler [3]. However, due to the difficulties related to sampling from the half-Cauchy distribution, there is currently no efficient and simple MCMC approach for the horseshoe+. Recently, Makalic and Schmidt [4] identified a novel scale mixture representation of the half-Cauchy distribution that was applied successfully to a specific form of Bayesian regression with half-Cauchy prior distributions called the horseshoe estimator [5,6]. This method outperformed the existing state-of-the-art implementations of the horseshoe, and given the favourable properties of the horseshoe+ relative to the horseshoe, it would be of great benefit if there was an efficient sampling algorithm for the horseshoe+.

This paper extends the work in [4] in two important ways: (i) we derive the first efficient Gibbs sampling scheme for the horseshoe+ estimator (see Sect. 2.1), and (ii) we exploit scale mixture representations to incorporate robust data models that naturally handle non-Gaussian data and are less sensitive to outliers (see Sect. 3). The resulting software implementation provides a powerful, flexible and robust tool for building prediction and classification models from potentially high-dimensional data and represents the state-of-the-art in Bayesian machine learning techniques.

2   Horseshoe+ Estimator

Following [4], we model the half-Cauchy distribution as a scale mixture of inverse gamma distributions. Specifically, let x and a be random variables such that

$$x^2|a \sim \mathcal{IG}(1/2, 1/a) \quad \text{and} \quad a \sim \mathcal{IG}(1/2, 1/A^2); \tag{9}$$

then $x \sim \mathcal{C}^+(0, A)$ [7], where $\mathcal{IG}(\cdot, \cdot)$ is the inverse gamma distribution (see Appendix A). The decomposition (9) may be used to represent the horseshoe+ prior distributions (6)–(8) through the following latent variable representation:

$$\lambda_j^2|\nu_j \sim \mathcal{IG}(1/2, 1/\nu_j),$$
$$\tau^2|\xi \sim \mathcal{IG}(1/2, 1/\xi),$$
$$\nu_j|\eta_j^2 \sim \mathcal{IG}(1/2, 1/\eta_j^2),$$
$$\eta_j^2|\phi_j \sim \mathcal{IG}(1/2, 1/\phi_j),$$
$$\phi_1, \ldots, \phi_p, \xi \sim \mathcal{IG}(1/2, 1).$$

where $(j = 1, \ldots, p)$. Although the above hierarchy introduces a number of additional latent variables, it leads to conjugate conditional posterior distributions for all parameters, greatly simplifying the resulting Gibbs sampling procedure.

2.1 Gibbs Sampling for the Horseshoe+

This section details the conditional posterior distributions for the horseshoe+ parameters required for the Gibbs sampler [3]. An advantage of the hierarchy (1)–(8) is that the conditional posterior distributions of the horseshoe+ do not depend on the choice of the sampling distribution of the data (1)–(3). Let

$$z_i = \begin{cases} (y_i^* - \frac{1}{2})/\omega_i^2 & \text{if } y_i \in \{0, 1\} \\ y_i & \text{otherwise} \end{cases}, \tag{10}$$
$$e_i = z_i - \mathbf{x}_i^{\mathrm{T}}\boldsymbol{\beta} - \beta_0, \tag{11}$$
$$\boldsymbol{\Omega} = \sigma^2 \operatorname{diag}(\omega_1^2, \ldots, \omega_n^2),$$

where $(y_1^*, \ldots, y_n^*)$ are latent variables defined in Sect. 3.3. The conditional posterior distribution for the intercept term $\beta_0 \in \mathbb{R}$ is the Gaussian distribution $\mathcal{N}(\tilde{\mu}, \tilde{\sigma}^2)$ where:

$$\tilde{\mu} = \left(\sum_{i=1}^n \frac{z_i - \mathbf{x}_i^{\mathrm{T}}\boldsymbol{\beta}}{\omega_i^2}\right)\left(\sum_{i=1}^n \frac{1}{\omega_i^2}\right)^{-1}, \qquad \tilde{\sigma}^2 = \sigma^2\left(\sum_{i=1}^n \frac{1}{\omega_i^2}\right)^{-1}$$

From the seminal paper by Lindley and Smith [8], the conditional posterior distribution for the regression coefficients $\boldsymbol{\beta} \in \mathbb{R}^p$ is the p-variate Gaussian distribution $\mathcal{N}_p(\tilde{\mu}, \mathbf{A}^{-1})$ where:

$$\tilde{\mu} = \mathbf{A}^{-1}\mathbf{X}^{\mathrm{T}}\boldsymbol{\Omega}^{-1}(\mathbf{z} - \beta_0 \mathbf{1}_n)$$
$$\mathbf{A} = \left(\mathbf{X}^{\mathrm{T}}\boldsymbol{\Omega}^{-1}\mathbf{X} + \boldsymbol{\Lambda}^{-1}\right)$$
$$\boldsymbol{\Lambda} = \tau^2\sigma^2 \operatorname{diag}(\lambda_1^2, \ldots, \lambda_p^2)$$

This paper uses Rue's algorithm [9] for efficient sampling from the multivariate Gaussian conditional posterior distribution of the regression coefficients when the sample size is greater than the number of predictors $(n > p)$. Rue's algorithm is based on Cholesky factorisation of the conditional posterior variance matrix and has cubic complexity in terms of the number of predictors p. An alternative approach to sampling multivariate Gaussian densities of this form was recently introduced by Cong et al. [10]. When the number of predictors is greater than the sample size $(p > n)$, we instead use the sampling algorithm by Bhattacharya et al. [11], which has linear complexity in p.

In the case of continuous data $y_i \in \mathbb{R}$, the conditional posterior distribution for $(\sigma^2 > 0)$ is the inverse gamma distribution $\mathcal{IG}(\tilde{\alpha}, \tilde{\beta})$ where:

$$\tilde{\alpha} = \frac{n+p}{2}, \qquad \tilde{\beta} = \frac{1}{2}\left(\sum_{i=1}^{n} \frac{e_i^2}{\omega_i^2} + \sum_{j=1}^{p} \frac{\beta_j^2}{\tau^2 \lambda_j^2}\right).$$

In the case of binary data, the noise variance parameter is simply set to $(\sigma^2 = 1)$ and does not require sampling. The conditional posterior densities for the remaining hyperparameters are:

$$\lambda_j^2 | \beta_j, \nu_j, \tau^2, \sigma^2 \sim \mathcal{IG}\left(1, \frac{1}{\nu_j} + \frac{\beta_j^2}{2\tau^2\sigma^2}\right),$$

$$\tau^2 | \boldsymbol{\beta}, \boldsymbol{\lambda}, \xi, \sigma^2 \sim \mathcal{IG}\left(\frac{p+1}{2}, \frac{1}{\xi} + \frac{1}{2\sigma^2}\sum_{j=1}^{p} \frac{\beta_j^2}{\lambda_j^2}\right),$$

$$\nu_j | \eta_j, \lambda_j \sim \mathcal{IG}\left(1, \frac{1}{\eta_j^2} + \frac{1}{\lambda_j^2}\right),$$

$$\xi | \tau^2 \sim \mathcal{IG}\left(1, 1 + \frac{1}{\tau^2}\right),$$

$$\eta_j^2 | \nu_j, \phi_j \sim \mathcal{IG}\left(1, \frac{1}{\nu_j} + \frac{1}{\phi_j}\right),$$

$$\phi_j | \eta_j \sim \mathcal{IG}\left(1, 1 + \frac{1}{\eta_j^2}\right).$$

Importantly, the conditional posterior distributions for all horseshoe+ hyperparameters are inverse gamma distributions for which computationally efficient sampling algorithms are readily available.

3   Robust Data Models

The decision to represent the sampling distribution of the data as a Gaussian scale mixture distribution naturally extends the data model (1)–(3) to a wide range of non-Gaussian distributions. For example, the Student-t and Laplace distributions can be represented in a scale mixture form. These distributions can be used to form hierarchical Bayesian estimators that are robust to data outliers and heavy-tailed errors. Additionally, Polson et al. [12] have recently extended this scale mixture representation to model discrete data through logistic regression (see Sect. 3.3) and negative binomial regression.

For the following robust linear regression models, we let $(z_i = y_i)$ as in Eq. (10). In the case of standard linear regression with Gaussian errors, the latent variables $(\omega_i^2 = 1)$ for all $i = (1, \ldots, n)$ and sampling of ω_i^2 is not required. The prior distributions and corresponding conditional posterior densities for regression with Laplace noise (see Sect. 3.1) and regression with Student-t noise (see Sect. 3.2) are discussed below.

3.1   Regression with Laplace Noise

It is well known that estimators using the Gaussian distribution to model errors can be negatively influenced by even a single outlying data point. This is due to the light tails of the Gaussian distribution, which are not able to capture large departures from the distribution mean. A popular alternative to the Gaussian distribution for modelling outliers is the Laplace distribution, which has heavier tails while still possessing finite mean and variance (see Appendix A.4). The Laplace distribution may be represented as a Gaussian-exponential scale mixture distribution where the scale parameters follow

$$\omega_i^2 \sim \text{Exp}(1), \quad (i = 1, \ldots, n),$$

and $\text{Exp}(1)$ denotes the exponential distribution with a mean of one. This choice of mixing distribution ensures that the scale parameter σ^2 is equal to the variance of the residuals $(e_i \in \mathbb{R})$ (see (11)), as in standard Gaussian regression.

Given the residuals $(e_1, \ldots, e_n)$ and the variance parameter σ^2, the conditional posterior distribution of $1/\omega_i^2$ is

$$\frac{1}{\omega_i^2} \mid y_i, \mathbf{x}_i, \boldsymbol{\beta}, \beta_0, \sigma^2 \sim \text{IGauss}\left(\left(\frac{2\sigma^2}{e_i^2} \right)^{\frac{1}{2}}, 2 \right)$$

where $\text{IGauss}(\mu, \lambda)$ denotes the inverse Gaussian density with mean $(\mu > 0)$ and shape parameter $(\lambda > 0)$ (see Appendix A.2).

3.2   Regression with Student-t Noise

The Student-t distribution is often used in Bayesian robust regression as an alternative to both the Gaussian and Laplace distributions. The Student-t distribution is parameterised by a location, a scale and a degrees of freedom parameter $(\delta > 0)$ (see Appendix A.3). When $(\delta = 1)$, the Student-t distribution is equal to the Cauchy distribution. For all finite values of $(\delta > 0)$ the Student-t distribution has heavier tails than the Gaussian distribution. For $(\delta < 6)$, the Student-t distribution has heavier tails than the Laplace distribution and results in estimators that are significantly more resistant to outliers in the data. One potential drawback of modelling the errors with the Student-t distribution is that it has infinite variance when $(\delta \leq 2)$.

The Student-t distribution with degrees of freedom $(\delta > 0)$ may be represented as a Gaussian-inverse gamma scale mixture distribution where

$$\omega_i^2 \sim \mathcal{IG}\left(\frac{\delta}{2}, \frac{\delta}{2}\right).$$

Given the residuals $(e_1, \ldots, e_n)$ and the scale parameter σ^2, the conditional posterior distribution of ω_i^2 is

$$\omega_i^2 \mid y_i, \mathbf{x}_i, \boldsymbol{\beta}, \beta_0, \sigma^2, \delta \sim \mathcal{IG}\left(\frac{\delta + 1}{2}, \frac{1}{2}\left(\frac{e_i^2}{\sigma^2} + \delta\right)\right).$$

When the degrees of freedom parameter $(\delta > 2)$, the variance of the residuals is related to σ^2 by

$$\mathrm{Var}(e_i) = \sigma^2\left(\frac{\delta}{\delta - 2}\right).$$

When $(\delta \leq 2)$, the parameter σ^2 can only be interpreted as a scale parameter.

3.3   Binary Data

When dealing with binary data, we model the relationship between the predictors $(\mathbf{x}_i \in \mathbb{R}^p)$ and the outcome variable $y_i \in \{0, 1\}$ $(i = 1, \ldots, n)$ using logistic regression. Directly sampling from the logistic regression model is difficult due to the analytic form of the likelihood function. Recently, Polson et al. [12] introduced a latent variable representation of the logistic likelihood function that is easily integrated into the hierarchy (1)–(8). Here, the logistic likelihood function is represented as a Gaussian scale mixture with a Pólya-gamma mixing density (see Appendix A.5).

Implementation of logistic regression within the hierarchy (1)–(8) requires only a minor change in the way the latent variables $(\omega_1, \ldots, \omega_n)$ and the scale parameter σ^2 are handled. In the case of logistic regression, we set $(y_i^* = y_i)$ for all $i = (1, \ldots, n)$ in (10), $(\sigma^2 = 1)$ and sample the latent variables $(\omega_1, \ldots, \omega_n)$ from

$$\frac{1}{\omega_i^2} \mid \mathbf{x}_i, \boldsymbol{\beta}, \beta_0 \sim \mathrm{PG}(1, \beta_0 + \mathbf{x}_i^{\mathrm{T}}\boldsymbol{\beta}),$$

where $\mathrm{PG}(\cdot, \cdot)$ is a Pólya-gamma distribution (see Appendix A.5). An algorithm for efficient sampling from the Pólya-gamma distribution was recently introduced by Windle et al. [13] and is employed in this paper.

Robust Logistic Regression. In real data, misrecording of the labels of the outcome variable or the values of the predictors is relatively common. For continuous outcomes, a standard approach to handle these data errors is to employm heavy-tailed, non-Gaussian error models to moderate the effects of erroneous data points. In the case binary outcome variables, the only possible mislabelling of the outcome is a transposition, a change from $(y_i = 0)$ to $(y_i = 1)$ or from $(y_i = 1)$ to $(y_i = 0)$. The effect of mislabelled binary data is to reduce the magnitude of the observed association, in contrast with the continuous case where data contamination usually results in artificially inflated estimates. For binary data, an interesting symmetry exists between the labels y_i and the predictors $\mathbf{x}_i$ in the sense that mislabelling an outcome variable is essentially equivalent to moving a predictor from one side of the decision plane to the other.

In this paper, we handle contaminated binary data by adapting the misclassification model introduced in [14] and later analysed by Copas [15] and Carrol and Pederson [16]. Let

$$\mathbb{P}(y_i = 1|y_i^* = 0, \gamma_0) = \gamma_0,$$
$$\mathbb{P}(y_i = 0|y_i^* = 0, \gamma_0) = (1 - \gamma_0),$$
$$\mathbb{P}(y_i = 0|y_i^* = 1, \gamma_1) = \gamma_1,$$
$$\mathbb{P}(y_i = 1|y_i^* = 1, \gamma_1) = (1 - \gamma_1),$$

where $(y_1^*, \ldots, y_n^*)$ are latent variables representing the unobserved, correctly specified values of the observed outcomes $(y_1, \ldots, y_n)$, γ_0 is the probability of a $(y_i^* = 0)$ being misrecorded as a $(y_i = 1)$, and γ_1 is the probability of a $(y_i^* = 1)$ being misrecorded as a $(y_i = 0)$.

Rather than fix the values of the hyperparameters γ_0 and γ_1 *a priori*, they are integrated into the hierarchy (1)–(8) and sampled along with all other parameters. To simplify sampling, γ_0 and γ_1 are assigned independent Beta(1/2,1/2) prior distributions. Bayesian estimation of this misspecification model involves sampling from the conditional posterior distributions of the latent variables $(y_1^*, \ldots, y_n^*)$

$$\mathbb{P}(y_i^* = j|y_i, \beta_0, \boldsymbol{\beta}) \propto \mathbb{P}(y_i|y_i^* = j, \gamma_j)\,\mathbb{P}(y_i = j|\beta_0, \boldsymbol{\beta}), \qquad (12)$$

where

$$\mathbb{P}(y_i = 1|\beta_0, \boldsymbol{\beta}) = \frac{1}{1 + \exp(-\beta_0 - \mathbf{x}_i^{\mathrm{T}}\boldsymbol{\beta})}$$

is the logistic function. Normalising (12) is straightforward given that the conditional posterior distributions of $(y_1^*, \ldots, y_n^*)$ are independent and y_i^* is a binary variable $(y_i^* \in \{0, 1\})$. The conditional posterior distributions of the hyperparameters (γ_0, γ_1) are

$$\gamma_k \mid (y_1^*, \ldots, y_n^*), (y_1, \ldots, y_n) \sim \text{Beta}(n_{0k} + 1/2, n_{1k} + 1/2),$$

where $(k = 1, 2)$ and

$$n_{jk} = \sum_{i=1}^{n} \mathbb{I}(y_i = j, y_i^* = k)$$

is the count of how many times we observe $(y_i = j)$ together with $(y_i^* = k)$.

The advantages of the Bayesian misspecification model over the frequentist approaches [15,16] is that the Bayesian model automatically provides posterior estimates and credible intervals for both hyperparameters (γ_0, γ_1). These hyperparameters can be interpreted as the estimated rate of transposition errors in a given data set. Furthermore, the posterior expectation

$$\mathbb{E}\left(\mathbb{I}(y_i \neq y_i^*)|y_1, \ldots, y_n\right)$$

can be interpreted as the estimated posterior probability of an individual data point being misrecorded which is a useful diagnostic when analysing real data.

4 Results and Discussion

This section compares our novel implementation of the horseshoe+ estimator against the horseshoe estimator implemented in [4]. As the key difference between these estimators is an extra level of Gibbs sampling required for the horseshoe+, the computational speed and memory usage of both implementations is expected to be approximately equivalent. Further, based on the theoretical analysis presented in [2] for the multiple means problem, the horseshoe+ estimator is expected to outperform the horseshoe estimator in terms of asymptotic prediction error (i.e., as the sample size $n \to \infty$). However, non-asymptotic comparison of the two estimators in Gaussian and non-Gaussian regression models has not been attempted to date. Consequently, we have chosen to compare the horseshoe and horseshoe+ estimators on their empirical prediction performance in finite sample regression tests.

4.1 Robust Linear Regression

The test procedure for comparing the two estimators is now described. For each test, we first generated a training set of predictors $(\mathbf{x}_1, \ldots, \mathbf{x}_n)$ where each $\mathbf{x}_i$ $(i = 1, \ldots, n)$ was generated from a zero mean p-variate Gaussian distribution $\mathbf{x}_i \sim \mathcal{N}_p(\mathbf{0}_p, \mathbf{I}_p)$ with the variance–covariance matrix equal to the identity matrix $\mathbf{I}_p$. The sample size of the training data was set to $(n = 100)$ in each test iteration. The training data $(\mathbf{y} \in \mathbb{R}^n)$ was generated from the model

$$y_i = \mathbf{x}_i^{\mathrm{T}} \boldsymbol{\beta}^* + \epsilon_i, \quad (i = 1, \ldots, n),$$

where $\boldsymbol{\beta}^* \in \mathbb{R}^p$ is the vector of true regression coefficients and the noise variable ϵ_i follows a zero-mean Gaussian, Laplace or Student-t distribution. The variance of the noise variables was chosen to control a pre-specified signal-to-noise (SNR) ratio for each test. The regression coefficients $\boldsymbol{\beta}^*$ were chosen to represent the ultra-sparse data model for which the horseshoe and horseshoe+ estimators were originally developed. In particular,

$$\boldsymbol{\beta}^* = (1, 1, \ldots, 1, 0, 0, \ldots, 0)^{\mathrm{T}},$$

where 5% of the entries of $\boldsymbol{\beta}^*$ were set to one, with the remaining 95% of the entries set to zero. The complete test procedure was repeated for 100 iterations

Table 1. Mean squared prediction errors for the horseshoe (HS) and horseshoe+ estimators computed over 100 test iterations (standard errors shown in parenthesis). For each test, the number of non-zero coefficients was set to 5% of the total number of predictors p. The sample sizes of the training and test data were set to $n = 100$ and $n = 10^5$ respectively.

p	SNR	Gaussian		Laplace		Student-t ($\delta = 5$)	
		HS	HS+	HS	HS+	HS	HS+
50	1	3.308 (0.18)	**3.273** (0.17)	3.197 (0.15)	**3.175** (0.14)	3.261 (0.18)	**3.229** (0.17)
	4	0.816 (0.03)	**0.810** (0.03)	0.792 (0.03)	**0.789** (0.02)	0.806 (0.04)	**0.801** (0.03)
	8	0.407 (0.02)	**0.405** (0.02)	0.396 (0.01)	**0.394** (0.01)	0.403 (0.02)	**0.400** (0.02)
100	1	6.393 (0.71)	**6.324** (0.75)	5.787 (0.50)	**5.704** (0.50)	5.951 (0.51)	**5.862** (0.53)
	4	1.447 (0.09)	**1.427** (0.09)	1.367 (0.05)	**1.354** (0.05)	1.398 (0.07)	**1.381** (0.07)
	8	0.719 (0.04)	**0.711** (0.04)	0.682 (0.03)	**0.676** (0.03)	0.697 (0.03)	**0.689** (0.03)
200	1	**16.61** (1.49)	16.75 (1.59)	**16.28** (1.36)	16.36 (1.45)	**16.22** (1.50)	16.37 (1.63)
	4	3.705 (0.60)	**3.498** (0.59)	3.379 (0.52)	**3.181** (0.42)	3.417 (0.73)	**3.272** (0.67)
	8	1.712 (0.22)	**1.631** (0.20)	1.565 (0.15)	**1.512** (0.13)	1.592 (0.21)	**1.540** (0.19)

for $p = \{50, 100, 200\}$ and SNR $= \{1, 4, 8\}$. The performance of the two estimators was measured using the mean squared prediction error metric computed on independently generated test data at the Rao-Blackwellised posterior mean. For each estimator, the posterior mean was computed from $1,000$ samples from the corresponding posterior distribution with a 'burnin' period of $1,000$ samples and a thinning level of 5. The sample size of the test data sets was set to $n = 10^5$. In the spirit of reproducible research, MATLAB code implementing both estimators and all simulations will be made available on the authors' web pages[1].

Table 1 presents the mean squared predictions errors for the horseshoe and horseshoe+ estimators. In the case of ($p = 200$) and (SNR $= 1$), the horseshoe estimator resulted in smaller prediction error in contrast to the new horseshoe+ estimator. However, the horseshoe+ estimator obtained improved prediction error for all other combinations of SNR and p that were examined. Importantly, the horseshoe+ estimator performed well even when the noise distribution was non-Gaussian. Consequently, the finite sample performance of the horseshoe+ estimator appears to be at least as good as the original horseshoe procedure in the robust linear regression setting.

4.2 Robust Logistic Regression

This section demonstrates the utility of the logistic regression misclassification model, discussed in Sect. 3.3, when applied to the analysis of a real data set. We consider the Pima Indians diabetes data set [17] collected by the National Institute of Diabetes and Digestive and Kidney Diseases and available for download from the UCI Machine Learning Repository. The data set comprises eight

---

[1] www.emakalic.org/blog and www.dschmidt.org.

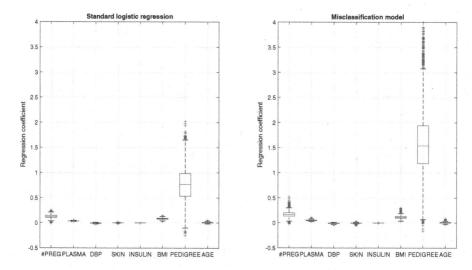

Fig. 1. Analysis of the Pima Indian diabetes data set with the Bayesian horseshoe+ estimator. The left Figure depicts the posterior estimates of the regression coefficients from the non-robust logistic regression model. The right Figure shows the regression coefficients estimated from the misclassification model.

$(p = 8)$ predictor variables and $(n = 768)$ observations collected from female patients of Pima Indian heritage who are at least 21 years of age. The aim is to build a prediction model of diabetes from the eight predictor variables. The original donor of the data set indicated that there are no missing values in the data. However, this appears to be erroneous because the data contains predictors with values that are biologically implausible. For example, the variable diastolic blood pressure contains entries with zero blood pressure, which is clearly incorrect.

The Pima Indians diabetes data set was analysed using the following two models: (i) standard (non-robust) logistic regression and (ii) the misclassification model (see Sect. 3.3). For both analyses, the Bayesian horseshoe+ estimator was used to estimate all unknown parameters. Box and whisker plots of the posterior estimates of the regression coefficients are shown in Fig. 1. It is clear that the observed associations under the misclassification model (Fig. 1, right) are inflated in contrast with standard logistic regression (Fig. 1, left). This is expected because the effect of ignoring data contamination in binary logistic regression is a reduction in the absolute magnitude of the regression coefficients. The posterior estimates of the rate of transposition errors in the data were found to be relatively large, $(\gamma_0 = 0.03)$ and $(\gamma_1 = 0.13)$. Further analysis of the unobserved latent variables $(y_1^*, \ldots, y_n^*)$ was then used to discover which particular observations are likely to be contaminated. A manual examination of the top 10 observations identified as highly likely to contain contaminated data was then performed. These top observations were found to be clearly erroneous because they contained biologically impossible predictor values, highlighting the usefulness of the robust logistic regression model.

A   Appendix

A.1   Inverse Gamma Distribution

The inverse gamma probability density function is given by

$$p(x|\alpha, \beta) = \frac{\beta^\alpha}{\Gamma(\alpha)} x^{-\alpha-1} \exp\left(-\frac{\beta}{x}\right), \quad (x > 0),$$

with shape parameter ($\alpha > 0$) and scale parameter ($\beta > 0$). The first two moments are

$$E(x) = \frac{\beta}{\alpha - 1}, \quad \mathrm{Var}(x) = \frac{\beta^2}{(\alpha - 1)^2(\alpha - 2)},$$

where the mean and variance only exist for ($\alpha > 1$) and ($\alpha > 2$) respectively.

A.2   Inverse Gaussian Distribution

The inverse Gaussian probability density function is given by

$$p(x|\mu, \lambda) = \left(\frac{\lambda}{2\pi x^3}\right)^{\frac{1}{2}} \exp\left(-\frac{\lambda(x - \mu)^2}{2\mu^2 x}\right),$$

for ($x > 0$), where ($\mu > 0$) is the mean and ($\lambda > 0$) is the shape parameter. The first two moments are

$$E(x) = \mu, \quad \mathrm{Var}(x) = \frac{\mu^3}{\lambda}.$$

A.3   Student-t Distribution

The Student-t distribution probability density function is given by

$$p(x|\mu, \sigma^2, \nu) = \frac{\Gamma\left(\frac{\nu+1}{2}\right)}{\Gamma\left(\frac{\nu}{2}\right)\sqrt{\pi\nu\sigma^2}} \left(1 + \frac{1}{\nu}\frac{(x - \mu)^2}{\sigma^2}\right)^{-\frac{\nu+1}{2}},$$

where ($x \in \mathbb{R}$), ($\mu \in \mathbb{R}$), ($\sigma^2 > 0$) and the degrees of freedom ($\nu > 0$). The first two moments are

$$E(x) = \mu, \quad (\nu > 1), \quad \mathrm{Var}(x) = \sigma^2\left(\frac{\nu}{\nu - 2}\right), \tag{13}$$

where the mean and variance only exist for ($\nu > 1$) and ($\nu > 2$) respectively.

A.4   Laplace Distribution

The probability density function of the Laplace distribution is

$$p(x|\mu, b) = \frac{1}{2b} \exp\left(-\frac{|x - \mu|}{b}\right),$$

where ($x \in \mathbb{R}$), ($\mu \in \mathbb{R}$) is the location parameter and ($b > 0$) is the scale parameter. The first two moments are

$$E(x) = \mu, \quad \mathrm{Var}(x) = 2b^2.$$

A.5   Pólya-Gamma Distribution

A random variable x follows a Pólya-gamma distribution [12], $x \sim \mathrm{PG}(b, c)$, if

$$x \stackrel{D}{=} \frac{1}{2\pi^2} \sum_{k=1}^{\infty} \frac{g_k}{(k - 1/2)^2 + c^2/(4\pi^2)},$$

where $g_k \sim \mathrm{Ga}(b, 1)$ are independent gamma random variables, $(b > 0)$ and $(c \in \mathbb{R})$ are the parameters and $\stackrel{D}{=}$ denotes equality in distribution. The first two moments of x are

$$\mathrm{E}(x) = \frac{b}{2c}\tanh\left(\frac{c}{2}\right), \quad \mathrm{Var}(x) = \frac{b}{4c^3}(\sinh(c) - c)\operatorname{sech}^2\left(\frac{c}{2}\right).$$

References

1. Andrews, D.F., Mallows, C.L.: Scale mixtures of normal distributions. J. R. Stat. Soc. (Ser. B) **36**(1), 99–102 (1974)
2. Bhadra, A., Datta, J., Polson, N.G., Willard, B.: The horseshoe+ estimator of ultra-sparse signals (2015). arXiv:1502.00560
3. Geman, S., Geman, D.: Stochastic relaxation, Gibbs distributions, and the Bayesian restoration images. IEEE Trans. Patttern Anal. Mach. **6**, 721–741 (1984)
4. Makalic, E., Schmidt, D.F.: A simple sampler for the horseshoe estimator. IEEE Signal Process. Lett. **23**(1), 179–182 (2016)
5. Carvalho, C.M., Polson, N.G., Scott, J.G.: The horseshoe estimator for sparse signals. Biometrika **97**(2), 465–480 (2010)
6. Polson, N.G., Scott, J.G.: Shrink globally, act locally: sparse Bayesian regularization and prediction. In: Bayesian Statistics. vol. 9 (2010)
7. Wand, M.P., Ormerod, J.T., Padoan, S.A., Fruhwirth, R.: Mean field variational Bayes for elaborate distributions. Bayesian Anal. **6**(4), 847–900 (2011)
8. Lindley, D.V., Smith, A.F.M.: Bayes estimates for the linear model. J. R. Stat. Soc. (Ser. B) **34**(1), 1–41 (1972)
9. Rue, H.: Fast sampling of Gaussian Markov random fields. J. R. Stat. Soc. (Ser. B) **63**(2), 325–338 (2001)
10. Cong, Y., Chen, B., Zhou, M.: Fast simulation of hyperplane-truncated multivariate normal distributions (2016)
11. Bhattacharya, A., Pati, D., Pillai, N.S., Dunson, D.B.: Dirichlet-Laplace priors for optimal shrinkage. J. Am. Stat. Assoc. **110**, 1479–1490 (2015)
12. Polson, N.G., Scott, J.G., Windle, J.: Bayesian inference for logistic models using Pólya-Gamma latent variables. J. Am. Stat. Assoc. **108**(504), 1339–1349 (2013)
13. Windle, J., Polson, N.G., Scott, J.G.: Sampling Pólya-Gamma random variates: alternate and approximate techniques (2014)
14. Ekholm, A., Palmgren, J.: Correction for misclassification using doubly sampled data. J. Official Stat. **3**(4), 419–429 (1987)
15. Copas, J.B.: Binary regression models for contaminated data. J. R. Stat. Soc. Ser. B (Methodol.) **50**(2), 225–265 (1988)
16. Carroll, R.J., Pederson, S.: On robustness in the logistic regression model. J. R. Stat. Soc. Ser. B (Methodol.) **55**(3), 693–706 (1993)
17. Lichman, M.: UCI machine learning repository (2013)

High Resolution Self-organizing Maps

Van Tuc Nguyen[✉], Markus Hagenbuchner, and Ah Chung Tsoi

School of Computing and Information Technology, University of Wollongong,
Wollongong, Australia
{vtn966,markus,act}@uow.edu.au

Abstract. Kohonen's self organizing feature map (SOM) provides a convenient way for visualizing high dimensional input features by projecting them onto a low dimensional display space. This map has an appealing characteristic: feature vectors close to one another in the high dimensional input space remain close to one another in the low dimensional display space. Owing to the computational requirements, the display space so far remains of relatively low resolutions. In this paper, we provide an implementation of the SOM by making use of the highly parallel architecture of a graphic processing unit to increase its computational speed to allow a substantial increase in the resolution while keeping the computation to within an acceptable wall clock time. Armed with such an implementation, we find that the high resolution SOM can display intricate details concerning the relationships among the input feature vectors. These details would be lost if a low resolution SOM was deployed. The capability of the high resolution SOM is demonstrated through an application to an artificially generated dataset, the policeman dataset. The dataset allows us to design intricate relationships among the input feature vectors.

Keywords: Self-organizing map · Artificial neural network · Clustering and projection · High resolution · Unsupervised learning

1 Introduction

The Self-Organising feature Map (SOM) [1] is popularly used for data visualization in the exploration stage of a data mining application. One of the key properties of a SOM is that it creates a topology-preserving mapping of a high dimensional input (feature) space onto a low dimensional discrete grid called the display space [1]. The grid can be of any dimension although a two-dimensional grid is most common to aid the visualization of the projections. Through such visualization, it helps the user in understanding any intricate relationships among the input vectors via exploration in the display space. Such visualization would often act as a prelude to further processing of the input data [1]. Each grid point is referred to as a *neuron*, characterized by a codebook vector of the same length as the input vectors. The number K of grid points defines the *resolution* of the SOM, and the SOM is said to consist of K neurons. The SOM training algorithm

© Springer International Publishing AG 2016
B.H. Kang and Q. Bai (Eds.): AI 2016, LNAI 9992, pp. 441–454, 2016.
DOI: 10.1007/978-3-319-50127-7_38

aims to order the codebook vectors located at grid points so that data points represented by high dimensional vectors which are similar in input space are mapped to nearby grid points. Once the ordered codebook vectors are obtained and have converged to a stable equilibrium [1], interesting and useful insights into the properties of input vectors can be made. A main problem with the SOM is that its mapping space is discrete, thus the quality of the mapping depends on the magnitude of K.

A SOM consisting of a very large number of neurons, i.e. the magnitude of K is relatively large, is called a High Resolution SOM (HRSOM). The reason for creating HRSOMs is to better visualize the macro as well as micro structures, indicating relationships existing among the input vectors [8,11]. HRSOMs allow more room to separate dissimilar input patterns, and are more suitable for datasets that exhibit complex relations among its vectors. In contrast, a low resolution SOM (LRSOM) where K is of a low magnitude, intricate and complex relationships among input vectors will be lost as the intricate relationships would merge into simpler structures.

The training algorithm of SOM scales linearly with the number of neurons and the size of the dataset [1]. The linear computational time complexity of the algorithm prevents it from training sufficiently large SOMs. This also prevents the construction of a display space which is approximately continuous, by having very large magnitude K value. To the best of our knowledge, nobody has yet succeeded in implementing a SOM with K in the order of millions. There were various attempts in improving the granularity of the display space but these were based on a hierarchical SOM structure [9,10] and by a social hierarchical structure: the tree SOM [6,7]. The basic idea behind these approaches is that the SOM adapts the topology of each hierarchical layer to the properties of the input vectors, starting with a very small SOM of grid size 1×1, then growing/enlarging the SOM in places where the quantization errors are high [6,7,9,10]. The approach of growing SOM online and only in locations where high quantization error occurs, gives the SOM a tree-like appearance [9,10]. The approach does deploy a number of additional neurons in order to separate the input vectors from different categories. However, there are two main drawbacks associated with this approach:

1. The SOM can only grow in restricted areas. Thus, the shape of clusters formed can be distorted so that they no longer reflect the size and shape of a cluster in the high dimensional input space.
2. The growth of the neuron number increases the computational demand. It hence remains difficult to solve problems involving datasets with complex relationships among input vectors. In contrast, the HRSOM introduced in this paper does not suffer from these drawbacks, as it can provide a display map with relatively large magnitude K.

An important observation is that the SOM is an Artificial Neural Network and is a massively parallel system. The SOM has traditionally been implemented on CPU systems thus limiting the degree of parallelism to the number of cores in a CPU. There have been various recent attempts in porting the SOM algorithm to GPU (Graphics Processing Unit) in order to take advantage of hundreds and thousands of cores, though each core would have access to relatively small fast onboard memory capacity, typically found on a GPU. It was claimed that this significantly reduces the execution time of the SOM algorithm [12–14][1]. Moreover, there is a trend to computing clusters which are equipped with multiple GPUs internally connected together, and which are much more powerful than CPU clusters in parallel and distributed applications [15,16]. The aims of this paper is to (a) describe techniques which further enhance the speed of a GPU implementation and (b) analyze the effects and results of using HRSOM in a dataset which contains intricate relationships among their input vectors.

The contributions of this paper are as follows: (a) an implementation of the SOM on a GPU which takes into account its architectural characteristics. To the best of our knowledge, the resulting GPU implementation is the fastest known implementation of the SOM algorithm. (b) Demonstrating that a HRSOM can display intricate and complex relationships among input vectors by applying the GPU SOM implementation to an artificially generated dataset with known intricate and complex relationships among its vectors. It is shown that the HRSOM is able to capture and display the intricate and complex relationships among its input vectors. To the best of our knowledge the demonstrations shown in this paper have not been seen before in the literature.

The rest of this paper is organized as follows: Sect. 2 presents the procedures for implementing the SOM algorithm on a GPU. Section 3 applies the HRSOMs to an artificially generated dataset, called a policeman dataset. This dataset is designed so that there will be intricate and complex relationships among its vectors. Conclusions are drawn in Sect. 5 with some indications of further research directions.

2 The High Resolution Self-organizing Map

The SOM algorithm [1] performs a nonlinear and topology preserving projection of the high dimensional input data (feature) space onto a discretized display space consisting of K neurons arranged in a d-dimensional grid. In this paper, without loss of generality, we assume that $d = 2$ and that the neurons are organized on a N x M grid. Each neuron i of the map is associated with an n-dimensional codebook vector $m_i = (m_{i1}, \ldots, m_{in})^T$, where T denotes the transpose operator. n is also the dimension of the input vectors. Neurons adjacent to neuron i belong to a neighbourhood denoted as N_i where the neighborhood relation is often hexagonal in shape [1]. The SOM training algorithm [1] can be presented in two steps:

Step 1 - Competitive step: An input vector u is randomly drawn from the input dataset. Its similarity to the codebook vectors is computed. Most commonly, it is to find the minimum Euclidean distance $\|u - m_i\|$ between u and

---

[1] Some of the papers do not provide sufficient detail for us to validate their claims.

the codebook vector of neuron i, m_i. The winning neuron r must satisfy the relationship $r = \arg\min_i \|u - m_i\|$.

Step 2 - Cooperative step: Related codebook vectors are adjusted. The best matching codebook vector m_r and its neighbours are moved closer to the input vector. The magnitude of the adjustment is controlled by the learning rate α and by the neighbourhood function $f(\Delta_{ir})$, where Δ_{ir} is the topological distance between the two codebook vectors m_r and m_i. The amount of change in the codebook vector m_i is computed as $\Delta_{m_i} = \alpha(t)f(\Delta_{ir})(m_i - u)$. The learning rate $\alpha(t)$ decreases with the training process. The neighbourhood function $f(.)$ controls the amount by which the codebooks of the neighbouring neurons are updated. Popular is a Gaussian neighbourhood function: $f(\Delta_{ir}) = \exp\left(-\frac{\|l_i - l_r\|^2}{2\sigma^2}\right)$, where σ is the spread or radius which controls the operating region of function $f(.)$. The two values l_r and l_i are respectively the location of the winning neuron, and the location of the i-th neuron on the map.

The two steps are repeated for each training sample and for a pre-defined number of iterations. While there is no proof of the convergence of the training algorithm [1], it is empirically found that the training algorithm always converges, as when the learning constant is reduced to a value close to 0 the training algorithm stops updating the codebook vectors.

2.1   GPU Acceleration of the SOM Algorithm

GPU Architecture: A GPU, sometimes called the visual processing unit, is a specialized graphical unit. The electronic circuit design of the GPU is for the particular purpose of rapidly manipulating and altering memory to accelerate the processing of data stored in a frame buffer, and to optimize the visualizing process on a display. GPUs are widely used in various hardware architectures such as embedded systems, personal computers and workstations. In a personal computer, a GPU can be integrated on a video card, or can be embedded on the motherboard. Modern computers allow us to attach one or more GPUs. The more recent GPU technology allows for efficient and general purpose massive parallel computations that is not limited to graphics processing.

In software programming, a GPU is used to extensively support computational problems that require high-performance and highly-parallel computations. The GPU is constructed from a number of *stream multiprocessors*. A multithreaded program is employed in parallel by allocating the number of threads equal to the number of processing blocks and the number of blocks is partitioned into a grid of blocks. All threads have access to the GPU's global memory which is also used as a communication channel between the multiprocessors and between GPU and CPU. Threads of an individual block have common access to another type of memory called shared memory that helps threads in a block to have access to each other's data. A thread from one block cannot access the shared memory of another block. It is important to note that there are several types of memory found on a GPU:

1. Register memory: Data stored in this memory is limited to be accessible to the thread that uses it and is destroyed when the thread terminates. The size of register memory is in the order of a few bytes.
2. Local memory: This has a similar role to register memory, however its size is in the order of kilobytes but it is slower than register memory.
3. Shared memory: Data stored in shared memory is accessible to all threads within the operating block. This type of memory allows for threads to share data and to communicate with one another. The shared memory lasts for the life-time of the associated block and is larger but slower than local memory.
4. Global memory: This is the most accessible type of memory. The memory is accessible by all threads and by the CPU host system. Its content lasts for the duration of the host allocation, and is in size in the order of gigabytes although it is the slowest of all.
5. Constant and texture memory: This is a read only memory and is used to store data that is not altered during code execution. This memory, too, can be gigabytes in size and it is faster than global memory.

CUDA is a parallel computing platform and is also the application programming interface (API) model. CUDA is created by NVIDIA allowing software developers to use a GPU for general purpose, so-called GPGPU code. In other words, CUDA is a software layer that gives direct instructions to the GPU for the execution of GPU kernels. The CUDA platform accommodates various programming languages such as C and C++. A multiprocessor adopts an unique architecture called SIMT (single-instruction, multiple-thread). For this reason, all parallel threads execute the same set of instructions but can operate on different data. CUDA introduces simple kernel calls to execute code on the GPU. This allows the CPU host system to define and call kernels on the GPU.

Porting the SOM Algorithm to GPU: The SOM algorithm implemented on GPU is optimized for dealing with problems which require significantly large processing maps. There are computations involving the large data buffering required to find the best matching unit and update the related codebook vectors. Most of the parts of the SOM algorithm for high resolution maps contain massive parallel computations which can be implemented on GPU. There are a number of kernels that we have used, namely (1) initializing the codebook vectors and assigning the map coordinates to reference variables; (2) calculating all the Euclidean distances between the current input vector and the codebook vectors; (3) a reduction kernel for finding the minimum distance; (4) listing the neighbouring nodes based on the radius value; (5) updating the codebook vectors of nodes that are related to the winning node. Each of these operations can thus be implemented as a GPU kernel function. In order to optimize these kernels, a number of strategies have been applied:

1. Reduce the number of host to device and device to host memory transfers. When transferring data from CPU to GPU and back, we use one continuous

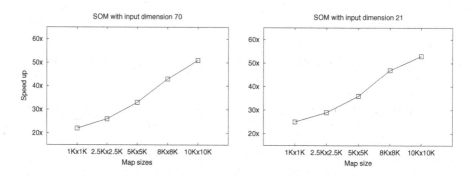

Fig. 1. GPU rate of speed-up depending on map size. 1K means 1000 and Ax means A times speed up when compared to the CPU.

memory block. For example, by concatenating all data samples into a single array there is only one instruction to transfer the data to the GPU. For large datasets this improves the transfer speed by more than 10 times when compared to transferring input samples separately. The input data and the coordinate reference variables are constant and hence they can be stored in the texture memory which allows much faster access to its contents.

2. For all kernel functions store run-time variables in shared memory or in local memory as much as possible, and keep kernels simple and small since the cost of kernel launches is negligible and since this improves speed because fewer registers are used. By using small kernels one can better utilize the registers, shared memory and constant memory because the memory resources are limited to each kernel.

3. The reduction kernel is a choke point of the algorithm. Its speed can be optimized by assuming that the number of threads is a power of two value. Hence the number of threads is chosen to suit the reduction kernel.

4. A stream multiprocessor can handle at most 2048 threads concurrently and can accommodate 16 active blocks. If one sets a block to contain 128 threads, the number of concurrently active blocks will be $2048/128 = 16$ (the maximum threshold). On the other hand a block size of 256 would also use the available computation resource. We found that setting the block size to any other value seems to be wasting the resources since there will exist a block which contain fewer threads than the others. We found that using blocks that contain 128 threads produces the best acceleration.

5. Prevent threads from diverging by ensuring that the conditional jumps branch equally for all threads. We implemented conditional branching based on a multiple of the wrap size and, in addition, we also unroll loops.

The speed improvement of the GPU implementation increases with the degree of parallelism (although our implementation is for a single GPU rather than for multi-GPU systems). This can be observed in Fig. 1 which shows that the rate of speed improvement increases with the size of a SOM. The speed comparison is relative to the Intel(R) Core(TM) i7-5960X CPU @ 3.00 GHz (extreme edition). That multiple-core CPU was the fastest consumer type CPU from Intel at the time of writing. Figure 1 shows that the GPU implementation can be 52 times faster than the fastest consumer CPU. The results were obtained by using GeForce GTX TITAN X (black edition) - one of the fastest GPUs for single precision computations. The results were obtained by using the compiler optimization flag -03 for both, the CPU and GPU version of the code. Moreover, the CPU version is based on Kohonen's SOM software package (som_pak) which implements *tricks* to accelerate code execution such as (a) not updating codebooks that are more than 3σ away from the winning codebook, and (b) breaking the loop early when computing the Euclidean distance. Theses tricks improve the execution speed of the CPU code by approximately three to 10 times depending on network size, and are not (and do not need to be) implemented in the GPU code.

It is difficult to make a comparison with others who ported the algorithm to a GPU since the others use different computation resources, do not specify the type of CPU and GPU used, do not specify whether or not optimization tricks were implemented, or do not provide the code. Nevertheless, we found that we achieved a maximum on the memory bandwidth on the GPU and hence are confident that our GPU implementation processes the data at the maximum possible rate. Moreover, the speed improvement observed matches the expected theoretical improvement of a massive parallel implementation which grows with $O(\log n)$ when compared to a sequential implementation which is $O(n)$.

3   Evaluation Methods

There are a number of standard measurements which are usually used to determine the quality of clusters with regard to the known input samples' categories. The micro purity and the macro purity are commonly used in the machine learning and data mining community. While the micro purity measures the overall clustering performance with respect to a given sample's classes, the macro purity calculates the average of the individual clustering performance for each class. Given a SOM mapping result of a n-class clustering problem, the number of samples in the dataset is denoted as A, and the number of samples in class k is denoted as A_k. The number of samples with the majority label in class k is denoted as a_k. The calculation of micro and macro purity is as follows:

$$\text{micro-purity} = \frac{\sum_{k=0}^{n} a_k}{A}, \qquad \text{macro-purity} = \frac{\sum_{k=0}^{n} \frac{a_k}{A_k}}{n}. \qquad (1)$$

We will also use a third evaluation method to compute the clustering quality. The quantity is computed based on a group of neurons on the SOM map. Given a node c on the map, a group is defined, which includes all nodes in the direct neighborhood to node c. In other words, all nodes in the group are directly connected to c. The evaluation performance is computed as follows. For every node and corresponding group we count the number of samples with the majority label for every class. The result obtained is divided by the number of samples in the whole dataset. We denote this evaluation as the *grouping index*. In practice, the local relation between neurons on the map is considered in this evaluation method, hence reducing the sparsity problem when one uses more number of neurons than the number of input samples.

4 Experiments

4.1 Problem Description

The policemen dataset [5] is an artificial dataset which is used for the purpose of assessing the clustering performance of HRSOMs. The dataset is created based on an attributed plex grammar for generating image databases [5]. It contains three categories of images: policeman, house, and sailing boat. The number of image categories can be increased arbitrarily, though at present, three categories are sufficient to demonstrate the visualization capabilities of the HRSOM. Each category of images contains N_c classes, dependent on the specific features associated with the image. Each generated image is encoded into a feature input vector which is the concatenation of the center of gravity coordinates of the image's parts. For example, a policemen image contains a hat, a head, a torso, two arms and two legs. A house has a roof, a chimney, a door and windows. A house with two windows is considered to be in a different class to a house with four windows. The dataset contains 15,000 artificial images, which are then described by 15,000 corresponding feature vectors. The maximum number of parts in an image is 14. The number of parts can also be increased again arbitrarily, though we find that 14 appears to be sufficient to demonstrate the visualization capabilities of the HRSOM. This means the data input dimension will be 28 as each element is a coordinate pair. For the images having fewer than 14 parts, the corresponding input vector is padded with zero. The distribution of the 12 classes in the dataset is shown in Table 1. It can be observed that the distribution of the data classes is considerably unbalanced. The largest class is more than 30 times larger than the smallest one.

Table 1. The distribution of the 12 classes in the policemen dataset.

Cls.ID	Description	Symbol	Sample
1	Policemen with raised left arm	+	2,520
2	Policemen with lowered left arm	×	2,480
3	Ships featuring two masts	▼	3,767
4	Ships with three masts	◇	1,233
5	Houses without windows	✳	84
6	Houses with 1 window in (LL) corner	□	227
7	Houses with 1 window in (UR) corner	■	251
8	Houses with 1 window in (UL) corner	○	241
9	Houses with 2 windows in (LL) and (UL)	●	658
10	Houses with 2 windows in (UL)) and (UR)	△	726
11	Houses with 2 windows in (LL) and (UR)	▲	663
12	Houses with 3 windows	▽	2,150

(LL) = lower left; (UR) = upper right; (UL) = upper left

4.2   The Cluster Forming Progress

A HRSOM of size 2500×2500 is trained using $\sigma = 600$, the number of training iterations is 120 and $\alpha = 0.6$. The progress by which the clusters formed is presented in Fig. 2. The mappings at iteration number 2, 40, 80, 90, 100, 110, 117 and 120 are shown from top left to bottom right, respectively. It can be observed that for about half of the training period, the clusters form relatively slowly. Separation of groups of samples commenced between the 70th and 80th iteration, and become well defined between the 100th to 120th epoch. The mapping result from the 115th iteration to the end of the training procedure is largely unchanged. Such a detailed clustering result has never before been seen for this dataset since the largest SOM applied thus far never exceeded the size of 256×256. The result presented here is useful since it not only provides an insight into the visually differences and similarities between sample classes, but

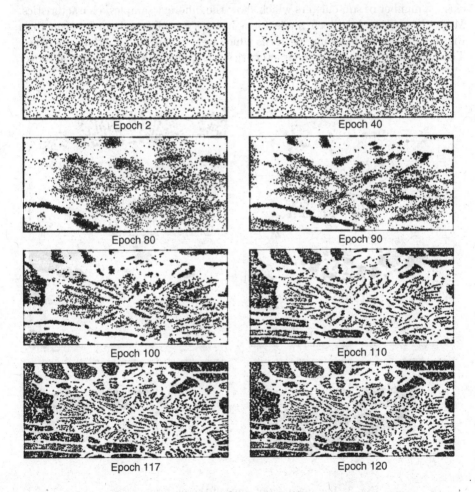

Fig. 2. The evolution of the mapping during the training procedure.

also supports the claim that the HRSOM can give a better visualization for a complex input space.

4.3   Closer View on Individual Clusters

The fully trained HRSOM in Fig. 2 shows well separated clusters although there exist pattern classes with a mapping which spreads across a larger region on the map; for example: the two classes denoted as the policemen with lowered left arm and the policemen with raised left arm. This indicates that the policemen classes contain a variety of sample features which are not very similar. Another reason could be that the two classes contains approximately ten times the number of samples of the other classes. A closer view of some of the individual clusters is provided by Fig. 3. Shown are the 4 clusters of houses, 2 clusters of policemen and 2 clusters of ships from top to bottom, respectively. Inside each cluster, there exists a number of sub-clusters which show the different samples' characteristics. More observation has been made for a house type class as can be seen from Fig. 4. This cluster has some patterns which were mapped closer than the other. These patterns have similar characteristics. Samples mapped far apart, however

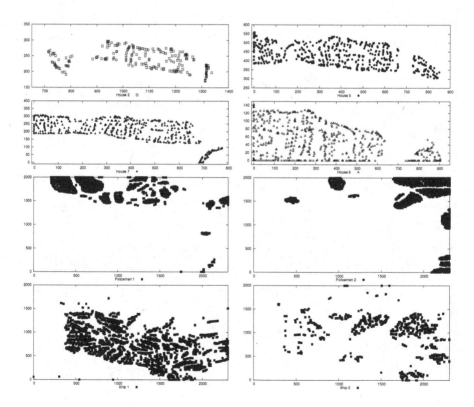

Fig. 3. The mapping of some of the pattern classes.

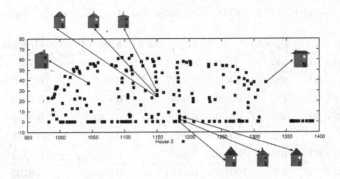

Fig. 4. The mappings of the samples in class "house one windows (UR)".

contain different features. It can be seen from the corresponding images of these samples, the patterns mapped near by, i.e. in the middle of the cluster, look very similar in the shape of the windows and the relative positions of chimneys. Further to the left of the figure, there are houses with chimneys located further on their right sides. On the other hand, further to the right of the figure, there are houses with chimneys located further on their left sides. Such observations were only possible due to the HRSOM and were not previously observed using lower resolution SOMs.

4.4 Comparing LRSOMs and HRSOMs

Table 2 compares the clustering performance when training SOMs with different map sizes for the policemen dataset. The map sizes are grouped into low resolution SOMs (LRSOMs) and HRSOMs. The former group includes SOMs of 400 neurons (80×50) to 75,000 neurons (300×250) while the latter contains SOMs with more than 100,000 neurons and up to 5,500,000 neurons (2500×2200). It should be noted that this is the first time that SOMs of such large size have been trained on complex clustering problems such as the policemen benchmark data. When training the SOMs we varied training parameters such as the learning rate α and number of training iterations as indicated in Table 2. The radius (σ) was adjusted to be about 40 % of the smaller side of the map although we varied σ to investigate the sensitivity of this parameter. Each experiment was repeated three times with different initializing conditions. The results shown are the average performance over the three runs. A number of interesting observations can be made from Table 2:

1. The performance of HRSOMs is generally much better than that of LRSOMs for all three assessment methods.
2. An almost perfect performance is obtained for the highest resolution SOM. This is interesting because the SOM is trained unsupervised but evaluated on using actual class labels. Only the largest SOMs offer sufficient mapping space to serve applications which require the separation of pattern instances in low

Table 2. A comparison of LRSOMs with HRSOMs when using the policemen dataset.

Map	Map size	Epoch	σ	α	Micro Purity	Macro Purity	Grouping Index
1	80×50	400	20	0.2	0.9351	0.8766	0.7334
2	80×50	400	22	0.5	0.9377	0.8877	0.7097
3	80×50	400	25	0.6	0.9385	0.8789	0.7435
4	100×80	400	35	0.5	0.9536	0.9200	0.7785
5	100×80	400	40	0.8	0.9532	0.9185	0.7776
6	100×80	400	37	0.3	0.9556	0.9192	0.7747
7	300×250	400	100	0.6	0.9773	0.9635	0.8702
8	300×250	400	115	0.4	0.9750	0.9641	0.8658
9	300×250	400	120	0.7	0.9717	0.9664	0.8668
10	1200×1000	200	400	0.1	0.9957	0.9894	0.9687
11	1200×1000	200	300	0.5	0.9970	0.9933	0.9745
12	1200×1000	200	500	0.9	0.9935	0.9876	0.9699
13	2300×2000	100	500	0.4	0.9823	0.9634	0.9712
14	2300×2000	150	500	0.4	0.9987	0.9982	0.9876
15	2300×2000	200	500	0.4	0.9997	0.9995	0.9965
16	2500×2200	80	600	0.6	0.9813	0.9524	0.9692
17	2500×2200	100	600	0.6	0.9993	0.9994	0.9964
18	2500×2200	120	600	0.6	1.0000	1.0000	0.9990

dimensional space while preserving the topology of the input data. HRSOMs are thus particularly well suited as a dimension reduction method while maintaining the information needed to separate pattern classes. A trained HRSOM is useful, for example, as a pre-processor in big data applications to reduce the dimensionality of a domain and speed up subsequent computations.

3. The grouping index experiences the most significant improvement among the three evaluation metrics. An improvement by approximate 27 % indicates the improvement in quality of the clusters.

4. HRSOMs are less sensitive to the choice of σ and the learning rate α. We attribute this observation to the fact that HRSOM offer a higher degree of freedom to the mappings of the data and hence do not rely on large α and large σ as is often needed for smaller SOMs (this is needed to allow the re-organization of mappings during the early stages of network training [1]).

The training time required ranged from less than 60 min for the smallest of the maps to 4 days and 9 h for the largest map. The training of the largest map would have taken nearly 8 months had it been executed on a state-of-the-art Intel CPU. Note also that the SOM only needs to be trained once and, when trained, the GPU version of the SOM can project data in $O(\log N)$ time, where N is the number of neurons. Provided a sufficiently large GPU is used the computational complexity of the HRSOM is thus independent to the number of samples that need to be projected as they can be processed independently and in parallel.

5 Conclusions

This paper demonstrated the properties of the HRSOM as a tool to visualize the intricate relationships among similar input vectors. It is shown that a suitable implementation on today's mainstream GPU technology allows for the training of SOMs with a granularity of larger than 6 million. Never before seen intricate details of the relationships among input vectors can be observed in the display space. These details would have been lost if the resolution of the SOM had been more limited. Our implementation of the SOM algorithm on a GPU is particularly efficient so that the limitation is now only the amount of available memory on the GPU. It is expected that the next generation of GPU would allow K to be in the order of tens of millions thus further reducing the difference between a discrete and a continuous mapping space. Owing to the relatively low cost of the GPU compared with CPU, one could envisage the HRSOM deployed as a visualization device in its own right.

We will make our software available to others on request, to promote the use of SOM as a visualization tool in data mining and big data applications, and so that they would not need to be encumbered with coding the SOM in CUDA themselves.

As a topic for future research, we suggest the implementation of SOM to run on a GPU cluster. This will allow the SOM to be deployed in big data applications, and in applications which require an even higher resolution of the mapping space.

References

1. Kohonen, T.: The self-organizing map. Proc. IEEE **78**(9), 1464–1480 (1990)
2. Scarselli, F., Gori, M., Tsoi, A.C., Hagenbuchner, M., Monfardini, G.: The graph neural network model. IEEE Trans. Neural Netw. **20**(1), 61–80 (2009)
3. Castillo, C., Donato, D., Becchetti, L., Boldi, P., Leonardi, S., Santini, M., Vigna, S.: A reference collection for web spam. SIGIR Forum **40**(2), 11–24 (2006)
4. Yahoo! Research. Web spam collections. barcelona.research.yahoo.net/webspam/datasets/
5. Hagenbuchner, M., Gori, M., Bunke, H., Tsoi, A.C., Irniger, C.: Using attributed plex grammars for the generation of image and graph databases. Pattern Recogn. Lett. **24**(8), 1081–1087 (2003)
6. Sauvage, V.: The T-SOM (Tree-SOM). In: Sattar, A. (ed.) AI 1997. LNCS, vol. 1342, pp. 389–397. Springer, Heidelberg (1997). doi:10.1007/3-540-63797-4_92
7. Samsonova, E.V., Kok, J.N., IJzerman, A.P.: TreeSOM: cluster analysis in the self-organizing map. Neural Netw. **19**(6), 935–949 (2006)
8. Forti, A., Foresti, G.L.: Growing hierarchical Tree SOM: an unsupervised neural network with dynamic topology. Neural netw. **19**(10), 1568–1580 (2006)

9. Bauer, H.U., Villmann, T.: Growing a hypercubical output space in a self-organizing feature map. IEEE Trans. Neural Netw. **8**(2), 218–226 (1997)

10. Villmann, T., Bauer, H.U.: Applications of the growing self-organizing map. Neurocomputing **21**(1), 91–100 (1998)

11. Skupin, A., Esperbé, A.: Towards high-resolution self-organizing maps of geographic features. Geograpchi visualization: concepts, tools and applications, 159–181 (2008)

12. Mathew, S., Joy, P.: Ultra fast SOM using CUDA. NeST-NVIDIA Center for GPU computing, hpc@ nestgroup. net (2010)

13. Xiao, Y., Leung, C.S., Ho, T.Y., Lam, P.M.: A GPU implementation for LBG and SOM training. Neural Comput. Appl. **20**(7), 1035–1042 (2011)

14. Wittek, P., Darányi, S.: A GPU-accelerated algorithm for self-organizing maps in a distributed environment. In: 20th European Symposium on Artificial Neural Networks, Computational Intelligence and Machine Learning, Bruges, Belgium (2012)

15. McConnell, S., Sturgeon, R., Henry, G., Mayne, A., Hurley, R.: Scalability of self-organizing maps on a GPU cluster using OpenCL and CUDA. J. Phys. Conf. Ser. **341**(1), 12–18 (2012)

16. Khan, S.Q., Ismail, M.A.: Design and implementation of parallel SOM model on GPGPU. In: 5th International Conference on Computer Science and Information Technology (CSIT), pp. 233–237. IEEE (2013)

Exceptional Contrast Set Mining: Moving Beyond the Deluge of the Obvious

Dang Nguyen[✉], Wei Luo, Dinh Phung, and Svetha Venkatesh

School of Information Technology, Centre for Pattern Recognition and Data
Analytics, Deakin University, Geelong, Australia
{d.nguyen,wei.luo,dinh.phung,svetha.venkatesh}@deakin.edu.au

Abstract. Data scientists, with access to fast growing data and computing power, constantly look for algorithms with greater detection power
to discover "novel" knowledge. But more often than not, their algorithms
give them too many outputs that are either highly speculative or simply confirming what the domain experts already know. To escape this
dilemma, we need algorithms that move beyond the obvious association
analyses and leverage domain analytic objectives (aka. KPIs) to look
for higher order connections. We propose a new technique Exceptional
Contrast Set Mining that first gathers a succinct collection of *affirmative* contrast sets based on the principle of redundant information elimination. Then it discovers *exceptional* contrast sets that contradict the
affirmative contrast sets. The algorithm has been successfully applied to
several analytic consulting projects. In particular, during an analysis of
a state-wide cancer registry, it discovered a surprising regional difference
in breast cancer screening.

1 Introduction

A task we face daily is to find meaningful differences among distinct groups. Are
there meaningful differences between private and government schools to justify
$30,000 extra cost per year [1]? Why are the hospital emergency departments
overcrowded on one day but not on others [2]? Such problems are formulated as
contrast set mining in the data mining community. The primary concern of contrast set mining is to reveal combinations of attributes and values whose distributions are different across groups [3]. The differences can be described using pairs
of contrasting conditional probabilities. For example, if social science researchers
compare gender groups, they might find that P(Occupation=engineer | Gen-
der=female)=15% while P(Occupation=engineer | Gender=male)=80%. In this
example, the contrast set is Occupation=engineer and two groups are Gen-
der=female and Gender=male. Mining contrast sets have many practical applications in different domains such as university admission [3] and folk music [4].

Many algorithms (including STUCCO [3] and CIGAR [5]) have been developed to tackle this important but difficult problem. With the emergency of

© Springer International Publishing AG 2016
B.H. Kang and Q. Bai (Eds.): AI 2016, LNAI 9992, pp. 455–468, 2016.
DOI: 10.1007/978-3-319-50127-7_39

big data, contrast set mining faces two challenges that have become more pronounced. First, real-world data analytic problems often have a large number of potential factors, making it very easy for an algorithm to output too many contrast sets, often with redundant information. Second, it is difficult to measure the value of a discovered contrast set. It is common to use some statistical significance measures to avoid patterns of speculative nature. But a statistically significant contrast set often just confirms some obvious prior knowledge, or detects incidental biases such as data collection artifacts. Just like other data mining problems, contrast set mining has to find a solution to balance the depth/coverage of mining and the succinctness/usefulness of the outputs.

We propose a new method Exceptional Contrast Set Mining (ECSM), an effective enhancement of contrast set mining. Its novelty lies in recognizing the high level of redundancy in traditional contrast set mining and then removal of these redundant patterns through statistical methods. It also distinguishes affirmative vs exceptional contrast set. Let us re-consider the above example. That affirmative contrast set expresses that men are more likely to work as engineers than women (a common knowledge). However, if we could find a contrast set that describes women are more likely work as engineers than men, then that exceptional contrast set is unexpected and reveals something new.

We formulate the algorithm and evaluate it on a diverse range of problems, that includes two real-world datasets: one concerns student performance in Portuguese language and the other one is related to breast cancer patients in a state cancer registry. The experimental results show that ECSM discovers relevant, succinct, and informative contrasts in large datasets. In particular, it discovered a surprising regional difference in breast cancer screening. That difference, we believe, can be attributed to the singer-songwriter and cancer survivor Kylie Minogue.

The rest of this paper is organized as follows. We briefly give the related background in Sect. 2. ECSM is described in Sect. 3 while the applications of ECSM to two real-world datasets are represented in Sect. 4. We conclude this paper and discuss directions for future work in Sect. 5.

2   Related Background

Data mining plays a crucial role in pattern discovery. One of main goals of data mining is to find important associations in data (e.g., correlations between risk factors and a disease). Frequent itemset mining has been proposed to achieve this purpose [6]. However, frequent itemset mining approaches are not suitable to identify major differences in data (e.g., distinguishing characteristics of two groups of patients who stay shortly and long in the hospital) because they do not enforce *consistent contrast* (i.e., using the same attributes to separate the group). Recently, contrast set mining has been proposed and developed to solve the shortcoming [3].

Contrast set mining was first introduced by Bay and Pazzani [7] with their STUCCO algorithm. STUCCO searched for contrast sets using an enumeration tree and used chi-square tests to check their significance. To perform multiple hypothesis tests, it employed a modified Bonferroni correction to control false positive errors. Since then, several algorithms adopted STUCCO's approach, including CIGAR [5]. However, both STUCCO and CIGAR often generated many contrast sets and they discovered common knowledge from data.

Some researches have attempted to reduce the number of results by finding maximal contrast sets [8] or correlated contrast sets [9]. These methods required many user-defined thresholds such as minimum frequency, mutual information, and all-confidence thresholds. In real-world applications, specifying many threshold values is a really challenging task for end users. Another important limitation of these methods is that they did not employ any approach for multiple testing to control false positive errors.

Class association rule mining [10] has been recently adapted for mining contrast sets [11]. However, this approach is more suitable for a predictive task than for a descriptive task aimed by traditional contrast set mining methods.

Exception rule mining aims at discovering unexpected association rules that express some correlations observed on few records and contradict commonsense association rules [12]. Note that our work is fundamentally different from exception rule mining. ECSM focuses on searching exceptional differences between groups (supervised learning) whereas exception rule mining focuses on finding exceptional associations between antecedent and consequent of rule (unsupervised learning).

3   Framework

3.1   Preliminaries

Let D be a dataset with n attributes $\mathcal{A} = \{A_1, A_2, ..., A_n\}$ and $|D|$ denotes the number of records where each record has an record identifier (RID). Let $G = \{g_1, g_2, ..., g_k\}$ be a list of groups (classes). Each attribute $A_i \in \mathcal{A}$ can take on values from its domain dom $(A_i) = \{a_{i1}, a_{i2}, ..., a_{im}\}$. The domain between attributes are distinct, i.e., dom $(A_i) \cap$ dom $(A_j) = \emptyset, \forall i \neq j$.

3.2   Contrast Set Mining

We briefly describe classic contrast set mining [3,7] in this section.

Definition 1. *A contrast set X is a conjunction of attribute-value pairs defined on groups g_1, g_2,..., g_k, such that no attribute occurs more than once.*

Let cover (X) be the set of $RIDs$ in D containing a contrast set X and cover (X, g_i) be the set of $RIDs$ in D containing both X and a group g_i.

Definition 2. *The support of a contrast set X w.r.t a group g_i, denoted by* sup (X, g_i), *is the percentage of records in g_i that contain X, that is,* sup $(X, g_i) = \frac{\text{cover}(X, g_i)}{|g_i|}$.

Table 1. Example dataset

RID	A	B	C	Group
1	a_1	b_1	c_1	g_1
2	a_1	b_2	c_1	g_2
3	a_2	b_2	c_1	g_2
4	a_3	b_3	c_1	g_1
5	a_3	b_1	c_2	g_2
6	a_3	b_3	c_1	g_1
7	a_1	b_3	c_2	g_1
8	a_2	b_2	c_2	g_2
9	a_1	b_3	c_2	g_3
10	a_3	b_1	c_1	g_3

A sample dataset is shown in Table 1. It contains ten records, three attributes (A, B, and C), and three groups (g_1, g_2, and g_3). The number of records belong to three groups g_1, g_2, and g_3 is 4, 4, and 2 respectively. For example, consider contrast set $X = \{(A, a_1)\}$. We have sup $(X, g_1) = \frac{2}{4} = 50\%$ because there are four records in group g_1, in which two records also contain $X = \{(A, a_1)\}$. Similarly, we also have sup $(X, g_2) = \frac{1}{4} = 25\%$ and sup $(X, g_3) = \frac{1}{2} = 50\%$.

Definition 3. *A contrast set X is significant if it satisfies $p_v(X) < \alpha$, where p_v refers to the p-value of X and α is the significance level.*

Definition 4. *A contrast set X is large if it satisfies* dev $(X) = \max_{ij} |$ sup $(X, g_i) -$ sup $(X, g_j)| \geq \delta$, *where δ is a user-defined threshold (called minimum support difference).*

The goal of contrast set mining is to find all contrast sets whose support differs meaningful across groups, meaning these contrast sets satisfy both Definitions 3 and 4.

Determination of whether a contrast set X is significant (i.e., X satisfies Definition 3): The significance of X is determined using chi-square test. We first compute 2-dimensional contingency table and chi-square value. The contingency table consists of 2 rows and k columns where the rows represent the truth of the contrast set and the columns represent the groups. The chi-square value is computed as follows:

$$\chi^2(X) = \sum_{i=1}^{2} \sum_{j=1}^{k} \frac{(O_{ij} - E_{ij})^2}{E_{ij}},$$

where O_{ij} is an observed value at the intersection of row i and column j and E_{ij} is an expected value calculated by $E_{ij} = (O_{i.} \times O_{.j})/|D|$, $O_{i.}$ is the total in row i, $O_{.j}$ is the total in column j, and $|D|$ is the total number of records.

We then compute p-value of X based on χ^2 value and the degrees of freedom df. Note that $df = k - 1$ where k is the number of columns of the contingency table. We can get the corresponding p-value by looking up an χ^2 distribution table. X is significant if the chi-square test rejects the *null* hypothesis that X and groups are *independent* (i.e., p-value of X is less than the significance level $\alpha = 0.05$).

Due to the large number of hypothesis tests, a contrast set mining method often suffers from a high false positive rate. STUCCO [3] uses a modified Bonferroni correction to limit the total Type I errors (false positives) for all chi-square tests to α. STUCCO used an enumeration tree to search for contrast sets. At the level i in the search tree, STUCCO computed a different α_i for the chi-square test. α_i is calculated as follows:

$$\alpha_i = \min\left((\alpha/2^i)/|C_i|, \alpha_{i-1}\right),$$

where $|C_i|$ is the number of candidates at level i. When we descend through the search tree, α_i is half that of α_{i-1}, causing the significance level to become more restrictive.

3.3   Redundant Contrast Set

A contrast set X is considered as *redundant* if one of the following conditions holds.

1. **Identical support** [3]: All its immediate subsets have the same support, that is, $\forall i \in [1, k]$, $\sup(X, g_i) = \sup(Y, g_i)$, for all $Y \subset X$ and $|Y| = |X| - 1$.
2. **Decreasing significance:** All its immediate subsets have strictly greater chi-square values, that is, $\chi^2(X) < \chi^2(Y)$, for all $Y \subset X$ and $|Y| = |X| - 1$.
3. **Independence** [3]: The observed distributions of X are not statistically different from its expected distributions (i.e., not statistically surprising), that is, $\forall i \in [1, k]$, $\sup(X, g_i) = \sup(X_1, g_i) \times \ldots \times \sup(X_h, g_i)$, for all $X_j \subset X$ and $|X_j| = |X| - 1$, $\forall j \in [1, h]$.

STUCCO [3] and CIGAR [5] only used two conditions (1) and (3) to prune redundant contrast sets whereas our proposed method (ECSM) also uses the condition (2) to seek a more succinct result.

3.4   Exceptional Contrast Set

In this section, we propose a new type of contrast set, called *exceptional* contrast set.

Definition 5. *The dominant group of a contrast set X, denoted by $g(X)$, is the group where X has its distribution being greater than its distributions in other groups, that is, $\exists i \in [1, k]$, $\sup(X, g_i) \geq \sup(X, g_j)$, for all $j \neq i$.*

A contrast set X is considered as *exceptional* if all the following conditions hold.

1. X and all its singleton subsets (i.e., subsets of size 1) are both large and significant.
2. Its dominant group is different from that of at least one of its singleton subsets.
3. Its chi-square value is greater than those of its singleton subsets that have the same dominant group.

More precisely, let X_1, X_2,..., X_h be singleton subsets of a contrast set X. X is exceptional iff X and X_l ($\forall l \in [1,h]$) are both large and significant and $\exists j \in [1,h]$, $g(X) \neq g(X_j)$ and if $g(X) = g(X_i)$ with $i \neq j$, then $\chi^2(X) \geq \chi^2(X_i)$. In this case, X is called exceptional contrast set while X_j is called affirmative contrast set.

3.5 Exceptional Contrast Set Mining

We propose an efficient method for mining exceptional contrast sets, called ECSM. It first finds large contrast sets. During the search process, it uses support and chi-square values to eliminate redundant contrast sets as described in Sect. 3.3. It then employs Benjamini and Hochberg's (BH) method [13] to generate significant contrast sets and control false positive errors. Finally, it explores exceptional contrast sets as described in Sect. 3.4.

Our algorithm for mining large contrast sets is described in Algorithm 1. It uses a tree structure for storing contrast sets. Each node in the tree contains a contrast set with metadata to quickly compute its support in each group. Unlike existing methods (STUCCO and CIGAR), our algorithm scans the dataset only once rather than scanning the dataset at each level of search tree.

Input: D - dataset, δ - min. support difference
Output: LS_1, LS_h - large contrast sets in D
$LS_1 \leftarrow \emptyset$;
begin
 foreach $A_i \in \mathcal{A}$ **do**
  foreach $X \in \text{dom}(A_i)$ **do**
   if dev $(X) \geq \delta$ **then**
    compute χ^2, p-value, $g(X)$;
    $LS_1 \leftarrow LS_1 \cup \{X\}$;
   end
  end
 end
 $LS_h \leftarrow$ CS-Miner(LS_1,δ);
end

Function $CS\text{-}Miner(LS_1,\delta)$
 $P_i \leftarrow \emptyset$;
 foreach $n_x \in LS_1$ **do**
  foreach $n_y \in LS_1$, with $y > x$ **do**
   $O \leftarrow n_x n_y$;
   if dev $(O) \geq \delta$ **then**
    if O is *not redundant* **then**
     compute χ^2, p-value, $g(O)$;
     $P_i \leftarrow P_i \cup \{O\}$;
    end
   end
  end
 end
 CS-Miner(P_i,δ);
end

Algorithm 1. Find-Large-CS(D,δ)

We discuss how we generate significant contrast sets and control false positive errors. False positive errors can be controlled based on two measures: *family-wise error rate* (FWER) and *false discovery rate* (FDR) [14]. STUCCO [3] used a modified Bonferroni correction to control FWER for all chi-square tests to α. This approach can be useful if the number of tests is small, but for a large number of hypothesis tests it can find very few significant contrast sets. We used a different approach to multiple testing; we did not control FWER but attempt to control FDR by using BH's method [13]. For contrast sets of cardinality (or size) 1, we compare their p-values with α and only keep significant contrast sets with p-values no larger than α. These contrast sets are then stored in CS_1. For contrast sets of size h ($h > 1$), we use BH's method as shown in Algorithm 2 to generate significant ones and store them in CS_h.

Input: LS_h - large contrast sets, α - significance level
Output: CS_h - significant contrast sets in LS_h
begin
    let $cs_1, ..., cs_M$ be M large contrast sets in LS_h;
    sort $cs_1, ..., cs_M$ in ascending order of p-value;
    let $p_1 \leq ... \leq p_M$ be corresponding p-values;
    define $L = \max \left\{ j : p_j \leq \frac{j \times \alpha}{M} \right\}$;
    return $cs_j, j = 1, ..., L$ (i.e., cs_j is significant);
end

Algorithm 2. BH-Method($LS_h, \alpha = 0.05$)

After obtaining two sets of large and significant contrast sets CS_1 and CS_h, we use Algorithm 3 to find exceptional contrast sets.

4 Experiments

We conducted experiments to validate our algorithm.

4.1 Experimental Setup

All experiments were performed on a computer with an Intel Core i7-5500U CPU at 3.00 GHz and 8 GB of RAM running OS Windows 8.1 (64-bit). The algorithms were coded in C# (.NET Framework 4.5.50938).

4.2 Datasets

The two datasets used in the experiments are Portuguese and Breast cancer.

Portuguese was downloaded from UCI Machine Learning Repository [15]. It contains student achievement on the Portuguese course of two secondary schools. The data attributes include student grades, demographic, social information and school related features. We used the final year grade $G3$ (issued at the third period) as the group attribute. Because $G3$ has a strong correlation with attributes $G1$ and $G2$ (the first and second period grades respective), we removed

Input: CS_1, CS_h - large and significant contrast sets
Output: ECS - exceptional contrast sets
$ECS \leftarrow \emptyset$;
begin
    foreach $X \in CS_h$ **do**
        foreach $X_i \subset X$ *and* $X_i \in CS_1$ **do**
            if $g(X) \neq g(X_i)$ **then**
                X is exceptional contrast set;
            else
                if $\chi^2(X) < \chi^2(X_i)$ **then**
                    $X \leftarrow null$;
                    break;
                end
            end
        end
        if $X \neq null$ **then**
            $ECS \leftarrow ECS \cup \{X\}$;
        end
    end
end

Algorithm 3. ECSM(CS_1,CS_h)

$G1$ and $G2$. All continuous attributes were discretized into equal intervals in which $G3$ was converted to *Fail* ($G3 < 10$) or *Pass* ($G3 \geq 10$).

Breast cancer consists of electrical medical records from 25,747 patients with breast cancer in a state cancer registry. The data attributes include patient demographic and diagnoses indicated by ICD-10 codes. All continuous attributes were discretized into equal intervals. We used the attribute *Area* (i.e., patient's living location is either *City* or *Outer region*) as the group attribute.

Table 2 represents the number of attributes, the number of groups, the group distribution, and the number of records in each dataset.

Table 2. Characteristics of the experimental datasets

Dataset	# attributes	# groups	Group dist. (%)	# records
Portuguese	30	2	(15.4, 84.6)	649
Breast cancer	13	2	(91.1, 8.9)	25,747

4.3   Parameter Settings

STUCCO and ECSM need only one parameter, the minimum support difference δ while CIGAR requires four parameters to be set, including the minimum support difference δ, minimum support β, minimum correlation λ, and minimum correlation difference γ. In our experiments, we used the same δ for all three methods but we did not set other parameters for CIGAR.

4.4   Evaluation Methods

We compared the performance of the methods by measuring their efficiency and evaluating the obtained contrast sets.

The efficiency of the algorithms was measured in terms of execution time (in second) while the quality of contrast sets was evaluated using different measures as follows.

Complexity measure: to evaluate the understandability of the knowledge extracted from the contrast sets. We used *The number of contrast sets* and *Average length of contrast sets*.

Generality measure: to quantify the generality of contrast sets. We used *Coverage* of a contrast set X, that is, $\text{Cov}(X) = \frac{\text{cover}(X)}{|D|}$ to compute the percentage of records covered by X [16].

Interest measure: to select and rank contrast sets regarding their potential interest to end users. We used χ^2 value and *dev* for this measure.

Quality measure: to measure the quality of contrast sets. We used *Sensitivity* of a contrast set X, that is, $\text{Sen}(X) = \frac{1}{k} \times \sum_{i=1}^{k} \sup(X, g_i)$ [16].

4.5   Execution Time Analysis

Figure 1 compares the execution time of ECSM with those of two state-of-the-art contrast set mining methods, namely STUCCO [3] and CIGAR [5], with various minimum support difference thresholds δ. The results show that ECSM was much faster than STUCCO and CIGAR. For example, consider the dataset Portuguese with $\delta = 12\%$ as shown in Fig. 1(a). The mining time of ECSM was only 1.837 (s) in comparison with 252.209 (s) and 208.830 (s) of STUCCO and CIGAR respectively. For this example, ECSM was 137 times faster than STUCCO and 114 times faster than CIGAR.

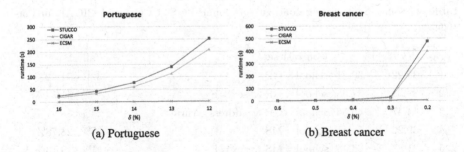

(a) Portuguese   (b) Breast cancer

Fig. 1. Execution times of STUCCO, CIGAR, and ECSM

4.6   Portuguese Dataset

Our goal is to find out what are major differences between two groups of students who failed and passed the Portuguese course. We assessed the quality of contrast sets generated by STUCCO, CIGAR, and ECSM in Portuguese.

We used $\delta = 10\%$ for all three algorithms. The summary of contrast sets generated by each algorithm is shown in Table 3.

It can be seen that contrast sets generated by STUCCO have the highest average coverage, χ^2, and *dev* while those generated by ECSM have the highest sensitivity. In addition, STUCCO and CIGAR generated numerous complicated contrast sets and most of them are not interesting. Note that STUCCO and CIGAR used two conditions (1) and (3) to remove redundancy as mentioned in Sect. 3.3. Table 4 lists out some examples of uninteresting contrast sets provided by both STUCCO and CIGAR. For example, consider the first three contrast sets in Table 4. Contrast set 1 describes that the students from school "Gabriel Pereira" are likely to pass the Portuguese course (71.22% vs 32%) and its specializations (contrast sets 2 and 3) add no new/interesting information.

On the contrary, ECSM generated a very small number of contrast sets. This can be explained by the fact that ECSM eliminated many redundant patterns and found exceptional patterns only. In additionally, the results of ECSM are interesting. Table 5 represents several exceptional contrast sets found by ECSM.

Table 3. Summary of contrast sets found in Portuguese

		STUCCO	CIGAR	ECSM
Complexity	# contrast sets	2,446	73,107	82
	avg. length	4.5	6.3	3.7
Generality	avg. coverage	20.64	15.24	18.34
Interest	avg. χ^2	39.35	13.80	19.31
	avg. *dev*	23.36	14.34	13.27
Quality	avg. sensitivity	18%	12%	21%

Table 4. Some uninteresting contrast sets found by STUCCO and CIGAR in Portuguese

No.	dev (%)	χ^2	Contrast set	Fail	Pass
1	39.22	58.41	school = GP	32%	71.22%
2	32.53	37.38	school = GP & sex = F	9%	41.53%
3	34.47	40.56	school = GP & address = rural	24%	58.47%
4	39.22	58.41	school = MS	68%	28.78%
5	21.87	23.43	school = MS & sex = F	41%	19.13%
6	18.34	21.03	school = MS & address = rural	32%	13.66%
7	39.71	103.84	past_failures = 0	51%	90.71%
8	33.46	37.38	past_failures = 0 & family_support = yes & take_higher_education = yes	21%	54.46%

Table 5. Some exceptional contrast sets found by ECSM in Portuguese

No.	dev (%)	χ^2	Contrast set	Fail	Pass	Type
1	16.13	10.41	address = rural	56%	72.13%	Affirmative
2	12.17	17.62	address = rural & sex = M & family_size = greater than 3 & study_time = less than 2 hours	18%	5.83%	Exceptional
3	11.35	15.92	address = rural & mother_edu = primary education & school_support = no & nursery = yes	17%	5.65%	Exceptional
4	13.58	7.63	alcohol_consume = very low	58%	71.58%	Affirmative
5	10.62	10.59	alcohol_consume = very low & address = urban & nursery = yes & internet_at_home = yes	19%	8.38%	Exceptional

Table 6. Redundancy eliminated by three algorithms in Portuguese

	STUCCO	CIGAR	ECSM
Identical sup.	229	4,483	30,786
Decreasing sig.	N/A	N/A	142,817
Independence	752	44,044	78

Table 6 summarizes the number of contrast sets removed by three algorithms in each type of redundancy as described in Sect. 3.3.

4.7   Breast Cancer Dataset

An Australian cancer control agency is interested in the impact of residential remoteness on cancer prevalence/incidence. The agency maintains a state-wide cancer registry; the data size is large but the data quality is not optimal. Contrast set mining was used for exploratory data analysis, to suggest hypotheses to bio-statisticians.

Our goal is to identify what are distinguishing characteristics of two groups of breast cancer patients who are living at cites and outer regions. We evaluated the quality of contrast sets generated by STUCCO, CIGAR, and ECSM in Breast cancer dataset. We used $\delta = 1\%$ for all three algorithms. Table 7 represents the summary of contrast sets generated by each algorithm.

Again, contrast sets found by STUCCO and ECSM have very high average coverage and sensitivity. STUCCO also obtained contrast sets with the highest average χ^2 value and *dev*. Regarding the complexity, CIGAR generated many contrast sets. The number of patterns found by STUCCO is few, however they are still not interesting. Table 8 describes several uninteresting contrast sets provided by STUCCO and CIGAR.

ECSM discovered the fewest contrast sets and they are both non-redundant and unexpected/interesting. Table 9 shows two exceptional contrast sets provided by ECSM.

In Australia, women receive free breast cancer screening from age 40. In Table 9, the first row confirms the well-accepted conjecture that women living in cities have better access to screening and hence may have slightly higher reported incidence rate. What is interesting is that rows 2 and 3 of Table 9 show that *only in 2005*, more breast cancers were detected for the first time (primary

Table 7. Summary of contrast sets found in Breast cancer

		STUCCO	CIGAR	ECSM
Complexity	# contrast sets	19	1,597	14
	avg. length	1.7	5.4	1.5
Generality	avg. coverage	17.75	8.84	15.59
Interest	avg. χ^2	15.54	10.51	9.56
	avg. *dev*	2.93	1.84	2.10
Quality	avg. sensitivity	18.27%	9.14%	15.53%

Table 8. Some uninteresting contrast sets found by STUCCO and CIGAR in Breast cancer

No.	dev (%)	χ^2	Contrast set	City	Outer region
1	5.65	28.23	Age = 60-79	37.96%	43.62%
2	4.84	22.65	Age = 60-79 & npri = 1	31.11%	35.95%
3	3.91	29.90	Age = 60-79 & npri = 1 & Site4 = C509	11.82%	15.73%
4	3.43	9.76	Age = 40-59	46.09%	42.66%
5	3.69	11.60	Age = 40-59 & npri = 1	41.47%	37.78%
6	1.04	7.93	Age = 40-59 & npri = 1 & Site4 = C508	2.87%	1.83%

Table 9. Two exceptional contrast sets found by ECSM in Breast cancer

No.	dev (%)	χ^2	Contrast set	City	Outer region	Type
1	3.43	9.89	Age = 40-59	46.09%	42.66%	Affirmative
2	1.03	11.01	Age = 40-59 & Sex = F & ydg = 2005 & npri = 1 & Site4 = C509	1.97%	3.01%	Exceptional
3	1.05	10.12	Age = 40-59 & Sex = F & ydg = 2005 & Site4 = C509	2.22%	3.27%	Exceptional

Table 10. Redundancy eliminated by three algorithms in Breast cancer

	STUCCO	CIGAR	ECSM
Identical sup.	2	574	5,148
Decreasing sig.	N/A	N/A	1,515
Independence	7	82	109

site without other accompanying cancers) in the outer regional areas. Further investigation reveals that this unexpected finding can be explained by the impact of the news coverage of Kylie Minogue's cancer in 2005 (related information can be found in [17]).

Table 10 summarizes the number of contrast sets removed by three algorithms in each type of redundancy.

5   Conclusion

We proposed ECSM, an effective enhancement of contrast set mining. Its novelty lies in the following aspects: (1) recognizing the high level of redundancy in traditional contrast set mining (as demonstrated in this paper) and application of disciplined statistical measures; (2) distinction of affirmative vs exceptional contrast sets; (3) leveraging succinct collection of affirmative contrast set to bootstrap exception discovery. Evaluation on a diverse range of problems suggests that ECSM constitutes an effective solution for discovering relevant, succinct, and informative contrasts in large datasets. Because the simplicity of the algorithm, it can be easily re-implemented and, if needed, customized to suit similar data analytical applications.

In the future, we plan to study theoretical aspects of ECSM including the impact of discovered exceptional contrast sets on classification, clustering, and outlier detection, and how to incorporate ECSM into more complex contrast set syntax to accommodate relational and temporal data.

Acknowledgment. This work is partially supported by the Telstra-Deakin Centre of Excellence in Big Data and Machine Learning.

References

1. Timna, J., Marc, M., Henrietta, C.: Primary-aged students in private schools perform only slightly better: NAPLAN. The Age Victoria, July 2015. http://goo.gl/hQ1q8V
2. Luo, W., Cao, J., Gallagher, M., Wiles, J.: Estimating the intensity of ward admission and its effect on emergency department access block. Stat. Med. **32**(15), 2681–2694 (2013)
3. Bay, S., Pazzani, M.: Detecting group differences: mining contrast sets. Data Min. Knowl. Disc. **5**(3), 213–246 (2001)

4. Neubarth, K., Conklin, D.: Contrast pattern mining in folk music analysis. In: Meredith, D. (ed.) Computational Music Analysis, pp. 393–424. Springer, New York (2016)

5. Hilderman, R., Peckham, T.: Statistical methodologies for mining potentially interesting contrast sets. In: Guillet, F.J., Hamilton, H.J. (eds.) Quality Measures in Data Mining, vol. 43, pp. 153–177. Springer, Heidelberg (2007)

6. Agrawal, R., Srikant, R.: Fast algorithms for mining association rules in large databases. In: Proceedings of the 20th International Conference on Very Large Data Bases, ser. VLDB 1994, pp. 487–499. Morgan Kaufmann Publishers Inc., San Francisco (1994)

7. Bay, S., Pazzani, M.: Detecting change in categorical data: mining contrast sets. In: The 5th ACM SIGKDD International Conference on Knowledge Discovery and Data Mining, pp. 302–306. ACM (1999)

8. Simeon, M., Hilderman, R.: COSINE: a vertical group difference approach to contrast set mining. In: Butz, C., Lingras, P. (eds.) AI 2011. LNCS (LNAI), vol. 6657, pp. 359–371. Springer, Heidelberg (2011). doi:10.1007/978-3-642-21043-3_43

9. Simeon, M., Hilderman, R., Hamilton, H.: Mining interesting correlated contrast sets. In: Bramer, M., Petridis, M. (eds.) Research and Development in Intelligent Systems XXIX, pp. 49–62. Springer, London (2012)

10. Nguyen, D., Nguyen, L.T., Vo, B., Hong, T.-P.: A novel method for constrained class association rule mining. Inf. Sci. **320**, 107–125 (2015)

11. Jabbar, M.S., Zaïane, O.R.: Learning statistically significant contrast sets. In: Khoury, R., Drummond, C. (eds.) AI 2016. LNCS (LNAI), vol. 9673, pp. 237–242. Springer, Heidelberg (2016). doi:10.1007/978-3-319-34111-8_29

12. Suzuki, E.: Autonomous discovery of reliable exception rules. In: KDD, vol. 97, pp. 159–176 (1997)

13. Benjamini, Y., Hochberg, Y.: Controlling the false discovery rate: a practical and powerful approach to multiple testing. J. R. Stat. Soc. Ser. B (Methodol.) **57**(1), 289–300 (1995)

14. Liu, G., Zhang, H., Wong, L.: Controlling false positives in association rule mining. Proc. VLDB Endow. **5**(2), 145–156 (2011)

15. Cortez, P., Silva, A.M.G.: Using data mining to predict secondary school student performance. In: Proceedings of 5th FUture BUsiness TEChnology Conference (FUBUTEC 2008), pp. 5–12. EUROSIS (2008)

16. Geng, L., Hamilton, H.: Interestingness measures for data mining: a survey. ACM Comput. Surv. (CSUR) **38**(3), 9 (2006)

17. Chapman, S., McLeod, K., Wakefield, M., Holding, S.: Impact of news of celebrity illness on breast cancer screening: Kylie Minogue's breast cancer diagnosis. Med. J. Aust. **183**(5), 247–250 (2005)

Smart Sampling: A Novel Unsupervised Boosting Approach for Outlier Detection

Mahsa Salehi[1]([✉]), Xuyun Zhang[4], James C. Bezdek[3],
and Christopher Leckie[2,3]

[1] IBM Research, Melbourne, VIC 3006, Australia
mahsalehi@au1.ibm.com
[2] National ICT Australia, Melbourne, VIC 3003, Australia
[3] Department of Computing and Information Systems, The University of Melbourne,
Melbourne, VIC 3010, Australia
{jbezdek,caleckie}@unimelb.edu.au
[4] Department of Electrical and Computer Engineering, University of Auckland,
Auckland 1010, New Zealand
xyzhanggz@gmail.com

Abstract. While various ensemble algorithms have been proposed for supervised ensembles or clustering ensembles, there are few ensemble based approaches for outlier detection. The main challenge in this context is the lack of knowledge about the accuracy of the outlier detectors. Hence, none of the proposed approaches focused on sequential boosting techniques. In this paper for the first time we propose a novel boosting algorithm for outlier detection called BSS, where we sequentially improve the accuracy of each ensemble detector in an unsupervised manner. We discuss the effectiveness of our approach in terms of bias-variance trade-off. Furthermore, an extended version of BSS (called DBSS) is proposed to introduce a novel source of diversity in outlier ensemble modeling. DBSS is used to analyze the effect of changing the input parameter of BSS on its detection accuracy. Our experimental results on both synthetic and real data sets demonstrate that our approaches outperform the two state-of-the-art outlier ensemble algorithms and benefit from bias reduction. In addition, our BSS approach is robust with respect to the changing input parameter. Since each detector in our proposed BSS/DBSS is only a subset of the whole dataset, our both techniques are well suited to application environments with limited memory processors (e.g., wireless sensor networks).

Keywords: Unsupervised boosting · Outlier detection · Ensemble analysis · Smart sampling

1 Introduction

Outliers (aka rare events or anomalies) are the set of data points that are inconsistent with the majority of the data set [1]. Outlier detection is an important

© Springer International Publishing AG 2016
B.H. Kang and Q. Bai (Eds.): AI 2016, LNAI 9992, pp. 469–481, 2016.
DOI: 10.1007/978-3-319-50127-7_40

component in many data mining applications. For example, in environmental monitoring in wireless sensor networks, outlier detection can be used for detecting faulty sensors or interesting patterns in the data set. Several outlier detection algorithms have been proposed in the literature and many surveys have categorized outlier detection algorithms [2,3]. Researchers traditionally have focused on statistical approaches and assume a priori an underlying distribution of normal data points. Another category of outlier detection (called distance based approaches) focus on the distances between data points and the k nearest neighbors (k-NNs) to evaluate the outlier scores of data points without assuming any underlying distribution. However, both categories of algorithms can be affected by either false positives or false negatives, which trigger the need for ensemble learning techniques.

Ensemble analysis is a technique that has been shown to improve the accuracy of many data mining tasks such as classification [4] and clustering [5]. Considering bias-variance trade-off, the accuracy improvement in ensemble classification comes from: (1) Variance reduction and (2) Bias reduction [6]. For example, subagging (aka subsampling) is an ensemble based technique in classification, where the parameters in subsamples of data are tuned during training phase using label information [4]. The idea behind this technique is that by aggregating multiple estimations, variance reduces in classification. Boosting [7] is another technique in classification which uses label information to sequentially reduce bias and improve the final classification accuracy. As clearly stated in a recent work [6], while ensemble techniques that reduce variance such as bagging [4] can be used in unsupervised outlier detection, proposing ensemble based techniques to reduce bias in outlier detection is challenging because we do not have label information as we have in classification. Hence, an open research problem in the literature is how to build effective outlier ensemble models by extracting knowledge from data in unsupervised scenarios where we have a lack of labels. In this paper we address this significant problem. We proposed a novel boosting algorithm for the problem of outlier detection without using any label information to improve the accuracy of the state-of-the-art outlier ensemble techniques. Since we proposed an effective *smart sampling* technique to increase detection accuracy, our algorithm is memory-efficient as well and is well-suited to memory-constrained environments. The contributions of this paper are as follows:

- A novel sequential boosting model for the problem of outlier detection is proposed based on smart sampling of the data set.
- An unsupervised approach is proposed to extract the ground truth from subsamples of the data set to increase the effectiveness of ensemble members.
- A novel and effective method for inducing diversity in ensemble detectors is proposed.
- The approaches are potentially scalable and can be applied to streaming data, where we can consider each incoming data chunk as a detector.

2   Related Work

Outlier detection algorithms can be categorized into *Model based* and *Distance based* approaches. In the first category a profile of the normal behavior of data points is usually generated based on an assumed distribution of the data points [8], while in the second category the distances between data points are used in assigning an outlier score to data points. In general distance based approaches outliers are detected by computing the distance of a point with respect to the other points in the data set [9]. In contrast, density based approaches assign outlier scores to data points with respect to their k nearest neighbors [10]. Since the LOF technique achieves good detection accuracy in non-homogeneous densities, it has become a popular approach and many variants of this technique have been proposed [11,12]. Finally, rather than considering distances, angle based approaches introduce a different perspective (variances of angles between data points) to detect outliers [13]. Several ensemble based models have been proposed for classification [4] and clustering problems [5]. In terms of outlier detection, Aggarwal [14] and Zimek et al. [15] surveyed the techniques in the literature. Recently, a different source of diversity has been proposed in [16], which randomly selects different subsamples of data and builds binary trees over them. This unsupervised algorithm is called an isolation forest (iForest) and randomly selects features in building the trees. Later in [17], random subsamples of a data set are introduced as a source of diversity. While [16,17] have investigated the problem of creating random samples as ensemble members for outlier detection, Kollios et al. [18] have proposed an approximate algorithm for clustering and outlier detection based on biased sampling. Later in [19] another sampling based outlier detection technique was presented. Rather than computing the nearest neighbors of each data point using the whole data set, they compute the neighbors of a data point with respect to a sample of the data set. In contrast to [19], Sugiyama and Borgwardt [20] claimed that using only one random sample of a data set can be both efficient and effective in most cases and their scheme outperforms a variety of outlier detection algorithms in their experimental results. But none of the three mentioned algorithms are benefiting from ensemble learning properties. Finally in [21] relevant chunks of history data are selected to detect outliers on streaming data sets.

The only two relevant state-of-the-art ensemble based methods are [17,20] which use *random* sampling techniques. To the best of our knowledge, no existing work has considered an *unsupervised* boosting technique to increase the detection accuracy in ensemble analysis of outliers.

3   Efficient Outlier Detection Ensemble Model

3.1   Problem Statement

Given a data set $P \in \Re^d$ with n data vectors, the goal is to assign a *local* outlier score $Score(P, p_j)$ to each data vector $p_j \in P$ with high accuracy using

an ensemble approach. Subsampling has proven to increase distance based outlier detection accuracy, by inducing diversity (using different subsamples) and
increasing the gap between the values that are assigned to outliers and inliers
[17]. However, in terms of increasing the gap between inliers and outliers, if the
ranking of the outlier scores remains the same, the final detection accuracy will
not change.

Specifically, let c and r be the center and radius of a hypersphere H in
d dimensional space $\Re^d$ where n data points are uniformly distributed in this
hypersphere. The *expected k-nearest neighbor (k-NN) distance* for Euclidean
data is [17]: $E\{d_k\} = \left(\dfrac{k}{n}\right)^{\frac{1}{d}} r$. Assume two hyperspheres H_o and H_i both with
radius r in a given data set, one with n_o data points and the other with n_i data
points, such that the data points are uniformly distributed in the hyperspheres
and $n_o < n_i$. Hence, H_o is less dense than H_i and reflects the outlier region,
while H_i reflects the inlier region in the data set. Using the above equation, the
expected k-NN distances for the points in H_o and H_i are $E\{d_k^o\} = \left(\dfrac{k}{n_o}\right)^{\frac{1}{d}} r$ and
$E\{d_k^i\} = \left(\dfrac{k}{n_i}\right)^{\frac{1}{d}} r$ respectively.

As mentioned in [17], after random subsampling of the data set by a ratio
m $(0 < m < 1)$, the expected k-NN distance for both regions in the data set
would increase. The change in the expected k-NN distance in H_o and H_i caused
by subsampling are $\Delta_o = f(m)\left(\dfrac{k}{n_o}\right)^{\frac{1}{d}} r$ and $\Delta_i = f(m)\left(\dfrac{k}{n_i}\right)^{\frac{1}{d}} r$ respectively,
where $f(m) = \dfrac{1 - m^{\frac{1}{d}}}{m^{\frac{1}{d}}}$. Therefore, $\Delta_o > \Delta_i$ which suggests that subsampling
increases the gap between outliers and inliers using the k-NN distance. However,
this change does not guarantee that the rank inversion between the outlier set
and inlier set is less probable and as a result the accuracy improves [6]. This
is because the ROC AUC (which is an accuracy measure for outlier detection)
just considers the rankings of the outlier scores, but not their values. Hence, in
terms of the bias-variance trade-off in ensemble analysis, subsampling techniques
lead to variance reduction, but not bias reduction [6]. We aim to address this
problem by proposing an effective approach to reduce bias in outlier ensembles
by manipulating ensemble members. In the following subsection we describe our
proposed approaches.

3.2   BSS: An Unsupervised Approach to Build Effective Ensembles Based on Smart Sampling

The BSS algorithm consists of three steps:

1. Constructing Ensembles: For a given data set P with n data points, we
randomly select a subset of the data set using a given ratio m $(0 < m < 1)$. As
a result, new subset of data with size mn is generated. Let s_j be this subset. We

refer to each s_j as an ensemble member in our algorithm. After constructing an ensemble member, we apply the next step of the algorithm on it.

2. Manipulating the Ensembles: For an ensemble member s_j, we further remove the points with high probability of outlierness to build more robust models of normal data. We compute the outlier scores of data points in subset s_j. Then we further decrease the density of subset s_j using the knowledge in the computed scores. In other words, we compute an *Outlier Removal Threshold* (*ORT*) as follows:

Definition 1. *The Outlier Removal Threshold (ORT$_s$) for a data set s is defined as:*

$$ORT_s = \mu(Y) \pm 3\sigma(Y) \tag{1}$$

where Y is a random variable that takes sample values from the set $\lambda = \{Score(s, p)|p \in s\}$, and $\mu(Y)$ and $\sigma(Y)$ are its mean and standard deviation, respectively. According to [22], the k-NN distance of data points follows the Gamma distribution. Also in [23], outlier scores are modeled by a mixture of Beta distribution components and in [24] Gaussian and Gamma distributions are considered for outlier scores. We have made a similar assumption about the distribution of the Score of data points in s and defined ORT$_s$ using three standard deviations.

Note the choice of $\pm$ in Eq. 1 depends on the algorithm by which the outlier scores are computed, as the outlier scores can be right-skewed or left-skewed. Also, here we can use any other removal outlier detection removal threshold. For instance, the third quartile or top q can be used as well.

Thereafter, we remove the set of data points Q_j from each subset s_j using its relevant threshold ORT_{s_j}, where Q_j is:

$$Q_j = \{q \,|\, q \in s_j \quad \text{and} \quad Score(s_j, q) > ORT_{s_j}\} \tag{2}$$

Removing such data points will further decrease the density of the highly probable outlier regions, while the inlier regions are kept intact. Hence, at the end of the second step of the BSS algorithm the number of data points in H_o and H_i on a subset s_j are $m'_j n_o$ and $m n_i$ respectively, where $m'_j < m$.

The output of this step is a *manipulated* subsets of the given data set. Although the removed data points may not be true outliers, removing them can give a more robust view of normal model of data set in comparison to random sampling.

After manipulating the ensemble detectors we remove the set Q_j from the whole dataset P, if P has at least $mn + |Q_j|$ number of data points, i.e., $P \leftarrow P \setminus Q_j$. The reason why we set this condition is that we want to make sure there are enough number of data points for the next round, so that we can select a subset with size mn from the updated dataset P.

Depending on the time requirements of the problem we may repeat the first and second steps of BSS to produce the desired number of manipulated subsets. Let $S = \{s_1, ..., s_c\}$ be the set of generated subsets, where c is the total number of subsets.

3. Voting: The next step in BSS is to compute the outlier score of each data point using the set S produced from previous steps. For the data point $p_j \in P$, the $Score(s_k, p_j)$ for all $s_k \in S$ are computed. The final outlier score for each data point is the average of the scores assigned using all of the updated subsets, i.e., $Score(P, p_j) = \frac{1}{c} \Sigma_{s_k \in S} Score(s_k, p_j)$.

As mentioned, since we have downsampled the data set P nonhomogeneously, the density of outlier regions will decrease by a higher ratio than inlier regions, and each individual detector thus becomes more effective. As a result, the consensus decision on the outlier scores of the data points will become more accurate than using an ensemble of random samples.

3.3   DBSS: An Unsupervised Approach to Induce Diversity in BSS

Another factor in the accuracy of an ensemble based algorithm is the diversity of its ensemble members. While the source of diversity in BSS is the different perspectives (samples) of the data set, DBSS proposes a novel source of diversity based on the Outlier Removal Threshold (ORT_s). This extension is proposed to study the effect of changing ORT on BSS detection accuracy.

In Definition 1 we consider three standard deviations of the distribution of outlier scores for defining a threshold to reduce the density of outlier regions. However, since we do not have the underlying distribution of the outlier scores, this is just an approximate estimation. By changing the threshold in each ensemble member, DBSS generates more diverse subsamples. In other words, DBSS constructs ensemble members similar to BSS. However, in the second step, each data subset s_j is manipulated using a different definition of the outlier removal threshold:

$$ORT_{s_j}^\eta = \mu(Y) \pm \eta\sigma(Y) \tag{3}$$

where η is a random number such that $2 < \eta < 3$. DBSS uses a different random η for each s_j to remove the highly probable outliers from the subsets. This approach not only proposes a new method of inducing diversity into the outlier ensembles, but also it is not limited to Eq. 3. Other definitions can potentially be used according to the nature of the data set and outlier definition. Since set of different values for η is produced in this algorithm, we can evaluate how the threshold parameter affects the final accuracy.

Finally, in the last step of DBSS all ensemble members can participate in the voting with similar or different weights, depending on the outlier removal threshold.

3.4   Time Complexity

In the first step of both the BSS and DBSS algorithms, given the total number of data points n, c subsets of size mn are randomly generated with sampling ratio $0 < m < 1$. The time complexity of the *Constructing step* is $O(n)$. In the second step, the outlier score of each ensemble member is evaluated. This step depends on the choice and implementation of the outlier detection algorithm.

If time complexity of a chosen outlier detection algorithm is $O(n^2)$, the total time complexity of the *Manipulating step* is $O(cm^2n^2)$. In the *Voting step* of the algorithm, the expected number of data points in each subset is reduced to $m'n^2$, where m' is the expected sampling ratio after step 2 of the algorithms and $m' < m$. Finally, the time complexity of both BSS and DBSS is $O(n^2)$ with respect to the number of data points, which is similar to the time complexity of the chosen outlier detection algorithm and random subsampling (RS) algorithm [17]. However, this can be affected by a higher choice of the constants c, m and as a result m'. Due to the potential for the BSS and DBSS algorithms to parallelize the ensemble computation, the time complexity can be further reduced to near linear.

Table 1. Data sets properties

	Dataset	# points	# dimensions	# classes	Outlier label
1	Synthetic-D	2000	Variable(5:5:50)	5	Add 5% uniform noise
2	Synthetic-M	5000	10	5	*Mahalanobis distance*
3	UCI Covertype	5743	10	7	'4'
4	UCI Vowels	1456	12	4	'1'
5	UCI Protein	5575	9	2	less than or equal to 0

4   Evaluation

In this section we evaluate the performance of BSS/DBSS with the base algorithm over all data points, and two state-of-the-art methods, i.e., [17], and the isolation forest algorithm [16]. These two algorithms are the most relevant approaches in the literature, as they use subsamples of a data set to construct outlier ensembles. Although [18] uses a biased sampling technique for outlier detection, this algorithm is not benefiting from ensemble learning properties and we are not using it for evaluation. The performance measure that we used is the ROC AUC for the outlier scores, which considers true positive rate (TPR) vs. false positive rate (FPR). The base algorithm that we used in our experiments is LOF [10], which is a well-established density based outlier detection algorithm. Note that BSS and DBSS are not dependent on a specific outlier detection algorithm and we can use them for any other technique, e.g., k-NN. All implementations were in Matlab R2015b and all experiments were done on a Core i7-2600 CPU 3.40 GHz running Windows 7.

4.1   Data Sets

In the experiments we have used two types of synthetic data sets consisting of 11 different synthetic data sets and 3 publicly available real data sets from the

UCI machine learning repository. We summarize the properties of the data sets in Table 1. For each data set we have assigned outlier labels in the following way. **(1) Synthetic-D**: We have generated 10 synthetic data sets in this group by changing the data dimensions from 5 to 50 while keeping the number of data points and the number of clusters fixed to evaluate the detection accuracy and scalability of our approaches. Also we add 5% outliers to all Synthetic-D data sets. **(2) Synthetic-M**: In this data set we use a different type of outlier to evaluate our algorithm. We used the *Mahalanobis distance* of each data point to its cluster center to label the data set. **(3) Covertype**: We have selected numerical features, i.e., 1 to 9 and 44. We select the labeled subset with the lowest number of data points and label them as outliers. The outlier rate is 2.23%. **(4) Protein**: We have chosen the first attribute as a label attribute and the values less than or equal to 0 are considered as outlier records. The outlier rate is 2.55%. **(5) Vowels**: We used the records in the training data set. The classes labels are 1, 6, 7 and 8. We randomly selected 50 data points from class 1 and labeled them as outliers. The outlier rate is 3.4%.

4.2   Outlier Detection Accuracy

In this subsection we compare our BSS and DBSS algorithms with the subsampling (we call it RS) algorithm [17]. We investigate how our boosting schema affects detection accuracy in details with every different sampling ratios. We change the sampling ratio m from 0.1 to 0.9 to investigate its impact on detection accuracy. Since the number of nearest neighbors (k) affects the detection accuracy as well, the authors in [17] change the number of nearest neighbors and then choose the results for which the base algorithm (LOF) performs better. Here we used the same technique to produce results. We have changed k to 2, 5 and 10 and find out that for all of the data sets in this paper (except Covertype) the AUC of LOF is highest when k equals 10. However, in the Covertype data set LOF performs better when k equals 5. Also, the number of ensemble members c in all experiments is set to 25, which is the same value used in the RS algorithm.

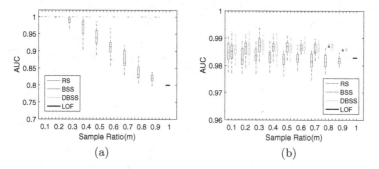

(a)   (b)

Fig. 1. Detection accuracy of BSS, DBSS and RS on: (a) Synthetic-D, (b) Synthetic-M (Color figure online)

Figure 1a shows the results on the first data set in the Synthetic-D group, where the data dimensionality is 5 and k is 10. The first nine values on the horizontal axis in Figs. 1a and b correspond to sampling ratio (m). The last value on the horizontal line shows the results of the base algorithm on the total number of data points in this data set. The red and orange boxplots are the AUC of BSS and DBSS respectively and the blue boxplots are the AUC of the RS algorithm. The source of variance in the boxplots is different ensemble detectors. In addition, the AUC of LOF is shown on the far right of the figure using all data points, where all BSS, DBSS and RS have the same AUC as LOF.

It is shown that the AUC of BSS and DBSS are 100% for smaller sampling ratios whereas RS AUC drops from $m = 0.2$. The AUC of LOF increases up to 20% by using BSS, DBSS in all sampling ratios. The results in this figure show that smart sampling in boosting for our both BSS and DBSS techniques increases, leading to higher detection accuracy than both the RS algorithm and LOF algorithm.

Figure 1b shows the results on the Synthetic-M data set. In this data set, both BSS and DBSS improve the AUC of LOF and RS algorithms. Note the nature of outliers in Synthetic-D and Synthetic-M datasets are different, and this impacts on the AUC differences in Fig. 1a and b. However, our approaches outperforms RSS and LOF in both synthetic data sets.

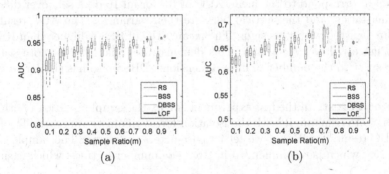

Fig. 2. Detection accuracy of BSS, DBSS and RS on: (a) Vowels, (b) Protein.

The results on the Vowels and Protein data sets are shown in Fig. 2a and b respectively. In both data sets, the AUC of BSS and DBSS are either almost equal to or greater than the RS algorithm. Also, in the Vowels dataset the AUC of BSS and DBSS for all sampling ratios are greater than or almost equal to the base algorithm and in the Protein dataset for all sampling ratio (except one) our algorithms outperform the base algorithm. However in this particular dataset all of the four algorithms have a very low detection accuracy. Note that the RS algorithm could not outperform the base algorithm in several sampling ratios.

The results on the Covertype data set are depicted in Fig. 3. The AUC results on this real data set show that BSS and DBSS clearly outperform both the LOF and RS algorithms for all sample ratios. Whereas RS algorithm is unable to outperform LOF in 5 sample ratios.

In conclusion, all of the above figures suggest that using smart sampling in our proposed boosting schema reduces bias in outlier detection, which leads to higher detection accuracy by approximately 6% in average without impacting the time complexity. Generally, both BSS and DBSS outperform the RS algorithm for the real life and synthetic datasets used in this paper. Also, except in one sampling ratio in Protein dataset, our both proposed approaches have higher/equal detection accuracy than the base algorithm, whereas RS cannot outperform the base algorithms in multiple sampling ratios.

BSS versus DBSS: As mentioned earlier in Sect. 3, the main aim of proposing DBSS algorithm is to evaluate the effect of changing the parameter ORT on the detection accuracy. In other words, the main difference between DBSS and BSS is that in BSS a constant ORT is used to construct ensemble detectors, whereas in DBSS this parameter is variable. All of the above boxplots suggest that DBSS has comparable accuracy with BSS. Hence, our boosting schema is robust to changes in the parameter ORT.

Scalability Comparison: In this experiment, we investigate the scalability of BSS with respect to data dimensionality. We use all 10 data sets in Synthetic-D with different data dimensions from 5 to 50. Figure 4 shows the result, where the circles correspond to the mean AUC of RS for all 10 data sets over different sampling ratios, and the vertical bars are their variances. The AUC results of BSS are depicted with plus signs. The mean AUC of BSS is higher than RS and the variance is lower than BSS. This demonstrates that BSS is more scalable with respect to the number of data dimensions. DBSS also has similar results.

Isolation Forest. In the last experiment, we fix the sampling ratio of both BSS and DBSS to compare it with the isolation forest (iForest) approach. To obtain the AUC results for iForest we set its parameters as follows: The sample size is set to 256, which is recommended in [16]. The number of trees, which is similar

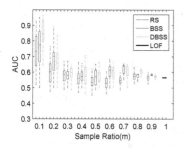

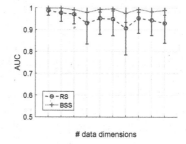

Fig. 3. Detection accuracy of BSS, DBSS and RS on Covertype

Fig. 4. Comparing the AUC mean and variance of BSS with RS

to the number of ensembles (c), is set to 25 to make the experimental results comparable. The height of the tree is set to 10 as in [16]). In BSS and DBSS we set m to 0.1 and the number of nearest neighbors k to 10. We also add the results of RS by fixing the sampling ratio $(m = 0.1)$.

Figure 5 shows the AUC results of the four algorithms over all synthetic and real data sets. The results show that the AUC of DBSS are either almost equal to or greater than the AUC of the RS algorithm in all data sets, whereas BSS has a higher or equal detection accuracy in most of the data sets. In terms of iForest, the AUC of BSS and DBSS are greater than or equal to iForest in all data sets except one (Protein). However, in this specific data set all of the four algorithms have a very low detection accuracy (below 70%).

The results show that DBSS outperforms both state-of-the-art algorithms in all data sets (except one for iForest). Although the results of DBSS and BSS are similar, DBSS performs slightly better, which again shows the robustness of our boosting schema to the threshold parameter. The results suggests that diversifying ensemble members and smart sampling in outlier ensembles are both promising techniques in outlier detection.

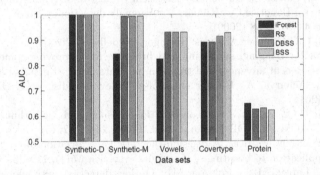

Fig. 5. Comparing AUC of BSS and DBSS with iForest and RS.

5 Conclusions

Proposing effective ensemble based techniques for unsupervised outlier detection is challenging due to the lack of label information in data sets. This motivates us to propose an ensemble learning algorithm to detect outliers in an unsupervised manner. Our second motivation comes from memory-constrained applications where the amount of memory on processors is limited and only a subset of data can be fit into memory of each processor. In this paper by considering the theoretical foundations of accuracy improvement in ensemble analysis (bias-variance trade-off), we proposed two novel sequential boosting models for outlier detection using smart sampling: (1) A new and effective approach is proposed to increase the accuracy of the ensemble members by extracting the ground truth in a new and effective unsupervised manner (BSS). (2) A new and unsupervised approach for inducing diversity in outlier ensembles is introduced (DBSS) to show the robustness of our proposed boosting schema with respect to changing a threshold parameter in BSS.

To the best of our knowledge there is no similar work tackling the problem of unsupervised boosting. Our approaches are promising due to the bias reduction in outlier detection. We have compared our approaches with two state-of-the-art ensemble based outlier detection algorithms (RS and iForest). Our empirical results on both synthetic and real data sets show the effectiveness of our approaches in terms of detection accuracy. In the data sets used in this paper the average accuracy improvement is 6% comparing to RS algorithm with the same time complexity, whereas it is 8% in comparison with iForest but with higher time complexity. However, due to the potential of the BSS and DBSS algorithms for parallelizing the ensemble computation, in the future we intend to investigate this problem to reduce the time complexity.

References

1. Barnett, V., Lewis, T.: Outliers in Statistical Data, vol. 3. Wiley, Hoboken (1994)
2. Chandola, V., Banerjee, A., Kumar, V.: Anomaly detection: a survey. ACM Comput. Surv. **41**(3), 1–58 (2009)
3. Aggarwal, C.C.: Outlier Analysis. Springer, Heidelberg (2013)
4. Buhlmann, P.: Bagging, subagging and bragging for improving some prediction algorithms. Recent advances and trends in nonparametric statistics (2003)
5. Ghosh, J., Acharya, A.: Cluster ensembles. Wiley Interdisc. Rev. DMKD **1**(4), 305–315 (2011)
6. Aggarwal, C.C., Sathe, S.: Theoretical foundations and algorithms for outlier ensembles. ACM SIGKDD Explor. Newsl. **17**(1), 24–47 (2015)
7. Freund, Y., Schapire, R.E.: A decision-theoretic generalization of on-line learning and an application to boosting. J. Comput. Syst. Sci. **55**(1), 119–139 (1997)
8. Yang, X., Latecki, L.J., Pokrajac, D.: Outlier detection with globally optimal exemplar-based GMM. In: SDM, pp. 145–154 (2009)
9. Knox, E.M., Ng, R.T.: Algorithms for mining distance-based outliers in large datasets. In: VLDB, pp. 392–403 (1998)
10. Breunig, M.M., Kriegel, H.P., Ng, R.T., Sander, J.: LOF: identifying density-based local outliers. SIGMOD **29**, 93–104 (2000)
11. Kriegel, H.P., Kröger, P., Schubert, E., Zimek, A.: LoOP: local outlier probabilities. In: CIKM, pp. 1649–1652 (2009)
12. Papadimitriou, S., Kitagawa, H., Gibbons, P.B., Faloutsos, C.: Loci: fast outlier detection using the local correlation integral. In: ICDE, pp. 315–326 (2003)
13. Pham, N., Pagh, R.: A near-linear time approximation algorithm for angle-based outlier detection in high-dimensional data. In: SIGKDD, pp. 877–885 (2012)
14. Aggarwal, C.C.: Outlier ensembles: position paper. SIGKDD Explor. Newsl. **14**(2), 49–58 (2013)
15. Zimek, A., Campello, R.J., Sander, J.: Ensembles for unsupervised outlier detection: challenges and research questions a position paper. SIGKDD Explor. Newsl. **15**(1), 11–22 (2014)
16. Liu, F.T., Ting, K.M., Zhou, Z.H.: Isolation-based anomaly detection. TKDD **6**(1), 3 (2012)

17. Zimek, A., Gaudet, M., Campello, R.J., Sander, J.: Subsampling for efficient and effective unsupervised outlier detection ensembles. In: SIGKDD, pp. 428–436 (2013)
18. Kollios, G., Gunopulos, D., Koudas, N., Berchtold, S.: Efficient biased sampling for approximate clustering and outlier detection in large data sets. TKDE **15**(5), 1170–1187 (2003)
19. Wu, M., Jermaine, C.: Outlier detection by sampling with accuracy guarantees. In: SIGKDD, pp. 767–772 (2006)
20. Sugiyama, M., Borgwardt, K.: Rapid distance-based outlier detection via sampling. In: NIPS, pp. 467–475 (2013)
21. Salehi, M., Leckie, C.A., Moshtaghi, M., Vaithianathan, T.: A relevance weighted ensemble model for anomaly detection in switching data streams. In: Tseng, V.S., Ho, T.B., Zhou, Z.-H., Chen, A.L.P., Kao, H.-Y. (eds.) PAKDD 2014. LNCS (LNAI), vol. 8444, pp. 461–473. Springer, Heidelberg (2014). doi:10.1007/978-3-319-06605-9_38
22. Dong, W., Wang, Z., Josephson, W., Charikar, M., Li, K.: Modeling LSH for performance tuning. In: CIKM, pp. 669–678 (2008)
23. Bouguessa, M.: Modeling outlier score distributions. In: Zhou, S., Zhang, S., Karypis, G. (eds.) ADMA 2012. LNCS (LNAI), vol. 7713, pp. 713–725. Springer, Heidelberg (2012). doi:10.1007/978-3-642-35527-1_59
24. Kriegel, H.P., Kröger, P., Schubert, E., Zimek, A.: Interpreting and unifying outlier scores. In: SIAM, pp. 13–24 (2011)

Approximating Message Lengths of Hierarchical Bayesian Models Using Posterior Sampling

Daniel F. Schmidt$^{(\boxtimes)}$, Enes Makalic, and John L. Hopper

Centre for Epidemiology and Biostatistics, The University of Melbourne,
Carlton, VIC 3053, Australia
{dschmidt,emakalic,j.hopper}@unimelb.edu.au

Abstract. Inference of complex hierarchical models is an increasingly common problem in modern Bayesian data analysis. Unfortunately, there are few computationally efficient and widely applicable methods for selecting between competing hierarchical models. In this paper we adapt ideas from the information theoretic minimum message length principle and propose a powerful yet simple model selection criteria for general hierarchical Bayesian models called MML-h. Computation of this criterion requires only that a set of samples from the posterior distribution be available. The flexibility of this new algorithm is demonstrated by a novel application to state-of-the-art Bayesian hierarchical regression estimation. Simulations show that the MML-h criterion is able to adaptively select between classic ridge regression and sparse horseshoe regression estimators, and the resulting procedure exhibits excellent robustness to the underlying structure of the regression coefficients.

1   Introduction

The Minimum Message Length (MML) [1] principle of inductive inference is a powerful, information theoretic, Bayesian framework for parameter estimation and model selection. The MML principle is based on the connection between statistical inference, algorithmic complexity and information theory, and has been extensively applied to a wide range of statistical and machine learning problems with great success. The basic idea is to quantify the fit of a model to an observed data string by the length of a decodable message, usually measured in *nits*, or base-e digits, that communicates both the model parameters as well as the length of the data string once compressed using the statistical properties of the model. This message length can be decomposed into two components:

$$I(\mathbf{y}^n, \boldsymbol{\theta}) = I(\boldsymbol{\theta}) + I(\mathbf{y}^n|\boldsymbol{\theta}),$$

where $I(\boldsymbol{\theta})$ is the length of the message required to describe the statistical model $\boldsymbol{\theta}$, and $I(\mathbf{y}^n|\boldsymbol{\theta})$ is the message length required to describe the data string $\mathbf{y}^n = (y_1, \ldots, y_n)$ using the properties of the nominated model $\boldsymbol{\theta}$. In this fashion, the message length naturally balances the ability of a model to fit a data string against the complexity of the model, and minimising the message length provides

© Springer International Publishing AG 2016
B.H. Kang and Q. Bai (Eds.): AI 2016, LNAI 9992, pp. 482–494, 2016.
DOI: 10.1007/978-3-319-50127-7_41

a natural framework for statistical inference. The MML principle is particularly attractive as both continuous model parameters, as well as discrete, structural parameters can be estimated by minimising the message length. A key idea in the MML principle is that estimates should be stated (coded) only to an accuracy warranted by the data. This leads to estimators that are invariant under one-to-one reparameterisations, a property not shared by most conventional Bayesian point estimators, and that gracefully handle many difficult statistical problems that cause significant issues for procedures such as maximum likelihood and Bayesian posterior mean and mode estimators.

Computing the exact message length is in general an NP problem; however, under suitable regularity conditions, the MML87 approximation [2] provides a simple formula to compute the message length of a statistical model $p(\mathbf{y}^n|\boldsymbol{\theta})$, characterised by continuous parameters $\boldsymbol{\theta}$ and associated prior distribution $\pi(\boldsymbol{\theta})$:

$$I_{87}(\mathbf{y}^n, \boldsymbol{\theta}) = -\log p(\mathbf{y}^n|\boldsymbol{\theta}) + \frac{1}{2} \log |\mathbf{J}(\boldsymbol{\theta})| - \log \pi(\boldsymbol{\theta}) + c(k), \qquad (1)$$

where $p(\mathbf{y}^n|\boldsymbol{\theta})$ is the likelihood of the data $\mathbf{y}^n$, $\mathbf{J}(\boldsymbol{\theta})$ is the Fisher information matrix, k is the number of free continuous model parameters and

$$c(k) = -\frac{k}{2} \log(2\pi) + \frac{1}{2} \log(k\pi) + \psi(1) \qquad (2)$$

are appropriate dimensionality constants. While the MML87 approximation has been applied to derive MML estimators for a wide range of statistical models, there are large classes of problems to which it does not immediately apply, particularly those in which the prior distributions are highly peaked or those in which the Fisher information matrix can be (near) singular.

1.1   Bayesian Hierarchical Models

Multi-level Bayesian hierarchical models in which the prior distribution $\pi(\boldsymbol{\theta})$ is decomposed into a chain of priors, each potentially depending on further hyper-parameters, are a class of models that cannot currently be easily handled within the MML framework. Hierarchical models are becoming increasingly common in machine learning because they can describe complex prior beliefs about the parameters $\boldsymbol{\theta}$, and can also be used as tools to generate estimators with desirable properties; examples of particular importance are the extensive number of Bayesian penalized regression procedures based on scale mixture priors such as the Bayesian horseshoe [3] which represent the state-of-the-art in penalized regression. The general Bayesian hierarchical model can be described by

$$\mathbf{y}^n|\boldsymbol{\theta} \sim p(\mathbf{y}^n|\boldsymbol{\theta})d\mathbf{y}^n,$$
$$\boldsymbol{\theta}|\boldsymbol{\alpha}_1 \sim \pi(\boldsymbol{\theta}|\boldsymbol{\alpha}_1)d\boldsymbol{\theta},$$
$$\boldsymbol{\alpha}_1|\boldsymbol{\alpha}_2 \sim \pi(\boldsymbol{\alpha}_1|\boldsymbol{\alpha}_2)d\boldsymbol{\alpha}_1,$$
$$\boldsymbol{\alpha}_2|\boldsymbol{\alpha}_3 \sim \pi(\boldsymbol{\alpha}_2|\boldsymbol{\alpha}_3)d\boldsymbol{\alpha}_2,$$
$$\cdots$$
$$\boldsymbol{\alpha}_q \sim \pi(\boldsymbol{\alpha}_q)d\boldsymbol{\alpha}_q,$$

where $(\boldsymbol{\alpha}_1, \ldots, \boldsymbol{\alpha}_q)$ are the hyperparameters of the prior hierarchy. While the original MML87 formula (1) cannot be applied directly to hierarchical models, Makalic and Schmidt [4] extended the formula to this class of problems, and subsequently applied the new approximation to the problem of shrinkage estimation in Gaussian regression models. However, the resulting hierarchical message length approximation depends crucially on modifying the key formulas to account for the highly peaked (non-uniform) prior distributions that can arise when the hyperparameters are estimated from data, rather than being specified *a priori*. As discussed in Sect. 2, this is only really plausible for specific conjugate likelihood-prior pairings, which greatly restricts the applicability of the method to general hierarchical structures.

1.2   Sampling Approaches to Inference of Hierarchical Models

The standard Bayesian approach to handling complex hierarchical models is to sample from the posterior distribution

$$p(\boldsymbol{\theta}, \boldsymbol{\alpha}_1, \ldots, \boldsymbol{\alpha}_q | \mathbf{y}^n) \propto p(\mathbf{y}^n | \boldsymbol{\theta}) \pi(\boldsymbol{\theta}, \boldsymbol{\alpha}_1, \ldots, \boldsymbol{\alpha}_q)$$

using Markov Chain Monte Carlo (MCMC) procedures such as Gibbs sampling. While such procedures allow the posterior distribution to be explored, they cannot directly be used to select between different model structures. Standard Bayesian model selection is usually based on the marginal probability of the data $\mathbf{y}^n$ under a given hierarchy, but this is notoriously difficult to compute. Chib's algorithm [5] provides a method for computing marginal probabilities directly from posterior samples, but it is slow, requiring multiple runs of the sampling chain. Further, it requires that all the conditional distributions be fully specified up to normalising constants. Given that this latter restriction is not required to efficiently sample from many posterior distributions, this represents a serious limitation. Other approaches include a proposal to use an empirical Laplace approximation of the posterior distribution [6], and an MML based approach called the message from Monte-Carlo formula [7], which appears to offer a way of efficiently computing message lengths from posterior samples. Unfortunately, it is not immediately clear how, or if, either of these procedures can be adapted to multi-level hierarchies.

What is required is a simple, computationally efficient procedure for performing Bayesian model selection of complex, hierarchical models using posterior samples drawn from a single run of a sampling chain. This paper proposes such a procedure based on the minimum message length principle.

1.3   Our Contribution: Model Selection of Hierarchical Models

Specifically, we propose a simple formula, called MML-h, which is motivated by the MML87 approximation and can be used to compute message lengths of complex Bayesian hierarchical models directly from a single set of posterior samples. In contrast to Chib's [5] algorithm, the method requires only that the *prior* distributions be exactly specified, a condition which is generally satisfied, and which dramatically expands the scope of the procedure.

We demonstrate the simple applicability of the procedure by using it to compute message lengths for two complex Bayesian penalised regression models, the ridge and horseshoe estimators, and show that it can be used to adaptively select between the different hierarchies depending on the properties of the observed data. To further underline the usefulness of our new procedure, we note that prior to this paper, computation of model selection scores for the horseshoe has, to the best of the authors' knowledge, not been undertaken anywhere in the literature due to the complexity of the hierarchy.

2   Curved Message Lengths for Conjugate-Priors

The MML87 approximation was derived under the assumption that the prior distribution, $\pi(\boldsymbol{\theta})$, is approximately uniform in a neighbourhood of the parameter estimates $\boldsymbol{\theta}$ determined by the accuracy to which the parameters are coded. In the case of Bayesian hierarchical models, where the hyperparameters that control the behaviour of the prior distributions are treated as free parameters, this assumption can be violated. In such settings, the approximation fails and minimisation of the resulting message length leads to degenerate estimates.

In the special case that the prior $\pi(\cdot)$ is *conjugate* with the likelihood $p(\cdot)$, C. S. Wallace suggested an ingenious correction to the usual MML87 formula for heavily curved priors that preserves the invariance property of the MML estimator. We discuss this procedure in some detail, as it directly motivates our approximation, and is unfortunately only briefly mentioned by Wallace ([1], pp. 235–236). Consider the two-level hierarchy

$$\mathbf{y}^n|\boldsymbol{\theta} \sim p(\mathbf{y}^n|\boldsymbol{\theta})d\mathbf{y}^n,$$
$$\boldsymbol{\theta}|\boldsymbol{\alpha} \sim \pi(\boldsymbol{\theta}|\boldsymbol{\alpha})d\boldsymbol{\theta},$$

where $\boldsymbol{\alpha}$ are specified hyperparameters controlling the behaviour of the prior distribution for $\boldsymbol{\theta}$. The key idea is to first propose some imaginary "prior data" $\mathbf{y}_0^m$ whose properties depend only on the prior hyperparameters $\boldsymbol{\alpha}$. It is then possible to view the prior $\pi(\boldsymbol{\theta}|\boldsymbol{\alpha})$ as a *posterior* distribution of this prior data $\mathbf{y}_0^m$ and some initial uninformative prior $\pi_0(\cdot)$ that does not depend on the hyperparameters $\boldsymbol{\alpha}$. Formally, we seek the decomposition

$$\pi(\boldsymbol{\theta}|\boldsymbol{\alpha}) = C_0\pi_0(\boldsymbol{\theta})p(\mathbf{y}_0^m|\boldsymbol{\theta}),$$

where $p(\mathbf{y}_0^m|\boldsymbol{\theta})$ is the likelihood of m imaginary prior samples, $\pi_0(\boldsymbol{\theta})$ is an uninformative prior and C_0 is a suitable normalisation constant not dependent on $\boldsymbol{\theta}$. The corrected Fisher information $\mathbf{J}^*(\boldsymbol{\theta})$ is then constructed from the combined likelihood $p(\mathbf{y}^n, \mathbf{y}_0^m|\boldsymbol{\theta}) = p(\mathbf{y}^n|\boldsymbol{\theta})p(\mathbf{y}_0^n|\boldsymbol{\theta})$, and the corrected MML87 approximation is simply

$$I_{87}^*(\mathbf{y}^n, \boldsymbol{\theta}) = -\log \pi(\boldsymbol{\theta}|\boldsymbol{\alpha}) + \frac{1}{2}\log|\mathbf{J}^*(\boldsymbol{\theta})| - \log p(\mathbf{y}^n|\boldsymbol{\theta}) + c(k),$$

$$= -\log \pi_0(\boldsymbol{\theta})C_0 + \frac{1}{2}\log|\mathbf{J}^*(\boldsymbol{\theta})| - \log p(\mathbf{y}^n, \mathbf{y}_0^m|\boldsymbol{\theta}) + c(k),$$

where $c(k)$ are terms depending only on k. This correction has the interesting interpretation of being based on the asymptotic lower bound of the inverse of the variance of the maximum likelihood estimator for the combined data $(\mathbf{y}^n, \mathbf{y}_0^m)$ rather than for the observed data $\mathbf{y}^n$ only, and in this sense, it naturally incorporates the effects of the curvature of the prior on the variance of the resulting estimator. While this is clearly a very neat solution, it does have two drawbacks. The first is that it can only be used if the prior is conjugate with the likelihood, a situation that will often not be the case in many Bayesian models. The second is that it cannot be applied when the prior distribution is itself a multi-level hierarchy with additional hyperparameters that require estimation.

2.1 Poisson-Exponential Hierarchy

We now demonstrate the above procedure on a simple, but commonly used, statistical model. Consider the following Bayesian hierarchy:

$$y_i|\lambda \sim \text{Poi}(\lambda),$$
$$\lambda|\alpha \sim \text{Exp}(\alpha),$$

where $\text{Poi}(\lambda)$ denotes a Poisson distribution with rate parameter λ, and $\text{Exp}(\alpha)$ is an exponential distribution with scale parameter α and probability density

$$p(\lambda|\alpha) = \frac{1}{\alpha} \exp\left(-\frac{\lambda}{\alpha}\right). \tag{3}$$

The hyperparameter α controls how concentrated the prior density is around $\lambda = 0$; for smaller α the curvature of the prior distribution becomes very large relative to the width of the coding quantum, and the usual condition of a "roughly uniform" prior under which MML87 was derived becomes increasingly violated as $\alpha \to 0$. As the exponential distribution is conjugate to the Poisson distribution, we can use the procedure discussed in Sect. 2 to derive an appropriately corrected Fisher information.

Let $\mathbf{y}^n = (y_1, \ldots, y^n)$ be n samples from an unknown Poisson distribution. The probability of $\mathbf{y}^n$ for a rate parameter λ is

$$p(\mathbf{y}^n|\lambda) = \frac{\lambda^{\bar{y}} \exp(-n\lambda)}{\prod_{i=1}^n \Gamma(y_i + 1)}, \tag{4}$$

where $\bar{y} = \sum_{i=1}^n y_i$ is the sufficient statistic. Following the procedure in Sect. 2 we can write the likelihood of the "prior data" $\mathbf{y}_0^m$ as

$$p(\mathbf{y}_0^m|\lambda) = \frac{\lambda^{\bar{y}_0} \exp(-m\lambda)}{K}.$$

Setting $m = 1/\alpha$, $\bar{y}_0 = 0$, $\pi_0(\lambda) \propto 1$ and $C_0 = K/\alpha$ yields

$$\pi(\lambda|\alpha) = \underbrace{(1)}_{\pi_0(\lambda)} \cdot \underbrace{\left(\frac{K}{\alpha}\right)}_{C_0} \cdot \underbrace{\left[\frac{\exp(-\lambda/\alpha)}{K}\right]}_{p(\mathbf{y}_0^m|\lambda)} = \frac{1}{\alpha} \exp\left(-\frac{\lambda}{\alpha}\right).$$

Thus, the effect of the prior is to augment the data with $1/\alpha$ additional samples. The augmented negative log-likelihood (up to constants) is then

$$l^*(\lambda) = -(\bar{y} + \bar{y}_0) \log \lambda + \lambda(n + 1/\alpha).$$

The corrected Fisher information, using the expectation $E\left[\bar{y} + \bar{y}_0\right] = \lambda(n+1/\alpha)$, is given by

$$J^*(\lambda) = \frac{n + 1/\alpha}{\lambda} \qquad (5)$$

The effect of the correction is to increase the sample size by $1/\alpha$; as $\alpha \to 0$ and the prior becomes highly curved, the Fisher information increases because the coding quantum must be correspondingly smaller.

3 Approximate Message Lengths from Posterior Samples

A well known property of the Fisher information matrix is that it provides a lower bound on the covariance of unbiased estimates. In particular, if $\hat{\boldsymbol{\theta}}(\mathbf{y}^n)$ is an unbiased estimator of $\boldsymbol{\theta}$ for some parametric model, then the Cramer-Rao lower bound states that

$$\text{Cov}(\hat{\boldsymbol{\theta}}(\mathbf{y}^n)) \geq \mathbf{J}^{-1}(\boldsymbol{\theta}).$$

Therefore, the accuracy to which parameters are encoded in the MML framework, which is determined by the determinant of the Fisher information matrix, is approximately inversely proportional to the (generalized) variance of the maximum likelihood estimates. The corrected Fisher information matrix $J^*(\boldsymbol{\theta})$ discussed in Sect. 2 can also be interpreted in a similar fashion by treating it as the Fisher information matrix for both real data and the additional Fisher information contributed by the prior distributions, so that the accuracy to which the parameters are encoded is approximately inversely proportional to the generalized variance of the resulting *maximum a posteriori* (MAP) estimates.

3.1 Approximating Message Lengths of Hierarchical Models

This observation motivates the main idea of this paper, which is to note that while the covariance matrix of the MAP estimates may be very difficult to compute, particularly for hierarchical models, we can approximate it by the covariance matrix of the posterior distribution $p(\boldsymbol{\theta}, \boldsymbol{\alpha}_1, \ldots, \boldsymbol{\alpha}_q | \mathbf{y}^n)$. This choice is particularly attractive because the posterior covariance automatically takes into account the effects of the prior distribution in a similar way to the curved conjugate-prior corrected Fisher information, and immediately extends to the complete set of parameters and hyperparameters in any hierarchy.

We propose to approximate the minimised message length of a complex, potentially non-conjugate hierarchical Bayesian model by

$$I(\mathbf{y}^n, \hat{\boldsymbol{\theta}}(\mathbf{y}^n)) \approx I_h(\mathbf{y}^n) = E\left[-\log p(\mathbf{y}^n|\boldsymbol{\theta}) - \log \pi(\boldsymbol{\theta}, \boldsymbol{\alpha}_1, \ldots, \boldsymbol{\alpha}_q) \,|\, \mathbf{y}^n\right]$$
$$-\frac{1}{2} \log |\text{Cov}(\boldsymbol{\theta}, \boldsymbol{\alpha}_1, \ldots, \boldsymbol{\alpha}_q \,|\, \mathbf{y}^n)| + c(k) - \frac{k_{\boldsymbol{\theta}}}{2}, \quad (6)$$

where the expectation is taken with respect to the posterior distribution, k is the total number of free parameters and hyperparameters in the model, $c(k)$ is given by (2), $k_{\boldsymbol{\theta}}$ is the dimensionality of $\boldsymbol{\theta}$ and $\hat{\boldsymbol{\theta}}(\mathbf{y}^n)$ are the estimates that minimise the exact message length. We call (6) the MML-h approximation. The last term in (6) accounts for the fact that the usual increase in the negative log-likelihood of $\mathbf{y}^n$ due to "rounding off" the model parameters $\boldsymbol{\theta}$ is already taken into account through the posterior expectation of the negative log-likelihood.

In general, analytical evaluation of (6) is a non-trivial proposition. However, as discussed in Sect. 1.2, a standard approach to Bayesian analysis is to characterise posterior distributions by pseudo-random samples generated from an MCMC sampling procedure. If a chain of samples drawn from the posterior distribution is available it is straightforward to approximate the posterior covariance matrix by the empirical posterior covariance matrix. Let

$$\boldsymbol{\theta}^{(j)}, \boldsymbol{\alpha}_1^{(j)}, \ldots, \boldsymbol{\alpha}_q^{(j)}, \ (j = 1, \ldots, m)$$

denote the chain of m parameter and hyperparámeter samples drawn from the posterior distribution $p(\boldsymbol{\theta}, \boldsymbol{\alpha}_1, \ldots, \boldsymbol{\alpha}_q | \mathbf{y}^n)$ of a Bayesian hierarchy. We can then approximate the minimised message length of the hierarchical model by the following simple formula:

$$I_h(\mathbf{y}^n) \approx \left(\frac{1}{m}\right) \sum_{j=1}^m \left[-\log p(\mathbf{y}^n | \boldsymbol{\theta}^{(j)}) - \log \pi \left(\boldsymbol{\theta}^{(j)}, \boldsymbol{\alpha}_1^{(j)}, \ldots, \boldsymbol{\alpha}_q^{(j)}\right) \right]$$
$$-\frac{1}{2} \log |\text{Cov}(\boldsymbol{\theta}^{(j)}, \boldsymbol{\alpha}_1^{(j)}, \ldots, \boldsymbol{\alpha}_q^{(j)})| + c(k) - \frac{k_{\boldsymbol{\theta}}}{2}, \tag{7}$$

where k is the total number of free parameters and hyperparameters in the complete hierarchy and $k_{\boldsymbol{\theta}}$ is the dimensionality of $\boldsymbol{\theta}$. To use the approximation (7) we require only the ability to sample from the posterior distribution, which makes it widely applicable. By taking m sufficiently large, we can approximate the formula (6) by (7) to any desired degree of accuracy.

3.2   Discussion

In comparison with the usual MML87 formula (1), the proposed MML-h approximation (7) uses the empirical covariance matrix of the posterior samples in place of the Fisher information matrix to determine the accuracy to which all parameters and hyperparameters are to be encoded. It also uses the posterior expected negative-log data-prior probabilities to approximate the maximised product of data and prior probabilities. Such an approximation is expected to work even if a hierarchy contains many hyperparameters, because the posterior variances of these hyperparameters will generally be large, and they will contribute very little to the total message length.

It is straightforward to establish that under suitable regularity conditions, as $n \to \infty$ the approximation (6) converges to the minimised MML87 message length in the case that the hierarchy contains no adjustable hyperparameters and the number of parameters remains fixed. Convergence in the general setting is more difficult to establish, and is a topic of future research.

The empirical covariance matrix can potentially be broken into a series of block-wise independent covariance matrices for each level of the hierarchy to improve numerical stability and reduce computational burden if the dimensionality of the complete parameter and hyperparameter space is large; in this case the determinant of the posterior covariance matrix can then be replaced by

$$|\text{Cov}(\boldsymbol{\theta}, \boldsymbol{\alpha}_1, \ldots, \boldsymbol{\alpha}_q \,|\, \mathbf{y}^n)| \approx |\text{Cov}(\boldsymbol{\theta}|\mathbf{y}^n)| \cdot \prod_{j=1}^{q} |\text{Cov}(\boldsymbol{\alpha}_j|\mathbf{y}^n)|.$$

Finally, it is important to note that while the approximation (6) allows us to assign message lengths to potentially complex hierarchical models, it is in general not invariant under reparameterisations, and the regular MML87 approximation should be preferred in those situations in which it can be applied. The lack of invariance is expected to be a minor issue, particularly for moderate to large sample sizes, as the differences in message lengths between parameterisations will generally be small, and will decrease with increasing sample size.

4   Example 1: Poisson-Exponential

As a simple example, we apply the MML-h approximation to the Poisson-exponential hierarchy discussed in Sect. 2.1. In this case, the posterior distribution of the Poisson parameter, and its associated posterior variance can be exactly computed, allowing for straightforward comparison of the MML-h message length (7) with the MML87 message length, without needing to approximate quantities via MCMC sampling. The resulting expressions are very close, which lends further confidence in the validity of the approximation (7) when applied to significantly more complex problems, such as those discussed in Sect. 5. Recall the Poisson-exponential hierarchy discussed in Sect. 2.1:

$$y_i|\lambda \sim \text{Poi}(\lambda),$$
$$\lambda|\alpha \sim \text{Exp}(\alpha).$$

Using the likelihood (4), the prior distribution (3) and the corrected Fisher information (5) derived in Sect. 2.1 in the MML87 formula (1), and minimising with respect to λ, yields the minimised MML87 message length

$$I_{87}(\mathbf{y}^n, \hat{\lambda}(\mathbf{y}^n)) = -(\bar{y}+1/2)\log\left(\frac{\bar{y}+1/2}{n+1/\alpha}\right) + (\bar{y}+1/2) + \sum_{i=1}^{n}\log\Gamma(y_i+1)$$

$$+ \log\alpha + \frac{1}{2}\log(n+1/\alpha) + c(1). \tag{8}$$

Under the specified hierarchy, the posterior distribution of λ is well known to be

$$\lambda\,|\,\mathbf{y}^n, \alpha \sim \text{Ga}(\bar{y}+1, n+1/\alpha),$$

where $\mathrm{Ga}(a, b)$ is a Gamma distribution with shape parameter a and inverse-scale parameter b. We can use this to compute the approximate minimised message length using approximation (6). The posterior variance is given by

$$\mathrm{Var}(\lambda \,|\, \mathbf{y}^n, \alpha) = \frac{\bar{y} + 1}{(n + 1/\alpha)^2}$$

and the required expectations are

$$\mathrm{E}\,[\lambda \,|\, \mathbf{y}^n, \alpha] = \frac{\bar{y} + 1}{n + 1/\alpha}, \quad \mathrm{E}\,[\log \lambda \,|\, \mathbf{y}^n, \alpha] = \psi(\bar{y} + 1) - \log(n + 1/\alpha).$$

Using the approximation $\psi(z) = \log(z) - 1/2/z + O(1/z^2)$, the MML-$h$ message length is given by

$$I_h(\mathbf{y}^n) = -(\bar{y} + 1/2) \log\left(\frac{\bar{y} + 1}{n + 1/\alpha}\right) + (\bar{y} + 1) + \sum_{i=1}^{n} \log \Gamma(y_i + 1)$$

$$+ \log \alpha + \frac{1}{2} \log(n + 1/\alpha) + c(1) + O(1/n). \tag{9}$$

Comparing (8) with (9) reveals close similarities between the two minimised message lengths. For finite n, the MML-h message length is slightly longer, and for large n the two message lengths converge.

5   Example 2: Ridge Versus Horseshoe Regression

Estimation of potentially high-dimensional regression models using Bayesian shrinkage techniques is an important and active area of research, with important applications in a wide range of areas such as statistical genetics. In this setting, the prior distribution expresses a belief about the relative magnitudes of the underlying regression coefficients. Consider the following local-global hierarchy for linear regression:

$$\mathbf{y}^n | \mathbf{X}, \boldsymbol{\beta}, \sigma^2 \sim \mathcal{N}_n(\mathbf{X}\boldsymbol{\beta}, \sigma^2 \mathbf{I}_n),$$
$$\beta_j | \lambda_j^2, \tau^2, \sigma^2 \sim \mathcal{N}(0, \lambda_j^2 \tau^2 \sigma^2),$$
$$\sigma^2 \sim \sigma^{-2} d\sigma^2, \tag{10}$$
$$\lambda_j \sim \pi(\lambda_j) d\lambda_j,$$
$$\tau \sim \mathcal{C}^+(0, 1),$$

where $\mathbf{X} \in \mathbb{R}^{n \times p}$ is a matrix of predictor variables (not necessarily full rank), $\mathcal{N}_k(\cdot, \cdot)$ is the k-variate Gaussian distribution, and $\mathcal{C}^+(0, 1)$ is the standard half-Cauchy distribution with probability density function

$$p(z) = \frac{2}{\pi(1 + z^2)}, \quad z > 0.$$

The parameter τ is a *global shrinkage* parameter that controls the overall level of regularisation applied to the estimated regression coefficients; in contrast, the λ_j hyperparameters are *local shrinkage* parameters that control the level of regularisation applied to individual regression coefficients. The choice of prior distributions $\pi(\lambda_j)$ associated with the local shrinkage parameters can lead to prior specifications with very different models of the underlying regression coefficients. The two priors that we consider are

$$\lambda_j \sim \begin{cases} \delta_1(\lambda_j)d\lambda_j & \text{(ridge regression)} \\ \mathcal{C}^+(0,1) & \text{(horseshoe regression)} \end{cases}$$

where $\delta_1(x)$ is a distribution with a point-mass at $x = 1$. In the case of ridge regression, there is no local shrinkage, which implies a belief that the underlying vector $\boldsymbol{\beta}$ is dense (most coefficients are non-zero), and that the majority of the coefficients have similar magnitude of association with the outcome y. In contrast, the powerful horseshoe hierarchy [3], which enjoys a number of favourable theoretical properties, allows each regression coefficient to be shrunk individually, and the half-Cauchy prior distribution over the λ_j hyperparameters implies a prior belief that the coefficients will either be close to zero, and unimportant, or large and relatively unaffected by shrinkage. These are two very different prior beliefs about $\boldsymbol{\beta}$, and both ridge regression and horseshoe estimation can yield excellent parameter estimates when applied in the appropriate setting.

An obvious question is whether we can use the data to decide which hierarchy we should be using to estimate the regression coefficients? In this example, we use the MML-h approximation to compute message lengths for both ridge and horseshoe models, and select the hierarchy that results in the shortest message length. This is an interesting and novel application of model selection to penalised regression that we believe has the potential to make significant improvements by *mitigating the weaknesses* of individual prior distributions, and is a topic for future research. This example also neatly highlights the ease of applicability of the MML-h approximation to highly complex hierarchical models such as the horseshoe, for which there are no software packages to compute model selection quantities such as marginal probabilities.

5.1 Message Lengths for Ridge and Horseshoe Regression

Sampling from the horseshoe, and ridge posterior distributions, can be efficiently done using the algorithm presented in [8]. The horseshoe and ridge hierarchies are nested, in the sense that the ridge is simply a special case of the horseshoe, and this makes computation of the approximate message lengths for both models relatively straightforward. In the case of ridge regression, we can simply set $(\lambda_j = 1)$ for all $j = (1,\ldots,p)$, and ignore these hyperparameters when computing (7). The covariance quantities used in the ridge regression case were

$$|\mathrm{Cov}(\boldsymbol{\beta},\tau,\sigma^2\,|\,\mathbf{y}^n)| = |\mathrm{Cov}(\boldsymbol{\beta}\,|\,\mathbf{y}^n)| \cdot \mathrm{Var}(\tau\,|\,\mathbf{y}^n) \cdot \mathrm{Var}(\sigma^2\,|\,\mathbf{y}^n).$$

In the case of the horseshoe, the covariance quantities were

$$|\mathrm{Cov}(\boldsymbol{\beta},\boldsymbol{\lambda},\tau,\sigma^2\,|\,\mathbf{y}^n)| = |\mathrm{Cov}(\boldsymbol{\beta},\tau,\sigma^2\,|\,\mathbf{y}^n)| \cdot |\mathrm{Cov}(\boldsymbol{\lambda}\,|\,\mathbf{y}^n)|.$$

Table 1. Median squared prediction errors, relative to ridge regression, for the horseshoe (HS), the model with smallest MML-h message length, and the MML-h weighted average of ridge and horseshoe regression, computed over 100 test iterations. For each test, the number of non-zero coefficients, out of $p = 20$, is given by p^*. The sample size of the training data was $n = 50$.

Coef. model	p^*	Pairwise correlation			Töplitz correlation		
		HS	MML-h	MML Avg	HS	MML-h	MML Avg
$\beta_j^* = 1$	1	0.071	0.071	0.071	0.067	0.067	0.067
	5	1.022	1.040	0.971	0.731	1.079	0.996
	10	1.924	1.051	1.104	1.379	1.000	1.001
	15	2.476	1.018	1.090	1.710	1.000	1.002
	20	2.886	1.049	1.144	2.112	1.000	1.000
$\mathcal{N}(0,1)$	1	0.075	0.075	0.075	0.093	0.093	0.093
	5	0.368	0.411	0.396	0.373	0.390	0.411
	10	0.788	0.960	0.953	0.778	0.932	0.932
	15	1.149	1.049	1.035	1.090	1.006	1.006
	20	1.323	1.074	1.049	1.348	1.021	1.020
$\mathcal{C}(0,1)$	1	0.082	0.082	0.082	0.060	0.060	0.060
	5	0.413	0.413	0.413	0.427	0.432	0.432
	10	0.485	0.617	0.630	0.467	0.550	0.550
	15	0.628	0.712	0.713	0.639	0.724	0.731
	20	0.589	0.674	0.652	0.691	0.767	0.734

We chose to use the block-diagonal covariance structure between $\boldsymbol{\beta}$ and $\boldsymbol{\lambda}$ because these parameters are largely uncorrelated. Both models have dimensionality $k_\theta = p + 1$. The total number of parameters and hyperparameters for the ridge model is $k = p + 2$, and for the horseshoe is $k = 2p + 2$.

5.2   Simulations

To test the ability of MML-h to discriminate between the horseshoe and ridge prior hierarchies we undertook a small simulation study. For each test iteration, ($p = 20$) covariates were generated from a multivariate normal distribution with zero mean, and either: (i) pair-wise correlations of $1/2$ between each covariate, or (ii) a Töplitz correlation structure with $\mathrm{corr}(X_i, X_j) = (1/2)^{|i-j|}$. Then, a true coefficient vector $\boldsymbol{\beta}^*$ was generated according to three different models: (i) $\beta_j^* \sim \delta_1(\beta_j^*)d\beta_j^*$ (i.e., $\beta_j^* = 1$), (ii) $\beta_j^* \sim \mathcal{N}(0,1)$ and (ii) $\beta_j^* \sim \mathcal{C}(0,1)$, where $\mathcal{C}(\cdot)$ denotes the Cauchy distribution. These models cover a wide range of true coefficient patterns. The first $p^* \leq 20$ coefficients were retained, and the remaining $(p - p^*)$ coefficients were set to zero to simulate different levels of sparsity.

Finally, ($n = 50$) datapoints were generated from the model $y_i = \mathbf{x}_i \boldsymbol{\beta}^* + \varepsilon_i$, where $\mathrm{Var}(\varepsilon_i)$ was chosen to attain a signal-to-noise ratio of five. Once the data was generated, $m = 2,000$ samples were drawn from the posterior distribution of the ridge and horseshoe model, and the coordinatewise medians of the samples were used as representative point estimates. The MML-h method was then applied, using the posterior samples, to estimate message lengths for the two different prior models and the prior model with the smaller message length was selected. Additionally, a posterior weighted average of the two coefficient estimates was produced, using the exponentiated negative message lengths as unnormalized weights. Finally, the expected squared prediction error for all four methods was calculated and recorded. This process was repeated 100 times for all combinations of sparsity level $p^* = \{1, 5, 10, 15, 20\}$, correlation structure and coefficient models. The results are presented in Table 1 as median prediction errors attained by the horseshoe, the model with smallest MML-h message length (ridge or horseshoe), and the MML-h weighted posterior average of the ridge and horseshoe models, all relative to the ridge regression model. Numbers less than one indicate a performance increase relative to ridge regression, while numbers greater than one indicate poorer performance relative to ridge regression.

5.3   Results and Discussion

The results demonstrate that for specific types of underlying true coefficients both ridge regression and horseshoe regression can have significant differences in performance relative to each other. For high to moderate sparsity, the horseshoe outperformed ridge regression, while for dense models, ridge regression generally performed better. The exception was when coefficients were generated from the Cauchy distribution. The heavy tails of this distribution lead to a mixture of large and small coefficients that ridge regression priors cannot adequately model.

In contrast, both MML-h based methods exhibit excellent *robustness* to the underlying structure. While they rarely achieve the outright smallest prediction errors, there are no situations in which they perform substantially worse than the best performing method. The most difficult situation in which to identify the best fitting hierarchy appears to be the grey region between dense and sparse, which is expected. These results clearly demonstrate that the MML-h message lengths, computed from only a small number of posterior samples, are able to discriminate well between complex prior hierarchies.

References

1. Wallace, C.S.: Statistical and Inductive Inference by Minimum Message Length. Information Science and Statistics, 1st edn. Springer, New York (2005)
2. Wallace, C.S., Freeman, P.R.: Estimation and inference by compact coding. J. R. Stat. Soc. (Ser. B) **49**(3), 240–252 (1987)
3. Carvalho, C.M., Polson, N.G., Scott, J.G.: The horseshoe estimator for sparse signals. Biometrika **97**(2), 465–480 (2010)
4. Makalic, E., Schmidt, D.F.: Minimum message length shrinkage estimation. Stat. Prob. Lett. **79**(9), 1155–1161 (2009)

494. D.F. Schmidt et al.

5. Chib, S.: Marginal likelihood from the Gibbs output. J. Am. Stat. Assoc. **90**(432), 1313–1321 (1995)
6. Lewis, S.M., Raftery, A.E.: Estimating Bayes`factors via posterior simulation with the Laplace-Metropolis estimator. J. Am. Stat. Assoc. **92**, 648–655 (1997)
7. Fitzgibbon, L.J., Dowe, D.L., Allison, L.: Univariate polynomial inference by Monte Carlo message length approximation. In: Proceedings of the Nineteenth International Conference on Machine Learning (ICML 2002), pp. 147–154 (2002)
8. Makalic, E., Schmidt, D.F.: A simple sampler for the horseshoe estimator. IEEE Signal Process. Lett. **23**(1), 179–182 (2016)

Kernel Embeddings of Longitudinal Data

Darren Shen[✉] and Fabio Ramos

School of Information Technologies, University of Sydney, Sydney, Australia
{darren.shen,fabio.ramos}@sydney.edu.au

Abstract. Longitudinal data is the repeated observations of individuals through time. They often exhibit rich statistical qualities, such as skew or multimodality, that are difficult to capture using traditional parametric methods. To tackle this, we build a non-parametric Markov transition model for longitudinal data. Our approach uses kernel mean embeddings to learn a transition model that can express complex statistical features. We also propose an approximate data subsampling technique based on kernel herding and random Fourier features that allows our method to scale to large longitudinal data sets. We demonstrate our approach on two real world data sets.

1 Introduction

The study of longitudinal data plays an important role in medicine and social sciences. Their defining characteristic is the repeated observation of outcomes for a group of individuals over time. Unlike cross-sectional studies, which only produce a single snapshot in time, longitudinal studies use repeated measurements to produce paths or trajectories through time. These trajectories are invaluable for studying how variables change over time. For example, a longitudinal study of the protein content of cow milk can reveal patterns about the effect of cow diet on milk over time.

Although longitudinal data can be very informative, they require careful statistical treatment because repeated observations from the same individual are often correlated. For example, the dosage of drugs given to patients undergoing medical treatment affects the severity of the disease, which in turn affects the dosage. Ignoring this correlation can lead to invalid inferences, so robust models for longitudinal data need to account for this effect.

One way to capture intra-subject correlation is to use a *transition model*, which describes how an observation relates to past observations. Suppose, for each individual i, we have a sequence of n_i observations $Y_{i1}, \ldots, Y_{in_i}$. A transition model assumes that the j-th observation Y_{ij} of individual i is a function of p previous observations $Y_{ij-1}, \ldots, Y_{ij-p}$ from the same individual. If we model the relationship with past observations as a probability distribution, then we obtain an order-p Markov model for each trajectory:

$$P(Y_{i1}, \ldots, Y_{in_i}) = P(Y_{i1}) \prod_{j=2}^{n_i} P(Y_{ij} \mid Y_{ij-1}, \ldots, Y_{ij-p}). \tag{1}$$

© Springer International Publishing AG 2016
B.H. Kang and Q. Bai (Eds.): AI 2016, LNAI 9992, pp. 495–506, 2016.
DOI: 10.1007/978-3-319-50127-7_42

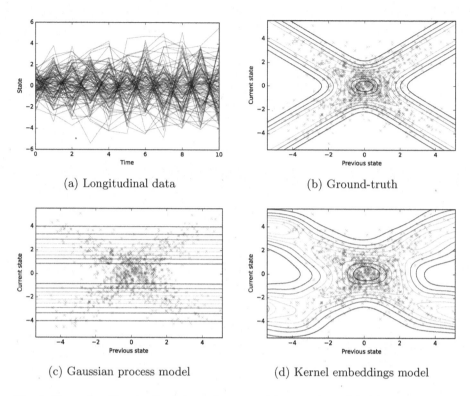

(a) Longitudinal data

(b) Ground-truth

(c) Gaussian process model

(d) Kernel embeddings model

Fig. 1. Example of longitudinal data where transitions are bimodal. (a) Spaghetti plot of the trajectories. (b) Ground-truth transition model $P(Y_{ij} \mid Y_{ij-1})$ for each value of Y_{ij} (y-axis) and Y_{ij-1} (x-axis). (c) Contour plot of transition model learned using a Gaussian process. (d) Contour plot of transition model learned using our method of kernel embeddings.

Traditional methods for learning the transition model $P(Y_{ij} \mid Y_{ij-1}, \ldots, Y_{ij-p})$ usually make parametric assumptions that restrict the expressiveness of the model. For instance, consider an order-1 Markov transition model with real-valued observations $Y_{ij} = \text{sign}(u_{ij})(0.9Y_{ij-1} + v_{ij})$, where $u_{ij}, v_{ij} \sim \mathcal{N}(0,1)$. Figure 1a shows 100 trajectories generated using this model. Because $\text{sign}(u_{ij})$ flips the sign of the state with 50% chance, the transition process is bimodal (see Fig. 1b). Figure 1c and d compares the transition model learned using a Gaussian process, with the one learned using our approach based on kernel mean embeddings. When we condition on a particular previous state Y_{ij-1} by slicing the contours vertically, we can see that the Gaussian process produces a normal distribution, which cannot account for the bimodality. In fact, the Gaussian process can only explain the bimodality by treating it as observation noise. On the other hand, each vertical slice in our method produces a distribution that faithfully reflects the characteristics of the data.

Our first contribution is a nonparametric Markov model for longitudinal data. By embedding observations into a reproducing kernel Hilbert space, we can learn the transition model from data without parametric assumptions, giving us freedom to model transitions with complex statistical features. We also propose an efficient data subsampling method based on a random Fourier features approximation of the method in [1]. The algorithm selects a representative subset of training points so that training on the subset gives comparable performance to training with the full data set. Our method finds m subsamples in $O(nmD)$ time and $O(nD)$ storage, where D is a parameter. This makes it possible to accurately model large longitudinal data sets without significant computational cost.

2   Related Work

Transition models applied to longitudinal data are typically parametric linear models (such as autoregressive processes). Although these methods can be easily interpreted, they are a poor fit for complex longitudinal data such as the trajectories of vehicles because of nonlinearity or multimodality in the transition process. Most nonparametric methods for longitudinal data such as splines and kernel density estimation[1] do not have strong theoretical guarantees when approximating probability distributions.

A related model is the state-observation model, where observations are generated by a sequence of hidden states. Our setting can be seen as a special case where the observations are noiseless. Several nonparametric methods based on kernel embeddings have been applied to these models [1,3]. In particular, [1] uses subsampling with kernel herding as a way to speed-up their method. The cost of this preprocessing step is, however, quadratic in the number of data points, which makes it infeasible to run on large data sets. Our method is a scalable alternative that focuses specifically on the noiseless observation case.

3   Kernel Embeddings

We briefly review kernel mean embeddings before applying it to longitudinal data. [5] offers a great exposition on the theory and its applications.

3.1   Kernel Mean

A kernel[2] $k_{\mathcal{X}} : \mathcal{X} \times \mathcal{X} \to \mathbb{R}$ defined on a measurable space $\mathcal{X}$ induces a unique Hilbert space $\mathcal{H}_{\mathcal{X}}$ of functions called a reproducing kernel Hilbert space (RKHS). It is thus named because its inner product $\langle \cdot, \cdot \rangle_{\mathcal{H}_{\mathcal{X}}}$ satisfies the *reproducing property* $f(x) = \langle f, k_{\mathcal{X}}(\cdot, x) \rangle_{\mathcal{H}_{\mathcal{X}}}$ for all $f \in \mathcal{H}_{\mathcal{X}}$ and $x \in \mathcal{X}$, where $k_{\mathcal{X}}(\cdot, x)$ is the function $k_{\mathcal{X}}$ with one of its parameters fixed at x.

---

[1] KDE is a closely related method, but we only use positive-definite kernels. Without this requirement, we lose all the theoretical benefits discussed in this paper.

[2] A *positive definite kernel* (or just a *kernel*) $k_{\mathcal{X}}$ defined on a measurable space $\mathcal{X}$ satisfies $\sum_{i=1}^{n} \sum_{j=1}^{n} c_i c_j k_{\mathcal{X}}(x_i, x_j) \geq 0$ for any $n \in \mathbb{N}$, $c_1, \ldots, c_n \in \mathbb{R}$, and $x_1, \ldots, x_n \in \mathcal{X}$.

Let X be a random variable on $\mathcal{X}$ with distribution $P(X)$. The framework of kernel embeddings revolves around mapping $P(X)$ to its corresponding *kernel mean* in $\mathcal{H}_\mathcal{X}$ using:

$$\mu[P(X)] = \mathbb{E}_X[k_\mathcal{X}(\cdot, x)] = \int_\mathcal{X} k_\mathcal{X}(\cdot, x)\, dP(X). \qquad (2)$$

A key advantage of this particular mapping is that it is injective for *characteristic* kernels [2]. In other words, all statistical features of the distribution are preserved by this mapping. The most commonly used characteristic kernel is the Gaussian kernel $k_\gamma(x, x') = \exp(-\gamma\|x - x'\|^2)$, where γ is a parameter. We will focus on the Gaussian kernel throughout this paper.

This formulation, however, requires $P(X)$ to be known explicitly, which is rare in practical problems. Suppose we only have samples from $P(X)$ and we wish to estimate $P(X)$. We can do this by estimating its kernel mean $\mu[P(X)]$ instead. Let $\{x_i\}_{i=1}^n$ be n samples drawn i.i.d. from $P(X)$, then the *empirical kernel mean* is defined to be the sample average,

$$\widehat{\mu}[P(X)] = \frac{1}{n}\sum_{i=1}^n k_\mathcal{X}(\cdot, x_i). \qquad (3)$$

If the Rademacher complexity of $P(X)$ and $\mathcal{H}_\mathcal{X}$ is bounded by $O(n^{-1/2})$, the distance $\|\widehat{\mu}[P(X)] - \mu[P(X)]\|$ in the RKHS $\mathcal{H}_\mathcal{X}$ converges to zero at a rate of $O(n^{-1/2})$ with high probability [2]. Since this convergence rate is independent of the dimensionality of $\mathcal{X}$, kernel embeddings works well in high dimensions.

3.2   Conditional Kernel Mean

We now focus on the kernel mean of conditional distributions, which forms the foundation of our method. Let X and Y be random variables over measurable spaces $\mathcal{X}$ and $\mathcal{Y}$, respectively. Given kernels $k_\mathcal{X}$ and $k_\mathcal{Y}$ on $\mathcal{X}$ and $\mathcal{Y}$, respectively, the embedding of $\mu[P(Y \mid x)]$ is defined as,

$$\mu[P(Y \mid x)] = \mathbb{E}_{Y|x}[k_\mathcal{Y}(\cdot, y)] = \int_\mathcal{Y} k_\mathcal{Y}(\cdot, y)\, dP(Y \mid x). \qquad (4)$$

Given i.i.d. samples $\{(x_i, y_i)\}_{i=1}^n$ from the joint distribution $P(X, Y)$, the empirical estimate of Eq. 4 is given by the weighted sum [3]:

$$\widehat{\mu}[P(Y \mid x)] = \sum_{i=1}^n w_i k_\mathcal{Y}(\cdot, y_i), \qquad (5)$$

where $w_i = ((K + n\epsilon_n I_n)^{-1}\mathbf{k}_x)_i$. K is the Gram matrix $(k_\mathcal{X}(x_i, x_j)) \in \mathbb{R}^{n \times n}$, I_n is the $n \times n$ identity matrix, $\mathbf{k}_x$ is the vector $(k_\mathcal{X}(x, x_i)) \in \mathbb{R}^n$, and ϵ_n is a regularisation parameter to reduce overfitting. The empirical kernel mean of marginal distributions from Eq. 3 is simply a special case of Eq. 5 with uniform weights $w_i = 1/n$. Indeed, non-uniform weights capture the effect of conditioning on $X = x$. When ϵ_n decreases at an appropriate rate as $n \to \infty$, this empirical embedding converges to the true conditional distribution, albeit at a slower rate than the marginal case [3].

4   Kernel Embeddings of Longitudinal Data

We now show how we can apply kernel embeddings specifically to longitudinal data. The mapping between kernel embeddings and our setting is summarised Table 1. As we can see, the mapping is straightforward, but there are some simplifying assumptions, which we discuss below:

1. We use an order-1 Markov model with each observations in $\mathbb{R}^d$. Hence, the model predicts the current observation Y given only the previous observation X. Since both X and Y represent observations, we can use the same space and kernel for both variables. We can extend our discussion to higher order Markov models by setting $\mathcal{X}$ to be the product space of multiple past observations and defining an appropriate kernel over that space.
2. We assume that transition probabilities are independent of time and the individual. This allows us to train a single transition model using all the data. We can relax this assumption by training separate transition models for different time measurements or different groups of individuals (e.g. separate models for male and female). This can reduce the variance of each model, but at the cost of less training data.
3. We focus on the Gaussian kernel $k_\gamma(x, x') = \exp(-\gamma\|x - x'\|^2)$ because it is widely used and has useful theoretical properties in the RKHS. Using other kernels is possible, but we lose some theoretical guarantees.

Table 1. Mapping between the general framework of kernel embeddings in Sect. 3 and our problem setting of longitudinal data.

Kernel embeddings	Our setting	Meaning in our setting
$X \in \mathcal{X}$	$X \in \mathbb{R}^d$	Previous observation
$Y \in \mathcal{Y}$	$Y \in \mathbb{R}^d$	Current observation
$\{(x_i, y_i)\}_{i=1}^n$	$\{(x_i, y_i)\}_{i=1}^n$	Transition model examples
$P(Y \mid X)$	$P(Y \mid X)$	Transition model
$k_\mathcal{X}(x, x')$	$k_\gamma(x, x') = \exp(-\gamma\|x - x'\|^2)$	Kernel on previous observation
$k_\mathcal{Y}(x, x')$	$k_\gamma(x, x') = \exp(-\gamma\|x - x'\|^2)$	Kernel on current observation

4.1   Embedding the Transition Model

To estimate the kernel mean of the transition model $P(Y \mid X)$, we first need to extract training samples from the trajectories. For each trajectory i, we can use consecutive observation pairs $\{(Y_{ij-1}, Y_{ij})\}_{j=2}^{n_i}$ as samples from the transition model. We can then combine the sample pairs from every trajectory to form the full set of training pairs which we denote as $\{(x_i, y_i)\}_{i=1}^n$. If we have t trajectories, then we have in total $n = \sum_{i=1}^t (n_i - 1)$ training examples. Substituting these n observation pairs into Eq. 5 gives us an estimate $\widehat{\mu}[P(Y \mid X)]$ of the transition model in the RKHS.

4.2   Estimation of Embedding Statistics

In order to use the embedding $\hat{\mu}[P(Y \mid X)]$ for probabilistic inference, we need to extract statistical information from it. Unfortunately, recovering the full conditional distribution that maps to the embedding is difficult (since it may not even exist). Typical methods assume that the distribution is a mixture and solve a quadratic program to find the mixture parameters [4]. These methods are time consuming and require the number of mixture components a-priori.

Fortunately, when we use Gaussian kernels, various statistics of an embedding in the form of Eq. 5 can be estimated by exploiting the reproducing property [6]. Notably, a consistent estimator of the density $p(y_0 \mid x)$ at $Y = y_0$ is given by,

$$p(y_0 \mid x) = \sum_{i=1}^{n} w_i J_{y_0,h}(y_i),\tag{6}$$

where $J_{y_0,h}(\cdot)$ is a Gaussian smoothing kernel centred at y_0 with bandwidth h:

$$J_{y_0,h}(y) = \frac{1}{\pi^{d/2} h^d} \exp(-\|y - y_0\|^2/h^2).\tag{7}$$

If any of the weights in Eq. 6 are negative, however, the density may be negative, so we may not be able to use this directly. We follow [4] by clipping negative weights at zero and renormalising them to get new weights $w_1^*, \ldots, w_n^*$ so that Eq. 6 forms a valid density:

$$w_i^* = \frac{\max(w_i, 0)}{\sum_{j=1}^{n} \max(w_j, 0)}.\tag{8}$$

Using these new weights may be justified as follows. Assume $y_1, \ldots, y_n$ are unique. As $h \to 0$, $J_{y_0,h}(y)$ becomes the Dirac delta $\delta(y_0 - y)$. When we set y_0 to be a training sample y_k, the right hand side of Eq. 6 reduces to just w_k. This implies that w_k is a consistent estimator of a density, which is nonnegative in the limit. Using a similar argument [1], we also can show that $\sum_{i=1}^{n} w_i$ is a consistent estimator of 1. Hence, in the limit of $n \to \infty$ and $h \to 0$, all weights are nonnegative and sum to 1, so the modified weights are the same as the true weights and we obtain the true density.

4.3   Summary of the Algorithm

The full algorithm takes a set of trajectories as input. We first extract ordered pairs of observations $\{(x_i, y_i)\}_{i=1}^{n}$ from the trajectories as samples from the transition process. We can then use Eq. 5 to get the weights of the predictive embedding conditioned on a given previous observation. To speed up prediction over many different previous observations, we can precompute $\rho = (K + n\epsilon_n I_n)^{-1}$ since it doesn't depend on the previous observation. Then, to make a prediction, we simply compute $\mathbf{k}_x$ and multiply with ρ to get the weights of the predictive embedding. We can then recover the density of the predictive distribution using Eq. 6 with the transformed weights from 8. We can tune the hyperparameters γ and ϵ_n using cross-validation. We set the bandwidth h to be $1/\gamma$. The complete pseudocode is given in Algorithm 1.

Algorithm 1. Learning and prediction of the transition model.

1: **function** LEARN-TRANSITION-MODEL(γ, $\{y_{1j}\}_{j=1}^{n_1}, \ldots, \{y_{tj}\}_{j=1}^{n_t}$)
2:     Extract training pairs $\{(x_i, y_i)\}_{i=1}^n$ from trajectories.
3:     Compute Gram matrix $K \in \mathbb{R}^{n \times n}$ where $K_{ij} = \exp(-\gamma\|x_i - x_j\|^2)$.
4:     **return** $\{(x_i, y_i)\}_{i=1}^n$ and $\rho = (K + n\epsilon_n I_n)^{-1} \in \mathbb{R}^{n \times n}$.
5: **end function**
6:
7: **function** PREDICT-TRANSITION-MODEL(γ, $\{(x_i, y_i)\}_{i=1}^n$, ρ, x)
8:     Compute $\mathbf{k}_x \in \mathbb{R}^n$ where the i-th entry is $\exp(-\gamma\|x - x_i\|^2)$.
9:     Compute prediction weights $\mathbf{w} = (w_1, \ldots, w_n) = \rho\mathbf{k}_x \in \mathbb{R}^n$.
10:     **return** new weights $w_i^* = \max(w_i, 0)/\sum_{j=1}^n \max(w_j, 0)$.
11: **end function**

5   Reducing the Computational Cost

The bottleneck of our method is computing the inverse $\rho = (K + n\epsilon I_n)^{-1}$, which takes $O(n^3)$ naively. We reduce this cost in two ways: a Nyström approximation of K, and a data subsampling method based on kernel herding [1].

5.1   Nyström Approximation

A common technique to compute ρ faster is to approximate K with a low rank matrix. The Nyström method first samples r data points $\widehat{x}_1, \ldots, \widehat{x}_r \in \mathcal{X}$ and then approximates K with the matrix $\tilde{K}_r = CW_r^\dagger C^\top$, where $C = k_\gamma(x_i, \widehat{x}_j) \in \mathbb{R}^{n \times r}$, $W_r = k_\mathcal{X}(\widehat{x}_i, \widehat{x}_j) \in \mathbb{R}^{r \times r}$ and $W_r^\dagger$ denotes the pseudo-inverse of W_r. The samples $\widehat{x}_1, \ldots, \widehat{x}_r$ can be selected in many ways. We found that running k-means on the data points and using the cluster centers as the samples worked very well. Once we compute $\tilde{K}$, we can use the Woodbury identity to approximate ρ as,

$$\rho = \frac{1}{n\epsilon_n} \left(I_n - C(n\epsilon_n I_r + W_r^\dagger C^\top C)^{-1} W_r^\dagger C^\top \right) \in \mathbb{R}^{n \times n}, \qquad (9)$$

where the matrix inversion in this expression is on a $r \times r$ matrix, rather than a $n \times n$ matrix. This reduces the cost of computing the weights to $O(n^2 r)$.

5.2   Subsampling with Kernel Herding

Even with the Nyström approximation, our algorithm still takes quadratic time in the number of data points, which makes it difficult to scale to large longitudinal data sets. A typical technique for reducing computational cost is to only train on a small random subset of the data. The problem of this approach, however, is that it is "too random", so the subset is unlikely to be representative of the original data.

A better approach is to pick representative data points that preserves the "information" in the original data. Since we are working in a RKHS, a natural definition of "information" is the joint embedding of the data in the RKHS [1].

Like before, let X and Y be random variables on $\mathbb{R}^d$ with a common Gaussian kernel $k_\gamma(\cdot, \cdot)$. The empirical joint embedding of samples $\{(x_i, y_i)\}_{i=1}^n$ from $P(X, Y)$ is given by,

$$\widehat{\mu}[P(X, Y)] = \frac{1}{n} \sum_{i=1}^n k^{(\times)}((\cdot, \cdot), (x_i, y_i)), \tag{10}$$

where $k^{(\times)}((\cdot, \cdot), (x_i, y_i))$ is the product $k_\gamma(\cdot, x_i) k_\gamma(\cdot, y_i)$. In other words, we wish to find a subset $\{(\bar{x}_p, \bar{y}_p)\}_{p=1}^m$ of the training data such that the empirical joint embedding of the subset is close to that of the original. Let $s_i = \sum_{j=1}^n k^{(\times)}((x_i, y_i), (x_j, y_j))$ and $\bar{s}_{ip} = \sum_{j=1}^{p-1} k^{(\times)}((x_i, y_i), (\bar{x}_j, \bar{y}_j))$. A greedy approach to pick each subsample was devised in [1]:

$$(\bar{x}_1, \bar{y}_1) = \underset{(x_i, y_i) \in \mathcal{D}}{\arg\max} \frac{1}{n} s_i \tag{11}$$

$$(\bar{x}_p, \bar{y}_p) = \underset{(x_i, y_i) \in \mathcal{D}}{\arg\max} \frac{1}{n} s_i - \frac{1}{p} \bar{s}_{ip}. \tag{12}$$

Intuitively, the term s_i favours samples that are representative of the full data set, while $\bar{s}_{ip}$ penalises samples that are too close to previous subsamples. Obtaining m subsamples takes $O(n^2 m)$ time. Unfortunately, this is not very useful since subsampling has the same cost as running Nyström on the full data.

The main bottleneck of this algorithm is computing s_i, which is a $O(n)$ summation. Using random Fourier features [7], we can approximate s_i in $O(D)$ time, where D is the number of random features. For a Gaussian kernel $k_\gamma(\cdot, \cdot)$, there exists a random feature map $\mathbf{z}_d : \mathbb{R}^d \to \mathbb{R}^D$ such that the dot product $\langle \mathbf{z}_d(x), \mathbf{z}_d(x') \rangle$ converges to $k_\gamma(x, x')$ when D is large:

$$\mathbf{z}_d(x) = \sqrt{\frac{2}{D}} [\cos(\boldsymbol{\omega}_1^\top x + b_1), \dots, \cos(\boldsymbol{\omega}_D^\top x + b_D)]^\top, \tag{13}$$

where $\boldsymbol{\omega}_1, \dots, \boldsymbol{\omega}_D \in \mathbb{R}^D$ are vectors with entries drawn from $\mathcal{N}(0, 2\gamma)$ and $b_1, \dots, b_D \in \mathbb{R}$ are drawn uniformly on $[0, 2\pi)$.

To approximate s_i, we need a random feature map for $k^{(\times)}$. Since this kernel is the product of two Gaussian kernels, $k^{(\times)}((x_i, y_i), (x_j, y_j))$ is just the Gaussian kernel $k_\gamma([x_i; y_i], [x_j; y_j])$ on $\mathbb{R}^{2d}$ where $[x_i; y_i] \in \mathbb{R}^{2d}$ is the concatenation of x_i and y_i. We can then use a neat trick to approximate s_i:

$$s_i = \sum_{j=1}^n k_\gamma([x_i; y_i], [x_j; y_j]) \tag{14}$$

$$\approx \sum_{j=1}^n \langle \mathbf{z}_{2d}([x_i; y_i]), \mathbf{z}_{2d}([x_j; y_j]) \rangle \tag{15}$$

$$= \langle \mathbf{z}_{2d}([x_i; y_i]), \sum_{j=1}^n \mathbf{z}_{2d}([x_j; y_j]) \rangle \tag{16}$$

$$\triangleq \langle \mathbf{z}^{(i)}, \mathbf{u} \rangle. \tag{17}$$

Algorithm 2. Kernel Herding Subsampling using Random Features.

1: **function** FAST-HERDING-SUBSAMPLE$(\gamma, \{(x_i, y_i)\}_{i=1}^{n})$
2:     Compute feature maps $\mathbf{z}^{(i)} \in \mathbb{R}^{D}$ for all i using Eq. 13.
3:     Compute the mean feature map: $\mathbf{u} = \sum_{i=1}^{n} \mathbf{z}^{(i)} \in \mathbb{R}^{D}$.
4:     Initialise $\bar{\mathbf{u}} \in \mathbb{R}^{D}$ be the zero vector.
5:     **for** $p = 1$ to m **do**
6:         $(\bar{x}_p, \bar{y}_p) = \underset{(x_i, y_i) \in \mathcal{D}}{\arg\max} \frac{1}{n} \langle \mathbf{z}^{(i)}, \mathbf{u} \rangle - \frac{1}{p} \langle \mathbf{z}^{(i)}, \bar{\mathbf{u}} \rangle$.
7:         Update $\bar{\mathbf{u}} \leftarrow \bar{\mathbf{u}} + \mathbf{z}_{2d}([\bar{x}_p; \bar{y}_p])$.
8:     **end for**
9:     **return** $\{(\bar{x}_p, \bar{y}_p)\}_{p=1}^{m}$.
10: **end function**

The key advantage of this approximation is that we only need to compute $\mathbf{z}^{(i)}$ and $\mathbf{u}$ once, so s_i can be computed with a $O(D)$ dot product instead of a $O(n)$ sum. The same idea applies to $\bar{s}_{ip}$. The pseudocode is given in Algorithm 2.

6   Experiments

We tested our algorithm in two real world data sets against other probabilistic methods that can learn a transition model. Both data sets have non-Gaussian transition models. To measure the performance of each algorithm, we first extracted pairs of observations from every trajectory and then performed 5-fold cross-validation on all the observation pairs. Hence, each algorithm uses a subset of the data to learn a transition model, which is then tested on unseen data.

For each test fold, we measured the mean negative log-likelihood (NLL) of the unseen data. For a set of test pairs $\{(x_i, y_i)\}_{i=1}^{n}$, the mean NLL is given by $-\frac{1}{n} \sum_{i=1}^{n} \log p(y_i \mid x_i)$. The lower this is, the more accurately the method generalises to test data.

6.1   NLSY97 PIAT Data

We extracted a subset of the National Longitudinal Study on Youths (NLSY97) data set containing annual results on a scholastic achievement test results called the Peabody Individual Achievement Test (PIAT). The data set contains test results for around 6000 individuals from 1997–2002. There were plenty of missing data, so the number of training pairs was only around 6000. This effect was prominent in the year 2002, where the number of data points was less than half of the previous year. We normalised the scores to be between 0 and 1.

We tested our kernel embeddings approach without any approximation (KME-EXACT). We also tested the approximate method (KME-APPROX) by first selecting $m = 200$ subsamples from kernel herding with $D = 50$ random features, and then selecting $r = 100$ K-means cluster centres for the Nyström approximation. We tuned the hyperparameters by maximising the likelihood from cross-validation. More specifically, for each value $\epsilon_n \in \{1.0, 0.1, 0.01\}$, we

Table 2. Cross-validation results for the PIAT data set. The 1st interval counts the % of test points that lie between the 0th and 25th percentile, and so on.

Method	NLL	1st (%)	2nd (%)	3rd (%)	4th (%)
AR(1)	-0.67 ± 0.04	17.6 ± 1.6	34.0 ± 0.9	28.5 ± 0.8	19.9 ± 1.1
GP	-0.70 ± 0.03	18.6 ± 2.1	34.8 ± 1.8	27.7 ± 1.0	19.0 ± 1.3
KME-Exact	-0.87 ± 0.03	22.1 ± 1.9	28.2 ± 1.4	27.4 ± 0.8	22.3 ± 1.2
KME-Approx	-0.74 ± 0.04	19.4 ± 2.3	32.7 ± 1.6	28.4 ± 1.9	19.5 ± 1.5

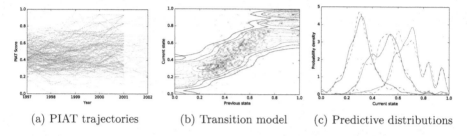

(a) PIAT trajectories   (b) Transition model   (c) Predictive distributions

Fig. 2. (a) Trajectories in the PIAT data set. (b) Contour of the transition model learned using our exact method on a 2000 data point subset of the PIAT data. (c) The predictive distribution conditioned on three different previous observations $X = 0.25, 0.5, 0.75$ (i.e. vertical slices of the contour). The solid and dashed line shows the densities of our predictive distribution using the exact and approximate methods, respectively.

used a 1D optimiser to tune γ. We found that this worked better than using 2D optimisation on both parameters with numerical gradients.

To compare, we trained an order-1 autoregressive model using ordinary least squares. We also trained a Gaussian process with a RBF kernel using the GPy library [9]. We optimised the hyperparameters (kernel variance and lengthscale, noise variance) by maximising the marginal likelihood of the data.

For each algorithm, we obtained their mean NLL on cross-validation as previously described. We also divided the predictive distributions of each method into four intervals that each have 25% of the total probability mass (i.e. the first interval is the mass between the 0th and 25th percentile, and so on). We then counted how many of the unseen points lie in each interval. The closer each of the four counts are as close to 25%, the better the predictive distribution matches the shape of the data. Table 2 reports the results for this data set.

In terms of log-likelihood, kernel mean embeddings outperformed AR(1) and GP. AR(1) performed worse than GP because it is a linear classifier, so it not as expressive as Gaussian processes which are nonlinear. KME-APPROX was comparable to GP, even though it used only 200 subsamples out of 6000 data points. Looking at the percentile interval results, we can see that both AR(1) and GP overconcentrated their mass in the middle because it is limited to a normal distribution. Our method had a similar concentration of mass, but to a lesser degree because it is not constrained by any parametric form (see Fig. 2c).

6.2   Edinburgh Pedestrian Data

This data set contains the 2D trajectories of pedestrians walking through an area in the University of Edinburgh [8]. We used trajectories from a single day (Aug 24), which had 664 trajectories and 76,260 data points. The movements of pedestrians are inherently Gaussian, which is not particularly interesting. Hence, for each trajectory i, we used observation pairs $\{(Y_{ij-k}, Y_{ij})\}_{j=k+1}^{n_i}$ that are k frames apart as data samples. This changed the task to modelling the position of a pedestrian after k frames, which exhibits more interesting statistical features like multimodality (see Fig. 3 for the effect of different k on our predictive distribution). We set $k = 10$, which gave us around 70,000 training points. Like the PIAT data set, we normalised the X and Y axes to be between 0 to 1.

Because the data set was so large, we could not use KME-EXACT due to memory constraints. We set $m = 500$, $D = 50$ and $r = 100$ for KME-APPROX. We also could not use exact GP inference, so we instead used sparse variational GP (SGP) from GPy with a RBF kernel and two outputs to obtain the 2D position. We used 100 inducing points to match the rank of the Nyström approximation in KME-APPROX, initialised using K-means cluster centres. We optimised both the hyperparameters and the inducing point locations. Table 3 shows the mean NLL results for 5-fold cross validation on this data set.

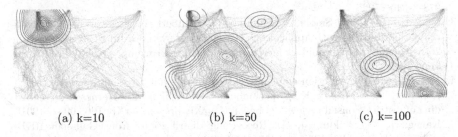

| (a) k=10 | (b) k=50 | (c) k=100 |

Fig. 3. Predictive distributions for the Pedestrian data set with different values of k. (a) Prediction for the location of a pedestrian $k = 10$ frames after, if the pedestrian started at the top left corner. (b) and (c) Predictions with larger k to predict farther into the future.

Table 3. Cross-validation results for the Pedestrian data set.

Method	NLL
KME-Approx	-6.00 ± 0.05
SGP	-1.44 ± 0.01

7   Conclusion

We described a nonparametric Markov model for longitudinal data where the transition model is learned using kernel mean embeddings. This allows us to model complex longitudinal data where the transition process exhibits statistical features such as skew and multimodality. We also proposed a random features approximation of the data subsampling method from [1], reducing the running time from $O(n^2m)$ to $O(nmD)$. This subsampling method can be used to speed up a wide variety of inference algorithms based on kernel mean embeddings.

For future work, we would like to incorporate explanatory variables by considering the distribution $P(Y_{ij} \mid Y_{ij-1}, x_{ij})$, where x_{ij} is a vector of variables that may influence the observation Y_{ij}. An interesting research direction would be to see if we can combine parametric models for explanatory variables with non-parametric transition models. We would also like to see the effect of using higher order Markov models on the performance of these methods.

References

1. Kanagawa, M., Nishiyama, Y., Gretton, A., Fukumizu, K.: Filtering with state-observation examples via kernel Monte Carlo filter. Neural Comput. **28**(2), 382–444 (2014)
2. Smola, A., Gretton, A., Song, L., Schölkopf, B.: A Hilbert space embedding for distributions. In: Hutter, M., Servedio, R.A., Takimoto, E. (eds.) ALT 2007. LNCS (LNAI), vol. 4754, pp. 13–31. Springer, Heidelberg (2007). doi:10.1007/978-3-540-75225-7_5
3. Song, L., Huang, J., Smola, A., Fukumizu, K.: Hilbert space embeddings of conditional distributions with applications to dynamical systems. In: Proceedings of the 26th Annual International Conference on Machine Learning, pp. 961–968. ACM, June 2009
4. McCalman, L.R.: Function embeddings for multi-modal Bayesian inference (2013)
5. Muandet, K., Fukumizu, K., Sriperumbudur, B., Schlkopf, B.: Kernel mean embedding of distributions: a review and beyonds. arXiv preprint arXiv:1605.09522 (2016)
6. Kanagawa, M., Fukumizu, K.: Recovering distributions from Gaussian RKHS embeddings. In: AISTATS, pp. 457–465 (2014)
7. Rahimi, A., Recht, B.: Random features for large-scale kernel machines. In: Advances in Neural Information Processing Systems, pp. 1177–1184 (2007)
8. Majecka, B.: Statistical models of pedestrian behaviour in the forum. Master's thesis, School of Informatics, University of Edinburgh (2009)
9. GPy: GPy: a Gaussian process framework in python. http://github.com/SheffieldML/GPy

Visual Analytical Tool for Higher Order k-Means Clustering for Trajectory Data Mining

Ye Wang, Kyungmi Lee, and Ickjai Lee[✉]

College of Business, Law and Governance, Information Technology Academy,
James Cook University, Cairns, QLD 4870, Australia
Ickjai.Lee@jcu.edu.au

Abstract. Trajectories are useful sources to understand moving objects and locations. Many trajectory data mining techniques have been researched in the past decade. Higher order information providing suggestions to what-if analysis when the best possible option is not feasible is of importance in dynamic and complex spatial environments. Despite of the importance of higher order information in trajectory data mining, it has received little attention in literature. This paper introduces new visualisation methods for determination of higher order k-means clustering for trajectory data mining. This paper proposes a radar chart-like visualisation for geometrical and directional higher order information and a k-means clustering technique for trajectory higher order information. This paper also demonstrates the usefulness of proposed visualisation methods and clustering technique with a case study using real world datasets.

Keywords: Higher order information · Visual analytics · Spatio-temporal data mining · Trajectory data mining

1 Introduction

Trajectories captured from moving entities in the geographical space are valuable sources of information for data mining. A trajectory can be represented as a set $P = \{p_1, p_2, ...p_n\}$ of continuous points, where each data point consists of a geospatial coordinate set and a timestamp such as $p_i = (x_i, y_i, t_i)$. Trajectory data contain not only information about the spatial and temporal aspects of moving entities, but also additional quantitative and qualitative attributes related to the moving entities where the movement takes place. A myriad of user-generated trajectory data are being collected with the advancement in Internet and low cost location based tracking and sensing technologies. Many consumer devices such as smart phones and wearables devices are now equipped with Global Positioning System (GPS) and many of these devices are also connected to the Internet enabling an automatic upload to online.

Various techniques [26] have been proposed for processing, managing, and mining trajectory data in a broad range of application domains, such application domains include surveillance analysis [10,13], animal movement studies [5],

© Springer International Publishing AG 2016
B.H. Kang and Q. Bai (Eds.): AI 2016, LNAI 9992, pp. 507–518, 2016.
DOI: 10.1007/978-3-319-50127-7_43

traffic analysis [3,24], sports performance analysis [16], supermarket shopping paths optimisation [17], and emergency management [8,9,21,23]. To deal with the complexity of trajectory dataset, researchers have been investigating different approaches and analytical tools for mining trajectory datasets in order to reveal useful information and knowledge efficiently from this large amount of trajectory datasets [27]. Many of these approaches for analysing trajectory datasets are based on data mining techniques such as DBSCAN [18], k-means/k-medoids [11], k-nearest neighborhood [4,23], and regions-of-interest [2]. Recently, a few visual analytic tools for trajectory data are proposed in literature [15,19,22]. Visual analytic tools allow users to explore and interact with the trajectory dataset with the application of visualisation and human cognitive capability to recognise patterns.

While there are increasing efforts devoted to trajectory analysis, research considering trajectory Higher Order Information (HOI) has received little attention in literature [6,25]. Many different types of valuable information can be derived from trajectory data to understand the underlying behaviours or activities that generate the trajectory data and to help make smart decisions. For instances, k-Nearest Neighbour (kNN) information and k-Order Region (kOR) information are the most popular research areas in HOI for trajectory data. HOI is of great importance in highly dynamic and complex environments. HOI for "what-if" analysis allows users to deal with the dynamicity and complexity of trajectory analysis especially when the best possible solution is unavailable/malfunctioning/fully booked. For instance, in emergency planning and management phases, HOI provides useful information to situations where more than k response centres are required to participate in the response and recovery phases.

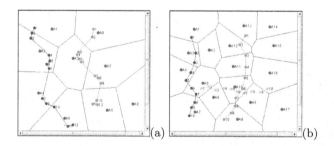

Fig. 1. Ordinary Voronoi diagrams: (a) Order-1 Voronoi diagram with 9 generators (labeled from A1 to A9), and 2 trajectories (12 timestamps); (b) Order-1 Voronoi diagram with 17 generators (labeled from A1 to A17), and 3 trajectories (10 timestamps). (Color figure online)

To find trends and similarity shared by different nodes in a large trajectory, one possible approach is to group similar nodes into clusters to reduce the complexity of the trajectory. Clustering in data mining is a process to group unlabelled objects into fewer classes of similar objects, called clusters, to simplify the data, providing summarised information of the objects and facilitating the discovery of hidden patterns [1]. Clustering is now an important area of application for a variety of fields including data mining, statistical data analysis, compression, and vector quantization. k-means clustering is popular for cluster analysis aiming to partition n observations into k clusters in which each observation belongs to a cluster with the nearest mean.

Consider an example shown in Fig. 1 which shows different Points of Interest (POI) datasets and trajectory datasets with Voronoi diagrams. Figure 1(a) displays P_1 9 POIs labelled from A1 to A9 in red, and 2 trajectories with 12 timestamps, T_1 is represented in green, T_2 is displayed in blue. In Fig. 1(b) P_2 shows 17 POIs labelled from A1 to A17 in red, and 3 trajectories with 10 timestamps, T_3 is represented in green, T_4 is displayed in blue, T_5 is presented in pink. For example, for T_1 in Fig. 1(a), using k-means clustering, the nodes can be classified into 4 clusters as the Voronoi diagram tessellates. Similarity, the nodes of T_2 can also be grouped into 4 clusters. However, in Fig. 1(b), T_3, T_4, and T_5, each node in the trajectories are quite evenly separated from each other. When only considering the node distance, using k-means clustering, no meaningful clusters could be identified. Even though trajectory nodes can be defined over different groups based on existing k-means method, HOI is not considered during the clustering process.

Research on trajectory data mining with HOI is limited. This study aims to introduce new geometrical and directional visualisation approaches and HOI k-means clustering method for analysing large trajectory datasets and HOI. These approaches are particularly efficient to provide new ways to represent geometrical HOI and directional HOI to provide semantic answers to problems such as top-k trajectory HOI. For example, during crisis like flooding or fire, information like the k nearest emergency response centres within a particular distance in a particular direction is very useful. The proposed visual analytical tool provides an implementation to combine different types of HOI, namely geometrical and directional HOI, as well as clustering data mining technique to simplify, analyse, and visualise trajectory datasets.

2   Preliminaries

kNN and kOR are the two most common HOI for analysing trajectory data. Given a point set S, a query point q and a positive interger k, the kNN query gives a point set kNNS(q) which is a subset of S such that $d(q, r) \leq d(q, p)$ holds for any point $r \in k$NNS(q) and for any point $p \in (S - k$NNS(q)$)$ where $d(q, r)$ denotes the distance between q and r. kNNQ is supplemented by the kth Order Region Query (kORQ) which returns a region R for a certain subset $P_i^{(k)}$ in P where the set of kNN for any location l in R is the same as $P_i^{(k)}$. kORQ can also be called kth-Nearest Point Diagram. kORQ associated with a given site is the set of points in the plane such that the site ranks number k in the ordering of the sites by distance from the point, and is a collection of one or more unconnected cells [14]. It is useful to establish a boundary on the number of cells, as their complexity can affect the accuracy of the final output of the kth order Voronoi diagram. kNNQ simply returns kth nearest neighbor, and typically returns quantitative geometrical and topological information. kNN has been widely used for classification, regression, "what-if" analysis. kOR returns kth nearest neighbor Voronoi regions. Typically, it returns qualitative geometrical and topological information. kOR has been extensively used for location optimisation, catchment analysis, market analysis, disaster management, and "what-if" analysis [7].

Both kNNQ and kORQ are modeled by Order-k Voronoi Diagram (OkVD) families which are generalisations of the Ordinary Voronoi Diagram (OVD). Unified Delaunay triangle based data structure [8] consists of a complete set of Order-k Delaunay triangles (from Order-0 to Order-(k-1)) for computing the OkVDs to be used in this study. The Delaunay triangulation is a dual graph of the OVD with edges connecting neighbouring points. It can be constructed by linking two adjacent Voronoi generators if they share a Voronoi edge together. For Order-0 triangles, no generator is on the circumcircle of each Delaunay triangle in the triangulation. However, there could be triangles whose circumcircles include a number of generators within them. Order-1 triangles are those triangles whose corresponding circumcircles include only one generator in it. Therefore, Order-k triangles are those ones whose corresponding circumcircles include k generators in it. A complete set of OkVDs could be drawn from this data structure, and the relationship between the data structure and OkVD is shown in Table 1. Based on these different OkVDs, we compute the geometrical and directional trajectory HOIs.

Table 1. The relationship between OkVDs and the unified data structure.

Order-k	OkVD
1	Order-0 triangle
2	Order-0 & Order-1 triangle
3	Order-1 & Order-2 triangle
4	Order-2 & Order-3 triangle
...	...
k	Order-(k-2) & Oder-(k-1) triangle

k-means clustering is one of the most common clustering approaches. It is a method of vector quantisation. The goal of clustering is to partition the data points into k clusters with each data point to one of the k clusters by solving the minimisation problem which minimises the *distance* from the data point to the centroid of the cluster. Euclidean distance is used in k-means clustering for the minimisation problem. This problem is NP-hard, a global minimum is not guaranteed. The proper number of clusters k to be used for clustering needs to be chosen by researchers carefully. An incorrect choice of k will invalidate the clustering results. A numerical way to find the best k is to try k-means clustering over a range of ks and observing the resulting sum of squares for each k until it is converging to a local minimum [12]. This study will utilise OkVDs generated from the unified data structure, and generate visual analytical tools to determine the number of k for clustering with HOI.

3   Algorithms

3.1   Higher Order Radar Chart (HORC)

In this section, we present a visualisation approach for displaying geometrical and directional HOI called Higher Order Radar Chart (HORC). Directional HOI (compass direction to generators) is visualised in 8 different directions: $dir = \{East, West, South, North, Southeast, Southwest, Northeast, Northwest\}$. For the geometrical HOI (distance from generators), the Min-Max normalisation is used to obtain a relative ratio for visualisation.

The Min-Max normalisation is one of the most widely used normalisation mthods that converts quantitative values into a certain range $[V_{newMax}, V_{newMin}]$. The Min-Max normalisation is computed as follows:

$$v\prime = \frac{v - V_{Min}}{V_{Max} - V_{Min}}(V_{newMax} - V_{newMin}) + V_{newMin}, \tag{1}$$

where $v\prime$ is a new normalised value, v is the old value before normalisation, V_{Max} is the maximum value whilst V_{Min} is the minimum value, V_{newMax} is the maximum value in the new range whilst V_{newMin} is the minimum value in the new range.

To create a HORC, first of all, geometrical information for each trajectory node $t_k \in T$ (original trajectory T for geometrical HORC) is calculated for that distance to POIs. Once the cardinality of each distance is calculated then it is normalised into [0,1] for relative visualisation. Then, directional information for each trajectory node $d_k \in D$ (original trajectory D for directional HORC) is determined for that direction from POIs.

Figure 2 shows HORC of the two different groups of datasets in Fig. 1. We take Fig. 1 datasets as examples to explain the proposed method. The Radar chart consists of 4 circles of difference sizes sharing the same center. The ratios of the radii of the circles are 1:2:3:4. We use this to measure the Min-Max normalisation and represent into this Radar chart. At the $East$ side of circle, we display the V_{Max} distance of the Min-Max normalisation to measure and compare with trajectories. The kNN POIs of each trajectory node show inside of these circles with different colors to represent POIs of distance (geometrical) and direction (directional) information.

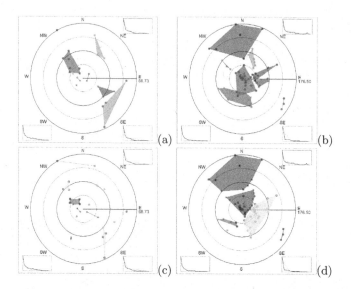

Fig. 2. HORCs: (a) O1VD HORC; (b) O3VD HORC for trajectory T_1 and T_2 based on P_1 with clusters determined for each trajectory; (c) O1VD HORC; (d) O3VD HORC for trajectory T_1 and T_2 based on P_1 with clusters determined for both trajectories together. (Color figure online)

3.2 Higher Order k-Means

In order to cluster quantitative top-k HOI for higher order trajectory data mining, Higher Order k-Means (HOk-Means) is introduced. HOk-Means is a k-Means approach to calculate an average distance up to k^{th} order HOI, first, each trajectory node finds k^{th} order POIs; then, HOk-Means calculates all k^{th} order POIs "*mean*" to find out the mean of top-k of this trajectory HOI. The higher order cluster centre $T\prime$ is defined as below:

$$T\prime = \sum_{i=1}^{n} \mathrm{mean}(\mathrm{kNN}(t_i)). \tag{2}$$

For example, given a trajectory $T = \{t_1, ..., t_n\}$ of n nodes and a set $P = \{p_1, ..., p_m\}$ of m generators, let $kNN(t_i)$ ($t_i \in T$) return a subset $P_i \subset P$ of kth order nearest neighbours, and find the "*mean*" of P_i, and let each trajectory node $t_i \in T$ find new $T\prime$ which includes centroids of all clusters. In the end, the Sum Squared Error (SSE) for all values of cluster number k is computed. The SSE is defined as the sum of the squared distance between each member of the cluster and its centroid. Mathematically it is defined as:

$$HOkMean = \sum_{i=1}^{k} \sum_{j=1}^{n} \left\| t_i^{(j)} - T\prime_j \right\|^2, \tag{3}$$

where $\left\| t_i^{(j)} - T\prime_j \right\|^2$ is a chosen distance measure between a trajectory data
node $t_i^{(j)}$ and the higher order cluster centre $T\prime_j$, which is an indicator of the
distance of the n data points from their corresponding higher order cluster centres. This method is not only minimising cluster's centroid, but also considering
HOI of each trajectory.

Figure 2(a) shows HORC based on the Order-1 Voronoi Diagram, HOI of
T_1 is displayed in green, HOI of T_2 is displayed in blue, P_1 is shown in red.
Figure 2(a) shows that based on O1VD, most of the POIs are very close to trajectory T_1, almost half of POIs distance are inside the innermost circle, that is
under 25% of 60.17 km. Generally, most of POIs are sitting at the east side of
trajectory T_1. Figure 2(b) displays O3VD HORC which considers the 3 nearest
POIs for each node. Using the proposed HOk-Means clustering method, considering each trajectory separately, 6 clusters are identified for T_1, and 3 clusters
are identified for T_2 based on the HOI and O1VD. Using O3VD, 5 clusters are
identified for T_1 and 4 clusters are identified for T_2. Figure 2(c) and (d) show the
results, when considering the HOI of both T_1 and T_2 together, 11 clusters are
identified based on O1VD, and 6 clusters are identified based on O3VD.

Figure 3(a) and (b) display the HORC and clustering results of P_2 and trajectories T_3, T_4, and T_5 based on O1VD, and Fig. 3(c) and (d) display the HORC
and clustering results based on O3VD. For this dataset, based on O1VD, 4 clusters are identified for each of the 3 trajectories when the trajectories are considered separately. Putting all the trajectories together, 8 clusters are identified.

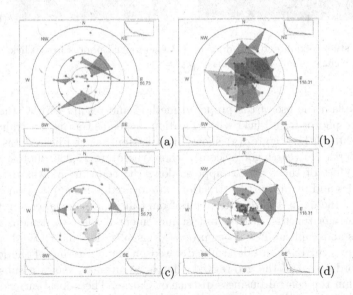

Fig. 3. HORCs: (a) O1VD HORC; (b) O3VD HORC for trajectory T_3, T_4, and T_5
based on P_2 with clusters determined for each trajectory; (c) O1VD HORC; (d) O3VD
HORC for trajectory T_3, T_4, and T_5 based on P_2 with clusters determined for all 3
trajectories together.

When using O3VD, 2 clusters are identified for T_3, 4 clusters for T_4, and 3 clusters for T_5. Again, when all 3 trajectories are considered based on O3VD, 9 clusters are identified.

On the corners of the HORC, the line charts for SSE for different k values. On the top right corner is the line chart for using original elbow method [20], on the bottom left is the line chart for using the HOk-Means method with all trajectories considered together. Finally, on the bottom right is the line chart for using the HOk-Means method with each trajectory considered separately.

4   Case Study

Flickr is a photo sharing website. It has more than 6 billion images uploaded and shared by their users. In this section, we demonstrate how the HORC and HOk-Means HOI can be used for real world datasets.

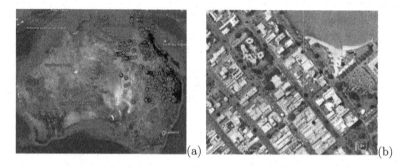

Fig. 4. A study region and real dataset: (a) Geo-tagged photos from Flickr; (b) Distribution of clusters in CBD of Cairns ($|P_3| = 14$).

We explain the use of the proposed method and results with a study using geo-tagged photos from Flickr where location tagged as Cairns, Australia, in the years of 2010-2011. Figure 4(a) shows the distribution of Flickr dataset using Google Earth 6.0 (earth.google.com), where there are 17,066 photos of Flickr uploads. Most of these photos appear along the coastline of Australia where major cities and main tourist attractions are located. This application will take a subset of the dataset with data points from Cairns, a regional city in Queensland, Australia together with a tourist movement dataset to demonstrate the usefulness and applicability of our proposed visualisation and HOk-Means clustering approach. POIs are determined based on the clustering of the locations (GPS coordinates) of the geo-tagged photos. Figure 4(b) shows a subset P of POIs within the central business district of Cairns. These locations can be set up as tourist information centres.

Figure 5(a) demonstrates higher order Voronoi diagrams of locations of 14 POIs P_3, and the O1VD with two photo-taker trajectories, whilst Fig. 5(b) displays O5VD. P_3 is a set of centroids determined for the Flickr dataset using

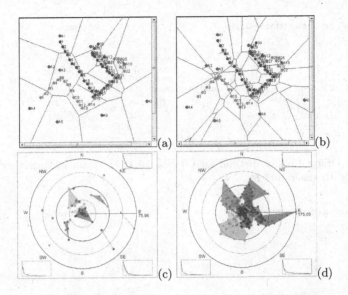

Fig. 5. (a) O1VD; (b) O5VD with two trajectory datasets; (c) HORC of O1VD (d) HORC of O5VD. Trajectory datasets T_6 and T_7 with generators P_3. (Color figure online)

k-means clustering of the photo records. The two trajectory datasets in Fig. 5 are tourist movement datasets collected every 30 s for 15 min from two different tourist sightseeing in Cairns. The data collected from tourist T_6 is represented in green and T_7 is shown in blue in Fig. 5. Let us assume P_3 represents best locations for setting up tourist information centres for tourists to seek help and advice when they are visiting the region.

Figure 5(c) and (d) show HORC and HOk-Means of the two trajectories. Figure 5(c) displays HORC in O1VD, T_6 is presented in green, T_7 is presented in blue. In Fig. 5(c), a HORC shows that the POIs are mostly located in the west side and very close to trajectory based on these two trajectories' O1VD information. However, when k increases to 5, as shown in Fig. 5(d), T_7 moves to the east side and north side. Obviously, most of the nearest POIs are located in the south west side at around half of V_{Max} distance. T_6 is located from west to east, and most of POIs are very close to the trajectory (less than 25% of V_{Max} distance).

The clustering results for order-1 information are displayed in Fig. 6(a) with original elbow method, and in Fig. 6(b) for the HO1-Means method. As we are only looking for the 1st nearest neighbour HOI, the result will be the same because it calculates only 1 nearest neighbour information. In Fig. 6, we can see that in the elbow chart showing the SSE after running k-means clustering for k going from 1 to 29. We see a pretty clear elbow at $k = 7$ for trajectory T_6 and $k = 6$ for trajectory T_7, indicating that 6 is the best number of clusters for T_6 and 7 is the best number of clusters for T_7.

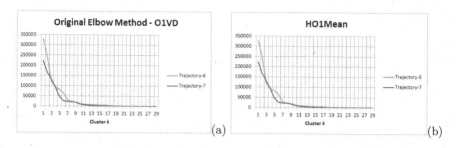

Fig. 6. (a) Original elbow method for O1VD; (b) HO1-Means of trajectory datasets T_6 and T_7 with generators P_3.

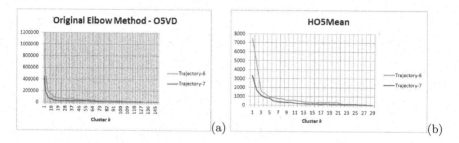

Fig. 7. (a) Original elbow method for O5VD; (b) HO5-Means of trajectory datasets T_6 and T_7 with generators P_3.

In another case, when we consider HOI for $k = 5$, Fig. 7(a) shows the original elbow method whilst Fig. 7(b) shows HO5-Means. Notice how the elbow chart for the original method does not have a clear elbow. It is a fairly smooth curve, and it is unclear what is the best value of cluster k. This result shows that the original method of k-means clustering does not work very well for trajectory HOI. On the other hand, Fig. 7(b) presents the results by HO5-Means, it suggests that the optimal number of clusters that the HO5-Means has selected for T_6 is $k = 5$ and for T_7 is $k = 6$. Therefore, the proposed HOk-means method is superior to the original method for selecting the best k for clustering of trajectory HOI.

5   Conclusion

Due to the complexity of trajectory datasets and related data analysis, there is a need to develop new and innovative ways to allow researchers and data analysts to reveal hidden information and knowledge embedded in trajectory datasets. Most of the existing research efforts focus on trajectory data mining of geometrical information, studies related to HOI for trajectory data require more attention. This paper proposes a new visual analytical approach to visualise HOI with regards to target generators, and a new trajectory clustering method with trajectory HOI. In particular, by combining geometrical HOI with directional

HOI and determining the optimal k, we propose a HORC and HOk-Means clustering. The chart is a new visual analytic tool to explore both the geometrical and directional HOI at the same time. The HOk-Means clustering helps to discover hidden groups in the dataset while considering HOI. Such chart and clustering are very useful for complex situations when both geometrical HOI and directional HOI have to be considered at the same time. An example for this situation could be during natural disasters like flooding and hurricane. Not only knowing the k-nearest emergency centres is important, should the direction of the centres also be considered as the water or wind may go from a particular direction to another direction. HORC and the HOk-Means clustering provide information to allow management to allocate different levels of resources to different clusters. Experimental results demonstrate promising outcomes that support the usefulness and applicability of the proposed methods. Future work includes further experiments with large trajectory datasets and extensions to multivariate HOI.

References

1. Berkhin, P.: A Survey of Clustering Data Mining Techniques, pp. 25–71. Springer, Berlin (2006)
2. Cai, G., Hio, C., Bermingham, L., Lee, K., Lee, I.: Sequential pattern mining of geo-tagged photos with an arbitrary regions-of-interest detection method. Expert Syst. Appl. **41**, 3514–3526 (2014)
3. Chu, D., Sheets, D., Zhao, Y., Wu, Y., Yang, J., Zheng, M., Chen, G., et al.: Visualizing hidden themes of taxi movement with semantic transformation. In: Pacific Visualization Symposium (PacificVis), pp. 137–144. IEEE (2014)
4. Güting, R.H., Behr, T., Xu, J.: Efficient k-nearest neighbor search on moving object trajectories. Int. J. Very Large Data Bases **19**(5), 687–714 (2010)
5. Handcock, R.N., Swain, D.L., Bishop-Hurley, G.J., Patison, K.P., Wark, T., Valencia, P., Corke, P., O'Neill, C.J.: Monitoring animal behaviour and environmental interactions using wireless sensor networks, GPS collars and satellite remote sensing. Sensors **9**(5), 3586–3603 (2009)
6. Beni, H.L., Mostafavi, M.A., Pouliot, J., Gavrilova, M.: Toward 3D spatial dynamic field simulation within GIS using kinetic Voronoi diagram and Delaunay tetrahedralization. Int. J. Geog. Inf. Sci. **25**(1), 25–50 (2011)
7. Keim, D., Andrienko, G., Fekete, J.D., Görg, C., Kohlhammer, J., Melançon, G.: Visual analytics: Definition, process, and challenges. Springer, Heidelberg (2008)
8. Lee, I., Lee, K.: A generic triangle-based data structure of the complete set of higher order Voronoi diagrams for emergency management. Comput. Environ. Urban Syst. **33**(2), 90–99 (2009)
9. Lee, I., Pershouse, R., Phillips, P., Christensen, C.: What-if emergency management system: a generalized voronoi diagram approach. In: Yang, C.C., Zeng, D., Chau, M., Chang, K., Yang, Q., Cheng, X., Wang, J., Wang, F.-Y., Chen, H. (eds.) PAISI 2007. LNCS, vol. 4430, pp. 58–69. Springer, Heidelberg (2007). doi:10.1007/978-3-540-71549-8_5
10. Lin, L., Lu, Y., Pan, Y., Chen, X.: Integrating graph partitioning and matching for trajectory analysis in video surveillance. IEEE Trans. Image Process. **21**(12), 4844–4857 (2012)

11. Liu, D., Lim, E.P., Ng, W.K.: Efficient k nearest neighbor queries on remote spatial databases using range estimation. In: Proceedings of the 14th International Conference on Scientific and Statistical Database Management, pp. 121–130 (2002)

12. Milligan, G.W., Cooper, M.C.: An examination of procedures for determining the number of clusters in a data set. Psychometrika **50**(2), 159–179 (1985)

13. Morris, B.T., Trivedi, M.M.: A survey of vision-based trajectory learning and analysis for surveillance. IEEE Trans. Circ. Syst. Video Technol. **18**(8), 1114–1127 (2008)

14. Okabe, A., Boots, B., Sugihara, K., Chiu, S.N.: Spatial Tessellations:Concepts and Applications of Voronoi Diagrams, vol. 501. Wiley, Hoboken (2009)

15. Palmer, J.D.: Using line and texture to visualize higher-order Voronoi diagram. In: Proceedings of the 3rd International Symposium on Voronoi Diagrams in Science and Engineering, pp. 166–172. IEEE Computer Society (2006)

16. Pingali, G., Opalach, A., Jean, Y., Carlbom, I.: Visualization of sports using motion trajectories: providing insights into performance, style, and strategy. In: Proceedings of the Conference on Visualization 2001, pp. 75–82. IEEE Computer Society (2001)

17. Popa, M.C., Rothkrantz, L.J., Shan, C., Gritti, T., Wiggers, P.: Semantic assessment of shopping behavior using trajectories, shopping related actions, and context information. Pattern Recogn. Lett. **34**(7), 809–819 (2013)

18. Sander, J., Ester, M., Kriegel, H.P., Xu, X.: Density-based clustering in spatial databases: the algorithm GBDSCAN and its applications. Data Min. Knowl. Disc. **2**(2), 169–194 (1998)

19. Telea, A., van Wijk, J.J.: Visualization of generalized voronoi diagrams. In: Ebert, D.S., Favre, J.M., Peikert, R. (eds.) Data Visualization 2001, pp. 165–174. Springer, Heidelberg (2001)

20. Tibshirani, R., Walther, G., Hastie, T.: Estimating the number of clusters in a data set via the gap distance. J. R. Stat. Soc. B **63**(2), 411–423 (2001)

21. Wang, Y., Lee, I.: Visual analytics of higher order information for trajectory datasets. Int. J. Environ. Earth Sci. Eng. **7**(12), 1587–1592 (2013)

22. Wang, Y., Lee, K., Lee, I.: Directional higher order information for spatio-temporal trajectory dataset. In: 2014 IEEE International Conference on Data Mining Workshop (ICDMW), pp. 35–42. IEEE (2014)

23. Wang, Y., Lee, K., Lee, I.: Visual analytics of topological higher order information for emergency management based on tourism trajectory datasets. Procedia Comput. Sci. **29**, 683–691 (2014)

24. Wang, Z., Lu, M., Yuan, X., Zhang, J., Van De Wetering, H.: Visual traffic jam analysis based on trajectory data. IEEE Trans. Vis. Comput. Graph. **19**(12), 2159–2168 (2013)

25. Xiong, X., Mokbel, M.F., Aref, W.G.: Sea-cnn: Scalable processing of continuous k-nearest neighbor queries in spatio-temporal databases. In: Proceedings of 21st International Conference on Data Engineering, ICDE 2005, pp. 643–654. IEEE (2005)

26. Zheng, Y.: Trajectory data mining: an overview. ACM Trans. Intell. Syst. Technol. (TIST) **6**(3), 29 (2015)

27. Zheng, Y., Zhou, X. (eds.): Computing with Spatial Trajectories. Springer, New York (2011)

A Framework for Mining Semantic-Level Tourist Movement Behaviours from Geo-tagged Photos

Guochen Cai, Kyungmi Lee, and Ickjai Lee[⊠]

College of Business, Law and Governance, Information Technology Academy,
James Cook University, Cairns, QLD 4870, Australia
ickjai.lee@jcu.edu.au

Abstract. This study investigates tourist movement patterns on the type of place semantic-level. We extract the semantic common movement patterns that a group of tourists have similar movement trajectories on the semantic level, and find out semantic trajectory patterns which are sequences of the type of place objects with transit time. Using real geo-tagged photos, we find out interesting common movement patterns and trajectory patterns. These results provide richer information and understanding of tourist movement behaviour on the type of place semantic-level.

Keywords: Semantics · Trajectory data mining · Movement behaviors · Geo-tagged photos

1 Introduction

Advances in Web 2.0/3.0 technologies promote a rapid increase of user-generated contents. These contents reflect their actual lives, ideas and activities. User-generated contents can be associated with location and time information through geo-tagging services the websites provide. These geo-referenced contents indicate tourists' movement patterns that help understand their behaviours.

This study aims to provide a framework to extract semantic-level tourist movement behaviours from geo-tagged contents. Previous studies mainly focus on spatial level movement behaviours by using only geometric features of trajectory data. These spatial level behaviours fail to reflect contextual semantic-level patterns. A semantic-level movement behaviour is a trajectory behaviour whose predicate bears on some contextual data [8] such as the type of place semantic information. This kind of behaviour provides detailed and fine-tuned information in the contextual semantic-level, and can help specific applications provide targeted services to tourists based on the semantic-level behaviours, like the type of place context semantics.

2 Literature Review

On-line user-generated contents are being constantly generated providing a data-rich environment for various tourist behaviour mining activities [5]. Especially,

© Springer International Publishing AG 2016
B.H. Kang and Q. Bai (Eds.): AI 2016, LNAI 9992, pp. 519–524, 2016.
DOI: 10.1007/978-3-319-50127-7_44

the social media data associated with geographic location and time information, that are annotated by geo-tagging services, are useful sources for spatio-temporal patterns of tourist movements.

One recent research is to find tourist spatial travel behaviours of location preferences. [7] discovers POI that a high density of tourists visits, and further finds out their visitation association rules. Another recent novel research is to learn tourist spatio-temporal dynamics from geo-tagged social media data. The geo-tagged data form tourist trajectories, spatio-temporal movements, by chronologically connecting all geo-tagged data. These trajectory data are then used to analyse tourist movement behaviours. [5] uncovers dynamic visitor traffic flow, inbound and outbound trajectories between cities. Another popular travel behaviour is the spatial movement pattern that a group of tourists share. [2] extracts tourist spatio-temporal sequential patterns whilst [9] mines tourist popular travel routes. Each route is a sequence of spatial region of areas. A popular travel route shows the common path that a group of tourists share.

A common drawback of these previous studies is that they only focus on spatial level travel behaviours that are based on geometric feature of spatial information. This spatial geometric feature only is not sufficient for some specific applications [8]. For instance, some applications require more meaningful background semantic information, like movement behaviours between different types of place, that the geometric feature merely cannot provide. The semantic-level behaviours provide more detailed and fine-tuned information about tourist movements on semantic-level that are valuable to domain experts like tourism industry and urban management.

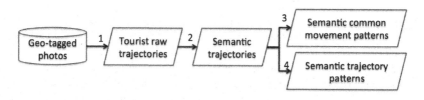

Fig. 1. Conceptual framework.

3   Framework and Approaches

Figure 1 shows our proposed framework. From geo-tagged photos, collected from a photo-sharing platform Flickr (https://www.flickr.com/) in this study, the framework firstly creates raw trajectories and then classifies them into tourist and non-tourist trajectories in Step 1 based on the time span of trajectory. The time span of a trajectory is calculated by using the time gap between the last photo and the first photo. We consider a photo-taker as a tourist if the time span of trajectory is less than 31 days. We use these tourist trajectories in this study. In Step 2, we enrich trajectories with semantics and transform trajectories into semantic trajectories. A semantic trajectory is a sequence of semantic annotations. Using the semantic Region-of-Interest (RoI) mining method [3], we detect semantic RoIs from trajectories. Each semantic RoI is annotated with a type of place semantics. We use this type of place as the basic semantics of travel to learn tourist movement behaviours on the type of place semantic-level. As a result, a basic semantic trajectory is a sequence of type of place annotations. In this study, we enrich basic semantic RoIs with more city, temporal and weather condition annotations by using geographic information database, visit time of RoI and weather observation database respectively, and RoIs become multidimensional. In particular, temporal semantic contains two features: date type and day time. Date type is the day of week, weekday and weekend, and day time is the time period of the day. Step 3 requires a clustering approach to group similar trajectories into the same common movement pattern. This framework adopts the EXTRACTDBCAN-Clustering method [1] to generate clusters from the ordering results and uses the SemT-OPTICS algorithm [3] to deal with enriched semantic trajectories. Since Steps 1–3 in our framework are based on [1,3], we recommend readers to refer to them for more details.

A semantic trajectory pattern is a sequence of visited objects with the transit time between two neighboring objects. Step 4 is to find these semantic trajectory patterns. The proposed method adopts $\mathcal{TAS}$ algorithm [4] which is a projection-based method built on the PrefixSpan method [6] designed for sequential patterns. $\mathcal{TAS}$ algorithm uses T-sequence data type instead of normal sequence used in projections. A T-sequence is a projected sequence enriched with an annotation sequence where the annotation sequence includes records of occurrences of the prefix in the original sequence. Our method adopts the T-sequence data type, but we use a progressive increase approach to calculate frequent interval time and semantic trajectory patterns. In addition, the proposed framework utilities arbitrary combination of dimensions when it generates trajectory patterns. Semantic trajectories are associated with four additional semantic dimensions. We find not only the trajectory patterns associated with a set of all four dimensions, but also patterns with subsets of four dimensions.

4 Experimental Results

4.1 Dataset

In this study, we use real geo-tagged photos collected from Flickr for Queensland area in Australia for a period between April 2014 and March 2015. We collected 64,733 photo data, and generate 1404 valid raw trajectories that the length is greater than 1. Then we select the trajectories with time span less than 31 days as tourist trajectories. Finally, we obtain 770 tourist trajectories. To enrich additional semantic annotations to trajectories, We use the Australia gazetteer data and cities 1000 dataset from GeoNames (http://www.geonames.org/) as our geographic information database. For the weather information database, we use the the observation stations database and daily weather observation database from Bureau of Meteorology Australia (http://www.bom.gov.au/climate).

For parameters, the methods we used to extract tourist travel behaviour of movement patterns require several parameters, respectively. It is a non-trivial problem to choose best values of parameters that produce meaningful and insightful RoIs and patterns. The following default values were chosen for this particular experiment for experimental purposes. For semantic trajectory generation, the semantic RoI mining method relies on the minimum support ($MinSup$) value for a cell to become a RoI and also on the size of cell ($CellSize$) that is used to partition the study region. We choose a value of 0.004 which means 0.4 Km for $cellSize$ and a value of 0.007 which means 0.7% for $MinSup$. For semantic trajectory patterns, the semantic trajectory pattern mining algorithm requires parameters $MinSup$ and time tolerance (tau) which is the acceptable range for a time interval. We use a value 2 days for tau. For SemT-OPTICS algorithm in semantic common movement pattern mining, we use all basic and additional semantics in this study. We select the basic semantics PLACE_TYPE as default compulsory dimension, and other four semantics, CITY, DAY_TYPE, DAY_TIME and WEATHER, as optional dimensions. A set of compulsory dimensions plays an important or crucial role in patterns. For simplicity, we set each optional dimension with an average weight value, which is used to compute the similarity between trajectories, a value of 0.25. We set values for ohter parameters element matching score threshold ($elematThreshold = 0.3$) and ratio threshold ($rThreshold = 0.3$) for our similarity method, basic parameters $minPts = 4$ and $epsilon = 1$ that OPTICS algorithm requires, and set a value for parameter $epsilon = 0.5$ for extractDBSCAN-clustering method.

We find 72 semantic RoIs. A semantic RoI is a spatial region with a type annotation indicating the type of the place in the region. Each type of place annotation is represented as a feature code defined in Geonames geographic database. We obtain 399 semantic trajectories and 204 valid semantic trajectories (length > 1). A sample of final semantic trajectory is presented as below.

$$< (HTL_{[GoldCoast][weekday][Clear][evening]}) (BCH_{[GoldCoast][weekday][Clear][dawn]}) >$$

4.2   Semantic Common Movement Patterns

In this section, we present some tourist semantic common movement patterns. Applying the SemT-OPTICS algorithm on the semantic trajectory dataset and extractDBSCAN method to ordering list, we found 5 clusters. We choose the first semantic trajectory cluster to represent the common movement pattern of that cluster. Table 1 shows two common movement patterns. The general form of pattern is a semantic trajectory. The first common movement pattern is a trajectory of starting at a hotel in Gold Coast and then going to a populated place and a park in Brisbane and visiting a hotel and a beach in Gold Coast. All of these visits are in clear weekday, but are at different day time that first four visits are at dawn in the day whilst the last visit is in the afternoon. This pattern shows a tourist common movement between Gold Coast and Brisbane which are nearby cities. Another common movement pattern is in Cairns that tourists start from a hotel in the morning and then go to the pier (where fleet

Table 1. Semantic common movement patterns (HTL: Hotel, RSTN: Railroad station, PPLX: Section of populated place, BCH: Beach, PRK: Park, PIER: Pier).

Cluster	Common movement pattern
cluster 1	$[(\text{HTL}_{[GoldCoast][weekday][Clear][dawn]}) \rightarrow$ $(\text{PPLX}_{[Brisbane][weekday][Clear][dawn]}) \rightarrow$ $(\text{PRK}_{[Brisbane][weekday][Clear][dawn]}) \rightarrow$ $(\text{HTL}_{[GoldCoast][weekday][Clear][dawn]}) \rightarrow$ $(\text{BCH}_{[GoldCoast][weekday][Clear][afternoon]})]$
cluster 2	$[(\text{HTL}_{[Cairns][weekday][Lightrain][morning]}) \rightarrow$ $(\text{PIER}_{[Cairns][weekday][Lightrain][morning]}) \rightarrow$ $(\text{HTL}_{[Cairns][weekday][Lightrain][evening]})]$

stations are) and back to hotel in the evening at last. This pattern supports a popular full-day tour to Great Barrier Reef (departing from the fleet station in the early morning and returning late in the afternoon).

4.3 Semantic Trajectory Patterns

We present several typical tourist travel behaviours of semantic trajectory patterns in this section including basic semantic trajectory patterns and multidimensional semantic trajectory patterns. A semantic trajectory pattern shows a sequence of visited types of place with transit time that occurs in a density of tourist trajectories. Multidimensional semantic trajectory patterns contain more semantics and provide much richer information about tourists frequent trajectory patterns. Due to the limitation of space, we list some multidimensional semantic trajectory patterns for the basic pattern PRK $\xrightarrow{[0,8]}$ HTL associated with various combinations of additional dimensions including dimension WEATHER and combinations of dimensions DAY_TYPE, CITY and WEATHER. We observe that additional semantics provide more information to basic patterns. For a combination of CITY dimension, we know this pattern is in Gympie, a regional town in the Wide Bay-Burnett region of Queensland. For combinations of dimensions DAY_TYPE, CITY and WEATHER together, we obtain more day type and weather information about the semantic trajectory pattern (Table 2).

Table 2. Multidimensional semantic trajectory patterns for basic pattern: PRK $\xrightarrow{[0,3]}$ HTL.

Combination of dimensions	Semantic trajectory patterns
WEATHER	$\text{PRK}_{[clear]} \xrightarrow{[0,8]} \text{HTL}_{[clear]}$
DAY_TYPE, CITY, WEATHER	$\text{PRK}_{[weekday][Gympie][clear]} \xrightarrow{[0,8]} \text{HTL}_{[weekday][Gympie][clear]}$

5   Conclusions

We presented a study of converting geo-tagged photos into trajectories and further into semantics enriched trajectories to extract semantic-level tourist movement patterns. The proposed framework enabled us to find out tourist common movement patterns on the type of place semantic-level with additional city, temporal and weather condition semantic information. In addition, we discover tourist semantic trajectory patterns, frequent visitation sequences of the type of place with transit time information. The trajectory patterns are also associated with various combinations of additional semantics that supply further insights into movements in different semantics contexts. Overall, these detected patterns are indicative of semantic-level tourist movement behaviours, and provide richer knowledge about travel movements on semantic-level.

References

1. Ankerst, M., Breunig, M.M., Kriegel, H.P., Sander, J.: Optics: ordering points to identify the clustering structure. ACM SIGMOD Rec. **28**(2), 49–60 (1999)
2. Bermingham, L., Lee, I.: Spatio-temporal sequential pattern mining for tourism sciences. Procedia Comput. Sci. **29**, 379–389 (2014)
3. Cai, G., Lee, K., Lee, I.: Discovering common semantic trajectories from geo-tagged social media. In: Fujita, H., Ali, M., Selamat, A., Sasaki, J., Kurematsu, M. (eds.) IEA/AIE 2016. LNCS (LNAI), vol. 9799, pp. 320–332. Springer, Heidelberg (2016). doi:10.1007/978-3-319-42007-3_27
4. Giannotti, F., Nanni, M., Pedreschi, D.: Efficient mining of temporally annotated sequences. In: SDM, pp. 348–359. SIAM (2006)
5. Girardin, F., Dal Fiore, F., Blat, J., Ratti, C.: Understanding of tourist dynamics from explicitly disclosed location information. In: Symposium on LBS and Telecartography, vol. 58. Citeseer (2007)
6. Han, J., Pei, J., Mortazavi-Asl, B., Pinto, H., Chen, Q., Dayal, U., Hsu, M.: Prefixspan: mining sequential patterns efficiently by prefix-projected pattern growth. In: proceedings of the 17th International Conference on Data Engineering, pp. 215–224 (2001)
7. Lee, I., Cai, G., Lee, K.: Mining points-of-interest association rules from geo-tagged photos. In: 2013 46th Hawaii International Conference on System Sciences (HICSS), pp. 1580–1588. IEEE (2013)
8. Parent, C., Spaccapietra, S., Renso, C., Andrienko, G., Andrienko, N., Bogorny, V., Damiani, M.L., Gkoulalas-Divanis, A., Macedo, J., Pelekis, N., et al.: Semantic trajectories modeling and analysis. ACM Comput. Surv. (CSUR) **45**(4), 42 (2013)
9. Zheng, Y.T., Zha, Z.J., Chua, T.S.: Mining travel patterns from geotagged photos. ACM Trans. Intell. Syst. Technol. (TIST) **3**(3), 56 (2012)

Learning High-Level Navigation Strategies via Inverse Reinforcement Learning: A Comparative Analysis

Michael Herman[1,2]([⊠]), Tobias Gindele[1], Jörg Wagner[1], Felix Schmitt[1], Christophe Quignon[1], and Wolfram Burgard[2]

[1] Robert Bosch GmbH, 70442 Stuttgart, Germany
{michael.herman,tobias.gindele,joerg.wagner3,felix.schmitt}@de.bosch.com,
christophe.quignon@inf.h-brs.de
[2] University of Freiburg, 79110 Freiburg, Germany
burgard@informatik.uni-freiburg.de

Abstract. With an increasing number of robots acting in populated environments, there is an emerging necessity for programming techniques that allow for efficient adjustment of the robot's behavior to new environments or tasks. A promising approach for teaching robots a certain behavior is Inverse Reinforcement Learning (IRL), which estimates the underlying reward function of a Markov Decision Process (MDP) from observed behavior of an expert. Recently, an approach called Simultaneous Estimation of Rewards and Dynamics (SERD) has been proposed, which extends IRL by simultaneously estimating the dynamics. The objective of this work is to compare classical IRL algorithms with SERD for learning high level navigation strategies in a realistic hallway navigation scenario solely from human expert demonstrations. We show that the theoretical advantages of SERD also pay off in practice by estimating better models of the dynamics and explaining the expert's demonstrations more accurately.

Keywords: Inverse Reinforcement Learning · Simultaneous Estimation of Rewards and Dynamics · Reinforcement Learning · Learning from demonstration

1 Introduction

The number of robots performing tasks in populated areas is constantly increasing. To ensure a fast and easy deployment of robots in various environments, it is necessary to provide simple programming approaches for non-experts to parameterize new behaviors or tasks. A class of approaches offering such possibilities is IRL [10], which specifies the problem of recovering the underlying reward function of an MDP from expert demonstrations. This corresponds to estimating the motivation of an expert, which then can be used to synthesize optimal policies even for new environments. Plenty of approaches have been

© Springer International Publishing AG 2016
B.H. Kang and Q. Bai (Eds.): AI 2016, LNAI 9992, pp. 525–534, 2016.
DOI: 10.1007/978-3-319-50127-7_45

proposed that solve the IRL problem under various assumptions, e.g. [1,9]. As humans are rarely able to provide optimal demonstrations, Ziebart et al. [11] proposed a probabilistic model of maximum (causal) entropy to capture stochastic behavior while matching feature expectations. Bloem and Bambos [2] extended the approach to infinite time horizons. The class of maximum (causal) entropy IRL algorithms has been successfully applied to a variety of problems, such as learning a priority-adaptive navigation [6] or learning to navigate in crowded environments [5]. Many existing approaches repeatedly apply a Reinforcement Learning (RL) algorithm as part of the IRL algorithm. Solving the RL problem typically requires the system dynamics to be known. To enable IRL under unknown transition models, model-free IRL algorithms have been proposed, such as [3,8]. These IRL variants either require expert demonstrations that are rich of observed transitions, additional samples from an arbitrary policy, or use heuristics. Since additional samples can be expensive and heuristics may tamper with the reward estimate, Herman et al. [7] proposed an approach called Simultaneous Estimation of Rewards and Dynamics (SERD), which exploits the fact that the expert's policy has been influenced by the reward function and the dynamics. Their evaluation has shown that SERD outperforms traditional approaches if expert demonstrations are used that stem from the same type of policy that is modeled. However, real human demonstrations may follow a different type of distribution. Therefore, it is unclear, whether SERD or IRL generalizes better in such cases.

In this paper, we investigate the performance of IRL approaches for robot programming tasks from human demonstrations. We compare SERD [7], Maximum Discounted Causal Entropy IRL [2] and Relative Entropy IRL [3] for learning high-level navigation strategies in a densely populated hallway scenario. Further, we assume that only human demonstration are available and both dynamics and rewards are unknown beforehand. The evaluation shows that SERD outperforms the evaluated IRL approaches in the training task and generalizes well to transfer tasks.

2   Fundamentals

A discounted, infinite horizon MDP is a tuple $M = \langle S, A, P(s'|s,a), \gamma, R, P(s_0) \rangle$, where S is the state space with states $s \in S$, A is the action space with actions $a \in A$, $P(s'|s,a)$ is the probability of a transition from s to s' when action a is applied, $\gamma \in [0,1)$ is a discount factor, $R : S \times A \to \mathbb{R}$ is a reward function assigning a scalar reward for picking action a in state s, and $P(s_0)$ is a start state probability distribution. A policy $\pi(s,a) = P(a|s)$ specifies the probability for an agent to choose action a in state s. The goal of an MDP is to find an optimal policy $\pi^*(s,a)$, which maximizes the expected, discounted, cumulated future reward. The Q-function $Q^\pi(s,a) = \mathbb{E}\left[\sum_{t=0}^\infty \gamma^t R(s_t, a_t)|s_0 = s, a_0 = a, \pi\right]$ specifies the expected, discounted, cumulated reward for starting in state s, picking action a and then following policy π. Furthermore, it satisfies the Bellman equations, and can therefore be updated iteratively to find the Q-function under the optimal policy $\pi^*(s,a)$:

$$Q_i(s,a) = \left[R(s,a) + \gamma \sum_{s' \in S} P(s'|s,a) \max_{a'} [Q_{i-1}(s',a')] \right]. \qquad (1)$$

Then, the optimal policy can be derived from the converged Q-function $Q_\infty(s,a)$ in Eq. (1) by choosing the action with the highest value $\pi^*(s) = \text{argmax}_a \, Q_\infty(s,a)$.

3 Learning Reward Functions from Human Demonstrations

IRL describes the problem of recovering the reward function from observed behavior of an expert which is acting according to some policy. The IRL problem is specified by the tuple $M \setminus R$ and either the expert's policy π or demonstrations $D = \{\tau_1, \tau_2, \ldots, \tau_N\}$ with τ being trajectories $\tau = \{(s_0^\tau, a_0^\tau), (s_1^\tau, a_1^\tau), \ldots, (s_{T_\tau}^\tau, a_{T_\tau}^\tau)\}$. Extracting the expert's reward function $R(s,a)$ corresponds to estimating his motivation or goal from observed behavior. Often, the reward is expressed as a linear combination of weighted features $R(s,a) = \boldsymbol{\theta}_R^\mathsf{T} \boldsymbol{f}$ with state- and action-dependent features $\boldsymbol{f} : S \times A \to \mathbb{R}^d$. Since humans are rarely able to demonstrate optimal behavior, IRL approaches are necessary that can cope with suboptimal or noisy demonstrations. Therefore, this section introduces IRL approaches that satisfy this requirement.

3.1 Maximum Discounted Causal Entropy IRL

The Maximum Discounted Causal Entropy IRL (MDCE IRL) approach by Bloem and Bambos [2] is an extension of [11] and models stochastic expert behavior, which makes it suitable for learning from human demonstrations. Bloem and Bambos [2] formulate a maximum causal entropy optimization problem under the constraint to match the feature expectation of the model $\boldsymbol{f}_\theta = \mathbb{E}\left[\sum_{t=0}^\infty \gamma^t \boldsymbol{f}(s_t^\tau, a_t^\tau) | s_0 = s, \pi_\theta, P(s'|s,a)\right]$ to the empirical one of the expert $\widetilde{\boldsymbol{f}} = \frac{1}{|D|} \sum_{\tau \in D} \sum_{t=0}^{T_\tau} \gamma^t \boldsymbol{f}(s_t^\tau, a_t^\tau)$. Their solution yields a policy that is based on a simplified stationary soft value iteration. The soft state-action value $Q_\theta(s,a)$ is recursively defined as $Q_\theta(s,a) = \boldsymbol{\theta}^T \boldsymbol{f}(s,a) + \gamma \sum_{s' \in S} P(s'|s,a) V_\theta(s')$ with $V_\theta(s) = \log\left(\sum_{a \in A} \exp(Q_\theta(s,a))\right)$. Similar to the Bellman equation, the soft value iteration is a fixed point equation [2]. After convergence, the soft state-action value $Q_\theta(s,a)$ is used to derive a stochastic policy:

$$\pi_\theta(s,a) = \frac{\exp(Q_\theta(s,a))}{\sum_{a' \in A} \exp(Q_\theta(s,a'))}. \qquad (2)$$

Estimating the expert's reward function corresponds to finding the feature weights under which the stochastic policy (2) yields the same feature expectation as the expert. According to [2], the feature weights can be estimated by gradient-based optimization, where the gradient is the difference between empirical feature counts of the expert demonstrations and the expected feature count under the current model: $\nabla_\theta L_\theta (D) = \tilde{f} - f_\theta$. It can be shown that the corresponding maximum likelihood problem that matches the stochastic policy in Eq. (2) to the expert demonstrations results in the same gradient, if the dynamics are known and the number of expert demonstrations approaches infinity.

3.2   Relative Entropy IRL

Boularias et al. [3] propose Relative Entropy IRL (REIRL), which is a model-free IRL approach and extends MaxEnt IRL [12] to unknown dynamics. They attempt to minimize the relative entropy between the learner's probability distribution of trajectories $P(\tau)$ and a baseline distribution $Q(\tau)$ under the constraint of matching feature counts. This corresponds to maximizing the entropy of $P(\tau)$ while matching features if the baseline distribution $Q(\tau)$ stems from a policy of maximum entropy, too. Boularias et al. propose a solution, which uses importance sampling from an arbitrary, known policy π_A. The parameters are updated according to the gradient

$$\frac{\partial g}{\partial \theta} = \tilde{f} - \frac{\sum_{\tau \in D_{\pi_A}} \frac{U(\tau)}{\Pi_A(\tau)} \exp\left(\theta^\mathsf{T} f_\tau\right) f_\tau}{\sum_{\tau \in D_{\pi_A}} \frac{U(\tau)}{\Pi_A(\tau)} \exp\left(\theta^\mathsf{T} f_\tau\right)} - \alpha \circ \epsilon \tag{3}$$

with the set of demonstrations D_{π_A} from an arbitrary policy π_A, the joint probability of actions under the baseline policy $U(\tau)$, the joint probability of actions under the arbitrary policy $\Pi_A(\tau)$, the cumulated, discounted feature counts of a single trajectory f_τ, as well as $\forall \alpha_i = \text{sign}\,\theta_i$, thresholds ϵ, and the Hadamard product $\circ$. However, we assume that only expert demonstrations are available and the dynamics are unknown. Therefore, the additional demonstrations D_{π_A} can only be generated artificially by sampling from a transition model that is naively trained from observed transitions.

3.3   Simultaneous Estimation of Rewards and Dynamics

In many real world problems the dynamics of the environment are not known. Furthermore, it is often not possible to estimate the dynamics directly from observed transitions, as demonstrations are biased towards states and actions with high expected, future rewards. To solve such problems, Herman et al. [7] propose an approach that simultaneously estimates the dynamics and the reward function, by exploiting the fact that policies are the result of both. Therefore, they introduce additional parameters θ_T of a parametric model of the dynamics $P_{\theta_T}(s'|s,a)$. The parameter vector $\theta = (\theta_R^\mathsf{T}, \theta_T^\mathsf{T})^\mathsf{T}$ is optimized by maximizing the log likelihood of the demonstrations. Herman et al. [7] propose a gradient-based solution given by:

$$\nabla_\theta L_\theta (D) = \sum_{\tau \in D} \sum_{t=0}^{T_\tau - 1} \left[\nabla_\theta \log \pi_\theta \left(s_t^\tau, a_t^\tau\right) + \nabla_\theta \log P_{\theta_T} \left(s_{t+1}^\tau | s_t^\tau, a_t^\tau\right) \right]. \tag{4}$$

For $\pi_\theta\left(s_t^\tau, a_t^\tau\right)$, the MDCE policy from Eq. (2) is used, since it has proven successful for learning from human demonstrations. Thus, the partial derivative of the log policy

$$\frac{\partial}{\partial\theta_i}\log\pi_\theta\left(s,a\right) = \frac{\partial}{\partial\theta_i}Q_\theta(s,a) - \mathbb{E}_{\pi_\theta(s,a')}\left[\frac{\partial}{\partial\theta_i}Q_\theta(s,a')\right] \tag{5}$$

is depending on the derivative of the soft Q-function. This gradient can be computed by solving the linear equation system in Eq. (6). Since it is a contraction, the soft Q-gradient can be approximated by repeatedly applying Eq. (6) to an arbitrary initial gradient.

$$\frac{\partial}{\partial\theta_i}Q_\theta(s,a) = \frac{\partial}{\partial\theta_i}\theta_R^\intercal f(s,a) + \gamma\sum_{s'\in S}\left[\left(\frac{\partial}{\partial\theta_i}P_{\theta_{T_A}}\left(s'|s,a\right)\right)V_\theta\left(s'\right)\right]$$
$$+\gamma\sum_{s'\in S}\left[P_{\theta_{T_A}}\left(s'|s,a\right)\cdot\mathbb{E}_{\pi_\theta(s',a')}\left[\frac{\partial}{\partial\theta_i}Q_\theta(s',a')\right]\right]. \tag{6}$$

4 Learning High-Level Navigation Strategies

Due to the complexity of planning problems in large, stochastic environments, motion planning is often decomposed into high-level strategies and local motion planning. High-level strategies typically provide global directions without taking into account all aspects of the dynamic environment, while local motion planners consider local dynamics and output control commands. The following section provides a discrete grid-based model for learning high-level navigation strategies. The resulting policy serves as a global strategy which incorporates the estimated reward function as well as the trained dynamics of the environment including the influence of humans.

4.1 Environment

A coarse discretization of a 2D map is used as a discrete state space for the high-level navigation strategy. Accordingly, the state is fully defined by the robots position. In each state the robot can pick from 5 different actions, which are moving into one of the four directions or waiting. We assume that the dynamics are invariant of the preferred movement direction and that the robot is only able to move to adjacent cells. Figure 1 illustrates the available successor states

Fig. 1. Transition model.

$\Psi = \{0, D, R, B, L\}$ (0: Staying, D: Preferred direction, R: Right, L: Left, B: Backwards) if the robot is located in state 0. The dynamics are modeled by parameters $\Theta = \{\theta_0, \theta_D, \theta_R, \theta_B, \theta_L\}$ of a Gibbs distribution:

$$P(\Psi = i) = \frac{\exp\left(\beta\theta_i\right)}{\sum_{j\in\Psi}\exp\left(\beta\theta_j\right)} \tag{7}$$

with β being a scaling parameter. The gradient of this transition model is given by

$$\frac{\partial}{\partial\theta_j}P(\Psi = i) = \begin{cases} \beta\dfrac{\exp(\beta\theta_i)\sum_{k\in\Psi\setminus i}\exp(\beta\theta_k)}{\left(\sum_{k\in\Psi}\exp(\beta\theta_k)\right)^2} & \text{if } i = j \\[2ex] -\beta\dfrac{\exp(\beta(\theta_i+\theta_j))}{\left(\sum_{k\in\Psi}\exp(\beta\theta_k)\right)^2} & \text{if } i \neq j \end{cases} \tag{8}$$

The local dynamics may differ depending on the state or action. Therefore, it may be necessary to introduce several local transition models. Our assumption is that only expert demonstrations are available and that the dynamics are unknown. However, many IRL approaches require the transition model to learn reward functions. Therefore, a transition model is initially estimated from expert demonstrations with an M-estimator.

4.2 Features

We use a set of features that allow all IRL algorithms to learn suitable rewards for explaining the expert demonstrations. The goal identifier $f_G \in \{0,1\}$ is 1 at the goal and 0 otherwise. The social norm feature $f_S \in \{-1,0,1\}$ is 1 if the agent waits or if he satisfies the social norm of walking on the right side of the hallway. If the agent walks into the opposite direction, the feature is -1 and 0 otherwise. The last feature is the normalized inverse distance to walls $f_D \in [0,1]$, since preferring one side of a hallway can also be explained by favoring small distances to walls.

5   Experimental Evaluation

In order to evaluate the IRL algorithms in a realistic scenario, we use a simulation of a densely populated hallway (Fig. 2a). We chose this scenario, as the transition model of the motion planner is stochastic and unknown, the resulting policy is non-trivial, and the discretized state is fully observable. In the simulation, the Social Force model [4] controls pedestrians to permanently walk through the hallway, satisfying a social norm to walk on the right side. In the evaluation, the estimates of the IRL approaches are analyzed with respect to the resulting policies and transition models. Therefore, it is separated in two parts: In the training task, the robot starts randomly in one of the blue grid cells (Fig. 2a), while in the transfer task he starts in the red ones. We implemented the scenario in ROS, using the movebase as a local motion planner. Therefore, the learned high-level strategy sends intermediate waypoints to the movebase. The environment dynamics consist of four different transition models: a transition model for moving with the human flow, a model for moving against the human flow, one for an orthogonal movement, and a transition model for choosing to stand still.

The demonstrations were provided by humans that were advised to guide the robot to the goal. To achieve this, the robot had to move through the hallway, cross an intersection, and enter the corridor. While the shortest way leads diagonally through the hallway, all humans operated the robot to stay in the human

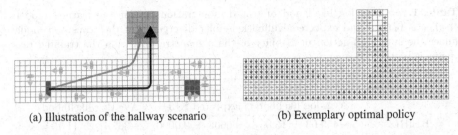

(a) Illustration of the hallway scenario   (b) Exemplary optimal policy

Fig. 2. (a) The robot's start positions are depicted in blue (training) and red (transfer). The green area is the goal zone. Simulated pedestrians are colored in green. Human experts typically guided the robot along the longer violet path and not along the shorter orange one. (b) An exemplary optimal policy which has been computed based on an estimate of MDCE SERD. The arrows indicate optimal actions. In cells without an arrow the optimum action is waiting. (Color figure online)

flow until they reached the intersection. Altogether, twenty training demonstrations were provided. For evaluation, ten additional demonstrations were recorded for each the training and the transfer task.

We compare MDCE IRL, REIRL, and MDCE SERD with respect to the quality of the estimates. MDCE IRL and REIRL need a transition model for learning the rewards. For SERD meaningful initial dynamics are beneficial. Therefore, dynamics are estimated with an M-estimator from observed transitions of the expert. Initial feature weights are uniformly sampled between $[-10, 10]$ before applying the IRL algorithms. Table 1 presents the average log likelihood of test demonstrations under the policy of the resulting estimates. MDCE SERD outperforms all other approaches in terms of the average log likelihood followed by MDCE IRL, which is trained via maximum likelihood estimation, while REIRL performed worse. This is probably caused by differing model assumptions of REIRL and errors in approximating the feature expectation via importance sampling. MDCE IRL and MDCE SERD optimize the log likelihood of the training demonstrations directly and, thus, cause the stochastic policy to match the observed behavior. Figure 2b illustrates an exemplary optimum policy based on an estimate of MDCE SERD. To reach the stated MDCE SERD log likelihood in Table 1, the feature weights have been initially estimated independently before simultaneously optimizing the dynamics. In addition, we implemented an early stopping criterion to prevent overfitting by using twenty percent of the training demonstrations for validation.

Figure 3(a) illustrates the empirical cumulative distribution function (CDF) of the times of arrival (TOA) of the resulting optimal policies. Since they are comparable, all approaches were able to find meaningful estimates, while MDCE SERD guides the robot a little bit faster to the goal. It is important to note that not all CDFs of the TOA are increasing up to 1, which indicates that some estimates do not model the desired behavior accurately enough to guide the robot to the goal area.

Table 1. Average log likelihood of test demonstrations under the learned models (higher is better) and expected Kullback Leibler divergence of the transition model under the state and action probability of the demonstrations from the training task (lower is better). ($\mu \pm \sigma$)

	Training task		Transfer task
	Avg. log likelihood	Expected KLD	Avg. log likelihood
REIRL	-459.03 ± 15.06	0.0663 ± 0.0000	-362.16 ± 11.15
MDCE IRL	-368.27 ± 2.39	0.0663 ± 0.0000	-300.88 ± 0.83
MDCE SERD	$\mathbf{-335.74 \pm 0.20}$	$\mathbf{0.0246 \pm 0.0002}$	$\mathbf{-269.79 \pm 0.44}$

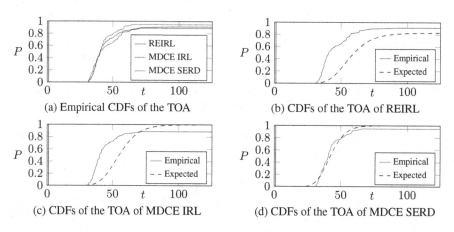

(a) Empirical CDFs of the TOA

(b) CDFs of the TOA of REIRL

(c) CDFs of the TOA of MDCE IRL

(d) CDFs of the TOA of MDCE SERD

Fig. 3. Expected and estimated cumulative distribution functions (CDF) of the time of arrival (TOA). If a CDF does not increase up to 1, the robot was not always able to reach the goal.

Table 1 shows the expected Kullback-Leibler divergence (KLD) of the true transition model (learned from all available demonstrations) to the estimated ones under the joint probability of states and actions of expert demonstrations. Hence, it specifies the accuracy of the transition models, while more likely transitions are more strongly weighted. The results of MDCE IRL and REIRL are equal, as their transition models are identically estimated, while MDCE SERD yields more accurate estimates. Figure 3(b)–(d) illustrate the CDFs of the expected time of arrival (TOA) and the empirical one, which is generated by running simulations of the resulting optimal policies. It can be seen that the empirical and the expected TOA CDFs differ strongly for MDCE IRL and REIRL, while the one of MDCE SERD is more accurate and very similar to the empirical one.

Finally, the estimates are evaluated in a transfer task, by estimating the log likelihood of ten demonstrations that were provided by an expert. Table 1 summarizes the results. Similar to the training task, MDCE SERD outperforms the other approaches closely followed by MDCE IRL, while REIRL yields worse results. This implies that accurate, generalizable models can be trained by MDCE SERD without overfitting.

6 Conclusions

In this paper, we compared IRL algorithms for learning under unknown dynamics, if only human expert demonstrations are available. Therefore, the reward function and the dynamics are estimated from demonstrations. The evaluation shows that both MDCE IRL and REIRL were able to accurately estimate the reward function. However, MDCE SERD shows improved performance in terms of accuracy of the estimated rewards and dynamics, since it incorporates the learning of the transition model into IRL. Therefore, the usage of SERD is beneficial if better transition models are useful. Such scenarios are, for example, high-level task planning problems, which need to accurately infer the time of arrival of robots. Future work could focus on feature learning for shared features between both the reward function and the dynamics.

References

1. Abbeel, P., Ng, A.Y.: Apprenticeship learning via inverse reinforcement learning. In: Proceedings of the Twenty-first International Conference on Machine Learning (2004)
2. Bloem, M., Bamos, N.: Infinite time horizon maximum causal entropy inverse reinforcement learning. In: IEEE 53rd Annual Conference on Decision and Control (2014)
3. Boularias, A., Kober, J., Peters, J.: Relative entropy inverse reinforcement learning. In: Proceedings of 14th International Conference on Artificial Intelligence and Statistics (2011)
4. Helbing, D., Molnar, P.: Social force model for pedestrian dynamics. Phys. Rev. E **51**(5), 4282 (1995)
5. Henry, P., Vollmer, C., Ferris, B., Fox, D.: Learning to navigate through crowded environments. In: IEEE International Conference on Robotics and Automation (2010)
6. Herman, M., Fischer, V., Gindele, T., Burgard, W.: Inverse reinforcement learning of behavioral models for online-adapting navigation strategies. In: IEEE ICRA (2015)
7. Herman, M., Gindele, T., Wagner, J., Schmitt, F., Burgard, W.: Inverse reinforcement learning with simultaneous estimation of rewards and dynamics. In: AISTATS (2016)
8. Klein, E., Geist, M., Piot, B., Pietquin, O.: Inverse reinforcement learning through structured classification. In: Proceedings of Advances in Neural Information Processing Systems (2012)
9. Neu, G., Szepesvári, C.: Apprenticeship learning using inverse reinforcement learning and gradient methods. In: UAI (2007)

10. Ng, A.Y., Russell, S.J.: Algorithms for inverse reinforcement learning. In: ICML (2000)
11. Ziebart, B.D., Bagnell, J.A., Dey, A.K.: Modeling interaction via the principle of maximum causal entropy. In: Proceedings of the International Conference on Machine Learning (2010)
12. Ziebart, B.D., Maas, A., Bagnell, J.A.D., Dey, A.: Maximum entropy inverse reinforcement learning. In: Proceeding of AAAI 2008 (2008)

Artificial Neural Network: Deep or Broad?
An Empirical Study

Nian Liu and Nayyar A. Zaidi[✉]

Clayton School of Business and Econometrics, Monash University,
Clayton, VIC 3800, Australia
sunnystliu@gmail.com, nayyar.zaidi@monash.edu

Abstract. Advent of *Deep Learning* and the emergence of *Big Data* has led to renewed interests in the study of Artificial Neural Networks (ANN). An ANN is a highly effective classifier that is capable of learning both linear and non-linear boundaries. The number of hidden layers and the number of nodes in each hidden layer (along with many other parameters) in an ANN, is considered to be a model selection problem. With success of deep learning especially on big datasets, there is a prevalent belief in machine learning community that a deep model (that is a model with many number of hidden layers) is preferable. However, this belies earlier theorems proved for ANN that only a single hidden layer (with multiple nodes) is capable of learning any arbitrary function, i.e., a shallow broad ANN. This raises the question of whether one should build a deep network or go for a broad network. In this paper, we do a systematic study of depth and breadth of an ANN in terms of its accuracy (0–1 Loss), bias, variance and convergence performance on 72 standard UCI datasets and we argue that broad ANN has better overall performance than deep ANN.

1 Introduction

The emergence of *Big Data* and success of deep learning on some problems have sparked a new interest in the study and evaluation of Artificial Neural Networks (ANN). ANN has a long history in the field of Artificial Intelligence. Earliest instance of ANN came in the form of Perceptron algorithm [1]. Perceptron was not capable of handling non-linear class boundaries and, therefore, interest in them remained fairly limited. Breakthrough was achieved by the invention of Backpropagation algorithm and multi-layer Perceptron that could model non-linear boundaries led to the golden era of ANN. However, results were not as good as one had expected. Attention of Artificial Intelligence (AI) community was soon diverted to more mathematically sound alternative such as Support Vector Machines (with convex objective function) and non-parametric models such as Decision Trees, Random Forest, etc.

With advancements in the computing infrastructure, access to much bigger training datasets and better (for example greedy layer-wised) training algorithms [2] – ANN demonstrated much successes in its application on structured

© Springer International Publishing AG 2016
B.H. Kang and Q. Bai (Eds.): AI 2016, LNAI 9992, pp. 535–541, 2016.
DOI: 10.1007/978-3-319-50127-7_46

problems[1] and, therefore, there has been a renewed interest in the study of ANN mostly under the umbrella term of *deep* learning. However, applicability of deep learning on problems where there is no structure in the input and studying its efficacy over other competing machine learning algorithms remains to be further investigated.

Deep refers to the fact that the network has to be very deep. Why is deep learning effective? We believe that this is due to the feature engineering capability of a deep network. For example, for image recognition, lower-level layers represent the pixels and middle-levels represent the edges made of these pixels and higher-levels represent the concepts made from these edges. An example is the convolutional neural network (CNN), where there is an explicit filtering stage (convolution), followed by max-polling and then a fully-connected network. In an CNN, an image is convolved with a kernel to incorporate the spatial information present in the image, resulting in higher-order features. This is followed by some sampling steps, for example, max-pooling to reduce the size of the output. There can be many convolution and max-pooling layers, which are followed by a fully-connected multi-layer perceptron. What about problems where we can not rely on interactions in the features? In other words, if we cannot rely on convolutional and max-pooling layers, will deep learning be as effective as it is on the structured problems? Well, we argue, that the success of deep learning can be attributed to the fact that it takes into account higher-order interactions among the features of the data. And when faced with extremely large quantities of training data, taking into account these higher-order interactions is beneficial [3]. Therefore, for un-structured problems, there are several directions one could take of which we mention only three in the following:

1. Take the quadratic, cubic or higher-order features and feed them to an ANN with no hidden layers, – shallow model, [4].
2. Train a multi-layer ANN with many hidden layers – deep model.
3. Train an ANN with single hidden layer but with many nodes – broad model.

We leave option one as an area of future research, and will focus on the second and third option. To the best of our knowledge, there are no comparative studies of deep and broad ANN in terms of their performance, bias and variance profiles and convergence analysis. We claim our main contributions in this paper are:

– We provide a comparison of 0–1 Loss, bias and variance of deep ANNs (with two, three, four and five hidden layers) and broad ANNs (with two, four, eight and ten nodes in one hidden layer). We show that on standard UCI datasets [5], broad ANNs leads to better results than deeper models.
– We compare the convergence analysis of deep ANNs and broad ANNs. We show that broad ANN have far superior convergence profiles than deeper models.

---

[1] These are mostly the problems in text, vision and NLP where there is a certain structure present in the input features. For example, deep learning performed extremely well on MNIST digit dataset (accuracy improved over 20%) when compared to typical machine learning algorithms.

2   A Simple Feed-Forward ANN

An Artificial Neural network (ANN) mimics how the neurons in a biological brain work. In the brain, each neuron receives input from other neurons. The effect and magnitude of the input is controlled by synaptic weight. The weight will change over time so that the brain can learn and perform tasks.

As this paper is focused on the empirical study between broad and deep ANN, any specific implementation of ANN can be used provided the implementation is held consistently across different model structures during comparison. As such, we have selected simple feed-forward ANN with the traditional back-propagation algorithm using stochastic gradient descent. All nodes in the ANN are sigmoid nodes.

3   Experimental Details

We compare and analyze the performance of deep and broad models on 72 natural domains from the UCI repository of machine learning datasets [5]. The list of datasets used in this work and the definition of bias and variance can be found in [7]. Each algorithm is tested on each dataset using 3 rounds of 2-fold cross validation. We report Win-Draw-Loss (W-D-L) results when comparing the 0–1 Loss, bias and variance of two models. A two-tail binomial sign test is used to determine the significance of the results. Results are considered significant if $p \leq 0.05$ and shown in bold. When discussing results, we denote all the datasets as *All*. We denote the following datasets as *Big* – Poker-hand, Covertype, Census-income, Localization, Connect-4, Shuttle, Adult, Letter-recog, Magic, Sign, Pendigits. Numeric features are discretized by using the Minimum Description Length (MDL) supervised discretization method [8].

4   Deep vs. Broad – Comparison in Terms of W-D-L

Let us start by comparing five broad models with five deep models in terms of their Win, Draw and Loss results on standard datasets. We denote broad models as NN2, NN4, NN6, NN8 and NN10. This denotes ANN with one hidden layer with two, four, six, eight and ten nodes in the hidden layer. We denote deep models as NN2, NN22, NN222, NN2222, NN22222. This denotes ANN with one hidden layer with two nodes, two hidden layers with two nodes each, three hidden layers with two nodes each, four hidden layers with two nodes each and five hidden layers with two nodes each, respectively. We also include an ANN with no hidden layer. This is denoted as NN0. This is actually similar to Logistic Regression except that it is using Mean-square-error instead of Conditional Log-Likelihood (CLL) [4,6].

Table 1. A comparison of Bias and Variance of broad models in terms of W-D-L on *All* datasets.

	vs. NN0		vs. NN2		vs. NN4		vs. NN6		vs. NN8	
	W-D-L	p	W-D-L	p	W-D-L	p	W-D-L	p	W-D-L	p
All Datasets - Bias										
NN2	35/3/34	1								
NN4	45/4/23	**0.010**	49/7/16	**<0.001**						
NN6	47/4/21	**0.002**	47/5/20	**0.001**	37/7/28	0.321				
NN8	48/3/21	**0.002**	44/5/23	**0.014**	37/7/28	0.321	36/11/25	0.200		
NN10	52/3/17	**<0.001**	47/5/20	**0.001**	41/9/22	**0.023**	43/10/19	**0.003**	40/15/17	**0.003**
All Datasets - Variance										
NN2	20/2/50	**<0.001**								
NN4	21/2/49	**0.001**	38/6/28	0.268						
NN6	27/3/42	0.091	43/7/22	**0.013**	40/8/24	0.060				
NN8	32/2/38	0.550	42/7/23	**0.025**	44/8/20	**0.004**	36/9/27	0.314		
NN10	30/3/39	0.336	42/7/23	**0.025**	43/9/20	**0.005**	34/13/25	0.298	33/10/29	0.704

Broad Models. Let us start with comparing the bias and variance of broad models (Table 1). A systematic trend in bias, as expected can be seen. A broader model is lower-biased than the less broader model. NN10 being lowest biased winning significantly in terms of W-D-L when compared to other lesser broad models. Variance results are slightly surprising. An NN10, though has (nonsignificant) higher variance than NN0, has significantly lower variance than NN2, NN4 and non-significantly lower variance than NN6 and NN8.

Table 2. A comparison of 0–1 Loss of broad models in terms of W-D-L on *All* and *Big* datasets.

	vs. NN0		vs. NN2		vs. NN4		vs. NN6		vs. NN8	
	W-D-L	p	W-D-L	p	W-D-L	p	W-D-L	p	W-D-L	p
All Datasets – 0–1 Loss										
NN2	27/2/43	0.072								
NN4	31/6/35	0.712	50/9/13	**<0.001**						
NN6	33/3/36	0.801	49/3/20	**<0.001**	45/7/20	**0.003**				
NN8	37/1/34	0.813	50/5/17	**<0.001**	44/8/20	**0.004**	31/14/27	0.694		
NN10	40/2/30	0.282	51/4/17	**<0.001**	49/5/18	**<0.001**	38/9/25	0.130	40/8/24	0.060
Big Datasets – 0–1 Loss										
NN2	6/0/6	1.226								
NN4	7/0/5	0.774	12/0/0	**0.011**						
NN6	7/0/5	0.774	12/0/0	**0.001**	11/0/1	**0.006**				
NN8	8/0/4	0.388	12/0/0	**<0.001**	9/0/3	0.146	8/0/4	0.388		
NN10	8/0/4	0.388	12/0/0	**<0.001**	10/0/2	**0.039**	9/0/3	0.146	9/0/3	0.146

Lower-bias of broad models translates into better 0–1 Loss and RMSE performance on *Big* datasets. This can be seen in Table 2. On *Big* datasets, NN10 leads to much better performance than NN8, NN6, NN4, NN2 and NN0. Some of the results are not significant statistically, but a general trend can be seen that a broader ANN leads to better performance than less broader ANN.

Deep Models. It can be seen from Table 3, unlike broad models, a deep model does not lead to low bias at all. In fact, bias increases as the model is made deeper. In terms of the bias, only NN2 wins to NN0 (win over one dataset), whereas all other deeper models lose to NN0. Similarly, all deeper models except NN22222 loses to NN0 in terms of variance. From Table 4, it can be seen that NN2 and NN22 are slightly competitive with NN0 in terms of 0–1 Loss on *Big* datasets – deeper models NN222, NN2222 and NN22222 are not very competitive with NN0.

Table 3. A comparison of Bias and Variance of deep models in terms of W-D-L on *All* datasets.

	vs. NN0		vs. NN2		vs. NN22		vs. NN222		vs. NN2222	
	W-D-L	p	W-D-L	p	W-D-L	p	W-D-L	p	W-D-L	p
All Datasets – Bias										
NN2	35/3/34	1								
NN22	30/3/39	0.336	28/4/40	0.182						
NN222	26/1/45	**0.032**	21/3/48	**0.002**	24/4/44	**0.021**				
NN2222	5/0/67	**<0.001**	3/1/68	**<0.001**	3/2/67	**<0.001**	4/9/59	**<0.001**		
NN22222	0/1/71	**<0.001**	0/1/71	**<0.001**	1/2/69	**<0.001**	1/9/62	**<0.001**	0/61/11	**<0.001**
All Datasets – Variance										
NN2	20/2/50	**<0.001**								
NN22	20/1/51	**<0.001**	27/6/39	0.175						
NN222	24/1/47	**0.009**	34/3/35	1	32/4/36	0.905				
NN2222	34/1/37	0.813	34/1/37	0.813	36/2/34	0.905	32/9/31	1		
NN22222	40/2/30	0.282	38/1/33	0.6353	39/2/31	0.403	35/9/28	0.450	8/61/3	0.227

Deep vs. Broad Models – Discussion. In Figs. 1 and 2, we compare the convergence profiles of broad and deep models respectively on sample datasets. Note, what we plot is the variation in mean-square-error on training data. Of course, the results are averaged over three rounds of two-fold cross validation. We do not compare broad and deep models directly, but instead used NN2 as the reference point between the two paradigms. As we expected, a broader model results in better convergence profile than less broader model. On most datasets, we found that NNk results in better convergence than NN(k-1). For deep models, it can be seen that most models are worst than NN2 (red line), where NN22 (green line) leads to the best convergence in most cases. We argue, that based on these results, broader models not only lead to better results, but also have better convergence profile. Of course, this property will lead to faster training and better efficiency overall.

Table 4. A comparison of 0–1 Loss of deep models in terms of W-D-L on *All* and *Big* datasets.

	vs. NN0		vs. NN2		vs. NN22		vs. NN222		vs. NN2222	
	W-D-L	p	W-D-L	p	W-D-L	p	W-D-L	p	W-D-L	p
All Datasets – 0–1 Loss										
NN2	27/2/43	0.072								
NN22	28/1/43	0.096	24/5/43	**0.027**						
NN222	24/1/47	**0.009**	25/5/42	**0.050**	28/3/41	0.148				
NN2222	7/0/65	**<0.001**	4/2/66	**<0.001**	4/2/66	**<0.001**	3/9/60	**<0.001**		
NN22222	7/1/64	**<0.001**	5/1/66	**<0.001**	4/2/66	**<0.001**	3/9/60	**<0.001**	1/61/10	**0.012**
Big Datasets – 0–1 Loss										
NN2	6/0/6	1.226								
NN22	5/0/7	0.774	4/0/8	0.388						
NN222	4/0/8	0.388	2/0/10	**0.039**	4/0/8	0.388				
NN2222	2/0/10	**0.039**	0/0/12	**<0.001**	1/0/11	**0.006**	1/1/10	**0.012**		
NN22222	1/1/10	**0.012**	0/0/12	**<0.001**	0/0/12	**<0.001**	0/1/11	**<0.001**	0/6/6	**0.031**

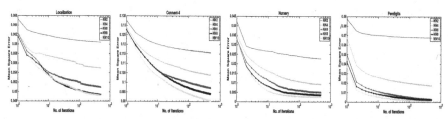

Fig. 1. Variation in Mean Square Error of NN2, NN4, NN6, NN8 and NN10 with increasing number of (optimization) iterations on sample datasets – localization, connect-4, nursery and pendigits. (Color figure online)

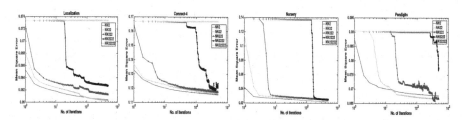

Fig. 2. Variation in Mean Square Error of NN2, NN22, NN222, NN2222 and NN22222 with increasing number of (optimization) iterations on sample datasets – localization, connect-4, nursery and pendigits. (Color figure online)

5   Conclusion and Future Works

The results we have presented in this paper are preliminary and warrants further investigation. There are several areas which we needs to be explored:

– The number of nodes in deeper model are set to two. Since the datasets in our experiments are not hugely big (biggest is the poker-hand with 1.17 million instances), we assumed that five hidden-layer neural network with two hidden nodes in each layer covers the essence of a deep network. Obviously, the number of nodes in each layer effects the performance and a systematic study is needed to assess the effect on performance.

- The maximum number of optimization iterations are set to 500. It can be the case that a deeper network just converge slowly, therefore, variations with more number of iterations needs to be tested.
- In this work, we have constrained ourselves to (feed-forward) multi-layer perceptrons – further experimentations needs to be conducted to check if results will hold on other ANN types such as Deep Belief Networks or Restricted Boltzmann machines, etc.

Acknowledgment. This research has been supported by the Australian Research Council (ARC) under grant DP140100087, and by the Asian Office of Aerospace Research and Development, Air Force Office of Scientific Research under contracts FA2386-15-1-4007 and FA2386-15-1-4017. Nian Liu was supported by Early Career Researcher seed grant (2015) by the Faculty of Information Technology, Monash University, Australia.

References

1. Rosenblatt, F.: The perceptron-a perceiving and recognizing automaton. Cornell Aeronautical Laboratory, Technical report 85-460-1 (1957)
2. Hinton, G.E., Osindero, S., Teh, Y.W.: A fast learning algorithm for deep belief nets. Neural Comput. **18**, 1527 (2006)
3. Zhang, X., LeCun, Y.: Text understanding from scratch. arXiv:1502.01710 (2015)
4. Zaidi, N.A., Petitjean, F., Webb, G.I.: Preconditioning an artificial neural network using Naive Bayes. In: Bailey, J., Khan, L., Washio, T., Dobbie, G., Huang, J.Z., Wang, R. (eds.) PAKDD 2016. LNCS (LNAI), vol. 9651, pp. 341–353. Springer, Heidelberg (2016). doi:10.1007/978-3-319-31753-3_28
5. Frank, A., Asuncion, A.: UCI machine learning repository (2010). http://archive.ics.uci.edu/ml
6. Zaidi, N.A., Carman, M.J., Cerquides, J., Webb, G.I.: Naive-Bayes inspired effective pre-conditioners for speeding-up logistic regression. In: IEEE International Conference on Data Mining, pp. 1097–1102 (2014)
7. Zaidi, N.A., Webb, G.I., Carman, M.J., Petitjean, F., Cerquides, J.: ALR^n: accelerated higher-order logistic regression. Mach. Learn. **104**(2), 151–194 (2016)
8. Fayyad, U.M., Irani, K.B.: On the handling of continuous-valued attributes in decision tree generation. Mach. Learn. **8**(1), 87–102 (1992)

An Empirically-Sourced Heuristic for Predetermining the Size of the Hidden Layer of a Multi-layer Perceptron for Large Datasets

Amanda Lunt[✉] and Shuxiang Xu

School of Engineering and ICT, University of Tasmania, Launceston, Australia
{Amanda.Lunt,Shuxiang.Xu}@utas.edu.au

Abstract. We recommend a guiding heuristic to locate a sufficiently-sized multilayer perceptron (MLP) for larger datasets. Expected to minimise the search scope, it is based on experimental research into the comparative performance of 14 existing approaches with global minimum ranges on 31 larger datasets. The most consistent performer was Baum's [1] equation that sets the number of hidden neurons equal to the square root of the number of training instances.

Keywords: Neural network · Multilayer Perceptron · Hidden layer size · Global minimum · Local minimum

1 Introduction

Trained under supervision, a 3-layer multilayer perceptron (MLP) will find 'hidden' relationships within a set of data by approximating continuous functions [2]. The trained network may then be used for prediction tasks on previously unseen data from the same domain, with the final configuration unique to each specific dataset. The size of the hidden middle layer, N^h, has a strong bearing on the prediction accuracy of the final model [3], yet the predominant technique to locating N^h is resizing through trial-and-error. Exhaustive search through a range of N^h becomes problematic with larger datasets, increasing demands on processor capacity and extending the time required for training. The usefulness of the heuristic proposed in this paper is in minimising the scope of the search to reach a suitably optimal network size. We note that a reasonable network architecture may not be limited to a single 'correct' configuration [4] so long as the underlying function can be learnt while retaining enough smallness to generalise [5].

A set of proposed mathematical relationships between N^h and the numbers of input neurons, N^i, output neurons, N^o, both fixed, and instances of the dataset used for training, N^{tr}, is summarised in Table 1. We used N^{tr} for our calculations rather than N^{TOT}, total number of instances, as it directly relates to the training process.

© Springer International Publishing AG 2016
B.H. Kang and Q. Bai (Eds.): AI 2016, LNAI 9992, pp. 542–547, 2016.
DOI: 10.1007/978-3-319-50127-7_47

Table 1. Fourteen ways to determine N^h. Approach number was attributed randomly.

Source	Equation	Approach number	Source	Equation	Approach number
[6, 7]	$N^h \geq 2N^i + 1$	(1)	[1, 8]	$N^h = \frac{N^{tr}}{N^i}$	(8)
[9] in [10]	$N^h \leq \frac{N^{tr}}{(N^i+1)}$	(2)	[1, 8]	$N^h = \frac{N^{tr}}{(N^i+N^o)}$	(9)
[1]	$N^h = \sqrt{N^{tr}}$	(3)	[11]	$N^h = 2N^i$	(10)
[12]	$N^h = \log(N^{tr})$	(4)	[13]	$N^h > N^o$	(11)
[14]	$N^h = \sqrt{N^i N^o}$	(5)	[15, 16]	$N^h \leq N^i - 1$	(12)
[17]	$N^h = \frac{(2N^i + 3)}{(N^i - 2)}$	(6)	[1, 16]	$N^h = \frac{N^i}{N^{tr}}$	(13)
[18]	$N^h = C \left(\frac{N^{tr}}{N^i \log N^{tr}} \right)^{1/2}$	(7)	[15]	$N^h \geq \frac{N^i}{3}$	(14)

1.1 Research Question

Which of the existing approaches can assist the search for a suitable number of neurons in the single hidden layer of a MLP for larger datasets?

2 Experiment

Our simple experiment investigates the performance of each approach when compared with global minimum benchmarks [19]. Thirty-one datasets with many attribute-target pairs or high dimensionality were sourced [20–22] (see Table 2).

A lower and upper limit to N^h was established for the training of each dataset based on calculations from the approaches in Table 1. We set lower bound at the calculation closest to 0, while upper bound was based on a sense of being able to train to that N^h, with flexibility to extend with working processor capacity. Where an approach takes the form of a lower or upper bound, the calculated N^h at the bound was used.

Weights were initialised randomly to represent prior knowledge [23]. Training, test and validation sets (70-15-15% of N^{TOT}) were also randomly generated for the best opportunity to locate the global minimum [24]. Each-sized network was trained 10 times with cross-validation, accounting for random influences [25]. We performed our experiment using MATLAB Neural Network Toolbox version 6 add-on's patternnet function with the scaled conjugate backpropagation algorithm [26, 27].

Table 2. Characteristics of 31 datasets. Most are from http://archive.ics.uci.edu/ml/datasets except (b) http://mldata.org and (c) http://osmot.cs.cornell.edu/kddcup. Larger sets were excluded as too slow to train with available resources.

Working title	N^i	N^o	N^{TOT}	Working title	N^i	N^o	N^{TOT}
(b) 2Norm	20	2	7400	PokerHand	10	10	25010
Abalone	8	29	4177	(c) ProteinHomology	74	1	145751
AdultIncome	14	2	48842	PubChem362	144	2	4279
Chess	6	18	28056	PubChem456	153	2	9982
Connect4	42	3	67557	PubChem687	153	2	33067
FirstOrderTheorem	51	6	6118	(c) QuantumPhysics	78	1	50000
Gisette	5	2	7000	Shuttle	9	7	58000
LandSat	36	6	6435	Skin	3	2	245057
LetterRecognition	16	26	20000	Spambase	57	1	4601
Madelon	50	2	2600	Thyroid	21	3	7200
MagicGamma	10	2	19020	WallFollowRobot2	2	4	5456
Musk2	16	2	6598	WallFollowRobot4	4	4	5456
Nomao	17	2	34465	WallFollowRobotFull	24	4	5456
Nursery	8	5	12960	Waveform	21	3	5000
OptDigits	64	10	5620	WineQuality	11	11	4898
PageBlocks	10	5	5473				

3   Results

The global minimum was located for each dataset at the N^h with the smallest averaged performance error from the mean of squared errors comparing the actual output against the desired output [25]. Approaches (3) and (1) calculated the global minimum in one case each, *WallFollRobot2* and *AdultIncome* respectively.

Not all approaches gave us a sensible calculation for N^h for every dataset. We obtained a result for all of the 31 datasets with approaches (4), (5) and (7) only. Table 3 demonstrates a combination of this raw count [A] and the count of datasets where

Table 3. An excerpt of the simple ranking of approaches according to relative usefulness, ordered from 'most' useful and truncated for brevity. [B] was scaled in the final column to indicate its relationship to the research question, with no impact on the final rank.

Approach number	Result count [A]	Comparison of means [B]	[A] + [B]	[A] + 2[B]
(5)	31	26	57	83
(7)	31	25	56	81
(3)	29	23	52	75
(1)	28	23	51	74
(12)	28	22	50	72
(4)	31	18	49	67
(14)	22	18	40	58

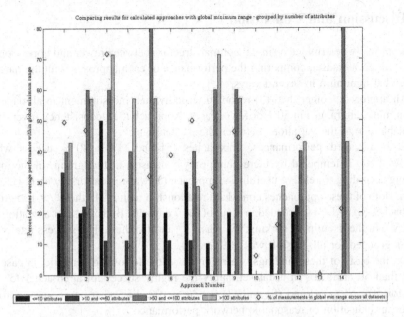

Fig. 1. Performance at calculated N^h compared with global minimum range over 31 datasets.

performance at the approach's calculated N^h intersects with the global minimum (95% CI) from a multiple comparison of means [B].

Figure 1 gives an overview of two further comparisons with the performance range at the global minimum N^h. Single diamonds are derived from the count of *individual* performance measures for an approach within the global minimum range over all datasets. You can clearly see the success of approach (3) $N^h = \sqrt{N^{tr}}$ in this, with occurrences 71.7% of times across all datasets.

The second set of comparisons is presented as bar graphs that have been separated into N^i groupings to allow for disparity in attribute dimensionality across the datasets: $N^i \leq 10$; $10 < N^i \leq 50$; $50 < N^i \leq 100$; and $N^i > 100$. This ratio is the per cent of times an *average* of the 10 performance measures at each pre-calculated N^h occurred within the range of performances recorded at the global minimum N^h, grouped by N^i. Approaches (5) and (14) were both highly successful in the $50 < N^i \leq 100$ group (4 out of 5 cases), with (3) and (8)'s average occurring within the global minimum range for 5 out of the 7 cases in the $N^i > 100$ group.

Also of note, (2), (3), (8) and (9)'s averages placed in the global minimum range for the $50 < N^i \leq 100$ group in 3 of the 5 cases. The results for the two groups where $N^i \leq 50$ (the remaining 19 datasets) were no better than 50%.

4 Discussion and Conclusion

We empirically determined a single, optimal structure between lower and upper bounds for N^h for each dataset, comparing the performance of each approach with the range at this global minimum in several ways.

All approaches other than (3) recorded an individual measurement in all datasets' global minimum ranges in 50% or fewer cases. Approach (3)'s consistency (over 71%) is notable due to the variations between the 31 datasets.

With averaged performances, approaches (5) and (14)'s 80% success where $50 < N^i \leq 100$ is tempered by there being only 5 datasets in that group. In the initial ranking according to relative usefulness, approach (5) was ranked first, with (14) lower down. Both of these approaches consider a relationship with N^i. In the $N^i > 100$ group, approaches (8) and (3) succeeded in 5 out of the 7 datasets. Both consider a relationship with N^{tr}. In the usefulness ranking, (8) was 11th and (3) third. The success rate in the results grouped for all $N^i \leq 50$ was 50% or less.

On the basis of these findings, we recommend the following heuristic: in cases of more than 50 attributes in a dataset, apply the highly successful approaches (5) and (14) for $50 < N^i \leq 100$ and (8) and (3) for $N^i > 100$. For other cases, use approach (3) for an indication of reasonable network performance.

References

1. Baum, E.B.: On the capabilities of multilayer perceptrons. J. Complex. **4**, 193–215 (1988)
2. Cybenko, G.: Approximation by superpositions of a sigmoidal function. Math. Control Signals Syst. **2**, 303–314 (1989)
3. Hornik, K.: Approximation capabilities of multilayer feedforward networks. Neural Netw. **4**, 251–257 (1991)
4. Zeng, X., Yeung, D.S.: Hidden neuron pruning of multilayer perceptrons using a quantified sensitivity measure. Neurocomputing **69**, 825–837 (2006)
5. Aran, O., Yildiz, O.T., Alpaydin, E.: An incremental framework based on cross-validation for estimating the architecture of a multilayer perceptron. Int. J. Pattern Recogn. Artif. Intell. **23**, 159–190 (2009)
6. Hecht-Nielsen, R.: Kolmogorov's mapping neural network existence theorem. In: Proceedings of IEEE First Annual International Conference on Neural Networks, pp. III-11–III-14. (1987)
7. Sprecher, D.A.: A universal mapping for kolmogorov's superposition theorem. Neural Netw. **6**, 1089–1094 (1993)
8. Barron, A.R.: Approximation and estimation bounds for artificial neural networks. Mach. Learn. **14**, 115–133 (1994)
9. Rogers, L.L., Dowla, F.U.: Optimization of groundwater remediation using artificial neural networks with parallel solute transport modeling. Water Resour. Res. **30**, 457–481 (1994)
10. Somaratne, S., Seneviratne, G., Coomaraswamy, U.: Prediction of soil organic carbon across different land-use patterns. Soil Sci. Soc. Am. J. **69**, 1580–1589 (2005)
11. Denker, J.S., Schwartz, D., Wittner, B., Solla, S., Howard, R., Jackel, L., Hopfield, J.: Large automatic learning, rule extraction and generalization. Complex Syst. **1**, 877–922 (1987)

12. Wanas, N.M., Auda, G.A., Kamel, M.S., Karray, F.O.: On the optimal number of hidden nodes in a neural network. In: IEEE Canadian Conference on Electrical and Computer Engineering 1998, vol. 2, pp. 918–921 (1998)
13. Gallinari, P., Thiria, S., Soulie, F.F.: Multilayer perceptrons and data analysis. In: IEEE International Conference on Neural Networks 1988, vol.391, pp. 391–399 (1988)
14. Shibata, K., Ikeda, Y.: Effect of number of hidden neurons on learning in large-scale layered neural networks. In: ICROS-SICE International Joint Conference 2009, pp. 5008–5013. SICE, Fukuoka International Congress Center, Japan (2009)
15. Arai, M.: Bounds on the number of hidden units in binary-valued three-layer neural networks. Neural Netw. 6, 855–860 (1993)
16. Huang, S.-C., Huang, Y.-F.: Bounds on the number of hidden neurons in multilayer perceptrons. IEEE Trans. Neural Netw. 2, 47–55 (1991)
17. Deepa, S.N., Sheela, K.G.: Estimation of number of hidden neurons in back propagation networks for wind speed prediction in renewable energy systems. Draft (2013)
18. Xu, S., Chen, L.: A novel approach for determining the optimal number of hidden layer neurons for FNNs and its application in data mining. In: Proceedings the 5th International Conference on Information Technology and Applications 23–26 June 2008, Cairns, Qld, pp. 683–686 (2008)
19. Gorman, R.P., Sejnowski, T.J.: Analysis of hidden units in a layered network trained to classify sonar targets. Neural Netw. 1, 75–89 (1988)
20. Bache, K., Lichman, M.: UCI Machine Learning Repository. School of Information and Computer Science, University of California, Irvine (2013)
21. Hoyer, P.O., Ong, C.S., Henschel, S., Braun, M.L., Sonnenburg, S.: IDA Benchmark Repository, vol. 0.1.6. ML Group, Berlin (2013)
22. ACM Special Interest Group on Knowledge Discovery and Data Mining: KDD Cup 2004: Particle physics; plus protein homology prediction. ACM (2004). http://www.kdd.org
23. Dayhoff, J.: Neural Network Architectures: An Introduction. International Thomson Computer Press, Boston (1996)
24. http://ulcar.uml.edu/~iag/CS/Intro-to-ANN.html
25. Flexer, A.: Statistical evaluation of neural network experiments: minimum requirements and current practice, pp. 1005–1008. The Austrian Research Institute for Artificial Intelligence, Schottengasse 3, A-1010 (1994)
26. Demuth, H., Beale, M., Hagan, M.: Neural Network Toolbox 6 User's Guide. The MathWorks Inc., Natick (2009)
27. Møller, M.F.: A scaled conjugate gradient algorithm for fast supervised learning. Neural Netw. 6, 525–533 (1993)

Distributed Genetic Algorithm on GraphX

Seemran Mishra[1], Young Choon Lee[2(✉)], and Abhaya Nayak[2]

[1] National Institute of Technology, Rourkela, India
[2] Macquarie University, Sydney, Australia
seemran.mishra1996@gmail.com,
{young.lee,abhaya.nayak}@mq.edu.au

Abstract. In this paper we explore the application of a recent breed of distributed systems, graph processing frameworks in particular, to solving complex research problems. These frameworks are designed to take full advantage of today's abundant resources with their inherent distributed computing functionalities. Abstraction of many technical details, such as networking and coordination of multiple compute nodes is a desirable feature provided by these graph-processing frameworks. While these frameworks are largely used to process and analyse the web graphs and social networks, their capacity is not limited to this direct application. This paper is based on design and implementation of a genetic algorithm (GA) using a graph processing tool, GraphX for the task scheduling problem as a case study. Our experimental results show that GraphX can significantly aid in devising distributed solutions for complex problems.

1   Introduction

Many research problems, primarily in computing related disciplines, increasingly face computational challenges and "Big data" issues. These problems are often too large for any single workstation computer to effectively deal with. In particular, many computational simulations, machine learning algorithms and image processing techniques require distributed processing/computing with a large number of processor cores and a massive amount of networked storage.

What has followed is the development of distributed systems, such as MapReduce, Hadoop, Apache Spark, Pregel and TensorFlow. These software systems in conjunction with the availability of abundant elastic and on-demand cloud resources have powered many large-scale data processing, graph processing and machine learning endeavours. These distributed systems are great tools for solving a wide range of research problems.

In this paper, we explore the application of GraphX[1] to solving research problems. In particular, we design and implement a genetic algorithm (GA) on GraphX for the task scheduling problem. GA fits well with Valiant's Bulk Synchronous Parallel (BSP) model [3], the inspiration behind Pregel. An instance

---

[1] GraphX (http://spark.apache.org/graphx/) is an open source implementation of the Google Pregel graph processing framework [1], provided as an API of Apache Spark [4].

© Springer International Publishing AG 2016
B.H. Kang and Q. Bai (Eds.): AI 2016, LNAI 9992, pp. 548–554, 2016.
DOI: 10.1007/978-3-319-50127-7_48

of the GA runs as a Pregel computation in each vertex/node of a directed graph in a distributed manner; hence we refer to our GA implementation as distributed GA or simply *dsGA*. Iterations/generations of GA can be seen as supersteps of Pregel. At the end of each superstep, a certain number of best chromosomes are exchanged among all GA instances as messages.

Our experimental results show the effectiveness of GraphX on leveraging distributed computing. In particular, dsGA provides comparable schedules in terms of schedule length to those generated by the "singleton" GA and HEFT (Heterogeneous Earliest Finish Time) [2]. More importantly, dsGA scales well as evidenced in its execution time being over 30% shorter than singleton GA when the population size is 48 or more.

Algorithm 1. Genetic Algorithm for task scheduling

1. Generate initial population P with N_c random chromosomes.
2. **while** termination criteria not satisfied **do**
3.   Evaluate P
4.   Select best chromosome P_{best}
5.   **foreach** crossover in N_{cr} **do**
6.   Select two random parents, $par1$ and $par2$, from P - P_{best}
7.   Select crossover point co.
8.   Perform crossover of $par1$ and $par2$ at co for $chi1$ and $chi2$
9.   **foreach** child in $chi1$ and $chi2$ **do**
10.   **foreach** mutation in N_{mut} **do**
11.   Randomly generate a chromosome
12.   **foreach** task in chi **do**
13.   Mutate assigned proc of chi to value in randomly gen. chrm.
14.   **if** better schedule is not obtained by mutation
15.   Modify assigned processor bit of chi to original value
16.   **end if**
17.   **end for**
18.   **end for**
19.   **end for**
20.   Remove two worst chromosomes C_{w1} and C_{w2} from P
21.   Insert child chromosomes $chi1$ and $chi2$ in place of C_{w1} and C_{w2} in P
22.   **end for**
23. **end while**
24. Evaluate P
25. Select the best chromosome

2   Distributed Genetic Algorithm

With the implementation of the genetic algorithm on task graphs we intend to acquire competitive schedules, the deciding factor being schedule length, within a reasonable amount of time. This algorithm (Algorithm 1) trails the usual steps of the classic genetic algorithm, namely initialization, selection, crossover, mutation and inheritance. Chromosomes, considered as the basic unit of GA, are implemented in form of schedules. Our input data is a basic task graph along with its attributes such as the precedence relation, task type, processor type and computation as well as communication cost. We prioritise tasks taking into account their computational costs (varying between different processors) along with their communication costs and precedence relationships.

Our distributed genetic algorithm (dsGA, Algorithm 2), specifically designed with GraphX, trails the classic GA (Algorithm 1). The design is in accordance with the Bulk Synchronous Parallel model and uses the pregel function of GraphX for graph processing. The graph is made up of n_v vertices that are interconnected by edges. The population, P, for each generation of GA is distributed among all the vertices with each vertex having P/n_v number of chromosomes. There are n_e edges which are required to pass messages among the vertices in order to determine the best schedule. Since each vertex is connected to every other vertex, the number of edges, n_e, is $n_v * n_{v-1}$ with all the edges being directed and initial value being true indicating that the message can be passed between them. The Pregel function of GraphX needs three functions namely 'Compute', 'Sendmsg' and 'Mergemsg' in order to execute.

Algorithm 2. GA for task scheduling in distributed environment

---

1. Generate n_v vertices.
2. Interconnect all the vertices to generate n_e edges and graph G
3. Assign P/n_v set of random schedules to all the vertices
4. Implement G.pregel(Compute, Sendmsg, Mergemsg, N_S, Initial message)
5. Select best chromosome P_{best} by evaluating P

Initially the same set of random schedules is allocated to each vertex. In order to improve performance, a schedule generated by HEFT algorithm is then added to the population of each vertex. The function named 'Compute' consists of a set of operations that needs to be executed on each vertex for each superstep. In our implementation we redefine each superstep to be one generation. Our 'Compute' function carries out selection, crossover and mutation operations which are completely similar to their counterparts in Algorithm 1. The best chromosome, P_{best}, is determined for every vertex. From the rest of the population, two parent chromosomes, $par1$ and $par2$, are selected randomly and the crossover operation is carried out to generate two child chromosomes, $chi1$ and $chi2$. Each child chromosome is further mutated P_{mut} times. The two child chromosomes, $chi1$ and $chi2$, then replace the two chromosomes, C_{w1} and C_{w2}, that have the worst schedule length. The above crossover operation is carried out N_{cr} times. Ultimately the best schedule from the modified population is selected and stored separately as M in order to be passed on as a message to other vertices. When the termination condition is satisfied for any vertex, that vertex becomes inactive and the values of the edges emerging from that vertex are reassigned to false.

The 'Sendmsg' function sends the best schedule stored in every vertex during each superstep to all other vertices. This function can be desirably modified with a dynamic termination condition, for instance, to render the vertex inactive after the three best schedule lengths are same. In case a vertex is inactive and it receives a message from another vertex, it becomes active again.

The 'Mergemsg' function evaluates which message is to be accepted by the vertex from the list of received messages. It compares all messages and selects the message having the least schedule length. The initial message which is a random schedule is passed to all vertices. While, in the case of statically decided number of generations, the number of supersteps, N_S, is equal to number of generations, N_{gen}, in the case of dynamic termination it is set to maximum integer value. After the termination is reached, the best schedules of all the vertices are compared to determine the best schedule.

3   Experiments

The performance of the distributed GA (dsGA) is evaluated primarily based on the Normalized Schedule Length (NSL), the ratio of the schedule length obtained by a particular algorithm to that obtained by HEFT. We have also compared the performance of dsGA with the singleton GA (or simply GA) based on both NSL and Normalized Execution Time (NET). The NET is defined as the ratio of the execution time obtained by dsGA or dsGA with HEFT (or dsGA^{HEFT}) to the execution time obtained by GA or GA with HEFT (GA^{HEFT}), respectively.

3.1   Experiment Settings

We have used both random task graphs and real-world task graphs, LU decomposition, Fast Fourier Transform and Laplace equation solver. There were 70 task graphs for each of these four types of task graph. The number of tasks in the task graph was varied between 15 to 255. Experiments were performed on seven variations of number of processors, i.e., 3, 5, 10, 15, 20, 30 and 50.

For the first four experiments, we compared NSL of HEFT, dsGA and dsGA^{HEFT} with varying numbers of processors. For each experiment, the task graphs were of varying sizes and their average was taken into account. The number of vertices for the implementation of dsGA was set as 6. The population size of P was kept constant as 30 for dsGA. The number of crossovers, N_{cr}, for each superstep or generation was fixed to be 10. The number of mutations N_{mut} for each child was set as 25. Instead of statically allocating the number of generations or number of supersteps N_S, dynamic termination criteria was given. The implementation rendered the vertex inactive when the best three schedules of that particular vertex had same schedule length. We provide four separate comparison graphs, one each for all types of task graphs. The task graph size was varied between 15 to 255 in all the four cases.

In the second set of experiments we have compared dsGA and dsGA^{HEFT} with the singleton GA and the singleton GA^{HEFT} with respect to varying population sizes. We first compared the NSL of HEFT, GA, dsGA, GA^{HEFT} and dsGA^{HEFT} followed by the NET comparison. The number of vertices for the distributed GA implementations (dsGA and dsGA^{HEFT}) was set to 6. The population size P was varied from 12 to 60 with an interval of 12 for the singleton GA implementations (GA and GA^{HEFT}). For a fair comparison we set the population size to an instance of dsGA in each vertex to range from 2 to 10; (note that we used 6 vertices in our dsGA implementation in GraphX). The number of crossovers, N_{cr}, for each superstep or generation was determined to be one-third of the population size. N_{mut}, which is the number of mutations for each child, was set to 25. The number of generations of supersteps N_S was statically allocated to be 30. For the smaller graphs, the distributed system gave higher execution time than the standalone implementation and for larger graphs the opposite was observed. Since the average was to be taken for all the graphs, in order to provide more pronounced results, we chose the larger task graphs with their size ranging from 125 to 255. This experiment was carried out for all the four kinds of task graphs and the average was taken for every instance of population.

3.2   Results

For the small numbers of processors (3, 5, 10 and 15 in Fig. 1, except in the LU case in Fig. 1(c)) the schedule length obtained from dsGA (without the schedule from HEFT in the initial population) tends to be comparable to the HEFT

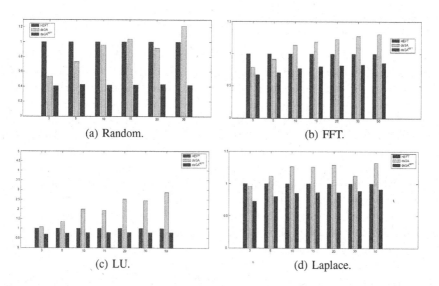

(a) Random.   (b) FFT.

(c) LU.   (d) Laplace.

Fig. 1. Normalised schedule length w.r.t. different task graphs and varying numbers of processors, 3, 5, 10, 15, 20 and 30 indicated on the x-axis.

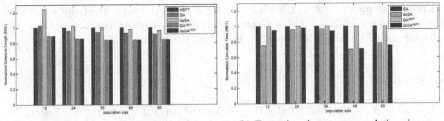

(a) Schedule length w.r.t. population sizes.   (b) Execution time w.r.t. population sizes.

Fig. 2. Performance w.r.t. different population sizes.

algorithm. On the other hand, schedule lengths deteriorated with increase in the number of processors, i.e., the increase of the search space. The small population size of 30 (5 per each vertex) in this set of experiments also contributed to such deterioration, as observed from the effect of population size on the schedule length in Fig. 2(a). In particular, schedule lengths from dsGA are equivalent/superior to those from HEFT for population sizes of 36, 48 and 60. However, in case of GA^{HEFT}, schedule lengths obtained for all types of processors were better than those obtained from HEFT. These results demonstrate that the implementation of dsGA for task scheduling yields comparable results and the best ones were obtained on adding the schedule from HEFT to GA.

Our main motivation was to compare the schedule quality and execution time of (singleton) GA and dsGA. From the results obtained (Fig. 2(b)), we observe that the execution time for dsGA was much lower than the execution time of GA for both the implementations, with and without HEFT. On increasing the population size, both implementations, with and without HEFT gave better schedule lengths as compared to the HEFT algorithm (Fig. 2(a)). Although schedule lengths of GA^{HEFT} were similar to those of $dsGA^{HEFT}$ regardless of different population sizes, execution times favour dsGA with an approx. 15% on average and over 30%, with population sizes of 48 and 60, shorter with dsGA. The main source of this performance gain is the power of distributed computing capability that GraphX inherently provides.

4   Conclusion

In this paper we have demonstrated the application of a recent graph processing framework to solving the task scheduling problem. Specifically, we implemented a distributed GA in GraphX. The design and implementation of our dsGA can be used as a reference implementation in many other problem domains. Our experimental results show that GraphX facilitates the distributed processing/computing of GA with performance superior to that with the singleton GA in terms of the quality of schedule and execution time. In particular, it is observed that dsGA can significantly outperform the singleton GA when the search space becomes large.

Acknowledgement. Young Choon Lee acknowledges the support of the Australian Research Council (ARC) Linkage Grant LP140100980. Abhaya Nayak's work is partially supported by ARC Discovery Project: DP150104133.

References

1. Malewicz, G., Austern, M.H., Bik, A.J., Dehnert, J.C., Horn, I., Leiser, N., Czajkowski, G.: Pregel: a system for large-scale graph processing. In: Proceedings of the International Conference on Management of Data (SIGMOD 2010), pp. 135–146 (2010)
2. Topcuouglu, H., Hariri, S., Wu, M.: Performance-effective and low-complexity task scheduling for heterogeneous computing. IEEE Trans. Parallel Distrib. Syst. **13**(3), 260–274 (2012)
3. Valiant, L.G.: A bridging model for parallel computation. Commun. ACM **33**(8), 103–111 (2012)
4. Zaharia, M., Chowdhury, M., Franklin, M.J., Shenker, S., Stoica, I.: Spark: cluster computing with working sets. In: Proceedings of the 2nd USENIX Conference on Hot Topics in Cloud Computing (HotCloud 2010) (2010)

Restricted Echo State Networks

Aaron Stockdill[✉] and Kourosh Neshatian

Department of Computer Science and Software Engineering,
University of Canterbury, Christchurch, New Zealand
aas75@uclive.ac.nz

Abstract. Echo state networks are a powerful type of reservoir neural network, but the reservoir is essentially unrestricted in its original formulation. Motivated by limitations in neuromorphic hardware, we remove combinations of the four sources of memory—leaking, loops, cycles, and discrete time—to determine how these influence the suitability of the reservoir. We show that loops and cycles can replicate each other, while discrete time is a necessity. The potential limitation of energy conservation is equivalent to limiting the spectral radius.

1 Introduction

Feed-forward neural networks are limited to learning "pure" functions. This excludes an entire class of problems which are time-dependent. Extending neural networks to handle this involves adding recurrency, and thus we have *recurrent neural networks*. These networks are capable of holding state, but it comes at the cost of them being difficult to train. Approaches to managing this complexity vary from unfolding the network sufficiently many steps to create a feed-forward network so that back-propagation can be used (this is called back-propagation through time [6]), or more advanced methods such as Kalman filtering [7].

A type of recurrent neural network is the *reservoir neural network* is becoming a common approach to learning. They are simple and fast to train, and as capable as other approaches [2]. A prominent model of reservoir neural network is the *echo state network* (ESN) described by Jaeger in 2001 [3]. Jaeger defines the ESN by a reservoir of neurons connected by synapses with weights that remain unaltered, in contrast to trained artificial neural networks, and restricts all the learning to a readout layer. Section 2 provides a description of ESNs.

An important component of a reservoir neural network is the eponymous reservoir. In the original definition, there are no restrictions on how the neurons connect—neurons can connect to any other, including themselves, and may form cycles. This gives the network powerful memory abilities, to the point that the past inputs entirely define the current state, after sufficient inputs. This is the *echo property* in ESNs. Because the structure of the reservoir has a strong influence on the expressive power of the ESN, this paper explores the effect of making structural changes: removing loops (wherein a neuron can output directly back into itself), cycles (as a sequence of nodes such that the output of a node eventually becomes input to itself), and the discrete time steps, and forcing a reservoir to be "conservative", where the outgoing weights for each neuron sum to one.

© Springer International Publishing AG 2016
B.H. Kang and Q. Bai (Eds.): AI 2016, LNAI 9992, pp. 555–560, 2016.
DOI: 10.1007/978-3-319-50127-7_49

Although software allows arbitrary reservoirs, this is not the case in hardware. Thus this work aims to detail how the structure of a reservoir can impact on learning potential, and thus inform and guide neuromorphic hardware development. Work from Sillin et al. and other researchers has shown great potential for reservoir-style machine learning on novel hardware [5]. Because physical laws govern the structure of the network when it must exist in a physical system, these restrictions we impose become relevant—even when they are not a problem encountered in a purely software environment.

2   Echo State Networks

An echo state network (ESN) is a recurrent neural network first described by Jaeger in 2001 [3] that uses a reservoir of neurons that do not need training. Instead, the training happens in a readout layer. A more complete summary of ESNs is available from Lukoševičius [4], but can be fully defined by

$$
\begin{aligned}
\mathbf{y}(t) &= \mathbf{W}^{\text{out}}[1; \mathbf{u}(t); \mathbf{x}(t)] \\
\mathbf{x}(t) &= (1 - \alpha) \times \mathbf{x}(t - 1) + \alpha \times \tanh\left(\mathbf{W}^{\text{in}}[1; \mathbf{u}(t)] + \mathbf{W}\mathbf{x}(t - 1)\right).
\end{aligned}
\tag{1}
$$

The readout layer $\mathbf{W}^{\text{out}}$ is the map from the input vectors $\mathbf{u}(t)$ to the output vectors $\mathbf{y}(t)$, based on the internal state of the network $\mathbf{x}(t)$. The operator $[\cdot; \cdot]$ denotes vertical vector concatenation. The constant α is the leaking rate, referring to the mixing between current and previous output.

The ESN's readout layer is typically trained using a linear model, leading to a simpler recurrent neural network than otherwise possible with techniques such as back-propagation through time, or extended Kalman filters. The ESN also retains all the expressiveness in the recurrent neural network, performing comparably to these other training methods [2]. The linear model is for the ESN author to choose, but a common and recommended choice we use here is ridge regression [4]. Thus our learning is solving the matrix equation

$$
\mathbf{W}^{\text{out}} = \mathbf{Y}^{\text{target}} \mathbf{X}^{\top} \left(\mathbf{X}\mathbf{X}^{\top} + \beta \mathbf{I}\right)^{-1}.
\tag{2}
$$

The matrix $\mathbf{Y}$ is the sequence of input vectors arranged horizontally, while the matrix $\mathbf{X}$ is the sequence of $[1; \mathbf{u}(t); \mathbf{x}(t)]$ vectors arranged horizontally.

The reservoir used by an ESN is the source of, and restriction on, its memory capacity. Normally, the only restriction placed on an ESN is the spectral radius, a measure of scaling performed by the reservoir. If the input could potentially contain a zero vector, then the spectral radius must be less than one [4]. Sometimes, the sparsity of the reservoir is also restricted, but this is for performance reasons. By placing further restrictions, we can reduce the 'power' of the ESN. Čerňanský and Makula demonstrated this in their work exploring the feed-forward ESN [1]. By removing both cycles and loops in the reservoir, they show an ESN becomes equivalent to a feed-forward neural network with inputs representing up to n steps back in the input history, where n is the number of neurons in the reservoir.

3   Exploring Reservoir Variations

We have identified four sources of memory in an ESN: leaking, cycles, loops, and the discrete time steps. Because leaking is inherently 'outside' the reservoir, this is a simple addition to any network. Any state read from the network can be stored to mix in with the next state before being sent to the readout layer. Thus we will not consider leaking any further.

Cycles are when there is a sequence of neurons $n_1 n_2 \ldots n_k n_1 n_2 \ldots$. Loops are an edge that connects neuron n_i back to itself. The discrete time steps are when the state of a neuron at time t receives information from its neighbours from time $t - 1$. This final property works in tandem with the first two to exploit the state of the network and provide the memory so vital in its power.

By removing these features, and combinations thereof, we potentially weaken the ESN, but in doing so make it more closely resemble the hardware implementations available. Two distinctions remain between the "fully-weakened" ESN and a hardware network. First, the hardware network is updating the connection weights while the network is running. Modern neuromorphic hardware is not composed of static resistors, but often some variation of memristive hardware, which has a dynamic resistance depending on the history of voltage or current. To explore the effect this might have, we introduce a *wobbling weights* matrix, which changes its weights based on the previous input across it, much like memristors. Second, the software reservoir is able to amplify and suppress energy arbitrarily, whereas hardware must conserve electricity. To remove this, we force the output weights of each neuron to sum to one.

3.1   Loops

Loops are a source of memory for the network. Because the neuron now has explicit access to its own state at time $t - 1$, it creates a type of weighted average, effectively giving each neuron total memory of past inputs. Cycles give the same effect, but the tighter effect of the loop is more easily emulated in hardware solutions by sensors and external voltage sources.

As shown by Čerňanský and Makula, removing both cycles and loops reduces an ESN to a feed-forward network with delayed-time inputs [1]. The memory of the network is limited by the longest chain. The network was still capable of solving the typical sorts of problems such as Mackey-Glass because the memory requirement is by convention set at 17 steps, and the reservoirs are trivially made larger than this. Removing just one of loops or cycles will not cause the same reduction in expressive power for temporal datasets. By removing only loops and not cycles, there is no immediate loss of power—any memory a loop supported is replicated with a cycle, but with a k-step delay, where k is the length of the shortest cycle through a neuron. Thus learning may slow, but not stop.

Consider a simple network of two neurons connected by a directed edge in both directions. If no loops are available, it is not immediately possible to mix the input to neuron i at time t, denoted $u_i(t)$, with $u_i(t - 1)$. But we can mix $u_i(t)$ with $u_i(t - 2)$. Thus the length of the cycle through neuron i is two, so there is a two-step delay in the network. In the meantime, neuron j is mixing $u_i(t-1), u_i(t-3), \dots$. The readout layer can mix both streams, thus mixing $u_i(t)$ for all t. This scales appropriately for cycles of length k.

3.2 Cycles

Cycles provide the network with 'infinite' memory. Removing cycles is an important research question, because hardware implementations are unable to recreate cycles. Kirchhoff's Voltage Law limits the amount of energy in a circuit, and forces conservation. That is, a junction is unable to amplify a signal, and so there cannot be cycles in the network. Having cycles would imply an infinite sequence of groups where the potential difference drops forever, leading to an impossible infinitely-descending structure:

$$V_1 > V_2 > \cdots > V_k > V_1 > \cdots \implies V_1 > V_1 \nleq \tag{3}$$

If cycles were to form, energy would cycle forever and become infinite, something not possible in a physical circuit.

Now, having removed cycles, infinite mixing of inputs is not available to every neuron, but infinite mixing of input at neuron n for input to neurons $m < n$ is available because we have not yet excluded loops. By modifying input to be repeated (i.e. $\mathbf{u}(t) \mapsto [\mathbf{u}(t); \mathbf{u}(t)]$), the inputs to neurons $m < n$ are also available at neurons $o > n$. Thus having loops can be made equivalent to having cycles, although particular mixes may not be available within the same number of time steps. This is important as a device which mixes the previous voltage across a memristor with the present voltage is conceivable.

As an illustrative example, consider a network that once contained the cycle of two neurons m and n such that $m \to n \to m$. Normally we could infinitely mix $u_m(t)$ with $u_n(t-2k-1)$ and $u_m(t-2k)$ for any natural number k, and vice versa, by allowing the inputs to cycle around each other. By removing cycles, such a structure is unavailable. Instead, we can simulate it with $m \to n \to m' \to n'$ such that $u_m(t) = u_{m'}(t)$ and $u_n(t) = u_{n'}(t)$, and every neuron also loops back into itself. It is now possible to mix $u_m(t)$ with $u_n(t-k)$ and $u_m(t-k-1)$ for any natural number k, as it now occurs further back in the network, and the cycle acts as an infinite internal delay mechanism for the input. This is a stronger guarantee than necessary, but does ensure the desired effect of cycles.

3.3 Conservation of Energy

The restriction of conservation of energy is not a restriction at all. It limits a network in the same manner as the spectral radius, the spectral radius being the largest absolute eigenvalue of the weights matrix. By ensuring that a neuron's outputs sum to one, we have effectively forced each column in the weights matrix to sum to one. The eigenvalues of matrix $\mathbf{W}$ are the same for $\mathbf{W}^\top$, so we can consider the matrix $\mathbf{W}^\top$ with row sums equal to 1. For some $\mathbf{v}$, we have

$$\mathbf{W}^\top \mathbf{v} = (\mathbf{w}_1^\top \mathbf{v}, \mathbf{w}_2^\top \mathbf{v}, \dots)^\top = (\mathbf{w}_1 \cdot \mathbf{v}, \mathbf{w}_2 \cdot \mathbf{v}, \dots)^\top \tag{4}$$

Algorithm 1. Propagate the input $\mathbf{u}(t)$ over the reservoir defined by $\mathbf{W}$

```
 1: procedure PROPAGATE(W, W^in, u(t))
 2:     v ← W^in u(t)
 3:     o ← (0, 0, ..., 0)^⊤
 4:     for all n ∈ toposort(W) do
 5:         s ← v_n
 6:         for all m ∈ predecessors(n) do        ▷ finds all nodes with edges into n
 7:             s ← s + o_m W_{n,m}                ▷ W_{n,m} is the weight from m to n
 8:         end for
 9:         o_n ← tanh(s)
10:     end for
11:     return o
12: end procedure
```

Given that $\mathbf{w}_i \cdot \mathbf{v} = \|\mathbf{w}_i\| \|\mathbf{v}\| \cos\theta$, $\|\mathbf{w}_i\| \leq 1$, and $-1 \leq \cos\theta \leq 1$, the largest absolute scaling possible by $\mathbf{W}^\top$ (and thus also by $\mathbf{W}$) is 1. Hence conservation of energy is equivalent to specifying a spectral radius of at most one. This is not an issue: ESNs are only guaranteed to work for spectral radii below one [3].

3.4 Discrete Time Steps

The discrete time nature of an ESN is the fundamental feature of its memory. This is also a difficult feature to replicate in hardware. Because a circuit will have the electricity pass through at significant fractions of the speed of light, no matter how rapidly we switch the input voltage, we are essentially saturating the network with the same signal millions of times before switching. Because of this speed disparity, a hardware network will essentially not contain discrete time steps, instead it will function more like a traditional feed-forward neural network, which we will call the one-hop reservoir, where the input $\mathbf{u}(t)$ is influencing the entire network at time t, but inputs $\mathbf{u}(s)$ from times $s < t$ are not in the network. The difference is now, there is no new information written to the network before the propagation is complete. Because of this distinct termination, the network is not allowed to have cycles or loops. Algorithm 1 outlines the propagation.

Because there is now no state in the network beyond the leaking rate, the network will be unable to learn any function requiring knowledge of previous time steps. Essentially, we remove the echo property. The state now depends solely on the random initial weights, not the history of previous inputs as required by an ESN. This network is now an untrained feed-forward neural network. Hence this reservoir is incapable of learning any of the time-series problems it was designed to solve. While there are potential applications for traditional machine learning, by training an equivalent neural network in software and 'burning in' the weights to hardware, this is not a suitable use for memristors—they will update their weight, and move away from their desired weight.

This comparison is not fair, because a network of memristors *does* maintain a state, because the weights *do* get updated. The question then becomes does the memristor's state act as a suitable substitute for the ESN's discrete time steps? The answer would seem to be no. By making the ESN have a 'wobbling' weights matrix to simulate the updating conductances of the memristors and switches, we handicap the readout layer by removing the underlying assumption of regression—for a given input x, there is a function $f(x)$ that we attempt to find. Because f is a function, each x uniquely maps to some y. By changing the weights matrix, we change the function we are trying to fit, and so prevent the linear regression from successfully fitting the training data.

4 Conclusion

We have presented four important sources of memory in an ESN: leaking, loops, cycles, and discrete time. While removing both loops and cycles is known to undermine the infinite memory capability in a reservoir, removing just one leaves the reservoir sufficiently powerful for all learning, assuming some modifications to inputs. The single most important feature of a reservoir neural network is the discrete time step nature of input propagation. Should this be unavailable, as is likely in hardware implementations, the reservoir will cease to function as an effective learner. The potential for self-updating hardware such as memristors does not provide the necessary memory to overcome the lack of discrete time propagation. A primary concern in hardware implementations of reservoir neural networks is the strict requirement of energy conservation. However we have shown this to be no more severe than spectral radius scaling.

References

1. Čerňanský, M., Makula, M.: Feed-forward echo state networks. In: Proceedings. 2005 IEEE International Joint Conference on Neural Networks 2005, vol. 3, pp. 1479–1482 (2005)
2. Čerňanský, M., Tiňo, P.: Comparison of echo state networks with simple recurrent networks and variable-length Markov models on symbolic sequences. In: Sá, J.M., Alexandre, L.A., Duch, W., Mandic, D. (eds.) ICANN 2007. LNCS, vol. 4668, pp. 618–627. Springer, Heidelberg (2007). doi:10.1007/978-3-540-74690-4_63
3. Jaeger, H.: The "echo state" approach to analysing and training recurrent neural networks. GMD Report (2001)
4. Lukoševičius, M.: A practical guide to applying echo state networks. In: Montavon, G., Orr, G.B., Müller, K.-R. (eds.) Neural Networks: Tricks of the Trade. LNCS, vol. 7700, 2nd edn, pp. 659–686. Springer, Heidelberg (2012). doi:10.1007/978-3-642-35289-8_36
5. Sillin, O.: H., Aguilera, R., Shieh, H.-H., Avizienis, A.V., Aono, M., Stieg, A.Z., Gimzewski, J.K.: A theoretical and experimental study of neuromorphic atomic switch networks for reservoir computing. Nanotechnology 24(38), 384004 (2013)
6. Werbos, P.J.: Backpropagation through time: what it does and how to do it. Proc. IEEE 78(10), 1550–1560 (1990)
7. Williams, R.J.: Training recurrent networks using the extended Kalman filter. In: International Joint Conference on Neural Networks 1992. IJCNN, vol. 4, pp. 241–246 (1992)

Feature-Aware Factorised Collaborative Filtering

Farhad Zafari[✉] and Irene Moser

Faculty of Science, Engineering and Technology, Swinburne University of Technology,
Melbourne, VIC 3122, Australia
{fzafari,imoser}@swin.edu.au

Abstract. In the area of electronic commerce, recommender systems have become more and more popular. The quality of recommendations depends on the quality of the preference model extracted by the recommender system. Recently, latent factor models based on probabilistic matrix factorisation have gained great attention in both industry and academia, owing to their superior accuracy over traditional recommender systems. Although latent factor models are very efficient, the latency of the features captured in these models impedes explaining the learnt model to the users. A lack of understanding of the latent features makes it difficult to decide on the optimal number of features to give as input to these models. Therefore, the model accuracy degrades when less relevant features are introduced into the model. To tackle this problem, in this paper we propose an extension to the basic matrix factorisation, so that the model takes into account the relevancy of the features beside their values. We test the accuracy of the proposed method on two benchmark datasets. The experiments show that the proposed method makes remarkable improvements over the basic method and some of the state of the art latent factor models.

Keywords: Recommender systems · Latent factor models · Probabilistic Matrix Factorisation · Feature-Aware Probabilistic Matrix Factorisation

1 Introduction

Recommender systems suggest items (movies, books, music, news, services, etc.) that appear most likely to interest a particular user. Matching users with the most appropriate items is key to enhancing user satisfaction and loyalty. Therefore, recommender systems have become the centre of attention for retailers, and many famous e-commerce leaders such as Amazon and Netflix have made recommender systems a salient part of their websites [1].

Typically, recommender systems are based on collaborative filtering (CF), in which the preferences of a user are predicted by collecting rating information from other similar users or items [2]. Among CF systems, Probablistic Matrix Factorisation (PMF) method is one of the most popular and widely researched and commonly employed latent factor[1] models [1]. In fact, latent factor models

---

[1] Throughout this paper, the terms *factor* and *feature* are used interchangeably.

© Springer International Publishing AG 2016
B.H. Kang and Q. Bai (Eds.): AI 2016, LNAI 9992, pp. 561–569, 2016.
DOI: 10.1007/978-3-319-50127-7_50

attempt to explain the ratings by characterising both items and users according to a number of latent features that are inferred from the ratings patterns. Latent factor models based on Probabilistic Matrix Factorisation have received much attention from both academia and industry because of their good prediction accuracy and therefore many recent studies have contributed extensions to the basic PMF by incorporating additional information. Despite their popularity and good accuracy, recommender systems based on latent factor models encounter some important problems in practical applications. One of the problems with latent factor models is how to distinguish between relevant and irrelevant features. In the current methods based on latent factor models, the number of features that capture the user preferences is assumed to be known in advance as one of parameters of the model. For example, it is assumed that the user's preferences over different movies can be explained by, say, 100 features, which are used to train a model. Since there are a limited number of features that explain most of the user's preferences, assuming too many item features leads to overfitting the model and degrades the accuracy. The degrading effect of training latent factor models with too many latent features has been emphasised by some researchers in the literature. For example, in order to find the features deemed most relevant by the users, Zhang et al. [3] conducted analytical experiments and showed that around 15 features would be sufficient to capture the user's preferences in two datasets. Therefore, the major research question that we are interested in this paper is: How many item features define the user preferences, and in particular, how can the system avoid the degrading effect of irrelevant features? The current work proposes a feature-aware extension of PMF to address this issue. The rest of the paper is organised as follows: The related work is introduced in Sect. 2. In Sect. 3, we first briefly introduce probabilistic matrix factorisation (Sect. 3.1), and then in Sect. 3.2 we introduce the proposed model, Feature-Aware PMF (FAPMF). In Sect. 4, we report on the experimental results of the proposed model, and finally we conclude the paper in Sect. 5, by summarising the main findings and giving the future directions of this work.

2   Related Work

Recommender systems can be broadly classified into content-based, and collaborative filtering (CF) systems. Content-based filtering, also referred to as cognitive filtering, originates in information retrieval and information filtering, and recommends items based on a comparison between the content of the items and a user profile. In these methods, the profile of a single user is used to predict the preferences of that user. The content of each item is represented as a set of terms or keywords, i.e. the words that occur in the item description. These methods analyse the content of the items that the user has liked before, build a user profile, and recommend items that have similar descriptions to the user profile. For example, if the user profile shows that they like a movie by a particular director, the movies directed by the same director are assumed to be of interest for the user as well. Content-based methods are more suitable for items that can be described in natural language.

CF approaches on the other hand predict the preferences of a user by collecting preference information from many users. These methods can be broadly classified into memory-based and model-based approaches. Memory- or instance-based learning methods predict the user preferences based on the preferences of other users or the similarity of the items. Item-based approaches in memory-based CF calculate the similarity between the items, and recommend the items similar to the items that the user has liked in the past. User-based approaches recommend items that have been liked by similar users [2].

Model-based CF learns the parameters of a model and only store those parameters. Algorithms in the category of model-based CF include clustering model, aspect models and latent factor models [2,4]. Latent factor models as an example of model-based collaborating filtering try to explain the ratings by characterising both users and items on a number of latent factors which are inferred from the rating patterns. Recently, latent factor models based on matrix factorisation have gained much popularity as they usually outperform traditional memory-based methods, and have achieved higher performance in some benchmark datasets [1]. Several matrix factorisation methods have been proposed in different problem settings, such as singular value decomposition [5], non-negative matrix factorisation [6], and probabilistic matrix factorisation [3,7]. Probabilistic matrix factorisation was originally proposed by Salakhutdinov and Mnih [7] and has become the foundation for a number of latent factor methods which incorporate additional information into the preference modelling. This paper shows that using the same amount of information as basic matrix factorisation, higher accuracies can be achieved by identifying and excluding irrelevant features (feature-awareness). To the best of our knowledge, the current work is the first attempt at incorporating feature awareness in the latent factor models.

3   Proposed Model

Since the proposed model expands on the popular probabilistic matrix factorisation [7], before introducing the proposed model, we first introduce probabilistic matrix factorisation.

3.1   Probabilistic Matrix Factorisation

In rating-based recommender systems, the observed ratings are represented by the rating matrix R, in which the element R_{ij} is the rating given by the user i

to the item j. Usually, R_{ij} is a 5-point integer, 1 point means very bad, and 5 points means excellent. Let $U \in \mathbb{R}^{N \times D}$ and $V \in \mathbb{R}^{M \times D}$ be latent user and item feature matrices, with vectors U_i and V_j representing user-specific and item-specific latent feature vectors respectively (N is the number of users, M is the number of items, and D is the number of item features). In probabilistic matrix factorisation, the log-posterior over the user and item latent feature matrices with rating matrix and fixed parameters is minimised. In other words:

$$argmin_{U,V} [\ln p(U, V | R, \sigma, \sigma_U, \sigma_V) = \ln p(R|U, V, \sigma) + \ln p(U|\sigma_U) + \ln p(V|\sigma_V) + C] \quad (1)$$

where C is a constant that is not dependent on U and V. σ_U, σ_V, and σ are standard deviations of matrix entries in U, V, and R respectively. Minimising the log-posterior probability in Eq. 1 is equivalent to minimising the error value in Eq. 2.

$$argmin_{U,V} [E = \frac{1}{2} \sum_{i=1}^{N} \sum_{j=1}^{M} I_{i,j} (R_{i,j} - \hat{R}_{i,j})^2 + \frac{\lambda_U}{2} \sum_{i=1}^{N} \|U_i\|_{Frob}^2 + \frac{\lambda_V}{2} \sum_{j=1}^{M} \|V_j\|_{Frob}^2] \quad (2)$$

where $\|.\|_{Frob}$ denotes the Frobenius norm, and $\lambda_U = \frac{\sigma^2}{\sigma_U^2}$ and $\lambda_V = \frac{\sigma^2}{\sigma_V^2}$ and $\hat{R}_{i,j} = U_i V_j^T$. $I_{i,j}$ is the indicator function that is equal to 1 if user i has rated item j and 0 otherwise. *Stochastic Gradient Descent* and *Alternating Least Squares* are usually employed to solve the optimisation problem in Eq. 2.

3.2   Feature-Aware Probabilistic Matrix Factorisation (FAPMF)

To address the problem of irrelevant features in matrix factorisation, we extend the basic matrix factorisation by incrementally adding features in the gradient descent process. The proposed method is abbreviated to FAPMF and is explained in the Algorithm 1. In Algorithm 1, first we get the learning ratios for each of the features (line 3). As studies have shown (e.g. [3,8]), only a limited number of factors are responsible for most of the observed user rating patterns. Therefore, we believe that learning the features incrementally with more learning iterations dedicated to first features would better capture the user rating patterns. With this approach, first we let the model learn the most important feature by iterating over the user ratings and estimating the matrices U and V with only one factor. In the next iteration, we add one more feature to the model (line 6) and go over another learning iteration (line 9). The ratio of the number of iterations when each feature is added, is determined by a *Lamé oval*. In a Lamé oval, the values of x and y are calculated according to the parametric function with $x = \sin(t)^{\frac{2}{r}}$ and $x = \cos(t)^{\frac{2}{r}}$.

```
Algorithm 1 FAPMF
 1: double TrainModel(Matrix U, Matrix V, double r, int maxIter, double λU, double λV,
    double γ)
 2: {
 3:   list< int> factorRatios ← GetFactorRatios(D, r);
 4:   int f ← 1;
 5:   totalNumFactors ← D;
 6:   for f ≺ totalNumFactors do
 7:     D ← f;
 8:     int l ← 1;
 9:     for l ≺ factorRatios.get( f) × totalNumFactors × maxIter do
10:       int loss ← 0;
11:       int lastloss ← 0;
12:       int i ← 1;
13:       for i ≺ N do
14:         int j ← 1;
15:         for j ≺ M do
16:           e_{i,j} ← R_{i,j} − U_i V_j^T;
17:           loss ← e_{i,j}^2;
18:           U_{i,f} ← U_{i,f} + γ·(e_{i,j}·V_{j,f} − λU·U_{i,f});
19:           V_{j,f} ← V_{j,f} + γ·(e_{i,j}·U_{i,f} − λV·V_{j,f});
20:           loss ← loss + λU × U_{i,f}^2 + λV × V_{j,f}^2;
21:           j ← j + 1;
22:         end for
23:         i ← i + 1;
24:       end for
25:       loss ← loss × 0.5;
26:       if |lastloss| > |loss| then
27:         γ ← γ × 1.005;
28:       else
29:         γ ← γ × 0.5;
30:       end if
31:       l ← l + 1;
32:       lastloss ← loss;
33:     end for
34:     f ← f + 1;
35:   end for
36: }
37: List< int> GetFactorRatios(int numFactors, double r)
38: {
39:   List< double> FactorRatios;
40:   int f ← 1;
41:   double sum ← 0;
42:   for f ≺ numFactors do
43:     double x ← \frac{r}{numFactors};
44:     FactorRatios.set( f, (1 − x^f)\frac{1}{f});
45:     sum ← sum + FactorRatios.get( f);
46:     f ← f + 1;
47:   end for
48:   f ← 1;
49:   for f ≺ numFactors do
50:     FactorRatios.set( f, \frac{FactorRatios.get( f)}{sum});
51:     f ← f + 1;
52:   end for
53:   return FactorRatios;
54: }
```

This model is a heuristic version of stochastic gradient descent search that is employed in probabilistic matrix factorisation. An optimal solution that uses all D factors is also an optimal solution for an individual factor (as well as a subset of D factors), provided that individual factor (or that subset of D factors) yields the same predictions as all D factors over the observed ratings. This means that using a heuristic approach, the factors can be learnt individually and incrementally. In Algorithm 1, we start the search process by only considering the group of parameters that belong to the first factor (line 9). By doing this, in fact we find the optimal parameter values ($U_{i,1}^*$ and $V_{j,1}^*$) for the first factor ($U_{i,1}$ and $V_{j,1}$). Then we move to the second factor and find the optimal solutions for this factor as well ($U_{i,2}^*$ and $V_{j,2}^*$), and repeat this process for all D factors. At the end of this process, we obtain U_i^* and V_j^* which, if it results in the same predicted ratings, is also an optimal solution for the original optimisation problem with D factors and $N \times M \times D$ parameters. To reduce the computational complexity, in this algorithm we ignore the strict limitation that the predictions of individual factors should exactly conform to the predictions of all D factors. The proof is omitted in this paper for the sake of brevity.

4   Experiments

In order to evaluate the effectiveness of the proposed method, we conducted a series of experiments on a number of benchmark recommendation datasets. To implement the proposed method, we use LibRec [9] which is an open source java library for recommender systems. Experiments were performed on the Movie-lens, and Filmtrust datasets. GroupLens Research have collected and the rating datasets from the Movielens web site. This dataset consists of 100000 ratings from 943 users on a total of 1,682 movies. The rating scale is [1,5], 5 for the most preferred movie, and 1 for the least preferred one, and each user in the dataset has rated at least 20 movies, and the density of the rating matrix is 6.30%. The Filmtrust dataset is a small dataset crawled from the Filmtrust website in June

2011. The dataset includes 35,497 ratings given by 1,508 users on 2,071 movies. The ratings are real values between 0.5 and 4, with lower values given for less favourable movies. The rating matrix is extremely sparse and only 1.14% of the ratings in the user rating matrix are known. In order to show the effectiveness of the proposed methods, we compared the results against the recommendation quality of three popular methods for discovering latent factor models, *Probabilistic Matrix Factorisation (PMF)* [7], *Bayesian Probabilisitic Matrix Factorisation (BPMF)*, [10] *Non-Negative Matrix Factorisation (NMF)* [6] In this paper, 80% of the ratings are randomly chosen for training and the remaining 20% are used for validation. We use 0.001 for both user and item regularisation parameters (λ_U and λ_V) and 0.01 for the learning rate (γ). We also set 200 as maximum number of learning iterations ($maxIter$ in Algorithm 1). Through experiments, we also noticed that using $\frac{1}{9}$ for r yields better accuracy for our model, so we used this value throughout our experiments. Each model training and test is repeated for 5 times to eliminate the randomness in the results and therefore to assure that the results are more reliable. Two standard and popular measures are used to measure and compare the performance of the models: Mean Absolute Error (MAE) and Root Means Square Error (RMSE).

4.1   Results

The evaluation results for the aforementioned latent factor models as well as the proposed model on the Filmtrust and Movielens datasets are shown in Tables 1a and b respectively.

From Table 1, we can see that our proposed model outperforms all the three latent factor models in almost all settings. The percentages on the right side of each table show the extent of improvement that the proposed model makes over the corresponding latent factor model. The proposed method yields extensive improvements over the basic probabilistic matrix factorisation in both datasets. We can also see that the accuracy of the proposed model in both datasets consistently decreases as the number of factors increases.

Table 1. Performance comparison of FAPMF in (a) Filmtrust and (b) Movielens datasets

(a)

Model	D	MAE Avg	MAE Stdev	RMSE Avg	RMSE Stdev	Impr. MAE	Impr. RMSE
FAPMF	50	0.6864	0.0073	0.9295	0.0120		
	100	0.6851	0.0023	0.9264	0.0026		
	150	0.6828	0.0109	0.9281	0.0115		
	200	0.6755	0.0064	0.9209	0.0090		
PMF	50	0.8522	0.0051	1.1331	0.0066	19.46	17.97
	100	0.8309	0.0039	1.1165	0.009	17.55	17.02
	150	0.854	0.0101	1.1479	0.0118	20.05	19.15
	200	0.8747	0.014	1.1779	0.0192	22.77	21.82
BPMF	50	0.7624	0.0112	1.0100	0.0120	9.97	7.97
	100	0.8019	0.0106	1.0666	0.0145	14.57	15.06
	150	0.8208	0.0101	1.0927	0.0134	16.81	15.06
	200	0.8481	0.0114	1.1327	0.0173	20.35	18.70
NMF	50	0.7407	0.0063	0.9905	0.0078	7.33	6.16
	100	0.7079	0.0038	0.9478	0.0051	3.22	2.26
	150	0.6916	0.0067	0.9295	0.009	1.27	0.15
	200	0.677	0.0092	0.9114	0.0095	0.22	-1.042
Average Improvement						12.80	11.53

(b)

Model	D	MAE Avg	MAE Stdev	RMSE Avg	RMSE Stdev	Impr. MAE	Impr. RMSE
FAPMF	50	0.7591	0.0023	0.9620	0.0022		
	100	0.7548	0.0047	0.9584	0.0054		
	150	0.7457	0.0043	0.9542	0.0047		
	200	0.7437	0.0035	0.9530	0.0047		
PMF	50	1.0106	0.0074	1.3062	0.0082	24.89	26.35
	100	0.8765	0.0029	1.1193	0.0029	13.89	14.38
	150	0.8190	0.0051	1.0424	0.0056	8.95	8.46
	200	0.8006	0.0043	1.0167	0.0063	7.11	6.27
BPMF	50	0.8046	0.0066	1.0375	0.0079	5.65	7.28
	100	0.8182	0.0067	1.0544	0.0083	7.75	9.11
	150	0.8254	0.0065	1.0611	0.0088	9.66	10.07
	200	0.8381	0.0072	1.0773	0.0095	11.27	11.54
NMF	50	0.8742	0.0048	1.1332	0.0049	13.17	15.1
	100	0.8694	0.0056	1.1250	0.0093	13.18	14.81
	150	0.8550	0.0051	1.1025	0.0054	12.79	13.45
	200	0.8392	0.0074	1.0807	0.0080	11.38	11.82
Average Improvement						11.64	12.39

Similarly, NMF also shows improvements in accuracy when the number of factors is increased. However, BPMF results in weaker accuracies when the number of factors is increased. This is probably because adding more features further complicates the solution space, by adding irrelevant features with less or even negative contribution to the accuracy. Therefore, the model fails to find the optimal solutions. In the proposed model, this problem is alleviated by learning the factors in an incremental way. In this model, first the model is trained with one factor for a specific number of iterations. Then a second factor is added and the training is done for less number of iterations. On the Filmtrust dataset, on average the proposed FAPMF method improves the MAE accuracy by 19.96%, 15.42%, and 3.01% relative to PMF, BPMF, and NMF respectively. The average improvement of FAPMF for RMSE measure is 19%, 13.72%, and 1.89% with respect to PMF, BPMF, and NMF respectively. In the case of the Movielens dataset, FAPMF improves 13.71%, 12.63%, and 8.58% over PMF, BPMF, and NMF respectively. The improvements for RMSE measure in this dataset are 13.87%, 13.8%, and 9.5%, respectively.

4.2   Effect of Parameter r in FAPMF

As explained in Sect. 3.2, FAPMF alleviates the negative effect of irrelevant features by training the model in a feature-aware manner. In this method, the first features are given more importance than the latter features, by doing more learning iterations over them. The parameter r in Algorithm 1 controls the number of iterations over each feature. If $r < 1$, the algorithm performs more iterations over the first features and few iterations over the later features. On the contrary, if $r > 1$, the training iterations would be equally distributed over all the features.

To demonstrate the effect of the parameter r on FAPMF's performance, we depict the change in RMSE with respect to learning iterations for each separate factor with two different values for r on the Filmtrust and Movielens datasets. We assume $r = 1$ and $r = 1000$, and we set the number of factors to 12. These two values for r result in different numbers of iterations for each factor. However, to be able to show both cases in one diagram, we set all iteration numbers to 400 and keep the accuracy fixed. To ensure that the results are not subject to randomness, each experiment is repeated 10 times, and the average errors of the iterations are used.

In Figs. 1a and b, the diagram on the top-left corner of each figure belongs to the first factor, and the diagram on the bottom-right corner of each figure belongs to the 12th factor. We can observe that when $r = 1000$, 200 iterations are used for all factors. However, when $r = 1$, the first factors receive more iterations. The deteriorating effect of later factors can be clearly seen in both datasets. We observe that in almost all factors, for a specific number of iterations the accuracy keeps improving before it deteriorates. Specifically we observe how with $r = 1$ the model avoids the degrading effect of irrelevant factors by limiting the number of learning iterations over that factor. Consequently, a lower value of parameter r in both datasets always results in better accuracies than a higher r value. We observe that as more factors are added, the gap in accuracy between

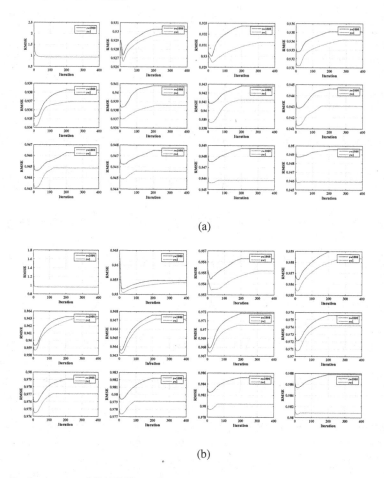

(a)

(b)

Fig. 1. Performance of FAPMF with respect to parameter r for each factor in (a) Filmtrust dataset and (b) Movielens dataset

the models with $r = 1$ and $r = 1000$ becomes larger. We also clearly see that the first factors reduce the error to a larger extent than the later factors. This actually means that the first factors leant by this model capture more of the rating patterns than the later factors and therefore they are more relevant.

5   Conclusion and Future Work

In this paper, we addressed the negative effect irrelevant features have in latent factor models, which has been observed by researchers. In order to tackle this problem, we proposed a feature-aware latent factor model based on probabilistic matrix factorisation. In the proposed method, the user preferences are learnt by incrementally iterating over each feature, and minimising the prediction error by using stochastic gradient descent. We used the Hein function to control the ratio of learning iterations over individual features. Using the Filmtrust and Movielens datasets with different numbers of features, the experiments showed that the proposed method considerably improves on basic PMF. The experiments also showed that the proposed method outperforms two other popular latent factor models based on Matrix Factorisation (BPMF and NMF).

Incorporating feature-awareness into the latent factor models has a great potential in improving the recommendation accuracy and explainability. In the future, we plan to develop more efficient and accurate feature-aware latent factor models. Furthermore, we are also interested in designing more efficient mathematical functions to estimate the training iterations (feature relevance) in FAPMF. We also plan to present a formal proof for the observation that an optimal solution with D factors includes optimal solutions for each individual factor (as well as any subset of D factors), provided that they result in the same predictions over observed ratings. This means that the solution space can be more easily explored by considering the factors individually and separately.

References

1. Koren, Y., Bell, R., Volinsky, C.: Matrix factorization techniques for recommender systems. Computer **42**(8), 30–37 (2009)
2. Ma, H., Yang, H., Lyu, M.R., King, I.: Sorec: social recommendation using probabilistic matrix factorization. In: Proceedings of the 17th ACM Conference on Information and Knowledge Management, pp. 931–940. ACM (2008)
3. Zhang, Y., Lai, G., Zhang, M., Zhang, Y., Liu, Y., Ma, S.: Explicit factor models for explainable recommendation based on phrase-level sentiment analysis. In: Proceedings of the 37th International ACM SIGIR Conference on Research & Development in Information Retrieval, pp. 83–92. ACM (2014)
4. Aghdam, M.H., Analoui, M., Kabiri, P.: A novel non-negative matrix factorization method for recommender systems. Appl. Math. Inf. Sci. **9**(5), 2721 (2015)
5. Sarwar, B., Karypis, G., Konstan, J., Riedl, J.: Application of dimensionality reduction in recommender system-a case study. Technical report, DTIC Document (2000)
6. Lee, D.D., Sebastian Seung, H.: Algorithms for non-negative matrix factorization. In: Proceedings Advances in Neural Information Processing Systems, pp. 556–562 (2001)
7. Salakhutdinov, R., Mnih, A.: Probabilistic matrix factorization. In: NIPS, vol. 20, pp. 1–8 (2011)
8. Jiang, M., Cui, P., Liu, R., Yang, Q., Wang, F., Zhu, W., Yang, S.: Social contextual recommendation. In: Proceedings of the 21st ACM International Conference on Information and Knowledge Management, pp. 45–54. ACM (2012)
9. Guo, G., Zhang, J., Sun, Z., Yorke-Smith, N.: Librec: a java library for recommender systems. In: Posters, Demos, Late-breaking Results and Workshop Proceedings of the 23rd International Conference on User Modeling, Adaptation and Personalization (2015)
10. Salakhutdinov, R., Mnih, A.: Bayesian probabilistic matrix factorization using Markov chain Monte Carlo. In: Proceedings of the 25th International Conference on Machine Learning, pp. 880–887. ACM (2008)

Social Intelligence

Mining Context Specific Inter-personalised Trust for Recommendation Generation in Preference Networks

Quan Bai, Weihua Li[✉], and Jing Jiang

Auckland University of Technology, Auckland, New Zealand
{quan.bai,weihua.li,jing.jiang}@aut.ac.nz

Abstract. This paper introduces a community-based approach to facilitate the generation of high-quality recommendations by leveraging the preferences of communities of similar users in preference networks. The proposed approach combines the idea of traditional recommendation systems and identification of network structures to explore context specific inter-personalised trust relationships among users. From the experimental results, we claim that the proposed approach can provide more accurate recommendations to individuals in a preference network.

Keywords: Community detection · Preference network · Recommender system

1 Introduction

In general, preference network represents the phenomenon that users have their own preferred items in a specific context, such as a social network; basically, it contains two types of elements, i.e., users and items [1]. Similar preference contributes to the trust relationships among the individuals, where trust refers to the level of belief established between two entities by considering past interactions in a certain context [2]. Whereas, the preference belief is normally subjective and conjuncted with certain context. Furthermore, trust can only be understood via observations and analysis as imperfect knowledge. Hence, it is difficult to explore certain objective behaviours of examined elements [2].

Recommender systems have emerged as an effective solution to the information overload problem [3]. In order to predict new likes or dislikes in preference networks, an automated recommendation approach is required for providing tailored and personalised information. However, traditional approaches, such as user-based collaborative filtering [4], item-based recommendation algorithms [5], only assume single and homogeneous trust relationships among the users, and evaluate item similarity from a simplistic world view. On the other side, community detection is widely applied to improve the accuracy for recommender systems, whereas, the available feedback ratings are ignored by many researchers [6,7]. Actually, this type of user-generated content is critical for perceiving users' preferences in a particular context.

© Springer International Publishing AG 2016
B.H. Kang and Q. Bai (Eds.): AI 2016, LNAI 9992, pp. 573–584, 2016.
DOI: 10.1007/978-3-319-50127-7_51

In this paper, we propose a community-based recommendation approach for preference networks, which is capable of covering the aforementioned research gaps. The proposed algorithms explore context specific inter-personalised trust by modelling massive transactional data and analysing network structures. Specifically, the context specific inter-personalised trust indicates multiple and heterogeneous trust relationships among individuals in terms of different contextual situations. In other words, a particular user may place trust to different individuals in terms of their multi-faceted interests. The approach is motivated by the intuition that, according to the rating history of users, a group of users share the similar feedback records for same items, as they have similar preference and criteria for items. It leverages the features of traditional recommendation systems and network structures to explore context specific inter-personalised trust relationships.

The rest of this paper is organised as follows. In Sect. 2, trust estimation protocol and formal definitions are given. In Sect. 3, the hierarchical community structures are elaborated and community-based recommendation algorithms are presented. In Sect. 4, experimental results are given to demonstrate the performance of the proposed model by comparing with some traditional recommendation systems. Finally, the conclusion is presented in Sect. 5.

2   Community-Based Trust Estimation Protocol

In this section, the trust estimation protocol is introduced, and the fundamental concepts are elaborated by giving formal definitions.

2.1   Trust Estimation Protocol

The protocol for community-based trust estimation approach is illustrated in Fig. 1. There are six modules in the protocol, i.e., *the Reply Module, the Interaction Record Database, the User Criteria Clustering Module, the Facet Object Set Generation Module, the Prediction Retrieval Module and the Trust Calculation Module*. In this section, we will introduce the overall process in general.

Reply Module tends to collect the user-item ratings and store the **Interaction Records** IR into Interaction Record Database. The objective of User Criteria Clustering Module is to cluster users into hierarchical communities according to the user-generated ratings. Similarly, Facet Object Set Generation Module aims to create object communities based on the hierarchical user criteria clustering tree generated from the User Criteria Clustering Module. Prediction Retrieval Module handles the **Item Enquires** IE about a particular item $IE.item_j$ that the user does not have previous interactions with, by searching all the related facet object sets. Next, Facet Object Set Generation Module transfers the facet object sets to the Trust Calculation Module, whose objective is to produce a quality prediction for $IE.item_j$ based on the preference of enquiring user $IE.u_i$.

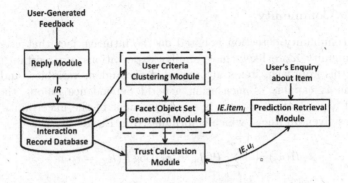

Fig. 1. Community-based trust estimation protocol

2.2 Formal Definition

In the current context, a preference network is comprised of an item set, i.e., $I = \{item_1, item_2, item_3, ..., item_n\}$, and a user set, i.e., $U = \{u_1, u_2, u_3, ..., u_m\}$. Each user may rate many different items, and every item can be rated by many users. Given a preference network having m users and n items, a $m \times n$ matrix R can be used to represent user-item ratings. Each entry $r_{m,n}$ in R denotes the feedback rating of item $item_n$ given by user u_m. $r_{m,n} = 0$, if u_m does not have any previous interactive experience with $item_n$.

Definition 1: Object Set O in a preference network is a set of objects. A particular object is represented as a two-tuple $o_{item_n}^{\tau} = <item_i, \tau_x>$, where $item_i \in I$, and τ_x denotes the rating value for $item_i$.

Once a pair of users, e.g., u_j and u_k, give a same rating τ_x to item $item_n$, the object $o_{item_n}^{\tau_x}$ is connected to both u_j and u_k. Thus, a preference network presents a bipartite pattern, consisting of two exclusive types of vertices representing users and the corresponding objects.

Definition 2: A Preference Network is a bipartite graph [8] represented as a three-tuple, i.e., $CG = <U, O, E>$, where U refers to the user set, O denotes the object set, and E indicates the edge set representing user-object interaction in CG, where $E = \{(u_j, o_{item_k}^{\tau_x})|u_j \in U, o_{item_k}^{\tau_x} \in O\}$.

Definition 3: Interaction Record IR refers to the interactive feedback to a specific item $item_j$ given by u_i, which is represented as a three-tuple, $IR = <u_i, item_j, o_{item_j}^{\tau_x}>$.

If u_i inquires the potential quality of $item_j$, and u_i lacks of interaction experience with $item_j$, the system assembles an **Item Enquiry** $IE = <u_i, item_j>$, which indicates user $IE.u_i$ enquires item $IE.item_j$.

3 Hierarchical Community Structure

In this section, a four-step trust mining algorithm is proposed to partition different types of elements into community structures.

3.1   User Community

The user community detection is based on the intuitive fact that users in the same community more likely have similar expectations of a certain group of items. In this approach, items are regarded as random variables, and mutual information is capable of measuring general dependence among them. The entropy of a user rating pattern is a measurement of the uncertainty in feedback values given on items, which is formulated in Eq. 1:

$$H(u_j) = -\sum_{i=1}^{n} P(R_{u_j} = r_{j,i}) \log P(R_{u_j} = r_{j,i}), \tag{1}$$

where n is the number of possible items which u_m rates. Higher entropy of users for item variables implies that their selection and rating pattern levels are more randomly distributed [9]. Mutual information describes the amount of common feedback ratings given by both users. Thus, the mutual information between user u_j and u_k is defined in Eq. 2:

$$I(u_j, u_k) = H(u_j) + H(u_k) - H(u_j, u_k). \tag{2}$$

The smaller $I(u_j, u_k)$, the greater difference between pair of user selection and rating patterns. However, mutual information is not bounded, and it would not be a suitable distance measurement for itself. Therefore, we transform the mutual information into a bounded mutual-information-based distance by normalizing it (See Eq. 3).

$$D(u_j, u_k) = 1 - \frac{I(u_j, u_k)}{\max(H(u_j), H(u_k))}. \tag{3}$$

In Eq. 3, $D(u_j, u_k)$ denotes the preference similarity between a pair of users. $D(u_j, u_k) = 0$, if identical users have the maximum possible selection and rating patterns, as well as the identical entropies, i.e., $H(u_j) = H(u_k) = I(u_j, u_k)$ [10]. Hence, given a user set with m users, an $m \times m$ mutual-information-based distance matrix can be calculated by using Eq. 3.

The user criteria clustering analysis algorithm is shown in Algorithm 1. In this algorithm, the inputs include user set U and user-item rating matrix R. While, the output is T, which denotes the hierarchical user criteria tree. Furthermore, $c_i.rating$ denotes the rating matrix for each cluster, and M_{i*j} denotes the entry of the mutual-information-based distance proximity matrix. $M_{|U|*|U|}$ is symmetric and the diagonal is zero.

In Algorithm 1, Line 1 initialises the leaf nodes of the user criteria cluster tree T by assigning each user into a cluster. Lines 2–8 aim to compute mutual-information-based distances among clusters. In Line 9, the closest pair of clusters are merged as a new cluster c_{temp}, while, a new cluster user set $c_{temp}.U$. c_{temp} and $c_{temp}.U$ are assigned as the latest internal node $T.Node_{K_T}$ and $T.Node_{K_T}.U$, respectively $T.Node_{c_i.ID}$ and $T.Node_{c_j.ID}$ are assigned as the left/right child node of the $T.Node_{K_T}$. Meanwhile, $T.Node_{K_T}$ becomes the parent node for these two nodes.

Algorithm 1. The User Criteria Clustering Analysis

Input: U, R
Output: T
1: $K_C = K_T = |U|$, $c_i.ID = i$, $c_i \leftarrow u_i$, $c_i.rating \leftarrow R_{u_i}$, $T.Node_i \leftarrow c_i$, $T.Node_i.left = T.Node_i.right = Null$
2: **while** $K_C > 1$ **do**
3:   **for** $\forall c_i \in C$ **do**
4:     **for** $\forall c_j \in C \wedge i \neq j$ **do**
5:       $M_{i*j} = D(c_i.rating, c_j.rating)$
6:       $(c_i, c_j) \leftarrow argmin(D(c_i.rating, c_j.rating))$
7:     **end for**
8:   **end for**
9:   $K_C = K_C - 1$, $K_T = K_T + 1$, $c_{temp} \leftarrow merge(c_i, c_j)$, $c_{temp}.U \leftarrow merge(c_i.U, c_j.U)$, $T.Node_{K_T} \leftarrow c_{temp}$, $T.Node_{K_T}.U \leftarrow c_{temp}.U$, $T.Node_{K_T}.left \leftarrow c_i$, $T.Node_{K_T}.right \leftarrow c_j$, $T.Node_{c_i.ID}.parent = T.Node_{c_j.ID}.parent = T.Node_{K_T}$, $c_i \leftarrow c_{temp}$, $C.remove(c_j)$, $c_i.ID = K_T$
10: **end while**
11: **return** T

3.2   Object Community

Recall that, in our approach, each item with a particular feedback rating is regarded as **an object**. Mathematically, an **Object Community** OC is a sub-graph of a preference network, which can be defined as a three-tuple, i.e., $OC =< U, O, E >$, where $OC.U \leftarrow T.Node.U$, $OC.O \leftarrow T.Node.parent.O$ and $OC.E = \{(u_i, o_{item_j}^{\tau_k}) | u_i \in OC.U, o_{item_j}^{\tau_k} \in OC.O\}$. The edge between user u_i and object $o_{item_j}^{\tau_k}$ is represented as $e_{(u_i, o_j^{\tau_k})}$. The weight of edge $w_{o_{item_j}^{\tau_k}}$ is formulated using Eq. 4, where $\deg(o_{item_j}^{\tau_k})$ denotes the degree of the corresponding object vertex.

$$w_{o_{item_j}^{\tau_k}} = \frac{1}{\deg(o_{item_j}^{\tau_k})} \tag{4}$$

The object communities are formed by leveraging network modularity approach [11]. Traditionally, modularity method starts off with each vertex representing a community which contains only one member, and then it calculates the changes of modularity to choose the largest of them [12]. However, in traditional method, the order of objects dramatically affects the computation time and efficiency [13]. In order to alleviate this issue, in our approach, we calculate the distance value $dv(o_{item_j}^{\tau_k})$ for each object $o_{item_j}^{\tau_k}$ belonging to the object community of its parent node $T.Node.parent.O$ by using Eqs. 5 and 6.

$$dv(o_{item_j}^{\tau_k}) = (\overbrace{\sum_{u_i \in T.Node.parent.U} e_{(u_i, o_{item_j}^{\tau_k})}}^{T.Node.parent} - \overbrace{\sum_{u_i \in T.Node.U} e_{(u_i, o_{item_j}^{\tau_k})}}^{T.Node}) \times w_{o_{item_j}^{\tau_k}} \tag{5}$$

$$e_{(u_i, o_{item_j}^{\tau_k})} = \begin{cases} 1 & \text{if the rating value which user } u_i \text{ gives item } item_j \text{ equals to } \tau_k \\ 0 & \text{otherwise} \end{cases} \tag{6}$$

Based on the decreasing order of the $dv(o_{item_j}^{\tau_k})$, the **Modularity Gain** $\Delta Q_{o_{item_j}^{\tau_k}}$ is formulated in Eq. 7, where the notations are explained as follows:

- m: The sum of the weights of all the edges in CG
- $m_{CG}(o_{item_j}^{T_x})$: The weight sum of the edge set $\{e_{(u_i,o_{item_j}^{T_k})}|o_{item_j}^{T_k} \in CG.O \setminus (u_i, o_{item_j}^{T_x}) \wedge u_i \in CG.U\}$
- $m_{OC}(o_{item_j}^{T_x})$: The weight sum of the edge set $\{e_{(u_i,o_{item_j}^{T_k})}|o_{item_j}^{T_k} \in CG.O \setminus (u_i, o_{item_j}^{T_x}) \wedge u_i \in OC.U\}$
- $l_{CG}(o_{item_j}^{T_k})$: The weight sum of the edge set $\{e_{(u_i,o_{item_j}^{T_k})}|u_i \in CG.U\}$
- $l_{OC}(o_{item_j}^{T_k})$: The weight sum of the edge set $\{e_{(u_i,o_{item_j}^{T_k})}|u_i \in OC.U\}$

$$\Delta Q_{o_{item_j}^{T_k}} = [\frac{m_{OC}(o_{item_j}^{T_x}) + 2 \cdot l_{OC}(o_{item_j}^{T_k})}{2m} - (\frac{m_{CG}(o_{item_j}^{T_x}) + l_{CG}(o_{item_j}^{T_k})}{2m})^2]$$
$$- [\frac{m_{OC}(o_{item_j}^{T_x})}{2m} - (\frac{m_{CG}(o_{item_j}^{T_x})}{2m})^2 - (\frac{l_{CG}(o_{item_j}^{T_k})}{2m})^2] \tag{7}$$

If $\Delta Q_{o_{item_j}^{T_k}}$ is positive, object $o_{item_j}^{T_k}$ is added into the object community of the current node of the tree $T.Node.O$ for which its gain is maximum. Otherwise, $o_{item_j}^{T_k}$ only stays in $T.Node.parent.O$. The modularity discrepancy $\Delta Q_{o_{item_j}^{T_k}}$ is expected to be as large as possible, so that $item_j$ is more likely to be rated as τ_k by users in the user community of the current node of the user criteria tree $T.Node.U$ than those outside the community. Furthermore, some objects are connected with limited users. If it is randomly distributed, these objects will be removed from higher object communities. On the other hand, such objects may be connected with particular user groups. Therefore, it is always being maintained in some object communities.

The hierarchical object community generation algorithm is demonstrated in Algorithm 2. Lines 1–11 aim to initialise the top object community based on the user criteria clustering tree. Lines 12–31 tend to generate the object community OC for each node of the user criteria cluster tree T. The output of the algorithm is the object community OC set, and each OC is assigned to the related node of the user criteria clustering tree T.

3.3   Facet Object Set

In the previous two steps, both user and object community are supposed to be figured out for each node of the user criteria clustering tree. The user community shares a common preference and accepts a similar criterion of items. Hence, the object community of this level implies a particular facet of the real-world. One important feature for hierarchical object community is that the lower level, the more significant correlations among objects. Too low levels of object community cannot include all the relevant objects. While, too high levels of object community may consist of too much noisy objects. Therefore, we narrow the scope of the object community to generate the corresponding facet object set, which implies the preference of a certain user community.

Let $FO = \{o_i|o_i \in O\}$ denote the facet object. The objects in a particular facet object are not only correlated with others, but also evaluated under

---

Algorithm 2. The Hierarchical Object Community Generation Algorithm

---

Input: $T, CG = <U, O, E>$
Output: $T, \{OC\}$
 1: $index = T.Node.size() - 1$
 2: $OC_{index}.U \leftarrow T.Node_{index}.U$
 3: $tempO1 \leftarrow CG.O$
 4: **for** $\forall o_j^{\tau x} \in tempO$ **do**
 5:     **for** $\forall u_i \in OC_{index}.U$ **do**
 6:         $sum = sum + e_{(u_i, o_j^{\tau x})}$

 7:     **end for**
 8:     **if** $(sum == 0)$ **then**
 9:         $tempO1.remove(o_j^{\tau x})$
10:     **end if**
11: **end for**
12: $OC_{index}.O \leftarrow tempO1, T.Node_{index}.O \leftarrow OC_{index}.O$
13: **for** $\forall T.Node_{index} \in T \wedge T.Node_{index} \neq Null$ **do**
14:     $tempO2 \leftarrow T.Node_{index}.parent.O$
15:     **for** $\forall o_j^{\tau x} \in tempO2$ **do**
16:         $tP = tC = 0$
17:         **for** $\forall u_i \in T.Node_{index}.parent.U$ **do**
18:             $tP = tP + e_{(u_i, o_j^{\tau x})}$

19:         **end for**
20:         **for** $\forall u_i \in T.Node_{index}.U$ **do**
21:             $tC = tC + e_{(u_i, o_j^{\tau x})}$

22:         **end for**
23:         $distanceValue_{o_j^{\tau x}} = (tP - tC) * w_{o_j^{\tau x}}$
24:         $distanceQue[].add(distanceValue_{o_j^{\tau x}}), sort(distanceQue[])$

25:     **end for**
26:     **for** $\forall distanceValue_{o_j^{\tau k}} \in distanceQue[]$ **do**
27:         calculate $\Delta Q_{o_j^{\tau k}}$
28:         **if** $(\Delta Q_{o_j^{\tau k}} < 0)$ **then**
29:             $tempO2.remove(o_{item_j}^{\tau k})$
30:         **end if**
31:     **end for**
32:     $OC_{index}.O \leftarrow tempO2, T.Node_{index}.O \leftarrow OC_{index}.O, OC_{index}.U \leftarrow T.Node_{index}.U$
33: **end for**
34: **return** OC, T

---

the same criteria by a group of users. In terms of each internal node $T.Node$ with child nodes $T.Node.left/T.Node.right$, users in the user community of left child node $T.Node.left.U$ also have interactions with part of objects belonging to the object community of right child node $T.Node.right.O$, and vice versa. Equation 8 defines the distance between two child nodes of current internal node. The community distance value $CDist(T.Node)$ is smaller if two objects in object communities of child nodes are more frequently and evenly connected with users in both two child user communities. It is necessary to specify a minimum acceptable threshold value, i.e., δ. If $CDist(T.Node) \geq \delta$, the contraction of facet object set will be terminated.

$CDist(T.Node)$

$$= \sqrt{\sum_{o^{\tau_k}_{item_j} \in T.Node.O} \left(\frac{\frac{\sum_{u_i \in T.Node.left.U} e_{(u_i, o^{\tau_k}_{item_j})}}{|T.Node.left.U|}}{\frac{\sum_{u_i \in T.Node.U} e_{(u_i, o^{\tau_k}_{item_j})}}{|T.Node.U|}} - \frac{\frac{\sum_{u_i \in T.Node.right.U} e_{(u_i, o^{\tau_k}_{item_j})}}{|T.Node.right.U|}}{\frac{\sum_{u_i \in T.Node.U} e_{(u_i, o^{\tau_k}_{item_j})}}{|T.Node.U|}} \right)^2}$$

with $T.Node$ outside. (8)

3.4 Context Specific Inter-personalised Trust Calculation

In terms of the inquired item $IE.item_j$ in particular enquirer, more than one facet object sets normally exist. Therefore, in order to make a more accurate prediction for enquirer $IE.u_i$, the system tends to compare the user's previous interaction records with the particular facet object sets related to inquired item, and then figure out the most trustable facet object set. Finally, the system suggests the most trustable item to the user. In our approach, the context specific inter-personalised trust value for particular facet object set is mainly determined by two factors: **Distance** and **Support**.

Distance represents the divergence between user's preference, R_{u_i} and facet object set, FO. It can be calculated by using Eq. 9.

$$Dist(u_i, FO_j) = \sqrt{\sum_{R_{u_i}.r_{i,k} \neq 0, o^{\tau_x}_k \in FO_j}^{R_{u_i}, FO_j} \left(\frac{R_{u_i}.r_{i,k} - \tau_x}{|u_i.ratedItemSet \cap FO_j.ItemSet|} \right)^2} \quad (9)$$

In Eq. 9, $|u_i.ratedItemSet \cap FO_j.ItemSet|$ denotes the number of items in facet object set $FO_j.ItemSet$ which are rated by u_i. While, $(R_{u_i}.r_{i,k} - \tau_x)$ calculates the difference between the rating given by u_i and τ_x implied by the object $o^{\tau_x}_{item_k}$ in facet object FO_j. $Dist(u_i, FO_j)$ is supposed to be small if objects in the facet object set FO_j are more appropriate for user's criteria about inquiry $item_k$.

Support is the ratio that each facet object set FO_j supports the rating history of user u_i, which is formulated in Eq. 10.

$$Support(u_i, FO_j) = \frac{|u_i.ratedItemSet \cap FO_j.ItemSet|}{|u_i.ratedItemSet \cup FO_j.ItemSet|} \quad (10)$$

By considering both distance and support, the context specific inter-personalised trust value is formulated in Eq. 11.

$$Trust(u_i, FO_j) = \frac{Support(u_i, FO_j)}{Dist(u_i, FO_j)} \quad (11)$$

4 Experiments

Experiments are conducted to analyse the performance of the community-based trust estimation approach. In the experiments, we compare the proposed approach with two memory-based collaborative filtering approaches, i.e. the user-based approach and the item-based approach, and one traditional data mining algorithms, i.e. K-Nearest Neighbour algorithm (KNN) [14].

4.1 Data Set

The real-world public dataset collected by Paolo Massa has been used for the experiments [15]. The dataset was crawled from *epinions*[1], which is a general consumer review website allowing users to share comments and reviews regarding various items, such as cars, books, music, etc. The ratings for each item ranges from 1 to 5. The dataset contains 195 users, 200 items and 5035 reviews.

A realistic collaborative filtering matrix probably contains millions of users and items. In practice, users give ratings to a few items only, and this results in a sparse matrix. The "sparseness" of a collaborative filtering matrix is defined as the percentage of empty cells [15]. Figure 2 demonstrates the number of users who created reviews. The X axis in Fig. 2 represents the user ID, while the Y axis indicates the item rating amount of the corresponding user. The sparseness of the dataset is around 87.1%, and more than 17% users rate no more than five items. The mean number of reviews is 25.82 with a standard deviation of 24.40, and the median is 19.

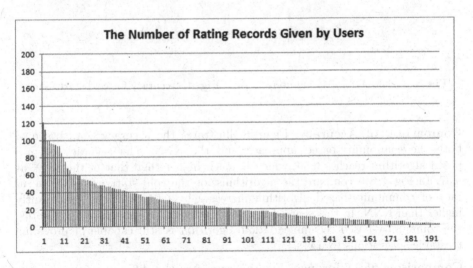

Fig. 2. Numbers of reviews rated by users with cold start users

4.2 Experimental Results

When a new user enters the system without any rating history, it is difficult to predict his or her preference since the user has never given any ratings before.

---

[1] www.epinions.com/.

We consider users with less than five rating records as "cold start users" [16]. The traditional collaborative filtering algorithms are usually unable to provide high quality recommendations for this group of users. Moreover, accurate predictions also create an incentive for such users to continue using the system. Therefore, we also compare algorithms' performances for "cold start users".

In the experiment, we mainly use two metrics, i.e., accuracy and difference, to compare the performance of the community-based trust estimation algorithm and the other three algorithms. Specifically, the accuracy signifies the percentage of potential quality prediction of items which are equal to the actual feedback rating values given by enquirers. However, neither user-based nor item-based approach can predict the exact rating values for required items. Hence, difference is adopted as another comparison metric. It measures the average distance between the actual and predicted rating values.

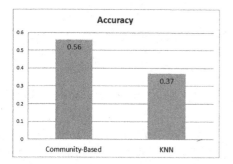
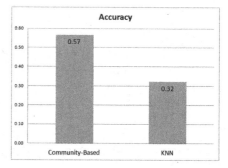

Fig. 3. Accuracy for existing users   **Fig. 4.** Accuracy for cold start users

Comparison of Accuracy. Figure 3 illustrates the accuracy comparison of both the community-based approach and the KNN. The accuracy of proposed algorithm reaches 0.56, which is much higher than that of the KNN at 0.37. In Fig. 4, we compare the algorithms for the cold start users, the accuracy of community-based algorithm increases by 0.01, which is significantly higher than KNN. In this sense, for a new user without enough rating records, the community-based recommendation algorithm is still capable of providing trustable suggestions to users.

Comparison of Difference. Figure 5 compares the difference values of the four algorithms. The community-based algorithm performs better than the other three algorithms, where the difference is approximate 0.72. Furthermore, as can be seen from Fig. 6 that, the difference of the community-based approach narrows to 0.6339 in terms of the "cold start users". However, the difference of the KNN and the item-based algorithm increased to above 1. Although the performance of the user-based recommendation algorithm performs better than KNN and item-based approach, the difference (0.8252) is still higher than the community-based trust estimation algorithm.

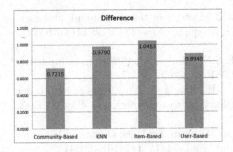
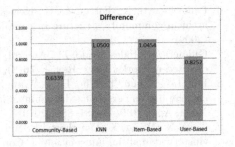

Fig. 5. Difference for existing users   **Fig. 6.** Difference for cold start users

5 Conclusion

In this paper, we proposed a community-based trust estimation approach to mine context specific inter-personalised trust in preference networks. In the approach, we organise the preference network as a set of more manageable interrelated communities. The approach mainly focuses on users with similar preference, and groups them into various user communities. Furthermore, object communities are partitioned to imply the interest and criterion of user communities for particular items. Finally, distance and support are considered in the approach to ascertain the most confident facet object set, and make the trustable quality prediction based on the rating value of the object about the enquired item $IE.item_j$ in this facet object set. From the experimental results, it can be seen that the community-based approach gives better performance than some other approaches in terms of both difference and accuracy, even under the "cold start users" situation. However, the community-based approach manages trust information in a centralized manner. In the future, we will extend the community-based mechanism to distributed environments.

References

1. Newman, M.E.J.: The structure and function of complex networks. SIAM Rev. **45**(2), 167–256 (2003)
2. Pitsilis, G., Marshall, L.: Trust as a Key to Improving Recommendation Systems. Springer, Heidelberg (2005)
3. Shah, L., Gaudani, H., Balani, P.: Survey on recommendation system. System **137**(7) (2016). http://www.ijcaonline.org/archives/volume137/number7/24291-2016908821
4. Luo, H., Niu, C., Shen, R., Ullrich, C.: A collaborative filtering framework based on both local user similarity and global user similarity. Mach. Learn. **72**(3), 231–245 (2008)
5. Li, D., Chen, C., Lv, Q., Shang, L., Zhao, Y., Tun, L., Ning, G.: An algorithm for efficient privacy-preserving item-based collaborative filtering. Fut. Gener. Comput. Syst. **55**, 311–320 (2016)

6. Lancichinetti, A., Fortunato, S., Kertész, J.: Detecting the overlapping and hierarchical community structure in complex networks. New J. Phys. **11**(3), 033015 (2009)

7. Guimerà, R., Sales-Pardo, M., Amaral, L.A.N.: Module identification in bipartite, directed networks. Phys. Rev. E **76**(3), 036102 (2007)

8. Liu, J.-G., Zhou, T., Che, H.-A., Wang, B.-H., Zhang, Y.-C.: Effects of high-order correlations on personalized recommendations for bipartite networks. Phys. A Stat. Mech. Appl. **389**(4), 881–886 (2010)

9. Zhou, X., Wang, X., Dougherty, E.R., Russ, D., Suh, E.: Gene clustering based on clusterwide mutual information. J. Comput. Biol. **11**(1), 147–161 (2004)

10. Dawy, Z., Hagenauer, J., Hanus, P., Mueller, J.C.: Mutual information based distance measures for classification and content recognition with applications to genetics. In: IEEE International Conference on Communications, ICC, vol. 2, pp. 820–824. IEEE (2005)

11. Dillen, N.B., Chakraborty, A.: Modularity-based community detection in fuzzy granular social networks. In: Satapathy, S.C., Bhatt, Y.C., Joshi, A., Mishra, D.K. (eds.) Proceedings of the International Congress on Information and Communication Technology. AISC, pp. 577–585. Springer, Singapore (2016)

12. Zhao, Z., Feng, S., Wang, Q., Huang, J.Z., Williams, G.J., Fan, J.: Topic oriented community detection through social objects and link analysis in social networks. Knowl. Based Syst. **26**, 164–173 (2012)

13. Vincent, D., Blondel, J.-L.G., Lambiotte, R., Lefebvre, E.: Fast unfolding of communities in large networks. J. Stat. Mech. Theory Exp. **2008**(10), P10008 (2008)

14. Larose, D.T.: k-nearest neighbor algorithm. In: Larose, D.T. (ed.) Discovering Knowledge in Data: An Introduction to Data Mining, pp. 90–106. Wiley, Hoboken (2005)

15. Massa, P., Bhattacharjee, B.: Using trust in recommender systems: an experimental analysis. In: Jensen, C., Poslad, S., Dimitrakos, T. (eds.) iTrust 2004. LNCS, vol. 2995, pp. 221–235. Springer, Heidelberg (2004). doi:10.1007/978-3-540-24747-0_17

16. Li, S., He, Y., Chang, E.Y., Wen, J.-R., Li, X.: Connecting social media to e-commerce: cold-start product recommendation using microblogging information. IEEE Trans. Knowl. Data Eng. **28**(5), 1147–1159 (2016)

Proactive Skill Posting in Referral Networks

Ashiqur R. KhudaBukhsh$^{(\boxtimes)}$, Jaime G. Carbonell, and Peter J. Jansen

Carnegie Mellon University, Pittsburgh, USA
{akhudabu,jgc,pjj}@cs.cmu.edu

Abstract. Distributed learning in expert referral networks is an emerging challenge in the intersection of Active Learning and Multi-Agent Reinforcement Learning, where experts—humans or automated agents—can either solve problems themselves or refer said problems to others with more appropriate expertise. Recent work demonstrated methods that can substantially improve the overall performance of a network and proposed a distributed referral-learning algorithm, DIEL (Distributed Interval Estimation Learning), for learning appropriate referral choices. This paper augments the learning setting with a proactive skill posting step where experts can report some of their top skills to their colleagues. We found that in this new learning setting with meaningful priors, a modified algorithm, proactive-DIEL, performed initially much better and reached its maximum performance sooner than DIEL on the same data set used previously. Empirical evaluations show that the learning algorithm is robust to random noise in an expert's estimation of her own expertise, and there is little advantage in misreporting skills when the rest of the experts report truthfully, i.e., the algorithm is near Bayesian-Nash incentive-compatible.

Keywords: Active learning · Referral network · Proactive skill posting

1 Introduction

Consider a network of experts with differing expertise, where any expert may receive a problem (aka a task or a query) and must decide whether to work on it or to refer the problem, and if so to which other expert. For instance, in a clinical network, a physician may diagnose and treat a patient or refer the patient to another physician whom she believes may have more appropriate knowledge, given the presenting symptoms. The referring physician may charge a referral fee and the receiving physician may charge a larger fee for diagnosing and treating the patient. Referral networks are common across other professions as well, such as members of large consultancy firms. If the experts are software agents, then the need for referral may be greater, given the likely narrower "expertise" typical of intelligent agents (including old-style expert systems). We can also envision a hybrid referral network comprising automated agents and possibly crowd-sourced human experts.

How does a network or how do individual experts in the network learn to refer effectively? Human referral networks are neither hardwired nor static. Potentially

© Springer International Publishing AG 2016
B.H. Kang and Q. Bai (Eds.): AI 2016, LNAI 9992, pp. 585–596, 2016.
DOI: 10.1007/978-3-319-50127-7_52

much larger networks of automated experts or hybrid networks with dynamic membership must likewise learn to refer with membership drift. One option is to maintain a global index and/or a "boss agent" telling all the others when to try and solve a problem or when to refer and to whom. In practice, however – in medicine, in academia, or in consulting companies – referrals occur to those one knows and trusts to do a good job. Hence a distributed learning setting, although it poses greater challenges to learning referrals, is a more realistic alternative.

To this end, a simple yet effective learning-to-refer method, dubbed DIEL (Distributed Interval Estimation Learning), has been proposed in [7]. The referral model assumes an initial sparse topology of a static referral graph where each expert knows a handful of colleagues so that $E \sim O(V)$ (E and V denote the number of edges and vertexes in the network, respectively). Learning consists of each expert improving its estimates of the ability of colleagues to solve different classes of problems. On a wide range of simulations with different network structures and parameters chosen to represent possible real-life scenarios, DIEL consistently outperformed greedy (Distributed Mean-Tracking, DMT) and random baselines, and Q-learning variants. However, all such experiments assumed an uninformative prior on the expertise of colleagues which may not correspond to a real-world setting. Moreover, in real life, we often see that experts clearly mention which type of tasks they are particularly good at and also often forge links to their colleagues via social networks. In turn, their colleagues may re-estimate their beliefs of expertise levels based on actual performance.

In this work, we augment the original DIEL expertise-learning setting with a local-network advertisement of expertise-by-topic by each expert in the network. Our primary contribution is this augmented learning setting and a distributed referral-learning algorithm, proactive-DIEL, that takes advertised priors from other colleagues into account. On the same data set used in [7], with an accurate estimate of true skill and truthful reporting proactive-DIEL substantially outperformed DIEL in the initial phase of learning. Additionally, we found that proactive-DIEL is robust to limited Gaussian noise in an expert's estimation of her own skills. Also, our experimental evaluations reveal that proactive-DIEL displays empirical evidence of being near Bayesian-Nash incentive-compatible, i.e., when all the other experts are truthful, lying about one's own level of expertise has little or no advantage.

The rest of the paper is organized as follows. We first illustrate the basic referral mechanism through a small example. After a discussion of related work, we present the list of assumptions on the referral network and expertise. Next, we describe our distributed learning algorithms and experimental setup, after which we present and discuss the results from our experiments. Finally, we end with some general conclusions and an outlook on future work.

2   Referral Mechanism

We illustrate the referral mechanism and its effectiveness with the simple graph of Fig. 1 (taken from [7]), which represents a five-expert network.

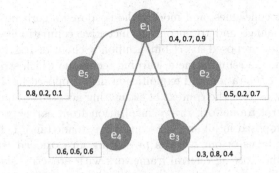

Fig. 1. A referral network with five experts.

The nodes of the graph are the experts, and the edges indicate that the experts 'know' each other, that is, they can send or receive referrals and communicate results. In the domain, three different topics (subdomains) can be distinguished – call them t_1, t_2, and t_3 – and the figures in brackets indicate an expert's expertise in each of these.

In this referral network, with a query belonging to t_2, if there was no referral, the client may consult first e_2 and then possibly e_5, leading to a probability of getting the correct answer of $0.2 + (1 - 0.2) \times 0.2 = 0.36$. With referrals, an expert handles a problem she knows how to answer, and otherwise if she had knowledge of all the other experts' expertise she could ask e_2 who would refer to e_3 for the best skill in t_2, leading to a solution probability of $0.2 + (1 - 0.2) \times 0.8 = 0.84$.

The referral mechanism consists of the following steps.

1. A user issues an *initial query* to an *initial expert*, chosen from a prior distribution, e.g. uniform: equiprobably among all experts, or based on proactive-social-net advertisement priors.
2. The initial expert examines the instance. If she is able to solve it, she returns the solution or label to the user.
3. If she is not able to solve the problem, she issues a *referral query* to a *referred expert*, i.e. colleague who may be better able to solve the problem. *Learning-to-refer* means improving the estimate of who is most likely to solve the problem and/or, which expert should best be sampled in order to learn their level of expertise on the topic of the problem.
4. If the referred expert succeeds, she communicates the solution to the initial expert, who in turn, communicates it to the user.

3   Related Work

In terms of the learning setting and referral learning algorithm, our primary basis for this work is [7] which first proposed a novel learning setting in the context of Active learning where experts are connected through a network and can refer instances to one another. We augment this learning setting by allowing advertisement to colleagues, and modify the proposed algorithm DIEL both for improved performance and to encourage incentive compatibility. The referral learning algorithm proposed in [7] builds upon a chain of research on interval estimation learning, a reinforcement learning technique which strives to strike a balance between exploration and exploitation first proposed in [4,5], which has been successfully used in the context of estimating accuracy of multiple labelers in [3]. The referral framework draws inspiration from earlier work in referral chaining, first proposed in [6] and subsequently extended in [11–14].

The primary focus of our work is presenting an extended learning setting by augmenting the existing referral framework with proactive skill posting and designing an appropriate distributed referral learning algorithm. However, in a real-world application, success will depend on preventing experts from misreporting their true skills as they advertise (e.g. to acquire more business). Among a large body of literature in truthful mechanism design [1,2,9,10] we highlight a few key differences with the *budgeted multi-armed bandit mechanism* motivated by crowdsourcing platforms presented in [2]. First, while *learning-to-refer* can be interpreted as a multi-armed bandit problem where each arm is a referral choice, our work deals with several such parallel multi-armed bandit problems in a distributed network setting. Also, in our setting each individual expert needs to learn appropriate referral choices based on topics and expertise estimates of colleagues on said topics. This is in contrast with homogeneous tasks, for example as considered in [2]. Finally, proactive-DIEL deals with partially available priors since experts are not bidding for all the topics because of a restricted advertisement budget (a factor [2] did not need to consider because of homogeneous tasks).

For similar reasons, our work differs from past work studying the communication between experts and non-experts in crowdsourcing. In [8], domain experts break down a complex task into simpler micro-tasks and actively supervise the non-expert crowd. In our approach, there is no such simplifying assumption: it is highly unlikely that one expert Pareto dominates other experts in a professional network across all topics. Instead, referral is bi-directional, and the main focus is on learning appropriate referral choices in a distributed manner.

4   Referral Networks

We follow the same notation, general setting for referral mechanism, and initial set of assumptions as in [7], which we describe here in further detail. Section 4.4 presents the additional assumptions and mechanisms for proactive skill posting first introduced in this paper.

4.1   Notation

- A *referral network* can be represented by a graph (V, E) of size k in which each vertex v_i corresponds to an expert e_i and each bidirectional edge $\langle v_i, v_j \rangle$ indicates a *referral link*.

- We call the set of experts linked to an expert e_i via a referral link, the *sub-network* of expert e_i.
- A *scenario* is a set of m queries $(q_1, \ldots, q_m)$ belonging to n topics $(t_1, \ldots, t_n)$, to be addressed by the k experts $(e_1, \ldots, e_k)$.
- As *expertise* for an expert-instance pair we simply take the probability that she can successfully solve the problem, i.e., $Expertise(e_i, q_j) = P(solve(e_i, q_j))$.

4.2 Initial Expert and Expertise Assumptions

Topic-wise distributional assumption: We take the expertise distribution for a given topic t to be a mixture of two truncated Gaussians (with parameters $\lambda = \{w_i^t, \mu_i^t, \sigma_i^t\}$ $i = 1, 2$.). One of them ($\mathcal{N}(\mu_2^t, \sigma_2^t)$) has a greater mean ($\mu_2^t > \mu_1^t$), smaller variance ($\sigma_2^t < \sigma_1^t$) and lower mixture weight ($w_2^t << w_1^t$). Intuitively, this represents the expertise of experts with specific training for the given topic, contrasted with the lower-level expertise of the layman population.

Instance-wise distributional assumption: We model the expertise of a given expert on instances under a topic by a truncated Gaussian distribution with small variance. i.e.,

$$Expertise(e_i, q_j) \sim \mathcal{N}(\mu_{topic_p, e_i}, \sigma_{topic_p, e_i}),$$
$$\forall q_j \in topic_p, \forall p, i : \sigma_{topic_p, e_i} \leq 0.2.$$

We further assume that an expert can accurately identify the topic of a query, that expertise does not change over time (but see future work in the conclusion), and that experts have no capacity constraints.

4.3 Network Assumption

The probability that a referral link exists between expert e_i and e_j is a function of how similar the two experts are, which we modeled as $P(ReferralLink(v_i, v_j)) = \tau + c\, Sim(e_i, e_j)$. As a similarity metric we used *cosine similarity of topic-means*. The parameter τ captures any extraneous reason two experts can be connected, e.g., same geolocation, common acquaintances, etc.

4.4 Proactive Skill Posting

An advertising unit is a tuple $\langle e_i, e_j, t_k, \mu_{t_k} \rangle$, where e_i is the *target expert*, e_j is the *advertising expert*, t_k is the topic and μ_{t_k} is e_j's (advertised) topical expertise.

We initially assume that experts can estimate their expertise (topic-means) accurately, and report truthfully. We assume each expert is allocated a budget of B advertising units, where B is twice the size of that expert's subnetwork. The advertising budget addresses the limited time that experts have to socialize with different colleagues and get to know each other's experience.

The advertising expert e_j reports to each target expert e_i in her subnetwork the two tuples $\langle e_i, e_j, t_{best}, \mu_{t_{best}} \rangle$ and $\langle e_i, e_j, t_{secondBest}, \mu_{t_{secondBest}} \rangle$, i.e., the top two topics in terms of the advertising expert's topic means. This is a one-time advertisement and happens right at the beginning of the simulation.

5   Distributed Referral Learning

In this section, we first present interval estimation learning [3] which is the primary building block of DIEL. Next, we outline the steps for DIEL and present the key ways in which proactive-DIEL differs.

5.1   Interval Estimation Learning

Action selection using Interval Estimation Learning (IEL) first estimates for each action a the upper confidence bound for the mean reward by

$$UI(a) = m(a) + t_{\frac{\alpha}{2}}^{(n(a)-1)} \frac{s(a)}{\sqrt{n(a)}} \tag{1}$$

where m(a) is the mean observed reward for a, $s(a)$ is the sample standard deviation of the reward, $n(a)$ is the number of observed samples from a, and $t_{\frac{\alpha}{2}}^{(n(a)-1)}$ is the critical value for the Student's t-distribution ($n(a) - 1$ degrees of freedom, $\frac{\alpha}{2}$ confidence level) (in our case, the action is the selection of a referred expert among possible choices in the subnetwork). Next, IEL selects the action with the highest upper confidence bound.

Input: A set of k experts e_1, e_2, ..., e_k. A set of n topics $topic_1$, $topic_2$, ..., $topic_n$. A $k \times k$ referral network.
Initialize rewards.
for $iter \leftarrow 1$ **to** $maxIter$ **do**
  Assign instance q to an initial expert e randomly
  if e fails to solve q **then**
    $topic \leftarrow getTopic(q)$
    $expectedReward \leftarrow 0$
    $bestExpert \leftarrow 0$
    for each expert e' in the subnetwork of e **do**
      if $expR_h(e', topic) \geq expectedReward$ **then**
        $bestExpert \leftarrow e'$
        $expectedReward \leftarrow expR_h(e', topic)$
      end
    end
  end
  $referredExpert \leftarrow bestExpert$
  if $referredExpert$ solves q **then**
    $update(reward(e, topic, referredExpert), 1)$
  else
    $update(reward(e, topic, referredExpert), 0)$
  end
end

Algorithm 1. DISTRIBUTED REFERRAL LEARNING, Q = 2

The intuition is that high mean selects for best performance, and high variance selects for unexplored expert capability on topic, thus optimizing for amortized performance, as variance decreases over time, and best mean is selected reliably among the top candidates. The parameter α weights exploration more when small and exploitation more when large. A partial parameter sweep confirmed that a value of $\alpha = 0.05$, settled on in [3], also worked well in [7].

5.2 Distributed Referral Learning

Algorithm 1 outlines the steps for expertise estimation in a distributed setting with single referral (a per-task query budget $Q = 2$). The function $expR_h(e', topic)$ estimates e''s topical expertise using heuristic h. DIEL (Distributed Interval Estimation Learning), and DMT (Distributed Mean-Tracking) differ in this heuristic, DIEL estimating reward by Eq. (1) and DMT by using the sample-mean.

One major challenge in the distributed setting is that there is no global visibility of rewards, i.e., $reward(e_i, topic_p, e_j)$ is only visible to expert e_i. When the task is solved successfully, $update$ assigns an additional reward of 1 to the referred expert (experts keep track of success and failure of experts they refer to by means of a sequence of 0s and 1s; here, we just mean that a "1" is appended to this sequence in the case of success).

Proactive-DIEL differs from DIEL in the following three key ways.

1. Emphasis on exploitation: The Student's t-distribution parameter has a large value for smaller n's and drops down towards 1 as n increases, thereby boosting exploration in early on in the learning. This is less important in our current setting where we start with a partially informative prior, hence we dropped this parameter in Eq. (1), leading to

$$UI(a) = m(a) + \frac{s(a)}{\sqrt{n(a)}} \qquad (2)$$

In fact, since we found that even the original DIEL performed better on the original data in [7] with this revised formula, we used it for subsequent performance comparisons in this paper.

2. Initialization: Rather than initializing DIEL sets $reward(e_i, t_k, e_j)$ for each i, j and k with a pair $(0, 1)$ in order to initialize mean and variance, as in DIEL, proactive-DIEL initializes $reward(e_i, t_k, e_j)$ for each advertisement unit $\langle e_i, e_j, t_k, \mu_{t_k} \rangle$ with two rewards of μ_{t_k}.

To initialize topics for which no advertisement units are available (recall that the budget was assumed to suffice for advertising an expert's top two skills only) we assumed it was safe and informative to initialize the rewards as if the expert's skill was the same as on her second best topic, that is, with two rewards of $\mu_{t_{secondBest}}$, effectively being an upper bound on the actual value.

3. Reward update function: When a referred expert e_j succeeds in solving a task on topic t_k, $update$ in proactive-DIEL's $update$ function, like DIEL's, assigns an additional reward of 1 to $reward(e_i, t_k, e_j)$.

When e_j fails, however, instead of always appending a 0 to $reward(e_i, t_k, e_j)$, proactive-DIEL, in the presence of an advertisement unit $\langle e_i, e_j, t_k, \mu_{t_k} \rangle$, appends a (negative) penalty P with probability μ_{t_k}. This way, over-reporting of skill leads to more frequent incurrence of the penalty.

In the absence of an advertisement unit, a penalty P is still appended, but with a probability equal to the sample mean of e_j observed by e_i on topic t_k. In our experiments, we set P to -0.35.

6   Experimental Setup

Since the performance of a referral network is sensitive to the topology of the network and expertise of experts on individual topics, we evaluated the performance of proactive-DIEL on the same wide range of scenarios considered in [7] by varying the parameters in the network. Table 1 lists the distributions from which the parameters are sampled. The data set consisted of 1000 scenarios, each with 100 experts, 10 topics and a referral network.

Following [7], we also report upper-bound performance of a network where every expert has access to an oracle that knows the true topic-mean (i.e., $mean(Expertise(e_i, q) : q \in topic_p) \, \forall i, p)$ of every expert-topic pair. Our measure of performance is the overall task accuracy of our multi-expert system.

Table 1. Parameters for synthetic data set.

Parameter	Description	Distribution
τ	$P(ReferralLink(v_i, v_j))$	Uniform(0.01, 0.1)
c	$= \tau + c \, Sim(e_i, e_j)$	Uniform(0.1, 0.2)
μ_1	Truncated mixture of two	Uniform(0, b)
μ_2	Gaussians for topics	Uniform(b,1) $b \in \{0.1, 0.2, 0.3, 0.4, 0.5\}$
σ_1		Uniform(0.2, 0.4)
σ_2		Uniform(0.05, 0.15)
w_2		$\mathcal{N}(0.03, 0.01)$, $w_2 \geq 0$

7   Results

In this section we present our three main results: (1) Access to (noisy) priors on their colleagues' expertise improves an expert's performance in both the DIEL and DMT algorithms, (2) in the augmented setting, even partial and possibly inaccurate information on the priors helps, and (3) misreporting skills confers little or no advantage when the other experts report truthfully.

7.1   DIEL and DMT with Informative Prior

We first show that informative priors on the means can be easily incorporated into DIEL and DMT and incorporating the priors is beneficial for both DIEL and DMT. Suppose every expert has access to an oracle than can estimate the true topic-mean of every other expert-topic pair within an error bound, i.e. $|\mu_{e_i,t_k} - \hat{\mu}_{e_i,t_k}| \leq \delta$. Unlike [7], instead of a 0 and a 1, all rewards $reward(e_i, t_k, e_j)$ are initialized with two rewards of $\hat{\mu}_{e_i,t_k}$. In Fig. 2, we see that even when δ is as high as 0.2, the performance of both DIEL and DMT substantially improved with uninformed DIEL still outperforming informed DMT at the later stage given enough samples.

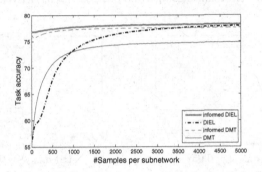

Fig. 2. DIEL and DMT with informative prior.

7.2   Performance of Proactive-DIEL

Next, we analyze the performance of DIEL with proactive skill posting. Figure 3 shows that even with a limited budget of $2 \times |subnetwork|$ (i.e., two advertisements per expert in the subnetwork), proactive-DIEL requires very few samples to reach a reasonably high overall network performance.

Even when experts post their skills truthfully, their self-estimates may be off. Here, we relax the accurate skill estimation assumption. Specifically, we considered $\hat{\mu} = \mu + \mathcal{N}(0, \sigma_{noise})$, where $\hat{\mu}$ is an expert's own estimate of her true topic-mean μ, and σ_{noise} is a small constant (we tried two different values for σ_{noise}, 0.05 and 0.1). Figure 4 compares the performance of proactive-DIEL with noisy estimates with the initial noise-free version and DIEL. Figure 4(a) shows that in the early stage, with a small amount of noise in estimation, eventually the same performance as the noise-free version is achieved while remaining strictly superior to DIEL throughout the entire course of simulation. With a larger value of noise, Fig. 4(b) shows that proactive-DIEL is slightly worse than the noise-free version, but nonetheless, outperforms DIEL.

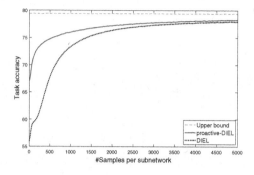

Fig. 3. Performance comparison of DIEL and proactive-DIEL.

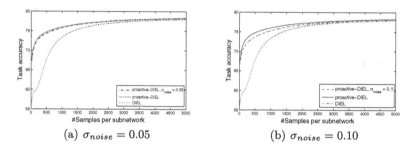

(a) $\sigma_{noise} = 0.05$　　　　　　　(b) $\sigma_{noise} = 0.10$

Fig. 4. proactive-DIEL with noisy skill estimation.

7.3   Bayesian-Nash Incentive-Compatibility

Our results show that proactive-DIEL is immune to a small amount of Gaussian noise in estimating topical expertise. Another important but somewhat orthogonal goal is to prevent deliberate misreporting, e.g. experts trying to get more business by overstating their skills. We treat the number of referrals received as a proxy for payment. Intuitively, proactive-DIEL's design discourages misreporting in the following two ways: For an advertised skill, if over-reported, a failed task would receive a higher skill-estimation penalty that it would otherwise. When a skill is under-reported, the expert will be selected less often, and hence it is naturally self-balancing.

Our empirical evaluation shows that in the steady-state, proactive-DIEL is fairly resilient to strategic lying, and is almost Bayesian-Nash incentive-compatible, i.e., there is little or no advantage for an individual expert in misreporting topical expertise when all other experts in the network report truthfully. We analyze the effect of misreporting the following way. First, we considered different combinations an expert can use while reporting their best and second-best skill (listed in Table 2). For a given strategy and scenario $scenario_i$, we first fix one expert, say e_l^i. Let $truthfulReferrals(e_l^i)$ denote the number of referrals received by e_l^i beyond a steady-state threshold (i.e., a referral gets counted if the

Table 2. Empirical analysis on Bayesian-Nash incentive-compatibility. Strategies where being truthful is no worse than being dishonest are highlighted in bold.

$\mu_{t_{best}}$	$\mu_{t_{secondBest}}$	Factor
Truthful	Overbid	0.99
Overbid	Truthful	**1.00**
Overbid	Overbid	0.97
Truthful	Underbid	**1.04**
Underbid	Truthful	**1.09**
Underbid	Underbid	**1.22**
Underbid	Overbid	**1.11**
Overbid	Underbid	**1.04**

initial expert has referred 1000 or more instances to her subnetwork) when e_l^i and all other experts report truthfully. Similarly, let $strategicReferrals(e_l^i)$ denote the number of referrals received by e_l^i beyond a steady-state threshold when e_l^i misreports while everyone else advertises truthfully. We compute the following factor:

$$\frac{\sum_{i=1}^{1000} truthfulReferrals(e_l^i)}{\sum_{i=1}^{1000} strategicReferrals(e_l^i)}$$ i.e., the ratio of the total number of $truthfulReferrals(e_l^i)$

to the total number of $strategicReferrals(e_l^i)$ across the entire data set. A value greater than 1 implies truthful reporting fetched more referrals than strategic lying. Table 2 shows that beyond the steady-state threshold, strategic misreporting is hardly beneficial and in most of the cases honest reporting of skills fetched more referrals.

8 Conclusion

In this work, we extended the referral-learning as proposed in [7] with a skill posting or advertising step, and revised the DIEL algorithm to (1) take advantage of informative priors and (2) include a mechanism to discourage cheating, such as skill over-reporting to get additional "business" by individual experts in a referral network.

Our new algorithm, proactive-DIEL, outperformed its predecessor convincingly, while nearly reaching empirical Bayesian Nash incentive compatibility.

We intend to extend these results in the following ways in the future.

- **Strategyproofness:** While misreporting was shown to be of little or no benefit when all other experts report truthfully, a stronger degree of incentive compatibility, strategyproofness, would require this to be the case no matter what other experts do. One future goal is to investigate what modifications to proactive-DIEL, or which conditions would ensure this.

- **Dynamic networks:** Our results showed an improved performance through informed priors in a static setting. Future research could compare the resilience of proactive-DIEL and DIEL in a dynamic setting, for instance when new experts join the network, or existing ones drop off.
- **Expertise drift:** In this work, we assumed expertise does not change with time. But it is conceivable that experts, for instance, improve with practice. Modifying proactive-DIEL to deal with time-varying expertise will be particularly challenging and may require designing an adaptive version of the reward mechanism.

References

1. Babaioff, M., Sharma, Y., Slivkins, A.: Characterizing truthful multi-armed bandit mechanisms. In: Proceedings of the 10th ACM Conference on Electronic Commerce, pp. 79–88. ACM (2009)
2. Biswas, A., Jain, S., Mandal, D., Narahari, Y.: A truthful budget feasible multi-armed bandit mechanism for crowdsourcing time critical tasks. In: Proceedings of the 2015 International Conference on Autonomous Agents and Multiagent Systems, pp. 1101–1109. International Foundation for Autonomous Agents and Multiagent Systems (2015)
3. Donmez, P., Carbonell, J.G., Schneider, J.: Efficiently learning the accuracy of labeling sources for selective sampling. In: Proceedings of KDD 2009, p. 259 (2009)
4. Kaelbling, L.P.: Learning in Embedded Systems. MIT Press, Cambridge (1993)
5. Kaelbling, L.P., Littman, M.L., Moore, A.P.: Reinforcement learning: a survey. J. Artif. Intell. Res. **4**, 237–285 (1996)
6. Kautz, H., Selman, B., Milewski, A.: Agent amplified communication, pp. 3–9 (1996)
7. KhudaBukhsh, A.R., Jansen, P.J., Carbonell, J.G.: Distributed learning in expert referral networks. In: European Conference on Artificial Intelligence (ECAI) 2016, pp. 1620–1621 (2016)
8. Nallapati, R., Peerreddy, S., Singhal, P.: Skierarchy: extending the power of crowdsourcing using a hierarchy of domain experts, crowd and machine learning. Technical report, DTIC Document (2012)
9. Tran-Thanh, L., Chapman, A., Rogers, A., Jennings, N.R.: Knapsack based optimal policies for budget-limited multi-armed bandits. arXiv preprint arXiv:1204.1909 (2012)
10. Tran-Thanh, L., Stein, S., Rogers, A., Jennings, N.R.: Efficient crowdsourcing of unknown experts using multi-armed bandits. In: European Conference on Artificial Intelligence, pp. 768–773 (2012)
11. Yolum, P., Singh, M.P.: Dynamic communities in referral networks. Web Intell. Agent Syst. **1**(2), 105–116 (2003)
12. Yu, B.: Emergence and evolution of agent-based referral networks. Ph.D. thesis, North Carolina State University (2002)
13. Yu, B., Singh, M.P.: Searching social networks. In: Proceedings of AAMAS 2003 (2003)
14. Yu, B., Venkatraman, M., Singh, M.P.: An adaptive social network for information access: theoretical and experimental results. Appl. Artif. Intell. **17**, 21–38 (2003)

Comprehensive Influence Propagation Modelling for Hybrid Social Network

Weihua Li[1(✉)], Quan Bai[1], and Minjie Zhang[2]

[1] Auckland University of Technology, Auckland, New Zealand
{weihua.li,quan.bai}@aut.ac.nz
[2] University of Wollongong, Wollongong, Australia
minjie@uow.edu.au

Abstract. The evolution of influencer marketing relies on a social phenomenon, i.e., influence diffusion. The modelling and analysis of influence propagation in social networks has been extensively investigated by both researchers and practitioners. Nearly all of the works in this field assume influence is driven by a single factor, e.g., friendship affiliation. However, influence spread through many other pathways, such as face-to-face interactions, phone calls, emails, or even through the reviews posted on web-pages. In this paper, we modelled the influence-diffusion space as a hybrid social network, where both direct and indirect influence are considered. Furthermore, a concrete implementation of hybrid social network, i.e., Comprehensive Influence Propagation model is articulated. The proposed model can be applied as an effective approach to tackle the multi-faceted influence diffusion problems in social networks. We also evaluated the proposed model in the influence maximization problem in different scenarios. Experimental results reveal that the proposed model can perform better than those considering a single aspect of influence.

Keywords: Hybrid Social Network · Indirect influence · Influence propagation · Influence maximization

1 Introduction

With the prevalence of social networks and spectacular growth of social media, people tend to spend more time on-line conducting social activities. Prior studies show that consumers perceive peers' influence from social networks as more trustworthy and persuasive than traditional media, such as radio and TV advertising [1]. Motivated by this background, how to analyse and model the influence diffusion in social networks has drawn great attention in the contemporary research field. Decision making applications, such as maximization of product adoption [2], are developed based on the influence propagation models. Specifically, the models are capable of estimating the spread of influence through the network topology, which can assist the business owners to make decisions on how to promote new products.

© Springer International Publishing AG 2016
B.H. Kang and Q. Bai (Eds.): AI 2016, LNAI 9992, pp. 597–608, 2016.
DOI: 10.1007/978-3-319-50127-7_53

Most research works investigate influence diffusion in social networks based on the existing network topological structure, where friendship-affiliation links represent influence propagation channels, and the strength of links is considered as the only factor affecting the influence propagation probability. Therefore, the assumption is friendship affiliation links are equivalent to influential links. However, this cannot hold in general, as they are naturally two different types of links coexisting in a social network.

Influence is a hybrid effect, which can be decomposed into multiple components focusing on different activities of human-beings [3]. Specifically, influence is presented as a mixed types of communications and interactions, such as perceiving information posted by the friends of on-line social networks, delivering messages or emails, conducting face-to-face discussions, reviewing the comments from web-pages, etc. Hence, any of these behaviours are capable of exerting influence and impacting individuals' decisions. However, most researchers ignore the multiple possible interactive diffusion channels. On the other side, in many situations, individuals are more likely getting influenced by the 'stimuli' left by others, especially in E-commerce domain. Feedbacks of a particular product, such as reviews, ratings, comments from previous buyers influence the purchasing decisions of others, even they are not adjacent neighbours and without any immediate interactions. Therefore, the influence is still capable of propagating through the users in the same context without explicit links, since they are affiliated implicitly via other features, such as similar preference and criteria for items. Obviously, this is an important feature to be considered in influence propagation modelling, but unfortunately, is ignored in most existing works.

In this paper, we model influence diffusion space as a Hybrid Social Network (HSN), which is formed by merging a number of homogeneous or heterogeneous networks representing possible influence propagation channels. Comprehensive Influence Propagation (CIP) model has been proposed to capture the multiple pathways of influence propagation. Furthermore, experiments have been conducted by making use of influence maximization [4,5] as a typical application. We evaluate the CIP in different scenarios, i.e., in the same network with divergent scale of initial negative influencers. Experimental results reveal that the proposed model can perform better than those considering a single aspect of influence.

The reminder of this paper is organized as follows. Section 2 introduces the general ideas of influence propagation in HSN. In Sect. 3, the CIP model has been elaborated. In Sect. 4, the CIP model has been evaluated in influence maximization problem, meanwhile, the experimental results are demonstrated. Finally, the paper concluded in Sect. 5.

2   Influence Propagation in Hybrid Social Networks

2.1   Influence Diffusion Models

Independent Cascade Model (ICM) and Linear Threshold Model (LTM) are two fundamental influence diffusion models which have been widely applied in many research works [5–7]. In both models, each node has two possible states, i.e., active and inactive. At the beginning, a limited set of influencers (nodes), i.e., **seed set**, are supposed to be selected as the initial active nodes, which attempt to propagate influence and affect the inactive neighbours at a certain probability. If any neighbour is activated, the state will be converted to active and it starts to propagate influence to its neighbours.

ICM is a non-deterministic diffusion model, where the receiver's state is not deterministically decided by the itself, but is affected and influenced with a pre-defined probability by the senders [8]. By contrast, LTM is a deterministic diffusion model, where each node is assigned a fixed threshold affecting the activation. In this paper, ICM is employed by extending its features, since individual's level of influence acceptance is not considered in this research work.

2.2 Direct and Indirect Influence

Influence diffusion is a sort of communication which concerns the spread of messages perceived as new ideas or innovations [9]. Thus, direct and indirect influence in this paper lay emphasis on the types of communicational channels. To be more specific, direct influence refers to the immediate interactions or message reciprocations among any users with explicit links, such as friendship affiliation.

The concept of indirect communicational influence stems from the ant and stigmergy algorithms, where ants interact each other and conduct group activities by leaving and sensing pheromones [10]. By tailoring this idea, in this paper, indirect influence describes a form of indirect communications among the users mediated by modifications of the environment. Specifically, some users are not connected explicitly, but they diffuse influences by leaving the messages, such as ratings, comments, reviews, beliefs, etc. Meanwhile, individuals are getting

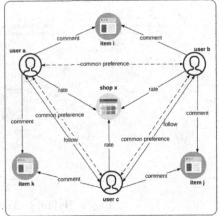

(a) Direct Influence Propagation   (b) Indirect Influence Propagation

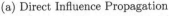

Fig. 1. Influence Diffusion in Social Networks

influenced by reading the information produced by their counterparts, since they locate in the same environment. Sometimes, the strength of indirect influence is even more prominent than direct influence, especially in sparse networks.

Figure 1 demonstrates two typical examples focusing on direct and indirect communicational influence. In Fig. 1a, three possible direct influence-diffusion channels exist. Whereas, other types of nodes, i.e., item and shop, are involved in Fig. 1b, where user a and b potentially influence each other via the messages delivered to the item and shop, though they are not connected explicitly in this social context. As for the other two pairs of users, i.e., user a and c, user c and b, they share both implicit and explicit influential links.

2.3 Hybrid Social Network

In this paper, HSN refers to an implicit heterogeneous user-centric network comprised of a number of social networks concerning possible direct and indirect influence propagation channels. It aims to model various influential relations among the individuals. Meanwhile, it implies the decomposition of influence effects, which gives high extensibility and flexibility. Specifically, when other available influential factors are added or the existing factors are changed, the model can be adapted easily by updating a particular facet.

Figure 2 demonstrates the generic model of HSN. Social networks inside the rectangle container represent different influence diffusion channels, and they are supposed to be extracted from the original heterogeneous social network. There are two types of network, the direct and indirect influence propagation network. The former is a homogeneous network, while the latter is a heterogeneous network, where the indirect influential relationships are established by via its corresponding object layer. Specifically, in this figure, $G_I^{a'}$ is object/item layer of network G_I^a. HSN G_{HSN} is constructed by merging all the social networks.

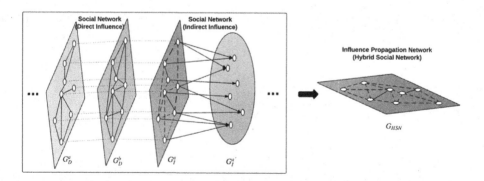

Fig. 2. Hybrid social network composition

3   Comprehensive Influence Propagation (CIP)

Comprehensive Influence Propagation (CIP) model is a concrete and simplified representations of influence propagation using HSN concept. CIP constructs HSN by extracting two types of networks, i.e., **direct influence-diffusion network** and **indirect influence-diffusion network**.

3.1   Formal Definitions

Definition 1: A Social Network in general refers to a graph $G = (V, E)$ containing a set of vertices $V = \{v_1, v_2, v_3, ..., v_n\}$ interconnected by edge set $E = \{e_1, e_2, e_3, ..., e_n\}$. An edge is represented by $e_{ij} = \{(v_i, v_j)|v_i, v_j \in V \wedge v_i \neq v_j\}$, and $w(e_{ij})$ denotes the weight of the corresponding edge e_{ij}. Recall that, HSN tends to model the users' direct and indirect influential relations, thus, two types of social network present in the current setting, i.e., direct influence-diffusion network G_D^n and indirect influence-diffusion network G_I^n, where n denotes the network index, $n \in \mathbb{N}$.

Definition 2: User is defined as a vertex $v_i, v_i \in V$ in any network $G_X^n = (V, E_X^n)$. Each user has a different set of neighbours in divergent networks. User v_i's neighbourhood in a particular graph G_X^n is represented as $\Gamma(v_i|G_X^n)$, where $(e_{ij}|G_X^n) \in E_X^n$, $v_j \in \Gamma(v_i|G_X^n)$. We regard v_i and v_j as adjacent neighbours in G_X^n if their tie strength $w(e_{ij}|G_X^n)$ is larger than threshold σ.

Another type of node, i.e., item also exists in the social context. **Item** is defined as a type of entity, such as product, shop, etc., that is visited by users. $I = \{i_1, ..., i_n\}$ represents the item set, where i_x denotes the xth item, $i_x \in I, x \in \mathbb{N}$. v_j has a **preference state** towards item i_x, i.e., $s_{jx}, s_{jx} \in \{PA, NA, IA\}$. $s_{jx} = PA$ implies that v_j shows a favour towards item i_x and tends to diffuse positive influence to its neighbours $\Gamma(v_j)$. Similarly, $s_{jx} = NA$ indicates that v_j expresses disfavour towards item i_x. While, IA refers to neutral opinion. The initial NA users are regarded as the **negative influencers**.

Definition 3: Direct Influence-Diffusion Network $G_D^n = (V, E_D^n)$ is a homogeneous network representing users V and their direct influential relations of nth layer, where the edges E_D^n presenting direct interactions are explicitly formed. Edge weight $w(e_{ij}|G_D^n)$ denotes the interaction frequency between users v_i and v_j in G_D^n. Equation 1 aims to calculate the edge weight in direct influence-diffusion network, where $f_i|G_D^n$ denotes user v_i's interactions with neighbours in G_D^n, and $max(f_i|G_D^n)$ refers to v_i's maximum interaction frequency G_D^n.

$$w(e_{ij}|G_D^n) = \frac{w(e_{ij}|G_D^n) - min(f_i|G_D^n)}{max(f_i|G_D^n) - min(f_i|G_D^n)} \cdot \frac{w(e_{ij}|G_D^n) - min(f_j|G_D^n)}{max(f_j|G_D^n) - min(f_j|G_D^n)} \quad (1)$$

Definition 4: Indirect Influence-Diffusion Network $G_I^n = (V, I, E_I^n, E_I^{n'})$ represents one of the indirect influence propagation spaces, where the users' relations E_I^n can be established implicitly based on the user-item interactions $E_I^{n'} = \{(v_m, i_n)|v_m \in V, i_n \in I\}$. Meanwhile, $\varphi(v_i, G_I^n)$ represents the interacted items of v_i in G_I^n. The edge weight $w(e_{ij}|G_I^n)$ denotes the implicit influential relations, such as similar preference, attention or criteria for items.

In this context, we assume $w(e_{ij}|G_I^n)$ is derived from the ratings to items. Specifically, r_{jx} indicates the rating value of i_x given by v_j, and $w(e_{ij}|G_I^n)$ is formulated using Jaccard index [11] by considering rating differences in Eq. 2.

$$w(e_{ij}|G_I^n) = (1 - \sqrt{\sum_{i_x \in \varphi(v_i, G_I^n) \cap \varphi(v_j, G_I^n)} (r_{ix} - r_{jx})^2}) \cdot \frac{|\varphi(v_i, G_I^n) \cap \varphi(v_j, G_I^n)|}{|\varphi(v_i, G_I^n) \cup \varphi(v_j, G_I^n)|}$$
(2)

Definition 5: Hybrid Social Network $G_{HSN} = (V, E_H^n)$ is an influence-diffusion network constructed by merging a collection of social networks in $G_D = \{G_D^1, G_D^2, G_D^3, ..., G_D^p\}$ and $G_I = \{G_I^1, G_I^2, G_I^3, ..., G_I^q\}$. Edge $e_{ij}|G_{HSN}$ denotes a comprehensive influence-diffusion channel between v_i and v_j, while the corresponding edge weight represents the possibility that influence propagates from one node to the other, which can be formulated in Eq. 3. $|G_X|$ denotes the cardinality of graph collection G_X.

$$w(e_{ij}|G_{HSN}) = 1 - \prod_{G_X^n \in \{G_D, G_I\}} \prod_{n=1}^{|G_X|} 1 - w(e_{ij}|G_X^n)$$
(3)

3.2   Influence Propagation Mechanism of CIP

The influence diffusion approach leveraged in the proposed model inherits and extends the key features of classic ICM, i.e., propagation and attenuation. The influence initiates from the **seed set**, i.e., activated nodes (both PA and NA). They transfer their influence through the correlation graph, whereas the power of this effect decreases when hopping further and further away from the activated nodes. Mathematically, $A = \{v_{a_1}, v_{a_2}, v_{a_3}, ..., v_{a_n}\}$ denotes an initial set of activated users, where $A \subset V, s_{v_{a_n}} \in \{PA, NA\}$. Furthermore, The classic IC model has been extended by accommodating the influence strength. In order to balance the influential capacity of both PA and NA nodes, we utilise the breadth-first influence diffusion algorithm, which is described in Algorithm 1.

In Algorithm 1, the input A_n is a set of activated vertices of level n and influence diffusion threshold σ_p. While, the output is global activated set A that is accumulated by each level of A_n. In each iteration, all the inactive neighbours of vertices in A_n are supposed to be activating candidates; the activated users including both PA and NA are added to S_n of current level. Lines 1–3 shows the termination criteria of this recursive algorithm. Line 5 aims to balance the positive and negative influence by randomizing the sequence of activated nodes set. Line 15 checks whether the propagation criteria meets. Similarly, Line 17 aims to determine if the neighbour v_j to be activated by considering **Influence Propagation Attenuation** (IPA). Line 19 updates the IPA after a successful propagation. Lines 24–26 demonstrate the increment of global activated vertices A. The algorithm is recursive, thus, S_n will be taken as input for the next level by invoking itself in Line 27.

Algorithm 1. Breadth-First Influence Propagation Algorithm

Input: A_n, G_{HSN}, σ_p
Output: A
1: **if** $A_n = \emptyset$ **then**
2:     Return
3: **else**
4:     Initialize $S_n := \emptyset$
5:     Shuffle the sequence of A
6: **end if**
7: **for** $\forall v_i \in A_n$ **do**
8:     **if** $NotRecursiveInoke$ **then**
9:         Initialize $v_i.IPA := 1$
10:     **end if**
11:     **for** $\forall v_j \in \Gamma(v_i|G_{HSN}) \wedge w(e_{ij}|G_{HSN}) \neq 0$ **do**
12:         **if** $s_j \neq IA$ **then**
13:             Next
14:         **end if**
15:         **if** $w(e_{ij}|G_{HSN}) \cdot v_i.IPA \geq \sigma_p$ **then**
16:             Generate a random decimal $d_r, 0 \leq d_r \leq 1$
17:             **if** $d_r \leq w(e_{ij}|G_{HSN}) \cdot v_i.IPA$ **then**
18:                 $s_{jx} := s_{ix}$
19:                 $v_j.IPA := v_i.IPA \cdot w(e_{ij}|G_{HSN})$
20:                 $S_n := S_n \cup \{v_j\}$
21:             **end if**
22:         **end if**
23:     **end for**
24:     **if** $S_n \neq \emptyset$ **then**
25:         $A := A \cup S_n$
26:     **end if**
27:     Invoke self and input S_n as variable
28: **end for**

3.3 CIP-Based Influence Maximization

A typical application of CIP is to tackle the influence maximization problem [4,5], where social network plays the medium for promoting a particular item. It aims to select a set of influential users with limited budgets to maximize the positive influence of a particular item in social networks [12]. The selection process is named as **seed selection**, and the selected set of users is **seed set**.

The CIP-based influence maximization problem tends to select seeds from a heterogeneous network having different types of nodes and links, where both positive and negative influence towards a particular item coexist, and they are capable of propagating through various channels. It aims to maximize the positive impact with different scale of negative influencers. Therefore, the evaluation metric of **influence effectiveness** is formulated in Eq. 4, where $|PA|$, $|NA|$ and $|V|$ represent the size of positive, negative and all the users respectively. The reward from positive nodes and penalization from the negative ones present in an asymmetric way by using a trade-off factor $\beta, \beta \in [0, 1]$, which is determined based on the specific business needs.

$$\xi(PA, NA) = \beta \cdot \frac{|PA|}{|V|} + (1 - \beta) \cdot (1 - \frac{|NA|}{|V|}) \tag{4}$$

There are a number of classic seed selection algorithms for influence maximization problem, such as greedy selection, rank-based selection and random selection [5, 7]. In the current setting, by considering multiple influence-diffusion channels, we utilise the following seed selection algorithms to evaluate the model performance in different scales of negative influencers.

- *Greedy Selection:* Obtain the maximum influence marginal gain in selecting each seed.
- *Rank-based Selection:* Rank users based on the node degree in a particular social network.
- *Random Selection:* Select users randomly.

4   Experiment and Analysis

In this section, experiment has been conducted to evaluate the proposed model in CIP-based influence maximization problem. Due to the current technical limitations and great privacy concerns, it is impossible to capture all the possible behaviours and relations of each individual. Therefore, we select two social networks for this experiment, i.e., the Trust Network (TNT) and Preference Network (PNT), presenting direct influence-diffusion network and indirect influence-diffusion network respectively.

4.1   Experiment Scenario

An organization intends to market a particular new product i_x via social networks. As the product has been introduced via other approaches, some users already have a prospective attitude towards i_x. Due to the limited budget, the organization plans to select a finite set of users as initial positive influencers, hoping that they can recommend i_x to their friend circles and spread positive influence in the network. We assume the possible negatives influencers are known. If any of them are selected as seeds, the preference state is not revised and they are not supposed to exert any positive influence on the neighbours.

4.2   Dataset

Movielens[1] dataset [13] has been used in our experiment. It is a stable benchmark dataset, which contains 1,000,209 anonymous ratings of approximately 3,900 movies made by 6,040 MovieLens users who joined MovieLens in 2000.

In order to reduce the computational cost, 500 users are selected randomly. Three social networks have been constructed based on the dataset, and the basic properties show in Table 1.

---

[1] https://grouplens.org/datasets/movielens.

Table 1. Properties of social networks

Networks	#Nodes	#Edges	Average degree	Average clustering coefficient
Hybrid Social Network (HSN)	500	3443	15.068	0.263
Preference Network (PNT)	500	4046	17.707	0.122
Trust Network (TNT)	500	5737	24.141	0.822

4.3 Experimental Results

Performance of the CIP is evaluated in influence maximization problem by using various traditional seed selection algorithms, where three rank-based approaches are based on the node degree of the corresponding social network. The evaluation metric, i.e., influence effectiveness, has been defined in Eq. 4. By considering both positive and negative influence, we assign $\beta = 0.5$ in our experiment. As influence diffusion and infection is a stochastic model, the values of influence effectiveness are averaged over 100 trials.

Selecting seed set from HSN implies identifying influencers by considering multiple influence-diffusion channels, i.e., the trust connectivity and user preference in the current setting. Whereas, selecting seed set from PNT or TNT indicates the consideration only covers a particular aspect of influence. Among all the seed selection algorithms, the greedy selection normally performs better than others though it is not scalable [6]. Therefore, we regard the seed set and its corresponding influence effectiveness produced by greedy algorithm as the optimal solution. In this experiment, seeds are selected from HSN using greedy selection, subsequently, they have been input into the TNT and PNT to evaluate the influence effectiveness against other algorithms. Next, we select seeds from TNT and PNT using greedy algorithm and apply them in HSN for evaluation.

Figures 3 and 4 demonstrate the influence effectiveness comparison in TNT and PNT respectively, where only 15 negative influencers present in this context.

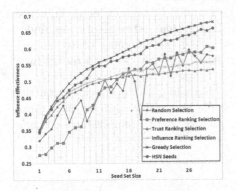

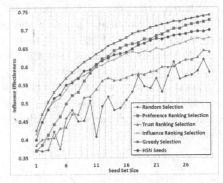

Fig. 3. Evaluation in TNT   **Fig. 4.** Evaluation in PNT

Table 2. Influence effectiveness difference comparison

Neg influencers size	15	15	15	30	30	30	50	50	50
Seed set size	10	20	30	10	20	30	10	20	30
TNT (random)	0.130	0.077	0.105	0.103	0.113	0.103	0.069	0.106	0.096
TNT (pref-rank)	0.120	0.073	0.080	0.105	0.076	0.081	0.074	0.074	0.087
TNT (trust-rank)	0.054	0.107	0.145	0.053	0.100	0.137	0.034	0.077	0.118
TNT (seeds from HSN)	**0.009**	**0.021**	**0.020**	**0.019**	**0.022**	**0.015**	**0.026**	**0.031**	**0.026**
PNT (random)	0.116	0.166	0.156	0.174	0.137	0.172	0.113	0.148	0.189
PNT (pref-rank)	0.040	**0.032**	**0.014**	0.040	**0.014**	**0.003**	0.028	**0.022**	**0.013**
PNT (trust-rank)	0.104	0.121	0.101	0.113	0.135	0.110	0.106	0.139	0.141
PNT (seeds from HSN)	**0.031**	0.035	0.041	0.021	0.024	0.021	**0.023**	0.026	0.030
HSN (random)	0.144	0.116	0.140	0.099	0.150	0.101	0.150	0.139	0.138
HSN (pref-rank)	0.105	0.049	0.042	0.097	**0.053**	0.057	0.082	0.057	0.053
HSN (trust-rank)	0.060	0.087	0.086	0.063	0.098	0.104	0.062	0.096	0.104
HSN (seeds from PNT)	0.048	0.049	0.052	0.058	0.062	0.067	**0.051**	**0.052**	**0.047**
HSN (seeds from TNT)	**0.031**	**0.044**	**0.038**	**0.051**	0.066	**0.048**	0.057	0.072	0.058

As we can observe from both figures, the seeds from HSN give more considerable performance, since the effectiveness is pretty close to the seeds produced by greedy algorithm, especially in TNT.

Moreover, we evaluate the proposed model further by comparing the influence effectiveness difference in various scenarios, where the "difference" refers to the influence effectiveness gap between the seed set selected by a particular algorithm and that produced by the greedy selection in the same network and scenario. Table 2 compares the effectiveness difference in different networks and scenarios. For example, TNT (random):0.130 refers to the influence effectiveness difference between random selection and greedy selection in TNT, when 15 negative influencers exist and the size of the seed set is 10. We also find that the optimal solution from HSN still performs well and even outperforms some classic heuristic-based algorithms in other social networks.

Whereas, the optimal solutions from TNT and PNT present higher influence effectiveness difference, though some are better than other algorithms under certain scenarios. In Figs. 5 and 6, by comparing the difference of exchanging the solutions, it is obvious that the seeds from HSN are closer to the optimal solutions in other social networks.

Based on the experiment and the above discussion, we can conclude that the proposed model can deliver more considerable and stable seed set than those considering a single aspect of influence, such as trust connectivity or user's preference.

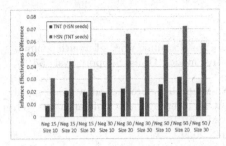 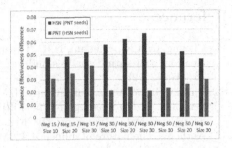

Fig. 5. Solution exchange: TNT and HSN   **Fig. 6.** Solution exchange: PNT and HSN

5   Conclusion and Future Work

In this paper, we proposed a novel approach, i.e., hybrid social network, to model the influence propagation in hybrid networks. We articulated the multi-faceted nature of influence, introduced the decomposition of influence effects, and defined direct/indirect influence. Furthermore, a concrete implementation of hybrid social network, i.e., CIP has been introduced, and validated in influence maximization problem. The experimental results reveal that seed set selected from HSN using greedy algorithm gives considerable and stable performance in other social networks, but not vice versa.

As we claimed that CIP is just one of the applications of hybrid social network. There are various potential directions to investigate influence diffusion by leveraging this generic approach, thus, the future work is set as follows.

– **Capture the dynamics of influence diffusion using HSN.** Hybrid social network implies the decomposition of influence effects, which gives high extensibility and flexibility. Specifically, when other available influential factors are added or the existing factors are changed, the model can be extended by granting adaptation capabilities, which is able to update a particular influence facet with the evolution of social networks.
– **Model the influence diffusion in a decentralised manner.** Influence agents can be allocated to each social network to monitor and manage the influential relations, while a central agent takes charge of the hybrid social network by communicating with the other agents.
– **Analyse major channels of influence diffusion.** The hybrid social network model can be extended by considering the impact factor in each influence facet. A particular influence can be diffused through various channels with different chances/possibilities. Further research works can be set to analyse the major channels for influences.

References

1. Cheung, C.M.K., Thadani, D.R.: The impact of electronic word-of-mouth communication: a literature analysis and integrative model. Decis. Support Syst. **54**(1), 461–470 (2012)
2. Bhagat, S., Goyal, A., Lakshmanan, L.V.S.: Maximizing product adoption in social networks. In: Proceedings of the Fifth ACM International Conference on Web Search and Data Mining, pp. 603–612. ACM (2012)
3. Anthony: The nature of influence, 02 March 2009
4. Domingos, P., Richardson, M.: Mining the network value of customers. In: Proceedings of the Seventh ACM SIGKDD International Conference on Knowledge Discovery and Data Mining, pp. 57–66. ACM (2001)
5. Kempe, D., Kleinberg, J., Tardos, É.: Maximizing the spread of influence through a social network. In: Proceedings of the Ninth ACM SIGKDD International Conference on Knowledge Discovery and Data Mining, pp. 137–146. ACM (2003)
6. Chen, W., Wang, Y., Yang, S.: Efficient influence maximization in social networks. In: Proceedings of the 15th ACM SIGKDD International Conference on Knowledge Discovery and Data Mining, pp. 199–208. ACM (2009)
7. Chen, W., Yuan, Y., Zhang, L.: Scalable influence maximization in social networks under the linear threshold model. In: IEEE 10th International Conference on Data Mining (ICDM), pp. 88–97. IEEE (2010)
8. Jiang, Y., Jiang, J.C.: Diffusion in social networks: a multiagent perspective. IEEE Trans. Syst. Man Cybern. Syst. **45**(2), 198–213 (2015)
9. Chen, Y.-C., Zhu, W.-Y., Peng, W.-C., Lee, W.-C., Lee, S.-Y.: CIM: community-based influence maximization in social networks. Acm Trans. Intell. Syst. Technol. (TIST) **5**(2), 25 (2014)
10. Dorigo, M., Bonabeau, E., Theraulaz, G.: Ant algorithms and stigmergy. Futur. Gener. Comput. Syst. **16**(8), 851–871 (2000)
11. Choi, S.-S., Cha, S.-H., Tappert, C.C.: A survey of binary similarity, distance measures. J. Syst. Cybern. Inf. **8**(1), 43–48 (2010)
12. Richardson, M., Domingos, P.: Mining knowledge-sharing sites for viral marketing. In: Proceedings of the Eighth ACM SIGKDD International Conference on Knowledge Discovery and Data Mining, pp. 61–70. ACM (2002)
13. Harper, F.M., Konstan, J.A.: The movielens datasets: history and context. ACM Trans. Interact. Intell. Syst. (TiiS) **5**(4), 1–19 (2015)

Multi-agent Planning with Collaborative Actions

Satyendra Singh Chouhan[(⊠)] and Rajdeep Niyogi

Department of Computer Science and Engineering,
Indian Institute of Technology Roorkee, Roorkee 247667, India
{satycdec,rajdpfec}@iitr.ac.in

Abstract. When multiple agents perform different actions simultaneously such that the combined effect of the actions achieve a given goal, then such actions are referred to as collaborative actions. Temporal planners can handle such actions by reasoning explicitly about time. However, algorithms for partial order planning can be used effectively for multi-agent domains with minor modifications without the requirement of explicit time. In this paper, we consider multi-agent planning (MAP) problems with such actions without reasoning explicitly about time. We give a new specification technique for handling collaborative actions in a PDDL-like language. We propose a new approach to solve such MAP problems. We have implemented and evaluated our approach on some benchmark planning problems. The results show the effectiveness of our approach.

Keywords: Multi-agent planning · Cooperative MAP · Collaborative actions

1 Introduction

Multi-agent system introduces the social aspect of planning between agents, where each agent performs its action with respect to actions of the other agents. Such actions are termed as collective actions (social actions). In multi-agent planning (MAP), agents may be cooperative or self interested. In cooperative multi-agent planning, agents cooperate with each other to achieve a common goal. There are different type of cooperation needed (results of collective actions) among the agents. Consider the following example.

Example 1. Multi-agent doorway domain. There are four rooms: *Room*1, *Room*2, *Room*3, and *Room*4; three agents *A*1, *A*2, and *A*3; boxes *B*1 and *B*2. The goal is to move all the boxes to the *Room*4 (Fig. 1). Box *B*1 is a light box and can be pickup by single agent individually however, two agents are needed to pickup the box *B*2. Rooms are connected through the doors. To move through a door an agent has to hold the door open so that other agent can pass through. Now, there are different types of cooperation among agents (results of various collective actions) needed in this domain such as:

Type 1. When agents are working in a common environment, the actions should not be conflicting. For example, two agents trying to pick up the box *B*1. In this case, only one of the agents would be able to pick up the box.

© Springer International Publishing AG 2016
B.H. Kang and Q. Bai (Eds.): AI 2016, LNAI 9992, pp. 609–620, 2016.
DOI: 10.1007/978-3-319-50127-7_54

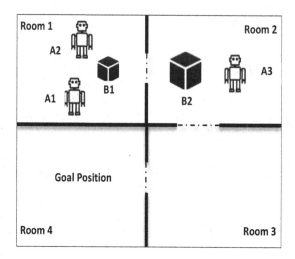

Fig. 1. Multi-agent doorway domain

Type 2. When multiple agents are performing the same action at the same time. For example, two agents are trying to pick up a heavy box.

Type 3. Another type of coordination would be, when different agents are performing different actions at the same point of time such that the combined outcome of the actions achieve the given goal. For example, two agents $A1$ and $A2$ are in $Room1$. Box $B1$ is $Room1$. The goal is to move $B1$ to $Room2$. $Room1$ and $Room2$ are connected through a door (d). Now, to move through the door (d) an agent has to hold the door open so that another agent (who might be carrying the box $B1$) can pass through. In this case, both actions $(hold\_door(A1,\ d)$ and $move(A2,\ d))$ can be termed as collaborative actions. These actions need to be performed simultaneously to achieve a given goal. There is a required temporal concurrency between these actions i.e., as long as an agent is holding the door the other agent can pass through.

There are various multi-agent planners that support cooperation of Type 1 [1–8]. There are few works for multi-agent planning problems with joint actions of Type 2 [9–12]. There are very few works in MAP that address the cooperation of Type 3.

In this paper, we focus on the planning problems that require cooperation among agents of Type 3. There is a long tradition of work in temporal planning to deal with such problems with required temporal concurrency. Temporal planners can handle concurrent actions with required temporal concurrency [13,14]. Temporal planners need to reason explicitly about time. However, algorithms for partial order planning can be used effectively for multi-agent domains with minor modifications without the requirement of explicit time. In this paper, we propose an approach that can solve such MAP problems (Type 3) without representation of explicit time. We identified two domains from IPC benchmark domains such that the multi-agent versions of these domains exhibit all the above three types of concurrency constraints between actions of the agents.

The rest of this paper is structured as follows. In Sect. 2 we discuss some related work. Section 3 presents the formal definitions and proposed work. Experimental results are given in Sect. 4. Conclusions are given in Sect. 5.

2   Related Work

Multi-agent planning domains can be classified into two types: loosely coupled and tightly coupled [15]. MAP problems that belong to loosely coupled can be solved by a single agent without cooperation of other agents (e.g., blocks-world domain). The tightly coupled planning problems requires cooperation among the agents (e.g., box-pushing domain).

In the recent year there are some state-of-the-art planners proposed that can solve loosely coupled and/or tightly coupled planning problems [1,2,16–18]. However, there are very few work suggested that focus on cooperative multi-agent planning with joint actions and/or collaborative actions. Now, we discuss some of the recent work that focus on such MAP problems.

In [9] a partial order planning algorithm is proposed that can handle concurrent interacting actions. It was a first attempt to handle concurrent actions without using time. A specification technique for concurrent actions based on the STRIPS language is given where the 'concurrent action' list has to be included in addition to the precondition and effect lists. In a multi-agent setting there typically arise two types of interaction of actions of different agents. First, consider the example, where an agent $Agent1$ can perform an action '$pickup(x)$' (to pick up an object x) only when no other agent (say, $Agent2$) is simultaneously making an attempt to pick up the object x i.e., perform '$pickup(x)$'. Second, consider the scenario of pushing a heavy box, where two agents are needed to perform the same action '$push$' simultaneously. The type of actions arising in the latter situation is referred to as joint actions in [9].

In [11], MAP planning problem is translated into temporal planning problem. The input MAP problem is modelled as a classical planning problem. In addition, concurrency constraints are mentioned on the $\langle action, object \rangle$ tuple. A concurrency constraints are represented as $\langle obj, act\_name, min, max \rangle$ where the interpretation of the concurrency constraints is that not more than max and at least min agents can simultaneously utilize object obj via the actions act_name.

For example, in MA-doorway domain a door (say $d1$) can only be used by one agent at a time for $open\_door$ action. Thus, the concurrency constraint will be $\langle d1, open\_door, 1, 1 \rangle$. However, from the same door more than one agent can pass through simultaneously. Therefore, the concurrency constraint for the $move$ action from $d1$ will be $\langle d1, move, 1, n \rangle$. Here, n be can be any number of agents. Similarly concurrency constraints for all the objects in the domain is specified in the input problem. For each action, the translation creates at most two new temporal actions. The translated problem is solved using an existing temporal planner. The details of translation are given in [11]. However, effectiveness of the given approach on the MAP problems with complex concurrency constraints is still an open question.

In [12], a new planning approach to MAP problems with concurrent actions is suggested. It suggested a translation technique that convert a multi-agent planning problem into a single agent planning problem. It specify the concurrency constraints on the ⟨action, object⟩ tuple in the domain. The single agent plan is generated using a classical planner. Thereafter, the single agent plan is compressed into to a multi-agent plan by a plan compression algorithm. The plan compression algorithm is based on dynamic programming. It iteratively merges the sequential actions to exhibits concurrency in the joint multi-agent plan.

In [10], MAFS algorithm is extended to supports joint actions. It specify a joint action by mentioning more than one agents in the action parameter list. The relevant precondition of all the agents is included in the parameter list of the joint action. All the agents jointly search the state space by exchanging relevant information (heuristic values) on a state to each other.

In all of the above works, multi-agent planning with concurrent interacting actions (joint actions) is considered where an action can be performed by multiple agents simultaneously. The approaches given in [11,12] will be useful when the agents are using (or trying to use) the same object at a same time (e.g., lifting a box together). However, it cannot handle other concurrency constraints that cannot be mentioned on objects, e.g., concurrency constraints in collaborative actions (Sect. 1). Thus, in this paper, we propose a new approach that can handle such type of MAP problems.

3   Proposed Work

In this section, we present formal definition of the multi-agent planning problem. Next, we discuss the specification of collaborative actions and planning approach to solve the MA-problem. We consider the Multi-agent planning as a MA-STRIPS problem given in [5]. Agents are cooperative and share a common goal.

Definition 1: MA-STRIPS problem for n agents is given by a quadruple $\pi = \langle P, A_{i=1 \ to \ n}, I, G \rangle$, where P is a finite set of propositions, $I \subseteq P$ represents the initial state, and $G \subseteq P$ is the goal state. A_i is the finite set of actions that the agent i can perform. Each action $a \in A = \cup A_i$ has the standard STRIPS syntax and semantics, that is, $a = pre(a), add(a), del(a)$ is given by its preconditions, add effects, and delete effects.

We consider that, an action template consists of an agent parameter that denotes the agent performing the action. An action having more than one agent in its parameter list represents that the action is performed by more than one agent–joint action.

3.1   Collaborative Actions

When different agents are performing different actions at the same point of time such that the combined outcome (synergistic effect) of the actions achieve a given goal, then such actions are referred to as collaborative actions.

The collaborative actions are special case of concurrent interacting actions such that actions must be performed concurrently (required concurrency). We discuss two domains from IPC (International Planning Competition) domains such that their multi-agent version of these domains exhibits all the types of concurrent actions (such as collaborative actions, joint actions).

1. Multi-agent (MA) - match cellar domain. This domain is a multi-agent version of match cellar domain from IPC benchmark domain (Temporal track). In classical match cellar domain, the goal is to mend fuses in a dark cellar (room). To mend a fuse requires the MEND_FUSE action for which there must be light for the entire duration of the action. The light can only be achieved by lighting a match (using LIGHT_MATCH action), which provides light only whilst it burns (i.e. for the duration of the action). To mend a fuse one must also have a hand free, the impact of which is that one can only fix one fuse at a time.

In the multi-agent version of the match cellar domain, the agents (electricians) must fix fuses by the light of matches as given above. The fuses are located at different rooms in a building, that the agents must navigate via corridors and lifts. A room can have more than one fuse. This navigation does not require concurrency. Since there is now more than one agent, more fuses can be fixed concurrently by the light of one match.

2. MA-doorway domain. This domain is a multi-agent version of turn and open domain from IPC benchmark domain (temporal track). In this domain, there are a number of agents, and a set of rooms. The rooms are connected through the doors. In order to open a door the agent must turn the doorknob and open the door at the same time. The set of room contains boxes of different sizes (heavy and light). Light boxes can be moved by single agents; however, to move a heavy box, two agents are needed. The goal is to find a plan to transport boxes from a given room to another. The characteristics and assumptions of this domain are:

Concurrent actions: This domain contains multiple agents that execute their actions simultaneously.

Concurrency constraints: There are constraints such that an action can (or must) be performed simultaneously. For example, in this domain, heavy boxes should be moved by two agents. In addition, if an agent (or agents) is carrying a box, then it cannot open a door; the door should be kept open by another agent. All the agents are cooperative in nature and they share a common goal.

3.2   Specification of Collaborative Actions in MAP

Actions in multi-agent planning raise two main issues: representation and planning. A basic syntax of action in PDDL is as follows:

:action $\langle action\_name \rangle$
:parameter $\langle parameter\_list \rangle$
:precondition $\langle precondition\_list \rangle$
:effect $\langle effect\_list \rangle$

In this syntax, parameter list contains all the objects involved in the action. The precondition list contains all the predicates that must be true at the time of performing the action. Effect list contains all the predicates that will be true or false, i.e.,

$$effect(\alpha) = add(\alpha) \cup del(\alpha)$$

where $add(\alpha)$ contains the predicates that will be true and $del(\alpha)$ contains the predicates that will be false after performing the action α.

Specification of Collaborative Actions in the Proposed Work. We extended the standard action template by adding a collaborative action list of collaborative actions, such that:

- collaborative action list contains the list of actions that provide temporal required concurrency for the action.
- An action may or may not have collaborative action list. For example moving through the corridor (MA-match cellar domain) does not require collaborative actions from other agents. Therefore, it will be an empty list for the *move* action.

Syntax: Action template with collaborative action list.

:action $\langle action\_name \rangle$
:parameter $\langle parameter\_list \rangle$
:precondition $\langle precondition\_list \rangle$
:collaborative $\langle collaborative\ action\_list \rangle$
:effect $\langle effect\_list \rangle$

For example, in MA-match cellar domain, syntax of *mend-fuse* action would be:

:action *Mend-fuse*
:parameter $(?a1 - agent\ ?fuse - fuse\ ?match - match)$
:precondition $(and(free\ ?a1)(lit\ match))$
:collaborative $(and(light - match\ ?a2 - Agent\ ?match - match)(not(= ?a1\ ?a2)))$
:effect $(mended\ ?fuse)$

Semantics: Collaborative actions are existentially quantified, that is one of the instance of collaborative action (grounded actions) must be performed concurrently with the main action. In the above example, light match action must be perform concurrently with *mend-fuse* action. There may exist more than one collaborative actions in the collaborative action_list. Then, all the actions in the collaborative action_list would be performed concurrently with the main action.

We require that the actions are performed by different agents; this is expressed by $not(= ?a1 ?a2)$ in the collaborative action list.

3.3 Planning Approach

Consider the $Mend - fuse$ action as shown above. The precondition of $Mend - fuse$ required temporal effect light that is provided by the action $light - match$. Thus, $light - match$ is a collaborative action for the $Mend - fuse$ action. However, unlike temporal planning, both actions are performed by different agents (Fig. 2). Thus handing this situation (required concurrency) is more complex in multi-agent planning because:

- Agents are real world entities.
- Both actions can be performed by different agents.
- An agent performing an action should be busy throughout the duration of action, i.e., the agent cannot perform two actions at the same time.

Handling Collaborative Actions. If an action a_i has a collaborative action a_j, then for each collaborative action, planning process adds two new actions: start-a_j and end-a_j. The start action contains the precondition and effects as follows:

precondition(start) = precondition(a_j) ∪ (free ?ag-agent)
effect(start) = effect(a_j) ∪ ¬(free ?ag-agent)

The end action contains the following precondition and effect:

precondition(end) = effect($start$) ∪ inv_pred(a_i)
effect(end) = ¬ effect(a_j) ∪ (free ?ag-agent)

The end action contains the 'inv_pred(a_i)' predicate (an invariant) as an extra precondition. This precondition is use to hold the order between the actions. For example, mend fuse action should be completed in between the 'start-light-match' and 'end-light-match' action i.e., the ordering constraints between the actions should be

$$\text{start-}a_j \prec a_i \prec \text{end-}a_j$$

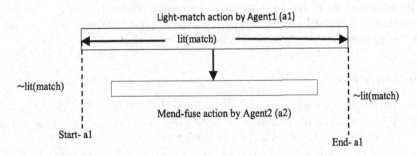

Fig. 2. Collaborative actions by different agents (example)

Here $\prec$ is a binary relation on the actions; it is irreflexive, antisymmetric, and transitive. So $\prec$ is a partial order.

The corresponding set of causal links will be:

$$\text{start-}a_j \xrightarrow{\text{c1}} a_i, \quad \text{start-}a_j \xrightarrow{\text{c2}} \text{end-}a_j, \quad \text{and} \quad a_i \xrightarrow{\text{c3}} \text{end-}a_j.$$

For example, start-light_match and end_light match action would be:

:action $start - light\_match$
:parameter $(?a1 - agent\ ?match - match)$
:precondition$(and(free\ ?a1)(unused\ ?match))$
:effect $(and(not(unused\ ?match)(lit\ ?match)$
$(not(free?ag))(lighting\ ?ag\ ?match))$

:action $end - light\_match$
:parameter $(?a1 - agent\ ?match - match)$
:precondition $(and(lit\ ?match)(lighting\ ?ag\ ?match)(done))$
:effect $(and\ (not\ (lit\ ?match))(not\ (lighting\ ?ag\ ?match))(free\ ?ag))$

Here, predicate '*done*' (similar to $inv\_pred(a_i)$ explained earlier) is used as an extra precondition. This precondition is used to preserve the ordering between the actions.

3.4   Planning Algorithm

We use the standard partial order planning (POP) algorithm [19] for planning purpose. A partial order plan is given as follows.

Definition 2: A partial order plan is a tuple $\rho = \langle A, S, \Gamma, C \rangle$ where, A is set of actions, S is the set of plan steps (instantiated actions), Γ is the set of ordering constraints between actions, and C is the set of causal links.

Here, if $\langle s_i, s_j \rangle \in \Gamma$ then s_i must precede the s_j in the plan. A tuple $\langle s_i, s_j, c \rangle \in C$ denotes a causal link (c) between s_i and s_j. It means that s_i provides the one or more precondition for step s_j.

In addition, we maintain a set of concurrency constraint (=). '=' is set of concurrent constraints on plan steps, i.e., if $\langle s_i, s_j \rangle \in\ =$ then steps s_i and s_j must be performed simultaneously (collaborative actions).

Planning Steps

1. *Initialization.* for each action $a_j \in \Lambda$, where $\Lambda = \cup A_i$ do the following;
   1.1 **if** a_j requires a collaborative action a_k then, create two new actions $start - a_k$ and $end - a_k$ (as per Sect. 3.3).
   - add a ordering constraint $\langle start - a_k,\ end - a_k \rangle$ in the set of ordering constraint Γ.

- add the concurrency constraint $\langle a_j, a_k \rangle$ in the set of concurrency constraint $(=)$.
- add $start - a_k$, $end - a_k$ into Λ' (initially $\Lambda' = \emptyset$).

1.2 add a_j into Λ'

2. *Planning.* Solve the modified planning problem (Λ') using a partial order planning algorithm (VHPOP) [20].
3. *Mapping.* Translate the plan obtained in step 2 into the multi-agent plan using the set of concurrency constraints $(=)$.

4   Implementation

In order to evaluate the performance of the propose approach, we performed experiments on two benchmark planning domains: MA-match cellar domains and MA-doorway domain. We have identified that the multi-agent versions of these planning domains have the scope of collaborative actions. Experiments were performed on the Intel core i7 with 3.40 GHz machine with 8 GB of RAM. The proposed approach uses an existing partial order planner (VHPOP) [20].

This section is organized as follows: first, we present the performance evaluation of proposed approach on above mentioned domain. Next, we compared the variants of pop algorithm having different search algorithms.

4.1   Experimental Results

Table 1 shows the performance evaluation of the proposed approach on two domains: MA-Match cellar domain and MA-doorway domain (Sect. 3).

Table 1. Experimental results on Multi-agent doorway domain and MA-match cellar domain

Sr .no.	No. of agents	MA-Match cellar domain		MA-doorway domain	
		Problem size (#fuses)	Time taken	Problem size (#boxes)	Time taken
1	2	5	95	5	16
2	3	8	226	8	138
3	5	15	1014	10	675
4	6	20	21095	12	1657
5	8	22	44684	15	53515
6	9	25	87090	17	258797
7	10	28	251626	18	795431
8	11	30	-	20	>1000000
9	12	32	-	25	-
10	15	35	-	30	-

legend (-) : Not solved

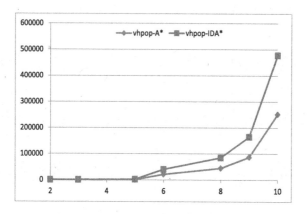

Fig. 3. Comparison of vhpop-A* with vhpop-IDA* (MA-Matchcellar domain)

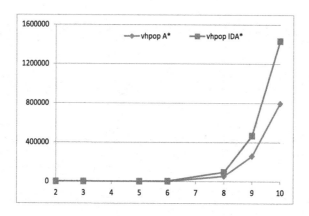

Fig. 4. Comparison of vhpop-A* with vhpop-IDA* (MA-Doorway domain)

In Table 1, Column 2 shows the number of agents in the problem instances. Column 3 and 5 shows the problem size with respect to the number of objects (fuses or boxes) in the problem instance. The experiments were run for ten times for each problem instance and the average running time is computed. Results show that the proposed approach is effective for problem instances with up to 10 agents.

Figures 3 and 4 present the comparison of variants of pop algorithm (with different search algorithms: A* and IDA*) on MA-match cellar domain and MA-doorway domain respectively. From the results given in Figs. 3 and 4 we observe that vhpop-A* is better than vhpop-IDA* on both the domains.

5   Conclusions

In this paper, we considered a class of multi-agent planning problems where agents perform collaborative actions to achieve a given goal. We are not aware of MAP approaches that can efficiently handle such problems. We identified some multi-agent planning domains (MA-match cellar domain and MA-doorway domain) where such actions are necessary. We gave a new specification technique for handling collaborative actions in a PDDL-like language. We proposed a partial order multi-agent planning algorithm to deal with collaborative actions without using explicit time, by modifying an existing partial order planning algorithm. Experimental results reveal that our approach is effective for small problem instances. As part of our ongoing/future work, we would like to extend our algorithm to handle joint actions.

Acknowledgements. The authors thank the anonymous reviewers of AI-2016 for their valuable comments and suggestions for improving the paper.

References

1. Competition of distributed and multiagent planners (CoDMAP). http://agents.fel. cvut.cz/codmap/. Accessed 30 April 2016
2. Štolba, M., Komenda, A.: MADLA: planning with distributed and local search. In: Competition of Distributed and Multi-Agent Planners (CoDMAP 2015), p. 21 (2015)
3. Borrajo, D.: Multi-agent planning by plan reuse. In: Proceedings of the International Conference on Autonomous Agents and Multi-Agent Systems (AAMAS), pp. 1141–1142 (2013)
4. Kvarnström, J.: Planning for loosely coupled agents using partial order forward-chaining. In: Proceedings of International Conference on Automated Planning and Scheduling (ICAPS) (2011)
5. Brafman, R., Domshlak, C.: From one to many: planning for loosely coupled multi-agent systems. In: Proceedings of International Conference on Automated Planning and Scheduling (ICAPS), pp. 28–35 (2008)
6. Nissim, R., Brafman, R.: Distributed heuristic forward search for multi-agent systems. arXiv preprint arXiv:1306.5858 (2013)
7. Torreño, A., Onaindia, E., Sapena, Ó.: FMAP: distributed cooperative multi-agent planning. Appl. Intell. **41**(2), 606–626 (2014)
8. Katz, M.J.: The generation and execution of plans for multiple agents. Comput. Artif. Intell. **12**(1), 5–35 (1993)
9. Boutilier, C., Brafman, R.: Partial-order planning with concurrent interacting actions. Journal of Artificial Intelligence Research **14**, 105–136 (2001)
10. Brafman, R., Zoran, U.: Distributed heuristic forward search for multi-agent systems. In: Proceedings of the 2nd ICAPS Distributed and Multi-Agent Planning workshop (ICAPS DMAP), pp. 1–6 (2014)
11. Crosby, M., Petrick, R.P.: Temporal multiagent planning with concurrent action constraints. In: ICAPS Workshop on Distributed and Multi-Agent Planning (DMAP) (2014)

12. Crosby, M., Jonsson, A., Rovatsos, M.: A single-agent approach to multiagent planning. In: 21st European Conference on Artificial Intelligence (ECAI 2014), vol. 263, pp. 237 (2014)
13. Do, M.B., Kambhampati, S.: SAPA: a multi-objective metric temporal planner. Journal of Artificial Intelligence Research **20**, 155–194 (2003)
14. Gerevini, A., Saetti, A., Serina, I.: Temporal planning with problems requiring concurrency through action graphs and local search. In: ICAPS, pp. 226–229 (2010)
15. Durfee, E.H.: Distributed problem solving and planning. In: Luck, M., Mařík, V., Štěpánková, O., Trappl, R. (eds.) ACAI 2001. LNCS (LNAI), vol. 2086, pp. 118–149. Springer, Heidelberg (2001). doi:10.1007/3-540-47745-4_6
16. Tozicka, J., Jakubuv, J., Komenda, A.: PSM-based planners description for CoDMAP competition. In: Proceedings of the Competition of Distributed and Multi-agent Planners (CoDMAP 2015), pp. 29–32 (2015)
17. Štolba, M., Komenda, A., Kovacs, D.L.: Competition of distributed and multiagent planners (CoDMAP). In: The International Planning Competition (WIPC 2015), p. 24 (2015)
18. Torreño, A., Onaindia, E., Sapena, Ó.: An approach to multi-agent planning with incomplete information. In: 20th European Conference on Artificial Intelligence (ECAI), pp. 762–767 (2012)
19. Weld, D.S.: An introduction to least commitment planning. AI Mag. **15**(4), 27–61 (1994)
20. Younes, H.L., Simmons, R.G.: VHPOP: versatile heuristic partial order planner. J. Artif. Intell. Res. **20**, 405–430 (2003)

Text mining and NLP

Open-Domain Question Answering Framework Using Wikipedia

Saleem Ameen, Hyunsuk Chung, Soyeon Caren Han,
and Byeong Ho Kang[✉]

School of Engineering and ICT, Tasmania 7005, Australia
{sameen, David. Chung, Soyeon. Han,
Byeong. Kang}@utas. edu. au

Abstract. This paper explores the feasibility of implementing a model for an open domain, automated question and answering framework that leverages Wikipedia's knowledgebase. While Wikipedia implicitly comprises answers to common questions, the disambiguation of natural language and the difficulty of developing an information retrieval process that produces answers with specificity present pertinent challenges. However, observational analysis suggests that it is possible to discount the syntactical and lexical structure of a sentence in contexts where questions contain a specific target entity (words that identify a person, location or organisation) and that correspondingly query a property related to it. To investigate this, we implemented an algorithmic process that extracted the target entity from the question using CRF based named entity recognition (NER) and utilised all remaining words as *potential* properties. Using DBPedia, an ontological database of Wikipedia's knowledge, we searched for the closest matching property that would produce an answer by applying standardised string matching algorithms including the Levenshtein distance, similar text and Dice's coefficient. Our experimental results illustrate that using Wikipedia as a knowledgebase produces high precision for questions that contain a singular unambiguous entity as the subject, but lowered accuracy for questions where the entity exists as part of the object.

Keywords: Open-domain · Question answering · Wikipedia

1 Introduction

The desire to produce a computational system that provides discrete answers to open ended questions with high precision and accuracy has been an area of academic inquiry for decades. While the advent of the World Wide Web and the prevalence of associated technologies such as search have presented an unmitigated opportunity for individuals to explore questions on dichotomous contexts, there remains pressing challenges – specificity, speed and relevance. There is a need for individuals to interact with information with greater ubiquity, efficiency and cohesion so that it reflects prevailing standards of social dogma. Question and answering (QA) systems have been a response to this problem but have operated with limited success in open domains. Despite the strides that research in natural language processing (NLP) has presented about the

© Springer International Publishing AG 2016
B.H. Kang and Q. Bai (Eds.): AI 2016, LNAI 9992, pp. 623–635, 2016.
DOI: 10.1007/978-3-319-50127-7_55

topological, ontological and semantic superstructures of expression; the problem of open-endedness has pertained due to the absence of a viable knowledgebase. Wikipedia, a perpetually expanding collation of information on diverse topics presents researchers with an opportunity to bridge the knowledge gap required to answer the dispersive distribution of questions presented within the open domain. This paper is therefore a preliminary investigation into the feasibility of utilising Wikipedia as a viable knowledgebase and through this study we explore the implications that this has for open-domain questions. One key tenet of Wikipedia that has strengthened its appeal is the prevalence of property-answer pairs for common characteristics related to people, locations and organisations. These properties can be extracted to present concise answers for factoid-based questions. We therefore place this as a constraint in our investigation, where we analyse entity specific questions, so that a constructive qualitative assessment about Wikipedia's knowledgebase is developed. Using named entity recognition (NER) to isolate the target entity in conjunction with standardised string matching algorithms such as the Levenshtein distance, similar text and Dice's coefficient, the efficacy of Wikipedia as the basis of a projection based question and answering framework is explored.

2   Related Works

There has been much research on Open-domain question answering (Open QA) for decades. Open QA systems require constructing a broad range of knowledge to achieve high accuracy [1]. Early systems focused on information retrieval approach [2]. Recently, large-scale knowledge base has been used to extract answer from a KB [3]. However, most of works require hand labeled on the logic that they applied to achieve high accuracy [4, 5]. A major research area in QA has been shifting to semantic parsing systems from small-sized and single-domain KBs to large-sized and multi-domain KBs [6–8]. Despite significant progress of constructing large-sized and multi-domain knowledgebase, major challenging problems of QA are still remained [9]. Recently, several QA systems proposed machine learning based systems that produce reasonable quality but still manual works are required such as designing lexicons, grammars and the knowledge base [4–6]. This paper investigates the feasibility of using large-sized multi-domain knowledgebase such as DBPedia.

3   Methodology

The objective of this paper was to investigate whether a framework that utilised Wikipedia's knowledgebase could be developed for an open domain question and answering system. We focused on examining the plausibility of answer retrieval for questions defined by two distinct characteristics: the existence of an entity and the prevalence of a related property. Our assumption was that if a question was comprised of a specific target entity, then that would be sufficient enough to query DBPedia's semantic knowledge graph by cross referencing the words embedded in the question against entity-specific keywords found within DBPedia using common string matching algorithms. This would enable an information retrieval process that could generate an answer to the user derived from the closest matching keyword. This process was implemented in stages using various methods: entity extraction using CRF based named entity recognition (NER); entity keyword generation using a SPARQL end-point to query DBPedia; property extrication using common string matching algorithms including the Levenshtein distance, similar text and Dice's coefficient; and answer representation through the implementation of a second SPARQL query using the entity and determined property. The processes implemented in this approach and the motivation for using it has been described in the following section.

3.1   Step 1: Entity Extraction – NER CRF Sequence Model

In the context of NER, the observation sequence is defined by the token sequence of the sentence. Consequently, the probability of the label sequence can be based on the independent features of the observation sequence without the model having to account for the distribution of those dependencies, which enables greater diversity among the input features and therefore greater precision in entity recognition. Implementing the model requires a training process by credible domain-relevant sources such as a dictionary or WordNet. In this study, the CRF model used has been investigated by the Stanford NLP lab and was trained with the CoNLL English training data based on the characteristics of four distinct classes (entity types): person, location, organisation and misc. In this experiment, word sequences that were labelled as either a person, location or organisation, were defined as the target entity (the subject of the question) and words that were labelled as 'misc', were defined as potential properties that the user is specifically querying about the target entity.

3.2   Step 2: Keyword Generation – DBPedia and SPARQL

The challenge associated with NLP within open domain question and answering is the underlying complexity of understanding the inherent meaning of the question that the user seeks to query. Even in situations where a singular subject target entity is successfully extracted using CRF-based NER, there remains the pressing problem of determining what the user query is actually asking about the target entity. Lexical and grammatical conventions suggest that it is reasonable to assume that questions marked by one entity subject, contains a distinct property that the user wants to ask about the entity within the remaining words of the question. This reduces the problem of understanding, to a simpler problem of property recognition. To put this into context, let us consider a scenario. Suppose a user wanted to ask 'Where was Barack Obama's place of birth?' NER would extract the target entity and subject of this question as 'Barack Obama', which places the remaining string of words 'where was place of birth' as potential candidate properties that the user wants to know about Barack Obama. Intuitively, it is clear that the property that the user wants to know about is Barack Obama's 'place of birth', however computationally determining this is difficult. In the aforementioned string, each word in isolation could be a property, each pair of words could be a property, or the entire string could be a property that the user wants information on and differentiation is a challenge. This is further exacerbated by the fact that there is no certainty that the answer to that property exists within Wikipedia's knowledgebase. The issue of property recognition is therefore twofold: 1. which word or word sequence should be selected as the target-entity property and 2. how do we determine whether the property selected has an answer?

Since the focus of our investigation was whether Wikipedia's knowledgebase alone was sufficient to answer questions that were characterised by one target entity, our SPARQL query extracted only the keyword properties that were defined by RDF data bound by Wikipedia's dataset and not the external RDF datasets linked to it. To contextualise this, suppose 'Barack Obama' was the target entity from the aforementioned example, DBPedia would return a list of all properties that were distinct to that entity based on its own algorithmic process that parses Wikipedia's dataset and labels each property accordingly. In our method, we extract all properties that are tagged with the RDF label 'dbp' or 'dbo', which represents property and ontology based characteristics respectively. The ontological based RDF label was introduced by DBPedia as a shallow, cross-domain ontology that was manually created to mitigate weaknesses in Wikipedia's infobox system related to data inconsistences, such as having different infoboxes for the same class, using different property names for the same property and not having clearly defined datatypes for property values (http://wiki.dbpedia.org/services-resources/ontology). DBPedia's ontology mappings produce a subsumption architecture of 685 classes that categorise 2,795 property types. Therefore, it was important to enact keyword generation that was based on both property and ontology labels, to produce a robust set of results that handled variation in the question asked by the user.

3.3   Step 3: Property Extraction – String Matching

Using a string matching algorithm, we wanted to select the closest matching keyword derived from the DBPedia data – data that is known to have an answer – against the potential candidate properties defined from the NER process. The following table illustrates how the data is represented prior to the string matching process:

Original Question: Where was Barack Obama's place of birth?
Target Entity: Barack Obama

Potential Candidate Properties*	Keywords (from DBPedia)**
Where	Abstract
Was	Birth date
Place	Birth place
Of	Office
Birth	Party
	Predecessor
	Religion
	Residence
	Spouse
	...

The premise of string matching algorithms is to accommodate for the following scenario: provided two strings, a text and a pattern, determine whether the pattern exists in the text. In this study, the 'text' is defined as the list of properties derived from the question and the 'pattern' is marked by each of the keywords generated from DBPedia. Given the fact that we have a many:many relationship between these two lists, we approach the problem in the following way:

```
For each potential candidate property
  -   For each keyword
        o   Compare the candidate property string with the keyword
            string
        o   Store the percentage match in a variable
        o   If the percentage match of the current keyword is greater
            than the percentage match of the previous keyword, update
            the state of the variable for the closest matching
            keyword
  -   Store the closest matching keyword of each property in a
      property match array with its corresponding percentage value
For each value in the property match array
  -   Sequentially compare the percentage match of each value
  -   Select the closest matching property as the keyword sequence
      with the highest similarity measure
```

3.4   Step 4: Answer Extraction – Using SPARQL and DBPedia

The final step in our approach was to use both the extracted target entity and the keyword with the highest matching similarity, as the two elements required to query DBPedia for an answer using the RDF triple construct. Our goal was to select the distinct object for the entity, which had the property identified. The output from the SPARQL query produces the resulting object answer. Given the fact that the keywords used for string matching were based from properties that DBPedia had in its database about the target entity, our approach prevents the problem of using a property that cannot inherently be queried.

4   Evaluation

The goal of this paper is to explore the efficacy of using Wikipedia as a knowledgebase for an automated open ended question and answering system. Given the underlying complexity associated with existing question and answering architectures, which place a heavy emphasis on NLP, we wanted to investigate whether a basic algorithmic approach that utilised Wikipedia was sufficient to produce answers for common questions.

4.1   Data Collection

For our study, it was necessary to collect a series of random open ended questions to evaluate the efficacy of our approach. The Text REtrieval Conference (TREC), which focuses on research in information retrieval provides a track of sample questions for

Table 1. Distribution of data according to entity characteristics

Entity characteristics	Number of questions
Person entities	29
Organisation entities	14
Location entities	59
Multiple entities (Same Type)	14
Multiple entities (Different Type)	16
No entity Recorded	68

use in systems that retrieve answers for open-domain, closed-class questions that are primarily fact based. TREC provides six possible datasets ranging from the TREC-8 (1999) to TREC (2004). To evaluate this study, we opted to use the TREC-8 dataset. The motivation behind this choice was driven by the fact that the answers supplied for this dataset were definitive in their representation with an emphasis on singular answers. By using the TREC-8 dataset, we established the view that it would enable us to produce a more robust representation about the efficacy of our system.

One of the assumptions that we defined in this study was that we were specifically interested in the category of questions that contained a distinct target entity. However, in the dataset, not all questions were defined according to this quality. Therefore, any question that contained more than one unique target entity or that had a non-existent entity, according to our entity categories, were omitted from this study. Table 1 shows the distribution of the questions relative to characteristics associated with their entities. Of the 200 questions, 102 questions were used for our evaluation.

4.2 Named Entity Recognition

Given the fact that our ability to generate answers was dependent on the correct classification of the target entity, we first evaluated the success of the CRF-based NER in correctly identifying and labelling the entities present. Of the 102 questions, 100 questions had been labelled according to its correct entity type, which illustrates high precision in CRF-based NER, producing an average accuracy rate of 98.04%. While the CRF-based NER was successful in identifying the entity within the question, we note certain limitations in utilising the extracted entity to produce queries for all questions. A qualitative exploration of the data illustrates how the nature of the question in terms of its semantics, colloquialisms and linguistic structure has an impact on whether the identified entity can indeed be defined as the 'target entity' of the question. We discuss these observations according to the various entity types and comment on a characteristic that was found to be universal to all three (Table 2).

4.2.1 NER for Entity Type 'Person'
One of the problems associated with questions that were related to people was the divergent ways that a particular name could be represented as a result of status or specificity. Each of these will be discussed with examples.

Table 2. The distribution of entities that were flagged as presenting issues in being the 'target'

	Person	Location	Organisation
Name not specific	37.93%	18.64%	35.71%
Name exists in object	10.34%	23.73%	14.29%
Incorrect Entity Type	3.45%	1.69%	0.00%

Status: In this context, status defines the construct of a name, which is led by the title the individual holds within society, rather than the name they occupy. In popularised culture, certain people of influence are known according to a title such as 'President', 'Queen', which truncates the full name of the person, in favour of the popularised form used to identify them. For example, one of the questions in the TREC-8 dataset was "Who was President Cleveland's wife?". In popular culture, the target entity here should be identified as "President Cleveland", however CRF-based NER identifies the name of the individual exclusively, independent of their title and thus the name 'Cleveland' is labelled and extracted. While it correctly identifies the entity type as a 'person', querying DBPedia according to the name 'Cleveland' produces ambiguity, as Cleveland could refer to innumerable individuals, including the location of Cleveland. This problem was compounded by the variety of contexts that this pertained to such as "When was Queen Victoria born?" In this question, the target entity required is 'Queen Victoria', however since NER focuses on the name, the word "Victoria" is extracted instead.

Specificity: An extension to the problem of status is the problem of specificity. It was noted that as an individual's popularity is propelled in society, it is reasonable to identify them with a truncated version of their name. For example, a question such as "Where was Lincoln assassinated?" results in the entity 'Lincoln' extracted. While the semantics of the question enables human judgement to recognise that this question refers to the president Abraham Lincoln, a computer struggles to make the same association. In isolation, the term Lincoln can be associated with a multitude of identities, which makes disambiguation a problem. Despite the apparent magnitude of this problem, the embedded ontological mapping found in DBPedia appears to support names whose truncation is sufficient enough to remain unique. For instance, in the question "When did Nixon die?", the term 'Nixon' is redirected to the president 'Richard Nixon', which places limited liability on issues associated with specificity. Thus, even though 37.9% of the questions related to people had problems with specificity; leveraging the ontology mapping of DBPedia, significantly reduced the error factor to 13.8%.

4.2.2 NER for Entity Type 'Location'

While the NER process was successful in identifying the target entity in several instances, it did not factor in the variation of syntactical and grammatical conventions where the entity identified was not necessarily the subject of the question, but in fact related to the object. To put this in context, consider the following question: "Which type of submarine was bought recently by South Korea?" Here, the subject of the question that is being asked about is the 'submarine' and the object that is related to it is 'South Korea'. NER in this case would extract the target entity as 'South Korea' and use it to direct the query. The issue with this, is that the properties related to South Korea are independent of the properties related to the submarine that was being queried about. This occurred 18.63% of the time, when analysing the cumulative distribution of the data. An interesting observation to note, was that 73.68% of these questions belonged to location based entities. We were unable to account for why the distribution skewed negatively towards location based entities, particularly given the uneven distribution of the sample size, however we were able to make speculative assertions. One speculation was the fact that it is more common to find questions that placed a location as a secondary component to a question as locations can often be used as descriptive terms to describe other objects, when compared to individuals or organisations.

4.2.3   NER for Entity Type 'Organisation'

Organisation based NER featured the best performance among the various entity types. The main discrepancy identified was the same issue of specificity that marked 'people' based questions. Just like people, organisations that exhibited exuberant levels of popularity over the course of their organisational lifecycle had their names truncated in colloquial discourse, which necessitated a level of intuition to discriminate. For example, in the question "Where is the Bulls basketball team based?", NER identifies the target entity correctly as 'Bulls'. However, this is ineffective in querying DBPedia for keywords or answers as the word 'bulls' is not a unique identifier. Instead, the terminology 'Chicago Bulls' is a more apt description to produce the required results. This occurred 14.3% of the time, which closely aligns to our 'people' based analysis. Leveraging DBPedia's ontology does not appear to be as important in organisation based queries as person based queries. One reason for this is because central to the identity of an organisation, is its name and thus, unlike people, it is rare for organisations to require the same level of semantic mapping.

4.3   Property Matching

Quintessential to the system's ability to answer open-ended questions is the efficacy in which it identifies the target property within the question. Our approach involved querying DBPedia for a list of keywords related to the target entity and finding the keyword with the highest percentage similarity when compared to the terms within the question (excluding the entity). To evaluate its performance, we used three different string matching approaches (levenshstein distance, similar text and Dice's coefficient) to determine whether there were any identifiable performance improvements that corresponded with the selected approach. Furthermore, we used three different approaches (full string, partial string, single word string) to segment the property list string within the original question to accommodate for scenarios where multiple words could define the composition of a particular keyword.

Table 3. Sample Data that illustrates a subset of questions that contain no property

Original question	Target entity	Property list
Where is Qatar?	Qatar	Where is
Where is Dartmouth College?	Dartmouth College	Where is
What did Richard Feynman say upon hearing he would receive the Nobel Prize in Physics?	Richard Feynman	What did say upon hearing he would receive the Nobel Prize in Physics
In what year did Joe DiMaggio compile his 56-game hitting streak?	Joe DiMaggio	In what year did compile his 56-game hitting streak
Who came up with the name El Nino?	El Nino	Who came up with the name

4.3.1   NER – Universal Issue: Entity as the Subject vs Entity as the Object

We first conducted a qualitative analysis of the property list that remained after the entity had been extracted from the original question to highlight distinguishing features. Our analysis primarily targeted the subset of questions that were independent of discrepancies derived in the entity selection process. The first observation we noted was that on average 38.9% of questions did not contain a clearly identifiable target property that was of relevance to the target entity; a necessary component to performing string comparison against a list of keywords.

The adverse implications of this can be best understood with reference to question one in Table 3, "Where is Qatar?" The target entity in this question is unmistakably identifiable as 'Qatar', which enables our algorithmic process to produce a set of keywords related to it. This yields a resultant list of 110 potential keywords that can be utilised using the approaches described in Sect. 4.3. The issue here is that for this to work, there must be a property that existed within the initial question that the keywords can liken a match to. However, the potential property strings that can be processed here are 'where', 'is', 'where is' – all of which are problematic in producing a reasonable match among the keyword list as they are not inherently the target property of the question, but instead are words that are reflective of 'location'. In this case, each string matching technique produced entirely divergent matches. The levenshtein distance metric produced 'timeZone' as a match, followed by similar text and Dice's coefficient that produced 'metricFirst' and 'legislature' respectively. This suggests that it is important to disregard matches that are below a particular threshold or alternatively, develop an approach that disambiguates key phrases by exploring a rule based approach. For instance, the word sequence 'Where is', which is often repeated among the questions, consistently references a location and can therefore be ontologically mapped to a more apt descriptive property. A second noteworthy observation pertains to questions that were compromised of multiple variables and therefore multiple properties. This subcategory of questions yields dispersive results as the key aspect of the question is chained in connection with a second element.

4.3.2 String Matching Approaches

For this evaluation, we wanted to determine whether using an algorithm that did not entail an underlying complex architecture would be sufficient in producing reasonable results. For this evaluation, we honed our analysis on the category of questions that aligned with our assumption and criteria. Out of the three string matching approaches that we implemented, as outlined in Sect. 4.3, there were significant performance differences between the approaches. Dice's coefficient illustrates that it produces stronger correlations between words of different form as shown in Fig. 1.

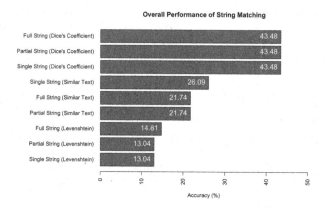

Fig. 1. Overall performance of string matching approaches in identifying the target keyword

4.3.3 Extending the Dataset: A Closer Examination

While the TREC-8 dataset was effective in providing a preliminary qualitative and quantitative evaluation of our approach, our findings suggested that the system was most effective in questions that aligned with specific criteria.

1. One specific target entity was identified
2. One relevant property related to the target entity existed
3. The property was related to the entity type as opposed to the entity itself.

Criteria three is perhaps the most significant observation that was extrapolated. While DBPedia produces a robust set of results related to the entity, the keywords that it isolates have a tendency to be specific to the type of entity that the entity belongs to, as opposed to the specifics of the entity itself. To expound on this notion, consider general qualities that a user would relate to the entity type 'person'. Such qualities include, but are not limited to, 'birthday', 'religion', 'education', 'spouse', among several others. By the same token, consider the general qualities associated with a company such as 'founder', 'founding date', 'net worth', 'product' and more. These qualities are a subset of intrinsic characteristics related to the labelled entity's type as opposed to the explicit characteristics or experiences of the extracted entity. With this in mind, we were interested to investigate how the system performed against a dataset that aligned with these three qualities. The motivation behind this stemmed from the fact that by identifying a sub-category of questions that the system performed

Table 4. Sample question template

Where was X Y?	Which X is Y in?
Who is X Y?	In what year was X Y?
When is X Y?	When was X Y?
What is X Y?	Provide me with information regarding the Y of X
How many Y does X have?	Who were the Y of X?
What Y does X have?	...

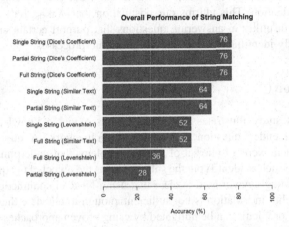

Fig. 2. Overall performance of string matching relative to guided question construction

efficiently with, it would better enable us to extend the system so that it aligns with the rules associated with various categories of questions. Given the profound difficulty of finding a credible data source that aligned with the characteristics of questions that we were interested in investigating; it was necessary for us to develop a random question data distribution that could be utilised to explore the implications of this system. To maximise the variability of the data, we generated question templates for common questions related to people, organisations and locations.

In Table 4, we document a series of question templates that accommodate for the various ways that basic questions that align with the stipulated criteria can be articulated. In these questions, X refers to the target entity and Y refers to the target property. Upon creating these question templates, we asked 50 students of diverse age and socio-ethnic backgrounds to produce a set of thirty questions distributed evenly between the respective entity type categories and place the name of a property that they deemed as relevant and related to the entity type in the position of Y as opposed to a property specific to the individual. While we considered randomly selecting properties from DBPedia's ontology network, we were interested in producing a dataset that was reflective of common questions that end users would be interested in querying about, guided by our question templates to ensure that our data considered different styles that questions could be asked. Upon collating the questions, we utilised NER and randomly selected a pool of 15 questions per category and re-evaluated the performance of the string matching algorithms (Fig. 2).

We observe a similar pattern in the re-sampled dataset as the TREC-8 dataset, with an expected higher precision. It appears that the differentiator between the algorithm's effectiveness stems primarily from the string matching approach more than the segmentation technique applied. This is particularly evident in the utilisation of Dice's coefficient, as we achieve the exact same performance across all three segmentation categories, in the same way that was identified in the original TREC-8 dataset. The drastic performance improvement in comparison to the TREC-8 dataset stems from the fact that we isolated the question structure to align with the subset of questions that were observed to produce consistent results through our preliminary qualitative and quantitative evaluation. This affirms our postulation that Wikipedia's knowledgebase can offer a useful utility in answering questions that purport certain criteria – criteria that can be easily identifiable using basic algorithmic processes.

5 Conclusion

Ultimately, this study illustrates that Wikipedia is a viable knowledgebase for an automated open ended question and answering framework. A qualitative analysis indicated that there were a finite set of distinct problems that underpinned the basis of natural language such as identifying the subject versus the object of the question coupled with disambiguating keywords. It is clear that using NER in conjunction with a string matching algorithm in isolation is not sufficient enough to alleviate these problems. In the future, these problems can be mitigated by using proven approaches in NLP such as word sense disambiguation and semantic labelling which can help render guided queries within Wikipedia's knowledgebase. We found that by focusing on questions that contained three attributes – a singular target entity and a single property related to the entity type – yielded substantive performance improvements. This validates our preliminary investigation and suggests that Wikipedia's knowledgebase performs with efficacy within the domain of open-ended questions. By compounding Wikipedia's knowledgebase with more advanced NLP features that address the problems observed in our qualitative analysis of the TREC-8 dataset, we have confidence that Wikipedia plays a vital role in addressing the issue of open-endedness in future QA systems.

Acknowledgement. This work was supported by the Industrial Strategic Technology Development Program, 10052955, Experiential Knowledge Platform Development Research for the Acquisition and Utilization of Field Expert Knowledge, funded by the Ministry of Trade, Industry & Energy (MI, Korea). This work was supported as part of the the the Office of Naval ResearchgrantN62909-16-1-2219.

References

1. Grosz, B.J., et al.: TEAM: an experiment in the design of transportable natural-language interfaces. Artif. Intell. **32**(2), 173–243 (1987)
2. Voorhees, E.M., Tice, D.M.: Building a question answering test collection. In: Proceedings of the 23rd Annual International ACM SIGIR Conference on Research and Development in Information Retrieval. ACM (2000)

3. Unger, C., et al.: Template-based question answering over RDF data. In: Proceedings of the 21st International Conference on World Wide Web. ACM (2012)
4. Kwiatkowski, T., et al.: Scaling semantic parsers with on-the-fly ontology matching. Association for Computational Linguistics (ACL) (2013)
5. Berant, J., et al.: Semantic parsing on freebase from question-answer Pairs. In: EMNLP (2013)
6. Cai, Q., Yates, A.: Large-scale semantic parsing via schema matching and Lexicon extension. ACL (1). Citeseer (2013)
7. Tsai, C., Yih, W., Burges, C.: Web-based question answering: revisiting AskMSR. Technical report MSR-TR-2015-20, Microsoft Research (2015)
8. Liang, P., Jordan, M.I., Klein, D.: Learning dependency-based compositional semantics. Comput. Linguist. **39**(2), 389–446 (2013)
9. Fader, A., Zettlemoyer, L., Etzioni, O.: Open question answering over curated and extracted knowledge bases. In: Proceedings of the 20th ACM SIGKDD International Conference on Knowledge Discovery and Data Mining. ACM (2014)

Predicting the Rank of Trending Topics

Dohyeong Kim[1], Soyeon Caren Han[2], Sungyoung Lee[1],
and Byeong Ho Kang[2(✉)]

[1] Department of Computer Engineering, Kyung Hee University,
Giheng-gu, Yongin, Korea
{dhkim,sylee}@oslab.khu.ac.kr
[2] School of Engineering and ICT, University of Tasmania,
Sandy Bay, Hobart, TAS 7005, Australia
{Soyeon.Han,Byeong.Kang}@utas.edu.au

Abstract. Trending topics is the most popular term list in the different web services, such as Twitter and Google. The changes in people's interest in a specific trending topic are reflected in the changes of its popularity rank (up, down, and unchanged). This paper proposes a temporal modelling framework for predicting rank change of trending topics, and delivers the real-time prediction service with only historical rank data. Historical rank data show that almost 70% of trending topics tend to disappear and reappear later. We handled those missing values, using deletion, dummy variable, mean substitution, and expectation maximization. On the other hand, it is necessary to select the optimal window size for the historical rank data. An optimal window size is selected based on the minimum length of topic disappearance in the same topic but with a different context. We examined our approach with four different machine-learning techniques using the twitter trending topics dataset, which is collected for 2 years. As an application, we implemented a trends prediction service, called TrendsForecast, applying our prediction model for Twitter trending topics in 10 different countries.

Keywords: Trending topic · Temporal prediction · Trends prediction

1 Introduction

By using different types of web-based services, such as search engines, social media, and Internet news aggregation sites, internet users can share and search information through the world. These services have caused a huge information-sharing paradigm shift by increasing personal information sharing and acquisition. This phenomenon, often called "the social data revolution", has resulted in the accumulation of unprecedented amounts of social data. This large amount of user created social data is like an untapped vein of gold in 21st century. Many information providers analyze their social data and provide a Trending Topics service, which displays the most popular terms that are discussed and searched within their community. Twitter collects their social data, extracts the

© Springer International Publishing AG 2016
B.H. Kang and Q. Bai (Eds.): AI 2016, LNAI 9992, pp. 636–647, 2016.
DOI: 10.1007/978-3-319-50127-7_56

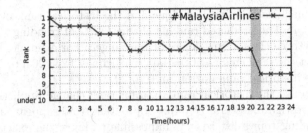

Fig. 1. The rank pattern of a trending topic

terms (including phrases and hash-tags) currently most often mentioned by their users, and publishes these on their site. Their list displays the top 10 trending topics of the moment, and displays these as part of the Twitter interface so all users can easily identify the current trending topics. Trending topics are estimated to reflect the real-world issues from the people's point of view. Over 85% of trending topics in Twitter are related to breaking news headlines, and the related tweets of each trending topic provides more detailed information of news and users' opinions [7].

The 'Trending Topics' list shows the top 10 trending topics in descending order of popularity. The lower the rank the higher the popularity, the higher the rank the lower the popularity. Based on the rank of a trending topic, it is possible to recognize the degree of current popularity of that topic. For example, on July 17th 2014, when a missile downed the Malaysia Airlines plane over Ukraine, all passengers were killed in the blast and it was breaking news around the world. During this time, the topic '#MalaysiaAirlines' appeared on the trending topics list. Figure 1 shows the hourly rank changes of the trending topic '#MalaysiaAirlines' in 24 h, which start from the point the topic first appeared on the trending topic list. You can see the trending topic has different hourly ranking changes based on people's interest change. The hourly ranking changes can be classified into three categories: up, down, and unchanged. In other words, this hourly ranking change represents the degree of change of popularity in that topic every hour, whether the people's interest in each trending topics is going up, down or staying unchanged. Predicting the trending topics hourly ranking change can be helpful to identify the influence of the topic in the near future. However, the 'Trending Topics' list displays only limited information, including the trending topic term, its rank, and updated date and time. We used only this available information to predict the future rank changes of trending topics. Therefore, the research aim of our study is to answer the following question: "can we predict the change of trending topics' popularity (up, down, and unchanged) and provide the real-time service?" In order to solve the problem, we proposed a temporal modeling framework using historical rank data and machine learning techniques. At time t, the problem, predicting the future rank change FRC of a trending topic T_x, can be expressed as follows:

$$FRC(T_x) = f[r_{t-n}, ..., r_{t-1}, r_t] \tag{1}$$

where f is a machine learning technique and the historical rank of n period is $[r_{t-n}, ..., r_{t-1}, r_t]$. The predicted ranking change FRC of trending topics can be classified into three classes: up, down, or unchanged. For example, let's assume that we predict the rank change of '#MalaysiaAirlines' from 20 h to 21 h (down), we can use its historical rank data from 0 to 20 h.

In order to use the historical rank data for rank change prediction, there is an issue to investigate. Several trending topics tend to disappear and reappear from the trending topics list so it is impossible to know the exact rank when it disappears. Figure 2 well represents the example of this topic disappearance issue with two different trending topics, including 'Ebola' and 'Black Friday'. It shows the 24 h rank pattern, which represents its rank from the point the topics was first detected on the 'Trending Topics' list. The topic 'Black Friday' is a seasonal trending topic. Twitter users are talking about what they will do on Black Friday. The topic appeared around the lowest ranks before the day, and disappeared for 5 h. It reappeared in the morning of that day. Another topic 'Ebola' relating to the deadly virus infection which raised concerns due to a resurgent epidemic in West Africa. In the initial three hours, the topic was about the first infection in Guinea and then it disappeared for 16 h. Following that, the topic reappeared and was about another infection in Liberia. Hence, these are the same topic term but each has a different context.

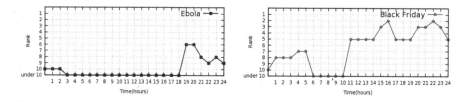

Fig. 2. Topic disappearance from the trending topic list

Historical rank data show that almost 70% of trending topics tend to disappear and reappear later. Therefore it is important to reflect this 'disappearance and reappearance' phenomenon in the prediction model, which is related to handling missing value and window size. First, it is necessary to handle the missing ranking value while the topic disappears. We applied four different missing value-handling approaches in Sect. 3.1, and identified the most successful approaches in trending topics rank prediction in Sect. 5.2. Secondly, as historical rank data is time-series data, it is necessary to select the optimal window size. Rather than choosing a random window size, we proposed a method to select the appropriate window size for predicting rank change of trending topics. As can be seen in Fig. 2, the context can change while the trending topic has disappeared. We need to find the minimum length of topic disappearance hours in the same topic with different contexts, and apply it to the window size.

2 Related Works

There are several studies aimed at predicting political events using social media. One of the most representative studies was applying tweets and news analysis for predicting election by Chung and Mustafaraj [4]. They collected all related tweets and news that contain the candidate's name and conducted sentiment analysis. UK general election was also forecasted by Franch [5]. The author used social data from various types of social media and those media are classified into its media level. The prediction performance was evaluated by using ARIMA(auto regressive integrated moving average), and it achieves 0.48 and 0.83% points off the real vote share. The data from social media have also received a lot of attention in economics field. Stock price is dynamically changed, which is based on real-world events and people's point of view. Therefore, social media could be the most effective resource to predict the stock price trends. Sprenger [11] considered twitter as a forum that leverages the wisdom of crowds for extracting the stock-related opinions. Bollen et al. [3] conducted sentimental analysis and examined how emotions can actually affect decision making in stock markets. The proposed approach gave 86.7% accuracy in daily prediction.

As people noticed that user data from social media sites well represents the people's interests, social media sites started to present the most discussed and searched topics, called trending topics [9]. Those trending topics services have received a great amount of attention. For example, 'Twitter Trending Topics', real-time event detection service provided by Twitter, shows the most often mentioned or posted short phrases, words, and hash-tags [2]. However, they show only the topic term and its rank with no detailed explanation. Hence, many researchers applied various summarisation and extraction approaches aimed at revealing the exact meaning of trending topics. Han and Chung [6] applied simple Term Frequency approach for extracting the representative keywords to disambiguate the approach. They also proved that the most successful approach to reveal the exact meaning of trending topics is simple Term Frequency, which is evaluated by 20 postgraduate students. Some researchers examined classifying trending topics. Lee et al. [8] classifies trending topics into 18 general categories by labeling and applying machine-learning techniques. Various types of prediction research in social media have been conducted. Nikolov and Shah [10] proposed a new algorithm for early detection of trending topics on Twitter. The performance achieved 95% accuracy. However, trending topics ranks and the prediction of rank changes has never been investigated before.

3 Trending Topic Rank Changes Model

The goal of this research is to predict the trend of trending topics rank change in the next hour. We propose a temporal modeling framework for predicting trending topics rank change using historical rank pattern and machine learning techniques. The proposed model can be described using the following equation:

$$FRC(T_x) = ML(PRP(T_x)) \tag{2}$$

In order to predict the next rank change FRC of a specific trending topic T_x, we used past rank pattern data (PRP) of the topic T_x. Then, machine learning techniques ML are applied for learning our model. Equation 3 describes the example of historical rank pattern PRP of a specific trending topic T_x at time t. It shows all historical rank patterns of a topic T_x in the specific period n. FRC represents the trends of the topic's ranking in the next hour. By comparing the current rank and the next-hour rank, the predicted rank change in the next hour will be one of three classes: up, down, and unchanged. For example, if the next-hour rank r_{t+1} is higher than the current rank r_t, the FRC will be 'down'.

$$PRP(T_x) = [r_{t-n}, ..., r_{t-1}, r_t] \tag{3}$$

$$FRC(T_x) = \begin{cases} up, & \text{if } r_t - r_{t+1} > 0 \\ down, & \text{if } r_t - r_{t+1} < 0 \\ unchanged, & \text{if } r_t - r_{t+1} = 0 \end{cases} \tag{4}$$

There are two main issues when we use the historical ranking data for our model: missing ranking handling and window size selection. Firstly, several trending topics tend to disappear and reappear from the 'Trending Topics' list. We specify how to handle missing rank values during the topic's disappearance. Secondly, the historical ranking patterns of trending topics are time-series data so it is crucial to select the appropriate window size for prediction. We propose an approach to select the optimal window size of our data. The detailed information of these approaches for those issues can be found in the following sections: (1) missing ranking handling and (2) window size selection.

3.1 Missing Ranking Handling

Since 'Trending Topics' list displays the top 10 trending topics of the moment, it displays the topics from rank1 to rank10. If the topic is suddenly out of the 'Trending Topic' list, it is impossible to recognize the exact ranking, whether the topic is ranked 11th or 50th. Manual inspection of the trending topics revealed that topic disappearance and reappearance is not limited to the type of topic. Various types of trending topics, including breaking news, persistent news (e.g. TV show or sport match) and hash tags seem to disappear and reappear randomly. We then analyzed how many trending topics actually disappear and reappears. Table 1 shows the percentage of trending topics that reappear and failed to reappear after the topic disappeared. The proportion of reappearing trending topics is almost 70%.

Table 1. The percentage of trending topics that reappeared or non-reappeared

	Reappearance	No-reappearance
Percentage	66.28%	34.82%

It is crucial to deal with missing rank data for our prediction model. We applied the following four missing value-handling approaches, including Pairwise deletion, Dummy Variable, Mean substitution, and Expectation maximization, reviewed by Allison [1]. The prediction results of these approaches will be discussed in the Sect. 5, 'Evaluation Result'.

3.2 Window Size Selection

The proposed temporal model uses historical trending topics and is learned using machine learning techniques, so it is important that sequences of the same window size should be used in training and testing. However, the primary difficulty is selecting an optimal window size for prediction using a good learning technique instead of trial and error. We analyze the actual trending topic ranking data on USA Twitter. According to the data analysis result, we found that the same topic terms are sometimes referring to different events, and this normally occurs when the time length of the topic disappearance exceeds a certain time. For example, Table 2 shows the example of analyzing the same trending topic '#MalaysiaAirlines' that is about two different events. The table displays the collected date and representative content of each topic. In 2014, there were two events related to Malaysia Airline: Firstly, the Malaysia airline flight MH370 disappeared on 8 March. The second referred to MH17 which is believed to have been downed by a missile in the eastern Ukraine on 17 July. The table shows that the same topic '#MalaysiaAirlines' are about two different events based on the collected date. Hence, before and after almost 4 months disappearance, the topic term '#MalaysiaAirlines' is separated into two different events.

We proposed the approach to identify the minimum length of topic disappearance that has different contexts by comparing the context similarity in two time-points (before-and-after the topic disappearance). As mentioned earlier, the trending topic terms consist of words, hash-tags or short phrases but it does not provide any description. It is almost impossible to recognize the exact meaning of a trending topic without extracting its detailed information. Hence, we proposed an approach to extract the representative contents for each trending topic and compare the context similarity in two time points (before-and-after the topic disappearance) if the topic disappeared at one point. The proposed approach is conducted as follows: (1) collect the trending topic and related tweets of the topic published less than 1 h ago, (2) preprocess the related tweets by removing stop words and (3) extract the representative 15 (fifteen) terms using term frequency

Table 2. The trending topic '#MalaysiaAirllines' with different events

Collected date	Extracted contents
2014/03/08	missing, flight, Malaysian, MH370, passenger, disappear, crash, pray, crew, lost, ocean, fail, safety, loss, airplane
2014/07/17	shot, down, missile, incident, kill, crash, attack, another, flight, victims, Malaysian, report, 259, explode

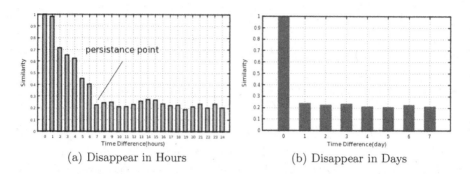

Fig. 3. The average of content similarity based on the topic disappearance time

(TF), and (4) calculate the cosine similarity of context of a specific trending topic at two different time-points (before and after the topic disappearance).

Figure 3 presents two sub-figures that show the result of context similarity based on the length of continuous disappearance; x-axis represents the length of topic continuous disappearance, and y-axis shows the cosine similarity rate (1 means exactly same and 0 is completely different). As you can see from the graphs, you can find that the context similarity is very low (0.2) if the topic continuously disappeared for over 7 h. Moreover, the similarity does not go down after 7 h, which is around 0.2. In other words, if a specific trending topic 'A' does not appear in the list for over 7 h and then reappears again, we can tell the first appeared topic 'A' and reappeared topic 'A' are talking about different contexts. Based on this result, the optimal window size would be the minimum length of continuous disappearance without different contexts in same topic term. The optimal size for U.S. twitter trending topic should be 7 (seven). The evaluation of prediction with different window size will be conducted in the evaluation result section.

4   Experimental Setup

In this section, we describe the collected data and applied machine learning techniques for evaluation of trending topics' rank change prediction. Algorithm 1 shows the whole algorithm for prediction. Based on Algorithm 1, we prepare the required data and machine learning techniques.

4.1   Evaluation Data

For the evaluation, we collected trending topic terms, related tweets and ranking patterns for those topics. By using the Twitter API, we crawled trending topics, related tweets and their ranks for two years (from 30th June, 2012 to 30th June, 2014) in different countries (USA, UK, and Australia). There are 57359 unique trending topics, and we trained this two years data for our evaluation.

The detailed data collection, including trending topic and related tweet, can be found in the below Algorithm 1.

In order to achieve the Eq. 2 in Sect. 3, the training data contains historical rank pattern as features, and the predefined future rank as class. The number of features are changing based on the optimal window size. For building the prediction model using our training data, we applied four machine learning techniques: Naive Bayes, Neural Networks, Support Vector Machines and Decision Trees.

Algorithm 1. Trending topics popularity prediction

1: Collect a trending topic T from Twitter with the rank r and the collected time(hour) h.

2: Put a topic T using search API and obtain the related tweets rt that are posted at the time $h - 1$ to h

3: Check whether the topic T appeared on the list before. If so, extract the representative words that describes the meaning of collected trending topics using Term Frequency. If not, the topic T is the new topic so skip to step 5.

4: Obtain all previous ranks PR of the trending topic T from the time(hour) $h - n + 1$ to h (n=window size).

5: Use this previous ranks PR as input data to the models trained by machine learning techniques

6: Predict the rank change FRC of the trending topic T will be up, down, or unchanged in the next hour.

---

5 Evaluation Result

We used 10-fold cross validation on two years of training data, which is described in Sect. 4.1. Based on this evaluation, we compare and summarize the prediction results with the proposed temporal model and different missing ranking handling approaches and window sizes.

5.1 Window Size Selection Examination

As we discussed in the Sect. 3.2, we proposed that the approach to selecting the optimal window size for trending topics' ranking change predictions. We found that optimal window size can be same as the minimum length of topic disappearance time that has same topic term with different meaning (see Sect. 3.2). We discovered the optimal window size for U.S twitter data can be 7(seven). We also applied our approach to the trending topic rank data from U.K. and Australia Twitter. Based on this examination, we found that the optimal window sizes for U.K. and Australia was 6(six) and 8(eight) respectively. We evaluate the prediction performance with those window sizes to examine whether the proposed approach selects the optimal window size of different data.

Table 3. U.S Trending topics ranking change prediction accuracies with different missing ranking handling approaches and window sizes

	Window Size	Missing Value	NB	NN	SVM	C4.5
(1)	5	Zero(0)	79.71%	88.20%	79.91%	88.74%
(2)	5	Lowest+1	80.11%	88.92%	80.82%	89.85%
(3)	5	Mean	75.10%	86.56%	77.29%	87.49%
(4)	5	Deletion	75.91%	85.42%	77.52%	85.74%
(5)	7	Zero(0)	83.91%	93.56%	85.36%	93.08%
(6)	7	Lowest+1	83.03%	93.68%	86.04%	94.01%
(7)	7	Mean	80.23%	91.06%	83.22%	92.91%
(8)	7	Deletion	82.93%	92.76%	83.93%	90.10%
(9)	9	Zero(0)	83.88%	92.53%	85.31%	93.00%
(10)	9	Lowest+1	83.00%	92.54%	85.61%	93.88%
(11)	9	Mean	80.34%	91.40%	83.29%	92.14%
(12)	9	Deletion	82.91%	90.92%	83.91%	90.11%

5.2   Prediction Evaluation

The experiments were designed to test the proposed model. We use the prediction performance as an indication of the suitability, which is obtained from four machine-learning techniques we discussed in the previous section. Each experiment result has different window sizes and different missing ranking handling techniques. Table 3 shows the prediction result of U.S. trending topics ranking changes with different window sizes (5,7,9) and four different missing handling techniques (Zero, Lowest+1, Mean, and Deletion). As mentioned in 'window size selection' section, we insist that the optimal window size for USA trending topics rank data can be size 7(seven). The experiment result shows that the prediction with size 7(seven) has the highest performance among 5, 7 and 9, which proves that our approach performs successfully. Since there is little difference in prediction accuracy of size 7 and 9, it is difficult to define whether 7 is better than 9. However, we can infer that if there is no difference between size 7 and 9, using size 7 is effective in performance, including data size and speed.

For missing ranking handling, missing value imputation with lowest+1 achieve the best prediction performance. This is because the other three approaches, mean, zero, and deletion, are not considered the nature of trending topics ranking. Among 4 machine learning approaches, C4.5 algorithm showed a higher performance than the others. Finally, we analyzed the performance of two more countries (U.K. and Australia) to make sure that our model performs well. As mentioned before, the optimal window sizes for two countries were size 6 and size 8 respectively, and it achieves the best performance (92.54%) with 6 instances and EM, and (80.13%) in 8 instances and EM.

Figure 4 presented that the example of rank change prediction in 24 h, which start from the point the topic first appeared on the trending topic list. Our prediction model forecasts the next-hour rank changes of trending topics using

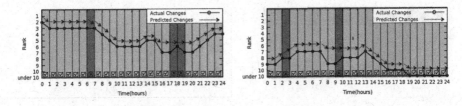

Fig. 4. Topic disappearance from the trending topic list

only historical data with proposed missing value and window size treatment approach.

5.3   Additional Feature

We put the additional features (topic features of the trending topic) into the training dataset in order to compare that of the historical rank data. We further classified the U.S. trending topics using the New York Times (NY times) classification service. As Trending topics are about real-time events, the traditional topic classification ontology cannot be applied. Unfortunately, if the category of a trending topic is extracted using a general document ontology, at any time, semantically related categories will also be extracted. There, we firstly search the trending topic term with the NY times classification service. We set the published time as the day that trending topic first emerged. Then we locate any related articles that were published with that term, on that day. Finally, the trending topics related categories are supplied by the NY Times service. Table 4 shows how U.S. trending topics are categorised.

Table 4. Topic distribution in U.S. trending topics

Topic	Entertainment	Sports	Politics	Fashion	World	Obituaries	Health	Technology
%	42%	28%	10%	6%	5%	4%	3%	2%

After we added this topic attribute into the training dataset, we learned the model with C4.5 decision tree, and the accuracy with topic attribute was 94.85%, which is slightly higher than that with only historical rank pattern (94.01%).

6   Application

Based on the evaluation result, we developed a trending topics popularity trends prediction system, called TrendsForecast[1], which represents the rank changes of trending topics in 10 different countries. The proposed system, TrendsForecast,

---

[1] TrendsForecast 2016 https://www.trendsforecast.net.

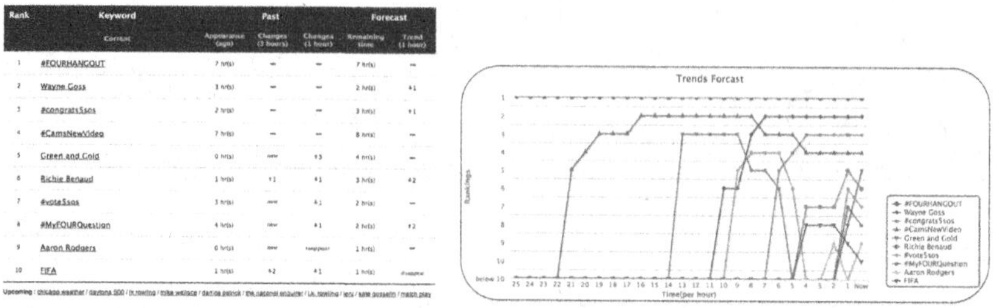

(a) Trending Topics Past and Future Rank Changes

(b) Historical Rank Changes for each trending topic

Fig. 5. Screenshot of TrendsForecast, trending topics rank change prediction system

applied C4.5 model with the best window size and missing value imputation method for each county. The table in the system, the first sub-figure in Fig. 5, presents not only future rank changes, but also various factors, including the time the trending topic first appeared, past rank changes in last 3 h and a hour. Therefore, users can see the popularity and importance of a specific trending topic by observing its historical rank changes, as well as its future rank changes. The second sub-figure in Fig. 5 displays the historical rank changes of top 10 trending topics in last 25 h.

7 Discussion

In this paper, we addressed trending topic rank prediction problems. Twitter Trending Topics service provides only the trending topic term and the rank of each trending keywords. Therefore, people may have question about whether any predication models using only this data can suggest any promising prediction results. This paper suggests a simple rank prediction that uses historical data with consideration of window size and missing value treatment. Surprisingly, our method achieved very significant performance (94.01% accuracy with C4.5 decision tree). On the one hand, this implicitly implies that the changing trends are the most important factors for rank prediction. On the other hand, it would be possible to improve performance of rank prediction. For example, we put additional information (topic feature) but it turns out that it would be very difficult to predict rank perfectly (100% accuracy), which is not because of algorithmic factors but because of trending topics' irregularly changing nature.

8 Conclusion

In this paper, we proposed a temporal modeling framework that predicts trending topics' hourly ranking change. We developed the learning procedure that can be used to construct models of trending topics ranking trends based on the historical trending topics ranking patterns. We also suggested the novel approaches to handle missing ranking and window size for trending topics rank change prediction. Rather than using complex features, we used historical ranking pattern and machine learning techniques, and it achieved the successful result (94.01%), and provide the real-time high performed prediction service.

Acknowledgment. This work was supported by the Industrial Strategic Technology Development Program, 10052955, Experiential Knowledge Platform Development Research for the Acquisition and Utilization of Field Expert Knowledge, funded by the Ministry of Trade, Industry Energy (MI, Korea).

References

1. Allison, P.D.: Missing Data. Sage, Thousand Oaks (2000)
2. Becker, H., Naaman, M., Gravano, L.: Beyond trending topics: real-world event identification on Twitter. In: ICWSM 2011, pp. 438–441 (2011)
3. Bollen, J., Mao, H., Zeng, X.: Twitter mood predicts the stock market. J. Comput. Sci. **2**(1), 1–8 (2011)
4. Chung, J.E., Mustafaraj, E.: Can collective sentiment expressed on Twitter predict political elections? In: AAAI, April 2011
5. Franch, F.: (Wisdom of the crowds) 2: 2010 UK election prediction with social media. J. Inf. Technol. Polit. **10**(1), 57–71 (2013)
6. Han, S.C., Chung, H.: Social issue gives you an opportunity: discovering the personalised relevance of social issues. In: Richards, D., Kang, B.H. (eds.) PKAW 2012. LNCS (LNAI), vol. 7457, pp. 272–284. Springer, Heidelberg (2012). doi:10. 1007/978-3-642-32541-0_24
7. Kwak, H., Lee, C., Park, H., Moon, S.: What is Twitter, a social network or a news media? In: Proceedings of the 19th International Conference on World Wide Web, pp. 591–600. ACM, April 2010
8. Lee, K., Palsetia, D., Narayanan, R., Patwary, M.M.A., Agrawal, A., Choudhary, A.: Twitter trending topic classification. In: 2011 IEEE 11th International Conference on Data Mining Workshops (ICDMW), pp. 251–258. IEEE, December 2011
9. Naaman, M., Becker, H., Gravano, L.: Hip and trendy: characterizing emerging trends on Twitter. J. Am. Soc. Inf. Sci. Technol. **62**(5), 902–918 (2011)
10. Nikolov, S., Shah, D.: A nonparametric method for early detection of trending topics. In: Proceedings of the Interdisciplinary Workshop on Information and Decision in Social Networks (WIDS 2012). MIT, November 2012
11. Sprenger, T.O.: TweetTrader.net: leveraging crowd wisdom in a stock microblogging forum. In: ICWSM, AAAI Press, May 2011

A Topic Transition Map for Query Expansion: A Semantic Analysis of Click-Through Data and Test Collections

Kyung-min Kim[1(✉)], Yuchul Jung[2], and Sung-Hyon Myaeng[1]

[1] Korea Advanced Institute of Science and Technology,
Daejeon, Republic of Korea
{kimdarwin,myaeng}@kaist.ac.kr
[2] Korea Institute of Science and Technology Information,
Daejeon, Republic of Korea
jyc77@kisti.re.kr

Abstract. Term mismatching between queries and documents has long been recognized as a key problem in information retrieval (IR). Based on our analysis of a large-scale web query log and relevant documents in standard test collections, we attempt to detect topic transitions between the topical categories of a query and those of relevant documents (or clicked pages) and create a Topic Transition Map (TTM) that captures how query topic categories are linked to those of relevant or clicked documents. TTM, a kind of click-graph at the semantic level, is then used for query expansion by suggesting the terms associated with the document categories strongly related to the query category. Unlike most other query expansion methods that attempt to either interpret the semantics of queries based on a thesaurus-like resource or use the content of a small number of relevant documents, our method proposes to retrieve documents in the semantic affinity of multiple categories of the documents relevant for the queries of a similar kind. Our experiments show that the proposed method is superior in effectiveness to other representative query expansion methods such as standard relevance feedback, pseudo relevance feedback, and thesaurus-based expansion of queries.

Keywords: Topic Transition Map · Semantic categorization of terms · Query expansion · Relevance feedback

1 Introduction

With millions of queries submitted to search engines every day, it is important to be able to mine the user interests hidden in the queries and their clicked web pages. We assume that the user interests and their relations to web pages are encoded in click-through data, and attempt to extract and represent user intentions in such a way that they can be used for search improvements. We attempt to ameliorate the term mismatching and term ambiguity problems that have long been recognized as critical problem for information retrieval (IR).

B.H. Kang and Q. Bai (Eds.): AI 2016, LNAI 9992, pp. 648–664, 2016.
DOI: 10.1007/978-3-319-50127-7_57

Queries and their clicked web pages separately or together have been analyzed in the past for different purposes [1–6]: query classification at a coarse-grain level such as informational, navigational, and transactional purposes; document classification using 14 categories at the top level of the Open Directory Project (ODP) hierarchy; and generation and analysis of click-graphs based on Query-URL log data but without topic-level content analysis. The click-graph approaches tried to analyze the complex structure of Query-clicked URLs [4, 5] and took an initial step toward improving query intent classifiers, especially for jobs and products [6]. A recent work [9] addressed the issue of intrinsic diversity of queries that often have little ambiguity in intent but seek the content covering a variety of aspects on a shared theme.

Motivated by the click-graph approaches [3–6], we analyze a large-scale click-through data and found that user queries and those shown in clicked pages often appear different from each other at a certain semantic level. More specifically, we observed transitions of user interests from the query side to the page side when user queries and clicked pages are represented with the semantic categories of the ODP taxonomy. The same kind of transitions was also detected between the topics (queries) and their relevant documents in a standard information retrieval dataset. We express the transitions from the query side to the documents (or page) side as a topic transition map (TTM).

TTM is constructed with a set of queries and the associated clicked pages or relevant documents, which are all mapped to the ODP hierarchy by a classifier. That is, topic transitions are captured at a semantic level, not between individual queries and documents as in other click graphs [3–5]. Once a TTM is available for a large click-through log or an IR test collection, a new query can be expanded based on which semantic category it belongs to; search terms are selected from the document categories in TTM, not the clicked or relevant documents.

This approach can be seen as an attempt to recognize the changes of user interest between two situations: at the time a query is entered and at the time retrieved documents are browsed and assessed. The changes may be attributed to different reasons: a real flow of information needs as new contents are discovered from the retrieved documents; to the under-specification of user queries, especially in web searching; or to the differences and mismatches in vocabulary. Regardless of the reasons, TTM has a potential to capture an average behavior of topic or interest transitions for a community of users or a set of queries to predict what semantic category would satisfy a user query. Taking the average behavior of topic transitions into consideration for searching is in the same spirit with the idea of using popularity of web pages for ranking in web search engines [7]. The difference lies in the time when user interests are captured: page creation time vs. search and browsing time.

2 Related Work

Some work has been done to categorize the needs of users with an intention to use the results for search engines. For example, the categorization proposed by [8] in which three classes (navigational, informational and transactional) was considered. In addition, [1] tried to identify user's interests in a web search engine based on a query log. Their analysis was made from two perspectives: the user's informational objects (informational/not information/ambiguous) and the topic categories (top 1st level of ODP hierarchy). However, the proposed categories are not at the topical level that is sufficient to represent specific users' needs or interests and hence inappropriate for handling the problem of term mismatches between queries and documents. A study [2] tried to make an approximation to the user intent using clicked-through data at a general level whereas a recent work [9] addressed the issue of intrinsic diversity of queries that often have little ambiguity in intent but seek the content covering a variety of aspects on a shared theme. However, these studies did not utilize the document sides together.

In the area of web query classification, a study [10] introduced a bridging classifier based on the ODP taxonomy and category names, which outperforms the best results of KDDCUP 2005 (http://www.acm.org/sidkdd/kddcup). This is a reasonable effort to classify web queries, which are short and ambiguous, into a set of target categories. Since a fine-grained topic classifier can provide a significant medium for identifying user interest, this line of work can lead to effective retrieval. However, there has been no attempt yet to relate it to the goal of improving search effectiveness, let alone the topic transition aspect. ODP taxonomy has been used to classify web documents into a deep hierarchy [12, 13]. Subtopic mining approaches such as [29] extract text fragments from different parts of retrieved documents, cluster fragments and thus merge similar ones, generate a subtopic for each cluster and finally rank and diversify the subtopics. The fine-grained classification approaches for web documents and the web query classification work led us to investigate on topic transitions between queries and relevant documents (or clicked documents).

There have been some attempts to elicit relations between queries and documents. The work in [4] studied a large click-graph with the goal of extracting the relations that are implicit in the actions of users submitting queries and clicking answers from query logs. They suggested a graph mining technique that can generate huge amounts of interesting relations, but focusing only on edges between queries and their visited URLs without an analysis at a semantic level. As a way to infer semantic query relations, the work in [11] tried to build a semantic query network based on high quality transaction data consisting of online searching, production viewing, and product buying activities from a large-scale query logs. It showed that extraction of relations by means of mapping user queries to higher dimensions is effective in determining related search terms.

As a concept-based query expansion method, the work in [15] employed association rules to mine query relations and built a query relation graphs for identifying salient concepts (or entities) through capturing transitive relations. Others proposed effective query expansion techniques that consider topics (or concepts) with the help of the ODP hierarchy [16, 17]. While their approaches are deemed semantic, they are limited to the query side. The closest to our work is the study of topic dynamics for pages visited by people in web search [26] to predict topic transitions using only the 1st level of ODP hierarchy.

A study considered both the query space and the document space together by computing probabilistic correlations between query terms and document terms based on query logs [18]. Its initial hypothesis is that the click-through information on search logs represents a clue for inferring relation between queries and documents chosen to be visited by users. A recent study on a latent semantic query suggestion method [19] used the rich information embedded in the query-click bipartite graph for recommending relevant queries. Although the concept of a latent semantic map (LSM) [27] was proposed to suggest better user queries and categorize search results, it didn't contain experiments.

While the ideas in [12–14, 26, 27] share some of our motivations and approaches, our method differs from theirs in that we attempt to find such query-document relations and topic transitions at different semantic levels of a huge hierarchy, apply them for query expansion, and validate both the notion of topic transition and its utility for query expansion through a series of experiments.

3   Topic Transition Map

Constructed from a set of queries and the associated clicked pages (or relevant documents[1]), TTM is intended to aggregate the interests of many users, which are expressed in relatively simple queries first and clicked (or relevant) documents later. In order to capture the commonality of interests of many users and their transitions at a semantic level, we use the ODP hierarchy (http://dmoz.org/) as the semantic medium. It has 16 root categories (nodes) at the top level, and the corresponding trees have varying numbers of branches at different levels and varying depths for the trees with the maximum being 36. Each category has a set of representative snippets, each of which has its own title, URL, and description. A query or document is mapped to the ODP category by a classifier that builds a centroid from the associated snippets. For our final TTM, however, we pruned some of the branches of the original ODP hierarchy to create a more compact one that suits our needs.

3.1   TTM Construction

Our TTM construction process is divided into three parts: adaptation of the original ODP hierarchy, query and document classification, and topic transition computation. The first part is to build a semantic hierarchy for our purpose, which is the basis for the topic and document classifier. Then the queries and documents are classified into the hierarchy, and the results are used to create a map between a query category (QC) and a document category (DC).

The original ODP hierarchy has page snippets for each category, but we crawled the snippets and web pages for the URL's in the Google version of ODP (https://web.archive.org/web/20080227072915/http://www.google.com/dirhp) to enrich our category representation. While the total number of the categories is about one half of a million, we chose to select those with more than 10 snippets, discarding the rest, because they are the basis for training the classifiers. As the categories at a low level tend to have a smaller number of snippets, the filtering process provides the effect of

---

[1] Hereafter we use *queries* and *relevant documents* without losing generality as we use the text in actual pages corresponding to the clicked URLs in our analysis.

Table 1. Numbers of categories after reconstruction

	Level 1	Level 2	Level 3	Level 4	Level 5	Level 6	Level 7	Level 8	...	All
#doc>10	13	387	2,495	4,720	4,399	2,739	1,504	512	...	16,930

pruning low level categories with too much specificity. In addition, we decided not to use sub-hierarchies such as 'regional' and 'reference' that contain links to other existing nodes and are deemed not suitable for representing semantic categories [12]. The resulting hierarchy has only 16,930 topic categories as in Table 1, which also shows the number of categories at each level.

To identify fine-grained topics of queries and documents in the collection based on the modified ODP taxonomy, we turn the problem into that of classifying them to the topic categories. While query and document classification for a hierarchy has been done for different purposes [10, 12, 13, 20], we chose to implement a nearest-neighbor classifier with the Rocchio's formula using only positive examples, which was used successfully in [20]. For each category, we calculate a centroid using the term features in the snippets, which is to be compared to document or query vectors based on a Cosine similarity measure. An advantage of using the nearest-neighbor classifier is that categories can be ranked so that top k topic categories are assigned to a query or a document where k determines the shape of TTM and hence affects the retrieval performance of query expansion.

For actual topic map construction, we build a bipartite graph $G = <Q, D, E>$ where Q and D represent a set of queries and a set of documents, respectively, and an edge $e \in E$ connects a query and its relevant or clicked document. For all $q_i \in Q$ and $d_j \in D$ for which an edge exists, we can generate a query category set $QC = \{qc_1, qc_2, qc_3, ..., qc_n\}$ and a relevant document category set $DC = \{dc_1, dc_2, dc_3, ..., dc_m\}$ based on the classification algorithm. As a result, we obtain a bipartite graph $G' = <QC, DC, E'>$ where an $e' \in E'$ is a weighted edge that connects a qc and dc. Such an edge is created only if there exists $e \in E$ such that it connects q_i and d_i whose categories are qc and dc, respectively.

The weight of an edge between a query category and a document category is computed by accumulating the relevance scores resulted from the query and document classifications. Given an edge $(x, y) \in E'$ and a function K that accepts a query or a document and returns a set of classes with a relevance score s higher than a threshold, the weight for the edge is computed as follows:

$$w_{xy} = \sum s(q_i) + \sum s(d_j)$$

for all q_i and d_j such that $x \in K(q_i)$ and $y \in K(d_j)$ and $(q_i, d_j) \in E$.

In effect, the weight for an edge between qc_i and dc_j is computed as the sum of all the relevance scores of the query and document pairs such that the categories of the query and the document include qc_i and dc_j, respectively, and the document in the pair is relevant to the query. In other words, the weight of an edge is reinforced linearly when a new (query, document) pair is detected for the edge in a click-through data or

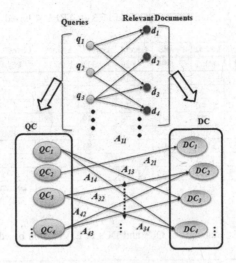

Fig. 1. Topic transition map construction

test collection. The bipartite graph can be represented as a matrix A where an element A_{ij} has the weight between qc_i and dc_j.

In Fig. 1, the initial query-document graph is converted into a category-category graph with A_{jk} indicating the weight of the topic flow from a query category (QC_j) to a document category (DC_k). Although a click graph model in a prior click graph study [3] is built by producing a probabilistic ranking of documents for a given query, the final TTM is constructed by computing the probabilities of individual topic transitions between them. The probability of reaching a document category from a query category can be computed by normalizing the transition weight with the sum of the weights on the transitions from the current query category:

$$P(dc_k|qc_j) = \frac{A_{jk}}{\sum_i A_{jk}}$$

The probability values computed as such are used in the final TTM.

3.2 Analysis of TTM

Based on the method described above, we separately built TTMs for the TREC-3/4 data and the MSN Live Search log-2006 data (LIVE-2006). The total numbers of queries and the associated documents are: 50 queries and 9,349 relevant documents (186.98 per query on average) out of 742,358 for TREC-3, 49 queries and 6,322 relevant documents (132.67 per query on average) out of 567,529 for TREC-4, and 1,502,387 queries (755,453 unique queries) and 370,907 clicked URL documents for LIVE-2006.

Table 2. Statistics of QCs and DCs in TREC-4 based TTM

TREC-4	Level1	Level2	Level3	Level4	Level5	Level6	Level7	Level8	All
ODP(doc>10)	13	387	2,495	4,720	4,399	2,739	1,504	512	16,930
#QC(S)	0	0	8	14	14	9	4	0	49
#QC(L)	1	15	44	81	62	25	15	2	245
#distinctQC(S)	0	0	8	14	13	9	4	0	48
#distinctQC(L)	1	14	41	78	58	23	15	2	232
#DC(S)	5	307	1,176	1,846	1,070	717	856	74	6,062
#DC(L)	57	1,694	6,584	9,193	5,694	3,468	3,055	512	30,310
#distinctDC(S)	4	75	294	382	264	151	78	25	1,277
#distinctDC(L)	6	168	763	1,102	743	436	225	57	3,519

L: Lenient, S: Strict

Table 3. Statistics of QCs and DCs in LIVE-2006 based TTM

LIVE-2006	Level1	Level2	Level3	Level4	Level5	Level6	Level7	Level8	All
ODP(doc>10)	13	387	2,495	4,720	4,399	2,739	1,504	512	16,930
#QC(S)	168	16,625	112,370	199,428	165,252	93,753	50,854	15,581	659,651
#QC(L)	1,258	93,296	576,352	960,448	794,994	469,658	235,971	73,623	3,233,123
#distinctQC(S)	12	384	2,475	4,612	4,280	2,663	1,456	496	16,535
#distinctQC(L)	13	387	2,495	4,719	4,398	2,738	1,503	511	16,924
#DC(S)	207	18,048	81,712	103,610	67,876	44,468	20,955	5,477	344,237
#DC(L)	1,968	93,516	400,000	506,446	352,394	224,707	103,047	28,559	1,720,875
#distinctDC(S)	12	375	2,398	4,396	3,950	2,462	1,310	444	15,494
#distinctDC(L)	13	387	2,494	4,704	4,382	2,732	1,497	509	16,878

L: Lenient, S: Strict

In LIVE-2006, the clicked documents whose URL is no longer valid and the associated queries were removed from the 10% sample of the entire data set.

To get basic statistics, we counted the number of topic categories at each level (in the ODP hierarchy), to which the queries and the documents were classified separately (i.e., QCs and DCs). Since top k categories for both queries and documents were selected in the TTM construction process, the statistics can vary depending on how many categories we consider for each query or document (i.e. the value of k). Two different counting methods are used in Tables 2 (TREC-4) and 3 (LIVE-2006): a lenient counting method (L) that takes top five QCs or DCs and a strict counting method (S) that takes only the top one QC or DC. TREC-3 is not shown because it is similar to TREC-4.

The numbers of distinct categories show that most heavily used topic categories are positioned at the levels 2 to 7 (TREC-3/4) and levels 3 to 7 (LIVE-2006) of the ODP hierarchy. We believe the categories at those levels are at the right granularity in representing the semantic categories of queries and documents, supporting our pruning process to create a more compact hierarchy. The first line in each table shows the number of the categories that have more than 10 snippets each.

Table 2 shows that 50 queries were mapped to 49 distinct categories because of their lengths and uniqueness in TREC-3 whereas the relevant documents were mapped to duplicated categories (from 7,522 to 1,426 in the strict case). The large reduction rate from 37,610 to 3,938 in the lenient case indicates that the documents across the top five categories are relatively homogeneous in their semantics, compared to the queries.

A unique aspect in Table 3 is that unlike TREC-3/4, the reduction ratio from the number QCs to the distinct QCs is high, from 650,651 to 16,535. It indicates that there are many duplicates in the query set and that many queries have the same semantic categories. On the other hand, the reduction ratio for relevant documents in LIVE-2006 is similar to that of TREC-3/4.

4 Query Expansion

4.1 Motivation

The inception of the notion of TTM was derived from our initial observation that there is a gap between the semantic categories of a query and those of the relevant documents. Figure 2 gives some examples for a mismatch between QC and DC resulting from the lenient counting method. It can be seen that except for Topic 215, where the QC is ranked as the third category on the DC side, the QC's are not listed at all on the DC sides. For Topic 215, the QC and the three DCs are reasonable categories in which some aspects of the information need can be found. For Topic 238, however, the QC is incorrect and so are the DCs with a possible exception of the second one. Topic 204 is the case where the DCs are far from the QC but all of the DCs have a potential to have relevant information.

These examples illustrate the possibility of retrieving relevant documents from the document topic categories different from the query topic category. First of all, the topic for a query can be misinterpreted or misrepresented by a query processing component of a search engine, resulting in very irrelevant documents. In this case, the document topic categories that can be reached through the transitions can compensate for the failure in the query analysis. This case is likely to occur with short queries. Second, relevant documents can be found not only in the category that is identical to the correctly identified query category but also in some other topic categories that can be reached with the transitions as in the case with Topic 215.

To validate the TTM idea beyond the anecdotal cases, we compared the semantic categories of the queries and those of their relevant documents in two collections: TREC-3 and TREC-4. The TREC-3/4 collections were used instead of LIVE-2006 because topic transitions are less likely to occur with longer queries, making our analysis conservative. Since some of the QCs were incorrectly classified, however, we used manually annotated query categories for the comparison and further experiments in order to avoid propagation of errors arising from the query classification phase.

Table 4 shows the matching ratio of a unique query category (top 1 QC) and document categories (DC) in the two collections when more than one category is allowed (up to five) for each document. It includes the distributions of two major topic flow types – Topic Retention (TR) and Topic Transition (TT). TR means that the

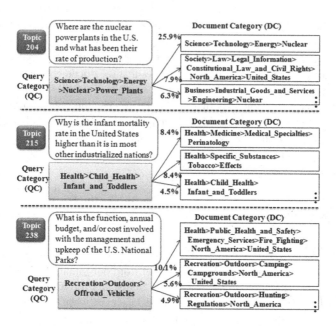

Fig. 2. Three Examples from TREC-4 based TTM

Table 4. Distributions of each topic flow type

	Topic Retention			Topic Transition		
	1D	2D	Full	1D	2D	Full
TREC-3	56%	37%	4%	35%	55%	
TREC-4	49%	27%	3%	43%	66%	

1D: 1st level; 2D: 2nd level; and Full: full depth

correctly identified QC is found in the DCs whereas TT means none of the DCs matches the QC.

The ratios vary depending on the hierarchy level at which comparisons are made. The higher (more general) the level, the more chance QCs and DCs match. When comparisons are made at the deepest level, there is only a slight chance that a QC match one of the DCs, over-emphasizing that there is a topic transition. When a QC ('A> B>C>...') and a DC ('A> C>D>...') are given, for example, the pair is considered topic retention at the 1st level, contributing to TR[1D], but counted as topic transition at the 2nd level (TT[2D]). This observation is taken into account when the topic transition phenomenon is used for query expansion.

The statistics support our premise that we need to capture topic transitions from QC to DC in a systematic way. This is related to the question of whether there is a potential for serendipitous encounters during web search [28]. While a user's interest may actually change from the query time to the search result browsing time as they encounter them, we take a less radical stance for generality: the topical categories of relevant documents are not necessarily the same as that of a query. As a result, a set of document categories associated with the category of a query in the past data can be used to predict the categories in which relevant documents can be found for the query.

4.2 TTM-Driven Query Expansion Method

Given a query and its category (QC), top k DCs based on the edge probabilities in TTM can be identified and used to select expansion terms. There are two different resources for each category, from which expansion terms can be extracted: snippets associated with a category and the documents in the relevance judgment collection, which have been classified to the category. Since snippets have been processed to extract the term features used in forming a centroid for classification, they are candidates for term expansion. Top n terms were used in our experiments. The documents from the relevance judgment collection, which belong to the category are used in a similar fashion to extract the expansion terms.

While it is conceivable to use the transition probabilities from QC to DC in computing the ranks of expansion terms from different DCs when more than one DC are used, we chose a simple term selection method in which term relevance scores used for centroid construction are employed. When a term occurs in more than one DC, we selected the highest score for ranking the expansion terms. Term weighting schemes for merging centroid vectors corresponding to the multiple DCs and for using transition probabilities are left for future research.

5 Experiments

The main goal of the experiments is two-fold: to find out how useful TTM is in query expansion for IR and to determine ways TTM-based query expansion can be used for retrieval performance improvements. For the first part, we compared our TTM-based expansion method against other feedback methods. For the second part, we ran experiments with different numbers of expansion terms and other variations such as the number of DCs used in TTM.

5.1 Data

While both TREC-3/4 and LIVE-2006 were used for our initial investigation on the existence of topic transitions and for actual construction of TTM, we conducted retrieval experiments on the TREC-4 data only. The real world data in LIVE-2006 helps showing that topic or interest transitions indeed occur in web search. However, we felt that the data would provide skeptical evaluation results since the clicked pages are not necessarily all relevant.

Instead, we chose to use TREC-4 because has not only full relevance judgments but also longer queries than web queries. The TREC-4 dataset consists of 49 official topics (7.5 words on average) and 567,529 documents (2.07 GB). The number of relevant documents for each query is 133 on average. We used only the title and description in the documents.

5.2   Experimental Design

We used Terrier v2.2.1 (http://ir.dcs.gla.ac.uk/terrier/) as our baseline retrieval system[2], which is based on the vector space model. Our baseline run is a regular retrieval without any feedback. It allows us to measure how much we can gain in retrieval performance by using TTM-based query expansion and other various feedback-based query expansion methods. For retrieval performance, we used standard measures: Interpolated precision/recall, MAP (Mean Average Precision), and P@n (precision at n documents). We ran five experiments: two other feedback-based term expansion methods, a semantic hierarchy based term expansion, and two TTM-based term expansion methods. Following are their descriptions.

- *Explicit relevance feedback (ERF)*. To simulate explicit relevance feedback provided by users, we used top k relevant documents available in the collection.
- *Pseudo Relevance Feedback (PRF)*. We used a standard pseudo or "blind" relevance feedback method where the top k documents in the ranked retrieval result are assumed relevant and used for feedback.
- *Topic Relevance Feedback (TRF)*. This feedback method uses the query category (QC) to which a given query is classified and manually judged as in Subsects. 3.1 and 4.1. Expansion terms are extracted from the documents determined to be relevant to the chosen query category. In effect, this method assumes that there is no topic transition for a given query. In our implementation, all the relevant documents are forced to be classified to the linked QC corresponding to the given query, without using the document category classifier. This process is done based on the assumption that relevant document always inherit the query's topic category. While the documents used for query expansion are the same as those used for PRF, the term extraction method is different; the semantic categories constructed with TTM and the term statistics in the documents in the categories play a critical role.
- *ODP-based Query Expansion (OQE)*. This case represents a thesaurus or concept hierarchy based term expansion. While past attempts using such resources like WordNet have not been successful in general, we wanted to see the effect of using ODP as the basis for term expansion. The snippets associated with the QC are used for expansion. For term ranking, the *df-icf* scores are used, too.
- *Topic Transition Relevance Feedback (TTRF)*. This is the expansion method we propose using TTM. As in the two previous cases (TRF and OQE), two resources can be used: the relevant documents or the ODP snippets associated with the DCs. The former is represented as 'TTRF' while the latter is represented as 'TTRF(ODP)' in Sect. 5.3. In addition, there are other variations in terms of the number of DCs to be considered, for which we show separate experimental results.

---

[2] We used the PL2 weighting model Terrier provides. Although we've tested other weighing scheme, such as, BM25, DFR_BM25, *tf-idf*, the PL2 showed the best performance on MAP and P@n measure when tested with TREC-4.

5.3   Experimental Results

For the legends in the graphs shown below, we use the following convention: [expansion method][the number of documents or topic categories used][the number of expanded terms]. For example, [ERF][4 docs][200] means the case where 200 terms were added from the explicit relevance feedback based query expansion method using the top 4 relevant documents. We ran a preliminary experiment to determine the best performing numbers of expansion terms for all the methods: 20 for TTRF, 120 for ERF, 120 for PRF, and 5 for TRF. We excluded OQE in the subsequent experiments because of its low performance.

Precision@n. To see the value of the proposed expansion method in a simulated web searching environment where only top n documents are often viewed by the users, we measured its retrieval performance using precision@n in comparison with the others. The TTRF method showed consistent superiority up to top 1000 retrieved documents as can be seen in Fig. 3. We believe that the high score terms selected from the top five DCs are effective in bringing relevant documents to the top n lists where n can be as large as 5.

Overall Performance Comparisons. Figure 4 compares the four different expansion methods and the baseline in terms of the traditional precision/recall. The experimental settings like the number of expansion terms and the number of categories used for the methods are different to make sure each reaches the best performance in our experiments. It's clear in the graph that TTRF has the best retrieval performance compared to the other expansion methods, followed by TRF, PRF, ERF, and TTRF (ODP).

To see the comparative results more clearly, we provide the following table where MAP and P@5 are shown. Two different parameter settings are shown for ERF, PRF, TRF, and TTRF because they gave slightly different relative superiority in MAP and P@n. Since the comparisons are among different feedback methods, the percent increases are based on the PRF method. The numbers show the highest value for each method with its optimal parameter setting.

Table 5 clearly shows that the overall performance increases with the order of TTRF > TRF > PRF (baseline) > ERF. In case of P@5 and P@10, the order is almost the same. The effect of using TTM over PRF is an increase of 27.31% in MAP, 33.80% in P@5 and 36.40% in P@10, respectively. This is a strong indication that recognizing and using topic transitions helps retrieval performance significantly.

Variations of TTRF. To better understand how best TTM can be used for query expansion, we compared different TTM-based term expansion strategies. The variations are mostly for the ways to select DCs, to combine the expansion terms. The other parameters like the number of terms were selected for the best condition for the case. In the order of the legends in the graph, they are:

- Case 1 ([TTRF(ODP)][6DC] [10]): ODP snippets within top 6 DCs were used and only top 10 terms were selected for expansion.
- Case 2 ([PRF][10docs][90]): It was included to be used as baseline as in Table 5.
- Case 3 ([TRF][50]): Top-50 terms were selected using TRF. It assumes the query topic is retained. Every relevant document is classified to the query category.

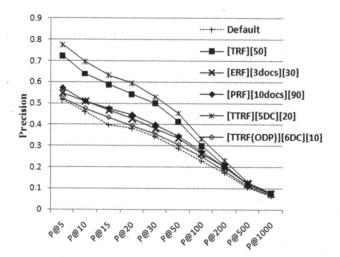

Fig. 3. Precision@n

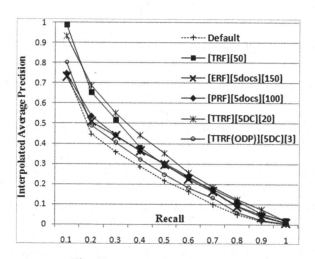

Fig. 4. Interpolated Precision-recall

- Case 4 ([TTRF][1DC] [5]): The first DC was used to obtain top five terms.
- Case 5 ([TTRF][3DC] [5]): The three DCs were combined to obtain top five terms.
- Case 6 ([TTRF][5DC] [20]): Top five DCs were combined to obtain top 20 terms.

As can be seen in Fig. 5, the best performance was obtained with Case 6 where top five DCs were used together. Because its performance was much better than Case 4 where only top DC was used, it indicates that the terms from lower-ranked DCs can help. However, the TTRF using ODP snippets for term expansion (Case 2) is only slightly better than the default, indicating that the snippets are not a good source for term expansion.

Table 5. Overall Performance Comparisons

Expansion method	MAP	P@5	P@10
[ERF]	[5docs][120]	[3docs] [20]	
	0.2407	0.5469	0.5041
	(−2.75%)	(−5.64%)	(−1.20%)
[PRF]	[5docs][120]	[10docs][90]	
(Baseline)	0.2475	0.5796	0.5102
[TRF]	[1DC][50]	[1DC][50]	
	0.2919	0.7224	0.6367
	(+17.94%)	(+24.64%)	(+24.79%)
[TTRF]	[5DC] [20]	[5DC] [20]	
	0.3151	0.7755	0.6959
	(+27.31%)	(+33.80%)	(+36.40%)

Cases 4 through 6 are considered a method of using a combination of topic retention and topic transition for query expansion. In other words, the improvements obtained in the cases are attributed to the effect of topic transition on selecting effective terms.

Discussion. Our experiments were conducted using a publically available search engine, Terrier, which has been used widely for research purposes. Nonetheless, the performances of the proposed method based on TREC-4 evaluation using MAP, P@5, and average precision are better than those of the highly ranked system (MAP: 0.2944 and P@5:0.5755) reported in evaluatIR.org[3]. However, it should be noted that our goal was not to achieve the state-of-the-art performance by tweaking the system parameters but to investigate on the values of the TTM and its use for query expansion in comparison with other well-known yet effective relevance feedback methods.

Known relevant documents in test collections are far from being perfect. As such, there should be additional relevant documents in the collection although a study showed that the imperfect nature of a test collection did not alter the relative rankings of participating IR systems [25]. However, the search results after relevance feedback is likely to be biased towards the average relevant documents used for the feedback. The proposed query expansion method based on the notion of topic transitions can alleviate the limitation since the different document categories are linked to the query side. Documents covering different aspects of the query topic can be brought in as a result of the feedback. An investigation on the characteristics of different sets of relevant documents associated with different DCs should lead to a method for expanding queries in multi-directional ways. We leave this line of research for the future.

The k-NN based query classifier we used in this work is not the best method for short query classification. The use of a k-NN based classifier is not to suggest the best performing query classification method, but just to identify the topic transition phenomenon between the queries and their relevant documents. We understand that

---

[3] The evaluteIR.org ALPHA web site provides useful information about IR test sets and systems for comparisons. Available at: https://web.archive.org/web/20150222083239/http://wice.csse.unimelb.edu.au:15000/evalweb/ireval/

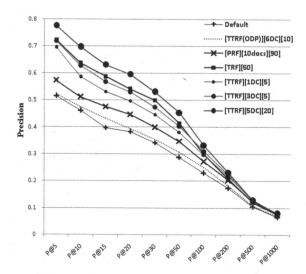

Fig. 5. Variations in TTM-based expansion strategies

classification accuracy would be an important issue if we were to use the proposed method for real applications involving short queries. Nonetheless, we feel it is important to recognize the topic transition phenomenon and attempt to find a way to utilize it. More fundamental is to understand the causes for topic transition. However, this paper focuses on how it can be used, leaving the "why" question for future research.

6   Conclusion and Future Work

In this paper, we introduced the notion of topic transitions and an approach to constructing Topic Transition Map (TTM) based on semantic analysis using the large-scale ODP hierarchy and a large-scale query log and TREC data. As a way to validate the idea and make a use of TTM, we proposed a new query expansion method. In a series of experiments using the TREC-4 collection, for which relevance judgements are available, we showed that our method is superior to other query expansion methods such as standard explicit relevance feedback, pseudo relevance feedback, and thesaurus-based term expansion. While the effect of query expansion based on the notion of topic transition would have a greater impact on web search, which can be studied with an appropriate web collection, we feel that the current work is a strong basis to warrant further studies.

With the new notion of TTM and its applicability shown in query expansion, there are several avenues to explore to fully exploit its value. First of all, it can be used to identify how user interests change and why users consider retrieved documents and pages interesting, not necessarily relevant. When a time factor is added, it should be also possible to see how interests of general users change at a semantic level over time. The second area is to identify user search goals, especially in terms of action-level clues (e.g., verb + related objects), by looking into both query and document sides at the same time. This can lead us to find an answer for what in the semantic content of query/document causes topic transitions. Findings along this line will help improve user satisfaction in web searching and advertising. The third area is to investigate on a specific method to make further improvements along term expansion. It should be possible to find a way to use different document categories for identifying different focal points of user interests and using them for improved search results. The fourth area of investigation is to apply the notion and the method to different domains, different data sets, and different users to build a generalized model for topic transitions.

Acknowledgments. This work was partly supported by Institute for Information & communications Technology Promotion(IITP) grant funded by the Korea government(MSIP) (No. R0101-15-0176, Development of Core Technology for Human-like Self-taught Learning based on a Symbolic Approach) and Industrial Strategic Technology Development Program grant funded by the Ministry of Trade, Industry & Energy (MI, Korea) (No. 10052955, Experiential Knowledge Platform Development Research for the Acquisition and Utilization of Field Expert Knowledge).

References

1. Baeza-Yates, R., Calderón-Benavides, L., González-Caro, C.: The intention behind web queries. In: Crestani, F., Ferragina, P., Sanderson, M. (eds.) SPIRE 2006. LNCS, vol. 4209, pp. 98–109. Springer, Heidelberg (2006). doi:10.1007/11880561_9
2. Nettleton, D.F., Calderon-Benavides, L., Baeza-Yates, R.: Analysis of web search engine clicked document. In: Proceedings of LA-Web 2006, pp. 209–219 (2006)
3. Craswell, N., Szummer, M.: Random walks on the click graph. In: Proceedings of ACM SIGIR 2007, pp. 239–246 (2007)
4. Baeza-Yates, R., Tiberi, A.: Extraction semantic relations from query logs. In: Proceedings of SIGKDD 2007, pp. 76–85 (2007)
5. Baeza-Yates, R.A., et al.: The anatomy of a large query graph. J. Phys. A Math. Theor. **41**, 1–13 (2008)
6. Li, X., Wang, Y.-Y., Acero, A.: Learning query intent from regularized click graph. In: Proceedings of ACM SIGIR 2008, pp. 339–346 (2008)
7. Cho, J., Roy, S.: Impact of search engines on page popularity. In: Proceedings of WWW 2004, pp. 20–29 (2004)
8. Broder, A.: A taxonomy of web search. ACM SIGIR Forum **36**(2), 3–10 (2002)
9. Raman, K., Bennett, P.N., Collins-Thompson, K.: Toward whole-session relevance: exploring intrinsic diversity in web search. In: Proceedings of SIGIR 2013, pp. 463–472 (2013)
10. Shen, D., Sun, J.-T., Yang, Q., Chen, Z.: Building bridges for web query classification. In: Proceedings of ACM SIGIR 2006, pp. 131–138 (2006)
11. Parikh, N., Sundaresan, N.: Inferring semantic query relations from collective user behavior. In: Proceedings of CIKM 2008, pp. 349–358 (2008)
12. Xue, G.-R., Xing, D., Yang, Q., Yu, Y.: Deep classification in large-scale text hierarchies. In: Proceedings of ACM SIGIR 2008, pp. 619–626 (2008)

13. Xing, D., Xue, G.-R., Yang, Q., Yu, Y.: Deep classifier: automatically categorizing search results into large-scale hierarchies. In: Proceedings of ACM WSDM 2008, pp. 139–148 (2008)
14. Wang, Q., et al.: Mining subtopics from text fragments for a web query. Inf. Retr. **16**, 484–503 (2013)
15. Fonseca, B.M., Golgher, P., Possas, B.: Concept-based interactive query expansion. In: Proceedings of CIKM 2005, pp. 696–703 (2005)
16. Chen, Y., Xue, G.-R., Yu, Y.: Advertising keyword suggestion based on concept hierarchy. In: Proceedings of ACM WSDM 2008, pp. 251–260 (2008)
17. Zhang, B., Du, Y., Li, H., Wang, Y.: Query expansion based on topics. In: Proceedings of FSDK 2008, pp 610–614 (2008)
18. Cui, H., Wen, J.-R., Nie, J.-Y., Ma, W.-Y.: Probabilistic query expansion using query logs. In: Proceedings of WWW 2002, pp. 325–332 (2002)
19. Ma, H., Yang, H., King, I., Lyu, M.R.: Learning latent semantic relations from click through data for query suggestion. In: Proceedings of ACM CIKM 2008, pp. 709–718 (2008)
20. Broder, A., Fontoura, M., Josifovsk, V., Riedel, L.: A semantic approach to contextual advertising. In: Proceedings of ACM SIGIR 2007, pp. 559–566 (2007)
21. Kaptein, R., Kamps, J.: Improving information access by relevance and topical feedback. In: Proceedings of the 2nd Workshop on Adaptive Information Retrieval (AIR 2008)
22. Rocchio, J.: Relevance feedback in information retrieval. In: The SMART Retrieval System: Experiments in Automatic Document Processing, pp. 313–323 (1971)
23. Fang, H., Zhai, C.: An exploration of axiomatic approaches to information retrieval. In: Proceedings of ACM SIGIR 2005, pp. 480–487 (2005)
24. Sebastiani, F.: Machine learning in automated text categorization. ACM Comput. Surv. **34** (1), 1–45 (2000)
25. Zobel, J.: How reliable are the results of large-scale information retrieval experiments? In: Proceedings of ACM SIGIR 1998, pp. 307–314 (1998)
26. Shen, X., Dumais, S., Horvitz, E.: Analysis of topic dynamics in web search. In: Proceedings of WWW 2005, pp. 1102–1103 (2005)
27. Kawamae, N., Suzuki, H., Mizuno, O.: Query and content suggestion based on latent interest and topic class. In: Proceedings of WWW 2004, pp. 350–351 (2004)
28. Andre, P., Teevan, J., Dumais, S.T.: From X-Rays to silly putty via uranus: serendipity and its role in web search. In: Proceedings of ACM SIGCHI 2009, pp. 2233–2036 (2009)
29. Broder, A., Fontoura, M., et al.: Robust classification of rare queries using web knowledge. In: Proceedings of ACM SIGIR 2007, pp. 231–238 (2007)

Unsupervised Keyphrase Extraction: Introducing New Kinds of Words to Keyphrases

Tho Thi Ngoc Le[✉], Minh Le Nguyen, and Akira Shimazu

Japan Advanced Institute of Science and Technology,
1-1 Asahidai, Nomi, Ishikawa 923-1292, Japan
{tho.le,nguyenml,shimazu}@jaist.ac.jp

Abstract. Current studies often extract keyphrases by collecting adjacent important adjectives and nouns. However, the statistics on four public corpora shows that about 15% of keyphrases contain other kinds of words. Even so, incorporating such kinds of words to the noun phrase patterns is not a solution to improve the extraction performance. In this work, we propose a solution to improve the extraction performance by involving new kinds of words to keyphrases. We have experimented on four public corpora to demonstrate that our proposal improve the performance of keyphrase extraction and new kinds of words are introduced to keyphrases. In addition, our proposal is also superior to the current unsupervised keyphrase extraction approaches.

Keywords: Automatic keyphrase extraction · Syntactic structure · Parse tree · Unsupervised approach

1 Introduction

Keyphrases are single-token or multi-token expressions that provide the essential information of a sentence or document. Automatic keyphrase extraction plays an important role in many applications of natural language processing (NLP). Many approaches have been proposed for extracting keyphrases automatically. Those approaches contain two common steps: (1) collecting as many tokens as possible for candidates which benefit keyphrase extraction; and (2) combining adjacent candidate tokens to obtain keyphrases using a fixed pattern of adjectives and nouns. Up to now, candidates for keyphrases are adjectives and nouns which are collected by many methods: applying linguistic knowledge (e.g. syntactic features like POS tags, NP chunks) and statistics (e.g. term frequency, inverse document frequency, n-grams) [1,4,13]; applying graph-based ranking technique [10]; or applying clustering technique [7,9]. An overview of approaches for automatic keyphrase extraction can be found in a survey by Hasan and Ng [3].

Since previous research applies a fixed pattern to extract keyphrases, i.e. adjacent adjectives and nouns, candidate tokens are restricted to a set of pre-specified words of only adjectives and nouns. This restriction causes a consequence that other kinds of words cannot be selected as candidates, and therefore

© Springer International Publishing AG 2016
B.H. Kang and Q. Bai (Eds.): AI 2016, LNAI 9992, pp. 665–671, 2016.
DOI: 10.1007/978-3-319-50127-7_58

never appear in keyphrases. For that reason, extraction still vary in a certain performance. Practically, not all of keyphrases are composed of only adjectives and nouns. Indeed, when shedding a light on the patterns of keyphrases in four public corpora, we found that there are approximately 15 % of keyphrases contain words other than adjectives and nouns. Unfortunately, the extraction performance is possibly decreased when we involve more kinds of words to the pattern of keyphrases since the extracted keyphrases are composed into incorrect grammar phrases. However, since other kinds of words do appear in keyphrases, there should be an investigation for a novel approach which tackling new words to keyphrase extraction.

In this work, we motivate to introduce words other than adjectives and nouns, which benefit the extraction performance, to keyphrases. We propose a novel approach to extract keyphrases by collecting noun phrases (NPs) as candidate keyphrases using syntactic information, i.e. chunks and constituent syntactic parse trees. Hence, the well-formedness of keyphrases are ensured by noun phrases from chunks and parse trees. In addition, words other than adjectives and nouns are also considered to keyphrases pattern if they appear in noun phrase candidates. We experimented keyphrase extraction on four public corpora and achieved very competitive performance. Compared to extraction using patterns and the whole chunks, our proposal takes advantage in performance while reserving the well-formedness of keyphrases and involving more kinds of words. We achieve better performance than the state-of-the-art on three corpora while we are still behind a supervised approach, which employs many features for machine learning. Therefore, we are able to conclude that our proposed approach is a competitive approach for unsupervised keyphrase extraction.

2   Corpora and Keyphrase Analysis

We consider four public corpora, namely DUC-2001 [12,14], Inspec [4], NUS [11], and SemEval-2010 [5], which are used for evaluating the extraction performance in previous studies. Some characteristics of these corpora have been analyzed in previous studies [2,5]. In this work, we examine two other characteristics in concern of our work and show them in Table 1. The characteristics in our concern are: the percentage of one-word keyphrases and the percentage of the types of keyphrase patterns. Each corpus has a different percentage of one-word keyphrases and the percentage of one-word keyphrases on four corpora is 19.5% on average. Since this is a significant percentage, a certain percentage of one-word keyphrases should be specified when extracting keyphrases.

When analyzing the patterns of keyphrases, we observe that not only adjectives and nouns appear in keyphrases, but other types of words also appear, such as: participles (*watermarking, ordering criteria, synthesized data*); adverbs (*highly nonlinear rule-based models, visually impaired people, partially ordered set*); cardinal numbers (*four main design patterns, category 5 hurricane, type II diabetes*).

The percentage of keyphrases for each type of keyphrase pattern is showed next to the number of keyphrases in Table 1. Note that, keyphrases in test set

Table 1. The characteristics of four public corpora of keyphrase extraction.

	Corpora			
	DUC-2001	Inspec	NUS	SemEval-2010
Type	News articles	Paper abstracts	Paper abstracts	Paper abstracts
# Documents for test	308	500	211	100
# Keys	2,484	4,913	2,327	1,482
# One-word keys	431 (17.4%)	659 (13.4%)	610 (26.2%)	309 (20.9%)
# Keys (adj+noun)	2,298 (92.5%)	4,221 (85.9%)	1,903 (81.8%)	(84.5%)
# Keys (w. participles)	53 (2.1%)	383 (7.8%)	206 (8.9%)	(7.2%)
# Keys (other patterns)	133 (5.4%)	309 (6.3%)	218 (9.4%)	(8.3%)
# Exist. keys in text	2,462 (99.1%)	3,826 (77.9%)	2,200 (94.5%)	(89.5%)
# Exist. keys (adj+noun)	2,277 (91.7%)	3,338 (68%)	1,837 (78.9%)	(80%)
# Exist. keys (w. participles)	53 (2.1%)	287 (5.8%)	178 (7.6%)	(3.2%)
# Exist. keys (other patterns)	132 (5.3%)	201 (4.1%)	185 (8%)	(6.3%)

of SemEval-2010 are provided as stemmed words, we examine the characteristic of keyphrases in training set instead. As shown in Table 1, the percentage of keyphrases which follow the patterns of adjectives and nouns is 86.2% on average. When looking closely to the annotated keyphrases, about 90% keyphrases actually exist in the text. The percentage of existing keyphrases which follow the patterns of adjectives and nouns is 79.6% on average. Consequently, the highest recall of extraction performance is less than 80% when involving only adjectives and nouns.

Among other types of words in keyphrases, verbs in forms of present and past participles have a considerable contribution to keyphrases. Therefore, in the following section, we examine whether involving participles as candidates in keyphrase patterns improves the extraction performance.

3 Analysis on Extracting Candidates Using Noun Phrase Patterns

This section describes the extracting process using patterns of noun phrases. Since approximately 86% of keyphrases are combinations of adjectives and nouns, to collect noun phrases, previous works often use a pattern to collapse adjacent adjectives and nouns. When examine the keyphrases pattern, we recognize that adjectives in forms of comparative and superlative also appear, e.g. *lower net income* and *nearest parent model*. Hence, the pattern for noun phrases is modified as the following regular expression

$$(JJ|JJR|JJS)*(NN|NNS|NNP|NNPS)+$$

As analyzed in Sect. 2, on average 6.5% of verbs in forms of present and past participles play the roles as adjectives and nouns in keyphrases. Hence, we involve them to the pattern of noun phrases and introduce another pattern for

Table 2. The performance of keyphrase extraction using patterns.

	DUC-2001			Inspec			NUS			SemEval-2010		
	Prec	Rec	F1	Prec	Rec	F1	Prec	Rec	F1	Prec	Rec	F1
Pattern(adj+noun)	21.3	**39.6**	**27.7**	35.4	44.9	39.6	15.7	20.2	**17.7**	21.0	20.7	**20.8**
Pattern(+participle)	20.8	38.8	27.1	34.0	44.6	38.6	14.9	19.6	16.9	19.4	19.4	19.4
TFIDF *n*-grams	-	-	27.0	-	-	36.3	-	-	6.6	14.9	15.3	15.1

noun phrases to examine whether including such participles improves extraction performance as following

$$(JJ| JJR| JJS| VBG| VBN)*(NN|NNS|NNP|NNPS| VBG)+$$

We assign weights to candidates and extract keyphrases as the work by Le et al. [6]. Experiments are run on four public corpora described in Sect. 2 and followed the evaluation criteria in SemEval-2010. We extract up to 15 highest weighted keyphrases for each document including one single keyphrases and 14 compound keyphrases. This technique is compared to a baseline [2,5], henceforth referred as *TFIDF n-grams* for convenience. In TFIDF n-grams, the top weighted *n*-grams of adjectives and nouns are extracted as keyphrases where the weight of a candidate is calculated by summing its constituent unigrams.

The performance of the proposed technique is presented in Table 2. Our technique achieves better performance than the TFIDF n-grams baseline for all corpora. Henceforth, we use these results as new baseline for keyphrase extraction in this work. When adding verbs in forms of present and past participles to the pattern of keyphrases, the performance decreases. The reason is that the participles which modify the meaning of noun phrases are confused with the verbs of sentences. For an example, considering the sentence *"Previous research has **indicated differing levels** of importance of perceived ease of use relative to other factors,"* the phrase *"indicated differing levels"* is wrongly extracted as a keyphrase since it satisfies the pattern of keyphrases. In fact, the participle *"indicated"* does not modify the meaning of noun phrase *"differing levels"* but it is a conjugation of verb in present participle tense.

Based on experimental results, we conclude that the performance of keyphrase extraction is not improved when involving present and past participles into noun phrase patterns. However, as keyphrases contain such parts-of-speech of words, an approach should be investigated to capture all possible words to keyphrases.

4   Extracting Candidates Using Syntactic Information

This section introduces a novel technique to improve the performance of keyphrase extraction by exploiting the syntactic information with two levels: *shallow (chunks)* and *deep (constituent parse trees)*. Our proposal to extract keyphrases using syntactic information is outlined as follows:

1. Parse all sentences in document for syntactic information;
2. Collect noun phrases as candidates;
3. Post-process candidates to make sure they are well-formed;
4. Assign weights to candidates to indicate their importance;
5. Collect the top weighted candidates as the keyphrases.

Weights of candidates are also computed as the way in work by Le et al. [6]. The post-processing step is to ensure that candidates are well-formed. This step removes the unnecessary tokens from the beginning and ending of candidates. Two ways are introduced to remove unnecessary tokens to ensure that:

- A Candidate begins with a token whose POS tag is JJ, JJR, JJS, NN, NNS, NNP, or NNPS; and ends with a token whose POS tag is NN, NNS, NNP, or NNPS.
- A Candidate begins with a token whose POS tag is JJ, JJR, JJS, NN, NNS, NNP, NNPS, VBG, or VBN; and ends with a token whose POS tag is NN, NNS, NNP, NNPS, or VBG.

In other words, we consider only adjectives and nouns in the first way while involving participles to the candidates in the second way.

Experiments and Evaluations

Illinois Chunker and Stanford CoreNLP tools are respectively employed to parse sentences into chunks and parse trees. Then, noun phrases are extracted using these syntactic information. After that, each noun phrase is post-processed to eliminate the punctuation, conjunctions and unnecessary tokens. In post-processing, each noun phrase is split at the position of punctuation or conjunctions (if any).

The experiments of our proposal are also run on four public corpora. The extraction performance using syntactic information is shown in Table 3 in comparison to two baselines: Pattern baseline (ref. Sect. 3) and Previous best F1 baseline. For DUC-2001 and NUS corpora, TFDIF n-grams yields the state-of-the-art performance [2]. For Inspec corpus, clustering approach [7] achieves highest F1-score. For SemEval-2010 corpus, HUMB [8], a supervised system, obtains the best performance.

Table 3. The performance of keyphrase extraction using syntactic information.

	DUC-2001			Inspec			NUS			SemEval-2010		
	Prec	Rec	F1	Prec	Rec	F1	Prec	Rec	F1	Prec	Rec	F1
Chunk(adj+noun)	21.3	39.6	27.7	38.6	45.3	**41.7**	17.5	21.7	**19.4**	22.7	21.5	22.1
Chunk(+participle)	21.4	39.8	**27.9**	38.1	46.1	**41.7**	17.2	21.8	19.2	23.4	22.7	**23.1**
Parse tree(adj+noun)	21.1	39.3	27.5	38.4	44.7	41.3	17.0	20.9	18.8	22.5	21.3	21.9
Parse tree(+participle)	21.1	39.3	27.5	37.6	44.8	40.9	16.8	21.0	18.7	21.9	21.1	21.5
Pattern(adj+noun)	21.3	39.6	27.7	35.4	44.9	39.6	15.7	20.2	17.7	21.0	20.7	20.8
Previous best F1	-	-	27.0	-	-	40.6	-	-	6.6	27.2	27.8	**27.5**

The results in Table 3 show that, in all corpora, our proposed approach beats the performance of Pattern baseline. In most of cases, the precision and recall scores are higher. When comparing to the previous best F1 scores, our proposal achieves the best performance on three corpora: DUC-2001, Inspec and NUS. On SemEval-2010 corpus, our approach still behind HUMB because this is a supervised method which exploits many features for machine learning: structure of the article, lexical cohesion of a sequence of words, TFIDF scores, and the frequency of the keyword in the global corpus.

When using the syntactic information for extracting candidates, we found that the recall is generally higher if participles are taken into account. In addition, other kinds of words, e.g. cardinal numbers, which occur in the middle of keyphrases are also included. For examples, keyphrases *modulo 2 residue class*, *category 5 hurricane* and *type II diabetes* are extracted by using syntactic information no matter participles are tackled or not. Even though words other than adjectives and nouns are involved, the syntactic information keeps the well-formedness of keyphrases. Therefore, both the recall and precision are improved.

5   Conclusions

We have demonstrated that keyphrases are not consistently the combination of adjectives and nouns. There are roughly 15% of keyphrases including other kinds of words such as participles, comparative/superlative adjectives and cardinal numbers. We believe that participles should be considered in keyphrase extraction since there is a recognizable percentage of keyphrases containing participles (6.5%). To improve the extraction performance and to take into account new kinds of words in keyphrases, we proposed to incorporate the syntactic information when extracting noun phrases as keyphrase candidates. We show the experimental results on four public corpora, in which performance has been improved and new kinds of words has been also introduced to the keyphrases.

Acknowledgments. This work was partly supported by JSPS KAKENHI Grant Number JP15K12094.

References

1. Frank, E., Paynter, G.W., Witten, I.H., Gutwin, C., Nevill-Manning, C.G.: Domain-specific keyphrase extraction. In: Proceedings of IJCAI 1999, pp. 668–673 (1999)
2. Hasan, K.S., Ng, V.: Conundrums in unsupervised keyphrase extraction: making sense of the state-of-the-art. In: Proceedings of COLING 2010: Posters, pp. 365–373 (2010)
3. Hasan, K.S., Ng, V.: Automatic keyphrase extraction: a survey. In: Proceedings of ACL 2014, pp. 1262–1273 (2014)
4. Hulth, A.: Improved automatic keyword extraction given more linguistic knowledge. In: Proceedings of EMNLP 2003, pp. 216–223 (2003)

5. Kim, S.N., Medelyan, O., Kan, M.Y., Baldwin, T.: SemEval-2010 Task 5: automatic keyphrase extraction from scientific articles. In: Proceedings of SemEval 2010, pp. 21–26 (2010)

6. Le, T.T.N., Nguyen, M.L., Shimazu, A.: Unsupervised keyword extraction for Japanese legal documents. In: Proceedings of JURIX 2013, pp. 97–106 (2013)

7. Liu, Z., Li, P., Zheng, Y., Sun, M.: Clustering to find exemplar terms for keyphrase extraction. In: Proceedings of EMNLP 2009, pp. 257–266 (2009)

8. Lopez, P., Romary, L.: HUMB: automatic key term extraction from scientific articles in GROBID. In: Proceedings of SemEval 2010, pp. 248–251 (2010)

9. Matsuo, Y., Ishizuka, M.: Keyword extraction from a single document using word co-occurrence statistical information. J. Artif. Intell. Tools **13**(1), 157–169 (2004)

10. Mihalcea, R., Tarau, P.: TextRank: bringing order into texts. In: Proceedings of EMNLP 2004, pp. 404–411 (2004)

11. Nguyen, T.D., Kan, M.-Y.: Keyphrase extraction in scientific publications. In: Goh, D.H.-L., Cao, T.H., Sølvberg, I.T., Rasmussen, E. (eds.) ICADL 2007. LNCS, vol. 4822, pp. 317–326. Springer, Heidelberg (2007). doi:10.1007/978-3-540-77094-7_41

12. Over, P.: Introduction to DUC-2001: an intrinsic evaluation of generic news text summarization systems. In: Proceedings of the International 2001 Document Understanding Conference (2001)

13. Turney, P.D.: Learning algorithms for keyphrase extraction. J. Inf. Retr. **2**(4), 303–336 (2000)

14. Wan, X., Xiao, J.: Single document keyphrase extraction using neighborhood knowledge. In: Proceedings of AAAI 2008, vol. 2, pp. 855–860 (2008)

Selected Papers from AI 2016 Doctoral Consortium

Shaping Interactive Marketing Communication (IMC) Through Social Media Analytics and Modelling

Pornpimon Kachamas[⊠]

Technopreneurship and Innovation Management Program,
Chulalongkorn University, Bangkok, Thailand
pornpimon.kac@student.chula.ac.th

Abstract. Social media marketing represents a dynamic field for intense research. However, existing researches have not evidently estimated the materiality of the information circulated on Social Media regarding the benefits of business entities. In this work, "Online consumer's behavior" will be analyzed using Artificial Intelligence (AI) for classificatory modelling through AISAS model by applying the family of Bayesian Classifiers. This study would help marketers in understanding key drivers of the perception towards their impact on the benefits, and enable users of such data to identify triggers that are worth monitoring and investing in.

Keywords: Machine learning · Data mining · Social media · AISAS · Social media materiality thailand and analytics

1 Introduction

A study from Zocial, Inc. [1] points out to the fact that Facebook users in Thailand express great interest in creating a fan page. The number of subscription continues to increase considerably and constantly. The country henceforth counts up to 35,000,000 accounts, about 34.62% rise comparing to the previous year, and claims the ninth position of country with most Facebook users. As online marketing becomes more and more accessible entailing larger internet traffic volumes, companies with online presence tend to explore ways that lead to understand properly the consumers' needs.

Nevertheless, there has not been any comprehensive analytical study on the precise drivers according to the AISAS Model, nor any statistical strength of such behavioral drivers on these metrics, namely Attention, Interest, Search, Action, and Share (AISAS) [2], in the online commercial space. AISAS is a consumption behavior model that takes contemporary Internet usage into consideration. It is therefore intrinsically interesting to treat with specific data and methods. In addition to finding the drivers of the above five AISAS attributes, this work aims to measure the significant impact of "Perception" (as defined by these attributes) on the benefits of the concerned Business entities. This is a notable contribution since it clarifies the presence (or lack of it in some circumstances) of measurable business value in undertaking research of Social Media data, in order to achieve a better financial performance.

© Springer International Publishing AG 2016
B.H. Kang and Q. Bai (Eds.): AI 2016, LNAI 9992, pp. 675–681, 2016.
DOI: 10.1007/978-3-319-50127-7_59

2   Related Work

Social network communication can be of great value in providing customers with satisfaction [3]. Online users enjoy social networking for various purposes. Due to the interactive nature of the online network, users can freely respond to the contents by way of comments, likes, or emoticons posting [4]. Social network communication can foster powerful product/brand awareness. That is because online communication is fast, convenient, and affordable [5]. Nowadays, most well-known retail corporations and individual entrepreneurs manage their products, images, and awareness by means of social network marketing [6]. Analyzing this Database using advanced analytics can be valuable – especially as potential customers usually express their opinions (like, comments, or emoticon posting) spontaneously and in real-time. The use of online marketing is unquestionably perceived as well-managed, modern and thus readily usable in increasing brand awareness and customer loyalty [7].

2.1   AISAS Model

Dentsu's AISAS Model is the framework in regards of the decision making process that online customers stimulate before the purchase transaction, i.e., attention, interest, search, action, and share [2]. The process can be reshuffled, interchanged, repeated, or skipped according to consumers' preferences. When a consumer shows interest in a product, a service, or an advertisement, he/she creates strong 'Attention' in such product that leads consequently to 'Interest' in such product. The consumer would do a search online for the potential comments about the product. The consumer then evaluates all related information before making the final purchase decision. In addition, peer or family member discussions on the interested product will be generally shared among them. Potential buyers will express their opinion for the final decision. Once a purchase is made, the consumer will become the opinion sharer on the online network. In order to set discriminatory models, each of these decision points should be clearly defined. For example, a precise trigger should be defined in order to categorize a certain pattern of online behavior as an instance of "Attention" / "Interest". This trigger will then be converted to logical code and run at the back-end to identify bloggers or online participants flowing in / out of the "Attentive" category. Historical online behavior will be used to create predictive attributes. These predictive attributes will then be tested through statistical techniques in order to evaluate their ability to predict the decision of accepting or purchasing a certain product, or any others metrics of interest that can affect the financial performance of the business entity (Table 1).

The above studies apply AISAS Model to analyze and explain behaviors of consumers using Facebook as a communication medium in expressing thoughts and opinions [13].

Table 1. Related research

Author	Detail
Rui Wan and Yongsheng Jin	This research establishes a first-order auto-regressive model describing the relationship between the followers' number and influence of microblogging [8]
Pei-Shan Wei and Hsi-Peng Lu	This study aims to compare the influence of celebrity endorsements to online customer reviews on female shopping behavior based on AIDMA and AISAS Models [9]
Ran Tang, Zhenji Zhang, Xiaolan Guan and Lida Wang	Based on AISAS, this study aims at real cased and data of government microblog information publication by using correlation analysis and regression analysis, and construct a quantitative model to measure the short-term effect of government microblog [10]
Du Zhiqin	This research studies the factors affecting the attitudes of university students towards WeChat marketing based on AISAS Model [11]
Hendriyani, Jessica Jane, Lenny Ceng	This research aims to confirm AISAS Model in the case of product (BB) that only uses Twitter as its promotion medium [12]

2.2 Naive Bayes Classifier

Bayesian Classifier, founded by Thomas Bayes [14], uses prior classification rule to predict the class of unknown events. Bayes' probability analysis applies prior probability to predict posterior outcome, by analyzing all the past occurrences of an event. Bayes' model employs joint probability distribution to demonstrate relationships among nodes. Each node uses the conditional probability in the parent nodes in which it predicts the probability status [15]. Basic Bayes Theorem calculates the posterior probability by employing the following probabilities [16].

$$P(c|x) = \frac{P(x|c)P(c)}{P(x)} \qquad (1)$$

Where

$P(c \mid x)$ = *Posterior Probability (likelihood of occurrence of the 'event' within profile 'x'*
$P(x \mid c)$ = *The likelihood of the occurrence of profile 'x' within all 'events'*
$P(c)$ = *Class Prior Probability of the event 'c' (e.g. Attention)*
$P(x)$ = *Predictor Prior Probability of obtaining the profile 'x' of an online user*

Specifically, the word frequency counts can be used to explore the state of satisfaction and willingness to buy of potential clients. Assuming strong dependencies of each variable, the formula above can be shown as below [17].

$$P(c_k|x_1,...,x_n) = \frac{1}{z}p(c_k)\prod_{i=1}^{n}p(x_i|c_k) \tag{2}$$

Where

$z = p(x)$ *is a scaling factor dependent only on* $x_1,...,x_n$

Many researches on this behavior have been conducted using various methods, namely Decision Tree, Neural Network, Support Vector Machine (SVM), and Naïve Bayesian Classifications. Studies show that Naïve Bayes classifier outperforms the SVM, and that it is the most promising method when external enriching is used through external knowledge base [16]. Studies also point out that Naïve Bayes, assuming independence, will achieve superior result [17].

The data mining on web analytics especially on comments and opinions is widely applied to many researches [18] analyzing comments, opinions, attitudes on line. This analysis is performed by separating comments into two categories: positive comments and negative comments. [19]. Using Naïve Bayesian method, the result comes up in form of a probability scale of positive or negative attitudes of the comments. Thus, the Bayesian equation presents clearly the strengths of Artificial Intelligence (AI) in an extremely legible and tractable manner, for ready use. Discriminative Accuracy of the Bayesian Classificatory equations would be evaluated by using metrics including Information Value (IV), Kolmogorov Smirnov Statistic (KS), ROC, and Gini. Predictive Accuracy would be assessed by using the Chi-Square tests and Brier's Score [20].

3 Design Science Research Process

This research follows the idea of Design Science Paradigm [21] which consists of seven steps in four stages [22] (Fig. 1).

3.1 Stage 1: Identifying Problems and Objectives

Social media marketing is a dynamic field of intense research. Online marketing managers need to analyze visitors' sentiments so as to understand emotions and feelings towards their brandings. However, existing researches have not evidently estimated the materiality of the information circulated on Social Media regarding the benefits of business entities. The objective of this research is to study and to apply classificatory modelling in order to:

1. Estimate relationship between sentimental patterns of users as expressed through the information posted online, and perception towards a category of business entities by means of Dentsu's AISAS Model framework.
2. Measure the materiality of "Perception" by connecting the specified component of "Perception" with the financial performance of the business entities.

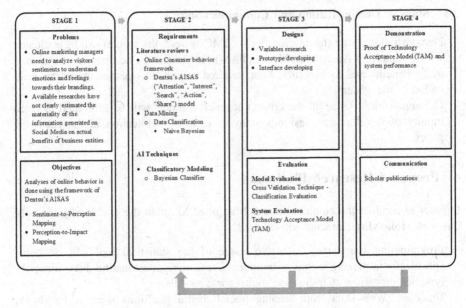

STAGE 1	STAGE 2	STAGE 3	STAGE 4
Problems	**Requirements**	**Designs**	**Demonstration**

• Online marketing managers need to analyze visitors' sentiments to understand emotions and feelings towards their brandings.
• Available researches have not clearly estimated the materiality of the information generated on Social Media on actual benefits of business entities
	Literature reviews
	• Online Consumer behavior framework
	  ○ Dentsu's AISAS ("Attention", "Interest", "Search", "Action", "Share") model
	• Data Mining
	  ○ Data Classification
	    ▪ Naïve Bayesian
	
	AI Techniques
		• Variables research
		• Prototype developing
		• Interface developing
			Proof of Technology Acceptance Model (TAM) and system performance

Objectives	• **Classificatory Modeling** ○ Bayesian Classifier	**Evaluation**	**Communication**
Analyses of online behavior is done using the framework of Dentsu's AISAS • Sentiment-to-Perception Mapping • Perception-to-Impact Mapping		**Model Evaluation** Cross Validation Technique - Classification Evaluation **System Evaluation** Technology Acceptance Model (TAM)	Scholar publications

Fig. 1. The four stages of the research project. (Adopted from Triepels et al., 2015).

3.2   Stage 2: Identifying Requirements

At this stage, the researcher holds review of related literature in favor of process, methods, and establishes the limitations from previous research in order to deal with the problem stated in stage 1. After reviewing all relevant articles, the researcher assures that AISAS Model is highly suitable to analyze and explain consumers' behaviors in expressing thoughts and opinions using Facebook as a communication medium [13]. Naïve Bayes, assuming independence, will achieve superior result than other methods [17]. This work will use Naïve Bayesian as a component in text mining technique along with the adoption of Dentsu's AISAS Model as a framework.

3.3   Stage3: Design and Evaluation

– Design: This work will be designed by variables research, prototype developing and interface developing
– Evaluation
   – Model evaluation: Cross-Validation by classification evaluation to test the model.
   – System evaluation: This stage is an initial evaluation of the system by using Technology Acceptance Model (TAM) [23] to collect feedback and improve system.

3.4   Stage 4: Demonstration and Communication

– Demonstration: After the development of IMC system, the system will be evaluated with Technology Acceptance Model (TAM) [23] to test the acceptance of system's usage benefits and its usability. Four hundred respondents participate in the evaluation of the system.
– Communication: Once all the criteria are met, a report will be submitted to community of development design system in view of publication in the academic papers.

4   Proposed Research Platforms

In order to conduct the empirical research required to attain the double objectives of this work, following elements will be used.

- Programming Language: Extensive usage of the statistical tool will be done, specifically in parsing strings of comments, text mining, estimating discriminatory-type Segmentations (Chi-Sq) and multivariate models.
- Database: Web Data from leading Social media platforms such as Facebook, Twitter will be used to attain Objective 1. At the same time, online activity data pertaining to mercantile websites (e.g. Amazon), will be needed in order to attain Objective 2 (or materiality measurement).

Activity data on mercantile websites will be combined with non-mercantile Social media data by using the IP addresses, such that an end-to-end behavioral pattern can be observed for each online user. This is one of the most crucial aspects of this work since the combined database will be used to evaluate the models used to measure the materiality of online behavior (dimensions within AISAS) on the financial performance of business entities.

5   Expected Results

This work serves as the focus on using Social media database in order to identify the drivers of customer interest (through the AISAS model) towards business entities, and to connect the text content on Social media to their financial performance. This study also discusses how to measure the returns of social media to business entities on investment and research.

References

1. Zocial Inc., May 2014. http://www.zocialinc.com/
2. Sugiyama, K.A.T.: The Dentsu way: 9 lessons for innovation in marketing from the world's leading advertising agency. McGraw-Hill, New York (2011)

3. Sashi, C.M.: Customer engagement, buyer-seller relationships, and social media. Manage. Decis. **50**(2), 253–272 (2012)
4. Viswanath, B., et al.: On the evolution of user interaction in Facebook. In: Proceedings of the 2nd ACM Workshop on Online Social Networks, pp. 37–42. ACM, Barcelona (2009)
5. Chu, S.-C.: Viral advertising in social media. J. Interact. Advertising **12**(1), 30–43 (2011)
6. Kim, D.H., et al.: Are you on timeline or news feed? the roles of facebook pages and construal level in increasing ad effectiveness. Comput. Hum. Behav. **57**, 312–320 (2016)
7. Mangold, W.G., Faulds, D.J.: Social media: the new hybrid element of the promotion mix. Bus. Horiz. **52**(4), 357–365 (2009)
8. Wang, R., Jin, Y.: An empirical study on the relationship between the followers' number and influence of microblogging. In: 2010 International Conference on E-Business and E-Government (ICEE) (2010)
9. Wei, P.-S., Lu, H.-P.: An examination of the celebrity endorsements and online customer reviews influence female consumers' shopping behavior. Comput. Hum. Behav. **29**(1), 193–201 (2013)
10. Tang, R., et al.: A study of short-term effect measurement for information publication in government microblog. Int. J. Hybrid Inf. Technol. **7**(1), 57–66 (2014)
11. Zhiqin, D.: Research into factors affecting the attitudes of university students towards WeChat marketing based on AISAS mode. In: 2015 IEEE International Conference on Electro/Information Technology (EIT). IEEE (2015)
12. Hendriyani, J.J., et al.: Online consumer behavior: confirming the AISAS model on twitter users
13. Wang, Y., Xu, P.: Micro-blogging marketing based on the characteristics of network consumer. In: 2012 International Conference on Engineering and Business Management (2012)
14. Press, S.J.: Bayesian Statistics: Principles, Models, and Applications. Wiley, Hoboken (1989)
15. Jensen, F.V.: An introduction to Bayesian Networks, vol. 210. UCL Press, London (1996)
16. Stuart, A., Ord, K.: Kendall's Advanced Theory of Statistics: Distribution Theory, vol. 1. Wiley, Hoboken (2009)
17. Murty, M.N., Devi, V.S.: Pattern Recognition: An Algorithmic Approach. Springer, Heidelberg (2011)
18. Dave, K., Lawrence, S., Pennock, D.M.: Mining the peanut gallery: opinion extraction and semantic classification of product reviews. In: Proceedings of the 12th International Conference on World Wide Web, pp. 519–528. ACM, Budapest (2003)
19. Chamlertwat, W., et al.: Discovering consumer insight from twitter via sentiment analysis. J. UCS **18**(8), 973–992 (2012)
20. Brier, G.W.: Verification of forecasts expressed in terms of probability. Mon. Weather Rev. **78**(1), 1–3 (1950)
21. Peffers, K., et al.: The design science research process: a model for producing and presenting information systems research. In: Proceedings of the First International Conference on Design Science Research in Information Systems and Technology (DESRIST 2006) (2006)
22. Triepels, R., Daniels, H.: Detecting shipping fraud in global supply chains using probabilistic trajectory classification (2015)
23. Davis, F.D.: Perceived usefulness, perceived ease of use, and user acceptance of information technology. MIS Q. **13**(3), 319–340 (1989)

Ingenious Product Form Co-design System for the Industry 4.0

Kittipong Sakornsathien[(⊠)]

Technopreneurship and Innovation Management Program,
Chulalongkorn University, Bangkok, Thailand
kittipong.s@student.chula.ac.th

Abstract. The Industry 4.0, when it comes into effect, will essentially transform supply chain management, action plans and business procedures [1]. However, a major challenge to the process of product design and development rests upon how flexible a production in a batch can be, and how to maintain the economic conditions of mass production in the same time [2]. This study aims to develop an automatic system that shows the potential of designing a product form by co-designing with the user. Artificial Intelligent techniques will be applied with Kansei engineering system in order to use as an ingenious product co-design system. First, supervised back propagation neural network (BPNN) will be co-operated with the genetic algorithm technique to optimize each design element value from Kansei engineering system. Then, the style and preference of each user will be used as a categorizing factor clustering the database into groups with K-mean technique. Each classifying cluster will use its own database in the system processing in order to obtain a set of precise design elements precisely based on the system. Moreover, the system acquires user's feedback as well as the preference cluster to revise its KE system formula. This project will apply the cross-validation as an unbiased model performance evaluation. The genuine use of this system will bring the benefit to the manufacturers by saving the lead time when their product is put on the market, and consequently ensure customers' satisfaction with the product form.

Keywords: Kansei engineering · Genetic algorithm · Product design · User preference · Emotional design · Industry 4.0 · Co-design

"Industry 4.0" is the term that outlines the vision of future manufacturing systems, which is driven by the interaction of a number of technological products that control their own production process. It will essentially transform the business procedures [1]. Product design process, Production plan and supply chain management are more or less influenced by this revolution. Recently, the concept that encourages consumers participation in the product development process engenders products that meet effectively consumers desire [2]. Furthermore, this concept guarantees the best practice model in product development [3]. However, due to the complexity and rapidly changing needs of consumers, [4] the design and the development of a product must necessarily take the time-to-market and the production flexibility into account. With the application of Industry 4.0, a major challenge of product design and development process then rises

© Springer International Publishing AG 2016
B.H. Kang and Q. Bai (Eds.): AI 2016, LNAI 9992, pp. 682–688, 2016.
DOI: 10.1007/978-3-319-50127-7_60

concerning the flexibility for the production while maintaining the economic conditions of mass production [5]. In order to do so, Innovative product design process requires tailored tools using the database over the process (Information Based), and performs with the network operation (Networked tools) that can process instantaneously (Real-time processing) [6]. However, the ability to develop a single product cannot be claimed as the overall key success of the product design.

To design successfully a product, the designers need to comprehend consumers' preferences, to hear the "voice of consumers", which consist of the key element that impacts the accomplishment of a new product. Years ago, many researchers emphasized on using consumers' opinions as a criterion for designing a product form, which are still one of the most important factors when consumers process the buying decision. There exists diverse possible techniques that can be used to survey the consumer's needs [6], Kansei engineering is the only prominent technique that converts emotion/feelings of customers perception towards products.

Kansei Engineering is defined as "translating technology of a consumer's feeling and image for a product into design elements" [7]. The semantic differential method is a general tool usually performed to elicit consumers' psychological feelings or preferences about a product in the emotional assessment for KE Technique [8]. Lately, Kansei Engineering approach is getting closer to the users through their more grounded association in the early steps of user experience's creation [9]. The system was applied and used in various industries such as in home and interior design [10], industrial products, material surface, and also many well-known product present on the market [11]. However, this impressive KE system could become more viable in the industry 4.0 era if we dispose of the system constraints, which will be explained in the next section.

1   Motivation

Our vision of a product form co-design system is the one that generates suggested product form for each customer by taking advantage of user's required "Emotion". Under the so-called vision, the ingenious system and the customer will act as a form designer of the product they want, in the same time the genuine product designer will take responsibility for design configuration in the system.

The innovative product design's process could decrease time-to-market period, and at the same time increase the chance of product sale. To attain the objective above, Product form Co-design aided system could be counted as a solution. The emotion that a customer expresses on the product becomes the main factor in this system. Kansei engineering is an internationally accepted technique. The technique helps to interpretation of emotions into designing elements for designers. On the other hand, there are some limitations to this system. The system may produce unstable results due to the massive quantity of input data [12], the period of usage [13], and the coverage of surveyed target group, all of which are used for the system formulation [14]. This project then attempts to answer to three research questions that would be a significant step for the system development.

- *How can we develop a KE system that is capable to categorize accurately various groups of users with different styles and preferences?*
- *How can we develop a KE system that is capable to diminish the impact of the outdated data while catching up with the trend?*
- *How can we develop a system that is capable to work along with the users as the product-form designing platform, and use the developed KE system that the users accept its usage benefits and its usability?*

2 Related Work

It cannot be denied that Kansei engineering serves as a grounded theory in emotional design engineering system. Researchers voluntarily make use of Kansei Engineering in term of application. In addition, Kansei Engineering offers a wide range of applications both in design domain and in service domain. According to Nomura, J. [15], a virtual space decision support system employing Kansei Engineering is applied for production and sales mainly in the kitchen business. Sato, N. et al. [16] extract the features of a movie using factor analysis from data of a Semantic Differential Gauge questionnaire, and then link the viewer's Kansei with the features using multiple linear regression analysis. As for Fu Guo et al. [13], the study uses a methodology of Kansei Engineering with Back Propagation neural networks and Genetics Algorithm to develop a Decision Support System (DSS) for industrial design, which function effectively with a non-designer. Moreover, Hsiao et al. [17] prove that the virtual model of optimum solutions in the study using genetic algorithm models produce results more rapidly and more accurately. Researchers endorse satisfactorily the success of the output from these two co-operated systems. In those studies, the fitness function in GA method can be defined by output valued from KE system. As a consequence, the output becomes the optimized design elements for the emotional needs.

As we can observe from previous research, a KE system's limitation about outdated-database is corrected with the technique called "Trend Card". This technique uses a set of trendy style pictures incorporated with the outdated database. The result asserts that the final product produced under the recommendation of the "Trend Card" with KE system is more modern, and receive more enthusiastic welcome in the market than those suggested from the KE system with outdated data [18]. Another limitation is the mismatching database issue. From the proved model, designing ideal products for positive responses, and inciting customers to buy the product depends on two factors. These factors are on the one hand individual tastes and preferences, on the other hand situational factors [19]. They have significance for developing the desired product form. The study conducted by Chuang et al. [14] examines the relationship between the preference perception of users preference towards mobile phones and their form design elements, and concludes that user's preference database serves as the styling benchmark to help designers in adopting a proper design and development perspective for the intended end users. Kansei Engineering system incorporating mismatching style and preference of user offered less precise output. As for Kuang and Jiang [20], they presents a new product platform design method based on Kansei engineering. Their study points out that the preference of clustered group was used as a platform functions applying to those users in the cluster. Hence, the proposed method could achieve the goal as a platform design aided system. Nevertheless, their method cannot solved the limitations of time and trend in the Kansei Engineering system.

3 The Proposed Model

The Methodology of this research follows the idea of Design Science Paradigm [21] which consists of seven steps and which is divided into four stages [22]. First of all, the paradigm begins with problems and objectives identification in the first stage. Following by the second stage, the objective is to identify requirements and limitations identified from the previous research in order to solve the problem stated in the first stage. The third stage is considered to be the most important of all. After coming up with the idea of proposed system, the designing and prototyping phase will be initiated. The model as well as the system will be evaluated before its real use in stage. Finally, the fourth stage consists of the demonstration and the communication of the finished system.

In this study, the first two stages have already been completed. The identified problem and the objective is set in order to answer three main research questions leading to the ingenious product form co-design system development. We come up with a proposed project framework shown in Fig. 1. The template of this model is adopted [23].

The main objective of this study is to reduce those limitations of Kansei Engineering system by reinforcing the system with the competence of neural networks that

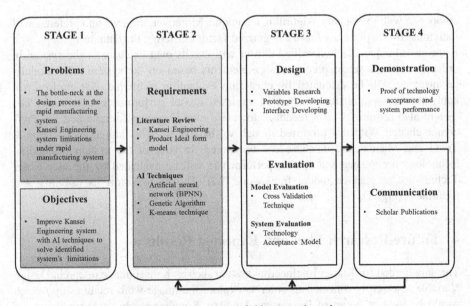

Fig. 1. The four stages of this research project.

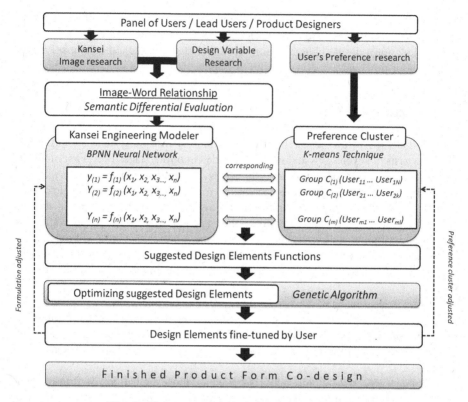

Fig. 2. The proposed design of system components.

co-operate with the Genetic Algorithm technique. Moreover, the style and preference of individual user will be used as a categorized factor clustering the data into groups with K-mean technique. Each classified cluster will use its own database in the system in order to obtain a set of precise design elements based on the system. The model's performance will be evaluated by clustering evaluation technique. The project will apply the cross-validation method as unbiased model performance evaluation. The system also acquires User's feedback to revise its KE system formula and the Preference cluster. With this proposed model, we believe that the limitations of Time & Trend and Mismatched database can be solved. In term of the system evaluation, technology acceptance and system performance will be evaluated by the user under Technology acceptance model framework [24]. This is to affirm its usability and usefulness (Fig. 2).

4   Future Research Plan and Expected Results

The data needed for Kansei Engineering system such as Kansei image research, Design Variable research is collected in order to explore the image-word relationship, which would be used as a part of the system database. Simultaneously, style preference of

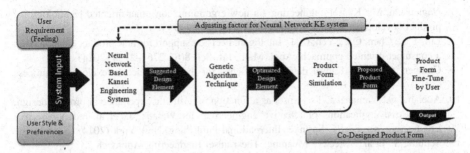

Fig. 3. Illustrated the expected system process

users will be collected and clustered, too. The information represents an input for the Neural Network based Kansei Engineering system. The Genetic Algorithm is finally combined into the system.

As for the results, this study expects to come up with the product-form design aided system that works with of Kansei Engineering technique and Genetics Algorithm method. The system is supposed to take diverse user's tastes and preferences into account as to cope with user's emotions. It is likely to be a useful system for many businesses in this upcoming industrial revolution, so-called Industry 4.0. The system is potentially capable to design automatically a product form by co-designing with the user. The real use of this aided system innovates the process of product development, and brings benefit to the manufacturers. Besides, it can save the lead time when their product is put on the market by encouraging the customer participation in the design. This will also lead to answer appropriately to their needs. And as a consequence, they can ensure customer satisfaction in the product form (Fig. 3).

References

1. Schmidt, R., et al.: Industry 4.0 - potentials for creating smart products: empirical research results. In: Abramowicz, W. (ed.) Business Information Systems, pp. 16–27. Springer International Publishing, Cham (2015)
2. Griffin, A.: PDMA research on new product development practices: updating trends and benchmarking best practices. J. Prod. Innov. Manage **14**, 429–458 (1997)
3. Cooper, G.R.: What seperates the winners from the losers and what drive success. In: Kahn, K.B. (ed.) The PDMA Handbook of New Product Development, pp. 3–34. Wiley, Hoboken (2013)
4. Yan, X.-T., Jiang, C., Eynard, B.: Advanced design and manufacture to gain a competitive edge: new manufacturing techniques and their role in improving enterprise performance, p. 891. Springer, London (2008)
5. Fogliatto, F.S., Da Silveira, G.J.C.: Mass customization engineering and managing global operations, in Springer series in advanced manufacturing, p. 1, online resource - xviii, 378 p. Springer, London (2011)
6. Lutters, E., et al.: Tools and techniques for product design. CIRP Ann. Manuf. Technol. **63**(2), 607–630 (2014)

7. Nagamachi, M.: Kansei engineering - a new ergonomic consumer-oriented technology for product development. Int. J. Ind. Ergon. **15**(1), 3–11 (1995)
8. Lin, Y.C., Chen, C.C., Yeh, C.H.: Intelligent decision support for new product development: a consumer-oriented approach. Appl. Math. Inf. Sci. **8**(6), 2761–2768 (2014)
9. Gentner, A., et al.: Kansei design approaches for the new concept development process (2012)
10. Adelabu, O., Yamanaka, T.: Kansei as a function of aesthetic experience in product design, in industrial applications of affective engineering. In: Watada, J., et al. (eds.) Affective Engineering, pp. 83–95. Springer International Publishing, New York (2014)
11. Schutte, S., et al.: Affective Meaning: The Kansei Engineering Approach
12. Nagamachi, M., Lokman, A.M.: Innovations of kansei engineering, in Industrial innovation series, p. 1, online resource -xiii, 140 p. CRC Press, Boca Raton (2011)
13. Guo, F., Ren, L., He, Z., Wang, H.: Decision support system for industrial designer based on kansei engineering. In: Rau, P.,L.,Patrick (ed.) IDGD 2011. LNCS, vol. 6775, pp. 47–54. Springer, Heidelberg (2011). doi:10.1007/978-3-642-21660-2_6
14. Chuang, M.C., Chang, C.C., Hsu, S.H.: Perceptual factors underlying user preferences toward product form of mobile phones. Int. J. Ind. Ergon. **27**(4), 247–258 (2001)
15. Nomura, J., et al.: Virtual space decision support system using Kansei engineering. In: Kunii, T.L., Luciani, A. (eds.) Cyberworlds. Springer, Tokyo (1998)
16. Sato, N., Anse, M., Tabe, T.: A method for constructing a movie-selection support system based on *Kansei* engineering. In: Smith, Michael, J., Salvendy, G. (eds.) Human Interface 2007. LNCS, vol. 4557, pp. 526–534. Springer, Heidelberg (2007). doi:10.1007/978-3-540-73345-4_60
17. Hsiao, S.W., Chiu, F.Y., Lu, S.H.: Product-form design model based on genetic algorithms. Int. J. Ind. Ergon. **40**(3), 237–246 (2010)
18. Lim, D., Bouchard, C., Aoussat, A.: Trends integration process as input data for Kansei engineering systems. In: Abraham, A., Dote, Y., Furuhashi, T., Köppen, M., Ohuchi, A., Ohsawa, Y. (eds.) Soft Computing as Transdisciplinary Science and Technology. Springer, Berlin (2005)
19. Bloch, P.H.: seeking the ideal form: product design and consumer response. J. Mark. **59**(3), 16–29 (1995)
20. Kuang, J., Jiang, P.: Product platform design for a product family based on Kansei engineering. J. Eng. Des. **20**(6), 589–607 (2009)
21. Hevner, A.R., et al.: Design science in Information Systems research. MIS Q. **28**(1), 75–105 (2004)
22. Peffers, K., et al.: The design science research process: a model for producing and presenting information systems research. In: Proceedings of the First International Conference on Design Science Research in Information Systems and Technology (DESRIST 2006) (2006)
23. Triepels, R., Daniels, H.: Detecting shipping fraud in global supply chains using probabilistic trajectory classification (2015)
24. Davis, F.D.: Perceived usefulness, perceived ease of use, and user acceptance of information technology. MIS Q. **13**(3), 319 (1989)

Selected Papers from SMA 2016

Data Lifecycle and Tagging for Internet of Things Applications

Sanghong Ahn[1], Hyeontaek Oh[1], Hwa Jong Kim[2],
and Jun Kyun Choi[1(✉)]

[1] School of Electrical Engineering, KAIST, Daejeon, Republic of Korea
{ancom21c,hyeontaek,jkchoi59}@kaist.ac.kr
[2] Department of Computer and Communications Engineering,
Kangwon National University, ChunCheon, Republic of Korea
hjkim@kangwon.ac.kr

Abstract. The number of sensors and the amount of data those sensors are collecting is increasing exponentially in the form of sensor networks collectively known as the Internet of Things (IoT). Using this large volume of IoT data in context-aware systems requires new preprocessing methods. This paper introduces an IoT data lifecycle, a proposed definition of an IoT-specific context tag, and required for an automated preprocessing method called data tagging before suggesting some uses of this IoT context tag and outlining possible directions for future research.

Keywords: Data tagging · Context-ware · Internet of Things

1 Introduction

The number of sensors and the amount of data those sensors are collecting is increasing exponentially in the form of sensor networks collectively known as the Internet of Things (IoT). According to Cisco, smart devices, the machines capable of sending and receiving data directly to and from other machines that compose these sensors networks, numbered over 600 million in 2015 and are expected to break 3 billion by 2020 [1]. These sensors generate data about things like the status of household appliances, the temperature, pressure, and quality of the air, and the location of individual users.

This data is used to understand contextual information relevant to individual users [2]. This contextual information can then be used to make decisions and perform actions to optimize predefined goals. In a simple example, if a smart thermometer registers a colder-than-desired temperature, a cloud-based smart home system may decide to turn the heater on.

In a more complicated example, imagine a user wears a wearable device to record his/her heart rate data while running. The device generates raw heartbeat data that is then written into a database. The database also ingests data about location from the user's phone; data about scheduled events from the user's calendar; and data about the weather from a weather information service. The application can then connect the relevant data points: the heart rate data indicates exercise; the location data indicates running; the calendar has a "marathon" event; and the weather service shows a "sunny day".

B.H. Kang and Q. Bai (Eds.): AI 2016, LNAI 9992, pp. 691–695, 2016.
DOI: 10.1007/978-3-319-50127-7_61

Together, the application knows the user is doing a cardio exercise (running) in the form of a marathon on a sunny day.

2 Needs and Requirements of IoT Data Tagging

IoT data is used to understand contextual information that can be used in context-aware computing systems. There are two levels of contextual information: primary contextual information and secondary contextual information. Imagine the rule "If humidity is greater than 70 percent, then it is humid." The humidity percentage is a primary context. The judgements "it is humid" and "it is not humid", which is made using the primary context, are secondary contexts. These secondary contexts are often used to make decisions.

The primary context may be interpreted as different secondary contexts depending on the application in question. For example, Fig. 1 shows how a single sensor, a heart rate monitor, can be used to measure both exercise intensity and psychological stress. Additionally, the interpretation may depend on other primary and secondary contexts. For example, a primary context of 25 degrees Celsius could lead to a "just right" secondary context in a room and a "too cold" secondary context in a greenhouse, depending on the primary context of which room the sensor is in.

Traditionally, these thresholds and interpretations are manually defined. This approach, however, will not scale as the number of IoT devices increases significantly. Methods that automatically generate metadata tags are needed to continue scaling contextual computing systems.

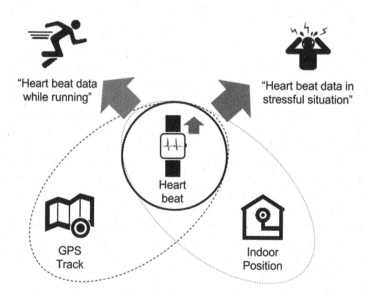

Fig. 1. An example of generating multiple secondary contexts from one primary context

3 Contextual Data Tagging in IoT Environment

This section describes the lifecycle of data in a standard IoT environment and proposes a contextual data tag for IoT data.

3.1 Data Lifecycle in IoT Environment

Conventional models of data lifecycle in IoT environment consist of four steps: data collecting, contextual awareness, data usage, and data post-processing. This paper expands this traditional definition by defining the data lifecycle model in Fig. 2 in a contextual perspective with data tagging and context sharing processes.

Data Source Profiling. Data source profiling involves generating a profile about the original source of the data that contains information like data type, data collecting period, location, owner, other related devices, etc. This profile is utilized in the later data tagging process.

Data Collecting. Data collecting process involves aggregating data from various sensors into a centralized data processing system.

Data Tagging. Data tagging involves generating metadata tags to associate with the raw data. The tags should be context-specific and used to help share, explore, and fuse data.

Context Awareness. Context awareness involves inferring contextual information from collected data. The specific method used to infer contextual information depends on the application.

Context Sharing. Context sharing involves sharing the context inferred from one set of data with other devices in the IoT sensor data.

Data Usage. Data usage involves analyzing the data and making decisions. The specific analysis and decision depends on the application.

Data Post-processing. Data post-processing involves dealing with data after it is used (e.g. stored for future use, published for open use, deleted).

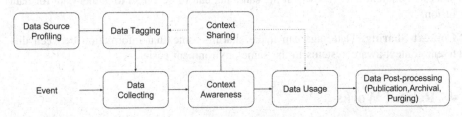

Fig. 2. Data lifecycle model with data tagging and context sharing processes

3.2   Context Data Tags in IoT

This paper uses the context model defined by Henricksen: "A context model identifies a concrete subset of the context that is realistically attainable from sensors, applications and users are able to be exploited in the execution of the task. The context model that is employed by a given context-aware application is usually explicitly specified by the application developer, but may evolve over time [3]."

The authors extend this definition by suggesting an attribute called a "context tag" that describes the current context inferred from a particular set of data. This IoT context tag has a unique identifier and a description that describes the context. The tag may contain spatiotemporal information, indexes of related entities, the current context, the direction of movement (like "increasing temperature" or "decreasing velocity"), and other information depending on the specific application. These tags often take the form of secondary contexts.

3.3   Requirements for IoT Data Tagging Mechanism

Analytics Speed. Data tagging should not slow down context generation and contextual reasoning.

Memory Usage. Different data types may share the same data tag. Individual datasets can have a large number of data tags. The system should try to have as little redundancy as possible to limit the amount of computer memory required.

Search Optimization. The purpose of data tags is to help users (including automated algorithms) search, discover, explore, and analyze data based on secondary contexts. The tags generated should make this as easy as possible.

3.4   Usages of Data Tags

Data Filtering and Preprocessing. Data tags can easily filter voluminous IoT data. Data providers can provide already tagged data to help data users reduce processing costs.

Data Extraction and Fusion. Data tags can accelerate the data search and discovery process. Different data types with the same tag can be searched to gather data for data fusion.

Context Sharing. Data tags can make sharing contextual information between different context-aware systems in the same environment easier.

4   Related Works

As the number of IoT sensors increases, the amount of generated data becomes increasingly unwieldy. Real-time data processing techniques, particularly models for real-time data processing middleware, have been widely studied [4, 5]. Proposed solutions focus on efficiently collecting data and creating context using semantic ontologies. However, these solutions do not explain how this context can be used and shared after it has been created.

Another relevant area of inquiry is the field of semantic annotation, which may help make data fusion more efficient [6, 7]. Previous works propose manual and semi-automatic knowledge-based construction methods based on a unified semantic representation model for annotation. Annotation techniques are still developing and require further study.

5 Conclusion

With the development of Internet of things and sensor networks, various types of data are being produced from sensors. In people's life and environments, more and more sensors are expected to detect location, measure temperature and air pressure, and record communication log. Processing huge amount of sensor data is a very important issue in IoT applications. This paper presented the needs and requirements of data tagging in IoT applications. Data tagging technique through proper data pre-processing provides solutions for traditional context-aware computing models by solving real-time data processing, data extraction and fusion and context sharing problems.

Acknowledgement. This work was supported by Institute for Information & communications Technology Promotion(IITP) grant funded by the Korea government(MSIP) (No: R-20160906-004163, Developing Bigdata Autotagging and Tag-based DaaS System).

References

1. Cisco: Cisco Visual Networking Index: Global Mobile Data Traffic Forecast Update, 2015-2020, White paper (2016)
2. Perera, C., Member, S., Zaslavsky, A., Christen, P., Georgakopoulos, D.: Context aware computing for the internet of things: a survey. IEEE Commun. Surv. Tuts. **16**(1), 414–454 (2014). IEEE, New York
3. Henricksen, K.: A framework for context-aware pervasive computing applications. Ph.D. thesis, The University of Queensland (2013)
4. Konstantinou, N., Solidakis, E., Zafeiropoulos, A., Stathopoulos, P., Mitrou, N.: A context-aware middleware for real-time semantic enrichment of distributed multimedia metadata. Multimedia Tools Appl. **46**(2), 425–461 (2010). Springer, Heidelberg
5. Spanos, D.-E., Stavrou, P., Mitrou, N., Konstantinou, N.: Sensorstream: a semantic real-time stream management system. Int. J. Ad Hoc Ubiq. Comput. **11**(2/3), 178–193 (2012). Inderscience, Geneva (2012)
6. Wu, Z., Xu, Y., Zhang, C., Yang, Y., Ji, Y.: Towards semantic web of things: from manual to semi-automatic semantic annotation on web of things. In: Wang, Yu., Yu, G., Zhang, Y., Han, Z., Wang, G. (eds.) BigCom 2016. LNCS, vol. 9784, pp. 295–308. Springer, Heidelberg (2016). doi:10.1007/978-3-319-42553-5_25
7. Xu, Y., Zhang, C., Ji, Y.: An upper-ontology-based approach for automatic construction of IoT ontology. Int. J. Distrib. Sens. Netw. **10**(4), 1–17 (2014)

A Content-Based Routing Scheme for Mobile Data Offloading in Pocket Switched Networks

Regin Cabacas[1] and In-Ho Ra[2(✉)]

[1] College of Information and Communications Technology,
West Visayas State University, Iloilo City, Philippines
rcabacas@wvsu.edu.ph
[2] School of Computer, Information and Communication Engineering,
Kunsan National University, Gunsan, South Korea
ihra@kunsan.ac.kr

Abstract. Existing cellular network infrastructure will likely be compromised in the future by the increase of mobile users and mobile data services. Continuous improvements of network infrastructure and several mobile data offloading strategies are promising solutions for cellular providers to cope with the demand increase of mobile data. However, these options require a lot of cost and time to implement. Few studies have been conducted to assess whether Pocket Switched Network (PSN) can be applied to support mobile data offloading. In this paper, we present a PSN mobile data offloading scheme that utilizes mobile users with available connectivity to deliver content-aware data to other mobile users. This paper also aims to evaluate whether PSN routing schemes can be applied to improve the current strategies in mobile data offloading. The simulation study exhibits promising results in terms of message delivery and latency reduction.

Keywords: Mobile data offloading · Pocket switched network · Content-based routing

1 Introduction

In recent years, cellular networks have begun to experience drastic problems in terms of keeping pace mobile data traffic demand. The ever-increasing number of mobile users, pervasiveness of mobile devices (due to improvement in features and inexpensiveness) and the eagerness of every user to be connected to the Internet are among the reasons of mobile data traffic explosion [1]. Mobile data offloading techniques such as accessing methods using Wi-Fi and femtocells [2] have been introduced to alleviate problems with mobile data traffic. In literature, cellular providers sought for other quick and promising alternative strategies to improve mobile data offloading like that of utilizing Delay Tolerant Networks [3, 4]. These approaches utilized any entity with mobility such as vehicles and mobile base stations to offload data to mobile users.

Huge portion of the mobile data traffic delivered by service providers do not have real time constraints. Thus, the delay tolerant nature of DTN promises great feasibility to support mobile data offloading. This paper specifically focuses on evaluating the

© Springer International Publishing AG 2016
B.H. Kang and Q. Bai (Eds.): AI 2016, LNAI 9992, pp. 696–701, 2016.
DOI: 10.1007/978-3-319-50127-7_62

possibility of offloading data services such as news, entertainment, weather and classified updates using opportunistic contacts among mobile users or the complementary network Pocket Switched Network (PSN). PSN is a solely human assisted network that use mobile users (with mobile devices like smartphones, PDA, etc.) to deliver content to other users using available short-range communication method such as Bluetooth, Wi-Fi and NFC. Furthermore, this paper introduces a content-based mobile data offloading and studies the possibility of utilizing a other PSN routing schemes to support mobile data offloading. The concept of social similarity interest of mobile users in the community and its temporal and spatial correlation has been integrated in the message transmission scheme.

The remainder of the paper is organized as follows; Sect. 2 discusses an overview of mobile data offloading and existing DTN researches on Mobile data offloading. In Sect. 3, the proposed scheme is presented in detail. Section 4 depicts the simulation settings and results. Lastly, Sect. 5 concludes this paper.

2 Background

The existing 3G or Long Term Evolution (LTE) networks currently are unlikely to provide as much bandwidth as in the wired Internet. Furthermore, the capacity shortage is becoming a serious challenge for cellular network providers. Thus, one of the efforts of cellular providers is to increase the physical capacity by reducing the cell size or by intelligent multiplexing of the shared radio medium [5] or produce better infrastructure conceptualizing 4G solutions and eventually 5G networks. However, these approaches still have fundamental limitations when the aggregate network demands exceed the physical capacity in the future. The use of complementary network technologies to deliver data that is commonly handled by established cellular networks is another cellular industry's strategy to mobile data offloading. With the increasing demand for mobile data brought about by the surging number of mobile users and proliferation of mobile services, several data offloading techniques have been implemented to support the underlying cellular network and cope with users mobile data needs. Wipro [6] enumerates six offloading options namely; Wi-Fi Hotspots, femto cells, LTE Small Cells, Direct tunnel, Internet Offload Gateway and M2 M Gateway. Among these, Wi-Fi and femto cells or an integration of both has been the popular option among telecommunication companies. However, these options has also some shortcómings like limited coverage, high installation and maintenance cost, which makes them not admirable to be implemented outdoor or mobile environment.

Recently, the area of Delay Tolerant Networks has been explored for applicability in mobile data offloading [7, 8]. This particular strategy offloads cellular traffic to mobile users by using other mobile users instead of retrieving duplicate data from base stations. Specifically, it exploits the opportunistic contacts between users to deliver same content to other users. Its end goal is to support mobile data offloading with little or no change to the current applications or underlying networks.

DTN in particular is comprised of a wide variety of complementary networks including Vehicular Ad hoc networks (VANET) and Pocket Switched networks (PSN). VANET have been introduced in [9, 10] and has presented several promising results for data offloading. PSN on the other hand, receives little attention among researches due to the fact that nodes in this kind of network have little capability in terms of communication range and buffer capacity unlike VANET. However, there are also possible applications of PSN such as content delivery which will be the focused of this paper.

3 Proposed Scheme

The proposed scheme takes advantage of the social interest similarities [11] (e.g., same profession, hobbies, etc.) of mobile users in a community to offload data. It aims to be an alternative support for other mobile data offloading strategies and reduce data traffic of often-subscribed mobile services (e.g. news, weather, entertainment updates and classified ads). Figure 1 illustrates the offloading strategy of the proposed scheme. First, the proposed scheme looks into the interest of every subscriber node in contact with a publisher node. A user (or a node) that has no interest with a particular content does not necessarily mean that it will not be better message forwarder. Therefore, when a subscriber node is not interested with the content currently in the publisher's buffer, the scheme identifies the relationship strength between the subscriber node and other nodes to decide if it receives the particular message.

To assess the capability of the node to be a forwarder, we use Freeman's degree centrality [12] shown in Eq. 1. This formula calculates for the relationship of every node to other nodes in the network, where N is the number of nodes, D is maximum degree and m is the number of edges.

$$C_d = (N * D - m)/((N - 1)(N - 2)) \tag{1}$$

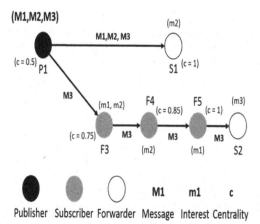

Fig. 1. Interest and centrality value-based mobile offloading decision rule

4   Simulation Results

The simulation was performed using the ONE (Opportunistic Network Environment) simulator. Different content-types were labeled as M1-M4. One (1) publisher node and three (3) subscriber node groups were simulated in the network emulating mobile users with walking speed and all equipped with Wi-Fi and Bluetooth interfaces. The 3 node groups have different number of interests such that 1 group has only 2 interest while the others has 3 and 4. The simulation time is set to 1 day where nodes move in the Helsinki map area using shortest path map based movement model. The following figures show the result of the simulation.

Figure 2 shows the total message size increases as the number of nodes in the network is also increased. This data shows that the denser the network, more messages will be offloaded even with only one publisher node given to the network. The result also shows a 500 MB total data size is offloaded with 150 users for 1 day. This means that with more various data content more data can be offloaded to users. Even if the proposed scheme focuses on offloading non-time sensitive data, message latency is an important factor to consider in data offloading. Latency suggests the time difference between the content is created and it was delivered. A data transferred on the latter time even with interest will not be valuable enough for the user. Figure 3 depicts the average latency of messages. As node is increasing, average latency decreases. A less than 3 h delay of delivery can be seen in all the scenarios and at most 1 h is achieved for 150 nodes in the simulation.

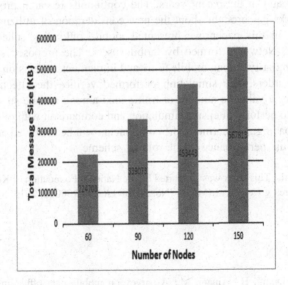

Fig. 2. Total offloaded data size

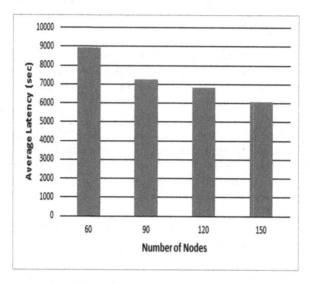

Fig. 3. Average message latency

5 Conclusion

Cellular providers have put significant effort in finding better solutions to cope with mobile data demand in the recent years. The continuous research progress in Delay Tolerant Networks has brought about the new consideration of utilizing it for mobile data offloading. In this paper, we presented mobile offloading scheme that utilize Pocket Switched Networks formed by mobile users. The proposed scheme utilized social interest similarity among mobile users and implements a decision rule to identify better data forwarders. The simulation performed verifies the effectiveness of this scheme to deliver a valuable amount of traffic load among mobile subscribers. In the future, we seek to perform intensive simulation and comparison with other schemes for mobile data offloading. Furthermore, the components will be reviewed and improved to further enhance the performance of the routing scheme.

Acknowledgement. This work was supported by the National Foundation of Korea (NSF) grant funded by the Korea government (MSIP) (No. 2016R1A2B40130002, 2013054460).

References

1. Aijaz, A., Aghvami, H., Amani, M.: A survey on mobile data offloading: technical and business perspectives. IEEE Wirel. Commun. **31**(2), 188–200 (2013)
2. Joe-Wong, C., Sen, S., Ha, S.: Offering supplementary network technologies: adoption behavior and offloading benefits. IEEE/ACM Trans. Netw. **23**(2), 355–368 (2015)

3. Li, Y., Qian, M., Jin, D., Hui, P., Wang, Z., Chen, S.: Multiple mobile data offloading through disruption tolerant networks. IEEE Trans. Mobile Comput. **13**(7), 1579–1596 (2014)
4. Izumikawa, H., Katto, J.: RoCNet: spatial mobile data offload with user-behavior prediction through delay tolerant networks. In: 2013 IEEE Wireless Communications and Networking Conference (WCNC), pp. 2196–2201 (2013)
5. Chapin, J.M., Lehr, W.H.: Mobile broadband growth, spectrum scarcity, and sustainable competition. In: Proceedings of the 39th Research Conference on Communication, Information and Internet Policy (2011)
6. Wipro Ltd. Website. https://www.wipro.com/documents/resource-center/Data_Offload_Approaches_for_Mobile_Operators.pdf
7. Go, Y., Moon, Y.G., Nam, G., Park, K.S.: A disruption-tolerant transmission protocol for practical mobile data offloading. In: Proceedings of the 3rd ACM International Workshop on Mobile Opportunistic Networks, pp. 48–56 (2012)
8. Mayer, C.P., Waldhorst, O.P.: Offloading infrastructure using delay tolerant networks and assurance of delivery. In: 2011 IFIP Wireless Days (WD), pp. 1–7 (2011)
9. El Mouna Zhioua, G., Zhang, J., Labiod, H., Tabbane, N., Tabbane, S.: VOPP: a VANET offloading potential prediction model. In: IEEE Wireless Communications and Networking Conference (WCNC), pp. 2408–2413, 6–9 April 2014
10. El Mouna Zhioua, G., Labiod, H., Tabbane, N., Tabbane, S.: Cellular content download through a vehicular network: I2 V link estimation. In: IEEE 81st Vehicular Technology Conference (VTC Spring), pp. 1–6 (2015)
11. Karbachi, G., Viana, A.C.: A content-based network coding to match social interest similarities in delay tolerant networks. In: ExtremeCom Workshop 2009, pp. 1–2 (2009)
12. Li, F., Wu, J.: LocalCom: a community-based epidemic forwarding scheme in disruption-tolerant networks. In: IEEE SECON, pp. 1–9 (2009)

Exploring the Use of Big Data Analytics for Improving Support to Students in Higher Education

Si Fan[1], Saurabh Garg[2], and Soonja Yeom[2(✉)]

[1] Faculty of Education, University of Tasmania, Hobart, Australia
Si.Fan@utas.edu.au
[2] School of Engineering and ICT, University of Tasmania, Hobart, Australia
{Saurabh.Garg,Soonja.Yeom}@utas.edu.au

Abstract. In the past two decades, with the globalisation of education, there has been a continuous increase in the diversity of students in Higher Education. This diversity form a basis for a culturally rich environment, although, the cultural and language differences and the diversity in teaching and learning styles also bring challenges. From a university's perspective, providing the maximum support to overcome these challenges and achieving maximised student engagement would be in its best interest. Recent advances in Big Data and increase in electronically available education data can help in achieving these aims. This paper reports the findings of a preliminary study which applies Big Data analysis methods to analyse education data gathered from learning management systems. The aims was to understand ways to improve student engagement and reduce student dropout. This paper documents the experience gained in this early exploration and preliminary analysis, and thereby provides background knowledge for reporting of data from the formal data collection stage which will be conducted at a later stage of research.

Keywords: Big data · Learning analytics · Higher education · Learning management system · Distributed system · Australian higher education

1 Introduction

Big Data is relevant to all organizations, including higher education institutions who produce large data sets and wish to benefit from the interpretation of these data. Through analysis of information flow, Big Data analytics looks for hidden threads, trends and patterns (Matteson 2013). Influential examples of Big Data applications include Google Flu Trends, which provides predictions and estimates of influenza activity for more than 25 countries, through aggregation of Google search queries (Pervaiz et al. 2012); and the research of Aslam et al. (2014), which reported on influenza-like illness rates through the collection and analysis of 159,802 tweets. While Big Data analytics is having a wide impact on public health and businesses' commercialization and marketing, its application remains limited in the field of education. Even in the few emerging studies on Big Data and learning analytics, the majority are carried out in America (e.g. Picciano 2012) and Canada (e.g. Ellaway et al. 2014).

© Springer International Publishing AG 2016
B.H. Kang and Q. Bai (Eds.): AI 2016, LNAI 9992, pp. 702–707, 2016.
DOI: 10.1007/978-3-319-50127-7_63

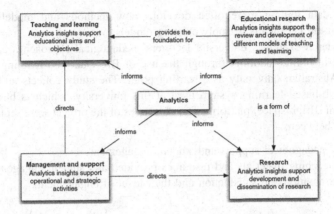

Fig. 1. One model of educational data analytics (Ellaway et al. 2014)

In the field of higher education, an increase in information flow can be seen through the wide adoption of web-based learning systems by educational institutions (e.g. MOOCs). These learning systems often feature a wide array of materials, which build on a large resource of publicly available data. These data may reveal hidden patterns to "predict student outcomes such as dropping out, needing extra help, or being capable of more demanding assignments" (West 2012, p. 2). Therefore, mining information from web-based learning systems can lead to insights regarding student performance and teacher pedagogical approaches (see Fig. 1).

In this paper, we give the details of our project to harness this insight from electronic education data available through learning management systems using Big Data technologies. In particular, this research focuses on extracting insights in relation to student behaviour in learning a higher education course, and factors that can affect their learning. This paper introduces the preliminary findings from a small data set which we intend to further validate using large data sets from different courses at this university.

2 Big Data Analysis for Australian Higher Education

Big Data is a concept that emerged with the rapid growth of web-based technologies and computer and mobile devices. It indicates "large pools of data that can be captured, communicated, aggregated, stored, and analysed" (Manyika et al. 2011, p. 4). Big Data in education is also gaining a growing attention from scholars (Eynon 2013). Data that are produced by university learning systems can be seen as one type of Big Data. Learning analytics is a term researchers often refer to when discussing the use of Big Data in the field of education. Analysis results can be generated either from build-in analytics functionalities in learning management systems, or through analysis of data gathered using data mining techniques. Researches (Clow 2013; Timms 2015) summarize the differences between learning analytics and educational data mining that, educational data mining has a greater emphasis on technical challenges, in that educational data mining more often develops new methods and models for data analysis, and the later tends to apply existing models.

This research examines university lecturers' pedagogical approaches and student engagement in online learning, through the use of Big Data and learning analytics, using one Australian university as a starting point. The study collects archived data from the web-based learning system used at this university, which is based on the Desire2Learn (Brightspace) platform. The objectives of the project were set to examine correlations between:

- lecturers' pedagogical support and student engagement;
- types of teaching materials and resources provided and student engagement; and
- student evaluation and satisfaction and their level of engagement.

3 Methodology

The research was approved by the Social Science Human Research Ethics Committee (HREC) Tasmania Network (reference H0016064). This project is being conducted in the following three stages.

Stage 1 of the project has been completed as a literature review. Relevant studies that have been conducted in higher education contexts using learning analytics have been reviewed and summarized.

Stage 2 of the project is a pilot study which will be conducted on approximately eight units from the disciplines of computing and education. This stage of the project will involve the data extraction from this specific learning management system and data cleaning. Archived data of student online discussions, lecturers' feedback, news items, unit materials and resources, and student engagement statistics generated by this learning management system are extracted, for the three year period of 2013-2015. After the initial data collection, invalid data not serving the aims of the study are filtered and removed.

Stage 3 involves the analysis of the collected data. To understand the students' interaction with learning materials, different features are identified. These features include: the number of times a student accesses a particular learning module, dates when the student visited the learning module, dates when the student submitted assignments, time when s/he accesses the assignment details, his/her achievement in terms of marks, and how frequently the student posted to the discussion board. To understand the teaching team's interaction, features were collected, such as: when teaching materials were uploaded, teaching strategies, number of assignments collected, and dates assignments are recorded. Data mining techniques such as clustering and classification are utilised to detect different patterns in the data. Big Data frameworks such as Hadoop will be utilised to analyse the data using machine learning techniques. Pearson's correlation coefficients for the association between student engagement and participation and other three variables: lecturers' pedagogical support and interactions; types of teaching materials and resources; and student evaluation and satisfaction, will be calculated in *R* Foundation for Statistical Computing.

4 Preliminary Study and Results

We conducted the preliminary study using Excel's statistical tool to guide what features we need to extract from future larger data sets and which factors we should study to generate insights. The small scale study consisted of two master level ICT units: UNIT1 (Web Development) and UNIT2 (Data Management Technology) offered at the chosen university. The enrolled cohorts in the two units overlap to a large extent. One of these units (UNIT1) is a lower level unit and the other (UNIT2) is the higher level unit. For the study, we used a set of 20 students in UNIT2, consisting of low achieving (PP = Pass), medium achieving (CR-DN = Credit or Distinction), and high achieving students (DN-HD = Distinction or High Distinction).

The aim is to study the behaviour or study pattern of the students in these units, thus we chose three factors: (1) frequency of access to each learning component; (2) date of access; and (3) their achievements in the units. To summarise and further analyse the data, we evaluated the minimum number of accesses, maximum number of accesses, and average number of accesses, in all the learning contents.

Figures 2 and 3 shows a summary of data. It can be seen clearly from the data, the study pattern of the students is similar in both units. This may indicate a high level of similarities in the patterns of teaching activities. At the same time, it can be noticed that on average the students in UNIT1 accessed learning materials less than those in UNIT2. This may indicate that more advanced units requires more effort from the students to understand the contents than lower level units. This can be evidenced further by the marks comparison between the two units. The data reveal that UNIT1 students got overall higher marks than in UNIT2. It is also interesting to notice that there are some high achieving students who enrolled in both units and achieved similar marks in both units. A difference can be seen in some medium achieving students who enrolled in both units. There is a one grade point difference in their achievement in the two units. This may be a reflection of the fundamental difference in the nature of these units. Both units have a portion on programming; however, UNIT2 has more difficult programming contents than UNIT1. Among low achieving students there is no clear

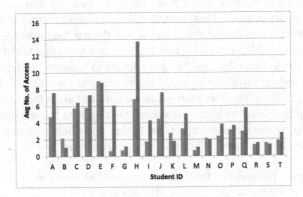

Fig. 2. Comparison Based on Number of Access (Blue: UNIT 1; Red: UNIT2) (Color figure online)

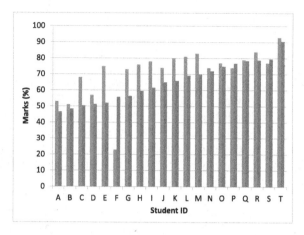

Fig. 3. Comparison Based on Student's Achievement (Blue: UNIT 1; Red: UNIT2) (Color figure online)

pattern, however, two of these students achieved almost 20% more marks in UNIT1. This may indicate that these students have specific interest in a particular area as included in UNIT1, due to which they were able to achieve significantly higher results.

5 Concluding Remarks

This paper presents the methodology and some preliminary analysis results from a small data set. Archived data were harvested from the learning management system used at one chosen Australian university. The aim was to study the innovative ways in which data management and analytics can benefit and contributes to Australian higher education. The preliminary study conducted using two master's level ICT units at one Australian university. The analysis validates that features, such as the number of visits to learning modules and their achievements in the units, are important indicators of users' interaction and learning in the unit. There are two key findings that emerged from this preliminary analysis. First, student interaction pattern remains the same from one unit to another. Second, students would access the teaching material more often if the unit is difficult or of higher level. These findings will be further tested and validated with larger data sets. Other features that are relevant to lecturers' pedagogical support and student engagement will also be added in later stage of the research. These features will be examined, using Big Data analytics techniques and skills that to date remain an unexplored area in Australian higher education. Starting with one university, the study will demonstrate relevance and impact for higher education nation wide, with the potential to contribute to online education at an international level.

Acknowledgement. This project is funded by the Research Enhancement Grant Scheme (REGS) at the University of Tasmania.

References

Aslam, A.A., Tsou, M.H., Spitzberg, B.H., An, L., Gawron, J.M., Gupta, D.K., Peddecord, K.M., Nagel, A.C., Allen, C., Yang, J.A., Lindsay, S.: The reliability of tweets as a supplementary method of seasonal influenza surveillance. J. Med. Internet Res. **16**(11) (2014)

Clow, D.: An overview of learning analytics. Teach. High. Educ. **18**(6), 683–695 (2013)

Ellaway, R.H., Pusic, M.V., Galbraith, R.M., Cameron, T.: Developing the role of big data and analytics in health professional education. Med. Teach. **36**(3), 216–222 (2014)

Eynon, R.: The rise of Big Data: what does it mean for education, technology, and media research? Learn. Media Technol. **38**(3), 237–240 (2013)

Manyika, J., Chui, M., Brown, B., Bughin, J., Dobbs, R., Roxburgh, C., Byers, A.H.: Big Data: The Next Frontier for Innovation, Competition, and Productivity. McKinsey Global Institute, US (2011)

Matteson, S.: Big data basic concepts and benefits explained. Web blog (2013). http://www.techrepublic.com/blog/big-data-analytics/

Pervaiz, F., Pervaiz, M., Rehman, N.A., Saif, U.: FluBreaks: early epidemic detection from Google flu trends. J. Med. Internet Res. **14**(5) (2012)

Picciano, A.G.: The evolution of big data and learning analytics in American higher education. J. Asynchronous Learn. Netw. **16**(3), 9–20 (2012)

Timms, M.: Big data in education: a guide for educators. Centre for Strategic Education Occasional Paper (No. 139), pp. 1–10 (2015)

West, D.M.: Big data for education: data mining, data analytics, and web dashboards. Governance Studies at Brookings, pp. 1–10 (2012)

DataCon: Easier Data Sharing, Exploration, and Fusion with Automatic Metadata Generation

Hwa Jong Kim[(⊠)]

Computer and Communications Engineering Department,
Kangwon National University, Chuncheon, Korea
hjkim3@gmail.com

Abstract. Data is now transforming the world. Individuals, organizations, companies and governments are rushing to build technologies that generate, manage and analyze ever-increasing amounts of data. The most valuable data applications often require fusing data from multiple sources. However, sharing data is a difficult and painful process. We propose a "DataCon" system that supports data sharing, exploration, and fusion between people using different software and different data formats. DataCons are data icons that contain structured metadata generated by Automatic Metadata Generation (AMG) algorithms. The Korean government is funding a three-year project with one million USD to develop a DataCon-based data sharing platform.

Keywords: Automatic Metadata Generation · Data fusion · Metadata · Tagging · Summarization · DataCon

1 Introduction

Data is transforming the world. Advances in hardware, software, and algorithms are revolutionizing domains from business to education to manufacturing. Existing technologies like urban planning tools and medical diagnosis systems are being reinvented by data. Emerging technologies like autonomous vehicles and augmented reality are made possible by data. Data is and will continue to influence, accelerate, and drive almost all human progress [1]. This has started a data arms race. Individuals, organizations, companies, and governments are rushing to build increasingly sophisticated technologies that generate, manage, and analyze ever-increasing amounts of data [2].

One of the key lessons from this arms race is that the most valuable data applications come from combining (fusing) data from many sources. However, because many people want to hoard their data in their respective organizational silos, it is difficult to discover, explore, and work with data cooperatively [3]. This is one of the key roadblocks to achieving a truly data-driven world [4, 5].

In 2016, the Korean government funded a three-year, $1 million USD project (the author of this paper is the principal investigator of the project) to develop algorithms that automatically generate standardized metadata for text, video, and IoT data along with a cloud-based system that allows data owners and data users to easily share,

© Springer International Publishing AG 2016
B.H. Kang and Q. Bai (Eds.): AI 2016, LNAI 9992, pp. 708–713, 2016.
DOI: 10.1007/978-3-319-50127-7_64

search, discover, access, explore, and analyze data from many sources. The project aims to develop a set of tools that automatically generate a "DataCon" object, a container for the metadata of raw data that can be shared and used like the icons on normal software files, through Automatic Metadata Generation (AMG). The AMG tools and the DataCon object will have an intuitive user interface and make it easier to share, exploration, and work with data. We hope this system will make sharing data practical and accelerate the development data-driven technologies.

Section 2 reviews related work. Section 3 describes our current approach to AMG. Section 4 contains our conclusions.

2   Background

2.1   Data Sharing Efforts

Many governments and institutes understand the value of sharing data openly. They operate and promote official data repository sites and encourage data applications that use open data. Most of these efforts focus on opening public data. However, only a few advanced nations have opened a significant portion of their data [6]. Additionally, governments only open a small portion of data they have. The compatibility of data formats is another issue for data sharing. For example, the US government lists 49 data formats on data.gov [7], but the Internet of Things (IoT) community alone has hundreds of data formats [8].

One of the key reason not many people share data is that there are many concerns, like privacy, with sharing raw data. Instead of sharing raw data, people often share metadata, data that describes the key aspects of the raw data, to people who may be interested in the full data. In many applications, the structure of metadata is based on the Dublin Core Metadata Initiative [9]. However, these metadata structures are far from ubiquitous, and few of the existing standards are designed for cross-domain applications that involve data fusion [10]. Metadata means many things to many people, from simple descriptions to detailed summarizations of the raw data. For example, tags are used widely to describe data. Tagging is an easy, convenient, and powerful tool that adds arbitrary supplementary information to the raw data that help users sort, search, and explore raw data. Specialized tags can be invented for things like privacy control [11, 12]. However, there is no commonly accepted method of tagging: the tags are often invented on the spot by the person responsible for managing the data. This applies to many other forms of metadata.

As a result, company and institutes often design their own metadata format, which makes sharing and understanding metadata difficult. Despite many efforts, there is no common way to share data using metadata. Sharing metadata is only the first step of sharing data. The metadata should be enough for the user to understand what the raw dataset contains. The next step is to access the raw data and begin the actual analysis.

2.2 Privacy Considerations

Data owners are often reluctant to open their data, even to close partners, often because of privacy concerns. For example, IoT devices generate sensitive data because many IoT applications are personalized to individual users [13]. The most valuable data-driven applications often involve personalization, which by definition involves generating sensitive data, so these privacy concerns will remain a key issue to many data owners. Any system that enables data sharing should contain safeguards that protect the privacy of individuals. In this paper, we propose an AMG scheme that can overcome this and many other challenges for safe data sharing.

3 Automatic Metadata Generation

3.1 Definition

AMG is performed by the data owner to produce a "DataCon" object, which is a generalized metadata structure that includes components such as summarization or tagging. DataCon objects are structured to save time when gathering, processing, and fusing data. The goal of AMG is not to replace the analytics performed by the data user. Instead, AMG should provide the analyst with an easy-to-understand overview of the data to help accelerate the data analytics process.

AMG helps analyze related data (e.g., analyzing multiple CCTV recordings simultaneously to find a person moving across the cameras) and cross-domain data sources (e.g. checking nearby public transportation and event schedules to guess why the person is moving in that direction). AMG helps data users decide how to preprocess the raw data; data owners promote their data on the data market for free or for a price; and data analysts find partners to perform mutually beneficial data fusion.

The meaning of "automatic" in the AMG is that the process should be formalized beforehand to produce a commonly acceptable DataCon format, even though the format is flexible and extendible by users.

3.2 DataCon Objects

DataCons contain components like summaries, indexes, and tags that convey information suggested by standards like Dublin Core [14] as well as additional information like how to access the data, how much of the data is available, and whatever else the data owner wishes to share. A typical DataCon would have the following items:

- author/contributor
- date/location
- title/description
- format/size/data type
- revision history
- how much of the data is available (open)
- link to the raw data (optional)

- link to a representative sample of the raw data (optional)
- information on how the summary is structured
- the summary

DataCon can also contain information specific to different data formats. For example:

Numerical Data: DataCons could include max, min, average, quartile values, and graphic data. It could also describe outliers and missing values.

Text Data: DataCons could include the estimated title of the document, the entities in the document, a text summary, captions of pictures.

Video Data: DataCons could include entities in the video, object descriptions, object appearance statistics, and video summarizations.

Sensor Data: DataCons could include contextual information, the purpose of sensing, and more.

3.3   Use of AMG with DataCon

Efficient Summarization: Summarization is widely used to explain unstructured data like text, speech, video, and system logs. Because the data is unstructured, the format of the summary must be flexible. For example, document summaries generate sentences that capture most of the meaning of documents like blogs, emails, news article, papers, and text documents. These summaries can help researchers choose which documents to read quickly [15], help analysts cluster similar documents, and increase reading comprehension by giving the reader a sense of what to expect. However, the actual length and content of the summary depends on the document in question. A summary of a long news article, for example, may be longer than a summary of a short email.

Videos can be summarized in a similar fashion by removing uninformative or boring scenes from movies. This enables things like automatically generated movie trailers, real-time analysis of video feeds, and descriptions of what events are occurring at what times [16].

However, most summarization methods are developed independently among institutes and companies, and do not allow for easy discovery, parsing, and fusion. Thereby most data users need to process the data again to extract another summary for further analysis. The AMG-based DataCons would help them to generate and use a commonly accessible summaries among data users.

Combined Data Exploration: Many data analytics solutions provide tools like graphical interface and visualization tools for exploratory analysis. Each solution, however, develops independent systems to store various types of raw data in a central server. When these systems cannot interpret new data types, or when those tools are not available, the analyst often has to gather many types of raw data and learn how to generate useful summary statistics on each individual data format. The AMG-based

DataCons would allow the analyst to easily examine data of many formats and suggest the best way of fusing that data together.

Set Level of Sharing: DataCon can indicate how much of the data is available for use based on the policy of the data owner. For example, we can define "openness levels" in the DataCon, which set an upper boundary to protect privacy, and a lower threshold to force specified data owners to open data beneficial to the public good. For example, cellphone location data might be freely available at one-hour intervals after all personally identifiable information has been scrubbed to help data users understand how people move, and this can be the lower bound of the data. By setting the lower bound through its DataCon, the data can have more chance to be open and used widely.

Support Data Fusion: The value of big data primarily comes from fusing data from a variety of sources. DataCon supports quick and ease exploration of data without the complicated and often resource-intensive downloading of raw data. Data markets and data portal services might use the DataCon instead of the raw data to help users decide what data is worth downloading fully. In many applications, DataCon could provide enough information for data fusion.

Data Discovery Platform: A cloud-based data discovery platform can be developed based on the DataCon. The platform would support searching and downloading data via DataCons, just like we use icons to find and use application softwares in a smartphone. DataCons cloud can be used as a solution for Data as a Service.

4 Conclusions

We propose Automatic Metadata Generation (AMG) and DataCon as a tool to share, open, and work with a variety of data while providing user-friendly options to mitigate concerns like privacy. The goal of AMG based DataCon is not to replace the analytics performed by the data user. Rather, it should provide the analyst with an overview of the data, suggest promising directions of inquiry, make further analysis more efficient, and increase the overall user-friendliness of working with data from a variety of sources. We are developing a pilot system of the proposed system over the next three years and expect to regularly release tools and platforms that help create a common framework for sharing data that accelerates the development of data-related technologies. In the pilot system, we are developing three AMG modules for text, video, and IOT data, and a testbed for DataCon cloud service.

Acknowledgements. This work was supported by Institute for Information & communications Technology Promotion (IITP) grant funded by the Korea government (MSIP) (No: R-20160906-004163, Developing Bigdata Autotagging and Tag-based DaaS System).

References

1. Chen, H., Chiang, R.H.L., Storey, V.C.: Business intelligence and analytics: from big data to big impact. MIS Q. **36**(4), 1165–1188 (2012)
2. Kim, G.-H., Trimi, S., Chung, J.-H.: Big-data applications in the government sector. Commun. ACM **57**(3), 78–85 (2014)
3. Fan, P.: Coping with the big data: convergence of communications, computing and storage. China Commun. **13**(9), 203–207 (2016)
4. Bauer, F., Kaltenböck, M.: Linked Open Data: The Essentials. Edition mono/monochrom, Vienna (2011)
5. Steinke, G.: Data privacy approaches from US and EU perspectives. Telemat. Inform. **19**(2), 193–200 (2002)
6. http://webfoundation.org/2015/10/seventeen-governments-adopt-the-new-international-open-data-charter/
7. http://catalog.data.gov/dataset#sec-res_format
8. Perrino, T., Howe, G., Sperling, A., Beardslee, W., Sandler, I., Shern, D., Pantin, H., Kaupert, S., Cano, N., Cruden, G., Bandiera, F., Brown, C.H.: Advancing science through collaborative data sharing and synthesis. Perspect. Psychol. Sci. **8**(4), 433–444 (2013)
9. DCMI. http://dublincore.org/
10. Shahi, A.: Activity-based data fusion for the automated progress tracking of construction projects, PhD thesis, presented to University of Waterloo, Ontario (2012)
11. Bar-Sinai, M., Sweeney, L., Crosas, M.: DataTags, data handling policy spaces and the tags language. In: Proceedings of the International Workshop on Privacy Engineering, pp. 1–8. IEEE Press, New York (2016)
12. Datatags. http://datatags.org/
13. Gubbi, J., Buyya, R., Marusic, S., Palaniswami, M.: Internet of Things (IoT): a vision, architectural elements, and future directions. Future Gener. Comput. Syst. **29**(7), 1645–1660 (2013)
14. Soundararajan, E., Meenachi, N.M., Sai Baba, M.: Semantic digital library - migration of Dublin core to RDF. In: International Conference on Signal and Image Processing (ICSIP), pp. 250–254 (2010)
15. Crangle, C.E.: Text summarization in data mining. In: Bustard, D., Liu, W., Sterritt, R. (eds.) Soft-Ware 2002. LNCS, vol. 2311, pp. 332–347. Springer, Heidelberg (2002). doi:10.1007/3-540-46019-5_24
16. Clarifai. https://www.clarifai.com/

An Empirical Evaluation of Job Classification Using Online Job Advertisements

Yang Sok Kim and Choong Kwon Lee[(✉)]

Department of Management Information Systems, Keimyung University,
1095 Dalgubeol-daero, Dalseo-Gu, Daegu 42061, Republic of Korea
{yagnsok.kim, cklee}@kmu.ac.kr

Abstract. Information system (IS) jobs have continually evolved over time along with the fast changes of information technology. Many scholars and organizations have tried to create various IS job classifications and have tried to classify job advertisements into specific job categories. However, we do not know how well the job classification represents real-world jobs. Since most companies post their job advertisements on the Web, it is valuable to use them for examining the limitation of the current IS job classification. By developing a crawling system to search job advertisements on the Web, we collected 139,573 job advertisements for about one year from eleven countries. We attempted to classify them to the pre-defined job categories suggested by Niederman et al. [1] using an exact term matching approach. We found that only small portion of the job advertisements were classified into the defined categories. This result implies that the current job classification scheme is not a sufficient tool to analyse IS jobs and a improved classification approach is needed for further job analysis.

Keywords: Job role · Job advertisement · Job classification · Job analysis

1 Introduction

Most countries have focused on developing IT industries in order to create new jobs and lower the rate of national unemployment because many studies have consistently shown the empirical evidences of how IT made a positive impact on the national economies. To understand jobs in IS area, many researchers have conducted job analysis. One of the fundamental steps in job analysis is to create the job classification. While some studies [2–5] choose specific jobs for job analysis, others [1] provide holistic classification for all jobs. Job classification is important in both cases. However, to answer whether or not the current job classification really reflects the real-world jobs is difficult because we should have sufficient job data and relevant analytic skills for assessing the appropriateness of job classification scheme. Toward this end, we developed a focused web crawling system which designed to collect large amount of online job postings and classified them into the job categories suggested by Niederman et al. [1] to see where the job advertisements fit into the categories.

© Springer International Publishing AG 2016
B.H. Kang and Q. Bai (Eds.): AI 2016, LNAI 9992, pp. 714–719, 2016.
DOI: 10.1007/978-3-319-50127-7_65

2   Related Studies

To analyze the changes of job trends in an industry, it is necessary to identify jobs in the market and classify them in a standardized approach. OECD has made efforts to develop a classification scheme at the level of IT industry since 2002 by dividing it into the two, manufacturing and service [6]. While IT services are further subdivided into three subsectors: wholesale of IT products; telecommunications; and computer software and services, these three major sub-sectors are further sub-divided into a number of industry categories. Public organizations and industry practitioners have attempted to extend the classification schemes of IT occupations because occupational structures reflects the environments of countries and change over time [7]. IT and Telecoms National Occupational Standards (NOS) [8] has been developed in parallel with and as part of the IT Professional Competency Model – e-skills Procom in UK. O*Net, which is sponsored by US Department of Labor/Employment and Training, also provides occupational classification scheme. It divides IT occupations into four broad categories - Information Support and Services; Interactive Media; Network Systems; and Programming and Software Development. Under these sub-categories, O*Net defines 58 occupations. In addition, private job agencies define their own job/occupation classification schemes reflecting current job market trends (e.g., Robert Half Technology [9]). Compared to ISCO-08, the classifications suggested by NOS, O*Net and Robert Half Technology have not much formalized. Researchers in academia have also suggested various classifications with regard to job skill analysis [7, 10–12]. Niederman et al. [1] have reviewed the co-evolution between the historical changes of IT and the job titles of IS. This research employed the classification scheme suggested by Niederman et al. [1].

3   Research Approach

Since the online advertisements are posted distributed ways, it is difficult to collect them manually. We developed a web crawling system which automatically collects online job advertisements. A total of 139,573 online job advertisements written in English were collected from 11 countries from March 2015 to March 2016. Table 1 summarizes data collection results by country. The countries that provided most job advertisements collected for this research were India (38.1%), England (19.0%), Ireland (13.4%), Canada (10.4%) and Australia (7.8%). It is necessary to define the job roles for further processing. Basically we used job role definition provided by O*Net. If a job role was not defined in O*Net, we referred to other job classification such as [8, 9]. In order to analyze how the jobs constitute the IS job market, we assigned each of online job advertisements into one or more job category. A job ad consists of job title, job metadata (e.g., Date Posted, Category, Location, Job Type, etc.) and job requirements.

The job titles of an ad are not exactly the same as the ones suggested by the classification. For example, the job classification uses 'Computer and IS Managers' to describe top level computer and IS managers, but few advertisements collected included such job title. Instead, terms such as 'CIO', 'chief information officer', 'CTO', and 'chief technology officer' were used as job titles in the advertisements collected.

Table 1. Data collection: Job advertisements

Country	Number of advertisements	Ratio
Australia	10,851	7.8%
Canada	14,467	10.4%
England	26,473	19.0%
Gulf	2,573	1.8%
Hong Kong	1,018	0.7%
India	53,125	38.1%
Indonesia	527	0.4%
Ireland	18,640	13.4%
Malaysia	3,725	2.7%
Singapore	7,600	5.4%
Thailand	574	0.4%
Total	139,573	100.0%

Table 2. An example (database administrators) of job titles in O*NET

Summary Report for: 15-1141.00 - Database Administrators
Administer, test, and implement computer databases, applying knowledge of database management systems. Coordinate changes to computer databases. May plan, coordinate, and implement security measures to safeguard computer databases
Sample of reported job titles: Database Administration Manager, Database Administrator (DBA), Database Analyst, Database Coordinator, Database Programmer, Information Systems Manager, Management Information Systems Director (MIS Director), Programmer Analyst, Systems Manager

For each occupation, as shown in Table 2, O*Net provides a list of job titles that could appear in job advertisements as well as the definition of the job.

Based on O*Net, we created a dictionary for job classification. Some dictionary words (e.g., IT Manager) were used in multiple jobs (e.g., Computer and IS Manager and IT Project Managers). This implies that a job ad can be classified into one or more jobs. We classified job advertisements into one of the jobs by assuring words in the title appeared in job advertisements matching with the dictionary. We used the exact pattern matching function, called REGEXP, provided by MySQL database management system.

4   Results

Overall 23.7% of all job advertisements (33,088) were classified into the defined job categories, but since some advertisements can be classified into one or more job categories, there are a total of 41,724 classifications as summarized in Table 3. The 'Software/Application Development' category accounted for the largest proportion (63.6%), followed by IT Service Professionals (Infrastructure) (17.2%), Project Management (11.3%), Enterprise Architecture (5.3%) categories. Jobs related to software

Table 3. Job ad classification results

Jobs & job categories	Classified jobs	%
1 Information Technology Management	405	1.0%
1.1 Computer and Information System Manager	405	1.0%
2 Project Management	4716	11.3%
2.1 Information Technology Project Managers	4716	11.3%
3 Software/Application Development	26518	63.6%
3.1 Computer System Analysts	5112	12.3%
3.2 Computer Programmers	8590	20.6%
3.3 Software Developers, Applications	9431	22.6%
3.4 Software QA Engineers and Testers	464	1.1%
3.5 Search Marketing Strategists	2	0.0%
3.6 Web Developers	2919	7.0%
4 IT Service Professionals (Infrastructure)	7190	17.2%
4.1 Database Administrators	2029	4.9%
4.2 Data Warehousing Specialist	25	0.1%
4.3 Network and Computer Systems Administers	3222	7.7%
4.4 Telecommunications Engineering Specialist	1	0.0%
4.5 Computer User Support Specialists	576	1.4%
4.6 Computer Network Support Specialists	1139	2.7%
4.7 Web Administrator		
4.8 Computer Operators	198	0.5%
5 Enterprise Architecture	2215	5.3%
5.1 Computer Systems Engineers/Architects	740	1.8%
5.2 Computer Network Architects	1185	2.8%
5.3 Database Architects	290	0.7%
6 IT Security	140	0.3%
6.1 Information Security Analysts	140	0.3%
7 Sourcing/Vendor Management	35	0.1%
7.1 Purchasing Managers	35	0.1%
8 IT Sales and Marketing Professionals	56	0.1%
8.1 Sales Representatives, Wholesale and Manufacturing, Technical and Scientific Products	56	0.1%
9 Emerging Roles for IT Professionals	449	1.1%
9.1 Business Intelligence Analysts	90	0.2%
9.2 Video Game Designers	26	0.1%
9.3 Multimedia Artists and Animators	333	0.8%
All Matched Jobs	41,724	100.0%

development, such as 'Computer Systems Analysts' (12.3%), 'Computer Programmers' (20.6%), 'Software Developers, Applications' (22.6%), 'Web Developer' (7.0%), and 'IT Project Managers' (11.3%) were most frequently advertised jobs in the market. Apart from these jobs, 'Network and Computer Systems Administrators' (7.7%) and 'Database Administrator' (4.9%) also contributed significant proportions of all advertised advertisements. Three emerging jobs ('Business Intelligence Analysts', 'Video Game Designer', and 'Multimedia Artists and Animators') accounted for only 1.1% of all advertised advertisements. Note that most frequently advertised jobs were similar to the jobs discussed in job skill analysis research.

5   Conclusion

Our research was based on the IT jobs and their classifications, job definition, job titles obtained from a famous site O*Net as well as academic references [1]. Our current results show that only 23.7% of the collected advertisements were classified by using 'exact' matching. This low classification rate might have come from various reasons. First, the current job classifications do not include some jobs defined O*Net by following Niederman et al. For example, the current job classifications do not include hardware related jobs and academic jobs. Second, the current job classification do not include newly evolving job titles appeared in the job market. For example, the jobs like user interface and user experience (UI/UX), and data analytics are frequently advertised in the market. Third, the exact matching approach makes too much restriction when classifying the advertisements. Even though the current approach of this research has these limitations, we assure that our research can be used as baseline for further studies. These limitations should be addressed in the future studies.

Acknowledgement. This research was supported by the Keimyung University New Faculty Research Grant of 2015 and 2016.

References

1. Niederman, F., Ferratt, T.W., Trauth, E.M.: On the co-evolution of information technology and information systems personnel. SIGMIS Database **47**(1), 29–50 (2016)
2. Lending, D., Dillon, T.W.: Identifying skills for entry-level IT consultants. In: Proceedings of the 2013 Annual Conference on Computers and People Research, Cincinnati, pp. 87–92. ACM (2013)
3. Ferrari, R., Madhavji, N.H., Wilding, M.: The impact of non-technical factors on Software Architecture. In: Proceedings of the 2009 ICSE Workshop on Leadership and Management in Software Architecture, pp. 32–36. IEEE Computer Society (2009)
4. Downey, J., Babar, M.A.: On identifying the skills needed for software architects. In: Proceedings of the First International Workshop on Leadership and Management in Software Architecture, Leipzig, pp. 1–6. ACM (2008)

5. Downey, J.: Systems architect and systems analyst: are these comparable roles? In: Proceedings of the 2006 ACM SIGMIS CPR Conference on Computer Personnel Research: Forty Four Years of Computer Personnel Research: Achievements, Challenges & the Future, Claremont, pp. 213–220. ACM (2006)
6. Crone, M.: Recent trends in ICT employment: benchmarking NI. In: Rogers, D. (ed.) Labour Market Bulletin, vol. 19, pp. 101–111 (2005)
7. Sabadash, A.: ICT employment statistics in europe: measurement methodology. European Commission Joint Research Centre Institute for Prospective Technological Studies (2012)
8. Morrow, C.: National Occupational Standards (NOS) IT and Telecoms (2009)
9. Robert Half Technology, RHT: Glossary of Job Descriptions for Information Technology (2016)
10. Todd, P.A., McKeen, J.D., Gallupe, R.B.: The evolution of IS Job Skills: a content analysis of IS job advertisements from 1970 to 1990. MIS Q. **19**(1), 1–27 (1995)
11. Hunter, D.: Occupations in information and communication technologies. options for upgrading the international standard classification of occupations. ILO Discussion Paper, April 2006
12. Donohue, P., Power, N.: Legacy job titles in IT: the search for clarity. In: Proceedings of the 50th Annual Conference on Computers and People Research, Milwaukee, pp. 5–10. ACM (2012)

Robust Text Detection in Natural Scene Images

Van Khien Pham and GueeSang Lee[(✉)]

Department of Electronics and Computer Engineering,
Chonnam National University, Kwangju, South Korea
poldpham@gmail.com, gslee@jnu.ac.kr

Abstract. Natural scenes of blurred text images are challenges to current text recognition field. In our paper, a novel method for text detection in natural scene image is suggested using edge detection, maximally stable extremal region (MSER) and tensor voting. Edge detection and MSER methods are combined to find the greatest character candidates from stable areas which are extracted from an input image. These text candidates are used to extract the text line information using tensor voting that creates normal vectors and curve saliency values in characters along the text lines. Therefore, the text line information is used to eliminate non-text areas. Our method is evaluated on the ICDAR2013 datasets and experiment results show that the proposed result is compared to the previous methods.

Keywords: MSER · Blurred images · Tensor voting · Text detection · Edge detection

1 Introduction

Text information in digital images is measured to be these important technologies in image processing, and computer vision. Natural scene images will be taken under various situations, which carry new challenges to accurately verify text areas. The background in natural scene images can be very complex such as the variation of the light intensity, different font styles, sizes, color and blur. Moreover, to detect text regions with high accuracy from original images, we must overcome these challenges.

The previous approaches of text detection need to be approximately divided into five main groups: connected component based, edge based, texture based, stroke based and learning based. Connected component based approaches [1, 2] can extract character candidates from input images followed by gathering character candidates into text by connected component analysis. The non-text regions are eliminated based on heuristic rules or classifiers. However, CC based approaches are hard to segment text connected components exactly without text location and scale information. Edge based approaches [3–5] are usually more efficient and simple in nature scene text extraction. These edges (eg. Canny, Sobel, Gradient…) of the text boundary are verified and collected, after that the non-text regions will be removed by some heuristic rules. Good performance is often found on scene images exhibiting strong edges. A major problem of edge based methods is hard to find under the effect of highlight and shadow. Texture based approaches are used on the observation which text regions have diverse textural

© Springer International Publishing AG 2016
B.H. Kang and Q. Bai (Eds.): AI 2016, LNAI 9992, pp. 720–725, 2016.
DOI: 10.1007/978-3-319-50127-7_66

properties from the background in original images. These methods [6–8] can verify text regions correctly even though images are complex background, However, computation time is very high and the final result is sensitive to text alignment direction. In stroke based approaches [9–11], character strokes offer robust features for text detection in input images. One feature which splits text from other elements of a natural scene is its almost constant stroke feature similar stroke width. These approaches on specific applications are very simple execution however complex backgrounds make text strokes hard to identify and segment. Learning based methods use some features to train a classifier (SVM [12], deep Convolutional Neural Network classifier [13–15]). The results of these methods show high precision, however the disadvantages of these methods are high processing time and need to train many images.

In our paper, a new method is proposed to overcome the difficulties with complex background images. We suggest a robust and precise Edge detection, MSER and tensor voting based on text detection method where these contributions are given. MSER algorithm [17–19] finds the best region detector due to its robustness against view point, and illumination changes, however MSER is sensitive to blurred image. Therefore, we need to combine edge detection and MSER to handle this problem. Besides that, tensor voting method [16] works well to identify text areas and remove non-text areas in complex backgrounds.

This paper is organized as the follow: Sect. 2 illustrates the proposed method. Section 3 shows the experiment method. Finally, Sect. 4 gives our conclusion.

2 Proposed Method

Firstly, edge and MSER detection are going to combine to extract good candidate text areas. Secondly, these center points of text candidate regions are used as input token for tensor voting. From the resulting tensors, we can detect and remove non-text connected components. Finally, text lines are generated based on center points of grouping connected character candidates.

2.1 MSER and Edge Detection

MSER has been known as one of the best region detectors with some cases as against view point, and illumination changes [19]. Text usually has different contrast to its background and relatively uniform intensity or color, therefore MSER is a good choice for text detection in natural scene images. It is detected using the basic CCs in the original image, and then too big and small regions will be eliminated at this process. MSER approach is able to detect text regions, even though the image is in low quality, scale and illumination. However, this method is sensitive with image blur, candidate text connected components overlap together. Therefore, we need to separate them; MSER and edge detection are used to solve this problem. The canny edge detection approach can create an edge map from the input image in the Fig. 2(c). Using canny edge detection, we can identify good character boundaries in the text regions.

2.2 Geometric Filtering

MSER and edge detection also extract many non-text regions that will really bother the text gathering procedure. Therefore, we need to eliminate the non-text regions while also maintaining the text regions. Actually, if we can effectively remove the non-text areas, the following text gathering and non-text removal can be very simple, even though some heuristic rules can remove false positive [20]. In the proposed method, we focus on the MSER and edge detection labeling process, that eliminates almost the non-text areas and identify text regions accurately in the Fig. 1(c).

(a)   (b)   (c)

Fig. 1. Candidate text regions process. (a) Original image, (b) MSER and Edge detection, (c) Text candidate regions

2.3 Tensor Voting Filtering

The input tokens corresponding candidate center points are primed by their information [21]. An input token that is an elementary curve with normal is denoted by a stick tensor with the largest eigenvector represents its normal vector. An un-oriented token is encoded by a ball tensor such as a separate pixel or a junction of curves. The tensor is represented by an ellipse. Intuitively, the ellipse's shape indicates the type of structure denoted and its size the saliency of this information.

Tensors have some smooth salient features (i.e., curves) that powerfully support each other. Each tensor votes its neighboring tensors with its information and also obtains votes from these tensors. The size, shape, the orientation and vote strength of this neighborhood are encapsulated in predefined kernels or voting fields.

2.4 Generating the Text Detection Result

The text candidate connected components are grouped by the text line information, to detect the text connected components. All connected components which have similar heights and are adjacent together in the horizontal orientation are gathered. After connecting candidate text regions, final text bounding boxes are shown as the Fig. 2(c). Those grouped connected components that either does not include tensor voting of text line are eliminated. Text line information is very important to improve text detection results.

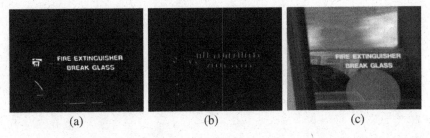

(a)   (b)   (c)

Fig. 2. Tensor voting process. (a) Input tokens, (b) tensor voting result, (c) text detection.

3   Experimental Results

In our method, we demonstrate the MSER, Edge detection, and Tensor voting algorithms. The evaluation of the proposed method and previous works are implemented on these images taken from 2013 ICDAR contest test images. ICDAR 2013 dataset is more interesting and the background areas are more complex than previous datasets.

Our method is compared with three main previous methods of Toan et al. [16], Chen et al. [19], and Huang et al. [14]. Chen's method gave the main idea using MSER, beside that Toan's method presented a new idea using tensor voting and Huang method used CNN to verify text and non-text regions. Figure 3 illustrates text detection results of some sample blur images from Toan, Chen, and our method. The experimental results show that the proposed approach can positively locate the text in several natural scene images as complex backgrounds with a lower false positive and negative compared to these above approaches in the Table 1.

Fig. 3. Some text detection results on blur images from Toan, Chen and our method

Table 1. Comparison of these methods from 2013 ICDAR Dataset.

Method	Precision	Recall	F-score
Chen [19]	74%	61%	66.9%
Toan [16]	82%	83%	82.5%
Huang [14]	88%	71%	78%
The proposed method	**89.3%**	**87.4%**	**88.3%**

The precision is represented as the number of accurate estimates given by the total estimate numbers. The recall is represented as the correct estimate numbers given by the total target numbers.

4 Conclusion

In our proposed method, robust text detection is suggested, which based on Edge detection, MSER, and Tensor Voting. MSER and Edge detection extract good candidate text areas. Besides that, tensor voting can remove almost non-text regions in the final step. Furthermore, tensor voting is very robust to noises, we can identify and remove connected components in noise regions or aligned vertically more efficiently. The experimental results are illustrated that the proposed method detects more text areas with lower false positives in complex natural scene images, which include blur, and low contrast scenes.

Acknowledgement. This research was supported by Basic Science Research Program through the National Research Foundation of Korea (NRF) funded by MEST (NRF- 2015R1D1A1A010 60172).

References

1. Jiang, R., Qi, F., Xu, L., Wu, G.: Using connected components' features to detect and segment text. J. Image Graph. **11**, 1653–1656 (2006)
2. Zhang, H., Liu, C., Yang, C., Ding, X., Wang, K.Q.: An improved scene text extraction method using conditional random field and optical character recognition. In: International Conference on Document Analysis and Recognition, ICDAR, pp. 708–712 (2011)
3. Liu, X., Samarabandu, J.: Multiscale edge-based text extraction from complex images. In: IEEE International Conference on Multimedia and Expo, ICME (2006)
4. Ye, Q., Jiao, J., Huang, J., Yu, H.: Text detection and restoration in natural scene images. J. Vis. Commun. Image Represent. **18**, 504–513 (2007)
5. Ezaki, N., Bulacu, M., Schomaker, L.: Text detection from natural scene images: towards a system for visually impaired persons. In: Proceedings of the 17th International Conference on Pattern Recognition, ICPR, vol. 2, pp. 683–686 (2004)
6. Zhou, G., Liu, Y., Meng, Q., Zhang, Y.: Detecting multi lingual text in natural scene. In: International Symposium on Access Spaces, ISAS, pp. 116–120 (2011)

7. Hanif, S.M., Prevost, L.: Text detection and localization in complex scene images using constrained adaboost algorithm. In: 10th International Conference on Document Analysis and Recognition, ICDAR, pp. 1–5 (2009)

8. Angadi, S.A., Kodabagi, M.M.: A texture based methodology for text region extraction from low resolution natural scene images. In: Advance Computing Conference, pp. 121–128 (2010)

9. Pan, Y.F., Zhu, Y., Sun, J., Naoi, S.: Improving scene text detection by scale adaptive segmentation and weighted CRF verification. In: International Conference on Document Analysis and Recognition, ICDAR, pp. 759–763 (2011)

10. Epshtein, B., Ofek, E., Wexler, Y.: Detecting text in natural scenes with stroke width transform. In: IEEE Conference on Computer Vision and Pattern Recognition, CVPR, pp. 2963–2970 (2010)

11. Yi, C., Tian, Y.L.: Text string detection from natural scenes by structure based partition and grouping. IEEE Trans. Image Proc. **20**, 2594–2605 (2011)

12. Gonzalez, A., et al.: Text location in complex images. In: IEEE 21st International Conference on Pattern Recognition, ICPR, pp. 617–620 (2012)

13. Krizhevsky, A., Sutskever, I., Hinton, G.E.: Image net classification with deep convolutional neural networks. In: Proceedings of NIPS, pp. 1106–1114 (2012)

14. Huang, W., Qiao, Y., Tang, X.: Robust scene text detection with convolution neural network induced MSER trees. In: Fleet, D., Pajdla, T., Schiele, B., Tuytelaars, T. (eds.) ECCV 2014, Part IV. LNCS, vol. 8692, pp. 497–511. Springer, Heidelberg (2014). doi:10.1007/978-3-319-10593-2_33

15. Jaderberg, M., Vedaldi, A., Zisserman, A.: Deep Features for Text Spotting. In: Fleet, D., Pajdla, T., Schiele, B., Tuytelaars, T. (eds.) ECCV 2014, Part IV. LNCS, vol. 8692, pp. 512–528. Springer, Heidelberg (2014). doi:10.1007/978-3-319-10593-2_34

16. Nguyen, T.D., Park, J., Lee, G.: Tensor voting based text localization in natural scene images. IEEE Sig. Process. Lett. **17**, 639–642 (2010)

17. Shi, C., Wang, C., Xiao, B., Zhang, Y., Gao, S.: Scene text detection using graph model built upon maximally stable extremal regions. Pattern Recogn. Lett. **34**, 107–116 (2013)

18. Yin, X.-C., Huang, K., Hao, H.-W.: Robust text detection in natural scene images. IEEE Trans. Pattern Anal. Mach. Intell. PAMI **36**, 970–983 (2013)

19. Chen, H., et al.: Robust text detection in natural images with edge-enhanced maximally stable extremal regions. In: 18th IEEE International Conference on Image Processing, ICIP (2011)

20. Li, Y., Lu, H.: Scene text detection via stroke width. In: IEEE 21st International Conference on Pattern Recognition, ICPR, pp. 681–684 (2012)

21. Nguyen, T.D., Lee, G.: Color image segmentation using tensor voting based color clustering. Pattern Recogn. Lett. **33**, 605–614 (2012)

Retraction Note to: Co-clustering for Dual Topic Models

Santosh Kumar, Xiaoying Gao, and Ian Welch

Retraction Note to:
Chapter "Co-clustering for Dual Topic Models" in:
B. H. Kang and Q. Bai (Eds.): *AI 2016:*
Advances in Artificial Intelligence, **LNAI 9992,**
https://doi.org/10.1007/978-3-319-50127-7_34

The Editors have retracted this conference paper [1] following an investigation by Victoria University of Wellington, for having significant overlap with a conference paper [2] by different authors. The latter [2] was submitted to a conference before the former [1]. Xiaoying Gao and Ian Welch agree to this retraction, Santosh Kumar does not agree to this retraction.

[1] Kumar, S., Gao, X., Welch, I.: Co-clustering for dual topic models. In: Kang, B.H., Bai, Q. (eds.) AI 2016. LNCS (LNAI), vol. 9992, pp. 390–402. Springer, Cham (2016). https://doi.org/10.1007/978-3-319-50127-7_34
[2] Rugeles, D., Zhao, K., Cong, G., Dash, M., Krishnaswamy, S.: Biclustering: an application of dual topic models. In: Proceedings of the 2017 SIAM International Conference on Data Mining, pp. 453–461. Society for Industrial and Applied Mathematics (2017).

The retracted version of this chapter can be found at
https://doi.org/10.1007/978-3-319-50127-7_34

Author Index

Printed in the United States
by Baker & Taylor Publisher Services